国外名校名著

化学工程的基本原理和计算方法

第九版

（英文原版影印）

[美] 戴维·M.希梅尔布劳　詹姆斯·B.里格斯　著

化学工业出版社

·北京·

内容简介

　　《化学工程的基本原理和计算方法》是由 David M. Himmelblau 与 James B. Riggs 所著的国际经典教材。本书第九版涵盖了化学工程、生物工程、石油工程和环境工程领域应用的基本原理和计算方法，系统阐述了物料与能量衡算、气体与多相平衡、非稳态系统及流程模拟应用，创新融合 MATLAB 与 Python 编程实践，将复杂概念转化为可执行代码，极大地提升了复杂计算的可操作性。本书超越传统教材的"知识灌输"模式，融入大量自测题与工业案例，辅以附录数据，内容编排强调工程化思维与系统化分析范式，实现理论推导与工程实践的高效衔接。

　　本书可作为化学工程、石油工程、环境工程及相关专业的本科生与研究生教材，也可为相关专业的工程师及科研人员提供方法论支持，是贯通基础理论与工业应用的权威指南。

　　北京市版权局著作权合同登记号：01-2025-1815

图书在版编目（CIP）数据

化学工程的基本原理和计算方法 = Basic Principles and Calculations in Chemical Engineering：第九版：英文 / （美）戴维·M. 希梅尔布劳（David M. Himmelblau），（美）詹姆斯·B. 里格斯（James B. Riggs）著. -- 北京：化学工业出版社，2025. 8. --（国外名校名著系列教材）. -- ISBN 978-7-122-48016-3

　　Ⅰ. TQ02

　　中国国家版本馆 CIP 数据核字第 202571CG80 号

责任编辑：任睿婷　徐雅妮　　　　装帧设计：刘丽华
责任校对：赵懿桐

出版发行：化学工业出版社
　　　　　（北京市东城区青年湖南街 13 号　邮政编码 100011）
印　　装：河北京平诚乾印刷有限公司
880mm×1230mm　1/16　印张 56½　字数 1421 千字
2025 年 8 月北京第 1 版第 1 次印刷

购书咨询：010-64518888　　　　售后服务：010-64518899
网　　址：http://www.cip.com.cn
凡购买本书，如有缺损质量问题，本社销售中心负责调换。

定　　价：259.00 元　　　　　　　版权所有　违者必究

推荐序

  《化学工程的基本原理和计算方法（第九版）》作为化学工程领域的经典著作，以其严谨的学科框架、创新的教学理念和与时俱进的实践工具，确立了其在工程教育领域的标杆地位。在化学工程的知识体系中，物料与能量衡算犹如树干，支撑着热力学、流体流动、传热、传质、反应器动力学、过程控制及过程设计等关键分支。因此，精研物料与能量衡算，既是贯通工程多尺度问题的认知基石，亦是构建系统化工程思维的范式根基。

  本书由 David M. Himmelblau 和 James B. Riggs 合著，是化学工程领域的经典教材，是原著的第九版。本书系统性地介绍了化学工程的核心原理与计算方法，以"物料与能量衡算"为脉络，构建了从宏观到微观的递进式学习路径，涵盖本科与研究生阶段的核心知识体系，并为未来的专业实践奠定坚实基础。本书采用问题导向的实践方法论，每一章均以"学习目标-原理剖析-解题策略-自测验证"为主线，使读者在循序渐进中完成知识内化与能力提升。本版教材创新性地引入 MATLAB 与 Python 编程实践，不仅契合现代工程对数字化工具的需求，更将复杂概念转化为可执行代码，极大提升了复杂计算的可操作性。本书超越传统教材的"知识灌输"模式，注重终身学习能力的塑造，以前沿的"学会如何学习"为理念，引导读者通过反思性习题建立批判性思维。

  本书不仅系统梳理了化学工程的核心知识体系，更通过跨学科视角与数字化技术的深度融合，为读者搭建了一座连接基础理论与工业实践的桥梁。本书既是一部授人以渔的方法论指南，也是一本面向未来的工程师成长手册。无论是初涉化工领域的学子，还是寻求知识更新的从业者，都能从本书中汲取养分——它不仅传授如何解题，更教会读者如何像工程师一样思考。期待这部经典之作继续引领新一代化工人才，在技术创新与可持续发展的征程中破浪前行。

天津大学/新疆大学教授

2025 年 3 月

前言

 本书作为一本导论，介绍了化学工程、生物工程、石油工程和环境工程领域中应用的原理与技术。尽管在过去三十年间，化学工程领域的学科范围有所扩大，但该研究领域的基本原理始终未变。本书介绍了化学工程本科与研究生学习以及化学工程专业实践所需的特定技能与基础知识。后续的化学工程课程学习将极大依赖于本课程中培养的技能，即解决抽象问题的能力以及对物料与能量衡算的应用能力。化学工程领域的学习可被视为一棵树，物料与能量衡算作为树干，而热力学、流体流动、传热、传质、反应器动力学、过程控制及过程设计等学科是从树干延伸出来的树枝。从这一角度出发，我们不难看出掌握本书内容的重要性。

 本书的主要目标是教你如何系统地构建并解决物料与能量衡算问题，甚至是使用本书所提供的方法系统地构建并解决所有类型的问题。此外，本书介绍了化学工程师所涉猎的各种过程，从炼油与化学工业，到生物工程、纳米工程及微电子产业。本书使用的分析方法主要基于宏观尺度（即把复杂系统看成一个统一的系统），之后的工程课程将教你如何构建微观尺度的物料与能量衡算，用于更完整地描述这些系统。事实上，要构建一个微观平衡，你只需将本书介绍的衡算方法应用到所关注过程内部的一个微小单元即可。

 本书结构安排如下：
- 第一篇引言：背景信息（第1-2章）。
- 第二篇物料衡算：如何构建与解决物料衡算（第3-5章）。
- 第三篇气体、蒸气与液体：如何描述气体与液体（第6-7章）。
- 第四篇能量衡算：如何构建与解决能量衡算（第8-9章）。
- 第五篇物料与能量综合衡算：如何使用宏观方法描述非稳态系统的行为（第10-11章）。
- 额外资料：第12-14章及附录。

 众所周知，学习需通过实践完成，即应用你所学习的知识。仅通过阅读与听讲不足以掌握本书中的信息与技能，为此，本书提供了多项资源以协助你完成这一过程，最重要的资源可能是书中每节末尾的自测题。这些自测问题与习题极具价值，通过回答它们并将你的答案与参考答案进行对比，你可以发现自己尚未完全理解的内容。

本版教材编写时融入了 MATLAB 与 Python 软件技术。书中提供了用于解线性方程组、确定线性方程组中独立方程数量、解单一非线性方程、对非线性数据集进行三次样条插值处理及积分初值常微分方程的 MATLAB 与 Python 代码。每种情况下，完成这些任务所使用的内置函数都会被单独呈现和描述。

我们衷心希望通过对本书及相关材料的学习，不仅能激励你持续追求成为一名化学工程师的目标，更能使你在追梦途中更加轻松。

Jim Riggs

得克萨斯州奥斯汀市

2021 年 11 月

作者简介

David M. Himmelblau 在得克萨斯大学化学工程系任教 42 年，是 Paul D. 和 Betty Robertson Meek 教授和美国 Petrofina 基金会百年荣誉退休教授。他于 1947 年获得麻省理工学院的学士学位，并于 1957 年获得华盛顿大学的博士学位。他是 11 本书籍和 200 多篇关于过程分析、故障检测与优化文章的作者，并曾担任 CACHE 公司（计算机辅助化学工程教育）的总裁，以及美国化学工程师学会（AIChE）的理事。他的著作《化学工程的基本原理和计算方法》被美国化学工程师学会评为该领域最重要的书籍之一。

James B. Riggs 于 1969 年获得得克萨斯大学奥斯汀分校的学士学位，1972 年获得该校的硕士学位。1977 年，他获得了加利福尼亚大学伯克利分校的博士学位。Riggs 博士在大学担任教授长达 30 年，其中前五年在西弗吉尼亚大学工作，其余时间则在得克萨斯理工大学任教。此外，他还拥有超过五年的工程经验，担任过多个职务。他的研究兴趣集中在先进过程控制和过程优化领域。在学术生涯中，他曾担任工业顾问，并创办了得克萨斯理工大学过程控制与优化联盟，领导该联盟长达 15 年。Riggs 博士还编写了几本广受欢迎的本科化学工程教材，包括《化学工程的计算方法》、《MATLAB 工程编程实战》和《化学与生物过程控制（第五版）》。

Vivek Utgikar 博士（审阅人）是爱达荷大学化学与生物工程系的教授。他还曾担任爱达荷大学工程学院核工程项目的主任以及负责科研与研究生教育的副院长。Utgikar 博士的教学经验丰富，讲授一系列基础与高级工程课程，包括传递现象、动力学、热力学、能源储存、电化学工程、氢能及乏燃料处置与管理等。他的研究兴趣涵盖先进能源系统、核燃料循环、多相系统建模以及生物修复。在加入爱达荷大学之前，Utgikar 博士曾在俄亥俄州辛辛那提市美国环境保护局国家风险管理研究实验室担任国家研究委员会副研究员。Utgikar 博士是一名注册工程师，拥有化工过程开发、设计相关工程经验，并获得辛辛那提大学的化学工程博士学位。他还拥有印度孟买大学的化学工程学士和硕士学位。他是培生公司出版的《化学工程基本概念与计算方法》与《可再生能源系统中的化学过程》两本书的作者。

致谢

我们要感谢许多曾经的老师、同事和学生，他们在本书的编写过程中，尤其是在本版的编写过程中，给予了直接或间接的帮助。特别感谢 C. L. Yaws 教授的关怀，同时与 Terry Ring 教授和 Clayton Radke 教授的讨论，对于本版内容的修订也起到了积极作用。使用这本教材的教师人数众多，他们提供的修正意见和建议不胜枚举，无法在此一一列名致谢。若您有任何评论或改进建议，我们将不胜感激。

Jim Riggs

出版方特别感谢 Vivek Utgikar 先生，他在本版的出版过程中提供了宝贵的帮助。

如何使用本书

欢迎阅读《化学工程的基本原理和计算方法（第九版）》。为了帮助您更高效地掌握本书所涵盖的化学工程基本原理与计算方法，除了主体内容之外，我们精心设计了多种学习辅助工具。希望您能够充分利用这些资源，提升学习效果。

学习辅助工具

1. 精选大量详细解析的示例，以阐释基本原理，加深对内容的理解。
2. 提供一套系统且统一的解题策略，适用于各类问题的求解过程。
3. 配备丰富的图表、示意图和草图，帮助您更直观地理解和巩固所学知识。
4. 每章开篇明确列出具体的学习目标。
5. 每节内容结束时设置自我评估测试，并附有详细答案，帮助您及时检测学习成果。
6. 每章结尾提供大量精选习题，并在附录 D 中给出参考答案。
7. 附录部分提供了与书中示例和习题相关的关键数据。
8. 每章均提供参考文献，供读者深入学习参考。
9. 每章末尾设有术语表，系统汇总关键概念与术语。

建议您先通读本书，熟悉并定位这些辅助工具，以便在后续学习过程中灵活运用。

良好的学习方法（学会如何学习）

"不能让所有人都穿同一双鞋。"

—— Publilius Syrus

从事学习心理学和教育心理学研究的专家指出，几乎所有人都通过实践与反思来学习，而不是通过被动地听别人讲授就能够掌握知识。"讲授不等于教学，听讲不等于学习。"真正的学习，需要通过动手实践来实现。

学习不仅仅是记忆

请不要将"记忆"简单等同于"学习"。单纯为了记住解题步骤而抄录或整理笔记，对真正理解物料衡算与能量衡算问题的解决方法帮助有限。只有通过反复练习，才能将知识真正内化，并运用于从未见过的新问题之中。

培养良好的学习习惯

您可能会发现，跳过正文直接去看公式或示例来解题，有时似乎能奏效，但从长远来看，这样的学习方式会带来挫败感。这种做法被称为"公式导向"学习方式，是一种非常糟糕的学习解决问题类课程的方法。采用这种方法，不仅难以举一反三，每道新题都会成为新的难题，同时也会错失本质相似问题之间的内在联系。

当然，适合不同人的有效学习方式（信息处理方式）是存在差异的，因此，您需要反思自己的学习过程，找到最适合自己的方法。有些同学擅长独立思考学习，有些则喜欢与同伴或导师讨论，边交流边学习。有人通过动手做实例掌握知识，有人则更偏好抽象概念的推演。通过图示和草图来辅助理解，通常对大多数人都是有效的。如果您发现自己容易对重复性的学习内容感到厌倦，也可以尝试参加一些学习风格测评，以了解最适合自己的有效学习模式。很多学生发现，这类测试不仅有趣，而且对改进学习方法大有裨益。

无论您的学习风格如何，下面将为您提供一些我们认为有效的学习建议，希望对您的学习有所帮助。

提升学习效果的建议

1. 合理安排学习时间，深入理解课程内容

本书的每一章需要大约三小时或更长时间来阅读、理解，并练习解决相关问题的技能。请在你的学习计划中预留出时间，确保在上课前已经阅读了相关材料。这样，您不仅可以在课堂上更好地理解教授讲解的内容，还能把知识的理解提升到更高的层次。虽然这种方法并非总能实现，但它无疑是最有效的学习方式之一。

2. 积极参与讨论，互助学习

如果你正在参加某个课程的学习，可以与一位或多位同学合作讨论和交流，但不要依赖别人完成作业。

3. 坚持每日学习，按时完成任务

每天坚持学习，按时完成预定任务，不要落后，因为每个新主题都是在前一个主题的基础上展开的。

4. 及时解决疑问，避免积压问题

遇到不理解的问题时，务必尽快找到答案。

5. 采用主动阅读法，提升理解和记忆

在阅读时，建议采取主动阅读的方式，每五到十分钟暂停一到两分钟，回顾总结刚才所学的内容，思考各部分之间的联系。如有必要，可以动手写下总结和要点，帮助梳理思路和加深理解。

有效使用本书的建议

如何才能最大限度地发挥本书的学习效果？以下是一些具体建议。

1. 了解目标，明确方向

在学习每一部分内容之前和之后，务必认真阅读该部分列出的学习目标，明确学习重点与需要掌握的知识。

2. 主动阅读，独立思考

阅读正文时，当遇到示例（例题）时，建议先遮住答案，尝试自己独立完成题目。对于那些通过阅读示例来学习的读者，也可以先阅读示例，再回过头来学习正文内容。

3. 及时练习，巩固所学

每完成一小节的学习后，请务必尝试解决该节末尾的自测题，并与参考答案进行核对，检验学习效果。每完成一章后，建议挑选几道章末练习题进行练习，进一步巩固和应用所学知识。

4. 坚持实践，深化理解

正如诺贝尔物理学奖得主 R. P. Feynman 所说："你对某件事一无所知，直到你付诸实践。"

5. 系统思考，规范解题

无论是使用计算器，还是借助计算机程序进行问题求解，都应采用系统化的方法来整理信息，以获得正确的解决方案。

本书的价值，就像一笔储蓄——你投入多少努力，最终将收获多少，并获得"利息"。

出版商说明

本书的出版是为了纪念 David M. Himmelblau 和 James B. Riggs，

以表彰他们在化学工程领域做出的杰出贡献。

单位换算表

力

	N	lb_f
N	1	0.2248
lb_f	4.448	1

能量

	J	cal	kWh	Btu	ft lb_f	hp h
J	1	0.2390	2.778×10^{-7}	9.478×10^{-4}	0.7376	3.725×10^{-7}
cal	4.184	1	1.162×10^{-6}	3.97×10^{-3}	3.086	1.558×10^{-6}
kWh	3.6×10^{-6}	8.606×10^{5}	1	3412.14	2.655×10^{6}	1.341
Btu	1055	252	2.930×10^{-4}	1	778.16	3.930×10^{-4}
ft lb_f	1.356	0.3241	3.766×10^{-7}	1.285×10^{-3}	1	5.051×10^{-7}
hp h	2.685×10^{6}	6.416×10^{5}	0.7455	2545	1.98×10^{6}	1

功率

	J s^{-1}	kW	ft lb_f s^{-1}	Btu s^{-1}	hp
J s^{-1}	1	10^{-3}	0.7376	9.478×10^{-4}	1.341×10^{-3}
kW	1000	1	737.56	0.9478	1.341
ft lb_f s^{-1}	1.356	1.356×10^{-3}	1	1.285×10^{-3}	1.818×10^{-3}
Btu s^{-1}	1055	1.055	778.16	1	1.415
hp	745.7	0.7457	550	0.7068	1

压力

	mm Hg	in.Hg	kPa	atm	bar	psia
mm Hg	1	0.03937	0.1333	1.316×10^{-3}	1.333×10^{-3}	0.01934
in.Hg	25.4	1	3.386	0.03342	0.03386	0.4912
kPa	7.502	0.2954	1	9.869×10^{-3}	0.01	0.1451
atm	760	29.92	101.3	1	1.013	14.696
bar	750.06	29.53	100	0.9869	1	14.5
psia	51.71	2.036	6.894	0.06805	0.06895	1

目　录

扫描二维码，查看电子版附录和电子版索引

PART I

INTRODUCTION

CHAPTER 1

Introduction to Chemical Engineering

Chapter Objectives

- Have a general knowledge of the evolution of chemical engineering.
- Understand what chemical engineering is and the types of jobs chemical engineers perform.
- Appreciate some of the issues associated with sustainability and green engineering.
- Understand the importance of ethics in the practice of engineering.

Introduction

Chemical engineers deal with processes that convert raw materials into useful products. Many times, these processes involve reactions followed by purification of the products, such as chemical reactions followed by concentration of the products, biological reactions followed by systems that recover and purify the products, or reactions and recovery of products on a nanometer scale. Overall, chemical engineers are process engineers—that is, chemical engineers deal with processes that produce a wide range of products.

1.1   A Brief History of Chemical Engineering

Chemical engineering evolved from the industrial applications of chemistry and separation science (i.e., the study of separating components from mixtures) primarily in the refining and chemical industry, which we refer to here as the **chemical process industries, CPI**. The first high-volume chemical process was implemented in 1823 in England for the production of soda ash, which was used to produce glass and soap.

In 1887, a British engineer, George E. Davis, presented a series of lectures on chemical engineering that summarized industrial practice in the chemical industry in Great Britain. These lectures stimulated interest in the United States and to some degree led to the formation of the first chemical engineering curriculum at MIT in 1888. Over the next 10 to 15 years, a number of US universities embraced the field of chemical engineering by offering curriculum in this area. In 1908, the American Institute of Chemical Engineers was formed and since has served to promote and represent the interests of the chemical engineering community.

Mechanical engineers understood the mechanical aspects of process operations, including fluid flow and heat transfer, but they did not have backgrounds in chemistry. Conversely, chemists understood chemistry and its ramifications but lacked the process skills. In addition, neither mechanical engineers nor chemists had backgrounds in separation science, which is critically important to the CPI. As a result, the study of chemical engineering evolved to meet these industrial needs.

The acceptance of the "horseless carriage," which began commercial production in the 1890s, created a demand for gasoline that ultimately fueled exploration for oil. In 1901, Patillo Higgins, a Texas geologist, and Anthony F. Lucas, a mining engineer, later to be known as "wildcatters," led a drilling operation that brought in the Spindletop Well just south of Beaumont, Texas. At the time, Spindletop produced more oil than all the other oil wells in the United States. Moreover, a whole generation of wildcatters was born, resulting in a dramatic increase in the domestic production of crude oil, which created a need for larger-scale, more modern approaches to crude refining. As a result, a market developed for engineers who could assist in the design and operation of processing plants for the CPI. The success of oil exploration was to some degree driven by the demand for gasoline for the automobile industry, and ultimately, it led to the widespread adoption of automobiles for the general population due to the resulting lower cost of gasoline.

These early industrial chemists/chemical engineers had few analytical tools available to them and largely depended on their physical intuition to perform their jobs as process engineers. Slide rules were used for performing calculations, and by the 1930s and 1940s, a number of nomographs were developed to assist them in the design and operation analysis of processes for the CPI. Nomographs are charts that provide a concise and convenient means to represent physical property data (e.g., boiling point temperatures or heat of vaporization) and can also be used to provide simplified solutions of complex equations (e.g., pressure drop for flow in a pipe). The availability of computing resources in the 1960s was the beginning of computer-based technology that is commonplace today. For example, since the 1970s, **computer-aided design (CAD) packages** have allowed engineers to design complete processes by specifying only a minimum of information; all the tedious and repetitive calculations are done by the computer in an extremely short period of time, allowing the design engineer to focus on the task of developing the best possible process design.

In 1959, Professors Bird, Stewart, and Lightfoot of the Department of Chemical Engineering at the University of Wisconsin published their textbook *Transport Phenomena* that covered fluid flow, heat transfer, and mass transfer. This book was widely adopted throughout the chemical engineering community and provided a much more mathematical and abstract analysis of these topics than had previously been used. The widespread use of this book ushered in a much more analytical approach to chemical engineering than the more empirical approach that preceded it.

During the period 1960–80, the CPI also made the transition from an industry based on innovation in which the profitability of a company depended to a large degree on developing new products and new processing approaches to a more mature commodity industry in which the financial success of a company depended on making their products using established technology more efficiently, resulting in less expensive products.

Globalization of the CPI markets began in the mid-1980s and led to increased competition. At the same time, development in computer hardware made it possible to apply process automation more easily and reliably than ever before. These automation projects provided improved product quality while increasing production rates and overall production efficiency with relatively little capital investment.

Beginning in the mid-1990s, new areas came on the scene that took advantage of the fundamental skills of chemical engineers, including the microelectronic industry, the pharmaceutical industry, the biomedical industry, and nanotechnology. Clearly, the analytical skills and the process training made chemical engineers ideal contributors to the development of the production operations for these industries. In the 1970s, more than 80% of graduating chemical engineers took jobs with the CPI and government. By 2000, that number had dropped to 50% due to increases in the number taking jobs with biotechnology companies, pharmaceutical/health care companies, and electronics and materials companies.

1.2   Types of Jobs Chemical Engineers Perform

Chemical engineers perform a wide range of jobs. Moreover, during your career, you are likely to have a number of different types of jobs. Following are the general types of jobs that chemical engineers perform:

- **Operations:** Operations engineers, or process engineers, are the first line of technical support for a processing plant. These engineers spend a lot of their time in the plant monitoring the operations and solving operational problems. When a serious technical problem occurs in the middle of the night, the operations engineer for that process is called in to resolve the problem. Many young chemical engineers start out as operations engineers for a few years so that they can become familiar with plant operations before they move to other assignments. This job also provides companies a view of how young engineers handle responsibility as well as how effectively they are able to work with others.

- **Technical sales:** Many products today are highly technical in nature and the consumer of these products often requires technical assistance to fully utilize them. Technical sales engineers provide that service as well as acquire new customers. Obviously, sales engineers need to be able to work effectively with their customers and to fully understand the technical issues associated with their company's products in order to maintain customer satisfaction.

- **Design:** Design is developing something new that meets a defined need and is used to develop new products and services, many times using teams of engineers. Design is a challenging endeavor because there is no limit to how many new ways something can be designed. Therefore, design requires creativity and experience. As a result, design teams often are made up of members with a wide range of experience and training. It is the design team's job to determine the best design for a product considering technical feasibility, economic viability, and the definition of the need for the end user.

- **Consulting:** Consulting companies specialize in specific areas of engineering—safety, design, control, and so on. When an operating company needs a consultant's expertise, it simply contracts with the consulting company for the needed services. Because consulting companies provide technical services on an as-needed basis, the company that hires a consultant does not have to employ an expert in a particular field as a full-time employee. Consulting companies often hire engineers who have many years of engineering experience in specific technical areas. Individuals also serve as industrial consultants after years of experience in industry, academia, or government laboratories.

- **Project management:** Project management engineers are similar to operations engineers in that they are called upon to provide a number of technical services for the day-to-day operation of a project (e.g., an expansion project for a process or the construction of a new process). Initially, these engineers are required to develop estimates of labor and material for the project, and this information is then used to receive approval for the project. The project management engineer is responsible for coordinating the project or a portion of the project when the project is approved. Coordinating the project requires working with a number of parties, e.g., the management team, the construction team, the suppliers, and the operations department in order to deliver a high-quality project on time and on budget.

- **Management: Corporate, operations, and technical:** Many companies use chemical engineers for their corporate management because the position requires technical knowledge. Engineers who move into corporate management usually have training in business or have attended an MBA program. They normally work their way up the management ladder from technical management and operations management positions. Corporate management directs the business at the corporate level and deals with issues such as the corporate image, identifying new business opportunities, and deciding how to handle economic downturns, all in an effort to improve the overall profitability of the corporation. Operations management deals with the day-to-day problems and opportunities associated with operating an industrial production facility. Technical management is concerned with managing engineers who deal with operations, research, and development.

- **Development:** Development teams work with design teams to apply various designs so that they can be further tested. During this phase, the real-world consequences of potential designs become apparent, and the development team is charged with solving these problems when possible. For a new process, a pilot-scale process can be constructed, operated, and monitored to evaluate the performance of the new process (e.g., to determine the activity and yield of a catalyst). In effect, development teams are asked to demonstrate whether a design concept is viable.

- **Research:** Research is the scientific investigation of physical systems using laboratory experiments and/or computer simulations. Fundamental, or "blue-sky," research studies the fundamental behavior of certain systems without regard to a specific industrial problem (e.g., studying the fundamental chemical reactions associated with a class of compounds). Industrial research is research aimed at solving an industrial problem (e.g., developing a new composite material that can be used in an industrial application). Whenever a technical issue has an important effect on society (e.g., developing green sources of energy), large amounts of government funding are usually offered to research-ers, who explore and propose ways to solve these problems.

- **University teaching:** Engineering professors typically have a PhD in engineering or a related field and divide their work effort between research, teaching, and service to the profession. Their research effort is based on fundamental studies of engineering systems, while their teaching relies on being able to effectively communicate abstract material and practical approaches to students in a way that the students can assimilate and apply this information. Engineering professors are evaluated for promotion and advancement on the basis of publications in peer-reviewed journals, their ability to develop research funding, their effectiveness as a teacher, and their contribution to the engineering profession. Being an engineering professor is a demanding profession because of the breadth of work the individual must perform, but it can be a very rewarding career to help young people along the path to becoming successful engineers.

1.3   Industries in Which Chemical Engineers Work

A chemical engineering education exposes the graduate to the fundamentals of process engineering and develops the graduate's ability to deal with complex problems. As such, chemical engineers are ideally suited to work in a wide variety of industries. Figure 1.1 shows the primary areas in which chemical engineers work.

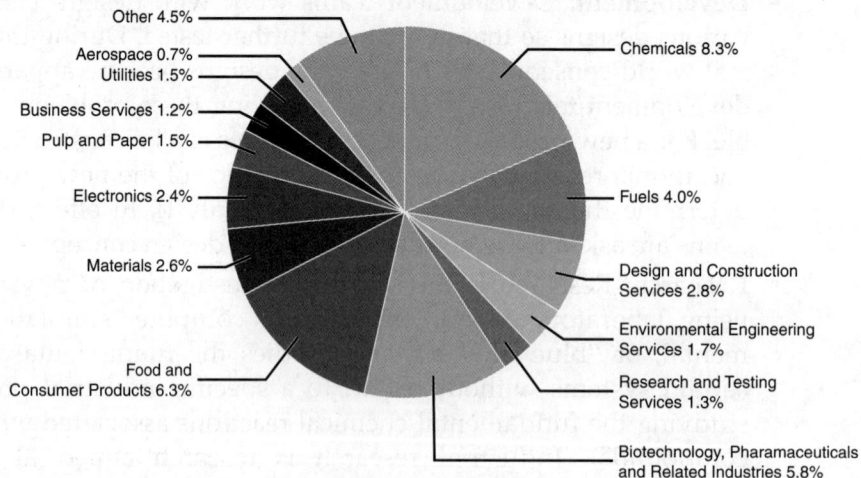

Figure 1.1   The primary areas in which chemical engineers work (Source: AIChE)

Following are brief descriptions of these primary CPI:

- **Refining:** Refining involves processing crude oil to produce fuels and lubrication oils for automobiles, trucks, and airplanes. In addition, refineries produce a wide range of chemical intermediate products, that is, products that serve as feed stocks for other processes (e.g., propylene for a process that produces polypropylene plastics). Refineries are typically large-scale processes having feed rates up to 600,000 barrels of crude oil per day.

- **Chemicals:** The chemical industry can be categorized as producing commodity, specialty, or fine chemical products. Commodity chemicals are high-volume products that are used as feed stocks for other chemical processes. For example, a petrochemical plant, which produces large volumes of chemical intermediates, is usually a part of most large-scale refineries. Specialty chemicals are relatively low-volume products that are often produced in batch operations. Examples of specialty chemicals include certain agricultural chemicals, paint pigments, and special-purpose solvents. Fine chemicals are commodity chemicals that are produced on a relatively small scale and are used as feed stocks for specialty chemicals.

- **Environmental:** Environmental engineers work to ensure that human health and nature's ecosystems are protected from emissions resulting from industry and human activity. Common efforts of environmental engineers include wastewater management, air and water quality control, waste disposal, processing or recycling waste streams, and documentation of these efforts (e.g., required reports to the Environmental Protection Agency [EPA] or environmental impact statements for proposed projects).

- **Equipment design and construction:** This category relates to the design and construction of new processes or expansion projects for existing processes. The design process involves determining the type and sequence of equipment and the sizing of this equipment, and this effort is typically performed by consulting companies that specialize in process design. Once the design is completed, chemical engineers are typically involved during the construction and startup of the process.

- **Pharmaceuticals and health care:** The pharmaceutical industry develops, produces, and markets synthetic drugs that can be administered to patients to treat or alleviate symptoms or protect against disease (e.g., vaccines). These products are usually produced using biological reactions.

- **Biotechnology:** Biotechnology uses living systems by harnessing cellular processes and biomolecular processes to develop products ranging from medicines to fuels to food. Although biotechnology is currently a developing research area, it has been used by humans for thousands of years to make or preserve foods (e.g., beer and wine, bread, and sauerkraut), to improve livestock and plants by selective breeding, and to improve agricultural soils by introducing bacteria that are able to fertilize crops and protect against attacks from insects.
- **Biomedical applications:** Biomedical engineers are involved with the application of engineering principles combined with knowledge of biology and human physiology for health care purposes. Application areas include diagnosis, monitoring, and therapy as well as the development of new technologies, such as artificial tissue and artificial organs.
- **Food production:** Chemical engineers work in a variety of ways for the "farm to fork" food industry, including agrochemicals and food processing, such as making potato chips, granola bars, candy, beer, or yogurt.
- **Government:** Chemical engineers work for certain government regulatory agencies [e.g., EPA or the Occupational, Safety and Health Administration (OSHA)] or government laboratories (e.g., Sandia National Laboratory and the National Institutes of Health).
- **Professional:** This category includes professions (e.g., patent lawyers and medical doctors) that require certification as well as specialized education (e.g., university professors).

1.4   Sustainability

A **sustainable product** is a product that meets the current needs without compromising the ability of future generations to meet their needs while protecting human health and the needs of society. Chemical engineers are ideally suited to evaluate the sustainability of a wide range of products and technologies, and therefore, we present an overview of sustainability and **green engineering** as an example of one of the ways that chemical engineering can contribute to society today and into the future.

Current estimates indicate that the societies of the world are consuming 50% more resources (e.g., energy, water, minerals, the ability to produce food) than the world can sustain (what the population of the world consumes in one year requires 18 months to replenish). The United States consumes natural resources at a rate that is more than 13 times the rate for the rest of the world. With the rapid growth in the economies of China and India, who make up almost 40% of the world's population, the world resource consumption rate is expected to accelerate.

In addition to preserving resources for the future, sustainable engineering involves protecting human health and the needs of society. That is, the effects of pollution and the impact on global warming are factors that also should be considered in any sustainable design project. As a result, a number of engineering professional groups are concerned that sustainability should be an integral part of future designs. That is, sustainability, or at least improved sustainability, should be an objective for engineering design work as opposed to basing design solely on minimum cost or maximum profit without regard to the impact on sustainability. The problem is that a sustainable design will, in general, cost more to implement than its nonsustainable counterpart. Therefore, the challenge for you as a future engineer is to develop new approaches that make sustainability as economically viable as possible.

As an example, consider a sustainable design for a building. The following features are aspects related to a sustainable design of a building:

- Nontoxic construction materials that can be produced from recycled materials using low-energy processing techniques
- Energy efficient design (e.g., low heat transfer rates to or from the building) using materials that require low amounts of energy to produce
- Renewable energy sources (solar panels, solar water heaters, etc.)
- High durability for the building, yielding a long service life; materials that develop character as they age
- Interior and exterior appearance as similar as possible to nature (i.e., producing a soothing environment for humans)
- Designing for a low total carbon footprint (i.e., the total carbon dioxide liberated during the production of the materials used in the building and the process of constructing the building)

- Using biomimicry (i.e., redesigning industrial processes along biological lines to produce building materials)
- Transferring ownership from an individual to a group of people, similar to car sharing
- Employing renewable materials that come from nearby sources

As you can see from this list, the design problem becomes more complicated when a more holistic approach to engineering is used, but on the other hand, this approach creates more opportunities for creative solutions.

1.4.1 Life-Cycle Analysis

A **life-cycle analysis** is a comprehensive method for developing a sustainable design (**green engineering**). A life-cycle analysis not only considers the effect of a product on the environment and on important resources but also considers all the steps used to produce a product and what happens to the product after its useful life has ended. Figure 1.2 shows a schematic example of a life-cycle analysis of a product.

As illustrated in Figure 1.2, raw materials are extracted from the earth, such as minerals and crude oil. These raw materials are refined into useable products, such as metals and chemical products, in a material processing operation. Next, these useable materials are used to manufacture parts of the final product and are assembled into the final product. Then, the product is used for its intended purpose for the life of the product. When the useful life of the product ends, the product must be disposed of and/or recycled. The recycling process recovers all or part of the product and returns the recovered material so that it can be used for other products in the future. Each of these steps, from raw material extraction to recycling materials, in general, requires the use of resources (e.g., energy) and has an environmental impact (e.g., generates pollutants), which is indicated by the two oppositely pointing arrows used in Figure 1.2. That is, each step in the life-cycle analysis generally requires resource consumption and results in pollution generation.

When a life-cycle analysis is used for a sustainable design, all the required resources and all the resulting loads on the environment are considered. Moreover, the effect of the design of the product on the ability of the product to be recycled at the end of life should be considered.

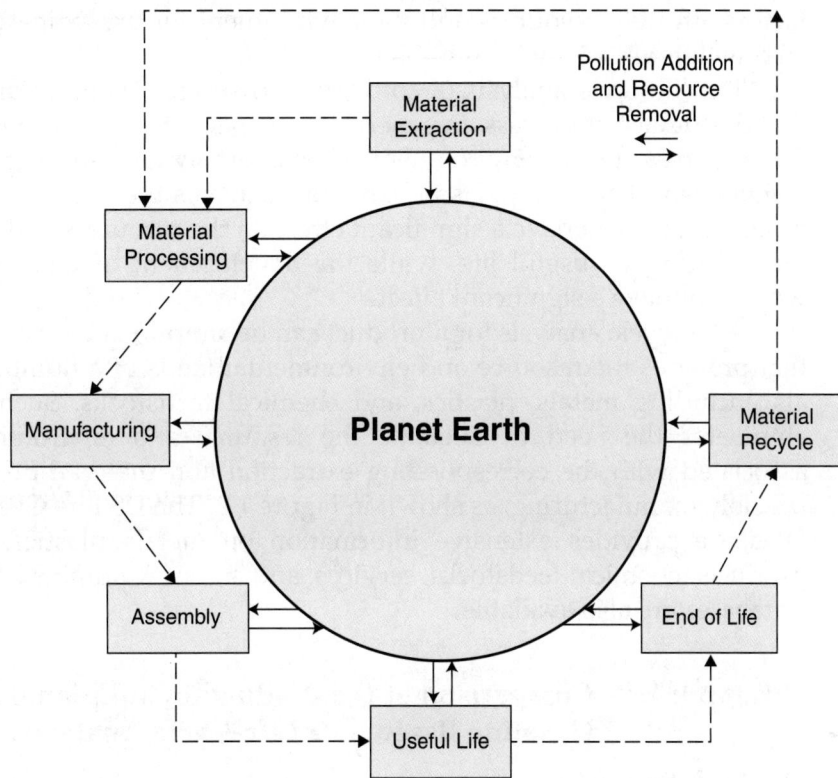

Figure 1.2   Schematic representation of a life-cycle analysis

Now consider how a life-cycle analysis would be applied to the four types of engineering designs: a device, a process, software, and services. The elements of a life-cycle analysis for a typical device are shown in Figure 1.2: The materials that compose the device must be extracted, refined, and manufactured into parts for the device. Then the parts are assembled into the device. And after the useful life is complete, material recycle can be used.

For a process, the elements of a life-cycle analysis closely follow the schematic in Figure 1.2. Moreover, the hardware elements used to implement a process (e.g., vessels, pumps, and processing equipment) are devices so that, with regard to the hardware of the process, the components of the life-cycle analysis are exactly the same as those used for a device. From an overall point of view, the pollution generated and the resources consumed during the useful life of a process would be expected to be the primary factors affecting resources and the environment far exceeding those associated with the hardware.

The life-cycle analysis of software and services is quite different from that of a device or process. In general, the impact of software and services on the resources and the environment is considerably less, although not always insignificant. For example, software that manages the operation of an automobile engine can have a significant effect on the resources and the environment during its useful life, while the development of the software itself would not have a significant effect.

A life-cycle analysis for a product can be simplified by using a database that provides the resource and environmental loads of a number of materials, including metals, plastics, and chemical feedstocks. Such a database eliminates the need to calculate the resource and environmental loads associated with the corresponding extraction and material processing and possibly manufacturing, as shown in Figure 1.2. The US Life Cycle Inventory Database provides extensive information on metals, plastics, agricultural products, chemical feedstocks, services, and so on. A number of commercial databases are also available.

Example 1.1   Comparison of the Production of Ethanol and Gasoline Based on a Life-Cycle Analysis

Problem Statement

Using a life-cycle analysis, compare ethanol (EtOH) from corn to gasoline as a transportation fuel with regard to greenhouse gas (GHG) generation.

Solution

Figure E1.1 shows schematics of the life-cycle analysis for the production of EtOH from corn and for the production of gasoline as a motor fuel. Note that for the production of EtOH, corn is produced by farming, which requires the use of fertilizers. The corn is used to produce EtOH using a fermentation and recovery process. The primary energy consumption and generation of GHG emissions occurs in the production of the fertilizer, growing the corn, and converting the corn to EtOH. In contrast, gasoline is produced by extracting crude oil from underground formations and refining the crude oil into gasoline and other products such as jet fuel, diesel, and lube oil. For gasoline, the primary energy consumption and generation of GHG emissions occurs during the crude oil production and refining, where refining is the largest contributor.

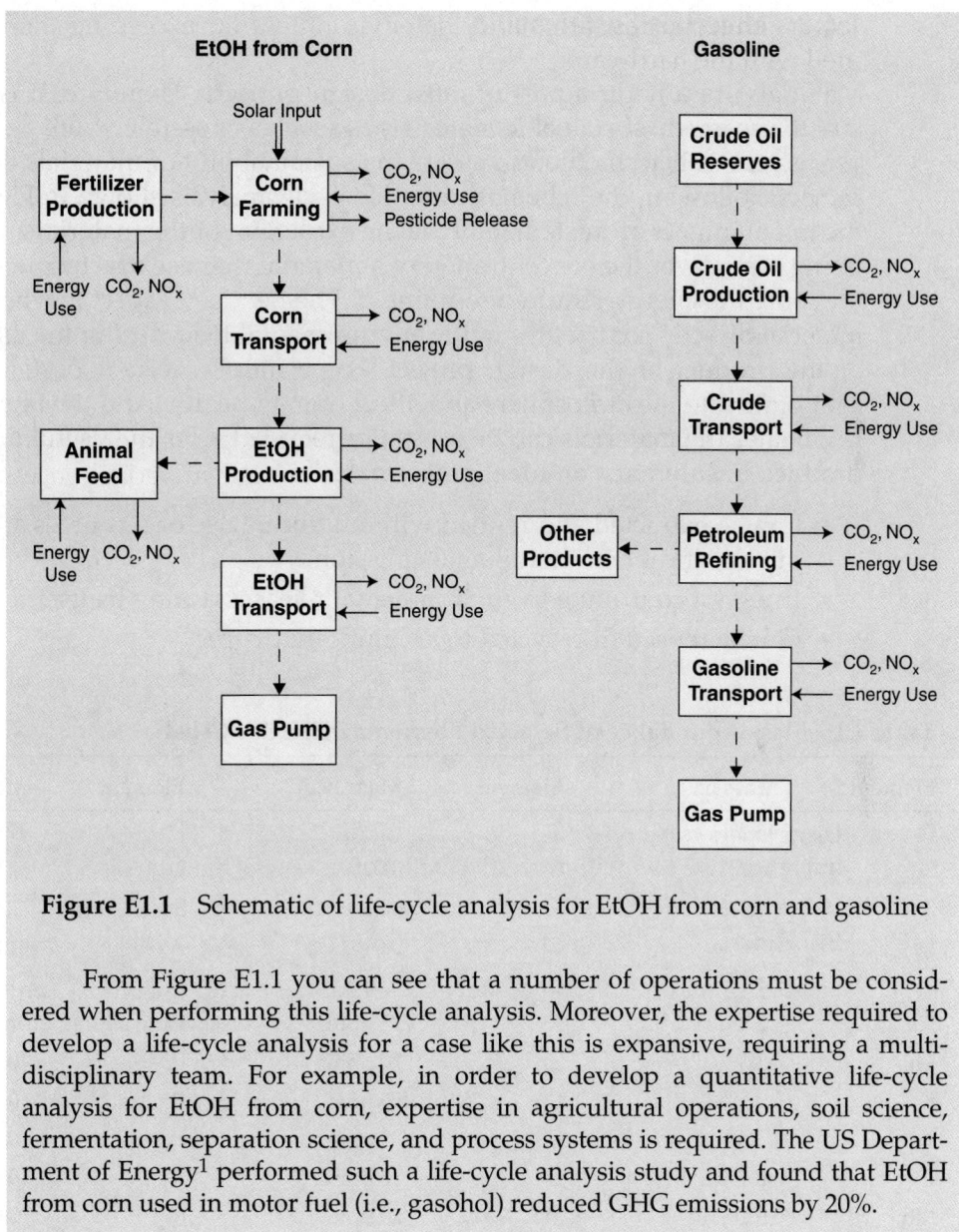

Figure E1.1　Schematic of life-cycle analysis for EtOH from corn and gasoline

From Figure E1.1 you can see that a number of operations must be considered when performing this life-cycle analysis. Moreover, the expertise required to develop a life-cycle analysis for a case like this is expansive, requiring a multidisciplinary team. For example, in order to develop a quantitative life-cycle analysis for EtOH from corn, expertise in agricultural operations, soil science, fermentation, separation science, and process systems is required. The US Department of Energy[1] performed such a life-cycle analysis study and found that EtOH from corn used in motor fuel (i.e., gasohol) reduced GHG emissions by 20%.

---

[1] J. Han, "Life-Cycle Analysis of Ethanol: Issues, Results and Case Simulations," Annual ACE Conference, Omaha, NE, August 15, 2015.

1.4.2   Materials Sustainability

Materials are a natural part of most design projects. Therefore, it is important to use green, sustainable materials as much as possible. A life-cycle analysis is an excellent method to assess the sustainability of materials used in a project. Following the schematic of a life-cycle analysis shown in Figure 1.2, the initial impact on sustainability is the extraction of the material and its refining. The lower the concentration of a material that is extracted, in general, the more the energy required to refine it. The next key aspect is whether the material directly contributes to the environmental load during the useful life of the product of the design project. For example, certain pesticides can evaporate into the atmosphere and affect human health. And the final aspect is whether the materials can be reused or recycled after the useful life of the product. In summary, **an ideal green material**

- Can be extracted and refined without undue use of resources and without significant environmental emissions.
- Does not contribute to environmental releases during its useful life.
- Can be reused or recycled to a significant degree.

Table 1.1   Mass Abundance of Selected Elements in Earth's Crust[2]

Element	Mass Fr.	Element	Mass Fr.	Element	Mass Fr.
O	46.4%	S	0.03%	Pb	12 ppm
Si	28.2%	C	0.02%	U	2.7 ppm
Al	8.2%	V	0.01%	Sn	2.0 ppm
Fe	5.6%	Cl	0.01%	As	1.8 ppm
Ca	4.1%	Cr	0.01%	Mo	1.5 ppm
Na	2.4%	Ni	75 ppm	W	1.5 ppm
Mg	2.3%	Zn	70 ppm	Bi	0.17 ppm
K	2.1%	Cu	55 ppm	Pd	0.15 ppm
Ti	0.6%	Co	25 ppm	Hg	0.08 ppm
P	0.1%	Li	20 ppm	Ag	0.07 ppm
Mn	0.1%	N	20 ppm	Pt	0.005 ppm
Fl	0.06%	Ga	15 ppm	Au	0.004 ppm

[2] S. R. Taylor, "Trace Element Abundances and the Chondritic Earth Model," *Geochimica et Cosmochimica Acta* 28, no. 12 (1964): 1989–98.

Table 1.1 lists the abundance of selected elements in the earth's crust. Even though gold makes up only 0.004 ppm of the total earth's crust, highly concentrated deposits of gold have been found, thus simplifying its recovery. In contrast, rare earth metals, some of which make up more of the earth's crust than gold, are found only in ores at relatively low concentrations. In terms of sustainability, abundance is only one factor. That is, the abundance and the net consumption from use together determine whether an adequate supply of a material is available. Listed in Table 1.2 are minerals that have been identified as having a limited supply based on their use in 1995.

Table 1.2   Elements with a Limited Supply[3]

Degree of Supply	Elements
Potentially highly limited	Ag, Au, Cu, As, Se, Te, Zn, Cd
Potentially limited	Co, Cr, Mo, Ni, Pb, Pt, Ir

Besides elements, the sustainability of materials that result from plant growth, such as crude oil and lumber, should be considered. For example, the production of lumber results in a removal of GHGs from the atmosphere even though the milling and transportation will generate some GHGs. Of course, the consumption of crude oil, in general, results in significant GHG emissions.

With regard to the availability and supply of a material, as the supply of a material decreases, the market price for that material tends to increase. For example, during the mid-1970s, a shortage of crude oil dramatically increased the price of crude oil and, as a result, the price of gasoline. This price increase for crude stimulated exploration for crude oil as well as conservation efforts. Therefore, by the early 1980s, there was an excess of crude on the market and the price of crude oil dropped dramatically.

Another important natural resource is phosphorus. Before the advent of modern farming practices, farmers used wastes (e.g., compost) to return phosphorus to their soil after their crops consumed it during their growth cycle. Today, most farmers use inorganic phosphate to fertilize their crops. Estimates predict that currently known reserves of phosphate (i.e., a source of phosphorus) will be exhausted in 80 years at the current consumption rate. What this means is that the current phosphate reserves that are easy to extract and refine will be exhausted. Even if new high-quality phosphate

---

[3] T. E. Graedel and R. R. Allenby, *Industrial Ecology*, Prentice Hall, 1995.

reserves are not identified, an expected increased price of phosphate should drive the processing of lower-quality phosphate reserves and the extraction of phosphorus from waste material. In addition, an increase in the cost of phosphate should also encourage farmers to use their fertilizer more efficiently.

With regard to the elements in Table 1.1 that have been identified as having a potentially limited supply, if the increased consumption of one of these elements begins to reduce its supply, the price of that element would be expected to increase. This increase in price would stimulate increased exploration for it and could possibly make ores that were previously uneconomical to refine financially viable for refining. In addition, the increased price of the material would be expected to increase the recovery from end-of-life products and recycling. During the design phase of a project, the use of a material with a potentially limited supply should be viewed as increasing the overall risk of the project. That is, a material with a potentially limited supply can be susceptible to significant price increases in the future, which could affect the economic viability of a project.

1.4.3   Environmental Releases and Toxicity

The release of chemicals during extraction, refining, use, or end of life can result in a significant environmental load affecting human health or the health of ecosystems.

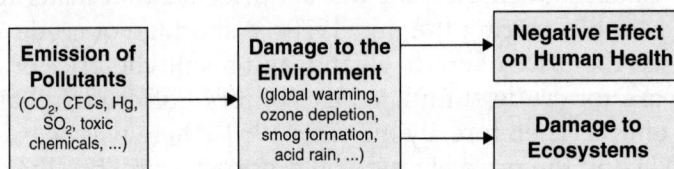

```
┌─────────────────┐     ┌─────────────────┐     ┌─────────────────┐
│  Emission of    │     │ Damage to the   │ ──► │ Negative Effect │
│  Pollutants     │ ──► │ Environment     │     │ on Human Health │
│ (CO₂, CFCs, Hg, │     │ (global warming,│     └─────────────────┘
│  SO₂, toxic     │     │ ozone depletion,│     ┌─────────────────┐
│  chemicals, ...)│     │ smog formation, │ ──► │ Damage to       │
└─────────────────┘     │ acid rain, ...) │     │ Ecosystems      │
                        └─────────────────┘     └─────────────────┘
```

Figure 1.3   Overall effect of the emission of pollutants

Figure 1.3 illustrates the connection between certain emissions (i.e., pollution) and human health and the health of ecosystems. Pollution results in global warming, ozone depletion, smog formation, acid rain, and so on. These in turn affect human health, cause damage to ecosystems, and disrupt human activities (e.g., increased damage from natural disasters). For example, consider the emission of chlorofluorocarbons (CFCs). CFCs damage ozone in the stratosphere, resulting in an increase in UVB radiation on the surface of the earth, which in turn increases skin cancer and the occurrence of cataracts and causes human immune system suppression, crop damage, and damage to marine life.[4]

Roughly 2000 new chemicals are introduced each year, and companies, government agencies, and the public need a method to evaluate the potential risks of these new chemicals. Evaluation is done by using screening tests, and in certain cases, extensive testing is required. That is, screening is used to identify which chemicals have the potential to be environmentally risky, and then extensive testing is used to evaluate those chemical to determine if they, in fact, represent a significant risk to human health and the health of ecosystems.

Table 1.3 outlines an approach used to screen chemicals for environmental risks. The first entry (dispersion and fate) has to do with the tendency of the chemical to accumulate in water, air, soil, and living organisms. The second entry (degradation rates) relates to how quickly the chemical is degraded in water, air, soil, and living organisms. The category "uptake by organisms" is related to specific factors that affect uptake and degradation in organisms while "uptake by humans" considers the rates at which the chemical is able to enter the human body and the rates at which the human body is able to expel it or degrade it into a nontoxic form. "Toxicity and other health effects" has to do with how the level of exposure to the chemical affects the health of organisms and humans.

Table 1.3   Chemical Properties Needed to Perform Environmental Risk Screenings[5]

Aspects Affecting the Environment	Relevant Properties
Dispersion and fate	Volatility, density, melting point, water solubility, soil sorption coefficient, effectiveness of wastewater treatment
Degradation rates in the environment	Atmospheric oxidation rate, aqueous hydrolysis rate, rate of degradation by sunlight, rate of microbial degradation, adsorption
Uptake by organisms	Volatility, the ability to dissolve in fat, molecular size, degradation rate in an organism
Uptake by humans	Transport through the skin, transport rates across lung membrane, degradation rates within the human body
Toxicity and other health effects	Dose–response relationships

---

[4] D. T. Allen and D. R. Shonnard, *Sustainable Engineering: Concepts, Design, and Case Studies,* Prentice Hall, 2012.

[5] After Allen and Shonnard, *Sustainable Engineering.*

If the screening process determines that a chemical could pose a significant risk to human health or to ecosystems, a detailed assessment of the impact of the chemical on the environment would be required. This assessment would involve testing with a range of laboratory animals with a range of exposure scenarios (e.g., exposure to a single large dose, exposure to multiple smaller doses, and continuous exposure to a low-dose level). The results of these studies may include the degree of reduction in life expectancy as well as the characteristics of the offspring of the laboratory animal.

1.4.4   Principles of Green Engineering

To effectively address the full range of design cases, it is necessary to have a general set of guidelines that, when applied, ensure a sustainable design. A number of guidelines for sustainable design have been developed.[6]

Following is an overview of the key factors identified here as the principles of green engineering, which are based on previous work in this area:

1. **Use energy and material inputs that are as inherently nonhazardous as possible.** Because material and energy input have such an important effect on the sustainability of a product, it is important to ensure that they are as nonhazardous as possible. When hazardous materials are used, special controls and planning for adverse conditions are required, which increases the cost and complexity of a design.

2. **Minimize wastes.** The generation of wastes creates special problems due to the difficulty and cost associated with dealing with wastes, especially hazardous wastes. Therefore, during the design phase, it is important to minimize waste generation. In certain cases, it is possible to find a use for a "waste product," such as using it as a feedstock for another process. For example, during the early days of crude oil refining, natural gas was considered a waste product and was burned into the atmosphere until it was determined that natural gas could be used for heating homes and businesses. In other cases, it may be possible to modify the chemistry of a reaction so that waste products are eliminated or at least significantly reduced.

3. **Minimize energy consumption.** In certain industries (e.g., the refining, petrochemical, and mineral purification industries), the energy usage for purification is a primary operating expense. In these cases, energy usage can be significantly reduced using heat integration, that is,

---

[6] Allen and Shonnard.

using waste heat (i.e., cooling that is required in the process) from one part of the process to provide heat to another part of the process.

4. **Minimize material usage.** Material usage can be minimized by designing the system or process so that the greatest possible conversion of the feed material to the product is obtained. Material use can also be minimized by extracting the feed components from waste streams so that they can be recycled to the process or using components from waste streams as a feedstock for other processes.

5. **Apply just-in-time manufacturing.** Just-in-time manufacturing is manufacturing that meets the demand for a product precisely when the product is needed, thus eliminating wastes and reducing the need for inventory. In terms of design, this means designing a system or process precisely for the expected demand and completing the project so that the product of the design is available only when the demand is present.

6. **Design for proper durability.** Proper durability means that a product is designed to last for the designed useful life and afterwards can be easily transformed into materials that can be recycled or reused or that can degrade into environmentally benign products.

7. **Design for recycling or reuse after end of life.** Materials can be recycled after end-of-life use by designing the product so that material recycling is simple and easy to implement. For example, scarce materials can be used in a way that facilitates their recovery for recycle. Components of a product can be designed so that they can be used in future generation devices (e.g., parts of a cell phone). Also, products can be designed for reuse (e.g., soft drink bottles).

8. **Use a life-cycle analysis to minimize the environmental impact of the project.** A life-cycle analysis is the most complete way to evaluate the impact of a project on the environment and natural resources. In this manner, the key areas of environmental load and resource depletion can be identified and addressed. For example, if the areas of material extraction and refining are the primary contributors to environmental load and resource depletion for a project, recycling would clearly be the best approach to improve the sustainability of the project.

9. **Use renewable sources of energy and materials.** The use of renewable sources of energy and materials reduces environmental impact and resource depletion. Solar panels are an example of a renewable energy source, and lumber is an example of a renewable material.

10. **Engage both communities and stakeholders in the project.** It is critically important to involve local communities from the conceptual stages through the completion of a design project when the project has any real or perceived impact on the local community. In addition, sustainable goals can sometimes be met by affecting a change in social behavior (e.g., the development of autonomous vehicles so that the use of vehicles is shared).

The principles of green engineering are really a checklist of factors that should be considered during the design process; otherwise, the design can be less sustainable than it could have been.

1.4.5   Optimization and Sustainable Engineering

Each of the principles of green engineering presented here should be applied in a balanced fashion. For example, strictly minimizing the energy consumption may result in an excess use of materials. That is, all the relevant factors, including the impact on the environmental load and ecosystems as well as the material and energy cost, should be considered when optimizing a sustainable design project (i.e., finding the optimal sustainable design).

The conventional approach to optimization of a design project neglecting the impact on the environment and society is to consider the costs associated with the end product of the design along with the expected income generation to identify the optimum design over the life of the product considering the time value of money.

When the impact on the environment and society are considered during the design process, the problem arises that the environmental and the societal impact are not easily represented on a monetary basis. Nevertheless, several different approaches are available for including the impact on the environment and society in the design process.

The most direct means of including the impact on resources and the environment is to follow government regulations. For this case, the government regulations would represent constraints on the design process that have to be satisfied for any valid design. For example, maximum SO_2 emissions are set by EPA regulations for coal-fired electric utilities. While there are certain cases where this approach is valid (e.g., the maximum safe chemical concentrations), it is not feasible to develop government regulations for the full range of factors that affect sustainability. For example, GHG emissions do not lend themselves to explicit limits.

Another approach is to rate products on the basis of their total impact on resources and the environment. For example, new home construction can be rated according to the total resource depletion and total pollution generation. Ratings such as a silver, gold, or platinum can be assigned to a new house based on sustainability and, of course, a platinum rating will command a higher price than a gold rating, which will command a higher price than a silver rating. The success of this approach depends on the judicious selection of the criterion to quality for each classification, taking into account the costs necessary to qualify for each classification and the resulting benefit to resources and the environment. However, not all design projects fit into this approach.

Another approach is to estimate the total cost to society of specific emissions. Total cost to society includes increases in medical costs, lost productivity, and a reduction in life expectancy. For example, the EPA has estimated the total cost to society for CO_2, CH_4, and NO_2 emissions. In this manner, the economic cost of GHG emissions can be considered directly during the optimization of a design project using a life-cycle analysis combined with values for the cost to society for the pollutants under consideration. This approach has generated considerable controversy and was removed from the EPA website in January 2017.

Self-Assessment Test

Questions

1. What is the difference between a conventional design and a sustainable design?
2. What is a life-cycle analysis, and how can it be used to develop sustainable designs?
3. How does sustainability affect the optimization of a design?

Answers

1. A conventional design is based on maximizing the profits of the design process without regard to the sustainability of the project. A sustainable design project also maximizes profits from the project, but it does so while producing a design that does not compromise the ability of future generations to meet their needs.
2. A life-cycle analysis is a thorough analysis of a process or product with regard to the resources consumed and the pollution generated considering everything from the materials used to the produce the product to the manufacturing process to the end-of-life of the product. It is used in a sustainable design in order to consider all the sources of resource consumption and pollution generated for a particular design.

3. Sustainable designs require a more comprehensive optimization analysis and generally result in more expensive designs compared to conventional designs because conventional designs neglect many factors considered by a sustainable design.

1.5   Ethics

Engineering ethics is a collection of moral principles applied in engineering practice. Put simply, engineering ethics is the rules of fair play that engineers operate under while serving the public, their employer, the client, and the profession. Engineering offers great potential for contributing to the public good, but at the same time, it can cause great harm if it is not applied correctly and ethically.

Around the turn of the twentieth century after the Industrial Revolution, engineers were playing a major role by contributing to manufacturing and the infrastructure for transportation. When structural failures caused by technical errors, construction problems, and ethical issues created major disasters [(e.g., Ashtabula River Railroad Disaster (1876), Tay Bridge Disaster (1879), Quebec Bridge Collapse (1907), and the Boston Molasses Flood (1919)], a number of the engineering societies adopted formal codes of ethics. These codes made it clear that engineers were responsible for protecting the safety of the public. In 1946, the National Society of Professional Engineers released the Canons of Ethics for Engineers and the Rules for Professional Conduct, which have evolved into the code of engineering ethics used today. Engineering ethics has become even more complicated today due to different cultural traditions encountered in global trade and when dealing with political corruption, environmental issues, and sustainability issues.

Arthur C. Little, who was a famous design engineer, once said, "Any sufficiently advanced technology is indistinguishable from magic." The engineering profession must thus use its magic wisely.

1.5.1   The Engineering Profession

A profession is a paid occupation that requires special education, training, or skills. Professions are known to evolve through a series of stages: the craft stage, the commercial stage, and the professional stage. The craft stage involves individuals who use common sense, intuition, and brute force to accomplish a task (e.g., building a bridge). As the demand for the task increases, the commercial stage develops and uses practitioners who use trial-and-error methods to improve the consistency and quality of the product. When science catches up with practice, the professional stage begins, combining scientific understanding with practice. As a result, professional engineering practitioners must be trained in scientific theory as well as engineering practice.

Significantly important to the engineering profession are professional engineering societies. Each of the major fields of engineering has a national society [(e.g., the American Institute of Chemical Engineers (AIChE)]. In addition, there are other engineering societies, including the Society of Women Engineers (SWE), the American Society for Engineering Education (ASEE), the National Society of Black Engineers (NSBE), Tau Beta Pi Engineering Honor Society (TBP), and the National Society of Professional Engineers (NSPE). Each of these societies represents a group of engineers and affords a means of interacting with other engineers who share similar backgrounds and interests.

The engineering profession is different from other professions, such as medical doctors, dentists, accountants, and lawyers, because these professionals deal largely with individuals, while engineers work primarily with organizations, such as companies or governmental agencies. The state of a profession depends on those professionals who are engaged in its practice. Moreover, the reputation and/or image of a profession can have a direct effect on the members of that profession. The future reputation of the engineering profession will depend on how technically well and ethically you perform your job as an engineer. Therefore, when you become an engineer, you accept the responsibility to improve or at least maintain the reputation of the engineering profession for those who will follow you.

1.5.2   Codes of Ethics

A number of ethics codes have been developed by engineering societies. Put simply, **engineering ethics boils down to being fair and honest while performing your duties for your employer and client but ultimately protecting the interests of the public.** That is, the public health, welfare, and safety supersede the interest of the employer and the client. If you are aware of a public safety issue, you are required by ethics codes to see that it is corrected or that the proper authorities are notified. In addition, engineers should work only on projects for which they have the necessary education or experience. Remember that whenever you sign your name to a document, you are confirming the accuracy and validity of it with your reputation.

The challenge associated with being ethical occurs when taking the right action costs you. For example, imagine if, based on the course syllabus, you earned a grade of C in a course, but you actually received an A in the course due to a clerical error. It is not ethical to ignore the error and keep the A. For another example, consider that you are working as an engineer and for the first time have been asked to lead a project. Therefore, this project is a launching point for your career. After the project was completed and deemed a huge success, you realize that the whole basis of the project violates a patent that your company does not hold. If you point out the patent infringement to your boss, who designed the project in the first place, this can sabotage your career.

A **conflict of interest** can create a problem for an organization even if no wrongdoing occurs. A conflict of interest can be defined as a set of circumstances that creates a risk that a professional judgment or decision may be affected by a secondary interest—for example, if you were a grader for homework in a class and had to grade your best friend in this class. As a result, **any potential conflict of interest should be disclosed to all parties by an engineer, and the parties should be left to decide if the potential conflict of interest is relevant.**

Plagiarism is the use of someone else's words, ideas, or other work (e.g., images, videos, and music) without referencing the original source. While many times plagiarism is not illegal, it is considered unethical in most organizations. Moreover, in universities, plagiarism is considered academic misconduct and can lead to disciplinary action. A number of software products are available to check for plagiarism, and many university professors routinely use them. Therefore, as an engineer, you should always be careful to reference the sources that you use for all work on which you place your name.

The primary portion of the fundamental canons of the code of ethics for the National Society of Professional Engineers follows:

I. Fundamental Canons

Engineers, in the fulfillment of their professional duties, shall:

1. Hold paramount the safety, health, and welfare of the public.
2. Perform services only in areas of their competence.
3. Issue public statements only in an objective and truthful manner.
4. Act for each employer or client as faithful agents or trustees.
5. Avoid deceptive acts.
6. Conduct themselves honorably, responsibly, ethically, and lawfully so as to enhance the honor, reputation, and usefulness of the profession.

Example 1.2   Copy of Exam
Case Description
Consider that you found a copy of the upcoming exam along with the solution in front of your professor's office door. Moreover, assume that you are currently failing this class and that a failing grade will result in your having to leave the university because you are on scholastic probation.

Analysis
While the fact that you are failing the class and on scholastic probation certainly increases the importance of your decision concerning the copy of the exam and solution, it does not impact the ethical issues of this case. The issue with regard to what to do with the exam is that it would not be fair to the other students in the class for you to use it to get a better grade on the exam. Sliding the material under your professor's door without looking at it further is probably the best option because taking it to the department office would likely embarrass your professor.

It is true that it would be quite difficult to do the right thing in this case given the compromised position that you find yourself in. This example makes the point that **it is much easier to make the right and ethical decision if you do not let yourself get into a compromised position.** Moreover, in this case, even if you were to use the exam and solution to pass this exam, you would not be able to stay in school unless you put the proper effort into your remaining classes.

Self-Assessment Test

Questions

1. Summarize what engineering ethics is.
2. What is a conflict of interest? Give an example.

Answers

1. Engineering ethics is being fair and honest while performing your duties to your employer and client, but ultimately protecting the interest of the public. In addition, as an engineer, you should always disclose any conflicts of interest to the affected parties. Moreover, you should only undertake work for which you have the appropriate background and experience to perform.
2. A conflict of interest is a set of circumstance that creates a risk that a professional judgment or decision may be affected by a secondary interest. If you, as a student, were asked to grade your own exam, that would be a conflict of interest.

Summary

- Chemical engineers are primarily process engineers and hold a variety of jobs for a wide range of industries.
- A sustainability analysis of a product or a process involves a wide range of considerations for a chemical engineer. The quantitative analysis of sustainability requires the use of detailed models, and the following chapters present some of the fundamentals used to develop these models.
- Ethics is about being honest and fair to all parties while protecting society. It is important for engineers to always act in an ethical fashion in order to maintain the reputation of the engineering profession for future generations.

Glossary

CAD   Computer-aided design packages (software) that perform design calculation.

CPI   The chemical process industries (e.g., refineries and chemical plants).

green engineering   Design based on minimizing the total impact of the design on the environment and important resources.

life-cycle analysis   A thorough approach for evaluating the impact of a design on the environment and important resources.

CHAPTER 2

Introductory Concepts

Chapter Objectives

- Understand SI and AE units and be able to make unit conversions.
- Apply units correctly when using equations.
- Understand the convention used to designate the accuracy of a number by the manner in which the number is written.
- Learn how to validate a solution.
- Properly use mass, moles, molecular weight, density, pressure, and flow rates to describe chemical engineering systems.

2.1 Units of Measure

An engineer must specify certain quantities in order to design something or to define the proper way to operate a system. For example, the design of a process would require drawings and specifications of each part of the process. The **dimensions** of all the elements of the process with their respective units must be specified accurately because this information will be used to ensure that each part of the process is accurately manufactured or selected. Process engineers use measurements to ensure that the process is operating properly. **Units of measure** are important because they provide a consistent means of precisely describing things. It has been determined that there are a total of seven fundamental measures of physical quantities:

- Length
- Time

- Mass
- Temperature
- Electric current
- Molecular amount
- Luminous intensity

These physical quantities and combinations of them can be used to describe the full range of physical characteristics. For example, your height (length) and your weight (mass) are listed on your driver's license to quantitatively describe you. Your car can be described in terms of its weight and the horsepower of its engine. **Extensive quantities,** such as mass and length, depend on the gross amount of material for their value. For example, for a container full of water, the larger the container, the greater the mass of the water in the container. Doubling the size of the container, however, does not change the temperature or density of the water. Quantities that are independent of the amount of material are referred to as **intensive quantities.**

Derived units are units based on combinations of fundamental units. For example, the velocity of an object can be expressed using length and time, such as feet per second or kilometers per hour. Other derived units, including units for force, energy, power and pressure, will be introduced when SI and American Engineering (AE) units are presented.

2.1.1 SI Units

SI unit system, formally known as Le Systeme d'Unites, is exclusively used in every industrialized nation in the world except the United States. SI units were developed to provide a simplified system of units that would serve as a standard for units throughout the world. Table 2.1 lists the **fundamental units** for the SI system of units. The SI system is a decimal-based system, and Table 2.2 lists the primary prefixes used for the SI systems; for example, mm is a millimeter (10^{-3} meters), Gg is a gigagram (10^9 grams), and µs is a microsecond (10^{-6} s). These prefixes are useful when displaying very large or very small quantities. For example, the distance from the earth to the moon is 384,400,000 m, or 384.4 Mm. This distance can also be conveniently expressed using scientific notation (i.e., 3.844×10^8 m). The centimeter–gram–second system **(CGS) system of units** is identical to the SI system except that grams replace kilograms and centimeters replace meters.

Table 2.3 lists some of the most common derived units for the SI system. **Force** (newtons) is mass times acceleration, or kg m s^{-2}, or kg m/s^2 [in this text, we use both negative exponents to designate units in the denominator (kg m s^{-2}) and a division sign (e.g., kg m/s^2)]. Energy (joules) is force times distance, or N m, and power (watts) is energy per time, or J s^{-1}.

Table 2.1   Fundamental Units for the SI System

Physical Quantity	Name	Symbol
Length	meter	m
Mass	kilogram	kg
Time	second, hour	s, h
Temperature	kelvin	K
Electrical current	ampere	A
Molecular amount	mole	mol, g mol, gmol
Luminous intensity	candela	cd

Table 2.2   SI Prefixes

Factor	Prefix	Symbol	Factor	Prefix	Symbol
10^{12}	tera	T	10^{-3}	milli	m
10^{9}	giga	G	10^{-6}	micro	μ
10^{6}	mega	M	10^{-9}	nano	n
10^{3}	kilo	K	10^{-12}	pico	p

Table 2.3   SI Derived Units

Physical Quantity	Name of SI Unit	Symbol	Definition
Area	square meter		m^2
Volume	cubic meter		m^3
Density	kilogram per cubic meter		$kg\ m^{-3}$
Velocity	meter per second		$m\ s^{-1}$
Acceleration	meter per second squared		$m\ s^{-2}$
Momentum	kilogram meter per second		$kg\ m\ s^{-1}$
Moles	kilogram moles	kgmol	$kg\ mole$
Concentration			$kg\ mol\ m^{-3}$
Temperature	degree Celsius	°C	$°C = K - 273.15$
Angle	radian	rad	$m\ m^{-1}$
Angular velocity			$rad\ s^{-1}$
Force	newton	N	$kg\ m\ s^{-2}$
Energy	joule	J	$kg\ m^2\ s^{-2}$

(Continues)

Table 2.3   **SI Derived Units (*Continued*)**

Physical Quantity	Name of SI Unit	Symbol	Definition
Power	watt	W	$\mathrm{kg\ m^2\ s^{-3}}$
Pressure or stress	pascal	Pa	$\mathrm{kg\ m^{-1}\ s^{-2}}$
Quantity of electricity	coulomb	C	$\mathrm{A\ s}$
Electrical voltage	volt	V	$\mathrm{kg\ m^2\ s^{-3}A^{-1}}$ or $\mathrm{W\ A^{-1}}$
Electrical resistance	ohm	Ω	$\mathrm{kg\ m^2\ s^{-3}A^{-2}}$ or $\mathrm{V\ A^{-1}}$
Frequency	hertz	Hz	$\mathrm{s^{-1}}$

2.1.2 American Engineering Units

The **AE system** of units originated from the British system of units and is based on the foot and the pound force. The fundamental units for the AE system are listed in Table 2.4, and the system is considered a gravitationally based system because it is based on defining force in terms of the gravitational field of the earth. A **pound force** is defined as the gravitational force of one **pound mass** at sea level and at a 45° latitude where the gravitational acceleration is 32.1740 ft/s^2. This can lead to some confusion in terminology. For example, weight, which is the gravitational force of an object, is used commonly to describe the mass of an object (e.g., the boy weighs 65 pounds or the bag of sugar weighs 5 pounds). The primary derived units for the AE system are listed in Table 2.5. Most engineers and industry in the United States use the AE units: feet, pounds, gallons, Btus, and horsepower. Also note that the AE system uses some of the fundamental and derived units from the SI system, such as amperes, radians, angular velocity, voltage, and frequency. A slug is a unit of mass that is sometimes used with the AE system of units: $1.0\ \mathrm{slug} = 1\ \mathrm{lb_f\ s^2\ ft^{-1}} = 32.174\ \mathrm{lb_m}$.

Example 2.1   The Variation of Weight with Elevation

Problem Statement

Determine the percentage change in the **weight** of a 150 lb$_\mathrm{m}$ person at an elevation of 5280 ft and on Mt. Everest (elevation is equal to 29,035 ft) compared to sea level.

Solution

The gravitational acceleration at sea level is 32.174 ft s^{-2}, at 5280 ft it is 32.158 ft s^{-2}, and at 29,035 ft it is 32.085 ft s^{-2}. First consider the force of gravity on Mt. Everest:

$$\text{On Mt. Everest:} \quad F = ma = \frac{150 \text{ lb}_\text{m}}{} \left|\frac{32.085 \text{ ft}}{\text{s}^2}\right| \frac{\text{s}^2 \text{ lb}_\text{f}}{32.174 \text{ ft}} = 149.6 \text{ lb}_\text{f}$$

$$\text{At an elevation of 5280 ft:} \quad F = ma = \frac{150 \text{ lb}_\text{m}}{} \left|\frac{32.158 \text{ ft}}{\text{s}^2}\right| \frac{\text{s}^2 \text{ lb}_\text{f}}{32.174 \text{ ft}} = 149.9 \text{ lb}_\text{f}$$

And of course, the force at sea level will be 150 lb$_\text{f}$. Note that only the gravitational acceleration changes during these calculations. Therefore, the percentage change in weight (i.e., the gravitational force) is the same as the percentage change in the gravitational acceleration. At an elevation of 5280 ft, the percentage decrease in weight is 0.05%, and on Mt. Everest, the percentage decrease is 0.27%. This example helps explain why the terminology of weight and mass are interchanged using the AE system when dealing with objects on the surface of the earth because numerically they are the same to a high degree of accuracy.

Table 2.4   Fundamental Units for the AE System

Physical Quantity	Name	Symbol
Length	foot, inch	ft, in.
Force	pound (force)	lb$_\text{f}$
Time	second	s
Temperature	degree fahrenheit, degree Rankine	°F, °R
Electrical current	ampere	A
Molecular amount	pound mole	lb-mol
Luminous intensity	candela	cd

It should be pointed out that pressure can be expressed as absolute pressure or as gauge pressure. The gauge pressure is the absolute pressure minus the atmospheric pressure; for example, if the absolute pressure is equal to 35 psia, then the gauge pressure would be 20.31 psig (i.e., 35 − 14.69). You will see in the next section that performing unit conversions within the SI system is much simpler and more direct than unit conversions within the AE system.

Table 2.5   AE Derived Units

Physical Quantity	Name of AE Unit	Symbol
Area	square feet, acre	ft^2, ac
Volume	cubic feet, gallons	ft^3, gal
Density	pounds (mass) per cubic foot	$lb_m\ ft^{-3}$
Velocity	feet per second	$ft\ s^{-1}$
Acceleration	feet per second squared	$ft\ s^{-2}$
Momentum	pounds (mass) feet per second	$lb_m\ ft\ s^{-1}$
Concentration	pounds moles per cubic foot	$lb\text{-mol}\ ft^{-3}$
Angle	radian	rad, $ft\ ft^{-1}$
Angular velocity	radians per second	$rad\ s^{-1}$
Mass	pound (mass)	lb_m
Energy	British thermal unit, foot pound (force)	Btu, $ft\ lb_f$
Power	horsepower	hp
Pressure	pound (force) per square inch	psi
Frequency	hertz	Hz
Quantity of electricity	coulomb	C
Electrical voltage	volt	V
Electrical resistance	ohm	Ω

Self-Assessment Test

Questions

1. List two intensive variables from the SI system of units.

2. Why is the AE system of units referred to as a gravitationally based system?

3. Why is the SI system of units referred to as a decimal system of units?

4. What is the SI derived units for density? For AE derived units? For CGS derived units?

Answers

1. SI intensive variables: temperature (K), pressure (kPa), density (kg m^{-3}), concentration (kgmol m^{-3}), angle (rad)

2. The AE system of units is referred to as a gravitationally based system because it is based on the pound force, which is defined in terms of the gravitational force.

3. The SI system of units is referred to as a decimal system because the decimal-based prefixes listed in Table 2.2 can be used with SI fundamental units (e.g., nm is a nanometer or 10^{-9} m).

4. SI: $kg\ m^{-3}$; AE: $lb_m\ ft^{-3}$; CGS: $g\ cm^{-3}$

2.2 Unit Conversions

Many times, the units of a quantity are not expressed in the units that you require. In that case, you must convert the units to the desired form. The tables listed on the inside cover of this text are quite useful for performing **unit conversions.** As an example, consider that you have a 2 horsepower (hp) electric motor and you want to know how many kilowatts (kW) the motor will produce. To make this unit conversion, use the fifth table on the inside cover, Power Equivalents, which is shown in Figure 2.1. Because the known quantity is expressed in horsepower, you first locate the row corresponding to hp in the leftmost column (marked with an arrow, →, in Figure 2.1). Next, locate the column corresponding to the required units (kW) in the top row of the table (also marked with an arrow). The cell intersecting the hp row and the kW column contains the desired conversion factor (i.e., 0.7457). Therefore, 2 hp is equal to 2×0.7457 kW, or 1.4914 kW.

Power Equivalents

	$J\ s^{-1}$	→ kW	$ft\ lb_f s^{-1}$	$Btu\ s^{-1}$	hp
$J\ s^{-1}$	1	10^{-3}	0.7376	9.478×10^{-4}	1.341×10^{-3}
kW	1000	1	737.56	0.9478	1.341
$ft\ lb_f s^{-1}$	1.356	1.356×10^{-3}	1	1.285×10^{-3}	1.818×10^{-3}
$Btu\ s^{-1}$	1.415	1415	778.16	1	1.415
→ hp	7457	<u>0.7457</u>	550	0.7068	1

Figure 2.1 Table of Power Equivalents

Another way to look at this unit conversion problem is to recognize that the conversion factor, 1 hp = 0.7457 kW, is an equality statement. As a result, 1 hp/0.7457 kW is equal to unity. Therefore, multiplying this unity ratio times the 2 hp does not change the result, just the units.

$$\frac{2\ \cancel{hp}}{} \left| \frac{0.7457\ kW}{1\ \cancel{hp}} \right. = 1.4914\ kW$$

Note that the hp from the unit conversion cancels the hp from the original quantity, yielding the desired results. When dealing with more complicated expressions requiring a number of unit conversion factors or applying units for equations (Section 2.3), writing out the full expression across the page and canceling the units is the recommended approach because you will be less likely to make a unit conversion error.

Now consider another example requiring two unit conversion factors. Convert the density 10 kg/m³ to lb_m/ft³. To perform this unit conversion, we require two unit conversion factors: kg (kilograms) to lb_m (pounds mass) and m³ (cubic meters) to ft³ (cubic feet). The first is obtained from the Mass Equivalent table and the second from the Volume Equivalent table on the inside cover. Applying these conversion factors and canceling units yields

$$\frac{10 \ \cancel{kg}}{\cancel{m^3}} \left| \frac{2.2046 \ lb_m}{1 \ \cancel{kg}} \right| \frac{\cancel{m^3}}{35.31 \ ft^3} = 0.6231 \ lb_m \ ft^{-3}$$

Remember that each conversion factor is equivalent to unity, and therefore, multiplying by them does not change the answer, only the units of the answer. That is, 0.6231 lb_m/ft³ is equivalent to 10 kg/m³. What if you applied the conversion factors **incorrectly?** That is,

$$\frac{10 \ kg}{m^3} \left| \frac{1 \ kg}{2.2046 \ lb_m} \right| \frac{35.31 \ ft^3}{m^3} = 160.2 \ \frac{kg^2 ft^3}{m^6 lb_m}$$

Here the units of mass (kg and lb_m) and units of volume (ft³ and m³) remain in the expression because they have not been canceled. Therefore, **if you cancel the units when applying conversion factors, you are less likely to make errors when applying conversion factors,** and thus, you minimize the errors associated with unit conversions. Note that the previous equation is a valid equality, but it does not have the desired final units.

Example 2.2   The Speed of Light

Problem Statement

The speed of light is approximately 300,000,000 m s⁻¹, or 3×10^8 m s⁻¹. Determine the speed of light in km per μs.

Solution

To solve this problem, we need to use the definitions of the prefixes for SI units (Table 2.2). That is, 1 km = 1000 m and 1 μs = 10^{-6} s. Performing the unit conversion yields

$$\frac{3 \times 10^8 \; \cancel{m}}{\cancel{s}} \left| \frac{1 \; km}{1000 \; \cancel{m}} \right| \frac{10^{-6} \; \cancel{s}}{\mu s} = 0.3 \; km \; \mu s^{-1}$$

Example 2.3   Unit Conversion for an Area

Problem Statement

For SI units, large areas are expressed in hectares (ha) where 1 ha = 10,000 m^2. In the United States, large areas are given in acres where 1 acre = 43,560 ft^2. Determine how many acres are in 45.3 hectares.

Solution

To solve this problem, we need some additional information: the number of square feet in a square meter. We use the number of feet in a meter from the Lengths Equivalents table (on the inside cover) and square it to get the necessary conversion factor. The conversion calculation becomes

$$\frac{45.3 \; \cancel{ha}}{} \left| \frac{10000 \; \cancel{m^2}}{1 \; \cancel{ha}} \right| \frac{(0.3048)^2 \; \cancel{ft^2}}{\cancel{m^2}} \left| \frac{1 \; ac}{43560 \; \cancel{ft^2}} \right. = 111.9 \; ac$$

This result agrees with the conversion factor 1 ha = 2.471 ac shown in the unit conversion tables.

2.2.1   Temperature Conversion

Temperature conversions can be different than the unit conversions presented so far because the reference points for temperatures in degrees Fahrenheit (°F) and degrees Celsius (°C) are different. For example, when converting from m to ft, both m and ft have the same value for zero length.

Therefore, converting from m to ft is accomplished by multiplying by a single factor (i.e., 3.2808). This is not true for converting a temperature in °F to °C because 0°F is not equal to 0°C. In addition, the size of a Δ°F is not the same as the size of a Δ°C. Note the designation of a temperature in this section is given as °C and °F, whereas a temperature change in the number of Fahrenheit degrees or Celsius degrees is indicated by Δ°F and Δ°C, respectively. Most texts do not use this convention, and it is not used in the remainder of the text except for this subsection. Therefore, in the future, you will have to examine how the temperature is used in order to determine if it is a specific temperature or a temperature difference.

The relation between the size of Δ°F and Δ°C is given by

$$\frac{\Delta °F}{\Delta °C} = \frac{180}{100} = \frac{9}{5} = 1.8 \tag{2.1}$$

because the temperature difference between the freezing point and the boiling point of water is 100 Δ°C (i.e., 100°C − 0°C) and 180 Δ°F (i.e., 212°F − 32°F).

In addition, the absolute temperature scales [i.e., Rankine degrees (°R) and kelvin (K)] are such that

$$\Delta K = \Delta °C \text{ and } \Delta °R = \Delta °F$$

That is, the **Rankine** and the **Fahrenheit** scales use the same size degree, whereas the **kelvin** and **Celsius** scales use the same size degree. Absolute temperatures are used in the ideal gas law and in other engineering equations. The relation between temperature in kelvin and degrees Celsius and between Rankine and Fahrenheit are given by

$$T(K) = T(°C) + 273.15$$
$$T(°R) = T(°F) + 459.67 \tag{2.2}$$

To convert from °C to °F, or vice versa, you can use the fact that water freezes at 0°C and at 32°F and the relative size of each degree. Therefore, the conversion from a temperature in degrees Celsius to degrees Fahrenheit is given by

$$T(°F) = 1.8\, T(°C) + 32 \tag{2.3}$$

And the conversion from a temperature in degrees Fahrenheit to degrees Celsius degrees is given by

$$T(°C) = [T(°F) − 32]/1.8 \tag{2.4}$$

From these equations, for a consistency check, note that at 0°C, T(°F) is equal to 32°F, and at 32°F, T(°C) is equal to 0°C. Figure 2.2 graphically demonstrates the Fahrenheit, Rankine, Celsius, and kelvin temperature scales.

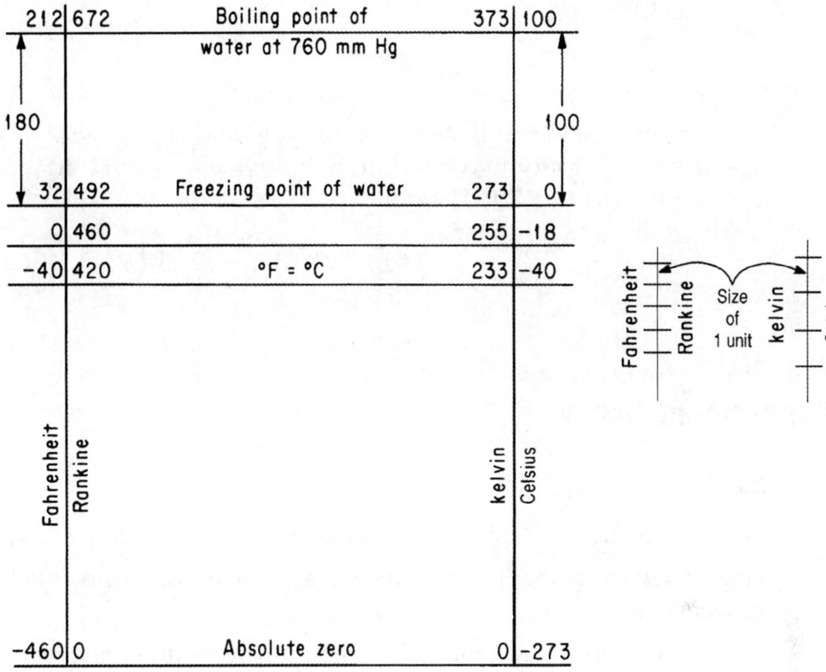

Figure 2.2   Temperature scales

Example 2.4   Temperature Conversion

Problem Statement

Convert 85°F to kelvin.

Solution

First, we convert the temperature to °C using Eq. 2.4.

$$T(°C) = [85 - 32]/1.8 = 29.44°C$$

Then it can be converted to kelvin using Eq. 2.2.

$$T(K) = T(°C) + 273.15 = 302.59 \text{ K}$$

Example 2.5   Temperature Conversion

Problem Statement

Convert 273.15 K to °R.

Solution

Because both Rankine and kelvin are absolute temperature scales, we can use a single conversion factor to convert from one to the other and that conversion factor is 5K = 9°R (i.e., Eq. 2.1) Therefore,

$$T(°R) = 9T(K)/5 = 491.67°R$$

Self-Assessment Test

Questions

1. How do you use the tables on the inside cover to obtain a conversion factor?

2. How does canceling units reduce the number of errors made during unit conversions?

3. Why are certain temperature conversion problems different from converting feet to meters?

Answers

1. Assume that you want to convert from units a to units b. First, locate the proper table depending on the types of units for a and b. Next, locate unit a in the left-hand column of the table and locate b in the row across the top of the table. Then, the conversion factor is determined by the intersection of the row for the unit of a and the column for the units of b. Finally, the conversion of a to b is accomplished by multiplying the number of units of a by the conversion factor to determine the amount in units of b.

2. By canceling the units in a unit conversion problem, you are able to make sure that you have correctly applied the unit conversion factors.

3. To convert from feet to meters, you simply multiply the number of feet by the conversion factor (0.3048). Conversely, if you want to convert from °C to °F, you could not simply multiply by a single factor because 0°C and 0°F are not equal. To convert from °C to °F, you can use the fact that water freezes at 0°C and at 32°F.

Problems

1. Convert 50 Btu gal^{-1} to SI units.

2. Convert 25°C to °R.

Answers

1. $\dfrac{50\ \text{Btu}}{\text{gal}}\left|\dfrac{1055\ \text{J}}{\text{Btu}}\right|\dfrac{\text{gal}}{3.875\times10^{-3}\ \text{m}^3}=1.361\times10^{7}\ \text{J m}^{-3}$

2. 25°C + 273.12 = 298.15 K; 298.15 K (9°R/5 K) = 536.67°R

2.3   Equations and Units

When applying engineering equations, it is essential to ensure that the units are applied properly. To illustrate the primary issues for units in equations, consider the following simple equation:

$$y = ax$$

First of all, the units on the left side of the equation must be the same as the units on the right side. Therefore, the product ax must have the same units as y. For example, if y has units of pressure in psi (lb$_f$ in^{-2}) and x has units of force (lb$_f$), a must have units of in^{-2}. Now let's add a constant term to the right side of this equation:

$$y = ax + b$$

For this case, b must have the same units as ax and y. What if y and ax had units of psi and b had units of °C? Then the result would be

$$\text{psi} = \text{psi} + °\text{C}$$

which is clearly *incorrect*. For this equation to be valid, b must have units of psi. That is, **an equation must be dimensionally homogenous (i.e., dimensional consistency)**—the quantities on both sides of the equal sign must have the same units, and terms that are added must also have the same units.

Now consider Newton's second law of motion (i.e., $F = ma$) for a mass of 2 kg under gravitational acceleration (9.8068 m s^{-2}) using SI units:

$$F = ma = \frac{2 \text{ kg}}{} \left| \frac{9.8068 \text{ m}}{\text{s}^2} \right| \frac{\text{N s}^2}{\text{kg m}} = 19.6136 \text{ N}$$

Note that the last conversion factor applied is simply the definition of a newton force. As a result, the product of mass and acceleration for SI units with mass in kg and acceleration in m s^{-2} yields force in N without any additional unit conversions. Now consider the same type of problem using the AE system. For a mass of 2 lb and gravitational acceleration,

$$F = ma = \frac{2 \text{ lb}_m}{} \left| \frac{32.174 \text{ ft}}{\text{s}^2} \right. = 64.348 \text{ lb}_m \text{ft s}^{-2}$$

Note that the units of this answer are not in lb$_f$, which is the unit of force in the AE system. To convert this answer into the proper AE units, we must apply the definition of force used. That is,

$$1 \text{ lb}_f = 1 \text{ lb}_m \times g = 32.174 \text{ lb}_m \text{ ft s}^{-2}$$

If we divide the left side into the right side of this equation, the result is a conversion factor known as g_c:

$$g_c = 32.174 \text{ lb}_m \text{ ft s}^{-2} \text{lb}_f^{-1}$$

Now if we apply g_c to the previous application of Newton's second law of motion in AE units,

$$F = \frac{ma}{g_c} = \frac{2 \text{ lb}_m}{} \left| \frac{32.174 \text{ ft}}{\text{s}^2} \right| \frac{\text{s}^2 \text{ lb}_f}{32.174 \text{ ft lb}_m} = 2 \text{ lb}_f$$

Note that if the local acceleration of gravity a were different than the standard gravitational acceleration, the resulting force would be slightly different. In certain cases, when using the AE system of units, g_c must be used to convert from lb$_m$ to lb$_f$, or vice versa.

Example 2.6   The van der Waals Equation

Problem Statement

The van der Waals equation is used to provide a more accurate approximation than the ideal gas law for high-density gases. The van der Waals equation is given by

$$\left(p + \frac{a}{V^2}\right)(V - b) = RT$$

Assuming that pressure is given in atmospheres (atm), the specific volume V is given in L gmol^{-1}, and temperature is given in K, determine the units for a and b and the value and units for R.

Solution

Because b is subtracted from V, b-must have the same units as V or L gmol^{-1}. The term a/V^2 must have units of atm. Therefore, a has units of L^2 atm gmol^{-2}. Finally, the value of R can be read out of the table on the back inside cover (i.e., $R = 0.08206$ atm L gmol^{-1}, K^{-1}) because of the units used for pressure, specific volume, and temperature.

Example 2.7   Reynolds Number

Problem Statement

The Reynolds number is a group of variables that are **dimensionless,** which can be used to calculate the frictional pressure drop losses for flow through a pipe and is given by

$$Re = \frac{\rho D v}{\mu}$$

where ρ is the density of the fluid (1.00 g cm^{-3}), D is the pipe diameter (2.00 in.), v is the linear velocity of the fluid (8.00 ft s^{-1}), and μ is the viscosity of the fluid (0.020 g cm^{-1} s^{-1}). Determine the numerical value of the Reynolds number for this case and demonstrate that it is, in fact, dimensionless.

(Continues)

Example 2.7   Reynolds Number (*Continued*)

Solution

Substituting into the equation for the Reynolds number and making the necessary unit conversions yields

$$\text{Re} = \frac{\rho D v}{\mu} = \frac{1\,\cancel{g}}{\cancel{cm}^3} \left| \frac{2.00\,\cancel{in.}}{} \right| \frac{8.00\,\cancel{ft}}{\cancel{s}} \left| \frac{\cancel{cm}\,\cancel{s}}{0.020\,\cancel{g}} \right| \frac{12\,\cancel{in.}}{1\,\cancel{ft}} \left| \frac{2.542^2\,\cancel{cm}^2}{1\,\cancel{in.}^2} \right| = 6.19 \times 10^4$$

Example 2.8   Kinetic Energy

Problem Statement

The kinetic energy of an object in motion is given by

$$KE = \frac{1}{2} m v^2$$

Determine the kinetic energy in Btu of an object with a mass of 100 lb_m at a velocity of 20.0 ft s^{-1}.

Solution

The kinetic energy can be calculated by using the above formula. Note that we have to use g_c in order to obtain the desired units for the solution because without g_c, the answer would contain units of mass (lb_m) and not force (lb_f).

$$KE = \frac{1}{2} m v^2 = \frac{1}{2} \left| \frac{100.\,\cancel{lb}m}{} \right| \frac{20.0^2\,\cancel{ft}^2}{\cancel{s}^2} \left| \frac{\cancel{s}^2\,\cancel{lb}_f}{32.174\,\cancel{lb}_m\,\cancel{ft}} \right| \frac{1.285\,10^{-3}\,\text{Btu}}{\cancel{lb}_f\,\cancel{ft}} = 827\,\text{Btu}$$

Example 2.9   Microchip Etching Rate

Problem Statement

Your handbook shows that microchip etching roughly follows the relation

$$d = 16.2 - 16.2 e^{-0.021t} \quad t < 200$$

where d is the depth of the etch in microns [micrometers (μm)] and t is the time of the etch in seconds. What are the units associated with the numbers 16.2 and 0.021? Convert this relation so that d becomes expressed in inches and t is specified in minutes.

Solution

After you inspect the equation that relates d as a function of t, you should be able to reach a decision about the units associated with each term on the right side of the equation. Based on the concept of dimensional consistency, both values of 16.2 must have the associated units of microns (μm). The exponential term must be dimensionless so that 0.021 must have units of s^{-1}. To carry out the specified unit conversion for this equation, look up suitable conversion factors inside the front cover of this book (i.e., convert 16.2 μm to inches and 0.021 s^{-1} to min^{-1}).

$$d(\text{in.}) = \frac{16.2\mu\text{m}}{} \left|\frac{1\text{m}}{10^6\,\mu\text{m}}\right| \frac{39.97\,\text{in.}}{1\text{m}} \left[1 - \exp\frac{-0.021}{\text{s}}\left|\frac{60\,\text{s}}{1\text{min}}\right|\frac{t(\text{min})}{}\right]$$

$$d(\text{in.}) = 6.38 \times 10^{-4}(1 - e^{-1.26t(\text{min})})$$

Example 2.10   Reaction Rate for a Bioreactor

Problem Statement

The growth rate for yeast used in the fermentation process for the production of ethanol (EtOH) from grains, such as corn, can be represented using Monod kinetics and is given by

$$\mu = \left[\frac{\mu_0}{1 + P/k_p}\right]\left[\frac{S}{k_s + S}\right]$$

where μ is the production rate of cells expressed in g-cells produced (g-cell)$^{-1}$h^{-1}, P is the EtOH concentration in g L^{-1}, and S is the glucose concentration in g L^{-1}. Determine the units used for μ_0, k_p, and k_s based on dimensional consistency.

Solution

From an overall analysis, you can see that the last term in the equation must be dimensionless. Therefore, the units for the first term must match the units for μ. In addition, the units for k_p are the same as P and the units for k_s are the same as S.

Example 2.11 Heat Capacity Equation

Problem Statement

Consider the following heat capacity equation:

$$C_p = a + bT$$

where C_p is given in Btu lb_m^{-1} °F^{-1}, T is the temperature (°F), and a and b are constants. Modify this equation so that temperature is given in kelvin and C_p is given in J kg^{-1} °C^{-1}.

Solution

First, note that a has units of Btu lb_m^{-1} °F^{-1}. Therefore, a must be converted to units of J kg^{-1} °C^{-1}:

$$\frac{a \text{ Btu}}{lb_m \text{ °F}} \left| \frac{lb_m}{0.4536 \text{ kg}} \right| \frac{1055 \text{ J}}{\text{Btu}} \left| \frac{9 \text{ °F}}{5 \text{ K}} \right. = 4186a$$

Likewise, the term bT has units of Btu lb_m^{-1} °F^{-1}, which must be converted to J kg^{-1} °C^{-1}. The specified temperature T must first be converted from kelvin to degrees Fahrenheit so that it can be substituted into the original equation to convert the temperature in the equation from °F to K. Note that this substitution does not change the units of the term bT. Therefore, T(°F) = 1.8T(K) − 459.67 can be substituted for the temperature term in bT before the units are converted to J kg^{-1} °C^{-1}:

$$\frac{b \text{ Btu}}{lb_m \text{ °F}} \left| \frac{lb_m}{0.4536 \text{ kg}} \right| \frac{1055 \text{ J}}{\text{Btu}} \left| \frac{9 \text{ °F}}{5 \text{ K}} \right| \frac{1.8T(K) - 459.67}{} =$$

$$4186b[1.8T(K) - 459.67] = 7535bT(K) - 1.924 \times 10^6 b$$

Therefore, the final form of the converted heat capacity equation is

$$C_p(\text{J kg}^{-1}\text{K}^{-1}) = 4186a + 7535bT(K) - 1.924 \times 10^6 b$$

(Note that, as discussed in section 2.2.1, ΔK = Δ°C; i.e., a change in temperature of 1 K is exactly equal to a change of 1 °C, and hence the units J kg^{-1}°C^{-1} and J kg^{-1} K^{-1} are equivalent.)

Self-Assessment Test

Questions

1. How do you ensure that units are being used properly in the application of an equation?
2. When is g_c used to convert the units of an equation?
3. What is a dimensionless group?

Answers

1. The units of each side of the equation must be equal, and the units of each term on the right side of the equation must be equal.
2. The conversion factor g_c is used to convert lb_m to lb_f, and vice versa.
3. A dimensionless group of variables is such that units of the variables that comprise the group cancel each other so that the resultant is a dimensionless number.

Problems

1. Consider the following equation: $y = ax + b$. If y has units of °F and x has units of lb_m h^{-1}, what units should a and b have?
2. Using the equation for potential energy ($PE = mgh$) and g_c, calculate the potential energy in Btus for a mass of 10.0 lb_m, a height h of 10.0 ft, and $g = 32.2$ ft s^{-2}.

Answers

1. The term b has units of °F, and term a has units of °F h lb_m^{-1}.

2. $$PE = mgh = \frac{10.0 \ \cancel{lb}_m}{} \left| \frac{32.2 \ \cancel{ft}}{\cancel{s}^2} \right| \frac{10.0 \ \cancel{ft}}{} \left| \frac{lb_f \ \cancel{s}^2}{32.2 \ \cancel{lb}_m \ \cancel{ft}} \right| \frac{Btu}{778.16 \ \cancel{lb}_f \ \cancel{ft}} = 0.129 \ Btu$$

2.4   Measurement Errors and Significant Figures

When I divide 5 by 3 using my calculator, I get an answer containing 10 digits (i.e., 1.666666666). Also, when I use my computer to solve a problem, I can get up to 15 digits in the answer. But when solving engineering problems, the data that you use are far less accurate than one part in 10^{10} or 10^{15}. Table 2.6 lists a number of industrial measurements and their typical uncertainty. For example, a temperature measured by a thermocouple has an uncertainty of about ±1°X; measured by an RTD, it has an uncertainty of ±0.1°C. Therefore, **the uncertainty in a measurement depends on the type of measurement and the device used to make the measurement.** You can understand this by considering how accurately you can determine the amount of gasoline in your gas tank from the gas gauge in your car or how accurately you can read the temperature from a thermometer. Note that the uncertainty is expressed as a percentage for most measurements, but for temperatures it is expressed in degrees.

Table 2.6   Typical Uncertainties for Industrial Measurements[1]

Measurement	Measured by	Uncertainty
Temperature	Thermocouple	±1C°
Temperature	Resistance temperature detector (RTD)	±0.1C°
Pressure	Pressure sensor	±0.1%
Mass flow rate	Orifice meter	±3–5%
Voltage	Volt meter	±0.1%
Solution pH	pH electrode	±0.1%

What are the sources of the errors listed in Table 2.6? Two sources of error are the background noise for the measuring instrument and the precision of the scale used to display the measurement, and these sources of error can be estimated by making repeated measurements under the same conditions. The ability to achieve consistent measurements is known as the **repeatability** and indicates the **precision** of the measurement. Other sources of error are **systematic errors,** which cause a consistent offset from the true readings. For an orifice meter, part of the ±3%–5% can be due to partial plugging of the pressure taps and/or a lack of calibration for the meter. For example, for an orifice meter with a partially plugged pressure tap, the measured flow rate would be consistently lower than if the pressure tap were not plugged. Therefore, the repeatability of a measurement is an indication of the precision of the instrument, while the difference between the true reading and the measurement indicates the **accuracy** of the instrument. The difference between precision and accuracy is illustrated in Figure 2.3.

---

[1] J. B. Riggs, M. N. Karim, and J. S. Alford, *Chemical and Bio-Process Control,* 5th ed. (Austin, TX: Ferret Publishing), 2020.

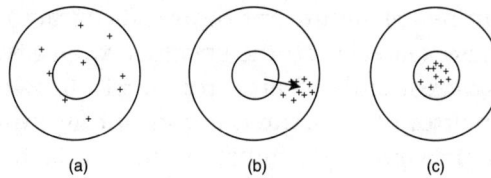

(a)   (b)   (c)

Figure 2.3   Targets that demonstrate the difference between accuracy and precision. (a) Neither accurate nor repeatable. (b) Repeatable but not accurate (the arrow indicates the bias error of the measurement). (c) Accurate and repeatable.

Because quantities used in engineering calculations have limited accuracy, we need a convenient means to specify the accuracy of the data that we are using or the accuracy of results calculated and the use of significant figures provides such a method. **The number of significant figures for a quantity is defined as the number of digits in a number except for zeros that are used only for the location of the decimal point.** For example, 0.0074 has only two significant figures because the zeros preceding 74 are used to locate the decimal point. If 0.0074 were expressed in scientific notation (i.e., 7.4×10^{-2}), it is clear that this number has only two significant figures. On the other hand, 800 could be an exact number or it could have only one or two significant figures. Scientific notation (8.00×10^2) can be used to clearly express 800 with three significant figures accuracy. Table 2.7 lists several examples of quantities and their number of significant figures.

Table 2.7   Significant Figures Examples

Quantity	Scientific Notation	No. of Significant Figures
8000	8.000×10^3	4
8000	8.00×10^3	3
8000	8.0×10^3	2
8000	8×10^3	1
0.01234	1.234×10^{-2}	4
0.3001	3.001×10^{-1}	4
55.64	5.564×10^1	4
0.003	3×10^{-3}	1
87.0	8.70×10^1	3

The number of significant figures (accuracy) roughly indicates the size of the error associated with the specified quantity. For example, one significant figure corresponds to an error of ±10%, two significant figures corresponds to an error of ±1%, three significant figures corresponds to an error of ±0.1%, and four significant figures corresponds to an error of ±0.01%. In this manner, the number of significant figures provides a rough estimate of the error associated with the quantity. Table 2.6 provides much more precise estimates of the error of the measurements listed by expressing the uncertainty explicitly. Applying the significant figures approach to the uncertainties listed in Table 2.6, the pressure, ampere, and pH readings provide results with three significant figures accuracy while an orifice meter provides less than two significant figures accuracy.

As a first approximation, when solving equations, the quantities with the smallest number of significant figures used to evaluate the equation can be used to set the accuracy of the solution (i.e., set the number of significant figures used to display the solution). That is, **the least accurate quantity used in the calculation of an equation determines the accuracy of the results of the equation.** For example, using Newton's second law of motion (i.e., $F = ma$), if the mass is 2.0 kg and the acceleration is 42.3 m s^{-2}, the resulting force F would be displayed as 84 N (i.e., with two significant figures because the mass was given with only two significant figures). The exception to this rule occurs when two numbers of approximately the same value are subtracted, which reduces the number of significant figures in the answer. For example, subtracting 1.232 from 1.244 yields 0.012. Though the two quantities used in this calculation both have four significant figures, the answer has only two significant figures. Conversely, if these numbers are added, the result will have four significant figures.

For a more accurate estimate of the uncertainty in the solution of an equation due to uncertainty in the inputs, you can vary the inputs according to their uncertainty and observe the variation in the result of the equation. That is, for each input, apply the positive and negative extreme of its value according to its uncertainty and observe the change in the result of the equation. For example, consider an input with a value of 0.8. The extreme values for this variable would be 0.8 + 0.08 and 0.8 − 0.08 because one significant figure accuracy corresponds to ±10% uncertainty. In this manner, you can directly assess the impact of each of the inputs on the result of the equation. The largest resulting uncertainty from an input would be used as the estimate for the uncertainty of the output from the equation.

Example 2.12   Microdissection of DNA

Nearly all living things contain DNA (which stands for deoxyribonucleic acid), namely, the molecule that stores **genetic information.** DNA consists of a series of nucleotides. Each nucleotide is denoted by the base it contains: adenine (A), cytosine (C), guanine (G), or thymine (T). The most famous form of DNA in a cell is composed of two very long backbones of sugar (S) phosphate (P) molecules forming two intertwined chains called a double helix that are tied together by base pairs, as shown in Figure E2.12. DNA can also take other forms not shown here.

The length of a segment of DNA is measured in the number of base pairs; 1 kb is 1000 base pairs (bp), and 3 kb = 1 μm. The sugar phosphate backbone is connected by successive combinations of A, T, G, and C.

A genome is one section of DNA in an enormously long sequence of A, C, G, and T.

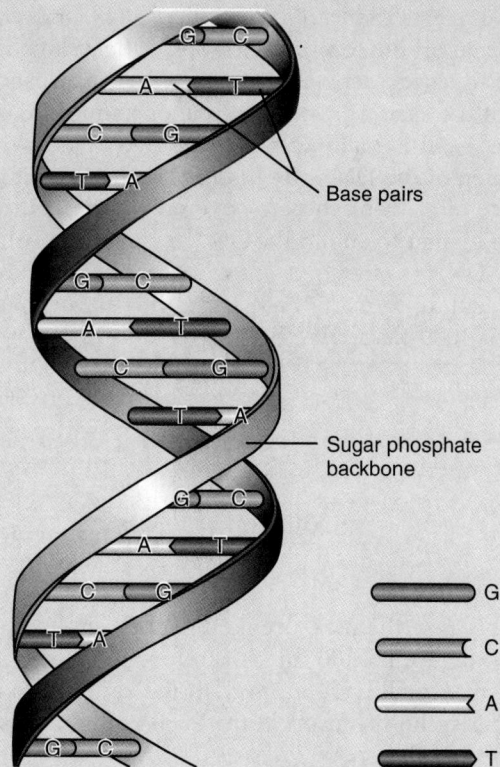

Figure E2.12   A three-dimensional representation of an arbitrary slice of an enormously long strand of DNA

(*Continues*)

Example 2.12 Microdissection of DNA (*Continued*)

Certain parts of the genome correspond to genes that carry the information needed to direct protein synthesis and replication. For a section of the genome to be a gene, the sequence of bases must begin with ATG or GTG and must end with TAA, TAG, or TGA. The length of the sequence must be an exact multiple of three. Protein synthesis is the production of the proteins needed by the cell for its activities and development. Replication is the process by which DNA copies itself for each descendant cell, passing on the information needed for protein synthesis. In most cellular organisms, DNA is organized on chromosomes located in the nucleus of a cell.

Problem Statement

A stretch-and-positioning technique on a carrier layer can be used for microdissection of an electrostatically positioned DNA strand. The procedure employs a glass substrate on the top of which a sacrificial layer, a DNA carrier layer, and a pair of electrodes are deposited. The DNA is electrostatically stretched and immobilized onto the carrier layer with one of its molecular ends aligned on the electrode edge. A cut is made through the two layers with a stylus as a knife at an aimed portion of the DNA. By dissolving the sacrificial layer, the DNA fragment on the piece of carrier can be recovered on a membrane filter. The carrier piece can then be melted to obtain the DNA fragment in solution.

If the DNA is stretched out to a length of 48 kb, and a cut is made with a width of 3 μm, how many base pairs (bp) should be reported in the fragment? Note: 1 kb is 1000 base pairs (bp), and 3 kb = 1 μm.

Solution

The conversion is

$$\frac{3\mu m}{} \left| \frac{3\,kb}{1\mu m} \right| \frac{1000\,bp}{1\,kb} = 9000\,bp$$

However, because the measurement of the number of molecules in a DNA fragment can be determined to three or four significant figures, and the 3 μm reported for the cut may well have more than the reported one significant figure if measured properly, the precision in the 9000 value may actually be better than indicated by the calculation.

Self-Assessment Test

Questions

1. What is the difference between precision and accuracy? Between precision and repeatability?

2. How do you determine the significant figures accuracy of a quantity listed in a reference source?

3. How do you determine the significant figures accuracy of a calculation when each of the numbers used in the calculation has a different number of significant figures?

Answers

1. Accuracy is related to how close the measurement is to the true value, and precision is related to how close multiple values of the same measurement are. Precision and repeatability are essentially the same thing, that is, how close multiple measurements of the same quantity are to each other.

2. The number of significant figures accuracy of measurement in a reference sources is equal to the number of significant figures reported in the reference.

3. The significant figures accuracy of a calculation is equal to the smallest number of significant figures for the values used in the calculation.

Problem

1. Indicate the number of significant figures accuracy for the following numbers: (a) 1000; (b)1000.0; (c) 1.00×10^3

Answers

1. (a) 1; (b) 5; (c) 3

2.5   Validation of Results

When solving either homework problems in school or problems encountered as a professional engineer, it is critically important that you ensure that your answers are accurate enough. By this we mean correct or close enough considering the problem requirements. Unfortunately, there is an almost unlimited number of ways to make errors when solving problems. Being able to eliminate errors when problem solving is an important attribute of a good engineer. Here are some suggestions to help you catch errors when problem solving:

- **Make sure your answers are reasonable.** Check to ensure that your answer seems reasonable according to your understanding of the physical system. For example, for a process plant, if you calculated a pipe diameter of 20 feet to transport 100 gal/min, you should know immediately that this is not a reasonable answer. As you gain engineering experience, you will be able to enhance your knowledge of what is a reasonable answer.

- **Check the details of your calculations.** Errors such as keying in the wrong number, confusing the proper decimal point of a number, reading an intermediate number incorrectly, transposing two numbers, and not being careful about the units are among the most common errors. Carefully repeat your calculations using your calculator or computer, perhaps in a different sequence. You can also simplify the numbers in your calculation so that you can calculate a rough approximate answer in your head or in the calculator. For example, the following exact calculation can be compared with an approximate calculation to confirm the validity of the more exact answer:

$$\frac{468}{0.0181} \left| \frac{6250}{2.95} \left(1 - \frac{3\pi}{9.5^3}\right) = 5.48 \times 10^7 \qquad \frac{5 \times 10^2}{2 \times 10^{-2}} \left| \frac{6 \times 10^3}{3 \times 10^0} = 5 \times 10^7 \right.$$

- **Review the solution procedure.** You should always review your problem solution immediately after it has been completed. Review the problem statement and problem specifications to ensure that you solved the correct problem. Also, make sure that the data used in the problem solution were correctly transferred or selected from a source of data. You should also review the assumptions that you used to make sure that they are reasonable and proper.

- **Back substitution.** To apply this approach, simply substitute your answer into the original equation to ensure that the equation is, in fact, satisfied.

It may seem that carrying out these steps to check your solution will cause extra work for you; it will. But if you realize how important it is to reliably develop accurate solutions now and in the future, you will appreciate the point of establishing good habits now.

Self-Assessment Test

Questions

1. What do you have to do to make sure that an answer to a problem is reasonable?

2. (a) What is the most common type of error in a problem solution, and (b) how can you correct it?

3. What benefit does reviewing a solution procedure provide?

Answers

1. Check to ensure that your answer seems reasonable according to your understanding of the physical system.

2. (a) Errors such as keying in the wrong number, confusing the proper decimal point of a number, reading an intermediate number incorrectly, transposing two numbers, or not being careful about the units are among the most common errors. (b) Carefully repeat your calculations using your calculator or computer, perhaps in a different sequence.

3. It reduces a wide variety of errors.

Problem

1. Develop an approximate solution for the following calculation of the volume in cubic feet:

$$\text{Volume} = \frac{\pi}{4} \left| \frac{(4\,\text{in.})^2}{} \right| \frac{10\,\text{ft}}{}$$

Answer

1. Assume π is equal to 3 and after canceling 4 from the numerator and the denominator, you are left with Volume $= 3 \times 4 \times 10 \times 120$. The actual solution is 125.7.

2.6   Mass, Moles, and Density

The quantity of material used in material and energy balances can be described by its mass or by the number of moles. When a material or energy balance is applied to a process in which reactions do not occur, these balances are usually applied using mass to represent the quantity of material involved. On the other hand, when reactions are occurring in a process, material is represented as moles because reactions occur on a molar basis. In addition, the density of material can be used to convert mass of a substance into its volume or the volume of a substance into its mass.

2.6.1   Choosing a Basis

Many times, when solving a problem, you have to choose a basis to work the problem. A **basis** is a reference chosen by you for the calculations you plan to make in a particular problem, and a proper choice of basis often can make a problem much easier to solve than a poor choice. The basis may be a period of time such as hours, or a given mass of material, or some other convenient quantity. To select a sound basis (which in many problems is predetermined for you but in some problems is not so clear), ask yourself the following three questions:

1. What do I have to start with (e.g., I have 100 lb of oil; I have 46 kg of fertilizer)?
2. What answer is called for (e.g., the amount of product produced per hour)?
3. What is the most convenient basis to use? (For example, if the composition of a given material is known in mole percent, then selecting 100 kg moles of the material as basis would make sense. On the other hand, if the composition of the material in terms of mass is known, then 100 kg of the material would be an appropriate basis.)

These questions and their answers will suggest suitable bases. Sometimes when several bases seem appropriate, you may find it is best to use a unit basis of 1 or 100 of something, for example, kilograms, hours, moles, or cubic feet. For liquids and solids in which a mass (weight) analysis applies, a convenient basis is often 1 or 100 lb or kg; similarly, because gas compositions are usually provided in terms of moles, 1 or 100 moles is often a good choice for a gas.

Always state the basis you have chosen for your calculations by writing it prominently on your calculation sheets or in the computer program used to solve the problem.

Example 2.13   Choosing a Basis

Problem Statement

The dehydration of the lower-molecular-weight alkanes can be carried out using a ceric oxide (CeO) catalyst. What are the mass fraction and mole fraction of Ce and O in the catalyst?

Problem Solution

Start the solution by selecting a basis. Because no specific amount of material is specified, the question "What do I have to start with?" does not help determine a basis. Neither does the question about the desired answer. Thus, selecting a convenient basis becomes the best choice. What do you know about CeO? You know from the formula that 1 mole of Ce is combined with 1 mole of O. Consequently, a basis of 1 kg mol (or 1 g mol, or 1 lb mol, etc.) would make sense. You can get the atomic weights for Ce and O from the back inside cover, and then you are prepared to calculate the respective masses of Ce and O in CeO. The calculations for the mole and mass fractions for Ce and O in CeO are presented in the following table:

Basis: 1 kg mole of CeO

Component	kg mol	Mole Fraction	Mol. Wt.	kg	Mass Fraction
Ce	1	0.50	140.12	140.12	0.8975
O	1	0.50	16.0	16.0	0.1025
Total	2	1.00	156.12	156.12	1.0000

You no doubt have heard the story of Ali Baba and the 40 thieves. Have you heard about Ali Baba and the 39 camels? Ali Baba gave his four sons 39 camels to be divided among them so that the oldest son got one half of the camels, the second son a quarter, the third an eighth, and the youngest a tenth. The four brothers were at a loss as to how they should divide the inheritance without killing camels until a stranger came riding along on his camel. He added his own camel to Ali Baba's 39 and then divided the 40 among the sons. The oldest son received 20; the second, 10; the third, 5; and the youngest, 4. One camel was left. The stranger mounted it, for it was his own, and rode off. Amazed, the four brothers watched him ride away. The oldest brother was the first to start calculating. Had his father not willed half of the camels to him? Twenty camels are obviously more than half of 39. One of the four sons must have received less than his due. But figure as they would, each found that he had more than his share. Cover the next few lines of text. What is the answer to the paradox?

After thinking over this problem, you will realize that the sum of 1/2, 1/4, 1/8, and 1/10 is not 1 but is 0.975. By adjusting (normalizing) the camel fractions (!) so that they total 1, the division of camels is validated:

	Camel Fractions	Normalizing		Correct Fractions		Distributed Camels (Integer)
	0.500	$\left(\dfrac{0.500}{0.975}\right)$	$=$	0.5128×39	$=$	20
	0.250	$\left(\dfrac{0.250}{0.975}\right)$	$=$	0.2564×39	$=$	10
	0.125	$\left(\dfrac{0.125}{0.975}\right)$	$=$	0.1282×39	$=$	5
	0.100	$\left(\dfrac{0.100}{0.975}\right)$	$=$	0.1026×39	$=$	4
Total	0.975	$\left(\dfrac{0.975}{0.975}\right)$	$=$	1.000	$=$	39

What we have done is to change the calculations from a basis total of 0.975 to a new basis of 1.000.

More frequently than you probably would like, you will have to change from your original selection of a basis in solving a problem to one or more different bases in order to put together the information needed to solve the entire problem. Consider the following example.

Example 2.14   Changing Bases

Problem Statement

Considering a gas containing O_2 (20%), N_2 (78%), and SO_2 (2%), find the composition of the gas on an SO_2-*free basis*, meaning gas without the SO_2 in it.

Solution

First choose a basis of 1 mol of gas (or 100 mol). Why? The composition for the gas is in mole percent. Next you should calculate the moles of each component, remove the SO_2, and adjust the basis for the calculations so that the gas becomes composed of only O_2 and N_2 with a percent composition totaling 100%:

Basis: 1.0 mol of gas

Components	Mol Fraction	Mol	Mol SO_2-Free	Mol Fraction SO_2-Free
O_2	0.20	0.20	0.20	0.20
N_2	0.78	0.78	0.78	0.80
SO_2	0.02	0.02		
	1.00	1.00	0.98	1.00

The round-off in the last column is appropriate given the original values for the mole fractions.

Self-Assessment Test

Questions

1. What are the three questions you should ask yourself when selecting a basis?
2. Why do you sometimes have to change bases during the solution of a problem?

Answers

1. See text.
2. For convenience or to simplify the calculations.

Problem

1. What would be good initial bases to select for solving problems 2.6.7b, 2.6.8, and 2.6.11 at the end of the chapter?

Answer

1. (a) 1 lb mol; (b) 100 kg mol; (c) 100 kg

2.6.2   The Mole and Molecular Weight

What is a mole? For our purposes, we will say that a **mole** is a certain amount of material corresponding to a specified number of molecules, atoms, electrons, or other specified types of particles.

In the SI system, a mole (which we will call a **gram mole,** i.e., g mol, to avoid confusing units) is composed of 6.022×10^{23} (Avogadro's number) molecules. However, for convenience in calculations and for clarity, we will make use of other specifications for moles, such as the **pound mole** (lb mol, composed of $6.022 \times 10^{23} \times 453.6$) molecules, the **kg mol** (kilomole, kmol, composed of 1000 moles), and so on. You will find that such nonconforming (to SI) definitions of the amount of material will help avoid excess details in many calculations.

One important calculation at which you should become skilled is converting the number of moles to mass and the mass to moles. To do this you make use of the **molecular weight**—*the mass per mole:*

$$\text{molecular weight (MW)} = \frac{\text{mass}}{\text{mole}} \qquad (2.5)$$

Based on this definition of molecular weight:

$$\text{g mol} = \frac{\text{mass in g}}{\text{molecular weight}}$$

$$\text{lb mol} = \frac{\text{mass in lb}}{\text{molecular weight}}$$

Therefore, from the definition of the molecular weight, you can calculate the mass knowing the number of moles or the number of moles knowing the mass. For historical reasons, the terms *atomic weight* and *molecular weight* are usually used instead of the more accurate terms *atomic mass* and *molecular mass.*

Example 2.15   Use of Molecular Weights to Convert Mass to Moles

Problem Statement

If a bucket holds 2.00 lb of NaOH:

  a. How many pound moles of NaOH does it contain?

  b. How many gram moles of NaOH does it contain?

Solution

You can convert pounds to pound moles, and then convert the values to the SI system of units. Look up the molecular weight of NaOH, or calculate it from the atomic weights. (It is 40.0.) Note that the molecular weight is used as a conversion factor in this calculation:

a. $\dfrac{2.00 \text{ lb NaOH}}{} \left| \dfrac{1 \text{ lb mol NaOH}}{40.0 \text{ lb NaOH}} = 0.050 \text{ lb mol NaOH}\right.$

b1. $\dfrac{2.00 \text{ lb NaOH}}{} \left| \dfrac{1 \text{ lb mol NaOH}}{40.0 \text{ lb NaOH}} \right| \dfrac{454 \text{ g mol}}{1 \text{ lb mol}} = 22.7 \text{ g mol}$

Check your answer by converting the 2.00 lb of NaOH to the SI system first and completing the conversion to gram moles:

b2. $\dfrac{2.00 \text{ lb NaOH}}{} \left| \dfrac{454 \text{ g}}{1 \text{ lb}} \right| \dfrac{1 \text{ g mol NaOH}}{40.0 \text{ g NaOH}} = 22.7 \text{ g mol}$

Example 2.16   Use of Molecular Weights to Convert Moles to Mass

Problem Statement

How many pounds of NaOH are in 7.50 g mol of NaOH?

(Continues)

Example 2.16   Use of Molecular Weights to Convert Moles to Mass (*Continued*)

Solution

This problem involves converting gram moles to pounds. From Example 2.15, the molecular weight of NaOH is 40.0:

$$\frac{7.50 \text{ g mol NaOH}}{} \left| \frac{1 \text{ lb mol}}{454 \text{ g mol}} \right| \frac{40.0 \text{ lb NaOH}}{1 \text{ lb mol NaOH}} = 0.66 \text{ lb NaOH}$$

Note the conversion between pound moles and gram moles was to proceed from SI to the AE system of units. Could you first convert 7.50 g mol of NaOH to grams of NaOH, and then use the conversion of 454g = 1 lb to get pounds of NaOH? Of course.

Values of the **molecular weights** (relative molar masses) are built up from the values of atomic weights based on a scale of the *relative* masses of the elements. The **atomic weight** of an element is the mass of an atom based on the scale that assigns a mass of exactly 12 to the carbon isotope ^{12}C. The value 12 is selected in this case because an atom of carbon 12 contains 6 protons and 6 neutrons for a total molecular weight of 12.

The back inside cover lists the atomic weights of the elements. On this scale of atomic weights, hydrogen is 1.008, carbon is 12.01, and so on. (In most of our calculations, we shall round these off to 1 and 12, respectively, for convenience). The atomic weights of these and other elements are not whole numbers because elements can appear in nature as a mixture of different isotopes. As an example, approximately 0.8% of hydrogen is deuterium (a hydrogen atom with one proton and one neutron); thus, the atomic weight is 1.008 instead of 1.000.

A **compound** is composed of more than one type of atom, and the molecular weight of the compound is nothing more than the sum of the weights of atoms of which it is composed. Thus, H_2O consists of 2 hydrogen atoms and 1 oxygen atom, and the molecular weight of water is (2) (1.008) + 16.000 = 18.016, or approximately 18.02.

You can compute **average molecular weights** for mixtures of constant composition even though they are not chemically bonded if their compositions are known accurately. Example 2.17 shows how to calculate the fictitious quantity called the average molecular weight of air. Of course, for a material such as fuel oil or coal whose composition may not be exactly known, you cannot determine an exact molecular weight, although you might estimate an approximate average molecular weight, which is good enough for most engineering calculations.

Example 2.17   Average Molecular Weight of Air

Problems Statement

Calculate the average molecular weight of air, assuming that air is 21% O_2 and 79% N_2.

Solution

Because the composition of air is given in mole percent, a basis of 1 g mol is chosen. The molecular weight of the N_2 is not actually 28.0 but 28.2 because the value of the molecular weight of the pseudo 79% N_2 is actually a combination of 78.084% N_2 and 0.934% Ar. The masses of the O_2 and pseudo N_2 are

Basis: 1 g mol of air

$$\text{Mass of } O_2 = \frac{1 \text{ g mol air}}{} \left| \frac{0.21 \text{ g mol } O_2}{\text{g mol air}} \right| \frac{32.00 \text{ g } O_2}{\text{g mol } O_2} = 6.72 \text{ g } O_2$$

$$\text{Mass of } N_2 = \frac{1 \text{ g mol air}}{} \left| \frac{0.79 \text{ g mol } N_2}{\text{g mol air}} \right| \frac{28.2 \text{ g } N_2}{\text{g mol } N_2} = 22.28 \text{ g } N_2$$

$$\text{Total} \qquad\qquad\qquad\qquad\qquad\qquad\qquad = 29.0 \text{ g air}$$

Therefore, the total mass of 1 g mol of air is equal to 29.0 g, which is called the average molecular weight of air. (Because we chose 1 g mol of air as the basis, the total mass calculated directly provides the average molecular weight of 29.0.)

Example 2.18   Average Molecular Weight

Problem Statement

Most processes for producing high-energy-content gas or gasoline from coal include some type of gasification step to make hydrogen or synthesis gas. Pressure gasification is preferred because of its greater yield of methane and higher rate of gasification.

Given that a 50.0 kg test run of gas averages 10.0% H_2, 40.0% CH_4, 30.0% CO, and 20.0% CO_2, what is the average molecular weight of the gas?

(Continues)

Example 2.18   Average Molecular Weight (*Continued*)

Solution

Let's choose a basis. The answer to question 1 is to select a basis of 50.0 kg of gas ("What do I have to start with?"), but is this choice a good basis? A little reflection shows that such a basis is of no use. You cannot multiply the given *mole percent* of this gas (remember that the composition of gases is given in mole percent unless otherwise stated) times kilograms and expect the result to mean anything. Try it, being sure to include the respective units. Thus, the next step is to choose a "convenient basis," which is, say, 100 kg mol of gas, and proceed as follows:

Basis: 100 kg mol or lb mol of gas

Set up a table such as the following to make a compact presentation of the calculations. You do not have to, but making individual computations for each component is inefficient and more prone to errors.

Component	Percent = kg mol or lb mol	Mol. Wt.	kg or lb
CO_2	20.0	44.0	880
CO	30.0	28.0	840
CH_4	40.0	16.04	642
H_2	10.0	2.02	20
Total	100.0		2382

$$\text{Average molecular weight} = \frac{2382 \text{ kg}}{100 \text{ kg mol}} = 23.8 \text{ kg/kg mol}$$

Check the solution by noting that an average molecular weight of 23.8 is reasonable because the molecular weights of the components range only from 2 to 44 and the answer is intermediate to these values.

To sum up, be sure to state the basis of your calculations so that you will keep clearly in mind their real nature and so that anyone checking your problem solution will be able to understand on what basis your calculations were performed.

Example 2.19   Average Molecular Weight of a Superconductor

Problem Statement

Since the discovery of superconductivity almost 100 years ago, scientists and engineers have speculated about how it can be used to improve the use of energy. Until recently, most applications were not economically viable because the niobium alloys used had to be cooled below 23 K by liquid He. However, in 1987, superconductivity in Y-Ba-Cu-O material was achieved at 90 K, a situation that permits the use of inexpensive liquid N_2 cooling.

What is the molecular weight of the cell of a superconductor material shown in Figure E2.19? (The figure represents one cell of a larger structure.)

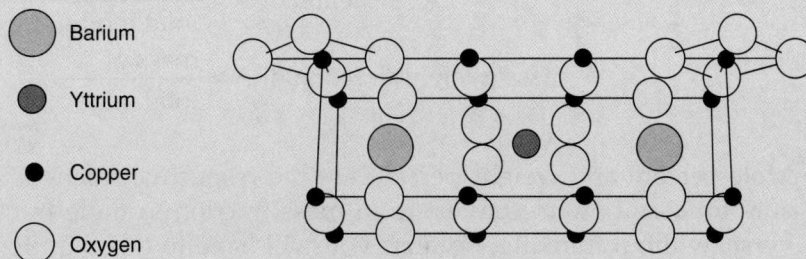

Figure E2.19   One cell of the Y-Ba-Cu-O superconductor

Solution

First look up the atomic weights of the elements from the table on the back inside cover. Assume that one cell is a molecule. By counting the atoms, you can find

Element	Number of Atoms	Atomic Weight (g)	Mass (g)
Ba	2	137.34	2(137.34)
Cu	16	63.546	16(63.546)
O	24	16.00	24(16.00)
Y	1	88.905	1(88.905)
		Total	1784.3

The molecular weight of the cell is 1784.3 atomic masses/1 molecule or 1784.3 g/g mol. Check your calculations and check your answer to ensure that it is reasonable.

Mole fraction is simply the number of moles of a particular substance in a mixture or solution divided by the total number of moles present in the mixture or solution. This definition holds for gases, liquids, and solids. Similarly, the **mass (weight) fraction** is nothing more than the **mass (weight)** of the substance divided by the total mass (weight) of all of the substances present in the mixture or solution. Although *mass fraction* is the correct term, by custom, ordinary engineering usage frequently employs the term **weight fraction.** These concepts can be expressed as

$$\text{mole fraction of A} = \frac{\text{moles of A}}{\text{total moles}}$$

$$\text{mass (weight) fraction of A} = \frac{\text{mass of A}}{\text{total mass}}$$

(2.6)

Mole percent and weight percent are the respective fractions times 100. Be sure to learn how to convert from mass fraction to mole fraction, and vice versa, without thinking, because you will have to do so quite often. **Unless otherwise specified, when a percentage or fraction is given for a gas, it is assumed that it refers to a mole percentage or a mole fraction. When a percentage or fraction is given for a liquid or a solid, it is assumed that it refers to a weight percentage or a weight fraction.**

Example 2.20   Conversion between Mass (Weight) Fraction and Mole Fraction

Problem Statement

An industrial-strength drain cleaner contains 5.00 kg of water and 5.00 kg of NaOH. What are the mass (weight) fraction and mole fraction of each component in the drain cleaner?

Solution

You are given the masses, so it is easy to calculate the mass fractions. From these values, you can then calculate the desired mole fractions. A convenient way to carry out the calculations in such conversion problems is to form a table, as shown here.

		Basis: 10.0 kg of total solution				

Component	kg	Weight Fraction	Mol. Wt.	kg mol	Mole Fraction	
H_2O	5.00	$\dfrac{5.00}{10.0} = 0.500$	18.0	0.278	$\dfrac{0.278}{0.403} = 0.69$	= 0.69
NaOH	5.00	$\dfrac{5.00}{10.00} = 0.500$	40.0	0.125	$\dfrac{0.125}{0.403} = 0.31$	= 0.31
Total	10.00	1.000		0.403	1.00	

Self-Assessment Test

Questions

1. Indicate whether the following statements are true or false:
   a. The pound mole is composed of 2.73×10^{26} molecules.
   b. The kilogram mole is composed of 6.023×10^{26} molecules.
   c. Molecular weight is the mass of a compound or element per mole.

2. What is the molecular weight of acetic acid (CH_3COOH)?

Answers

1. (a) T; (b) T; (c) T
2. 60.05

Problems

1. Convert the following:
   a. 120 g mol of NaCl to grams
   b. 120 g of NaCl to gram moles
   c. 120 lb mol of NaCl to pounds
   d. 120 lb of NaCl to pound moles

2. Convert 39.8 kg of NaCl per 100 kg of water to kilogram moles of NaCl per kilogram mole of water.

3. How many pound moles of $NaNO_3$ are there in 100 lb?

4. Commercial sulfuric acid is 98% H_2SO_4 and 2% H_2O. What is the mole ratio of H_2SO_4 to H_2O?

5. A solid compound contains 50% sulfur and 50% oxygen. Is the empirical formula of the compound (a) SO, (b) SO_2, (c) SO_3, or (d) SO_4?

6. A gas mixture contains 40 lb of O_2, 25 lb of SO_2, and 30 lb of SO_3. What is the composition of the mixture?

Answers

1. (a) 7010 g; (b) 2.05 g mol; (c) 7010 lb; (d) 2.05 lb mol

2. 0.123 kg mol NaCl/kg mol H_2O

3. 1.177 lb mol

4. 9

5. SO_2

6. O_2 0.62; SO_2 0.19; SO_3 0.19

2.6.3   Density and Specific Gravity

In ancient times, counterfeit gold objects were identified by determining the ratio of the weight to the volume of water displaced by the object (which is a way to measure the density of the material of the object) and comparing it to that of an object known to be made of gold.

A striking example of quick thinking by an engineer who made use of the concept of density was reported by P. K. N. Paniker in the June 15, 1970, issue of *Chemical Engineering*:

> The bottom outlet nozzle of a full lube-oil storage tank kept at a temperature of about 80°C suddenly sprang a gushing leak as the nozzle flange became loose. Because of the high temperature of the oil, it was impossible for anyone to go near the tank and repair the leak to prevent further loss.
>
> After a moment of anxiety, we noticed that the engineer in charge rushed to his office to summon fire department personnel and instruct them to run a hose from the nearest fire hydrant to the top of the storage tank. Within minutes, what gushed out from the leak was hot water instead of valuable oil. Some time later, as the entering cold water lowered the oil temperature, it was possible to make repairs.

Density (we use the Greek symbol ρ) is the ratio of mass per unit volume, such as kg/m^3 or lb_m/ft^3:

$$\rho = \text{density} = \frac{\text{mass}}{\text{volume}} = \frac{m}{V} \qquad (2.7)$$

Density has both a numerical value and units. Densities for liquids and solids do not change significantly at ordinary conditions with pressure, but they can change significantly with temperature for certain compounds if the temperature change is large enough, as shown in Figure 2.4. Note that between 0°C and 70°C, the density of water is relatively constant at 1.0 g/cm^3. In contrast, for the same temperature range, the density of NH_3 changes by approximately 30%. Usually, we ignore the effect of temperature on liquid density unless the density of the material is especially sensitive to temperature or the change in the temperature is particularly large.

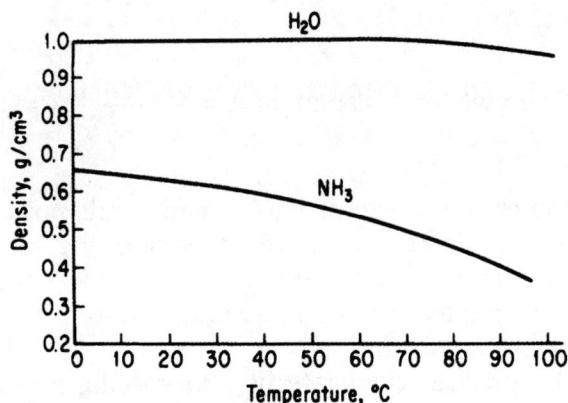

Figure 2.4   Densities of liquid H_2O and NH_3 as a function of temperature

Specific volume (we use the symbol $\hat{V}$) is the inverse of density, such as cm^3/g or ft^3/lb:

$$\hat{V} = \text{specific volume} = \frac{\text{volume}}{\text{mass}} = \frac{V}{m} \tag{2.8}$$

Because density is the ratio of mass to volume, it can be used to calculate the mass given the volume or calculate the volume knowing the mass. For example, given that the density of n-propyl alcohol is 0.804 g/cm^3, what would be the volume of 90.0 g of the alcohol? The calculation is

$$\frac{90.0\,g}{}\left|\frac{1\,cm^3}{0.804\,g} = 112\,cm^3\right.$$

Some quantities related to density are molar density (ρ/MW) and molar volume (MW/ρ). By analogy, in a packed bed of solid particles containing void spaces, the bulk density is

$$\rho_B = \text{bulk density} = \frac{\text{total mass of solids}}{\text{total empty bed volume}}$$

Now let's turn to specific gravity. **Specific gravity** is the ratio of the density of a substance to the density of a reference material. In symbols for compound A:

$$\text{specific gravity of A} = \text{sp.gr. of A} = \frac{(\text{g}/\text{cm}^3)_A}{(\text{g}/\text{cm}^3)_{\text{ref}}} = \frac{(\text{kg}/\text{m}^3)_A}{(\text{kg}/\text{m}^3)_{\text{ref}}} = \frac{(\text{lb}/\text{ft}^3)_A}{(\text{lb}/\text{ft}^3)_{\text{ref}}}$$

The reference substance for liquids and solids normally is water. Thus, the specific gravity is the ratio of the density of the substance of interest to the density of water, namely, 1.000 g/cm^3, 1000 kg/m^3, or 62.43 lb/ft^3 at 4°C. The specific gravity of gases frequently is referred to air but may be referred to other gases.

To be precise when referring to specific gravity, the data should be accompanied by both the temperature of the substance of interest and the temperature at which the reference density is measured. Thus, for solids and liquids, the notation

$$\text{sp.gr.} = 0.73\frac{20°}{4°} = 0.73$$

can be interpreted as follows: The specific gravity when the solution is at 20°C and the reference substance (implicitly water) is at 4°C is 0.73. In case the temperatures for which the specific gravity is stated are unknown, assume ambient temperature for the substance and 4°C for the water. Because the density of water at 4°C is very close to 1.0000 g/cm^3, in units of grams per cubic centimeter, the numerical values of the specific gravity and the density are essentially equal when the density is expressed in units of g/cm^3.

Note that the units of specific gravity as used here clarify the calculations and that the calculation of density from the specific gravity can often be done in your head. Because densities in the AE system are expressed in pounds per cubic foot, and the density of water is about 62.4 lb/ft^3, you can see that the specific gravity and density values are not numerically equal in that system. Densities of several liquids as function of temperature and pressure are available from the thermophysical properties database hosted by NIST - National Institute of Standards and Technology.

Example 2.21   Calculation of Density Given the Specific Gravity

Problem Statement

If penicillin has a specific gravity of 1.41, what is the density in (a) g/cm^3, (b) lb_m/ft^3, and (c) kg/m^3?

Solution

Start with the specific gravity to get the density via a reference substance. No temperatures are cited for the penicillin (P) or the reference compound (presumed to be water); hence, for simplicity, we assume that the penicillin is at room temperature (22°C) and that the reference material is water at 4°C. Therefore, the reference density is 62.4 lb/ft^3 or 1.00×10^3 kg/m^3 (1.00 g/cm^3).

a. $$\frac{1.41\dfrac{g\,P}{cm^3}}{1.00\dfrac{g\,H_2O}{cm^3}}\left|1.00\dfrac{g\,H_2O}{cm^3}\right. = 1.41\dfrac{g\,P}{cm^3}$$

b. $$\frac{1.41\dfrac{lb_m\,P}{ft^3}}{1.00\dfrac{lb_m\,H_2O}{ft^3}}\left|62.4\dfrac{lb_m\,H_2O}{ft^3}\right. = 88.0\dfrac{lb_m\,P}{ft^3}$$

c. $$\frac{1.41\,g\,P}{cm^3}\left|\left(\frac{100\,cm}{1\,m}\right)^3\right|\frac{1\,kg}{1000\,g} = 1.41\times10^3\dfrac{kg\,P}{m^3}$$

You should become acquainted with the fact that, in the petroleum industry, the specific gravity of petroleum products is often reported in terms of a hydrometer scale called °**API.** The equations that relate the API scale to density, and vice versa, are

$$°\text{API} = \frac{141.5}{\text{sp.gr.} \dfrac{60°\text{F}}{60°\text{F}}} - 131.5 \ \ (\text{API gravity}) \tag{2.9}$$

or

$$\text{sp.gr.} \frac{60°}{60°} = \frac{141.5}{°\text{API} + 131.5} \tag{2.10}$$

The volume and therefore the density of petroleum products vary with temperature, and the petroleum industry has established 60°F as the standard temperature for specific gravity and API gravity.

Example 2.22   Application of Specific Gravity to Calculate Mass and Moles

Problem Statement

In the production of a drug having a molecular weight of 192, the exit stream from the reactor containing water and the drug flows at the rate of 10.5 L/min. The drug concentration is 41.2% (in water), and the specific gravity of the solution is 1.024. Calculate the concentration of the drug (in kilograms per liter) in the exit stream, and the flow rate of the drug in kilogram moles per minute.

Solution

Read the problem carefully because this example is more complicated than the previous examples. You have a problem with some known properties, including specific gravity. The strategy for the solution is to use the specific gravity to get the density, from which you can calculate the moles per unit volume.

For the first part of the problem, you want to transform the mass fraction of 0.412 into mass per liter of the drug. Take 1.000 kg of the exit solution as a basis because the mass fraction of the drug in the product is specified in the problem statement.

Basis: 1.000 kg solution

How do you get mass of drug per volume of solution (the density) from the given data, which are in terms of the fraction of the drug (0.412)? Use the given specific gravity of the solution. Calculate the density of the solution as follows:

$$\text{density of solution} = (\text{sp.gr.})\,(\text{density of reference})$$

$$\text{density of solution} = \frac{1.024\,\dfrac{\text{g soln}}{\text{cm}^3\,\text{soln}}}{1.000\,\dfrac{\text{g H}_2\text{O}}{\text{cm}^3\,\text{H}_2\text{O}}}\left|\,1.000\,\dfrac{\text{g H}_2\text{O}}{\text{cm}^3\,\text{H}_2\text{O}}\right. = 1.024\,\dfrac{\text{g soln}}{\text{cm}^3\,\text{soln}}$$

The detail of the calculation of the density of the solution showing the units may seem excessive but is presented to make the calculation clear. Next, convert the amount of drug in 1.000 kg of solution to mass of drug per volume of solution using the density previously calculated, recognizing that there is 0.412 kg of the drug for the basis of 1.000 kg of solution.

$$\frac{0.412\ \text{kg drug}}{1.000\ \text{kg soln}}\left|\frac{1.024\ \text{g soln}}{1\ \text{cm}^3\ \text{soln}}\right|\frac{1\ \text{kg soln}}{10^3\ \text{g soln}}\left|\frac{1000\ \text{cm}^3\ \text{soln}}{1\ \text{L soln}}\right. = 0.422\ \text{kg drug/L soln}$$

Note that a distinction is drawn between properties of the solution (e.g., g soln, L soln) and the mass of the drug to prevent confusion in the cancelation of units. To get the flow rate, take a different basis, namely, 1 min.

$$\text{Basis: 1 min} = 10.5\ \text{L of solution}$$

Convert the selected volume to mass and then to moles using the information previously calculated:

$$\frac{10.5\ \text{L soln}}{1\ \text{min}}\left|\frac{0.422\ \text{kg drug}}{1\ \text{L soln}}\right|\frac{1\ \text{kg mol drug}}{192\ \text{kg drug}} = 0.0231\ \text{kg mol/min}$$

How might you check your answers?

Self-Assessment Test

Questions

1. Indicate whether the following statements are true or false:
   a. The inverse of the density is the specific volume.
   b. The density of a substance has the units of the mass per unit volume.
   c. The density of water is less than the density of mercury.

2. A cubic centimeter of mercury has a mass of 13.6 g. What is the density of mercury?

3. For liquid HCN, a handbook gives sp. gr. 10°C/4°C = 1.2675. What does this statement mean?

4. Indicate whether the following statements are true or false:
   a. The density and specific gravity of mercury are the same.
   b. Specific gravity is the ratio of two densities.
   c. If you are given the value of a reference density, you can determine the density of a substance of interest by multiplying by the specific gravity.
   d. The specific gravity is a dimensionless quantity.

5. If you put a glass bottle of beer in the freezer overnight, the bottle will burst. What does that tell you about how the density of beer is affected by temperature?

Answers

1. (a) T; (b) T; (c) T

2. 13.6 g/cm^3

3. The statement means that the density at 10°C of liquid HCN is 1.2675 times the density of water at 4°C.

4. (a) F—the units differ; (b) T; (c) T; (d) T

5. The specific volume of beer increases as the temperature is decreased. Therefore, the density decreases as the temperature decreases.

Problems

1. The density of a material is 2 kg/m^3. What is its specific volume?

2. If you add 50 g of sugar to 500 mL of water, how do you calculate the density of the sugar solution?

3. For ethanol, a handbook gives sp. gr. 60°F = 0.79389. What is the density of ethanol at 60°F?

4. The specific gravity of steel is 7.9. What is the volume in cubic feet of a steel ingot weighing 4000 lb?

5. A solution in water contains 1.704 kg of HNO_3/kg H_2O, and the solution has a specific gravity of 1.382 at 20°C. How many kilograms of HNO_3 per cubic meter of solution at 20°C are there?

Answers

1. 0.5 m^3/kg

2. Measure the mass of the water (should be about 500 g) and add it to 50 g of sugar. Measure the volume of the solution and divide the total mass by the volume.

3. 0.79389 g/cm^3 (assuming the density of water is also at 60°F)
4. 8.11 ft^3
5. 871 kg HNO$_3$/m^3 solution

2.7   Process Variables

As pointed out earlier, chemical engineers are process engineers, and this section addresses how processes are described in terms of commonly used process variables (e.g., temperatures, pressures, compositions, and flow rates). A **process** is a system that takes feed material and converts it into products on either a continuous or a batch basis. Examples include a crude unit of a refinery that uses crude oil as feed and converts it into gasoline, diesel fuel, and other products on a continuous basis; the fermentation process in an ethanol plant uses glucose made from corn as a feedstock and produces ethanol using fermentation yeast in a batch process to produce an aqueous solution of ethanol; silicon wafers are processed using lithography to produce microprocessors that are used in computers and cell phones; and potatoes are the feedstock for processes that make potato chips.

The remainder of this text addresses how to apply material and energy balances to processes. Material and energy balances are the foundation of process models, which are used to design or analyze the operation of processes. The design of a process determines the processing units and the size of each element of the process using process models so that the specified quality and production rate of products is attained. Process models also are used to analyze the operation of an existing process. For example, process models can be used to solve problems with the operation of an existing process (i.e., **process troubleshooting,** e.g., modifying the operation of a process so that it no longer produces off-specification products) or to evaluate ways to increase the production rate of an existing process typically requiring relatively minor modifications to the process (i.e., **debottlenecking**).

The following material in this section will present process variables that are commonly used to describe the operation of processes using material and energy balances. Various sensors are used to measure specific process variables of a process and the precision of the measurement depends to a large extent on the type of sensor used. In general, two different types of sensors are used on processes: **field-mounted sensors,** which have to be read while in the process, and **board-mounted sensors,** which can be read from the control room because their readings are electronically transmitted from the process to the control room. Table 2.8 lists the repeatability based on the proper operation of a number of board-mounted process sensors used in the CPI and biotechnology industries. That is, the repeatability is the normal variation in the measurement of a process variable. The sensors shown in Table 2.8 are described and discussed in the remainder of this section.

Table 2.8   Repeatability of Various Process Sensors

Process Measurement	Sensor Type	Repeatability
Temperature	Thermocouple	±1°C
	RTDs	±0.1°C
Pressure	Differential pressure (DP) cell	±0.1%
Level in a vessel	DP level indicator	±1%
Flow rate	Orifice meter	±0.3% − ±1%
	Magnet flow meter	±0.1%
	Vortex shedding meter	±0.2%
	Coriolis meter	±0.1% − ±0.5%
Solution Ph	pH electrode	±0.1 pH units
Composition	Dissolved O_2 electrode	±0.1% − ±0.5%
	Turbidity meter	±2.5%

Source: J. B. Riggs, M. N. Karim, and J. S. Alford, *Chemical and Bio-Process Control*, 5th ed. (Austin, TX: Ferret Publishing, 2020), 83.

2.7.1 Temperature

The temperature of a material is directly related to the kinetic energy of the atoms present in the material. Temperature measurements are typically based on a relative scale defined by the conditions under which water freezes and boils. Mercury-in-bulb thermometers are based on the fact that mercury expands when heated, and they are calibrated knowing that water boils at 100°C (212°F) at 1 atmosphere pressure.

Many times, temperature is an important process variable that requires close observation and control because it has a very strong effect on chemical and biochemical reactions. In general, the reaction rate of a chemical reaction will double for a 10°C increase in reaction temperature. In contrast, for biological reactions, an increase in temperature will also usually cause an increase in the reaction rate, but above a certain temperature, the elements of the biological reaction (i.e., enzymes and cells) become denatured, causing the reaction rate to decrease. Temperature measurements are also important for a number of processes other than reactors. For example, the temperatures of trays in a distillation column can be used to estimate the purity of products produced by the column. And the temperature of molten polymer in an extruder can have a dramatic effect on its operation.

Industrial Temperature Sensors. The temperature sensors most commonly used by industry are thermocouples, resistance temperature detectors (RTDs), and optimal pyrometers. Thermocouples are based on the fact that two metal junctions at different temperature will generate a voltage proportional to the temperature difference. RTDs are based on the observation that the electrical resistance of certain metals is a strong function of temperature. Thermocouples are less expensive and more rugged than RTDs, but RTDs are much more accurate. Therefore, for important temperatures, such as reactor temperatures or distillation tray temperatures, RTDs are normally used, while thermocouples are used for less important temperature points. Optimal pyrometers are used to measure very high temperature (700°C to 4000°C), such as the temperature of the tubes on the inside wall of a furnace.

2.7.2 Pressure and Hydrostatic Head

In Florentine Italy in the seventeenth century, well diggers observed that when they used suction pumps, water would not rise more than about 10 m (32.8 ft). In 1642, they came to the famous Galileo for help, but he did not want to be bothered. As an alternative, they sought the help of Torricelli. He learned from experiments that water was not being pulled up by the vacuum but rather was being pushed up by the local air pressure. Therefore, the maximum height that water could be raised from a well using a suction pump at the surface depended on the atmospheric pressure.

Pressure is defined as the normal (perpendicular) force per unit area. In the SI system, the force is expressed in newtons and the area in square meters; then the pressure is N/m^2 or pascal (Pa). (The value of a pascal is so small that the kilopascal, kPa, is a more convenient unit of pressure.) In the AE system, the force is the pounds force and the area used is square inches (lb_f/in^2).

What are other units for pressure? Look at Table 2.9 for some of the most common ones. Note that each pressure unit is expressed as the equivalent of 1 standard atmosphere. Keep in mind that the pounds referred to in psi are pounds force, not pounds mass.

Table 2.9 Convenient Conversion Factors for Pressure

Pressure Units	Conversion Factor
bar	1.013 bar = 1 atm
kPa	101.3 kPa = 1 atm
Torr	760 Torr = 1 atm
mm Hg	760 mm Hg = 1 atm
in. Hg	29.92 in. Hg = 1 atm
ft H_2O	33.94 ft H_2O = 1 atm
in. H_2O	407 in. H_2O = 1 atm
psi	14.69 psi = 1 atm

Examine Figure 2.5. Pressure is exerted on the top of the mercury in the cylinder by the atmosphere. The pressure at the bottom of the column of mercury is equal to the pressure exerted by the mercury plus that of the atmosphere on the mercury.

The pressure at the bottom of the **static** (nonmoving) column of mercury (also known as the hydrostatic pressure) exerted on the sealing plate is

$$p = \frac{F}{A} = \frac{mg}{A} + p_0 = \frac{mgh}{Ah} + p_0 = \frac{mgh}{V} + p_0 = \rho g h + p_0 \qquad (2.11)$$

where the first term after the pressure p is the definition of pressure, the second term is the combination of atmospheric pressure and the pressure change due to the column of liquid, and the third shows how h can be added to the numerator and denominator to get the volume in the denominator. In the fourth term, volume is substituted for area times height, and in the fifth term, the density is substituted for mass divided by volume. The notation used is as follows:

p = pressure at the bottom of the column of fluid
F = force
A = area
ρ = density of fluid
g = acceleration of gravity
h = height of the fluid column
p_0 = pressure at the top of the column of fluid

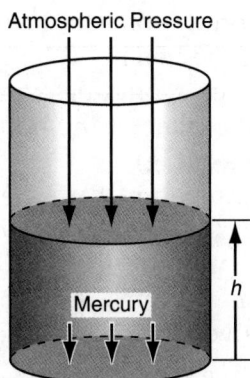

Figure 2.5 Pressure is the normal force per unit area. Arrows show the force exerted on the respective areas.

You can calculate the force exerted at the bottom of a static fluid by applying Equation (2.11). For example, suppose that the cylinder of fluid in Figure 2.5 is a column of mercury with a cross-sectional area of $1\ cm^2$ and is 50 cm high. From Table E.1 in Appendix E, you can find that the specific gravity of mercury at 20°C, and hence the density of the Hg, is $13.55\ g/cm^3$. Thus, the force exerted by the mercury alone on the $1\ cm^2$ section of the bottom plate by the column of mercury is

$$F = mg = \rho V g = \frac{13.55\ g}{cm^3} \left| \frac{980\ cm}{s^2} \right| 50\ cm \left| 1\ cm^2 \right| \frac{1\ kg}{1000\ g} \left| \frac{1\ m}{100\ cm} \right| \frac{1\ (N)(s^2)}{1\ (kg)(m)}$$

$$= 6.64\ N$$

The pressure on the section of the plate covered by the mercury is the force per unit area of the mercury *plus* the pressure (p_0) of the atmosphere that is over the mercury:

$$p = \frac{F}{A} + p_0 = \frac{6.64\ N}{1\ cm^2} \left(\frac{100\ cm}{1\ m} \right)^2 \frac{(1\ m^2)(1\ Pa)}{1\ N} \left| \frac{1\ kPa}{1000\ Pa} \right| + p_0 = 66.4\ kPa + p_0$$

If we had started with units in the AE system, the pressure would be computed as follows [the density of mercury is $(13.55)(62.4)\ lb_m/ft^3 = 845.5\ lb_m/ft^3$]:

$$p = \rho g V + p_0 = \frac{845.5\ lb_m}{1\ ft^3} \left| \frac{32.2\ ft}{s^2} \right| 50\ cm \left| \frac{1\ in.}{2.54\ cm} \right| \frac{1\ ft}{12\ in.} \left| \frac{1\ (s)^2(lb_f)}{32.174\ (ft)(lb_m)} \right| + p_0$$

$$= 1388\frac{lb_f}{ft^2} + p_0$$

Sometimes in engineering practice, a liquid column is referred to as *head of liquid,* the head being the height of the column of liquid. Thus, the pressure of the column of mercury could be expressed simply as 50 cm Hg, and the pressure on the sealing plate at the bottom of the column would be 50 cm Hg + p_0 (in centimeters of Hg).

Pressure, like temperature, can be expressed in either absolute **(psia)** or relative scales. Rather than using the word *relative,* the relative pressure is usually called **gauge pressure (psig)**. The atmospheric pressure is nothing more than the barometric pressure. The relationship between gauge and absolute pressure is given by the following expression:

$$p_{\text{absolute}} = p_{\text{gauge}} + p_{\text{atmospheric}} \tag{2.12}$$

Another term with which you should become familiar is **vacuum.** When you measure the pressure in *inches of mercury vacuum,* you are reversing the direction of measurement from the reference pressure, the atmospheric pressure, and toward zero absolute pressure; that is,

$$p_{\text{vacuum}} = p_{\text{atmospheric}} - p_{\text{absolute}} \tag{2.13}$$

As the vacuum value increases, the value of the absolute pressure being measured decreases. What is the maximum value of a vacuum measurement? A pressure that is only slightly below atmospheric pressure may sometimes be expressed as a *draft* in inches of water, as, for example, in the air supply to a furnace or a water cooling tower.

Example 2.23   Vacuum Pressure

Problem Statement

Express a vacuum pressure of 10.00 psia in terms of cm of Hg absolute.

Solution

A vacuum pressure of 10.00 psia corresponds to an absolute pressure of 4.69 psia (i.e., 14.69 − 10.00). Then, using the conversion factors in Table 2.9 yields

$$\frac{4.69 \text{ psia}}{} \left| \frac{760 \text{ mm Hg}}{14.69 \text{ psia}} \right| \frac{1 \text{ cm Hg}}{10 \text{ mm Hg}} = 26.26 \text{ cm Hg}$$

Here is an excerpt, along with a photograph (Figure 2.6), from a journal article[2] to remind you about vacuum.

Tanks are fragile. An egg can withstand more pressure than a tank. How was a vacuum created inside the vessel? As water was drained from the column, the vent to let in air was plugged up, and the resulting pressure difference between inside and out caused the stripper to fail.

Figure 2.6   A failed tank. (Image courtesy of "CCPS Process Safety Beacon," AIcHE)

Figure 2.7 illustrates the relationships among the pressure concepts. Note that the vertical scale is exaggerated for illustrative purposes. The dashed line illustrates the atmospheric **(barometric)** pressure, which changes from time to time. The solid horizontal line is the standard atmosphere. Point 1 in the figure denotes a pressure of 19.3 psia, that is, an absolute pressure referred to zero absolute pressure, or 4.6 psig (i.e., 19.3 psia − 14.7 psia) referred to the barometric pressure; point 2 is zero pressure; any point on the heavy line such as point 3 is a point corresponding to the standard pressure of 1 atm; point 4 illustrates a negative relative pressure, that is, a pressure less than atmospheric pressure. This latter type of measurement is described in Example 2.25 as a vacuum measurement. Point 5 indicates negative relative pressure that is above the standard atmospheric pressure.

---

[2] Roy E. Sanders, "Don't Become Another Victim of Vacuum," *Chemical Engineering Progress*, 89 (1993): 54–57.

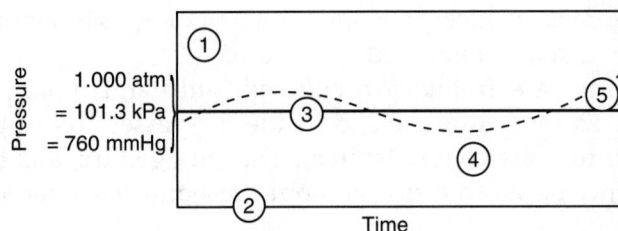

Figure 2.7   Pressure terminology

Please be careful not confuse the standard atmosphere with atmospheric pressure. The **standard atmosphere** is defined as the pressure (in a standard gravitational field) equivalent to 1 atm or 760 mm Hg at 0°C or other equivalent value and is fixed, whereas atmospheric pressure is variable and must be obtained from a barometric measurement or equivalent each time you need it.

You can easily convert from one pressure unit to another by using the relation between a pair of different pressure units as they respectively relate to the standard atmospheres so as to form a conversion factor. For example, let us convert 35 psia to inches of mercury and also to kilopascals by using the values in Table 2.9:

$$\frac{35 \text{ psia}}{} \left| \frac{29.92 \text{ in. Hg}}{14.7 \text{ psia}} = 71.24 \text{ in Hg}\right.$$

$$\frac{35 \text{ psia}}{} \left| \frac{101.3 \text{ kPa}}{14.7 \text{ psia}} = 241 \text{ kPa}\right.$$

Example 2.24   Pressure Conversion

Problem Statement

The pressure gauge on a tank of CO_2 used to fill soda-water bottles reads 51.0 psi. At the same time, the barometer reads 28.0 in. Hg. What is the absolute pressure in the tank in psia? See Figure E2.24.

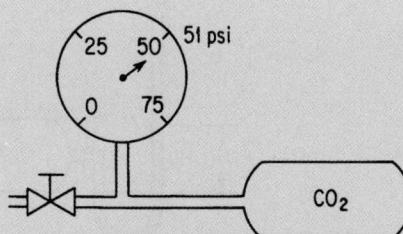

Figure E2.24

Solution

The first thing to do is to read the problem. You want to calculate a pressure using convenient conversion factors. Examine Figure E2.24. The system is the tank plus the line to the gauge. All of the necessary data are known except whether the pressure gauge reads absolute or gauge pressure. What do you think? It is more likely that the pressure gauge is reading psig, not psia, because the gauge has zero pressure marked on the gauge. Because the absolute pressure is the sum of the gauge pressure and the atmospheric (barometric) pressure expressed in the same units, you have to make the units the same in each term before adding or subtracting. Let's use psia. Start the calculations by changing the atmospheric pressure to psia:

$$\frac{28.0 \text{ in. Hg}}{} \left| \frac{14.7 \text{ psia}}{29.92 \text{ in. Hg}} = 13.8 \text{ psia} \right.$$

The absolute pressure in the tank is $51.0 + 13.8 = 64.8$ psia.

Example 2.24 identifies an important issue. **Whether relative or absolute pressure is measured in a pressure-measuring device depends on the nature of the instrument used to make the measurements.** For example, an open-end **manometer** (Figure 2.8a) would measure a gauge pressure because the reference pressure is the pressure of the atmosphere at the open end of the manometer. On the other hand, closing off the open end of the manometer (Figure 2.8b) and creating a vacuum in that end results in a measurement against a complete vacuum and hence is reported as an **absolute pressure.** In Figure 2.8b, if the pressure of the nitrogen in the tank is atmospheric pressure, what will the manometer read approximately?

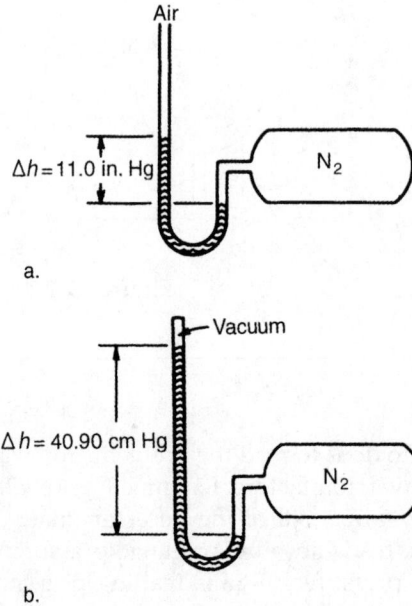

Figure 2.8 (a) Open-end manometer showing a pressure above atmospheric pressure in the tank; (b) manometer measuring absolute pressure in the tank

Sometimes you have to use common sense as to the units of the pressure being measured if just the abbreviation *psi* is used without the *a* or the *g* being appended for pressure. For any pressure unit, be certain to carefully specify whether it is gauge or absolute, although you rarely find people doing so. For example, state "300 kPa absolute" or "12 cm Hg gauge" rather than just 300 kPa or 12 cm Hg, as the latter two without the absolute or gauge notation can on occasion cause confusion.

Example 2.25   Vacuum Pressure Reading

Problem Statement

Small animals such as mice can live at reduced air pressures down to 20 kPa absolute (although not comfortably). In a test, a mercury manometer attached to a tank, as shown in Figure E2.25, reads 64.5 cm Hg, and the barometer reads 100 kPa. Will the mice survive?

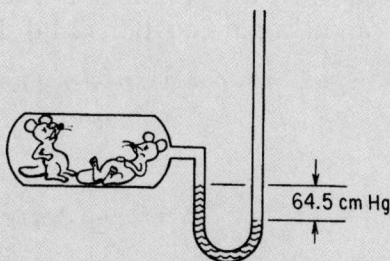

Figure E2.25

Solution

First read the problem. You are expected to realize from the figure that the tank is below atmospheric pressure. How? Because the left leg of the manometer is higher than the right leg, which is open to the atmosphere. Consequently, to get the absolute pressure, you subtract the 64.5 cm Hg from the barometer reading.

We ignore any temperature corrections to the mercury density for temperature and also ignore the gas density above the manometer fluid because it is so much less than the density of mercury. Then, because the vacuum reading on the tank is 64.5 cm Hg below atmospheric, the absolute pressure in the tank is

$$p_{\text{absolute}} = p_{\text{atmospheric}} - p_{\text{vacuum}} = 100 \text{ kPa} - \frac{64.5 \text{ cm Hg}}{} \left| \frac{101.3 \text{ kPa}}{76.0 \text{ cm Hg}} \right.$$

$$= 100 - 86 = 14 \text{ kPa absolute}$$

The mice will likely not survive.

Have you noted in the discussion and examples so far that we have said we can ignore the gas in the manometer tube above the measurement fluid? Is this OK? Let's see. Examine Figure 2.9, which illustrates the section of a U-tube involving three fluids.

When the columns of fluids are static (it may take some time for the dynamic fluctuations to die out!), the relationship among ρ_1, ρ_2, and the heights of the various columns of fluid is as follows. Pick a reference level for measuring pressure, such as the bottom of d_1. (If you pick the very bottom of the U-tube, instead of d_1, the left-hand and right-hand distances up to d_1 are equal, and consequently the pressures exerted by the right and left legs of the U-tube will cancel out in Equation (2.14). Try it and see.)

$$p_1 + \rho_1 d_1 g = p_2 + \rho_2 g d_2 + \rho_3 g d_3 \qquad (2.14)$$

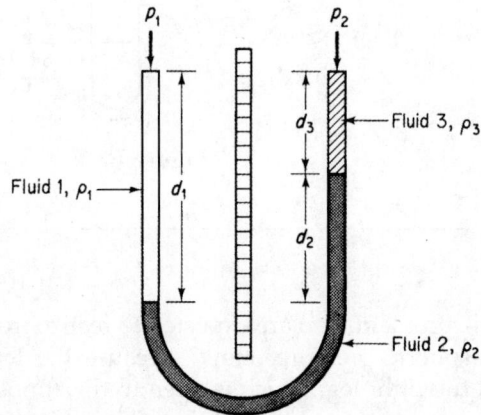

Figure 2.9   Manometer with three fluids

If fluids 1 and 3 are gases and fluid 2 is mercury, because the density of a gas is so much less than that of mercury, you can ignore the terms involving the gases in Equation (2.14) for practical applications.

However, if fluids 1 and 3 are liquids and fluid 2 is a nonmiscible fluid, the density of fluids 1 and 3 cannot be neglected in Equation (2.14). As the densities of fluids 1 and 3 approach that of fluid 2, what happens to the value of fluid level 2 (d_2) in Figure 2.9 for a given pressure difference $p_1 - p_2$?

Can you show for the case in which $\rho_1 = \rho_3 = \rho$ that the manometer expression reduces to the well-known differential manometer equation?

$$p_1 - p_2 = (\rho_2 - \rho) g d_2 \qquad (2.15)$$

As you may know, a flowing fluid experiences a **pressure drop** when it passes through a restriction such as the orifice in the pipe shown in Figure 2.10. The pressure difference can be measured with any instrument connected to the pressure taps, such as a manometer as illustrated in Figure 2.10, which can be used to measure the fluid flow rate in the pipe. Note that the manometer fluid reading is static if the flow rate in the pipe is constant.

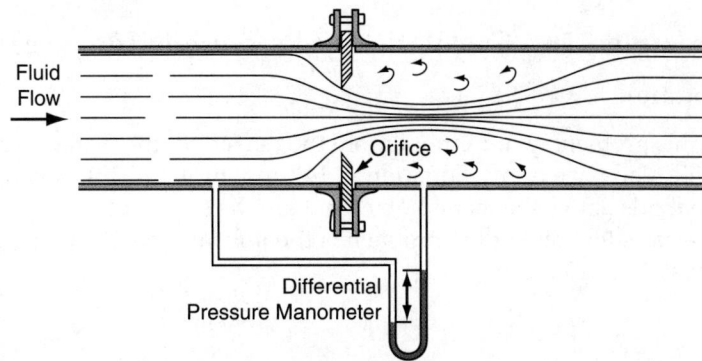

Figure 2.10   Concentric orifice used to restrict flow and measure the fluid flow rate with the aid of a manometer

Example 2.26   Calculation of Pressure Differences

Problem Statement

In measuring the flow of fluid in a pipeline through an orifice such as that in Figure E2.26 with a manometer used to determine the pressure difference across the orifice plate, the flow rate can be calibrated with the observed pressure drop (difference). Calculate the pressure drop $(p_1 - p_2)$ in pascals for the steady manometer reading in Figure E2.26.

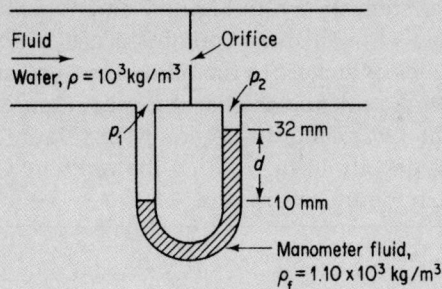

Figure E2.26

(Continues)

Example 2.26   Calculation of Pressure Differences (*Continued*)

Solution

In this problem, you cannot ignore the density of the water above the manometer fluid. Thus, we apply Equation (2.14), or simpler (2.15), because the densities of the fluids above the manometer fluid are the same in both legs of the manometer. The basis for solving the problem is the information given in Figure E2.26. Apply Equation (2.15):

$$p_1 - p_2 = \left(\rho_f - \rho\right)gd$$

$$= \frac{(1.10 - 1.00)10^3 \text{ kg}}{m^3} \left| \frac{9.807 \text{ m}}{s^2} \right| \frac{(22)\left(10^{-3}\right)m}{} \left| \frac{1 \text{ (N)}\left(s^2\right)}{(kg)(m)} \right| \frac{1(Pa)\left(m^2\right)}{1 \text{ (N)}}$$

$$= 21.6 \text{ Pa}$$

Check your answer. How much error would occur if you ignored the density of the flowing fluid?

Example 2.27   Unknown Liquid in a Manometer

Problem Statement

Consider the system shown in Figure 2.8a and 2.8b. In Figure 2.8a, a manometer is attached to a vessel, which contains a gas, and in Figure 2.8b, a manometer with a closed end is subject to the barometric pressure. Both manometers use the same unknown liquid, and the barometric pressure is equal to 765 mm Hg. If the readings of the two manometers are 30 in (Figure 2.8a) and 80 in (Figure 2.8b), determine the density in lb_m/ft^3 for the unknown fluid and the pressure in the vessel shown in Figure 2.8a in psia.

Solution

Because the barometric pressure and the height difference of the legs in the manometer are known, the density of the unknown liquid can be calculated, assuming that the density of the unknown liquid is much larger than the density of air by rearranging Equation (2.15):

$$\rho = \frac{p - p_0}{gh} = \frac{(765 - 0)\text{mm Hg}}{80 \text{ in}} \left| \frac{s^2}{32.174 \text{ ft}} \right| \frac{14.69 \text{ lb}_f}{760 \text{ mm Hg in}^2} \left| \frac{32.174 \text{ lb}_m \text{ ft}}{lb_f \text{ s}^2} \right| \frac{12^3 \text{ in}^3}{ft^3}$$

$$= 319 \text{ lb}_m \text{ ft}^{-3}$$

Now the pressure in the vessel can be calculated using Equation (2.11) again:

$$p = \rho g h + p_0 = \frac{319 \text{ lb}_\text{m}}{\text{ft}^3} \left| \frac{32.174 \text{ ft}}{\text{s}^2} \right| \frac{30 \text{ in}}{} \left| \frac{\text{lb}_\text{f} \text{ s}^2}{32.174 \text{ lb}_\text{m} \text{ ft}} \right| \frac{\text{ft}^3}{12^3 \text{ in}^3} + 14.79$$

$$= 5.54 + 14.79 = 20.33 \text{ psia}$$

It is interesting to note that you could also calculate the gauge pressure in the vessel directly by using the ratio of the manometer readings times the barometric pressure and converting to psig:

$$p - p_0 = \frac{30 \text{ in}}{80 \text{ in}} \left| \frac{765 \text{ mm Hg}}{} \right| \frac{14.69 \text{ psi}}{760 \text{ mm Hg}} = 5.54 \text{ psig}$$

Industrial Pressure Sensors. The most common board-mounted pressure sensor in the process industries is the differential pressure (DP) cell, which usually uses a balanced bar that is deflected on the basis of the differential pressure between two compartments. Because the two compartments of the DP cell are in contact with opposite sides of the diaphragm, the higher pressure side will exert more force on the balance bar (Figure 2.11). A precision forcing motor is used to maintain the balance bar in a balanced position in spite of the force due to the high pressure acting on the diaphragm. As a result, the differential pressure between the compartments is proportional to the force used to balance the bar. If one of the compartments is open to the atmosphere, the DP cell will measure the gauge pressure of the process. DP cells are also used to measure the pressure drop across an orifice plate in order to measure the flow rate or to measure the hydrostatic head as a level sensor.

Field-mounted pressure sensors include manometers, which have already been described, and the Bourdon pressure gauge (Figure 2.12). The Bourdon gauge is based on the fact that as the pressure being measured increases, it causes the C-shaped Bourdon tube to slightly straighten, which in turn moves the gauge needle to a higher pressure reading. The Bourdon pressure gauge measures the gauge pressure because it is the pressure above atmospheric pressure that causes the C-shaped tube to straighten.

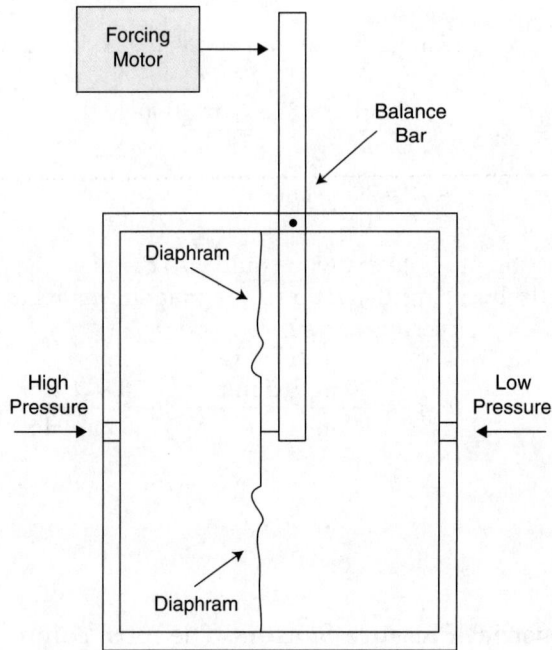

Figure 2.11   Schematic of a DP cell that uses a balance bar

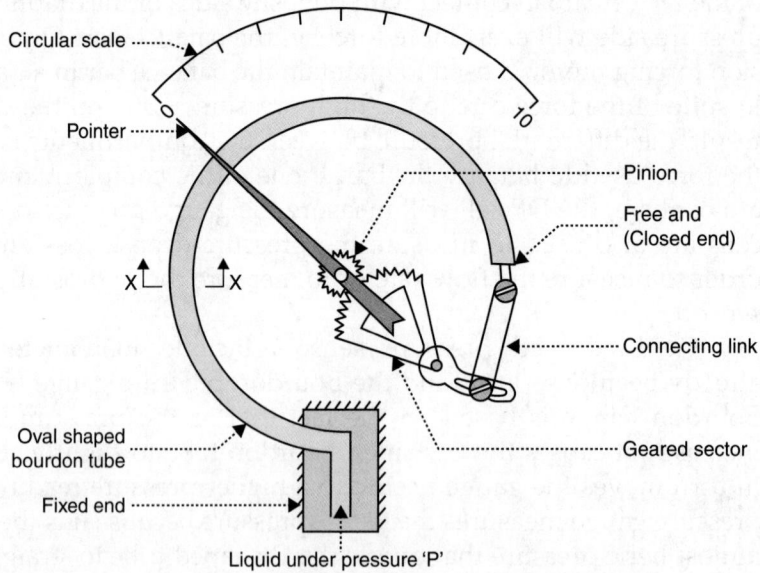

Figure 2.12   Schematic of a C-type Bourdon pressure gauge

Self-Assessment Test

Questions

1. Indicate whether the following statements are true or false:
   a. Atmospheric pressure is the pressure of the air surrounding us and changes from day to day.
   b. The standard atmosphere is a constant reference atmosphere equal to 1.000 atm or the equivalent pressure in other units.
   c. Absolute pressure is measured relative to a vacuum.
   d. Gauge pressure is measured in a positive direction relative to atmospheric pressure.
   e. Vacuum and draft pressures are measured in a positive direction from atmospheric pressure.
   f. You can convert from one type of pressure measurement to another using the standard atmosphere.
   g. A manometer measures the pressure difference in terms of the height of a fluid(s) in the manometer tube.
   h. Air flows in a pipeline, and the manometer containing Hg that is set up as illustrated in Figure 2.10 shows a differential pressure of 14.2 mm Hg. You can ignore the effect of the density of air on the height of the columns of mercury.

2. What is the equation to convert vacuum pressure to absolute pressure?

3. Can a pressure have a value lower than that of a complete vacuum?

Answers

1. All are true.

2. Atmospheric pressure − vacuum pressure = absolute pressure

3. No

Problems

1. Convert a pressure of 800 mm Hg to the following units:
   a. psia
   b. kPa
   c. atm
   d. ft H_2O

2. Your textbook cites five types of pressures: atmospheric, barometric, gauge, absolute, and vacuum pressure.
   a. What kind of pressure is measured in Figure SAT2.7.2 P2 a?
   b. In Figure SAT2.7.2P2 b?

c. What would be the reading in Figure SAT 2.7.2P2 c assuming that the pressure and temperature inside and outside the helium tank are the same as in parts a and b?

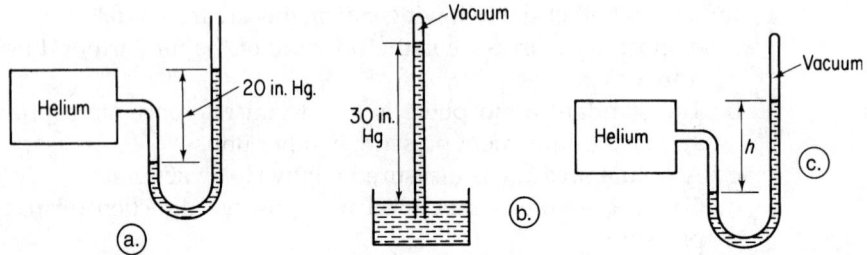

Figure SAT2.7.2 P2

3. An evaporator shows a reading of 40 kPa vacuum. What is the absolute pressure in the evaporator in kilopascals?

4. A U-tube manometer filled with mercury is connected between two points in a pipeline. If the manometer reading is 26 mm Hg, calculate the pressure difference in kilopascals between the points when (a) water is flowing through the pipeline, and (b) air at atmospheric pressure and 20°C with a density of 1.20 kg/m³ is flowing in the pipeline.

Answers

1. (a) 15.5; (b) 106.6; (c) 1.052; (d) 35.6

2. (a) Gauge pressure; (b) barometric pressure; (c) absolute pressure (50 in Hg)

3. In the absence of a barometric pressure, assume 101.3 kPa (1 atm). The absolute pressure is 61.3 kPa.

4. The Hg is static. (a) 3.21 kPa; (b) 3.47 kPa.

2.7.3   Levels

The liquid levels in vessels in a process plant must be monitored to ensure that the vessel neither overfills nor runs dry during operation. Such vessels are used to maintain inventory within a process and absorb flow rate fluctuations entering the vessel. In addition, many reactors maintain the desired reaction in the liquid phase in the reactor; therefore, the liquid level in a reactor will directly affect the production rates of products. For each of these cases, the level of liquid in the vessel should be measured continuously.

The most common level sensor in the process industries is based on measuring the hydrostatic head using a DP cell because the height of the liquid level is proportional to the pressure difference between the pressure at the top of the vessel and the pressure at the bottom of the vessel. This works well as long as there is a large difference in the density of the light and heavy phases (e.g., vapor and liquid, respectively). Rearranging Equation (2.15) yields

$$h = \frac{\Delta p}{\rho g} \tag{2.16}$$

where h is the measure height of the liquid level, Δp is the pressure difference between the upper and lower taps, ρ is the density of the liquid, and g is the gravitational constant.

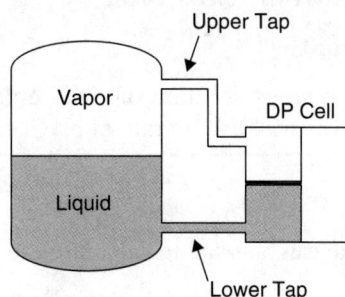

Figure 2.13   Schematic of a typical differential level sensor

2.7.4   Flow Rate

In the process industries, process streams are normally delivered to or removed from a process in pipes. The **flow rate** of a process stream is the rate at which material is transported through a carrying pipe. In this book, we usually use an overlay dot to denote a rate except for the volumetric flow rate F. The **mass flow rate** ($\dot{m}$) of a process stream is the mass (m) transported through a pipe per unit time (t):

$$\dot{m} = \frac{m}{t} \qquad [=] \text{ mass/time}$$

The **molar flow rate** ($\dot{n}$) of a process stream is the moles (n) of a substance transported through a pipe per unit time:

$$\dot{n} = \frac{n}{t} \qquad [=] \text{ moles/time}$$

The **volumetric flow rate** (F) of a process stream is the volume (V) transported through a line per unit time:

$$F = \frac{V}{t} \qquad [=] \text{ volume/time}$$

Example 2.28   Molar flow rate.

Problem Statement

A 20 wt. percent aqueous solution of NaCl enters a process at the rate of 20,000 lb_m/h. Determine the molar flow rate of NaCl entering the process.

Solution

This mass flow rate can be used to calculate the molar flow rate of NaCl by

$$\dot{n}_{NaCl} = \frac{20,000 \text{ lb}_m \text{ soln}}{\text{h}} \left| \frac{0.20 \text{ lb}_m \text{ NaCl}}{\text{lb}_m \text{ soln}} \right| \frac{\text{lb mol NaCl}}{58.44 \text{ lb}_m \text{ NaCl}} = 68 \text{ lb mol/h}$$

    Industrial Flow Rate Sensors. Magnetic, vortex shedding, and orifice meters are board-mounted flow rate sensors. Magnetic flow meters, which are low-pressure drop flow sensors, are based on measuring the voltage generated by an electrically conducting fluid passing through a magnetic field (e.g., municipal water). Vortex shedding flow meters are based on inserting a blunt object inside the pipe and measuring the downstream pulses created by the blunt object, which correlate to the flow rate. Orifice meters, which are similar to the one shown in Figure 2.10 except that a DP cell is used to measure the pressure drop across the orifice instead of a manometer, are the most commonly used flow sensor because they are less expensive and more generally applicable than magnetic and vortex shedding flow meters.

    A Coriolis meter, which is also a board-mounted flow rate sensor, passes the fluid through a U-shaped tube, and the resulting angular deflection in the tube is directly proportional to the mass flow rate. Coriolis meters are widely used in the biotechnology industries because they use smooth flow conduits that are relatively easy to clean and keep sterile and are generally low-maintenance sensors. Rotameters, which are field-mounted flow rate sensors, also are used in the biotechnology industries and use a float suspended in a tapered glass tube. As the flow rate increases, the float settles higher in the tapered tube.

Self-Assessment Test

Problems

1. Forty gallons per minute of a hydrocarbon fuel having a specific gravity of 0.91 flows into a tank truck with a load limit of 40,000 lb of fuel. How long will it take to fill the tank in the truck to its load limit?

2. Pure chlorine enters a process. By measurement, it is found that 2.4 kg of chlorine pass into the process every 3.1 min. Calculate the molar flow rate of the chlorine in kilogram moles per hour.

Answers

1. 132 min

2. 0.654 kg mol/hr

2.7.5   Concentration

Concentration designates the amount of a component (solute) in a mixture divided by the total amount of the mixture. The amount of the component of interest is usually expressed in terms of the mass or moles of the component, whereas the amount of the mixture can be expressed as the corresponding volume or mass of the mixture. Some common examples that you will encounter are

- **Mass per unit volume** (i.e., **mass concentration**), such as lb_m of solute/ ft^3 of solution, g of solute/L, lb_m of solute/bbl, kg of solute/m^3.
- **Moles per unit volume** (i.e., **molar concentration**), such as lb mol of solute/ft^3 of solution, g mol of solute/L, g mol of solute/cm^3.
- **Mass (weight) fraction,** the ratio of the mass of a component to the total mass of the mixture, a fraction (or a percent).

- **Mole fraction,** the ratio of the moles of a component to the total moles of the mixture, a fraction (or a percent).
- **Parts per million (ppm) and parts per billion (ppb),** a method of expressing the concentration of extremely dilute solutions; ppm is equivalent to a mass (weight) ratio for solids and liquids. It is a mole ratio for gases.
- **Parts per million by volume (ppmv) and parts per billion by volume (ppbv),** the ratio of the volume of the solute per volume of the mixture (usually used only for gases).

Other expressions for concentration from chemistry with which you should be familiar are molality (g mol solute/kg solvent), molarity (g mol/L), and normality (equivalents/L). Note that concentrations expressed in terms of mass (mass per unit volume, mass fraction, ppm) are referred to as **mass concentrations,** and those in terms of moles, as **molar concentrations** (e.g., moles per unit volume, mole fraction).

Example 2.29   Nitrogen Requirements for the Growth of Cells

Problem Statement

In normal living cells, the nitrogen requirement for the cells is provided from protein metabolism (i.e., consumption of protein in the cells). When cells are grown commercially, such as in the pharmaceutical industry, $(NH_4)_2SO_4$ is usually used as the source of nitrogen. Determine the amount of $(NH_4)_2SO_4$ consumed in a fermentation medium in which the final cell concentration is 35 g/L in a 500 L volume of fermentation medium. Assume that the cells contain 9 wt% N and that $(NH_4)_2SO_4$ is the only nitrogen source.

Solution

Basis: 500 L of solution containing 35 g/L

$$\frac{500 \text{ L}}{} \left| \frac{35 \text{ g cell}}{\text{L}} \right| \frac{0.09 \text{ g N}}{\text{g cell}} \left| \frac{\text{g mol}}{14 \text{ g N}} \right| \frac{1 \text{ g mol } (NH_4)_2 SO_4}{2 \text{ g mol N}} \left| \frac{132 \text{ g } (NH_4)_2 SO_4}{\text{g mol } (NH_4)_2 SO_4} \right.$$

$$= 7425 \text{ g } (NH_4)_2 SO_4$$

Here is a list of typical measures of concentration given in the set of guidelines by which the Environmental Protection Agency defines the extreme levels at which the five most common air pollutants could harm people if they are exposed to these levels for the stated periods of exposure:

1. **Sulfur dioxide:** 365 µg/m^3 averaged over a 24 hr period
2. **Particulate matter** (10 µm or smaller): 150 µg/m^3 averaged over a 24 hr period
3. **Carbon monoxide:** 10 mg/m^3 (9 ppm) when averaged over an 8 hr period; 40 mg/m^3 (35 ppm) when averaged over 1 hr
4. **Nitrogen dioxide:** 100 µg/m^3 averaged over 1 yr
5. **Ozone:** 0.12 ppm measured over 1 hr

Note that the gas concentrations are mostly mass/volume except for the ppm.

Example 2.30   Use of ppm

Problem Statement

The current OSHA 8 hr limit for HCN in air is 10.0 ppm. A lethal dose of HCN in air is (from the *Merck Index*) 300 mg/kg of air at room temperature. How many milligrams of HCN per kilogram of air is 10.0 ppm? What fraction of the lethal dose is 10.0 ppm?

Solution

In this problem, you have to convert ppm in a gas (a mole ratio, remember!) to a mass ratio.

Basis: 1 kg of the air-HCN mixture

We can treat the 10.0 ppm as 10.0 g mol HCN/10^6 g mol air because the amount of HCN is so small when added to the air in the denominator of the ratio.
    The 10.0 ppm is

$$\frac{10.0 \text{ g mol HCN}}{10^6 \text{ (air + HCN) g mol}} = \frac{10.0 \text{ g mol HCN}}{10^6 \text{ g mol air}}$$

(Continues)

Example 2.30   Use of ppm (*Continued*)

Next, get the molecular weight (MW) of HCN so that it can be used to convert moles of HCN to mass of HCN; the MW = 27.03. Then

$$\frac{10.0 \text{ g mol HCN}}{10^6 \text{ g mol air}} \left| \frac{27.03 \text{ g HCN}}{1 \text{ g mol HCN}} \right| \frac{1 \text{ g mol air}}{29 \text{ g air}} \left| \frac{1000 \text{ mg HCN}}{1 \text{ g HCN}} \right| \frac{1000 \text{ g air}}{1 \text{ kg air}}$$

$$= 9.32 \text{ mg HCN/kg air}$$

$$\frac{9.32}{300} = 0.031$$

Does this answer seem reasonable? At least it is not greater than 1!

Industrial Composition Sensors. A wide variety of composition analyzers are used in the process industries. Gas chromatographs (GCs), which are widely used in refineries and petrochemical plants, separate components on the basis of different affinities for absorption onto a packed chromatographic column. The accuracy and the associated sample processing time vary from one application to another. High-pressure liquid chromatography (HPLC) is used extensively in the biotechnology field and is similar to a GC in that it uses the variation in mobility of components to separate them from a liquid mixture using a chromatographic column.

The biotechnology industry uses a number of special-purpose analyzers, including dissolved oxygen (DO) sensors, pH sensors, and turbidity meters. Many bioreactors use oxygen as a reactant, and DO sensors, which are electrodes, are used to ensure that the oxygen concentration is in the desired range. pH has a significant effect on most bioprocesses, and pH sensors, which are also electrodes, are used to control or monitor the system pH. A schematic of a general electrode is shown in Figure 2.14. The glass membrane on the lower end of the electrode allows certain ions from the bioreactor to enter into the fill solution. The electrode reading is the voltage difference between the measuring electrode in the fill solution and the AgCl reference electrode. Turbidity is the measure of the amount of light scattering caused by suspended solids in a sample. The concentration of cells in a sample from a fermentator correlates strongly with cell concentration if the reaction mixture is clear except for cells.

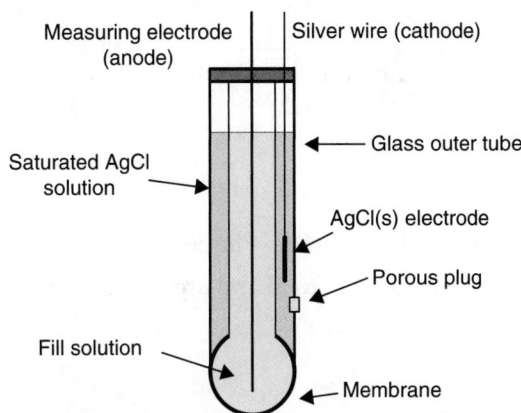

Figure 2.14   Cross section of an electrode

Self-Assessment Test

Questions

1. Do parts per million denote a concentration that is a mole ratio?

2. Does the concentration of a component in a mixture depend on the amount of the mixture?

3. Pick the correct answer. How many ppm are there in 1 ppb?
   a. 1000
   b. 100
   c. 1
   d. 0.1
   e. 0.01
   f. 0.001

4. Is 50 ppm five times greater than 10 ppm?

5. A mixture is reported as 15% water and 85% ethanol. Should the percentages be deemed to be by mass, mole, or volume?

6. In a recent EPA inventory of 20 greenhouse gases that are emitted in the United States, carbon dioxide constituted 5.1 million gigagrams (Gg), which was about 70% of all of the U.S. greenhouse gas emissions. Fossil fuel combustion in the electric utility sector contributed about 34% of all of the carbon dioxide emissions; and the transportation, industrial, and residential-commercial sectors accounted for 34%, 21%, and 11% of the total, respectively. Are the last four percentages mole percent or mass percent?

Answers

1. For gases, but not for liquids and solids
2. No
3. 0.001
4. Yes
5. Mass percentages because it is a liquid
6. They are percentages of a total and are dimensionless.

Problems

1. How many milligrams per liter are equivalent to a 1.2% solution of a substance in water?
2. If a membrane filter yields a count of 69 fecal coliform (FC) colonies from 5 mL of well water, what should be the reported FC concentration?
3. The danger point in breathing sulfur dioxide for humans is 2620 $\mu g/m^3$. How many ppm is this value?

Answers

1. Basis: 1 L of solution: $\dfrac{1\,\text{L}}{}\left|\dfrac{1000\,\text{g}}{\text{L}}\right|\dfrac{12\,\text{g Substrate}}{1000\,\text{g}}\left|\dfrac{1000\,\text{mg Substrate}}{\text{g Substrate}}\right. = 12,000\ \text{mg/L}$
2. Report 1400 FC/100 mL
3. 1 ppm

Glossary

absolute pressure   Pressure relative to a complete vacuum.

accuracy   A measure of how closely a measurement approaches the true value.

AE system   The American Engineering system of units based on the foot, pound force, and second.

API   Scale used to report specific gravity of petroleum compounds.

atomic weight   Mass of an atom based on ^{12}C being exactly 12.

average molecular weight   A pseudo molecular weight computed by dividing the mass in a mixture or solution by the number of moles in the mixture or solution.

barometric pressure   Pressure measured by a barometer; the same as absolute pressure.

basis   The reference material or time selected to use in making the calculations in a problem.

board-mounted sensor   A sensor that can be read from the control room

Celsius (°C)   Relative temperature scale based on zero degrees being the freezing point of an air-water mixture.

CGS   A system of units identical to SI Units except that centimeters are used instead of meters and grams are used instead of kilograms.

changing bases   Shifting the basis in a problem from one value to another for convenience in the calculations.

compound   A species composed of more than one element chemically bound together.

concentration   The quantity of a solute per unit volume or per a specified amount in a mixture.

debottlenecking   The activity of increasing the throughput for a process using the minimum changes to the process.

density   Mass per unit volume of a compound; molar density is the number of moles divided by the total volume.

derived units   A unit of measure that is based on combinations of fundamental units.

dimensional consistency   Each term in an equation must have the same set of net dimensions.

dimensionless group   A collection of variables or parameters that has no net dimensions (units).

dimensions   The basic concepts of measurement, such as length or time.

extensive quantities   Quantities that depend on the size of the system being considered.

Fahrenheit (°F)   Relative temperature scale with 32 degrees being the freezing point of an air-water mixture.

field-mounted sensor   A sensor that can be read only on the process.

flow rate   Amount of mass, moles, or volume of a material passing through a pipe or system per unit time.

force   A derived unit for the mass times the acceleration.

fundamental units   Units that can be measured independently.

gauge pressure   Pressure measured above atmospheric pressure.

g_c   A conversion factor used to convert between lb_m and lb_f.

gram mole 6.022×10^{23} molecules.

intensive quantities   Quantities that do not depend on the size of the system being considered.

kelvin (K)   Absolute temperature scale based on zero degree being the lowest possible temperature we believe can exist.

kg mol   1000 moles.

mass   A basic dimension for the amount of material.

mass concentration   A unit of concentration based on the mass of a component.

mass flow rate   The mass per unit time flowing through a pipe.

mass fraction   The ratio of the mass of a component to the total mass of a collection of material or a solution.

molar concentration   A unit of concentration based on the moles of a component.

molar flow rate   The total moles of material per unit time flowing through a pipe

mole   Amount of a substance containing 6.022×10^{23} entities.

molecular weight   Mass of a compound per mole.

mole fraction   Moles of a particular compound in a mixture or solution divided by the total number of moles present.

nondimensional group   See **dimensionless group**.

parts per billion (ppb)   Concentration expressed in terms of parts of the component of interest per billion parts of the mixture.

parts per million (ppm)   Concentration expressed in terms of parts of the component of interest per million parts of the mixture.

pound force   The unit of force in the AE system.

pound mass   The unit of mass in the AE system.

pound mole $6.022 \times 10^{23} \times 453.6$ molecules.

precision   How closely a number of measurements of the same quantity are grouped.

pressure   The normal force per unit area that a fluid exerts on a surface.

process   Take feed material and convert it into products.

process troubleshooting   The activity of solving a process problem.

psia   Pounds force absolute per square inch.

Rankine (°R)   Absolute temperature scale related to degrees Fahrenheit based on zero degrees being the lowest possible temperature we believe can exist.

relative error   Fraction or percent error for a number.

repeatability   An indication of how precise a measurements is.

SI system   The internationally recognized system of units based on the meter, kilogram, and second.

solution   Homogeneous mixture of two or more substances.

specific gravity   Ratio of the density of a compound to the density of a reference compound.

specific volume   Inverse of the density (volume per unit mass).

standard atmosphere   The pressure in a standard gravitational field equivalent to 760 (exactly) mm Hg.

systematic errors   Cause a consistent error from the true reading.

units conversion   Change of units from one set to another.

unit of measure   A means of precisely and explicitly describing things.

vacuum   A pressure less than atmospheric (but reported as a positive number).

volumetric flow rate   The volume of material per time flowing through a pipe.

weight   A force opposite to the force required to support a mass (usually in a gravitational field).

weight fraction   The historical term for mass fraction.

Problems

Section 2.1   Units of Measure

*2.1.1   The following questions will measure your SIQ.

    a. Which is (are) a correct SI symbol(s)?

        (1) nm   (2) °K

        (3) sec   (4) N/mm

    b. Which is (are) consistent with SI usage?

        (1) MN/m^2 (2) GHz/s

        (3) $kJ/[(s)\,(m^3)]$ (4) °C/M/s

    c. Atmospheric pressure is about

        (1) 100 Pa   (2) 100 kPa

        (3) 10 MPa   (4) 1 GPa

    d. The watt is

        (1) 1 joule per second   (2) Equal to $1\,(kg)\,(m^2)\,/s^3$

        (3) The unit for all types of power   (4) All of above

    e. Which height and mass are those of a petite woman?

        (1) 1.50 m, 45 kg   (2) 2.00 m, 95 kg

        (3) 1.50 m, 75 kg   (4) 1.80 m, 60 kg

    f. Which is a recommended room temperature in winter?

        (1) 15°C   (2) 20°C

        (3) 28°C   (4) 45°C

    g. The temperature 0°C is defined as
       (1) 273.15°K   (2) Absolute zero
       (3) 273.15 K   (4) The freezing point of water

    h. What force may be needed to lift a heavy suitcase?
       (1) 24 N   (2) 250 N
       (3) 25 kN   (4) 250 kN

**2.1.2 Two scales are shown in Figure P2.1.2, a balance (a) and a spring scale (b). In the balance, calibrated weights are placed in one pan to balance the object to be weighed in the other pan. In the spring scale, the object to be weighed is placed on the pan and a spring is compressed that moves a dial on a scale in kilograms. State for each device whether it directly measures mass or weight. Underline your answer. State in *one* sentence for each the reason for your answer.

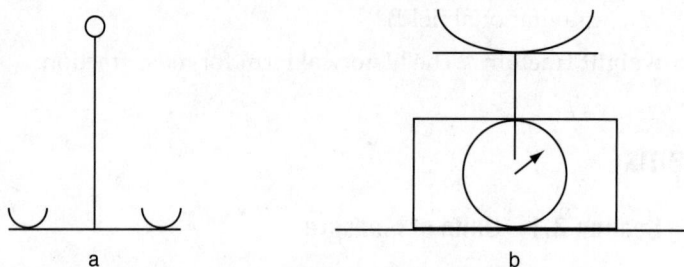

Figure P2.1.2

Section 2.2   Unit Conversions

  2.2.1 Convert the following to SI units using the correct number of significant figures:
    a. 2.5 miles
  *b. 52.66 Btu
    c. 5.00 hp
    d. 50. gal min^{-1}
    e. 22.5 lb$_f$ in^{-2}
    f. 45.6 slugs s^{-1}
  *g. 17.0 hp hr
    h. 35.9 gal s^{-1} ft^{-2}
    i. 357 °F
  *j. 7.6 lb$_m$ ft^{-3}

2.2.2 Make the following unit conversions using the correct number of significant figures:

    *a. Convert 6.000 ft to micrometers.

    b. Convert 100.0 km/h to in. μs^{-1}.

    c. Convert 500K to degrees Fahrenheit.

    d. Convert 1.00×10^6 Btu/h to kilowatts.

    e. Convert 7.5 kg m^{-3} to ounces per bushel (you may have to look up the needed conversion factors).

    *f. Convert 15.367 lbm ft^2 s^{-2} to Btu.

    g. Convert 0.2433 kg m^{-1} s^{-2} to psi (lbf $in.^{-2}$).

    h. Convert 10.1 A V to kilowatts.

    *i. Convert 32.17 ft s^{-2} to mm h^{-2}.

    j. Convert 0.779 lb_m ft^2 s^{-3} to kW.

2.2.3 Estimate how many gallons of gasoline in billions of gallons per year would be saved if everyone in the United Statves drove 1000 miles less for a year. For the year 2014, it was estimated that 136.8 billion gallons of gasoline were consumed in the United States and that the overall average number of miles driven per person per year is 13,500 miles.

2.2.4 Determine the monthly power bill to operate an air-conditioning unit if the unit delivers 2000 Btu h^{-1}. Assume that the air-conditioning unit has an energy efficiency ratio (EER, i.e., the ratio of Btus of cooling divided by W h of electrical energy used to drive the cooling unit) equal to 11.2 Btu W^{-1} h^{-1} and that the cost of electrical energy is $0.14 per kW h.

2.2.5 The astronomical unit (AU) is the average distance from the earth to the sun (9.29×10^7 miles). Determine how many AUs are between the earth and the nearest star (Proxima Centauri), which is 4.24 light years (ly) away. A light year is the distance that light travels in one year, and the speed of light is 2.99792×10^8 m s^{-1}.

*2.2.6 The astronomical unit parsec (pc) is used to express distance (1 pc = 3.0867×10^{16} m) and is based on the observed change in location of a distant star due to the orbit of the earth about the sun. Determine how many parsecs are between the earth and the brightest star in the night sky (Sirius), which is 8.6 light years (ly) away. A light year is the distance that light travels in one year, and the speed of light is 2.99792×10^8 m s^{-1}.

2.2.7 A large petroleum refinery can process 400,000 barrels of crude oil per day. Determine the crude feed rate in pounds per hour for this large refinery if the density of crude is 0.85 g cm^{-3}. A barrel of crude contains 42 gal.

2.2.8 Consider a rectangular kitchen sink (18 in. × 15 in. × 8 in.) with a plug in the drain. If it takes 297 seconds to fully fill the sink using the sink faucet, what is the flow rate of water from the faucet in gallons per minute?

2.2.9   You are driving an economy car across Canada, and you just filled your gas tank with 40 L of gasoline after traveling 480 km for a fuel efficiency of 12 km/L. What is the fuel efficiency in mi/gal?

2.2.10   During your trip across Canada, you see that the cost of gasoline is Can\$1.29/L. If the exchange rate is 1 Can\$ = US\$0.79, determine the price of gasoline in Canada in US\$/gal.

2.2.11   The ideal gas constant is 8.314 kPa m^3 kgmol^{-1} K^{-1}. Convert this gas constant into units of atm ft^3 lb mol^{-1} °R^{-1}.

2.2.12   Estimate the rainwater collection in gallons that would result from a 3.0 in. rain and a collection roof with a footprint of 3500 ft^2.

*2.2.13   Carry out the following conversions:
    a. How many m^3 are there in 1.00 (mile)^3?
    b. How many gal/min correspond to 1.00 ft^3/s?

2.2.14   Convert
    *a. 0.04 g/[(min) (m^3)] to lb$_m$/[(hr) (ft^3)]
    *b. 2 L/s to ft^3/day

    **c. $\dfrac{6(\text{in})(\text{cm}^2)}{(\text{yr})(\text{s})(\text{lb}_m)(\text{ft}^2)}$ to all SI units

*2.2.15   Convert the following:
    a. 60.0 mi/h to ft/s
    b. 50.0 lb/in.^{2} to kg/m^2
    c. 6.20 cm/hr^2 to nm/s^2

**2.2.16   A technical publication describes a new model 20 hp Stirling (air cycle) engine that drives a 68 kW generator. Is this possible?

**2.2.17   Your boss announced that the speed of the company Boeing 737 is to be cut from 525 mi/hr to 475 mi/hr to "conserve fuel," thus cutting consumption from 2200 gal/hr to 2000 gal/hr. How many gallons are saved in a 1000 mi trip?

**2.2.18   From *Parade* magazine (by Marilyn Vos Savant, August 31, 1997, p. 8): "Can you help with this problem? Suppose it takes one man 5 hours to paint a house, and it takes another man 3 hours to paint the same house. If the two men work together, how many hours would it take them? This is driving me nuts."

**2.2.19   In the AE system of units, the viscosity can have the units of (lb$_f$) (h)/ft^2, while in a handbook the units are (g)/[(cm)(s)]. Convert a viscosity of 20.0 (g)/[(m)(s)] to the given AE units.

**2.2.20   Thermal conductivity in the AE system of units is

$$k = \frac{\text{Btu}}{(\text{h})(\text{ft}^2)(^\circ\text{F}/\text{ft})}$$

Determine the conversion factor to convert from AE units to the following

$$\frac{\text{kJ}}{(\text{d})(\text{m}^2)(^\circ\text{C}/\text{cm})}$$

***2.2.21   Water is flowing through a 2-in.-diameter pipe with a velocity of 3 ft/s.
  a. What is the kinetic energy of the water in (ft) $(\text{lb}_f)/\text{lb}_m$?
  b. What is the flow rate in gallons per minute?

****2.2.22   Consider water pumped at a rate of 75 gal/min through a pipe that undergoes a 100 ft elevation increase by a 2 hp pump. The rate of energy input from the pump that goes into heating the water is approximately equal to the rate of energy input from the pump minus the rate of potential energy generated by pumping the water up the elevation change ($m'\,gh$, where m' is the mass flow rate of water and h is the elevation change). Estimate the rate of energy input for heating the water for this case in British thermal units per hour.

**2.2.23   What is meant by a scale that shows a weight of 21.3 kg?

***2.2.24   A tractor pulls a load with a force equal to 800 lb (4.0 kN) with a velocity of 300 ft/min (1.5 m/s). What is the power required using the given AE system data? The SI data?

***2.2.25   What is the kinetic energy of a vehicle with a mass of 2300 kg moving at the rate of 10.0 ft/s in British thermal units?

***2.2.26   A pallet of boxes weighing 10 tons is dropped from a lift truck from a height of 10 ft. The maximum velocity the pallet attains before hitting the ground is 6 ft/s. How much kinetic energy does the pallet have in (ft) (lb_f) at this velocity?

***2.2.27   Calculate the protein elongation (formation) rate per mRNA per minute based on the following data:
  a. One protein molecule is produced from x amino acid molecules.
  b. The protein (polypeptide) chain elongation rate per active ribosome uses about 1200 amino acids/min.
  c. One active ribosome is equivalent to 264 ribonucleotides.
  d. $3x$ ribonucleotides equal each mRNA.

Messenger RNA (mRNA) is a copy of the information carried by a gene in DNA and is involved in protein synthesis.

Section 2.2.1   Temperature Conversion

*2.2.28   "Japan, U.S. Aim for Better Methanol-Powered Cars" reads the headline in the *Wall Street Journal*. Japan and the United States plan to join in developing technology to improve cars that run on methanol, a fuel that causes less air pollution than gasoline. An unspecified number of researchers from Japanese companies will work with the EPA to develop a methanol car that will start in temperatures as low as −10°C. What is this temperature in degrees Rankine, kelvin, and Fahrenheit?

*2.2.29   Can negative temperature measurements exist?

***2.2.30   The heat capacity C_p of acetic acid in $J/[(g\ mol)(K)]$ can be calculated from the equation

$$C_p = 8.41 + 2.4346 \times 10^{-5}\,T$$

where T is in kelvin. Convert the equation so that T can be introduced into the equation in degrees Rankine instead of kelvin. Keep the units of C_p the same.

*2.2.31   Convert the following temperatures to the requested units:
   a. 10°C to °F
   b. 10°C to °R
   c. −25°F to K
   d. 150K to °R

**2.2.32   In a report on the record low temperatures in Antarctica, *Chemical and Engineering News* said at one point that "the mercury dropped to −76°C." In what sense is that possible? Mercury freezes at −39°C.

Section 2.3   Equations and Units

*2.3.1   The flow rate over a weir (i.e., a vertical plate that is perpendicular to the flow direction placed at the bottom of an open channel) is used to estimate the volumetric flow rate in an open channel. For an open rectangular channel, the equation for the volumetric flow rate (Q) in gallons per minute (gpm) in terms of the width of the open rectangular channel (W) in inches and the height of the flow over the weir (h) in inches is given by

$$Q = 4.8\,Wh^{1.5}$$

Determine the units for the constant 4.8 in this equation.

2.3.2   Assume that you want to use the ideal gas law ($pV = nRT$) with temperature in K, the number of moles n in gmol, the pressure in $lb_f\ in^{-2}$ (psi), and the volume in gal. Determine the gas constant R for this case.

2.3.3 When a shaft is subjected to a twisting torque, the resulting angle of the twist θ is given by

$$\theta = \frac{TL}{MS}$$

where θ is in radians, T is the applied torque in N m, L is the length of the shaft in m, M is the polar moment of inertia of the shaft, and S is the shear modulus of the material comprising the shaft in N m^{-2}. Determine the units of M based on dimensional consistency.

2.3.4 Hoover Dam, which was completed in 1935 during the Great Depression, was built to control flood waters, provide a source of irrigation water, and generate hydroelectric power. Hoover Dam dams the Colorado River and forms Lake Mead, the largest reservoir in the United States based on volume of water retained. On average, 4.3×10^6 gallons per min (gpm) are used to produce hydroelectric power, and the elevation used to generate the hydroelectric energy is 180 m. The equation that relates hydroelectric power (P) to the elevation change (h) and the mass flow rate of the water ($\dot{m}$) is

$$P = \eta \dot{m} g h$$

where η is the efficiency for converting potential energy into electrical energy (0.9), and g is the gravitational acceleration (9.8067 m s^{-1}). Determine the average electric power generation for Hoover Dam in units of TW h y^{-1}.

2.3.5 Consider the gasoline storage tank for a gas station. If the cylindrical storage tank is 15 ft in diameter with a length of 30 ft, determine how many cars can be filled from this tank if on average 15 gallons of gasoline are pumped into each car. The density of gasoline is approximately 0.73 g cm^{-3}.

*2.3.6 When an object (e.g., an automobile, a baseball, or an airplane) is traveling through air with a velocity v, a force F opposes the direction of the velocity due to air resistance. The drag coefficient CD is used to describe air resistance and is given by the following equation

$$C_D = \frac{2F}{\rho v^2 A}$$

where F is in lb$_f$, ρ is the density of air in lb$_m$ ft^{-3}, v is in ft s^{-1}, and A is the frontal area of the object in ft^2. Determine the units for C_D. Also, determine what results if g_c is used in this equation.

2.3.7   The gravitational force (F) between two objects of mass m and M is given by

$$F = G\frac{Mm}{r^2}$$

where F is in N, M and m are in kg, r is the separation between the two objects in m, and G is the gravitational constant. Determine the units for G. This equation is used to determine the properties of an orbit of a planet about a star.

*2.3.8   The density of a certain liquid is given an equation of the following form:

$$\rho = (A + BT)e^{CP}$$

where ρ = density (g/cm^3), T = temperature (°C), and P = pressure (atm). For this equation to be dimensionally consistent, what are the units of A, B, and C?

*2.3.9   Explain in detail whether the following equation for flow over a rectangular weir is dimensionally consistent. (This is the modified Francis formula.)

$$q = 0.415(L - 0.2h_0)h_0^{1.5}\sqrt{2g}$$

where q = volumetric flow rate (ft^3/s), L = crest height (ft), h_0 = weir head (ft), and g = acceleration of gravity (32.2 ft/s^2).

**2.3.10   In an article on measuring flows from pipes, the author calculated $q = 80.8$ m^3/s using the formula

$$q = CA_1\sqrt{\frac{2gV(p_1 - p_2)}{1 - (A_1/A_2)^2}}$$

where q = volumetric flow rate (m^3/s), C = dimensionless coefficient (0.6), A_1 = cross-sectional area 1 (m^2), A_2 = cross-sectional area 2 (m^2), V = specific volume (10^{-3} m^3/kg), P = pressure, $p_1 - p_2 = 50$ kPa, and g = acceleration of gravity (9.80 m/s).

Was the calculation correct? (Answer yes or no and explain briefly the reasoning underlying your answer.)

***2.3.11   Leaking oil tanks have become such an environmental problem that the federal government has implemented a number of rules to reduce the problem. A flow rate from a leak from a small hole in a tank can be predicted from the following relation:

$$Q = 0.61S\sqrt{\frac{2\Delta p}{\rho}}$$

where Q is the leakage rate (gal/min), S is the cross-sectional area of the hole causing the leak (in.^{2}), Δp is the pressure drop between the inside of the tank opposite the leak and the atmospheric pressure (psi), and ρ is the fluid density (lb/ft^3).

To test the tank, the vapor space is pressurized with N_2 to a pressure of 23 psig. If the tank is filled with gasoline (sp. gr. = 0.703) to the height of 73 in. and the hole is 1/4 in. in diameter, what is the value of Q (in cubic feet per hour)?

***2.3.12** A relation for a dimensionless variable called the compressibility (z), which is used to describe the pressure-volume-temperature behavior for real gases, is

$$z = 1 + B\rho + C\rho^2 + D\rho^3$$

where ρ is the density in g mol/cm^3. What are the units of B, C, and D? Convert the coefficients in the equation for z so that the density can be introduced into the equation in the units of lb$_m$/ft^3, thus:

$$z = 1 + B^* \rho{*} + C^*(\rho^*)^2 + D^*(\rho^*)^3$$

ρ^* is in lb$_m$/ft^3. Give the units for B^*, C^*, and D^*, and give the equations that relate B^* to B, C^* to C, and D^* to D.

2.3.13 The velocity in a pipe in turbulent flow is expressed by the following equation:

$$u = k \left(\frac{\tau}{\rho} \right)^{1/2}$$

where τ is the shear stress in N/m^2 at the pipe wall, ρ is the density of the fluid in kg/m^3, u is the velocity in m/s, and k is a coefficient.

You are asked to modify the equation so that the shear stress can be introduced in the units of τ, which are lb$_f$/ft^2, and the density ρ' for which the units are lb/ft^3, so that the velocity u comes out in the units of ft/s. Show all calculations, and give the final equation in terms of u, τ, and ρ' so a reader will know that AE units are involved in the equation.

2.3.14 In 1916, Nusselt derived a theoretical relation for predicting the coefficient of heat transfer between a pure saturated vapor and a colder surface:

$$h = 0.943 \left(\frac{k^3 \rho^2 g \lambda}{L \mu \Delta T} \right)^{1/4}$$

where

h is the mean heat transfer coefficient, Btu/[(hr) (ft^2) (°F)]

k is the thermal conductivity, Btu/[(hr) (ft) (°F)]

ρ is the density in lb/ft^3

g is the acceleration of gravity, 4.17×10^8 ft/(hr)^2

λ is the enthalpy change of evaporation in Btu/lb

L is the length of tube in ft

μ is the viscosity in lb$_m$/[(hr) (ft)]

ΔT is a temperature difference in °F

What are the units of the constant 0.943?

****2.3.15   The efficiency of cell growth in a substrate in a biotechnology process was given in a report as

$$\eta = \frac{Y_{x/s}^c \gamma_b \Delta H_b^c / e^-}{\Delta H_{out}}$$

In the notation table,

η is the energetic efficiency of cell metabolism (energy/energy)

$Y_{x/s}^c$ is the cell yield, carbon basis (cells produced/substrate consumed)

Y_b is the degree of reductance of biomass (available electron equivalents/g mole carbon, such as 4.24_e^- equiv./mol cell carbon)

$\Delta H_b^c / e^-$ is the biomass heat of combustion (energy/available electron equiv.)

ΔH_{cat} is the available energy from catabolism (energy/mole substrate carbon)

Is there a missing conversion factor? If so, what would it be? The author claims that the units in the numerator of the equation are (mol cell carbon/mol substrate carbon) (mol available e^-/mol cell carbon) (heat of combustion/mol available e^-). Is this correct?

****2.3.16   The Antoine equation, which is an empirical equation, is used to model the effect of temperature on the vapor pressure of a pure component. The Antoine equation is given by

$$\ln p^* = A - \frac{B}{C+T}$$

where p^* is the vapor pressure, T is the absolute temperature, and A, B, and C are empirical constants specific to the pure component and the units used for the vapor pressure and temperature. Determine under what conditions this equation will be dimensionally consistent.

**2.3.17 A letter to the editor says: "An error in units was made in the article 'Designing Airlift Loop Fermenters.' Equation (4) is not correct."

$$\Delta p = 4 f \rho \left[\frac{v^2}{2g} (L/D) \right] \qquad (4)$$

Is the author of the letter correct (i.e., is f dimensionless)?

***2.3.18 Heat capacities are usually given in terms of polynomial functions of temperature. The equation for carbon dioxide is

$$C_p = 8.4448 + 0.5757 \times 10^{-2}\, T - 0.2159 \times 10^{-5}\, T^2 + 0.3059 \times 10^{-9}\, T^3$$

where T is in °F and C_p is in Btu/[(lb mol)(°F)]. Convert the equation so that T can be in °C and C_p will be in J/[(g mol)(K)].

Section 2.4   Measurement Errors and Significant Figures

2.4.1 For each of the following numbers, express the number in scientific notation and indicate its number of significant figures.
   a. 12.4
   b. 0.0023
   *c. 100
   d. 22,340
   e. 0.20001
   *f. 20.00
   g. 0.009
   h. 0.00230
   i. 778
   *j. 778.0

2.4.2 For each of the numbers listed in Problem 2.4.1, indicate the percent uncertainty indicated by the number of significant figures.

2.4.3 For each of the following numbers, express the number in scientific notation and indicate its number of significant figures.
   a. 0.0011
   b. 5000
   c. 2.3001
   d. 88770
   e. 88770.0
   f. 435,400
   g. 100.10

h. 0.000561

i. 1.000

j. 10

2.4.4 For each of the numbers listed in Problem 2.4.3, indicate the percent uncertainty indicated by the number of significant figures.

*2.4.5 If you subtract 1191 cm from 1201 cm, each number with four significant figures, does the answer of 10 cm have two or four (10.00) significant figures?

*2.4.6 What is the sum of the following numbers to the correct number of significant numbers?

3.1472

32.05

1234

8.9426

0.0032

9.00

**2.4.7 Suppose you make the following sequence of measurements for the segments in laying out a compressed air line:

4.61 m

210.0 m

0.500 m

What should be the reported total length of the air line?

**2.4.8 Given that the width of a rectangular duct is 27.81 cm, and the height is 20.49 cm, what is the area of the duct with the proper number of significant figures?

**2.4.9 Multiply 762 by 6.3 to get 4800.60 on your calculator. How many significant figures exist in the product, and what should the rounded answer be?

***2.4.10 Suppose you multiply 3.84 times 0.36 to get 1.3824. Evaluate the maximum relative error in (a) each number and (b) the product. If you add the relative errors in the two numbers, is the sum the same as the relative error in their product?

Section 2.5   Validation of Results

2.5.1 For each of the following calculations, estimate the solution without using your calculator. Then using your calculator, determine the result and compare.

a. $(5.95)(4.04)/3.12$

b. $1.456 \times 10^3 - 3.26 \times 10^1$

c. $4\pi(1.96^3)/3$

d. $8\pi(16.7/0.02 - 27.3)/3$

e. $88.8(\cos(2.3/1035) + 1.035)/2.045$

Section 2.6   Mass, Moles, and Density

Section 2.6.1   Choosing a Basis

2.6.1   Read each of the following problems and select a suitable basis for solving each one. Do not solve the problems.

**a.   You have 130 kg of gas of the following composition: 40% N_2, 30% CO_2, and 30% CH_4 in a tank. What is the average molecular weight of the gas?

**b.   You have 25 lb of a gas of the following composition: 80%, 10%, 10%. What is the average molecular weight of the mixture? What is the weight (mass) fraction of each of the components in the mixture?

****c.   The proximate and ultimate analysis of coal is given in the following table. What is the composition of the volatile combustible material (VCM)? Present your answer in the form of the mass percent of each element in the VCM.

Proximate Analysis (%)		Ultimate Analysis (%)	
Moisture	3.2	Carbon	79.90
Volatile combustible material	21.0	Hydrogen	4.85
Fixed carbon	69.3	Sulfur	0.69
Ash	6.5	Nitrogen	1.30
		Ash	6.50
		Oxygen	6.76
Total	100.0	Total	100.00

*d.   A fuel gas is reported to analyze, on a mole basis, 20% methane, 5% ethane, and the remainder CO_2. Calculate the analysis of the fuel gas on a mass percentage basis.

***e.   A gas mixture consists of three components: argon, B, and C. The following analysis of this mixture is given: 40.0 mol % argon, 18.75 mass % B, 20.0 mol % C. The molecular weight of argon is 40, and the molecular weight of C is 50. Find (1) the molecular weight of B and (2) the average molecular weight of the mixture.

***2.6.2   Two engineers are calculating the average molecular weight of a gas mixture containing oxygen and other gases. One of them uses the correct molecular weight of 32 for oxygen and determines the average molecular weight as 39.2. The other uses an incorrect value of 16 and determines the average molecular weight as 32.8. This is the only error in the calculations. What is the percentage of oxygen in the mixture expressed as mole percent? Choose a basis to solve the problem, but do not solve the problem.

**2.6.3   Choose a basis for the following problem: Chlorine usage at a water treatment plant averages 134.2 lb/day. The average flow rate of water leaving the plant is 10.7 million gal/day. What is the average chlorine concentration in the treatment water leaving the plant (assuming no reaction of the chlorine), expressed in milligrams per liter?

Section 2.6.2   The Mole and Molecular Weight

*2.6.4   Convert the following:
a. 4 g mol of $MgCl_2$ to g
b. 2 lb mol of C_3H_8 to g
c. 16 g of N_2 to lb mol
d. 3 lb of C_2H_6O to g mol

*2.6.5   How many pounds are there in each of the following?
a. 16.1 lb mol of pure HCl
b. 19.4 lb mol of KCl
c. 11.9 g mol of $NaNO_3$
d. 164 g mol of SiO_2

***2.6.6   A solid compound was found to contain 42.11% C, 51.46% O, and 6.43% H. Its molecular weight was about 341. What is the formula for the compound?

**2.6.7   The structural formulas in Figure P2.6.7 are for vitamins.
a. How many pounds are contained in 2.00 g mol of vitamin A?
b. How many grams are contained in 1.00 lb mol of vitamin C?

Vitamin	Structural formula	Dietary sources	Deficiency symptoms
Vitamin A	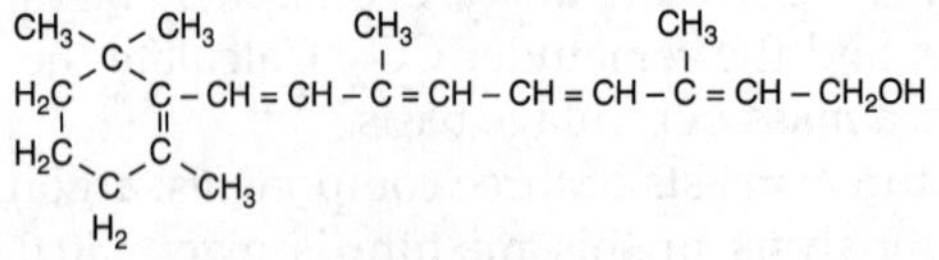 CH_3 CH_3 ... CH_3 ... CH_3 $C-CH=CH-C=CH-CH=CH-C=CH-CH_2OH$ Retinol	Fish liver oils, liver, eggs, fish, butter, cheese, milk; a precursor, b-carotene, is present in green vegetables, carrots, tomatoes, squash	Night blindness, eye inflammation
Ascorbic acid (vitamin C)	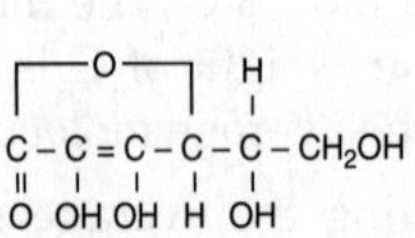 $C-C=C-C-C-CH_2OH$ O OH OH H OH	Citrus fruit, tomatoes, green peppers, strawberries, potatoes	Scurvy

Figure P2.6.7

*2.6.8   A sample has a specific volume of 5.2 m^3/kg and a molar volume of 1160 m^3/kg mol. Determine the molecular weight of the material.

**2.6.9   Prepare an expression that converts mass (weight) fraction (ω) to mole fraction (x), and another expression for the conversion of mole fraction to mass fraction, for a binary mixture.

*2.6.10   You have 100 kg of gas of the following composition:

$$CH_4 \quad 30\%$$
$$H_2 \quad 10\%$$
$$N_2 \quad 60\%$$

What is the average molecular weight of this gas?

*2.6.11   You analyze the gas in 100 kg of gas in a tank at atmospheric pressure and find the following:

$$CO_2 \quad 19.3\%$$
$$O_2 \quad 6.5\%$$
$$H_2O \quad 2.1\%$$
$$N_2 \quad 72.1\%$$

What is the average molecular weight of the gas?

**2.6.12   Suppose you are required to make an analysis of 317 lb of combustion gas and find it has the following composition:

$$CO_2 \quad 60\%$$
$$CO \quad 10\%$$
$$N_2 \quad 30\%$$

What is the average molecular weight of this gas in the AE system of units?

**2.6.13   Consider 10.0 kg mole of benzene (C_6H_6). How many of the following are present in this quantity of benzene: (a) molecules of benzene; (b) lb mol benzene; (c) lb$_m$ benzene; (d) g mol benzene; (e) g mol H; (f) g mol C?

Section 2.6.3   Density and Specific Gravity

*2.6.14   A solid has a specific gravity of 5.25. Estimate the volume in ft^3 of 100 lb$_m$ of it.

*2.6.15   You are asked to decide what size containers to use to ship 1000 lb of cottonseed oil of specific gravity equal to 0.926. What would be the minimum size drum expressed in gallons?

*2.6.16   The density of a certain solution is 8.80 lb/gal at 80°F. How many cubic feet will be occupied by 10,010 lb of this solution at 80°F?

**2.6.17   Which of the three sets of containers in Figure P2.6.17 represents respectively 1 mol of lead (Pb), 1 mol of zinc (Zn), and 1 mol of carbon (C)?

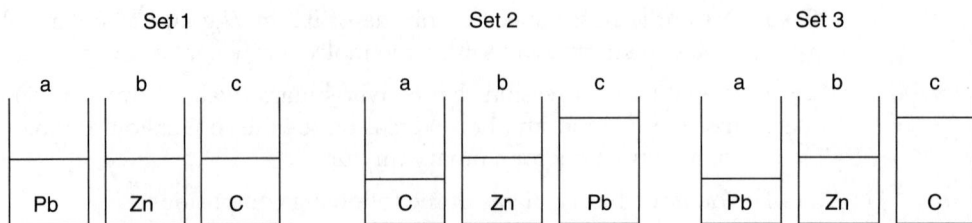

Set 1   Set 2   Set 3

a   b   c   a   b   c   a   b   c

Pb   Zn   C   C   Zn   Pb   Pb   Zn   C

Figure P2.6.17

Section 2.7   Process Variables

Section 2.7.2   Pressure and Hydrostatic Head

**2.7.1   From the newspaper:

> BROWNSVILLE, TX. Lightning or excessive standing water on the roof of a clothes store are emerging as the leading causes suspected in the building's collapse. Mayor Ignacio Garza said early possibilities include excessive weight caused by standing water on the 19-year-old building's roof. Up to six inches of rain fell here in less than six hours.

Flat-roof buildings are a popular architectural style in dry climates because of the economy of materials of construction. However, during the rainy season, water may pool on the roof decks so that structural considerations for the added weight must be taken into account. If 15 cm of water accumulates on a 10 m by 10 m area during a heavy rainstorm, determine (a) the total added weight from the standing water the building must support and (b) the force of the water on the roof in psi.

***2.7.2   A problem with concrete wastewater treatment tanks set belowground was realized when the water table rose and an empty tank floated out of the ground. This buoyancy problem was overcome by installing a check valve in the wall of the tank so that if the water table rose high enough to float the tank, it would fill with water. If the density of concrete is 2080 kg/m^3, determine the maximum height at which the valve should be installed to prevent a buoyant force from raising a rectangular tank with inside dimensions of 30 m by 27 m that is 5 m deep. The walls and floor have a uniform thickness of 200 mm.

**2.7.3   A centrifugal pump is to be used to pump water from a lake to a storage tank that is 148 ft above the surface of the lake. The pumping rate is to be 25.0 gal/min, and the water temperature is 60°F. The pump on hand can develop a pressure of 50.0 psig when it is pumping at a rate of 25.0 gal/min. (Neglect pipe friction, kinetic energy effects, and factors involving pump efficiency.)

a. How high (in feet) can the pump raise the water at this flow rate and temperature?

b. Is this pump suitable for the intended service?

**2.7.4 A manufacturer of large tanks calculates the mass of fluid in the tank by taking the pressure measurement at the bottom of the tank in psig and then multiplying that value by the area of the tank in square inches. Can this procedure be correct?

**2.7.5 Suppose that a submarine inadvertently sinks to the bottom of the ocean at a depth of 1000 m. It is proposed to lower a diving bell to the submarine and attempt to enter the conning tower. What must the minimum air pressure be in the diving bell at the level of the submarine to prevent water from entering the bell when the opening valve at the bottom is cracked open slightly? Give your answer in absolute kilopascals. Assume that seawater has a constant density of 1.024 g/cm^3.

**2.7.6 A pressure gauge on a welder's tank gives a reading of 22.4 psig. The barometric pressure is 28.6 in. Hg. Calculate the absolute pressure in the tank in (a) lb/ft^2, (b) in. Hg, (c) N/m^2, and (d) ft water.

**2.7.7 Person A says they calculated from a formula that the pressure at the top of Pikes Peak is 9.75 psia. Person B says that it is 504 mm Hg because they looked it up in a table. Which person is right?

***2.7.8 The floor of a cylindrical water tank was distorted into 7 in. bulges because of the settling of improperly stabilized soil under the tank floor. However, several consulting engineers restored the damaged tank to use by placing plastic skirts around the bottom of the tank wall and devising an air flotation system to move it to an adjacent location. The tank was 30.5 m in diameter and 13.1 m deep. The top, bottom, and sides of the tank were made of 9.35-mm-thick welded steel sheets. The density of the steel is 7.86 g/cm^3.

a. What was the gauge pressure in kilopascals of the water at the bottom of the tank when it was completely full of water?

b. What would the air pressure have to be in kilopascals beneath the empty tank in order to just raise it up for movement?

***2.7.9 Examine Figure P2.7.9. Oil 1 (density = 0.91g/cm^3) flows in a pipe, and the flow rate is measured via a mercury (density = 13.546 g/cm^3) manometer. If the difference in height of the two legs of the manometer is 0.78 in., what is the corresponding pressure difference between points A and B in mm Hg? At which point, A or B, is the pressure higher? The temperature is 60°F.

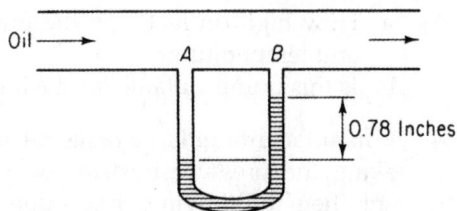

Figure P2.7.9

Section 2.7.4   Flow Rate

**2.7.10   Benzene is flowing through a 4 in. schedule 40 pipe with an internal diameter of 4.026 in. at an average velocity of 5.00 ft/s. Determine the mass flow rate in lb_m/h and the molar flow rate in lb mol/h.

**2.7.11   A source of methane (i.e., 96% CH_4; 4% N_2) enters a process at the rate of 25 m^3/h. Determine the flow rate of methane in kg mol/h. Because the pressure of this stream is moderate, assume the ideal gas law.

***2.7.12   Coal with the following composition is fed to a furnace at the rate of 10,000 lb_m/h:

Carbon	78.2%
Hydrogen	4.5%
Oxygen	7.2%
Nitrogen	1.4%
Sulfur	3.0%
Ash	7.7%

In addition, water is present with the coal in the amount of 3.75 lb_m water/lbm coal. Determine the molar feed rate of each element excluding the ash but including the water.

**2.7.13   The feed to a reactor that produces ammonia (NH_3) by the reaction $N_2 + 3H_3 \rightarrow NH_3$ contains 20% N_2 and 80% H_2 at a feed rate of 10,000 kg/h. Determine the feed rate of N_2 in kg/h for this process.

Section 2.7.5   Concentration

*2.7.14   Calculate the mass and mole fractions of the respective components in $NaClO_3$.

*2.7.15   The specific gravity of a solution of KOH at 15°C is 1.0824 and contains 0.813 lb KOH per gal of solution. What are the mass fractions of KOH and H_2O in the solution?

*2.7.16   You purchase a tank with a volume of 2.1 ft^3. You pump the tank out and add first 20 lb of CO_2 gas and then 10 lb of N_2 gas. What is the analysis of the gas mixture in the tank?

*2.7.17 How many ppb are there in 1 ppm? Does the system of units affect your answer? Does it make any difference if the material for which the ppb are measured is a gas, liquid, or solid?

**2.7.18 The following table lists values of Fe, Cu, and Pb in Christmas wrapping paper for two different brands. Convert the ppm to mass fractions on a paper-free basis.

	Concentration	ppm	
	Fe	Cu	Pb
Brand A	1310	2000	2750
Brand B	350	50	5

2.7.19 Harbor sediments in the New Bedford, Massachusetts, area contain PCBs at levels up to 190,000 ppm, according to a report prepared by Grant Weaver of the Massachusetts Coastal Zone Management Office [*Environ. Sci. Technol.*, **16, No. 9, 491A (1982)]. What is the concentration of PCBs in percent? Does this seem reasonable?

**2.7.20 The analysis of a biomass sample gave

Element	% Dry Weight of Cell
C	50.2
O	20.1
N	14.0
H	8.2
P	3.0
Other	4.5

This compound gives a ratio of 10.5 g cells/mol ATP synthesized in the metabolic reaction to form the cells. Approximately how many moles of C are in the cells per mole of ATP?

***2.7.21 A radioactive tracer-labeled microorganism (MMM) decomposes to NN as follows:

$$MMM \ (s) \rightarrow NN \ (s) + 3CO2(g)$$

If the $CO_2(g)$ yields 2×10^7 dpm (disintegrations per minute) in a detection device, how many μCi (microcuries) is this? How many cpm (counts per minute) will be noted if the counting device is 80% efficient in counting disintegrations? Data: 1 curie $= 3 \times 10^{10}$ dps (disintegrations per second).

***2.7.22   Several alternative compounds have been added to gasoline, including methanol, ethanol, and methyl tert-butyl ether (MTBE), to increase the oxygen content of gasoline in order to reduce the formation of CO on combustion. Unfortunately, MTBE has been found in groundwater at concentrations sufficient to cause concern. Persistence of a compound in water can be evaluated from its half-life, $t_{1/2}$, that is, the time for one half of the compound to leave the system of interest. The half-life depends on the conditions in the system, of course, but for environmental evaluation can be approximated by

$$t_{1/2} = \frac{\ln(2)}{k[OH^-]}$$

where $[OH^-]$ is the concentration of hydroxyl radical in the system that for this problem of the contamination of water is equal to 1.5×10^6 molecules/cm^3. The values of k determined from experiment are

	k cm^3/ [molecule(s)]
Methanol	0.15×10^{-12}
Ethanol	1×10^{-12}
MTBE	0.60×10^{-12}

Calculate the half-life of each of the three compounds, and order them according to their persistence.

***2.7.23   The National Institute for Occupational Safety and Health (NIOSH) sets standards for CCl_4 in air at 12.6 mg/m^3 of air (a time-weighted average over 40 hr). The CCl_4 found in a sample is 4800 ppb. Does the sample exceed the NIOSH standard? Be careful!

***2.7.24   The following table shows the annual inputs of phosphorus to Lake Erie:

A	Short Tons/Yr
Source	
Lake Huron	2240
Land drainage	6740
Municipal waste	19,090
Industrial waste	2030
Total of sources	30,100
Outflow	4500
Retained	25,600

  a. Convert the retained phosphorus to concentration in micrograms per liter, assuming that Lake Erie contains 1.2×10^{14} gal of water and that the average phosphorus retention time is 2.60 yr.
  b. What percentage of the input comes from municipal water?
  c. What percentage of the input comes from detergents, assuming they represent 70% of the municipal waste?
  d. If 10 ppb of phosphorus triggers nuisance algal blooms, as has been reported in some documents, would removing 30% of the phosphorus in the municipal waste and all the phosphorus in the industrial waste be effective in reducing the eutrophication (i.e., the unwanted algal blooms) in Lake Erie?
  e. Would removing all of the phosphate in detergents help?

**2.7.25 A gas contains 350 ppm of H_2S in CO_2. If the gas is liquefied, what is the weight fraction of H_2S?

**2.7.26 Sulfur trioxide (SO_3) can be absorbed in sulfuric acid solution to form a more concentrated sulfuric acid. If the gas to be absorbed contains 55% SO_3, 41% N_2, 3% SO_2, and 1% O_2:
  a. How many parts per million of O_2 are there in the gas?
  b. What is the composition of the gas on an N_2-free basis?

****2.7.27 Twenty-seven pounds (27 lb) of chlorine gas is used for treating 750,000 gal of water each day. The chlorine used up by the microorganisms in the water is measured to be 2.6 mg/L. What is the residual (excess) chlorine concentration in the treated water?

***2.7.28 A newspaper report says the FDA found 13–20 ppb of acrylonitrile in a soft-drink bottle, and if this is correct, it amounts to only 1 molecule of acrylonitrile per bottle. Is this statement correct?

***2.7.29 Several studies of global warming indicate that the concentration of CO_2 in the atmosphere is increasing by roughly 1% per year. Do we have to worry about the decrease in the oxygen concentration also?

PART II
MATERIAL BALANCES

CHAPTER 3

Material Balances

Chapter Objectives

- Understand the connection between a schematic of a process and the actual process.
- Be able to apply the material balance equation to single-unit processes in which reactions do not occur using the recommended step-by-step procedure.
- Understand how to apply a degree-of-freedom analysis to a material balance problem.
- Be able to solve a set of linear material balance equations using either MATLAB or Python.

Introductory Example

Let's begin with an example. Consider the rotary dryer shown in Figure 3.1. A wet granular polymer is fed into the dryer on one end, and hot dry air is fed into the dryer on the other end. The body of the dryer rotates, drops the wet polymer through the passing air, and moves the wet polymer through the dryer. Therefore, the water on the polymer will evaporate into the air and exit the process with the air on one end of the dryer, and dried polymer will exit the other end of the dryer. The material balances for this process indicate simply that (1) the amount of water that enters the process with the polymer leaves the process with the air, and (2) the amount of polymer that entered the process as wet polymer exits the process as dried polymer. This analysis is based on assuming **steady-state operation** (i.e., the feed streams, the exit streams and the conditions inside the process do not change with time).

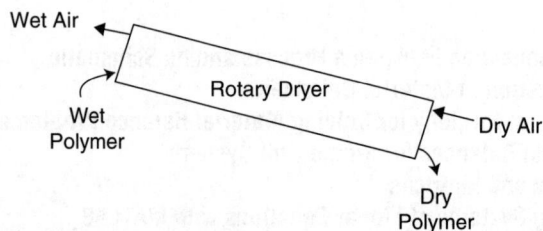

Figure 3.1   Schematic of a rotary dryer

You know from common sense that if you put a six-pack of beer into the refrigerator, you cannot remove more than six beers before putting more beer into the refrigerator. Also, if you drink two beers from the six-pack, you know that four beers remain. Both of these examples are applications of the **law of conservation of mass**, which states that mass cannot be created or destroyed.

If we apply the law of conservation of mass to a steady-state process, we arrive at the result that what enters the process must exit the process during a specified period of time, that is,

$$\text{Mass in} = \text{Mass out} \tag{3.1}$$

This relationship applies to any steady-state process. Because in this chapter we do not consider processes in which chemical reactions occur, this equation applies to each compound or element that enters the process as well as to the total mass. Chapter 4 addresses how to apply material balances for processes in which chemical reactions occur.

In this chapter, we deal with processes that are represented with material flowing into the process and material flowing out of the process (e.g., see the introductory example). Applying Equation (3.1) to a flow system for a convenient period of time (e.g., 1 hr, 1 min) yields the following result:

$$(\dot{m}_i)_{\text{in}} = (\dot{m}_i)_{\text{out}} \tag{3.2}$$

which states that the flow rate of each component entering the process is equal to the flow rate of that component leaving the process assuming steady-state operation. Note that if you choose a unit of time as your basis, Equation (3.2) becomes Equation (3.1). Equations (3.1) and (3.2) are forms of material balances that we will be applying throughout this chapter to steady-state systems. The remaining material in this chapter will demonstrate how to apply these equations to steady-state continuous systems and to batch operations.

3.1   The Connection between a Process and Its Schematic

For our purposes in this section, we consider a portion of the ethylene cracking process. Ethylene is polymerized to produce high-density polyethylene, which is the most common plastic comprising about one third of the total volume of all plastics produced. In addition, ethylene gas is used to a much lesser extent to ripen a number of fruits, including bananas and pears.

The feed to the ethylene cracking process, which is typically a light hydrocarbon feed such as natural gas, is passed through tubes mounted inside a furnace (Figure 3.2) in order to provide the energy necessary to crack the feed into a number of light hydrocarbon products, including ethylene (C_2H_4). After exiting the cracking furnaces, the cracked products are separated and purified by a series of distillation columns (Figure 3.2). Note that this simplified schematic does not reflect the complexity of the actual plant. A large number of pieces of ancillary equipment, such as pumps, surge vessels, lines delivering utilities to the process, valves, heat exchangers, and pipes are packed around and connected to these columns. In addition, the sensor readings from these columns are electronically transferred to the control computer, located in the control room, and the control computer maintains operation of the process by sending electronic signals to the control valves. The landscape of the process (and all chemical plants) is dominated by this ancillary equipment that is often not shown in the simplified schematics. The point is that these are very complex systems with numerous components.

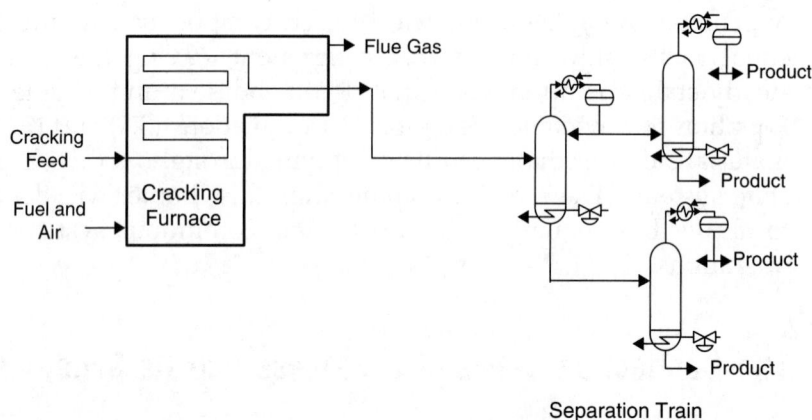

Figure 3.2 A general representation of an ethylene cracking process

If we neglect much of the detail (i.e., pumps, valves, and many of the lines) and focus on the overall function of a distillation column, we can develop a schematic of a single distillation column, as shown in Figure 3.3. The steam to the reboiler vaporizes a portion of the liquid in the bottom of the column (V), and the overhead condenser condenses the vapor in the top of the column and stores it in the accumulator. The vapor from the reboiler flows up the column, passing through trays, and the liquid from the accumulator flows down the column (L), contacting the vapor on each of the internal trays, thus concentrating the more volatile components in the vapor and the less volatile components in the liquid. Note that only the elements that affect material and energy balances are included in this figure, that is, the feed (F) and feed composition (z), the products (D and B) and their compositions (y and x, respectively), the steam to the reboiler, and the refrigerant to the condenser.

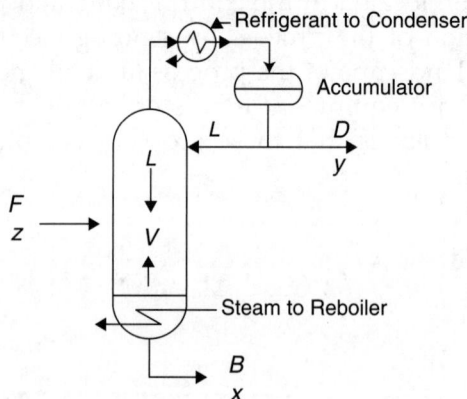

Figure 3.3 Schematic of a distillation column

We can draw a **system boundary** around any part or the whole of this schematic. For example, by drawing a boundary around the accumulator and the exit lines (boundary A in Figure 3.4), the result is shown in Figure 3.5, and it indicates that what enters the identified system equals what leaves that system. That is, the condensed liquid that enters the accumulator is equal to the sum of the distillate rate (D) and the reflux (L) that leaves the system.

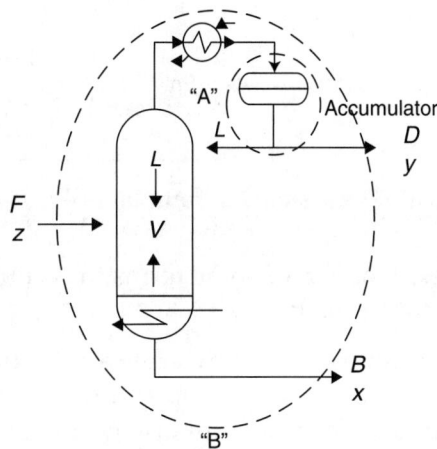

Figure 3.4   Boundaries applied to the schematic of a distillation column

On the other hand, if we draw a boundary around the entire column (boundary B) and consider only the material balance aspects of the system, the result is represented by the schematic shown in Figure 3.6. Note that neither the steam nor the refrigerant comes into contact with the components in the feed stream that are being separated by the column. That is, steam condenses in the reboiler, transferring heat to the material in the bottom of the column, and the refrigerant absorbs heat from the overhead vapors, but neither mixes with the material inside the column. From Figure 3.6, all of the details of the internal operation of the distillation column (i.e., the generation of vapor by the reboiler, condensation of the overhead vapor by the condenser, and the vapor/liquid contacting on each tray in the column) are neglected, and only the overall material balance behavior of the process is considered. Many times, this overall material balance behavior (i.e., a macroscopic model) is all that is needed to represent a process. Therefore, you can analyze the entire process or any portion of the process by simply applying the proper system boundary. This chapter focuses on applying material balances using schematics similar to Figures 3.5 and 3.6; that is, emphasis is placed on the overall behavior of the process and not the internal details.

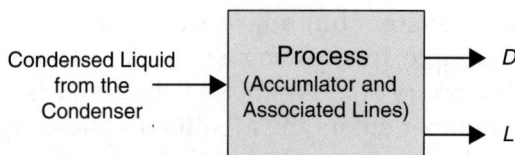

Condensed Liquid
from the
Condenser

Process
(Accumlator and
Associated Lines)

D

L

Figure 3.5   The system for boundary A

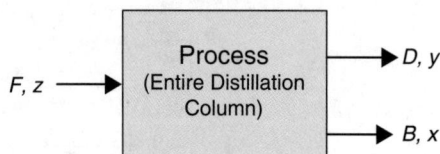

F, z

Process
(Entire Distillation
Column)

D, y

B, x

Figure 3.6   The system for boundary B

Now for the explanation of some terms that you will encounter frequently in the remainder of the text:

System: By **system** we mean any arbitrary portion of or a whole process you want to consider for analysis. You can define a system such as a reactor, a section of a pipe, or an entire refinery by stating in words what the system is. Or you can define the limits of the system by drawing the **system boundary**, which is a line that encloses the portion of the process that you want to analyze. The boundary could coincide with the outside of a piece of equipment or some section inside the equipment. Now, let's focus on some important characteristics associated with systems.

Closed system or process: Figure 3.7a shows a two-dimensional view of a three-dimensional vessel holding 1000 kg of H_2O. Note that material neither enters nor leaves the vessel; that is, no material crosses the system boundary. Figure 3.7a represents a **closed system**. Changes can take place inside the closed system, but no mass exchange occurs with the surroundings. A **batch reactor** is an example of a closed system, that is, after the reactants are added to the reactor, no material is added or removed from the reactor while the reactions occur.

Batch process: A batch process is a closed process in which a specific amount of material is processed to produce a product in a batchwise procedure. Making a cake is an example of a batch process because the ingredients are mixed and baked to produce a single cake.

Open system or process: Next, let us assume that, in an experiment, you add water to the tank shown in Figure 3.7a at the rate of 100 kg/min and withdraw water at the rate of 100 kg/min, as indicated in Figure 3.7b. Figure 3.7b is an example of an open system (also called a *flow system*) because the material crosses the system boundary. Figures 3.1, 3.2, and 3.3 are also examples of open systems.

Continuous process: A continuous process has one or more streams continuously entering the system and one or more streams continuously leaving the system. A continuous process is an open system.

Macroscopic balances: Macroscopic balances are concerned with only the overall behavior of the process under consideration and not the details of the internal operation of the system. For example, a macroscopic balance on a distillation column is concerned about the overall separation of the light and heavy components (i.e., the compositions of the streams leaving the column) and not the distribution of component concentrations inside the column (e.g., system B in Figure 3.4 and Figure 3.6).

Steady-state system or process: What does *steady state* mean? As pointed out earlier, a variable, such as an amount of material or a property of the material, is in the steady state if the value of that variable is invariant (does not change) with time. If you look at Figure 3.7b, you will note that the flow rates in and out are just water and presumably constant and equal, respectively; hence, the system is in the steady state, that is, the feed, products, and contents of the system remain constant with time.

Unsteady-state or transient system or process: The conditions in an unsteady-state process change with time. If material is added to a system, but no material is removed, such a system would be an unsteady-state process because the amount of material in the system would increase with time. A batch reactor is another example of an unsteady-state process because as the reaction proceeds in the reactor, the concentration of the components in the reactor would change with time. Chapter 11 addresses unsteady-state material and energy balances.

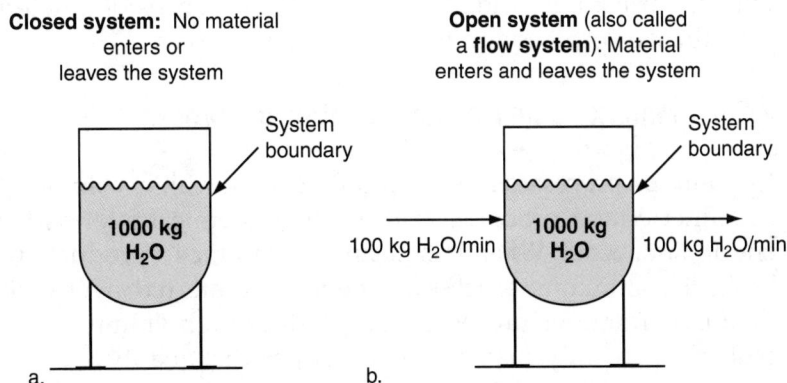

Closed system: No material enters or leaves the system

Open system (also called a **flow system**): Material enters and leaves the system

1000 kg H_2O

System boundary

a.

100 kg H_2O/min   1000 kg H_2O 100 kg H_2O/min

System boundary

b.

Figure 3.7   Comparison of (a) closed and (b) open systems

Self-Assessment Test

Questions

1. Does a process schematic contain all the details of a process? If not, what does it contain?

2. What constitutes a steady-state process?

3. What is an open system?

Answers

1. A process schematic does not contain all the details of the process. It contains only the elements of the process relevant to the overall behavior of the process.

2. A steady-state process is invariant with respect to time. That is, the streams entering and leaving the process as well as the conditions inside the process are invariant with respect to time.

3. An open system has material either entering or exiting the system boundary.

3.2   Introduction to Material Balances

This section applies Equations (3.1) and (3.2) to two very simple material balance problems to introduce you to how to apply material balances to continuous and batch processes. By formulating and solving simple material balance problems in this section, you should be better able to understand the detailed material balance solution procedure that is presented in the next section. Note that Equations (3.1) and (3.2) are macroscopic material balances because they consider only what enters and leaves the process and not what is happening inside the process. We first consider material balances for a continuous process and then for a batch process.

3.2.1   Material Balances for a Continuous Process

Continuous processes are commonly used for high-volume operations, such as refineries and processes that produce chemical intermediates (e.g., ethylene and benzene). When it is desired to produce a product on a high-volume basis, a continuous process is generally preferred over a batch operation. That is, continuous processes can produce high-volume products at a lower cost and, generally, maintain a higher product quality.

Now consider the mixer shown in Figure 3.8, a process that mixes a dilute stream of NaOH with a smaller flow rate of a concentrated stream to increase the concentration of NaOH in the dilute solution. Note that there are two ways to represent this process: (a) using compositions and (b) using component flow rates. Each approach has its advantages depending on the particulars of a problem. Both of these approaches provide the same information, just in a different form (i.e., the mass flow rate is simply the composition times the total stream flow rate, and the composition is simply the ratio of the component mass flow rate to the total stream flow rate).

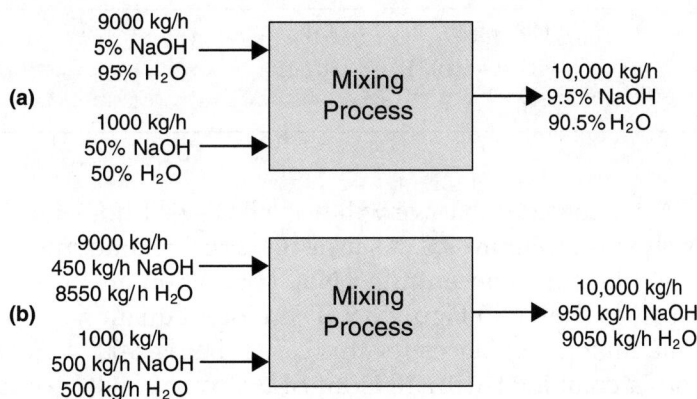

Figure 3.8 Two ways in which a process can be represented (a) in terms of compositions; (b) in terms of component flow rates

Using the values of the component flow rates listed in Figure 3.8, the total and **component balances** for mass and moles can be written using Equation (3.2). Note that if the tank is well mixed, the concentration of a component in the output stream will be the same as the concentration of the component inside the tank during mixing, an assumption frequently made that is relatively accurate if adequate mixing is used. Assuming steady-state operation of this process and applying Equation (3.2) yields the results shown in Table 3.1.

Table 3.1   Application of Equation (3.2) to a Mixer in Terms of kg/h

Component	In (kg/h)	=	Out (kg/h)
NaOH	450 + 500	=	950
H_2O	8550 + 500	=	9050
Total	10,000	=	10,000

Note that Equation (3.2) is satisfied for each component, and the total is based on the compositions provided in Figure 3.8 as shown in Table 3.1.

Next, we show the application of Equation (3.2) in terms of moles for Figure 3.8. We can convert the kilograms shown in Table 3.1 to kilogram moles by dividing each compound by its respective molecular weight (NaOH = 40.00 and H_2O = 18.02). The results are shown in Table 3.2.

Table 3.2   **Application of Equation (3.2) to a Mixer in Terms of kg mol/h**

Component	In (kg mol/h)	=	Out (kg mol/h)
NaOH	$(450 + 500)/40.00 = 23.75$	=	$950/40.00 = 23.75$
H_2O	$(8550 + 500)/18.02 = 502.25$	=	$9050/18.02 = 502.25$
Total	526.0	=	526.0

Now consider that we will use a basis of 1 hour for the continuous process shown in Figure 3.8. As a result, specific amounts of the dilute NaOH solution and the concentrated NaOH solution are combined in the mixing process, as shown in Figure 3.9. Therefore, Equation (3.1) can be used to formulate material balances for this case. That is, 9000 kg of a weak NaOH solution is combined with 1000 kg of a strong NaOH solution. Note that the only difference between this case and the previous case (i.e., Figure 3.8) is that this case considers specific amounts of each stream in units of kilograms, while the previous case considered flow rates of each stream in units of kg/h, but the amounts and concentrations are exactly the same. As a result, the application of Equation (3.1), as shown in Tables 3.1 and 3.2, will be exactly the same for the latter case except that the units are changed from kg/h to kg or kg mol/h to kg mol, respectively.

In summary, **for continuous processes, Equation (3.2) can be applied directly or by using a time interval basis, and Equation (3.1) can be applied in terms of either mass or moles.**

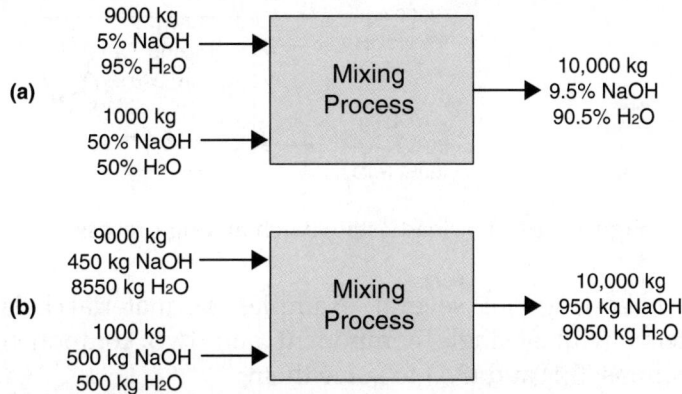

Figure 3.9 Combining specific amounts of material (a) in terms of compositions; (b) in terms of masses of components

3.2.2   Material Balances for Batch Process

Batch operations involve processing a fixed amount of material to produce a product. In order to use Equation (3.1) for a batch process, we must consider the system from a steady-state point of view. That is, initially the system is assumed to be empty. Then the feed material is added. Next, the feed is processed to produce the product. And finally, the product material is removed from the system. Consequently, the system behaves as a steady-state process because the system is empty to begin with and is empty after the processing is complete. Therefore, Equation (3.1) can be applied as a macroscopic material balance for a batch system.

If we consider the batch mixing of 9000 kg of dilute NaOH with 1000 kg of concentrated NaOH, the system shown in Figure 3.9 applies directly. Therefore, **using a basis of a time interval for material balances on continuous process is equivalent to applying material balances to a batch process.**

Now consider the batch process shown in Figure 3.10. This representation of the mixing process presented earlier shows the feed streams being added to the mixing tank. Therefore, the system would be an unsteady-state process because there is accumulation of material inside the mixing tank, so Equation (3.1) would not be applicable. Unsteady-state material balances are addressed in Chapter 11. An unsteady-state analysis of this system will yield the same result as applying Equation (3.1), although the analysis is more complicated than applying Equation (3.1), as shown in Figure 3.9a.

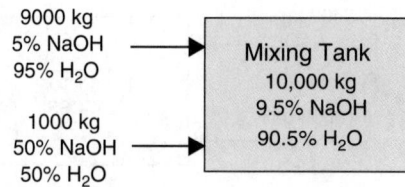

Figure 3.10 Unsteady state batch mixing process

Following are several examples of material balance problems for systems with a single component and two components. We will apply Equations (3.1) and (3.2) to solve them.

Example 3.1   Material Balance on a Lake

Problem Statement

Water balances on a lake can be used to evaluate the effect of groundwater infiltration, evaporation, or precipitation on the lake. Prepare a steady-state water balance, in symbols, for a lake, including the physical processes indicated in Figure E3.1 (all symbols are in mass flow rates). Express your answer as the outlet river flow (R_2) as a function of the other variables.

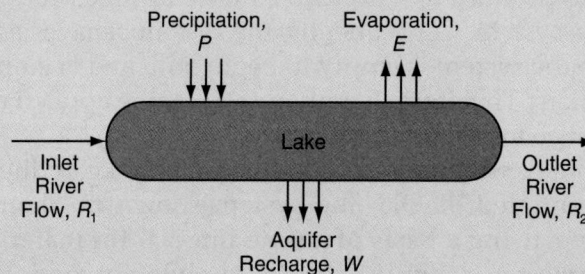

Figure E3.1

Solution

Applying Equation (3.2), assuming that the streams are in flow rates for water, a balance on this process yields

$$R_1 + P = R_2 + E + W$$

Solving for R_2 yields

$$R_2 = R_1 + P - E - W$$

Note that this equation can be applied based on the flow rates (i.e., a continuous process) or the amounts added or removed (i.e., a batch process).

Example 3.2   Rotary Dryer

Problem Statement

Return to the introductory example for this chapter. Assume that a wet polymer containing 30 wt % water enters the dryer at a rate of 5000 kg/h and that 10 kg of hot air is charged to the opposite end of the rotary dryer for each kilogram of wet polymer added to the dryer. Determine the weight fraction of water in the air exiting the dryer if the polymer is completely dry when it exits the dryer.

Solution

The statement that 10 kg of air are added for each kilogram of wet polymer sets the air addition rate at 50,000 kg/h (i.e., 10 kg air/kg wet polymer × 5000 kg wet polymer/h = 50,000 kg air/h). Figure E3.2 shows a schematic of this process along with the problem specification data. Note that this is a continuous process.

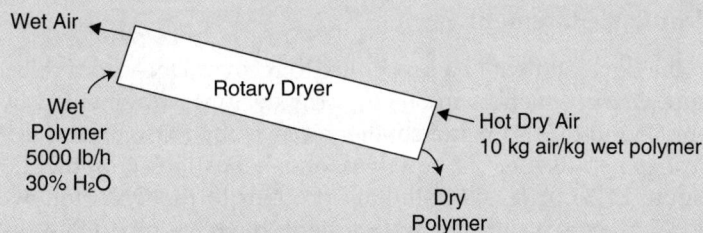

Figure E3.2

First, we apply a material balance on the polymer using Equation (3.2); that is, the amount of polymer that enters the dryer is equal to the amount of polymer leaving the dryer:

$$0.7(5,000 \text{ kg/h}) = \left(\dot{m}_{\text{polymer}} \right)_{\text{Out}} = 3500 \text{ kg/h}$$

(Continues)

Example 3.2   Rotary Dryer (*Continued*)

Note that both the polymer and the water are referred to as a **tie components** because they enter the process in only one stream and exit the process in only one stream. Therefore, a simple proportional relationship results between these two streams for both a polymer and a water balance (i.e., for the tie components). Next, we use Equation (3.2) to apply a material balance on the water to determine how much water will leave with the air:

$$0.3(5,000 \text{ kg/h}) = (\dot{m}_{water})_{Out} = 1500 \text{ kg/h}$$

Next, the material balance for air:

$$50,000 \text{ kg/h} = (\dot{m}_{air})_{Out}$$

Finally, the weight fraction of water in the air is equal to the flow rate of water divided by the sum of the flow rate of water plus the flow rate of air:

$$w_{water} = \frac{1500 \text{ kg/h}}{1500 \text{ kg/h} + 50,000 \text{ kg/h}} = 0.0291$$

Example 3.3   Efficiency of Recovery of DNA

Problem Statement

In the development of a procedure to recover DNA from cells and tissue, 20 μg of pure DNA sequences in 500 μg of water were fragmented by the application of ultrasound to 500 bp and smaller sizes. [A bp (base pair) is 0.34 nm (along the helical axis); 10.4 bp is equal to one helical turn in the DNA molecule.] See Figure E3.3a. After cross-linking proteins to the DNA followed by several additional processing and separation steps, the remaining DNA was precipitated from solution, cleaned, and dried, yielding 12.0 μg of DNA. What fraction of the DNA was lost in the processing steps?

   In nearly all living organisms, the DNA, which stands for deoxyribonucleic acid, is the molecule that stores genetic information. It forms long linear molecules of two intertwined chains called a double helix tied together as shown in Figure E3.3a.

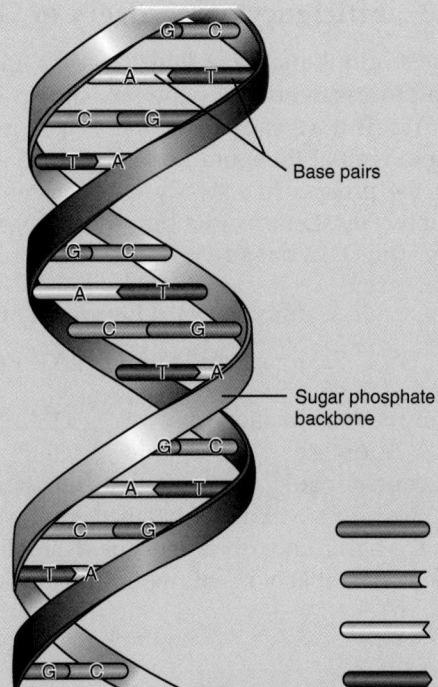

Figure E3.3a   Three-dimensional representation of a strand of DNA

Solution

This is an easy problem, but it illustrates the analysis needed to solve material balance problems. A review of the problem indicates that you can make a DNA balance using Equation (3.1) because the amounts of DNA and not the flow rates are specified. Therefore, this system is represented as a batch process. Do you have to worry about the water? No, because the processing involves water, and no information is given about that water. Figure E3.3b shows the given information.

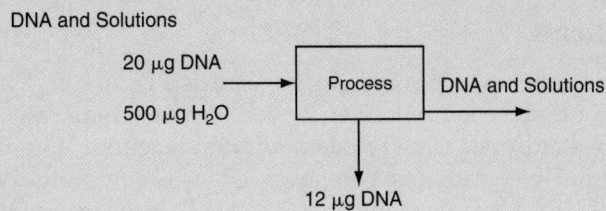

Figure E3.3b

(*Continues*)

Example 3.3   Efficiency of Recovery of DNA (*Continued*)

The first thing to do is pick a system, keeping in mind Equation (3.1). Let the system be the process denoted by the box shown in Figure E3.3b. Is the process an open one? Yes. Is it a steady-state one? If you assume that no DNA existed in the process at the initial time and none remained at the final time, as seems reasonable, then the process is a steady-state process. Does "what comes in must come out" apply? Yes. Let x be the output of DNA (the unknown) that is lost in the processing. The DNA balance is on the basis of 20 μg of DNA:

$$\text{Input} \qquad \text{Output} \qquad \text{Output}$$

$$20 \; \mu g \; DNA = 12 \; \mu g \; DNA + x \; \mu g \; DNA$$

The solution of the mass balance is $x = 8$ μg DNA, so that the fraction lost in the processing is 8/20, or 0.4.

    You, of course, could first calculate the fraction of the DNA recovered, 12/20, or 0.6. How does a material balance enter into the solution then? In effect, 1 μg becomes the basis, and x becomes the desired fraction. The material balance then is $1 = 0.6 + x$ so that x is still equal to 0.4. Could you choose 12 μg as the basis? Yes.

Example 3.4   Concentration of Cells Using a Centrifuge

Problem Statement

Centrifuges are used to separate particles in the range of 0.1 to 100 μg in diameter from a liquid using centrifugal force. Yeast cells are recovered from a broth (a liquid mixture containing cells) using a tubular centrifuge (a cylindrical system rotating about a cylindrical axis). Determine the amount of the cell-free discharge per hour if 1000 L/hr are fed to the centrifuge. The feed contains 500 mg cells/L, and the product stream contains 50 wt % cells. Assume that the feed has a density of 1 g/cm^3 and that there are no cells in the broth discharge from the centrifuge.

Solution

Several different types of centrifuges exist. Figure E3.4 implies continuous feed and continuous outputs; hence, you can conclude that the process involves a steady-state, open (flow) system without reaction. Note that this is a continuous process. Two components are involved: cells and broth. What should you take as a basis? Take a convenient basis of 1000 L/h for the feed to the centrifuge. Let P be the desired product and D be the discharge, both in grams/h.

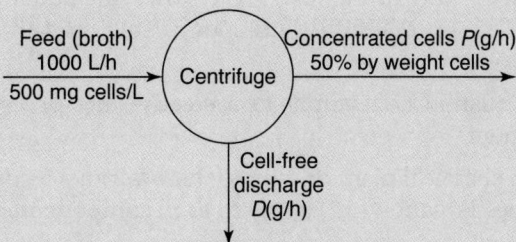

Feed (broth)
1000 L/h
500 mg cells/L

Centrifuge

Concentrated cells P(g/h)
50% by weight cells

Cell-free
discharge
D(g/h)

Figure E3.4

The material balances [total and broth component using Eq. (3.2)] are "what goes in must come out." Let us make a cell balance followed by a fluid balance because cells are a tie component.

Cell balance:

$$\frac{1000 \text{ L feed}}{\text{h}} \left| \frac{500 \text{ mg cells}}{1 \text{ L feed}} \right| \frac{1 \text{ g}}{1000 \text{ mg}} = \frac{0.50 \text{ g cells}}{1 \text{ g } P} \left| \frac{P \text{ g}}{\text{h}} \right. \quad P = 1000 \text{ g/h}$$

Fluid balance:

Using the calculated value of P in the fluid balance yields

$$\frac{1000 \text{ L}}{} \left| \frac{1000 \text{ cm}^3}{1 \text{ L}} \right| \frac{1 \text{ g fluid}}{1 \text{ cm}^3} = \frac{1000 \text{ g } P}{} \left| \frac{0.50 \text{ g fluid}}{1 \text{ g } P} \right. + D \text{ g fluid}$$

$$D = (10^6 - 500) \text{ g fluid/h}$$

Self-Assessment Test

Questions

1. How does a material balance relate to the concept of the conservation of mass?

2. In an automobile engine, as the valve opens to a cylinder, the piston moves down and air enters the cylinder. Fuel follows and is burned. Thereafter, the combustion gases are discharged as the piston moves up. On a very short timescale, say, a few microseconds, would the cylinder be considered an open or closed system? Repeat for a timescale of several seconds.

3. Without looking at the text, write down the equation that represents a material balance for (a) a steady-state open system and (b) a steady-state batch mixing system.

4. Does Equation (3.2) apply to a steady-state process involving more than one component?

5. When a chemical plant or refinery uses various feeds and produces various products, does Equation (3.2) apply to each component in the plant?

Answers

1. The conservation of mass focuses on the invariance of material in an all-encompassing system, whereas a material balance focuses on ensuring that the flows in and out of a more limited system and the material in the system are equated.

2. On a very short timescale when all of the valves are closed, the system behaves as a closed system. On a longer timescale, it is an open system.

3. See text.

4. Yes, based on two conditions: (1) no reactions take place; (2) the process represents a steady-state system.

5. Yes, but it neglects chemical reaction.

Problems

1. Draw a sketch of the following processes, and place a dashed line appropriately to designate the system boundary:
   a. A teakettle
   b. A hearth
   c. A swimming pool

2. Classify the following processes as open, closed, neither, or both:
   a. Oil storage tank at a refinery
   b. Flush tank on a toilet
   c. Catalytic converter on an automobile
   d. Fermentation vessel

3. As an example of a system, consider a bottle of beer. Pick a system.
   a. What is *in* the system?
   b. What is outside the system?
   c. Is the system open or closed?

4. A plant discharges 4000 gal/min of treated wastewater that contains 0.25 mg/L of PCBs (polychlorinated biphenyls) into a river that contains no measurable PCBs upstream of the discharge. If the river flow rate is 1500 ft^3/s after the discharged water has thoroughly mixed with the river water, what is the concentration of PCBs in the river in milligrams per liter?

Answers

1. A system that you pick will be somewhat arbitrary, as will be the time interval for analysis, but (a) and (c) can be closed systems (ignoring evaporation) and (b) open.

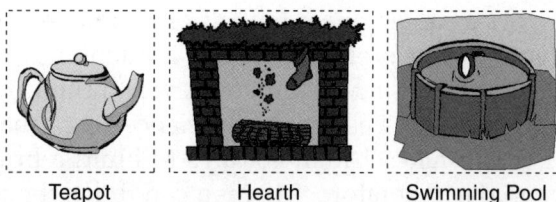

| Teapot | Hearth | Swimming Pool |

2. (a) Closed when no material is added or removed; (b) open (flow) when it is being flushed; (c) open (flow) while the car is running; (d) closed during the fermentation step

3. (a) The beer inside the bottle; (b) everything outside the bottle; (c) closed

4. PCB balance: 1000 mg/min = (4000 L/min + 1500*7.48*60 L/min) x_{PCB}; $x_{PCB} = 1.49 \times 10^{-3}$ mg/L

3.3   A General Strategy for Solving Material Balance Problems

Most of the literature on problem solving views a "problem" as a gap between some initial information (the initial state or problem statement) and the desired information (the final state or the answer to the problem). Problem solving is the activity of finding a path between these two states.

You will find as you go through this book that routine substitution of data into an appropriate equation will not be adequate to solve material (and later energy) balances other than the most trivial ones. You can, of course, formulate your own strategy for solving problems, and everyone has a different viewpoint. But adoption of the well-tested general strategy presented in this chapter has been found to significantly ease the difficulty students have when they encounter problems that are not exactly the same as those presented as examples and homework in this book—for example, problems in industrial practice. After all, the problems in this book are only samples, and simple ones at that, of the myriad problems that exist or could be formulated. Even if you pick your individual problem-solving technique, you will find the following steps to be a handy check on your work and a help if you get stuck.

An orderly method of analyzing problems and presenting their solutions represents training in logical thinking that is of considerably greater value than the mere knowledge of how to solve a particular type of problem. Understanding how to approach these problems from a logical viewpoint will help you to develop those fundamentals of thinking that will assist you in your work as an engineer long after you have read this material.

When solving problems, either academic or industrial, you should always use "engineering judgment" even though much of your training to date treats problems as an exact science (e.g., mathematics). For instance, suppose that it takes 1 man 10 days to build a brick wall; then 10 men can finish it in l day. Therefore, 240 men can finish the wall in 1 hr, 14,400 can do the job in 1 min, and with 864,000 men the wall will be up before a single brick is in place! Your password to success is to use some common sense in problem solving and always maintain a mental picture of the system that you are analyzing. Do not allow a problem to become abstract and unrelated to physical behavior.

You do not have to follow the steps in the following list in any particular sequence or formally employ every one of them. You can go back several steps and repeat steps at will. You can consolidate steps. As you might expect, when you work on solving a problem, you will experience false starts, encounter extensive preliminary calculations, suspend work for higher-priority tasks, look for missing links, and make foolish mistakes. The strategy outlined here is designed to focus your attention on the main path rather than the detours.

1. Read and understand the problem statement.
This means **read the problem carefully** so that you know what is given and what is to be accomplished. Rephrase the problem to make sure you understand it.

Here is a question to answer: How many months have 30 days? Now you may remember the mnemonic "30 days hath September . . ." and give the answer as 4, but is that what the question concerns—how many months have exactly 30 days? Or, does the question ask how many months have at least 30 days (the answer being 11)? Individuals reading the same problem frequently have different perspectives. If the streets in your town are numbered consecutively from 1 to 24, and you are asked by a stranger what street comes after 6th Street, you would be likely to respond 7th Street, whereas the stranger, if facing the opposite direction, would more likely be interested in 5th Street.

Be sure to decide if a problem is a simple or complex calculation and involves a steady-state or unsteady-state process, and when your calculations are completed, state your conclusion somewhere on your calculation sheet or computer printout, say, at the end or the beginning.

Example 3.5   Understanding the Problem

Problem Statement

A train is approaching a station at 105 cm/s. A man in one car is walking forward at 30 cm/s relative to the seats. He is eating a foot-long hot dog, which is entering his mouth at the rate of 2 cm/s. An ant on the hot dog is running away from the man's mouth at 1 cm/s. How fast is the ant approaching the station? Cover the solution, and try to determine what the problem requests before peeking.

Solution

As you read the problem, make sure you understand how each piece of information meshes with the others. Would you agree that the following is the correct analysis?

A superficial analysis would ignore the hot dog length but would calculate 105 + 30 − 2 + 1 = 134 cm/s for the answer. However, on more careful reading, it becomes clear that the problem states that the ant is moving away from the man's mouth at the rate of 1 cm/s. Because the man's mouth is moving toward the station at the rate of 135 cm/s, the ant is moving toward the station at the rate of 136 cm/s.

2. Draw a sketch of the process and specify the system boundary.

It is always good practice to begin solving a problem by drawing a sketch of the process or physical system. You do not have to be an artist to make a sketch. A simple box or circle drawn by hand to denote the system boundary with some arrows to designate flows of material will be fine. You can also state what the system is in words or with a label. Figure 3.11 illustrates some examples.

Figure 3.11c adequately represents Figure 3.11a, and Figure 3.11d represents Figure 3.11b because the internal details do not normally affect the application of Equations (3.1) and (3.2). The diagram itself will indicate if the system is open or closed.

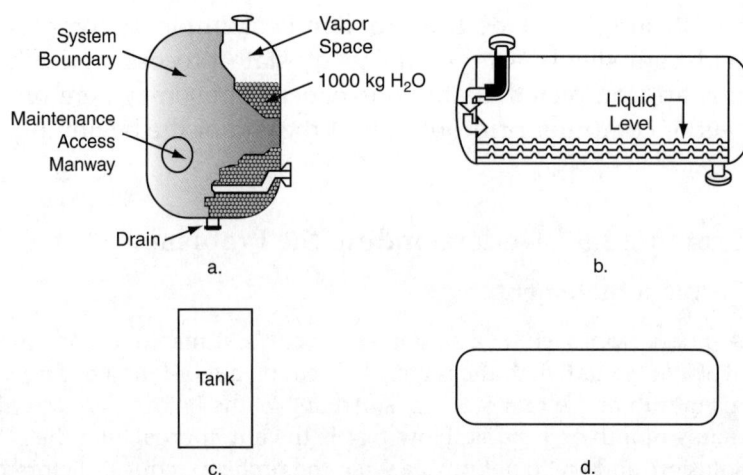

Figure 3.11   Examples of sketches used to represent process equipment

3. Place labels (symbols, numbers, and units) on the diagram for all of the known flows, materials, and compositions.

By putting data on the diagram, you will avoid having to look back at the problem statement repeatedly and will also be able to clarify what data are missing. For the unknown flows, materials, and compositions, insert symbols and units. Add any other useful relations or information. What kinds of information might you place on the diagram? Some specific examples are

- Stream flow rates (e.g., $\dot{F} = 100 \text{ kg} / \text{min}$)
- Compositions of each stream (e.g., $x_{H_2O} = 0.40$)
- Given flow ratios (e.g., $F/R = 0.7$)
- Given identities (e.g., $F = P$)
- Yields (e.g., $Y \text{ kg} / X \text{ kg} = 0.63$)
- Efficiency (e.g., 40%)
- Specifications for a variable or a constraint (e.g., $x < 1.00$)
- Conversion (e.g., 78%)
- Equilibrium relationships (e.g., $y/x = 2.7$)
- Molecular weights (e.g., MW = 129.8)

How much data should you place on the diagram? Enough to help solve the problem and be able to interpret the answer. Values of variables that are zero because they are not present in the problem can be ignored. If your diagram becomes too crowded with data, make a separate table and key it to the diagram. Be sure to include the units associated with the flows and other material when you write the numbers on your diagram or in a table. Remember that units make a difference!

Some of the essential data may be missing from the problem statement. If you do not know the value of a variable to put on the figure, you can substitute a symbol such as F_1 for an unknown flow or ω_1 for a mass fraction. The substitution of a symbol for a number will focus your attention on searching for the appropriate information needed to solve the problem.

Example 3.6   Placing the Known Information on the Diagram

Problem Statement

A continuous mixer mixes NaOH with H_2O to produce an aqueous solution of NaOH. The problem is to determine the composition and flow rate of the product if the flow rate of NaOH is 1000 kg/hr and the ratio of the flow rate of the H_2O to the product solution is 0.9. Draw a sketch of the process and put the data and unknown variables on the sketch with appropriate labels. We will use this example in subsequent illustrations of the proposed strategy.

Solution

Because no contrary information is provided about the composition of the H_2O and NaOH streams, we will assume that they are 100% H_2O and NaOH, respectively. Look at Figure E3.6 for a typical way the data might be put on a diagram.

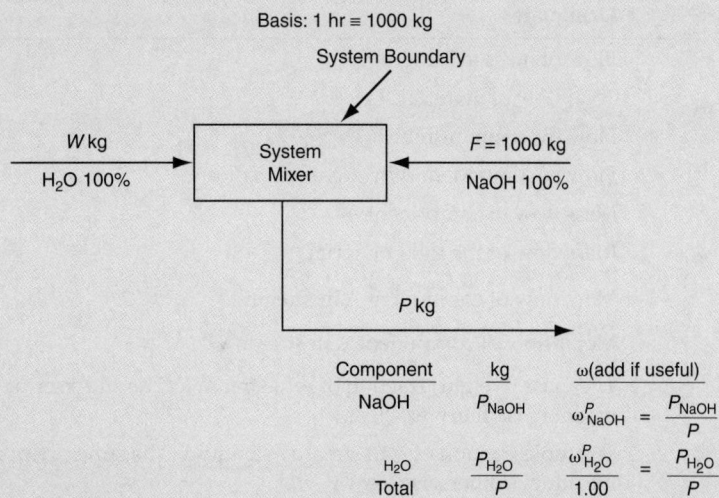

Component	kg	ω(add if useful)	
NaOH	P_{NaOH}	ω^P_{NaOH}	$= \dfrac{P_{NaOH}}{P}$
H_2O	P_{H_2O}	$\omega^P_{H_2O}$	$= \dfrac{P_{H_2O}}{P}$
Total	P	1.00	$\dfrac{P_{H_2O}}{P}$

Figure E3.6

(Continues)

Example 3.6 Placing the Known Information on the Diagram (*Continued*)

Note that the composition of the product stream is listed along with the symbols for unknown flows. Could you have listed the mass fractions instead of or in addition to the mass flows? Of course. Because you know the ratio $W/P = 0.9$, why not add that ratio to the diagram at some convenient place?

You will find it convenient to use a consistent set of algebraic symbols to represent the variables whose values are unknown (called the **unknowns**) in a problem. In this book, we frequently use mnemonic letters to represent the flow of material, both mass and moles, with the appropriate units attached or inferred, as illustrated in Figure E3.6.

When useful, employ m for the flow of mass and n for the flow of moles with appropriate subscripts and/or superscripts to make the meaning crystal clear. Table 3.3 lists some examples. In specific problems, pick obvious or mnemonic letters such as W for water and P for product to avoid confusion. If you run out of suitable letters of the alphabet, you can always insert superscripts to distinguish between streams, such as F^1 from F^2, or label streams as F^1 and F^2. Letters for flow *rates* should have overlay dots imposed.

Table 3.3 Some Examples of the Symbols Used in This Book

Symbol	Designates
F kg	Flow of mass in kilograms
F_{Total} or F_{Tot}	Total flow of material[*]
F^1	Flow in stream number 1[*]
F_A lb	Flow of component A in stream F in pounds
m_A	Mass flow of component A[*]
m_{Total} or m_{Tot}	Mass flow of the total material[*]
$m_A^{F^1}$	Mass flow of component A in stream F^{1}[*]
n_A^W	Molar flow of component A in stream W[*]
ω_A^F	The mass (weight) fraction of A in stream F (The superscript is not required if the meaning is otherwise clear.)
x_A^F	The mole fraction of A in stream F, a liquid (The superscript is not required if the meaning is otherwise clear.)
y_A^F	The mole fraction of A in stream F, a gas

[*]Units not specified but inferred from the problem statement.

4. Obtain any data you know are needed to solve the problem but are missing.

An evaporator cost $34,700. How much did it cost per pound? Clearly, something is missing from the problem statement.

When you review a problem, you may immediately notice that some essential detail is missing in the problem statement, such as a physical property (molecular weight, density, etc.). You can look up the values in physical properties databases available on the internet, in reference books, and in many other places. Or, some value may be missing, but you can calculate the value in your head. For example, you are given a stream flow that contains just two components; one is H_2O and the other, NaOH. You are given the concentration of the NaOH as 22%. There is no point in writing a symbol on the diagram for the unknown concentration of water. Just calculate the value of 78% in your head and put that value on the diagram.

5. Choose a basis.

We discussed the topic of basis in Chapter 2, where we suggested three ways of selecting a basis:

1. What do I have?
2. What do I want to find?
3. What is convenient?

Although picking a basis is listed in Step 5 in the proposed strategy, frequently you know what basis to pick immediately after reading the problem statement and can enter the value on your process diagram at that time. Although the basis we chose for the problem stated in Example 3.6 was a feed rate of 1000 L/h, you could have chosen a similar basis by choosing 1 hour as the basis. If you were to pick some other basis, it would not be as convenient.

Be sure to write the word *Basis* on your calculation page, and enter the value and associated units so that you and anyone who reads the page can later (weeks or months later) know what you did. Choosing a basis should eliminate at least one unknown.

6. Determine the number of variables whose values are unknown (the unknowns).

Note that frequently you will find it convenient to combine Steps 6, 7, and 8 as an aggregate to save space, but here we explain each step separately to focus on the details of the thought process that should occur as you proceed with the solution of a problem.

Determination of the number of unknowns in a problem is somewhat subjective. No unique number exists. Different views of what is known and not known yield different counts (e.g., whether to count a composition as unknown if it is a binary and the composition of the other component is known or whether to count a stream as unknown if it is obvious what its value is). The general objective in solving problems by hand is to reduce the number of simultaneous equations that have to be solved by assigning known values to as many variables as possible at the start of the count. Also, it is sensible to assign values to variables that you can calculate in your head. For example, if an input stream F is assigned a value of 100 kg because that value was selected as the basis, and you know that the input contains 60% NaCl and the other component is KCl, you can easily calculate that 60 kg of NaCl and 40 kg of KCl enter the system. If you plan to fill in a dialog box for a solution by a computer, you may be able to place all of the assigned values first in the proper cells and place any other known facts in the set of simultaneous equations without making any preliminary simple calculations. Also, you can omit from consideration values of variables that are zero.

The basic idea in Step 6 is to reduce the unknowns to as few variables as possible based on the problem specifications plus calculations that you can carry out in your head so that you have to solve a minimum number of simultaneous equations to complete the solution. In the problem stated in Example 3.6 from which Figure E3.6 was prepared, how many unknowns exist? There are nine variables, but you can assign values to all but four. We do not know the values of the following variables: W, P, P_{NaOH}, and $P_{\text{H}_2\text{O}}$, or, alternatively, W, P, ω_{NaOH}, and $\omega_{\text{H}_2\text{O}}$. In light of the necessary conditions stated in the next step, Step 7, you should be thinking about assembling four independent equations to solve the mixing problem.

7. **Determine the number of independent equations and carry out a degree-of-freedom analysis.**

IMPORTANT COMMENT

Before proceeding with Step 7, we need to call to your attention an important point from mathematics related to solving equations. Steps 6 and 7 focus on determining whether you can solve a set of equations formulated for a material balance problem. For simple problems, if you omit Steps 6 and 7 and proceed directly to Step 8 (writing equations), you probably will not be bothered by skipping the steps. However, for complicated problems, you can easily run into trouble if you neglect them. Computer-based process simulators take great care to make sure that the equations you formulate indeed can be solved.

What does solving a material balance problem mean? For our purposes, it means finding a unique answer to a problem. If the material balances you write are linear independent equations, as will be the vast majority of the equations you write, you are guaranteed to get a unique answer if the following necessary condition is fulfilled:

The number of variables whose values are unknown equals the number of independent equations you formulate to solve a problem. To check the sufficient conditions for this guarantee, you can calculate the rank of the coefficient matrix using MATLAB or Python (see Sections 3.6 and 3.7).

In Step 7, you want to *preview* the compilation of equations you plan to use to solve a problem, making sure that you have an appropriate number of independent equations. Step 8 pertains to actually formulating the equations. Steps 7 and 8 are frequently merged. What kinds of equations should you consider?

a. The **material balances** themselves

   You can write as many independent material balance equations as there are species involved in the system. In the specific case of the problem stated in Example 3.6, you have two species, NaOH and H_2O, and thus can write two independent material balance equations. If for the problem posed in Example 3.6 you write three material balances:

   • One for the NaOH
   • One for the H_2O
   • One total balance (the sum of the two component balances)

   only two of the three equations are independent. You can use any combination of two of the three in solving the problem.

b. The **basis** (if not already assigned in Step 6).

c. **Explicit relations** specified (i.e., the **specifications**) in the problem statement such as $W/P = 0.9$ stated in Example 3.6, or specified relations among the variables given in the problem statement, if not used in Step 6.

d. **Implicit relations,** particularly the sum of the mass or mole fractions in a stream being unity, or, alternatively, the sum of the amounts of each of the component materials equalling the total material. In Example 3.6 you have

$$\omega_{NaOH}^{P} + \omega_{H_2O}^{P} = 1$$

or, multiplying both sides of the equation by P, you get the equivalent equation

$$P_{NaOH} + P_{H_2O} = P$$

Frequently Asked Questions

What does the term *independent equation* mean? You know that if you add two independent equations together to get a third equation, the set of three equations is said to be *not independent;* they are said to be *dependent*. Only two of the equations are said to be independent because you can add or subtract any two of them to get the third equation. Figures 3.12 and 3.13 illustrate some examples of independent and dependent equations. **The rank of the coefficient matrix of a system of linear equations indicates the number of independent equations.** Sections 3.6 and 3.7 demonstrate how to use MATLAB or Python to determine the rank of a set of linear equations.

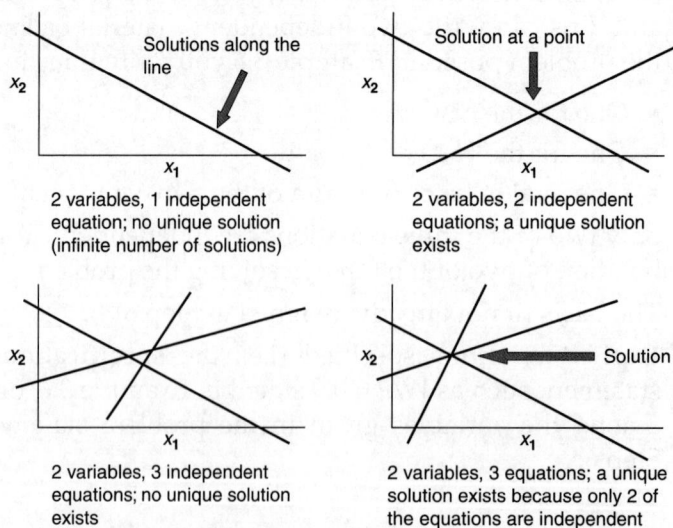

2 variables, 1 independent equation; no unique solution (infinite number of solutions)

2 variables, 2 independent equations; a unique solution exists

2 variables, 3 independent equations; no unique solution exists

2 variables, 3 equations; a unique solution exists because only 2 of the equations are independent

Figure 3.12 Further illustrations of independent and nonindependent equations

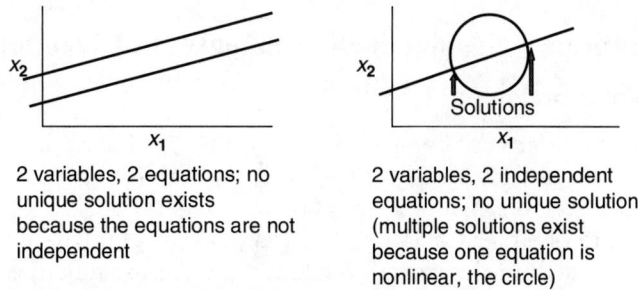

Figure 3.13   Illustrations of independent and dependent equations.

Once you have determined the number of unknowns and independent equations, it helps to carry out an analysis called a **degree-of-freedom analysis** to determine whether a problem is solvable. The difference is called the degrees of freedom available to the designer to specify flow rates, equipment sizes, and so on. You calculate the number of degrees of freedom (N_D) as follows, using the number of unknowns (N_U) and the number of independent equations (N_E):

$$N_D = N_U - N_E$$

When you calculate the number of degrees of freedom (N_D), you can ascertain what the solvability of a problem is. Three outcomes occur:

Case	N_D	Classification for Solution
$N_U = N_E$	0	**Exactly specified** (**determined**); a **unique solution** exists.
$N_U > N_E$	> 0	**Underspecified** (**determined**); more independent equations required.
$N_U < N_E$	< 0	**Overspecified** (**determined**); in general, no solution exists unless some constraints are eliminated or some additional unknowns are included in the problem.

For the problem in Example 3.6:

    From Step 6: $N_U = 4$
    From Step 7: $N_E = 4$

so that

$$N_D = N_U - N_D = \ 4 - 4 = 0$$

and a unique solution exists for the problem.

Example 3.7   Analysis of the Degrees of Freedom

Problem Statement

A gas cylinder containing CH_4, C_2H_6, and N_2 has to be prepared containing a mole ratio of CH_4 to C_2H_6 of 1.5 to 1. Available to prepare the mixture are (1) a cylinder containing a mixture of 80% N_2 and 20% CH_4, (2) a cylinder containing a mixture of 90% N_2 and 10% C_2H_6, and (3) a cylinder containing pure N_2, What is the number of degrees of freedom, that is, the number of independent specifications that must be made, so that you can determine the relative contributions from each cylinder to get the desired composition in the cylinder with the three components?

Solution

A sketch of the process greatly helps in the analysis of the degrees of freedom. Look at Figure E3.7. No specific amount of gas is required to be prepared; only the relative contribution from each cylinder is needed. Consequently, you can take as a convenient basis any value of the unknowns, although picking one of the Fs makes the most sense. Pick $F_1 = 100$ mol as the basis. (Did you contemplate using mass as the basis for a gas stream?)

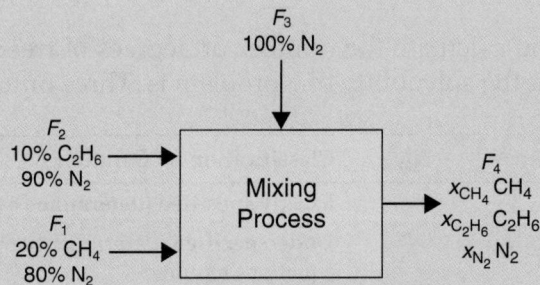

Figure E3.7

First count the number of variables, ignoring the ones whose value is zero. Do you get six (F_2, F_3, F_4, plus the three compositions of F_4)? Look at the following list. The next step is to determine the number of unknowns by assigning all of the known values to their respective variables, values that you can easily find from the problem statement and other sources or can easily calculate in your head. You can assign values to six variables. The question marks designate the unknowns.

$$n_{F_1}^{CH_4} = 20 \text{ specified } [(100)(0.20)] \quad n_{F_2}^{N_2} = 0.90 F_2 \text{ specified}$$

$$n_{F_1}^{N_2} = 80 \text{ specified } [(100)(0.80)] \quad n_{F_2}^{C_2H_6} = 0.10 F_2 \text{ specified}$$

$$F_1 = 100 \text{ used sum of } n_i \text{ in } F_1 \qquad F_2 = ?$$

$$n_{F_3}^{N_2} = (1.00) F_3 \qquad F_3 = ?$$

$$n_{F_4}^{C_2H_6} = ? \qquad n_{F_4}^{CH_4} = ?$$

$$n_{F_4}^{N_2} = ? \qquad F_4 = ?$$

Each one of the assignments is equivalent to one equation. The count of the question marks is six. Can you find any other obvious values to assign? If not, what independent equations can you involve to solve the problem?

> Three (3) species material balances: CH_4, C_2H_6, and N_2
> One (1) specified ratio: moles of CH_4 to C_2H_6 equal 1.5
> One (1) implicit equation: sum of the component molar quantities for product stream is equal to the total product flow rate.

Therefore, a total of five independent equations can be written for this problem. Thus, $6 - 5 = 1$ degree of freedom. Keep in mind that you must be careful when using equations to formulate a set of independent equations. Because based on the problem statement, N_2 does not need to be added to produce the desired product, you could set F_3 equal to zero. Would the equations be solvable then?

Example 3.8   Analysis of the Degrees of Freedom

Problem Statement

Examine Figure E3.8, which labels each of the components and streams in a process (say, centrifugation or dielectrophoresis) to separate living cells (superscript a for alive) from dead cells (superscript d) in water (superscript W). If the values of the mass fractions, $x_F^W, x_P^W, x_F^a, x_P^d$, as well as F, are prespecified (known), how many degrees of freedom remain that can be specified for the process? What values for the unknowns could be specified? All units are in mass.

(*Continues*)

Example 3.8   Analysis of the Degrees of Freedom (*Continued*)

Solution

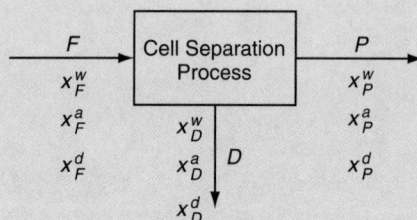

Figure E3.8

Steps 1–5

See Figure E3.8. The basis is F.

Step 6

Number of unknowns. Each stream has four labels; hence (4) (3) = 12 total variables exist, of which five values are prespecified so that seven unknowns exist:

$$D, P, x_F^d, x_P^a, x_D^W, x_D^a, x_D^d$$

Step 7

Number of independent equations needed: 7

Material balances: You can write four material balances, three component and one total, of which three are independent.

Sum of mass fractions equations: You can write three sum of mass fractions, one for each stream that is independent.

Thus, 7 – 6 = 1 degree of freedom exists.

In regard to picking the variable to be specified, you have to be careful. Do not pick a value for an unknown that will be redundant information or render one or more of the equations you have selected in Step 7 to become inadvertently dependent. For example, if you specify the value of x_F^d, because you have already counted the relation $(x_F^w + x_F^a + x_F^d) = 1$ as one of the independent equations, specification of x_F^d will not add any new information to the problem.

A comment: At the start of the analysis in Step 6, if you wanted to, you could have calculated the values of x_F^d and x_P^a by applying the respective sum of mass fraction equations in your head and reducing the number of unknowns by two and the independent equations by two (because of using two respective sum of mass fraction equations).

8. Write down the equations to be solved in terms of the knowns and unknowns.

Once you have concluded from the degree-of-freedom analysis that you can solve a problem, you are well prepared to write down the equations to be solved (if you have not already done so as part of Step 7). Bear in mind that some formulations of the equations are easier to solve by hand, and even by using a computer, than others. In particular, **you should attempt to write linear equations rather than nonlinear ones.** Recall that the product of variables, or the ratios of variables, or a logarithm or exponent of a variable, and so on, in an equation causes the equation to be nonlinear.

In many instances, you can easily transform a nonlinear equation to a linear one. For instance, in the problem posed in Example 3.6, one constraint given was that $W/P = 0.9$, a nonlinear equation. If you multiply both sides of the equation by P, you obtain a linear equation, $W = 0.9P$.

Another example of the judicious formulation of equations that we mentioned previously occurs in the choice of using a mass or mole flow, such as m or n, versus using the product of $\omega_{H_2O}^{P}$, the mass fraction of water in P, times P as two variables:

$$m_{H_2O} = \omega_{H_2O}^{P}P \quad \text{or} \quad n_{H_2O} = y_{H_2O}^{P}P$$

If you use the product $\omega_{H_2O}^{P}P$ for m_{H_2O} in the material balance for water, instead of having a linear equation for the water balance

$$F(0) + W(1.000) = m_{H_2O}$$

you would have

$$F(0) + W(1.000) = \omega_{H_2O}^{P}P$$

a nonlinear equation (which is why we did not use the product). Therefore, **if the composition and flow rate of a stream are unknown, apply the material balances using component flow rates (i.e., m or n) as unknowns.**

With these ideas in mind, you can formulate the set of equations to be used to solve the problem in Example 3.6. First, introduce the five specifications into the two material balances and into the summation of moles in P (or its equivalent, the summation of mass fractions).

Then you will obtain a set of four independent equations in four unknowns which were identified in Step 6. The basis is still the feed rate ($F = 1000$ kg/h) and the process has been assumed to be steady state. Recall from

Section 3.1 that in such circumstances a material balance simplifies to Equation (3.2).

NaOH balance: $1000 = P_{NaOH}$ or $1000 - P_{NaOH} = 0$ (1)

H_2O balance: $W = P_{H_2O}$ or $W - P_{H_2O} = 0$ (2)

Given ratio: $W = 0.9P$ or $W - 0.9P = 0$ (3)

Sum of components in P: $P_{NaOH} + P_{H_2O} = P$ or $P_{NaOH} + P_{H_2O} - P = 0$ (4)

Could you substitute the total mass balance $1000 + W = P$ for one of the two component mass balances? Of course. In fact, you could calculate P by solving just two equations:

Total balance: $1000 + W = P$

Given ratio: $W = 0.9P$

Substitute the second equation into the first equation and solve for P.

You can conclude that the symbols you select in writing the equations and the particular equations you select to solve a problem do make a difference and require some thought. With practice and experience in solving problems, this issue should resolve itself for you.

9. Solve the equations and calculate the quantities asked for in the problem. Industrial-scale problems may involve thousands of equations. Clearly, in such cases, efficient numerical procedures for the solution of the set of equations are essential. Process simulators exist to carry out the task on a computer as explained in Chapter 14. Because most of the problems used in this text have been selected for the purpose of communicating ideas, you will find that their solutions will involve only a small set of equations and can usually be solved for one unknown at a time using a sequential solution procedure. You can solve two or three equations by successive elimination of unknowns from the equations. For a larger set of equations or for nonlinear equations, use a computer program such as MATLAB (Section 3.6), Python (Section 3.7), Excel, or Mathcad. You will save time and effort by so doing.

Learn to be efficient at problem solving. For example, when given data in the AE system of units, say, pounds, do not first convert the data to the SI system, say, kilograms, solve the problem and then convert your results back to the AE system of units. The procedure will work, but it is quite inefficient and introduces unnecessary opportunities for numerical errors to occur.

Select a precedence order for solving the equations you write. One choice of an order can be more effective than another. We showed in Step 8 how the choice of the total balance plus the ratio $W/P = 0.9$ led to two coupled equations that could easily be solved by substitution for P and then W to get

$$P = 10,000$$

$$W = 9000$$

From these two values, you can calculate the amount of H_2O and NaOH in the product:

$$\text{NaOH balance: } P_{\text{NaOH}} = 1000 \text{ kg/h}$$

$$\text{From the}\begin{cases}\text{NaOH balance} \\ \text{H}_2\text{O balance}\end{cases}\text{you get}\begin{cases}P_{\text{NaOH}}=1000 \text{ kg/h} \\ P_{\text{H}_2\text{O}}=9000 \text{ kg/h}\end{cases}$$

so that

$$\omega^P_{\text{NaOH}} = \frac{1000 \text{ kg/h NaOH}}{10,000 \text{ kg/h total}} = 0.1$$

$$\omega^P_{\text{H}_2\text{O}} = \frac{9000 \text{ kg/h H}_2\text{O}}{10,000 \text{ kg/h total}} = 0.9$$

Examine the set of four equations listed in Step 8. Can you find a shorter or easier sequence of calculations to get a solution for the problem?

10. Check your answer(s).

Everyone makes mistakes. What distinguishes good engineers is that they are able to find their mistakes before they submit their work. Refer back to Chapter 2 for several ways to validate your solution. Good engineers use their accumulated knowledge as a primary tool to make sure that the results obtained for a problem (and the data used in the problem) are reasonable. Mass fractions should fall between zero and one. Flow rates normally should be nonnegative.

To the list of validation techniques that appeared in Chapter 2, we add one more very useful one: After solving a problem, use a redundant equation (i.e., an equation not used in the solution) to check your values. In the problem in Example 3.6 that we have been analyzing, one of the three material balances is redundant (not independent), as we pointed out several times. Suppose you solved the problem using the NaOH and H_2O balances.

Then the total balance would have been a redundant balance and could be used to check the answers:

$$P_{NaOH} + P_{H_2O} = P$$

Insert the numbers as a consistency check:

$$1000 + 9000 = 10{,}000$$

Here is a summary of the 10 steps for solving material balance problems:

1. Read and understand the problem statement.
2. Draw a sketch of the process and specify the system boundary.
3. Place labels for unknown variables and values for known variables on the sketch.
4. Obtain any missing needed data.
5. Choose a basis.
6. Determine the number of unknowns.
7. Determine the number of independent equations and carry out a degree-of-freedom analysis.
8. Write down the equations to be solved.
9. Solve the equations and calculate the quantities asked for.
10. Check your answer(s).

Self-Assessment Test

Questions

1. What does the concept "solution of a material balance problem" mean?
2. How many values of unknown variables can you compute (a) from one independent material balance; (b) from three; (c) from four material balances, three of which are independent?
3. What does the concept "independent equations" mean?
4. If you want to solve a set of independent equations that contain fewer unknown variables than equations (the overspecified problem), how should you proceed with the solution?
5. What is the major category of implicit constraints (equations) you encounter in material balance problems?

6. If you want to solve a set of independent equations that contain more unknown variables than equations (the underspecified problem), what must you do to proceed with the solution?

Answers

1. A solution means a (possibly unique) set of values for the unknowns in a problem that satisfies the equations formulated in the problem.

2. (a) one; (b) three; (c) three

3. Linear equations are independent if the vectors formed from the row coefficient in the equation are independent. For nonlinear equations, no simple definition exists.

4. Delete nonpertinent equations, or find additional variables not included in the analysis.

5. The sum of the mass or mole fraction in a stream or inside a system is unity.

6. Obtain more equations or specifications or delete variables of negligible importance.

Problems

1. A water solution containing 10% acetic acid is added to a water solution containing 30% acetic acid flowing at the rate of 20 kg/min. The product P of the combination leaves at the rate of 100 kg/min. What is the composition of P? For this process:
   a. Determine how many independent balances can be written.
   b. List the names of the balances.
   c. Determine how many unknown variables can be solved for.
   d. List their names and symbols.
   e. Determine the composition of P.

2. Can you solve these three material balances for F, D, and P?

$$0.1F + 0.3D = 0.2P$$
$$0.9F + 0.7D = 0.8P$$
$$F + D = P$$

3. How many values of the concentrations and flow rates in the process shown in Figure SAT3.3 P3 are unknown? List them. The streams contain two components, 1 and 2.

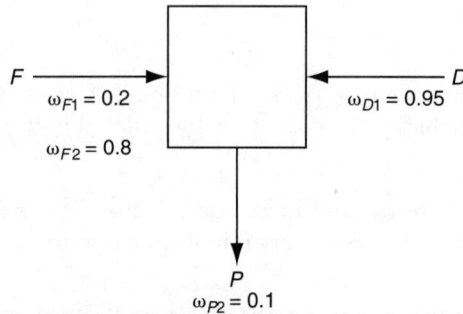

Figure SAT3.3 P3

4. How many material balances are needed to solve problem 3? Is the number the same as the number of unknown variables? Explain.

Answers

1. (a) two; (b) two of these three: acetic acid, water, total; (c) two; (d) feed of the10% solution (say, F_1) and mass fraction ω of the acetic acid in P; (e) 14% acetic acid and 86% water

2. Not for a unique solution because only two of the equations are independent.

3. $F, D, P, \omega_{D_2}, \omega_{P_1}$

4. Three unknowns exist. Because only two independent material balances can be written for the problem, one value of $F, D,$ or P must be specified to obtain a solution. Note that specifying values of ω_{D2} or ω_{P1} will not help.

3.4   Material Balances for Single Unit Systems

In the previous section, you read about solving material balance problems that do not involve chemical reaction. Can you apply these ideas now? You can hone your skills by going through the applications presented in this section, first covering up the solution and then comparing your solution with the one in the text. If you can solve each problem without difficulty, congratulations! If you cannot, analyze places where you had trouble. If you read the solutions without first trying to solve the problems, you will deprive yourself of the learning activity needed to improve your capabilities.

The use of material balances in a process allows you to calculate the values of the flows of species in the streams that enter and leave the plant equipment. You may want to find out how much of each raw material is used and how much of each product (along with some wastes) is produced by the plant. We present a wide range of problems in this section to demonstrate that no matter what the process is, the problem-solving strategy presented in the previous section can be effective for it. Remember that if the process involves rates of flow (i.e., a continuous process), you should apply Equation (3.2), and if the problem involves processing specific amounts of material (i.e., a batch process or a continuous process with a specified time interval), use Equation (3.1).

Example 3.9   Extraction of Streptomycin from a Fermentation Broth

Problems Statement

Streptomycin is used as an antibiotic to fight bacterial diseases in humans as well as to control bacteria, fungi, and algae in crops. First, an inoculate is prepared by placing spores of strains of *Streptomyces griseus* in a medium to establish a culture with a high biomass after it is scaled up to "straw" size to a seed vessel in several steps. The culture from the seed vessel is then introduced into a fermentation tank, which operates at 28°C and a pH of about 7.8 with the nutrients of glucose (the carbon source) and soybean meal (the nitrogen source). High agitation and aeration are needed. After fermentation, the biomass is separated from the liquid, and streptomycin is recovered by adsorption on activated charcoal followed by extraction with an organic solvent in a continuous extraction process. If we ignore the details of the process and consider just the overall extraction process, Figure E3.9 shows the net result.

Determine the mass fraction of streptomycin in the exit organic solvent based on the data in Figure E3.9, assuming that no water exits with the solvent and no solvent exits with the aqueous solution. Assume that the density of the aqueous solution is 1 g/cm^3 and the density of the organic solvent is 0.6 g/cm^3.

(Continues)

Example 3.9   Extraction of Streptomycin from a Fermentation Broth (*Continued*)

Solution

Step 1

Figure E3.9 indicates that the process is an open (flow), steady-state process without reaction (i.e., a continuous process). Assume because of the very low concentration of streptomycin in the aqueous and organic fluids that the volumetric flow rates of the entering fluids equal the volumetric flow rates of the respective exit fluids.

Steps 2–4

All of the data come from Figure E3.9.

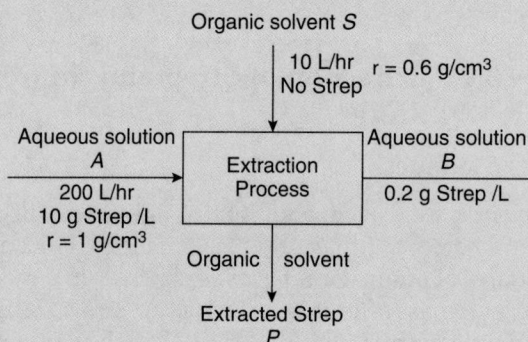

Figure E3.9

Steps 5–7

The basis is $A = 200$ L/h; $S = 10$ L/h.

The degree-of-freedom analysis is as follows:

From an analysis of Figure E3.9, you can see that there is a total of eight variables for this problem (i.e., four streams with two components in each stream). You want to reduce the eight variables to as few unknowns as possible. You can assign values to the following variables from the given data (the mass of a component is designated by m with appropriate superscripts and subscripts). Some of the values can be obtained by using the personal computer on top of your neck. Let A and B denote the input and output aqueous streams, respectively, and S and P denote the input and output organic streams, respectively.

$$A = 200\,\text{L/h} \qquad B = 200\,\text{L/h} \qquad S = 10\,\text{L/h} \qquad P = 10\,\text{L/h}$$

$$m_{\text{Strep.}}^{\text{in}} = 2000\,\text{g/h} \qquad m_{\text{Strep.}}^{\text{out}} = 40\,\text{g/h} \qquad m_{\text{Strep.}}^{\text{in}} = 0\,\text{g/h} \qquad m_{\text{Strep.}}^{\text{out}} = ?\,\text{g/h}$$

$$m_{\text{water}}^{\text{in}} = 2 \times 10^5\,\text{g/h} \quad m_{\text{water}}^{\text{out}} = 2 \times 10^5\,\text{g/h} \quad m_{\text{solvent}}^{\text{in}} = 6000\,\text{g/h} \quad m_{\text{solvent}}^{\text{out}} = 6000\,\text{g/h}$$

Number of unknowns: 1

Number of independent equations needed: 1

Note that we used one material balance for the water and one for the solvent in assigning values to variables. What independent balance is left? The material balance for the streptomycin.

Steps 8 and 9

What equation should you use for the streptomycin balance? This is a continuous process; therefore, you can use Equation (3.2):

$$\overset{\textbf{In}}{\frac{200\text{ L of }A}{\text{h}}\bigg|\frac{10\text{ g Strep.}}{1\text{ L of }A} + \frac{10\text{ L of }S}{\text{h}}\bigg|\frac{0\text{ g Strep.}}{1\text{ L of }S}} = \overset{\textbf{Out}}{\frac{200\text{ L of }A}{\text{h}}\bigg|\frac{0.2\text{ g Strep.}}{1\text{ L of }A} + m_{\text{Strep.}}^{\text{out}}}\ \text{g/h}$$

$$m_{\text{Strep.}}^{\text{out}} = 196\text{ g Strep./h}$$

To get the grams of Strep. per gram of solvent, you need to divide the mass flow rate of Strep. by the volumetric flow rate of S and convert the volume of S to mass. Use the specified density of the solvent:

$$\frac{196\text{ g Strep.}}{\text{h}}\bigg|\frac{\text{h}}{10\text{ L of }S}\bigg|\frac{1\text{ L of }S}{1000\text{ cm}^3\text{ of }S}\bigg|\frac{1\text{ cm}^3\text{ of }S}{0.6\text{ g of }S} = 0.0328\ \text{g Strep./g of }S$$

$$\text{The mass fraction of Streptomycin} = \frac{0.0328}{1 + 0.0328} = 0.0318$$

Could we have used a basis of 1 hour? Yes, but then you would use Equation (3.1) because you would be dealing with specific amounts of material instead of flow rates.

Example 3.10   Separation of Gases Using a Membrane

Problem Statement

Membranes represent a relatively new technology for the commercial separation of gases. One use that has attracted attention is the separation of nitrogen and oxygen from air. Figure E3.10a illustrates a nanoporous membrane, which is made by coating a very thin layer of polymer on a porous graphite supporting layer.

What is the composition of the waste stream if the waste stream amounts to 80% of the input stream?

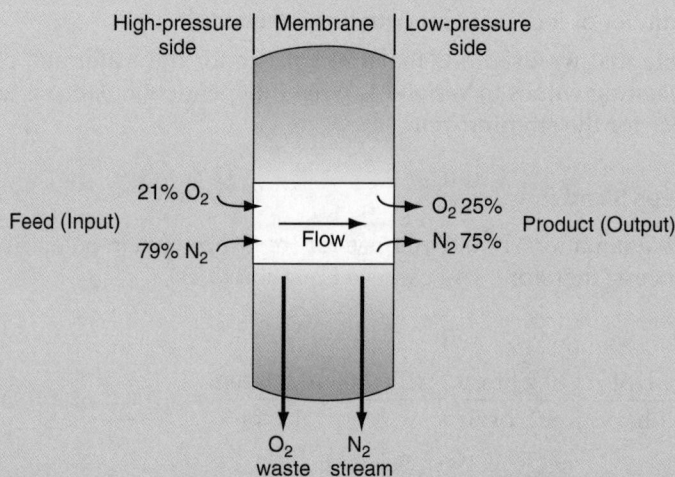

Figure E3.10a

Solution

Step 1

This is an open, steady-state process without chemical reaction. The system is the membrane, as depicted in Figure E3.10a. Let y_{O_2} be the mole fraction of oxygen in the waste stream W, as depicted in Figure E3.10a; let y_{N_2} be the mole fraction of nitrogen in the waste stream; and let n_{O_2} and n_{N_2} be the respective moles in each stream.

Steps 2–4

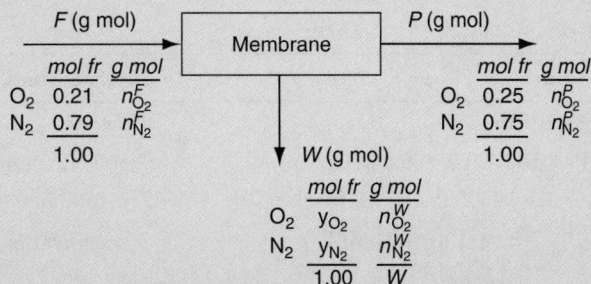

Figure E3.10b

All of the data and symbols have been placed in Figure E3.10b.

Step 5

Pick a convenient basis. The problem does not ask for the actual flow of mass or moles, just for the molar composition of W; hence only relative values have to be calculated. Let

$$\text{Basis: } F = 100 \text{ mol}$$

Note that this system is represented as a batch process. A degree-of-freedom analysis comes next.

Steps 6–8

To avoid forming nonlinear equations [by including terms such as $\left(y_{O_2}^W\right)(W)$] because both W and its composition are unknown, let's use as variables the number of moles (n) rather than the mole fractions (y). The total number of variables is nine, but it would be foolish to involve nine unknowns, particularly when you can reduce the number of unknowns at the start by using information in the problem statement to assign values to variables. How many preliminary substitutions and calculations you want to make depends on the information in the problem, of course. The following table is based on the application of Equation (3.1):

(Continues)

Example 3.10   Separation of Gases Using a Membrane (*Continued*)

Value	Information Used
$F = 100 \text{ mol/h}$	The basis
$W = 0.80 \times 100 = 80 \text{ mol/h}$	A specification
$P = F - W = 100 - 80 = 20 \text{ mol/h}$	The total material balance
$n_{O_2}^F = 0.21(100) = 21 \text{ mol/h}$	A specification plus the basis
$n_{N_2}^F = 0.79(100) = 79 \text{ mol/h}$	A specification plus the basis

Note the use of a material balance that involves only one unknown to get P. Now that P has been calculated, we can use another set of specifications to make two more ad hoc calculations:

$$n_{O_2}^P = 0.25P = 0.25(20) = 5.0 \text{ mol/h}$$

$$n_{N_2}^P = 0.75P = 0.75(20) = 15 \text{ mol/h}$$

The remaining unknowns are then

$$n_{O_2}^W \text{ and } n_{N_2}^W$$

Thus, we need to involve two more pieces of information given in the problem statement. Let's look at the process specifications. Are there any unused specifications? We have used five of the five process specifications in the preliminary calculations. Of the two species material balances, how many are independent? Only one, because we used the total material balance previously. Let's use the oxygen balance:

$$\frac{\text{In}}{21} = \frac{\text{Out in } P}{5} + \frac{\text{Out in } W}{n_{O_2}^W}$$

Solving for the amount of O_2 in the waste stream yields

$$n_{O_2}^W = 21 - 5.0 = 16 \text{ mol/h}$$

What other independent equation might be used? An implicit equation! The sum of the mole fractions in W, or the equivalent, the sum of the moles of the species in W, is an independent equation:

$$W = n_{O_2}^W + n_{N_2}^W = 80 = 16 + n_{N_2}^W$$

Solving for the amount of nitrogen in the waste product yields

$$n_{N_2}^W = 80 - 16 = 64 \, \text{mol/h}$$

Are the sums of the mole fractions in F and P independent equations? No, because the information in these two relations is redundant with the specifications that have been previously used.

The result of this analysis is that the values of all of the variables have been determined without solving any simultaneous equations by using a sequential solution procedure! No residual independent information exists, and the degrees of freedom are zero.

Step 9

The composition of the waste stream is

$$y_{O_2}^W = \frac{n_{O_2}^W}{W} = \frac{16}{80} = 0.20 \quad y_{N_2}^W = \frac{n_{N_2}^W}{W} = \frac{64}{80} = 0.80$$

Step 10

Check your results. You can use a redundant equation as a check, that is, one not used previously. For example, let's use the N_2 balance:

$$n_{N_2}^W + n_{N_2}^P = n_{N_2}^F$$
$$64 + 15 = 79 \quad \text{OK}$$

Be careful when formulating and simplifying the equations to be solved to make sure that you use only independent equations.

In the next problem, we give an example of distillation. Distillation is the most commonly used process for separating components in the refining and petrochemical industries, and it is based on the separation that results from vaporizing a liquid (see Chapter 7). As pointed out earlier, the counter-current contacting of the vapor and the liquid throughout the column concentrates the light components in the overhead product and the heavier components in the bottom product.

Example 3.11   Analysis for a Continuous Distillation Column

Problem Statement

A new manufacturer of ethyl alcohol (ethanol, denoted as EtOH for gasohol) is having a bit of difficulty with a distillation column. The process is shown in Figure E3.11. They think too much alcohol is lost in the bottoms (waste). Calculate the composition of the bottoms and the mass of the alcohol lost in the bottoms based on the data shown in Figure E3.11 that were collected in 1 hr of operation. Finally, determine the percentage of the EtOH entering the column that is lost in the waste stream.

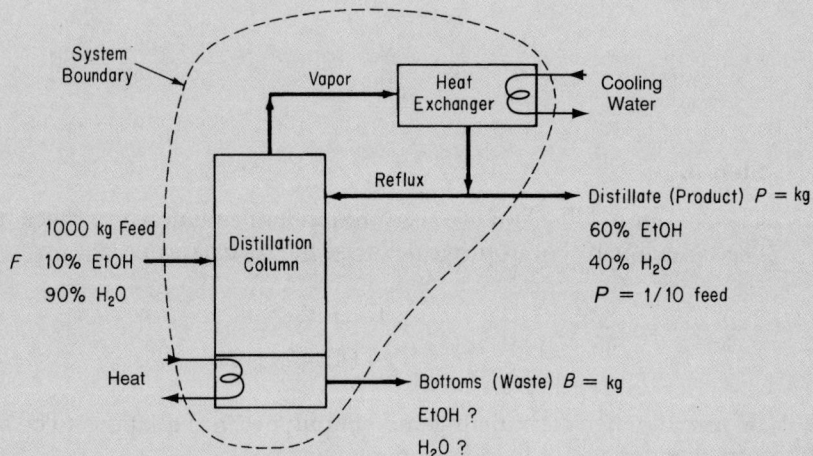

Figure E3.11   Schematic of a distillation column that recovers ethanol

Solution

Steps 1–4

Although the distillation process shown in Figure E3.11 is composed of more than one unit of equipment, you can select a system composed of all of the equipment included inside the system boundary as one lump. Consequently, you can ignore all of the internal streams for this problem. Let m designate the mass of a component. Clearly the process is an open system, and we assume it is in the steady state. No reaction occurs. The cooling water enters and leaves without mixing with the components being separated and can be ignored for the material balances for the system. In addition, the heat added at the bottom of the column does not involve mass entering or leaving the system and can also be ignored for the material balances.

All of the symbols and known data have been placed on Figure E3.11. This is an open, steady-state process based on processing a specific amount of feed, so you can apply Equation (3.1).

Step 5

Select as the basis the given feed:

$$\text{Basis: } F = 1000 \text{ kg of feed}$$

Therefore, this system is being represented as a continuous process with a specified time interval.

Steps 6 and 7

The next step is to carry out a degree-of-freedom analysis. Let m with appropriate superscripts and subscripts denote mass in kilograms. From Figure E3.11 you should be able to locate the following variables:

$$m^F_{\text{EtOH}}, m^F_{\text{H}_2\text{O}}, m^P_{\text{EtOH}}, m^P_{\text{H}_2\text{O}}, m^B_{\text{EtOH}}, m^B_{\text{H}_2\text{O}} \; F, P, B$$

Start the analysis by assigning known values to each variable insofar as is possible.

(Note that this basis allows the direct determination of P.) You are given that P is 10% of F, so $P = 0.1(1000) = 100$ kg.

(*Continues*)

Example 3.11   Analysis for a Continuous Distillation Column (*Continued*)

From the information in Figure E3.11:

$$m_{EtOH}^{F} = 1000(0.10) = 100$$

$$m_{H_2O}^{F} = 1000(0.90) = 900$$

$$m_{EtOH}^{P} = 0.60P = (0.6)(100) = 60\,kg$$

$$m_{H_2O}^{P} = 0.40P = (0.40)(100) = 40\,kg$$

$$P = m_{H_2O}^{P} + m_{EtOH}^{P} = 40\,kg + 60\,kg = 100\,kg$$

Thus, values have been assigned to six variables, leaving three unknowns: $m_{EtOH}^{B}, m_{H_2O}^{B}, B)$. What three independent equations would you suggest using to solve for the remaining unknowns? The usual categories to select from are

Material balances: EtOH, H_2O and total (but only two are independent)

Implicit equations: $\sum m_i^B = B$ or $\sum \omega_i^B = 1$

Steps 8 and 9

Let's use the EtOH balance to determine m_{EtOH}^{B}, which yields a value of 40 kg. Then, by using the implicit equation for B, $m_{H_2O}^{B}$ is equal to 860 kg. These results and the results for the mass fractions are shown in the following table.

Balance	kg Feed in	kg Distillate out	kg Bottoms out	Mass Fraction in B
EtOH balance	0.10(1000)	0.60(100)	40	0.044
H₂O balance	0.90(1000)	0.40(100)	860	0.956
Total $\sum m_i^B = B$			900	1.000

After all of the unknowns are determined for this problem, the percentage of EtOH can be calculated directly by

$$\text{Percentage of EtOH lost in } B = \frac{\text{EtOH in } B}{\text{EtOH in feed}} = \frac{40}{100} \times 100\% = 40\%$$

Step 10

As a check, let's use a redundant equation, the total balance: $B = 1000 - 100 = 900$ kg.

$$m_{EtOH}^B + m_{H_2O}^B = B \text{ or } \omega_{EtOH}^B + \omega_{H_2O}^B = 1$$

Examine the last two columns of the preceding table. Note that this example was solved using Equation (3.1) because the basis of 1 hour was chosen, resulting in a specific amount of material processed by the column. We could have used Equation (3.2) if we had chosen a basis of the feed rate equal to 1000 kg/h.

The next example represents an open system, but one that can be viewed as either unsteady state or steady state depending on how the process is actually carried out.

Example 3.12   Mixing of Battery (Sulfuric) Acid

Problem Statement

You are asked to prepare a batch of 18.63% battery acid as follows: A mixing vessel of old weak battery acid (H_2SO_4) solution contains 12.43% H_2SO_4 (the remainder is pure water). If 200 kg of 77.7% H_2SO_4 is added to the vessel, and the final solution is to be 18.63% H_2SO_4, how many kilograms of battery acid have been made? See Figure E3.12.

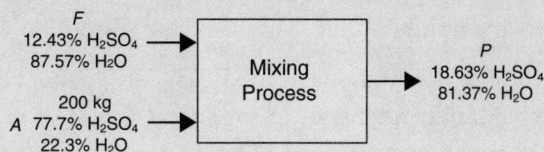

F
12.43% H_2SO_4
87.57% H_2O

200 kg
A 77.7% H_2SO_4
22.3% H_2O

Mixing Process

P
18.63% H_2SO_4
81.37% H_2O

Figure E3.12

Solution

Steps 1–4

All of the values of the compositions are known and have been placed on Figure E3.12. No reaction occurs. Should the process be treated as an unsteady-state

(Continues)

Example 3.12   Mixing of Battery (Sulfuric) Acid (*Continued*)

process or a steady-state process? This is a mixing process in which a specific amount of material is combined (i.e., a batch process) so it is a steady-state system, and Equation (3.1) can be applied:

$$in = out$$

Step 5

Take 200 kg of 77.7% strong acid A as the basis for convenience (the only quantity you know). Therefore, this system is represented as a batch process.

Steps 6 and 7

The analysis of the degrees of freedom is analogous to the ones carried out for the previous examples. Because the compositions of each stream are known, we use compositions times stream flow rates for each component instead of representing the component balances in terms of component flow rate. Why this choice? Because the resulting number of unknowns is smaller and the problem is simpler to solve.

The number of variables is three:

$$A, F, P$$

Let's assign known values to their respective variables:

$$A = 200 \text{ kg}$$

Therefore, the number of unknowns is two (F and P).

> Number of equations needed: 2
> What independent equations can you use?
> Material balances: 2 H_2SO_4, H_2O, and total (two are independent)
> The sum of the mass fractions in each stream conveys no new information.
> The degrees of freedom are zero because there are two independent equations and two unknowns.

Step 8

Let's insert the assigned values into the mass balances along with the specifications for the four mass fractions. By doing so, we can calculate F and P by solving just two simple simultaneous equations. The balances are in kilograms.

	Final		Initial				
H_2SO_4	P(0.1863)	–	F(0.1243)	=	200(0.777)	–	0
H_2O	P(0.8137)	–	F(0.8757)	=	200(0.223)	–	0
Total	P	–	F	=	200	–	0

Step 9

Because the equations are linear and only two independent equations exist, you can take the total mass balance, solve it for F, and substitute for F in the H_2SO_4 balance to calculate P and get

$$P = 2110 \text{ kg acid}$$

$$F = 1910 \text{ kg acid}$$

Step 10

You can check the answer using the H_2O balance. Does the H_2O balance?

Example 3.13   Separation Using a Chromatographic Column

Problem Statement

A chromatographic column can be used to separate two or more compounds. Figure E3.13a portrays the main features of such a column, which can be operated in either a horizontal or a vertical position. Figure E3.13a illustrates how the bands of separation occur as the injected mixture of compounds passes through the column at different points in time. Note that the columns in Figure E3.13a indicate the distribution of the sample at different times, with time increasing from the left to the right. The component that is more strongly absorbed on the packing in the column falls behind the more weakly adsorbed. Some compounds with appropriate packing in a sufficiently long column can be completely separated. A typical small-scale laboratory column might be from 5 mm to 5 cm in diameter and 10 cm long with a filled fraction of packing of 0.62 of the column volume. The smaller the column diameter, the smaller the scale of the packing used in the column. For biomaterials in laboratories, the columns tend use column diameters on the smaller end of the range.

(Continues)

Example 3.13   Separation Using a Chromatographic Column (*Continued*)

In the purification of a protein called bovine serum albumin (BSA) from NaCl (a process called *desalting*), 44.2 g of BSA and 97.7 g of NaCl in 500.0 g of H_2O are injected into a packed initially empty column (see Figure E3.13b). If the exit product P is collected in one container, after a while, you find that you have collected 386.3 g total that contain 302.7 g of H_2O, the remainder being BSA and NaCl. How much BSA and NaCl remained in the column? The ratio of the mass of BSA to the mass of NaCl in P is 0.78.

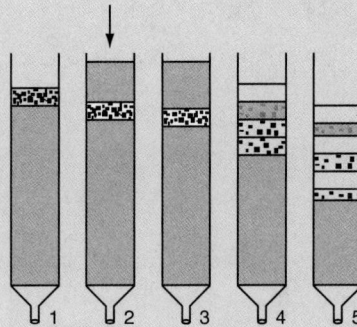

Figure E3.13a

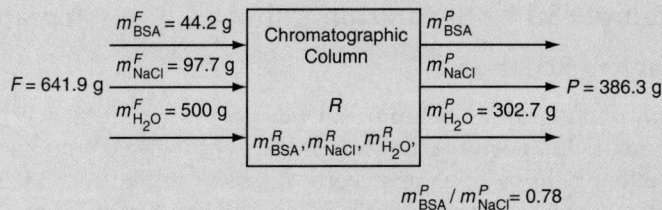

Figure E3.13b

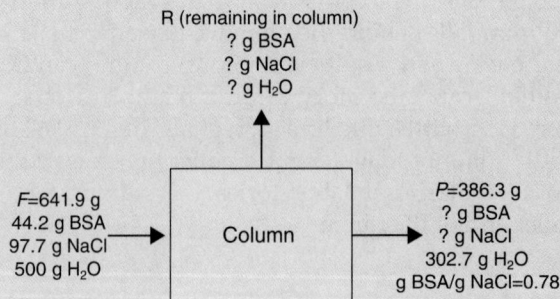

Figure E3.13c

Solution

Steps 1–4

All of the known data have been placed on Figure E3.13c along with the symbols for the masses of each compound. The process in Figure E3.13c is represented as a batch process without reaction. Equation (3.1) can be applied because the problem statement involves specific amounts of material and is not specified in terms of flow rates.

Step 5

$$\text{Basis: } F = 641.9 \text{ g (feed to the column)}$$

Therefore, this system is represented as a batch process.

Steps 6 and 7

If you examine Figure E3.13c, you can count the number of variables. Those whose values are not given are $m_{BSA}^P, m_{NaCl}^P, m_{BSA}^R, m_{NaCl}^R, m_{H_2O}^R, R.$

Number of unknowns after all of the known values are assigned their respective variables: 6
Number of independent equations needed: 6
Independent material balances: 3 (BSA, NaCl, H_2O)
Specification: 1 ($m_{BSA}^P / m_{Nacl}^P = 0.78$)
Implicit equations: 2($\sum m_i^P = 386.3$, $\sum m_i^R = R$; $\sum m_i^F = F$ is redundant)
Therefore, the degrees of freedom are zero.

Steps 8 and 9

At this point, the set of unknowns can be reduced by some simple calculations using equations involving just one unknown, if you want. For example, you can see from Figure E3.13c that a water balance is such an equation:

$$\left(H_2O \text{ balance:}   500 - 302.7 = m_{H_2O}^R   \text{ then } m_{H_2O}^R = 197.3g\right)$$

You could use the total balance next to get R, but instead let's solve two simultaneous equations to get the values of the unknowns in P. Remember, if you use the total balance at this stage, one of the two remaining available component balances will be redundant.

$$\sum m_i^P = P \text{ together with the specified ratio}(m_{BSA}^P / m_{Nacl}^P = 0.78)$$

$$\left. \begin{array}{l} m_{BSA}^P + m_{Nacl}^P + 302.7 = 386.3 \\ m_{BSA} = 0.78 m_{Nacl} \end{array} \right\} \begin{array}{l} m_{BSA}^P = 36.7\,g \\ m_{Nacl}^P = 47.0\,g \end{array}$$

(Continues)

Example 3.13   Separation Using a Chromatographic Column (*Continued*)

With the values of these two variables known, you can now solve an equation involving one unknown:

$$\text{BSA balance:} \quad 44.2 - 36.7 = m_{BSA}^{R} \quad m_{BSA}^{R} = 7.5\,g$$

followed by an NaCl balance:

$$97.7 - 47.0 = m_{NaCl}^{R} \quad m_{NaCl}^{R} = 50.7\,g$$

All that is left to do is to use the implicit equation $\sum m_i^R = R$ to get R:

$$R = m_{BSA}^{R} + m_{NaCl}^{R} + m_{H_2O}^{R} = 7.5 + 50.7 + 197.3 = 255.5\,g$$

Step 10

Check using the redundant total balance:

$$R = F - P \quad R = 641.9 - 386.3 = 255.6$$

The difference is due to round-off.

Example 3.14   Drying

Problem Statement

Fish can be turned into fish meal, and the fish meal can be used as animal feed to produce meat or used directly as food for human beings. The direct use of fish meal significantly increases the efficiency of the food chain. However, fish-protein concentrate, primarily for aesthetic reasons, is used mainly as a supplementary protein food. As such, it competes with soy and other oilseed proteins.

In the processing of the fish, after the oil is extracted, the fish cake is dried in rotary drum dryers, finely ground, and packed. The resulting product contains 65% protein. In a given batch of fish cake that contains 80% water (the remainder is dry cake denoted by BDC for "bone dry cake"), 100 kg of water are removed, and it is found that the fish cake is then 40% water. Calculate the weight of the fish cake originally put into the dryer. Figure E3.14 is a diagram of the process.

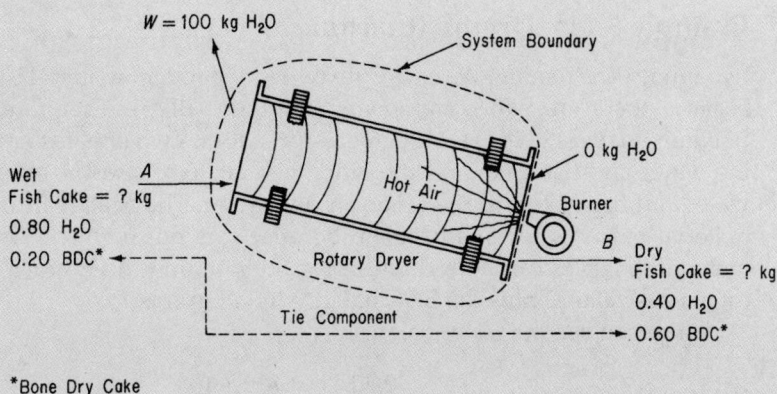

Figure E3.14

Solution

We will abbreviate the solution.

Steps 1–4

The process is a batch process without reaction. Equation (3.1) can be applied because the problem deals with specific amounts of water and fish cake. The system is the dryer. The relation between BDC in the wet and dry fish cake creates a special status for the BDC known as a **tie component** because the BDC enters the process in only one stream and leaves the process unchanged in only one stream. Thus, a tie component allows you to write a material balance for the unique component expressed as a fixed ratio of the streams containing the tie component. An example of the application of a tie component is shown in the following steps.

Step 5

Take a basis of what is given.

$$\text{Basis: 100 kg of water evaporated} = W$$

Steps 6–9

Because the compositions of each stream (A, B, and W) are known, A, B, and W are the only unknown variables, but because W was selected as the basis, only

(Continues)

Example 3.14   Drying (*Continued*)

two unknowns remain. We can write two independent material balance equations because there are two components (H_2O and BDC). Therefore, the degree-of-freedom analysis gives zero degrees of freedom. Two streams enter, one containing water and fish cake and the other, just air. Two streams exit, one containing water and fish cake and the other, air and water. The heated air is not used in the material balances for this problem because it is not required. Two independent material balances can be written using the notation in Figure E3.14. We use the total mass balance plus the BDC balance in kilograms.

You can use the water balance

$$0.80A = 0.40B + 100$$

as a check on the calculations.

$$
\left.
\begin{array}{lll}
 & \text{In} & \text{Out} \\
\text{Total balance} & A & = B + W = B + 100 \\
\text{BDC balance} & 0.20A & = 0.608
\end{array}
\right\} \text{mass balances}
$$

The solution is obtained by substituting the second equation into the first equation to eliminate B, and solving for A yields

$$A = 150 \text{ kg initial wet fish cake}$$

Applying a material balance for the tie component (BDC) allows you to calculate the ratio of the two streams involving the tie component, that is,

$$\frac{A\,\text{kg}}{B\,\text{kg}} = \frac{0.60}{0.20} = 3 \quad \text{or} \quad \frac{3B\,\text{kg}}{A\,\text{kg}} = 1$$

Step 10

Check via water balance:

$$\overset{?}{0.80(150) = 0.40(150)(1/3) + 100}$$
$$120 = 120 \qquad\qquad \text{OK}$$

Example 3.15   Hemodialysis

Problem Statement

Hemodialysis is the most common method used to treat advanced and permanent kidney failure. When your kidneys fail, harmful wastes build up in your body, your blood pressure may rise, and your body may retain excess fluid and may not make enough red blood cells. In hemodialysis, your blood flows through a device with a special filter that removes urea and preserves the water balance and the serum proteins in the blood.

The dialyzer itself (refer to Figure E3.15a) is a large canister containing thousands of small fibers through which the blood passes.

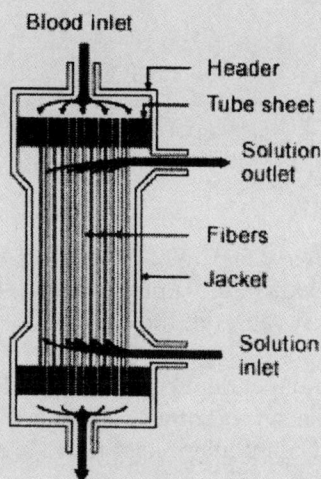

Figure E3.15a

Dialysis solution, the cleansing solution, is pumped around these fibers. The fibers allow wastes and extra fluids to pass from your blood into the solution that carries them away.

This example focuses on the plasma components in streams S (solvent) and B (blood): water, uric acid (UR), creatinine (CR), urea (U), P, K, and Na. You can ignore the initial filling of the dialyzer because the treatment lasts for an interval of two or three hours. Given the measurements obtained from one treatment, as shown in Figure E3.15b, calculate the grams per liter of each component of the plasma in the outlet solution.

(Continues)

Example 3.15 Hemodialysis (*Continued*)

Solution

This is an open, steady-state system.

Step 1

$$\text{Basis: 1 min}$$

Therefore, this is a continuous process, but because a time period was chosen as the basis, Equation (3.1) will be applied similar to a batch process.

Steps 2–4

The data have been inserted on Figure E3.15b. You can ignore the effect of the components of the plasma on the density of the solution for this problem. The entering solution can be assumed to be essentially water.

Steps 6 and 7

Number of unknowns: Figure E3.15b shows quite a few variables, but let's count only those variables whose values are not specified in the figure. Only eight unknowns exist, seven values of the components in the exit stream of water and the total exit flow.

    Number of independent equations needed: 8

    You can make seven component balances and use the implicit equation of the sum of the exit component mass flows to get the total flow out (in grams)

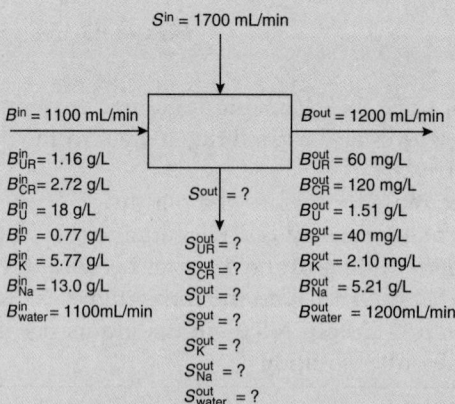

$S^{in} = 1700$ mL/min

$B^{in} = 1100$ mL/min

$B^{in}_{UR} = 1.16$ g/L
$B^{in}_{CR} = 2.72$ g/L
$B^{in}_{U} = 18$ g/L
$B^{in}_{P} = 0.77$ g/L
$B^{in}_{K} = 5.77$ g/L
$B^{in}_{Na} = 13.0$ g/L
$B^{in}_{water} = 1100$ mL/min

$S^{out} = ?$

$S^{out}_{UR} = ?$
$S^{out}_{CR} = ?$
$S^{out}_{U} = ?$
$S^{out}_{P} = ?$
$S^{out}_{K} = ?$
$S^{out}_{Na} = ?$
$S^{out}_{water} = ?$

$B^{out} = 1200$ mL/min

$B^{out}_{UR} = 60$ mg/L
$B^{out}_{CR} = 120$ mg/L
$B^{out}_{U} = 1.51$ g/L
$B^{out}_{P} = 40$ mg/L
$B^{out}_{K} = 2.10$ mg/L
$B^{out}_{Na} = 5.21$ g/L
$B^{out}_{water} = 1200$ mL/min

Figure E3.15b

Steps 8 and 9

The water balance in grams, assuming that 1 mL is equivalent to 1 g (a very convenient assumption), is

$$1100 + 1700 = 1200 + S_{\text{water}}^{\text{out}} \quad \text{hence} \quad S_{\text{water}}^{\text{out}} = 1600g$$

$$\text{or } 1.6 \text{ L}$$

The other component balances in grams are

$$\frac{g/L}{}$$

UR: $\quad 1.1(1.16) + 0 = 1.2(0.060) + 1.6S_{\text{UR}}^{\text{Out}} \qquad S_{\text{UR}}^{\text{Out}} = 0.75$

CR: $\quad 1.1(2.72) + 0 = 1.2(0.120) + 1.6S_{\text{CR}}^{\text{Out}} \qquad S_{\text{CR}}^{\text{Out}} = 1.78$

U: $\quad\;\; 1.1(18) + 0 = 1.2(1.51) + 1.6\;\; S_{\text{U}}^{\text{Out}} \qquad S_{\text{U}}^{\text{Out}} = 11.2$

P: $\quad\; 1.1(0.77) + 0 = 1.2(0.040) + 1.6S_{\text{P}}^{\text{Out}} \qquad S_{\text{P}}^{\text{Out}} = 0.50$

K: $\quad\; 1.1(5.77) + 0 = 1.2(0.120) + 16S_{\text{K}}^{\text{Out}} \qquad S_{\text{K}}^{\text{Out}} = 3.8$

Na: $\quad 1.1(13.0) + 0 = 1.2(3.21) + 1.6\;\; S_{\text{Na}}^{\text{Out}} \qquad S_{\text{Na}}^{\text{Out}} = 6.53$

Step 10

As a consistency check, evaluate the overall mass balance equation.

Frequently Asked Questions

1. All of the examples presented in this chapter have involved only a small set of equations to solve. If you have to solve a large set of equations, some of which may be redundant, how can you tell if the set of equations you select to solve is a set of *independent equations*? You can determine if the set is independent for linear equations by determining the rank of the coefficient matrix of the set of equations, which is equivalent to the number of independent linear equations. Computer programs such as MATLAB (Section 3.6), Python (Section 3.7), Mathcad, Polymath, and so on, provide a convenient way for you to determine the rank of the coefficient matrix without having to carry out the intermediate details of the calculations. Introduce the data into the computer program, and the output from the computer will

provide you with some diagnostics if a solution is not obtained. For example, if the equations are not independent, MATLAB returns the message "Warning: Matrix is singular to the working precision," which means that the rank of the coefficient matrix of the set of linear equations is less than the number of equations (i.e., one or more of the equations is not independent). Alternatively, you can directly apply built-in functions in MATLAB and Python to calculate the rank of the coefficient matrix for the set of linear equations.

2. What should you do if the computer solution you obtain by solving a set of equations gives you a negative value for one or more of the unknowns? One possibility to examine is that you inadvertently reversed the sign of a term in a material balance, say, from + to −. Another possibility is that you forgot to include an essential term(s) so that a zero was entered into the coefficient set for the equations rather than the proper number.

3. In industry, material balances rarely balance when Equations (3.1) or (3.2) are used with process measurements to check process performance. You want to determine what is called *closure*, namely, that the error between "in" and "out" is acceptable. The flow rates and measured compositions for all of the streams entering and exiting a process unit are substituted into the appropriate material balance equations. Ideally, in the steady state in the absence of reaction, the amount (mass) of each component entering a process or group of processes should equal the amount of that component leaving the system. The lack of closure for material balances in industrial process occurs for several reasons:

   a. The process is rarely operating in the steady state. Industrial processes are almost always in a state of flux, rarely approaching steady-state behavior.

   b. The flow and composition measurements have a variety of errors associated with them. Sensor readings include noise (variations in the measurement due to more or less random variations in the readings that do not correspond to changes in the process). Sensor readings can be inaccurate for a wide variety of reasons such as degradation of calibration, poor installment, wrong kind of sensor, dirt and plugging, failure of electronic components, and so on.

   c. A component of interest may be generated or consumed inside the process by reactions that the process engineer has not considered.

As a result, material balance closure to within ±5% for material balances for most industrial processes is considered reasonable. If special care and equipment are employed, discrepancies can be reduced somewhat.

Self-Assessment Test

Questions

1. Indicate whether the following statements are true or false:
   a. The most difficult part of solving material balance problems is the collection and formulation of the data specifying the compositions of the streams into and out of the system and of the material inside the system.
   b. All open processes involving two components with three streams involve zero degrees of freedom.
   c. Certain unsteady-state process problems can be analyzed and solved as steady-state process problems.
   d. If a flow rate is given in kilograms per minute, you should convert it to kilogram moles per minute.
2. Under what circumstances do equations or specifications become redundant?

Answers

1. (a) T; (b) F; (c) T; (d) F
2. When they are not independent

Problems

1. Assume that the conditions for a mixture of air and methane are such that it is able to ignite when the mole percent of air is between 7% and 12%. A mixture of air and methane containing 10% methane is flowing at a rate of 10,000 lb_m/h. Determine how air in lb_m/h should be added to the air/methane stream so that it will lower the mixture concentration of methane to 6%, which is safely below the ignition limit.
2. A cereal product containing 55% water is made at the rate of 500 kg/hr. You need to dry the product so that it contains only 30% water. How much water has to be evaporated per hour?
3. A cellulose solution contains 5.2% cellulose by weight in water. How many kilograms of 1.2% solution are required to dilute 100 kg of the 5.2% solution to 4.2%?

Answers

1. Methane is a tie component: $1000 = 0.06P$; $P = 16,667$; overall balance: $A = 6667$ lb_m/h
2. Cereal is a tie component: $0.45(500) = 225 = 0.7P$; $P = 321$; overall balance: $W = 500 - 321 = 179$ kg/hr
3. Cellulose balance: $5.2 + 0.012D = 0.042P$; overall balance: $100 + D = P$; $D = 33.3$ kg

3.5   Vectors and Matrices

A **matrix** is a rectangular array of numbers, symbols, or functions. For example,

$$\begin{bmatrix} 1 & -2 & 4 \\ -1 & 3 & 5 \\ 4 & 2 & -3 \end{bmatrix} \qquad \begin{bmatrix} x^2 & y-2 & 2 \\ 1-z & z^{1/2} & 2 \end{bmatrix} \qquad \begin{bmatrix} a_{11} & a_{12} \\ a_{21} & a_{22} \end{bmatrix}$$

are all matrices. Most of the matrices that we encounter are square matrices; that is, they have the same number of rows as columns, like the first and third cases in the previous examples.

A **vector** is a special case of a matrix: A column vector has a single column with a number of rows—for example,

$$\begin{pmatrix} 2 \\ 1 \\ 3 \end{pmatrix}$$

A row vector has a single row with a number of columns, such as (2 1 3). Unless otherwise specified, a vector is usually considered a column vector.

Two matrices are said to be equal; that is,

$$\mathbf{A} = \mathbf{B}$$

if the corresponding elements are equal (i.e., when $a_{ij} = b_{ij}$, for all values of i and j). Therefore, for two matrices to be equal, they must both have the same number of rows and columns.

Matrix addition is applicable only to matrices that have the same number of rows and columns. For example, for

$$\mathbf{A} + \mathbf{B} = \mathbf{C}$$

A, B, and **C** must have the same number of rows and columns. Also, this equation indicates that

$$a_{ij} + b_{ij} = c_{ij}$$

for all values of i and j. Therefore, matrix addition provides a concise means of representing a large number of parallel addition operations.

Matrix multiplication involves row and column multiplication, whereas matrix addition involves only element addition. Consider the following example:

$$AB = C$$

where

$$A = \begin{pmatrix} a_{11} & a_{12} \\ a_{21} & a_{22} \end{pmatrix} \qquad B = \begin{pmatrix} b_{11} & b_{12} \\ b_{21} & b_{22} \end{pmatrix} \qquad C = \begin{pmatrix} c_{11} & c_{12} \\ c_{21} & c_{22} \end{pmatrix}$$

By matrix multiplication, the first element of **C**, c_{11}, is given by the scalar product of the first row of **A** and the first column of **B**:

$$c_{11} = \begin{pmatrix} a_{11} & a_{12} \end{pmatrix} \begin{pmatrix} b_{11} \\ b_{21} \end{pmatrix} = a_{11}b_{11} + a_{12}b_{21}$$

Likewise, c_{21} is formed by the scalar product of the second row of **A** and the first column of **B**. In general, c_{ij} is formed by the product of the ith row of **A** and the jth column of **B**. Therefore, **for the multiplication of two matrices, the number of columns of the first matrix of the product must be equal to the number of rows of the second matrix.**

Matrix multiplication provides a very simple, concise means of representing a system of linear equations. Consider the following matrix equation

$$Ax = b$$

where

$$A = \begin{pmatrix} a_{11} & a_{12} & a_{13} \\ a_{21} & a_{22} & a_{23} \\ a_{31} & a_{32} & a_{33} \end{pmatrix} \qquad x = \begin{pmatrix} x_1 \\ x_2 \\ x_3 \end{pmatrix} \qquad b = \begin{pmatrix} b_1 \\ b_2 \\ b_3 \end{pmatrix}$$

Performing the matrix multiplication yields the following set of linear equations.

$$a_{11}x_1 + a_{12}x_2 + a_{13}x_3 = b_1$$
$$a_{21}x_1 + a_{22}x_2 + a_{23}x_3 = b_2$$
$$a_{31}x_1 + a_{32}x_2 + a_{33}x_3 = b_3$$

The matrix **A** is called the **coefficient matrix.** Therefore, a large system of linear equations can be expressed compactly using a coefficient matrix **A** and a **constant vector b.** In addition, this form lends itself quite readily to generalized computer-based solution algorithms. The **rank** of the **A** matrix is equal to the number of independent equations represented by matrix **A** and vector **b.**

Self-Assessment Test

Questions

1. How are vectors and matrices alike? How are they different?
2. How is the multiplication of two matrices performed?

Answers

1. Vectors and matrices are both arrays of numbers, strings, and functions. A vector has a single column or a single row, while a matrix has a number of both rows and columns.
2. For the product matrix **C**, c_{ij} is formed by the product of the ith row of **A** and the jth column of **B**.

3.6   Solving Systems of Linear Equations with MATLAB

In the remainder of this text, certain examples and problems will result in a system of linear equations that does not lend itself to a solution by hand. For these cases, MATLAB offers a built-in function that you can easily use to develop a solution. To use this function, you simply use the coefficient matrix and the constant vector for the system of linear equations with the built-in function.

The most direct way to use MATLAB to solve a system of linear equations is to use the following function:

$$x = A \backslash b$$

where **x** is a vector containing the solution to the set of linear equations, **A** is the coefficient matrix, and **b** is the constant vector for the system of linear equations being solved. Therefore, using function **x=A\b** in your MATLAB code is like calling a built-in function that uses **A** and **b** to define the system of linear equations that it analyzes and solves.

In addition, function **rank()** can be applied to the coefficient matrix to determine the number of independent equations (i.e., k) corresponding to matrix **A**:

k=rank(A)

To apply function **x=A\b**, you must be able to create the matrix **A** and the column vector **b**. Column vectors can also be entered by using a semicolon (;) between elements, or you can convert a row vector into a column vector by using the transpose function (i.e., an apostrophe added to the name of the row vector) as shown here on the right.

```
>>vector=[1;-2;3;-4]          >> vector=[1,-2,3,-4]
vector=                       >> Column_vector=vector'
     1                        Column_vector =
    -2                             1
     3                            -2
    -4                             3
                                  -4
```

The easiest way to generate a matrix is to specify the elements of each row, followed by a semicolon. Note that the elements of each row can be separated by either a space or a comma.

```
>> Matrix=[1,2,3;4,5,6]       >> Matrix=[1 2 3;4 5 6;7 8 9]
Matrix =                      Matrix =
     1     2     3                 1     2     3
     4     5     6                 4     5     6
                                   7     8     9
```

Example 3.16   Application of MATLAB's Built-In Linear Equation Solver

Problem Statement

Apply the built-in function **A\b** to solve the following set of linear equations:

$$3x_1 - 2x_2 + 3x_3 = 8$$
$$4x_1 + 2x_2 - 2x_3 = 2$$
$$x_1 + x_2 + 2x_3 = 9$$

(Continues)

Example 3.16   Application of MATLAB's Built-In Linear Equation Solver (*Continued*)

Solution

Following is the MATLAB code used to solve this problem.

MATLAB Solution for Example 3.16

```
%%%%%%%%%%%%%%%%%%%%%%%%%%%%%%%%%%%%%%%%%%%%%%%%%%%%%%%%%%%%%%%%
%
%                          NOMENCLATURE
%
% A – the coefficient matrix for the set of linear equations
% b – the constant vector for the set of linear equations
% x – the unknowns in the set of linear equations
%
%
%%%%%%%%%%%%%%%%%%%%%%%%%%%%%%%%%%%%%%%%%%%%%%%%%%%%%%%%%%%%%%%%

%                            PROGRAM
function Ex3_15_LinEqnSolver
clear; clc;
A=zeros(3); b(1:3)=0; b=b';      % Insert zeros for the elements of A
                                 %   and b
A=[3,-2,3; 4,2,-2; 1,1,2];       % Insert non-zero elements for A
k=rank(A)
b=[8;2;9];                       % Inset non-zero values for b
x=A\b;                           % Apply function A\b
fprintf('x1 =%8.4f  x2=%8.4f  x3 =%8.4f  k=%2d \n',x(1),x(2),x(3),k)
end

%                          PROGRAM END

%%%%%%%%%%%%%%%%%%%%%%%%%%%%%%%%%%%%%%%%%%%%%%%%%%%%%%%%%%%%%%%%

x1 =  1.0000   x2 =  2.0000  x3 =  3.0000    k=3
```

Description of Program and Results. After the elements of **A** and **b** are initialized at zero, the nonzero elements are added. Function `rank()` is used to determine the rank of the coefficient matrix to ensure that the equations are all independent. Then the `A\b` function is used to obtain the solution of the set of three linear equations. Note that the element values for vector **b** must be inserted as a column vector. If the third equation from the previous set of equations is used for the second and third equations (i.e., only two independent equations), the following results are obtained:

```
Warning: Matrix is singular to working precision.
> In Ex12_6_LinEqnSolver at 20
x1 =    -Inf   x2=    -Inf   x3 =    Inf
```

Example 3.17   Using MATLAB to Solve Example 3.13

Problem Statement

Apply the built-in function `A\b` to solve the set of linear equations resulting from Example 3.13.

Solution

Listed here is the complete set of balances that can be used to solve Example 3.13:

$$\text{BSA Balance:} \qquad m_{\text{BSA}}^R + m_{\text{BSA}}^P = 44.2$$

$$\text{NaCl Balance:} \qquad m_{\text{NaCl}}^R + m_{\text{NaCl}}^P = 97.7$$

$$\text{H}_2\text{O Balance:} \qquad m_{\text{H}_2\text{O}}^R = 500 - 386.3 = 197.3$$

$$\text{Specification:} \qquad m_{\text{BSA}}^P - 0.78 m_{\text{NaCl}}^P = 0$$

$$\text{Implicit Relation:} \quad R - m_{\text{BSA}}^R - m_{\text{NaCl}}^R - m_{\text{H}_2\text{O}}^R = 0$$

$$\text{Implicit Relation:} \quad m_{\text{BSA}}^P + m_{\text{NaCl}}^P = 386.3 - 302.7 = 83.6$$

Note that there are six equations and six unknowns; therefore, the degree of freedom is equal to zero. Defining the unknowns:

$$x_1 = R;\ x_2 = m_{\text{BSA}}^R;\ x_3 = m_{\text{BSA}}^P;\ x_4 = m_{\text{NaCl}}^R;\ x_5 = m_{\text{NaCl}}^P;\ x_6 = m_{\text{H}_2\text{O}}^R$$

(*Continues*)

Example 3.17   Using MATLAB to Solve Example 3.13 (*Continued*)

The following set of linear equations results:

$$x_2 + x_3 = 44.2$$
$$x_4 + x_5 = 97.7$$
$$x_6 = 113.7$$
$$x_3 - 0.78x_5 = 0$$
$$x_1 - x_2 - x_4 - x_6 = 0$$
$$x_3 + x_5 = 83.6$$

Then the coefficient matrix **A** and the constant vector **b** based on the definition of the unknowns and this sequence of equations become

$$\mathbf{A} = \begin{pmatrix} 0 & 1 & 1 & 0 & 0 & 0 \\ 0 & 0 & 0 & 1 & 1 & 0 \\ 0 & 0 & 0 & 0 & 0 & 1 \\ 0 & 0 & 1 & 0 & -0.78 & 0 \\ 1 & -1 & 0 & -1 & 0 & -1 \\ 0 & 0 & 1 & 0 & 1 & 0 \end{pmatrix} \quad \mathbf{b} = \begin{pmatrix} 44.2 \\ 97.7 \\ 113.7 \\ 0 \\ 0 \\ 83.6 \end{pmatrix}$$

Now that **A** and **b** have been formulated, we can directly solve this problem using MATLAB:

MATLAB Solution for Example 3.17

```
%%%%%%%%%%%%%%%%%%%%%%%%%%%%%%%%%%%%%%%%%%%%%%%%%%%%%%%%%%%%%%%%%%%%
%
%                        NOMENCLATURE

%
%  A - the coefficient matrix for the set of linear equations
%  b - the constant vector for the set of linear equations
%  x - the unknowns in the set of linear equations
%
%
%%%%%%%%%%%%%%%%%%%%%%%%%%%%%%%%%%%%%%%%%%%%%%%%%%%%%%%%%%%%%%%%%%%%
```

```
%                              PROGRAM
function Ex3_16_LinEqnSolver
clear; clc;
A=zeros(6); b(1:6)=0; b=b';     % Insert zeros for the elements of A
                                %   and b
                                % Insert elements for A
A=[0,1,1,0,0,0;0,0,0,1,1,0;0,0,0,0,0,1;0,0,1,0,-0.78,0;
1,-1,0,-1,0,-1;0,0,1,0,1,0];
k=rank(A)                       % Determine rank of A
b=[44.2;97.7;113.7;0;0;83.6];   % Inset non-zero values for b
x=A\b;                          % Apply function A\b
fprintf('x1 =%8.4f   x2=%8.4f   x3 =%8.4f   \n',x(1),x(2),x(3))
fprintf('x4 =%8.4f   x5=%8.4f   x6 =%8.4f   k=%2d\n',x(4),x(5),x(6),k)
end

%                              PROGRAM END
%%%%%%%%%%%%%%%%%%%%%%%%%%%%%%%%%%%%%%%%%%%%%%%%%%%%%%%%%%%%%%%%%%%%%%

x1 =256.6000   x2 =  7.5663   x3 = 36.6337
X4 = 50.7337   x5 = 46.9663   x6 =197.3000      k=6
```

Description of Program and Results. After the elements of **A** and **b** are initialized at zero, the nonzero elements are added. Then the **A\b** function is used to obtain the solution of the set of six linear equations. The rank of the coefficient matrix is calculated to ensure that this is a set of independent equations. Note that the element values for vector **b** must be inserted as a column vector. Developing the **A** matrix from the system of linear equations in terms of the unknowns is a tedious process and one in which it is easy to make errors.

Self-Assessment Test

Questions

1. What MATLAB function is used to solve a system of linear equations?
2. How do you define a matrix in MATLAB?

Answers

1. The **x=A\b** function
2. The rows are defined with either spaces or commas between the elements, and each row is ended with a semicolon. The entire expression is inside a set of square brackets.

3.7   Solving Systems of Linear Equations with Python

In the remainder of this text, certain examples and problems will result in a system of linear equations that does not lend itself to a solution by hand. For these cases, Python offers a built-in function that you can easily use to develop a solution. To use this function, you simply use the coefficient matrix and the constant vector for the system of linear equations with the built-in function.

The built-in function for solving systems of linear equations is

```
x=numpy.linalg.solve(A, b)
```

where **x** is a vector containing the solution to the set of linear equations, **A** is the coefficient matrix, and **b** is the constant vector for the system of linear equations being solved. Remember that you must import this function before you use it in a Python code.

In addition, function **rank()** can be applied to the coefficient matrix to determine the number of independent equations (i.e., **k**) corresponding to matrix **A**:

```
k=numpy.linalg.rank(A)
```

To apply function **solve(A,b)**, you must be able to create the matrix **A** and the column vector **b**. You can simply enter the elements of a vector directly into the call to np.array() by separating each element with a comma and enclosing the entire vector within square brackets (i.e., [ ]):

```
In[3]: vector=np.array([1, 2, 5, 10, 50, 90])
In[4]: vector
Out[4]: array([1, 2, 5, 10, 50, 90])
```

The easiest way **to generate a matrix is use function np.array( ) to specify the elements of each row separated by commas and enclosed by square brackets, and then the rows enclosed by square brackets are separated by**

commas, and the entire expression is enclosed by square brackets. The following matrix

$$\begin{bmatrix} 1 & 2 & 3 \\ 4 & 5 & 6 \end{bmatrix}$$

can be generated in Python using the following code by directly inserting the elements using square brackets around each row, separating the rows by a comma, and placing a square bracket around the entire array:

```
In[1] : Matrix=np.array([[1, 2, 3], [4, 5, 6]])
In[2] : Matrix
Out[2]: array([[1, 2, 3],
               [4, 5, 6]])
```

Example 3.18   Application of Python's Built-In Linear Equation Solver

Problem Statement

Apply the built-in function `numpy.linalg.solve` to solve the following set of linear equations:

$$3x_1 - 2x_2 + 3x_3 = 8$$
$$4x_1 + 2x_2 - 2x_3 = 2$$
$$x_1 + x_2 + 2x_3 = 9$$

Solution

Following is the Python code used to solve this problem.

Python Code for Example 3.18

```
Ex3_18 Linear Eqns.py
##################################################################
#                       NOMENCLATURE
#
# A - the coefficient matrix for the set of linear equations
# b - the constant vector for the set of linear equations
```

(Continues)

Example 3.18   Application of Python's Built-In Linear Equation Solver (*Continued*)

```
#  x – the vector of unknowns in the set of linear equations
#
#
################################################################################
#                          PROGRAM
Import numpy as np
Import numpy.linalg
A=np.array([[3, -2, 3], [4, 2, -2], [1, 1, 2]])    # Specify the coefficient
                                                     matrix A
k=numpy.linalg.rank(A)                             # Determine the rank of
                                                     the coefficient matrix
b=np.array([8, 2, 9])                              # Define the constant
                                                     vector b
#
# Apply function numpy.linalg.solve for the solution
#
xsol=numpy.linalg.solve(A,b)

#                          PROGRAM END
################################################################################
```

IPython console:

```
In[1]: runfile(…
In[2]: xsol, k
Out[2]: array([1., 2., 3.]), 3
```

Description of Program and Results. The elements of **A** and **b** are first specified in array form. Then function **numpy.linalg.solve** is used to obtain the solution of the set of three linear equations. If the third equation from the previous set of equations is used for both the second and third equations (i.e., only two independent equations), the following results are obtained:

```
In[1]: runfile(…
LinAlgError: Singular matrix
```

Example 3.19   Using Python to Solve Example 3.13

Problem Statement

Apply the function `numpy.linalg.solve` to solve the set of linear equations resulting from Example 3.13.

Solution

Listed here is the complete set of balances that can be used to solve Example 3.13:

BSA Balance:	$m_{BSA}^R + m_{BSA}^P = 44.2$
NaCl Balance:	$m_{NaCl}^R + m_{NaCl}^P = 97.7$
H$_2$O Balance:	$m_{H_2O}^R = 500 - 386.3 = 113.3$
Specification:	$m_{BSA}^P - 0.78 m_{NaCl}^P = 0$
Implicit Relation:	$R - m_{BSA}^R - m_{NaCl}^R - m_{H_2O}^R = 0$
Implicit Relation:	$m_{BSA}^P + m_{NaCl}^P = 386.3 - 302.7 = 83.6$

Note that there are six equations and six unknowns; therefore, the degree of freedom is equal to zero. Defining the unknowns:

$$x_1 = R; \ x_2 = m_{BSA}^R; \ x_3 = m_{BSA}^P; \ x_4 = m_{NaCl}^R; \ x_5 = m_{NaCl}^P; \ x_6 = m_{H_2O}^R$$

The following set of linear equations results:

$$x_2 + x_3 = 44.2$$
$$x_4 + x_5 = 97.7$$
$$x_6 = 113.7$$
$$x_3 - 0.78x_5 = 0$$
$$x_1 - x_2 - x_4 - x_6 = 0$$
$$x_3 + x_5 = 83.6$$

Then the coefficient matrix **A** and the constant vector **b** based on the definition of the unknowns and on this sequence of equations become

$$\mathbf{A} = \begin{pmatrix} 0 & 1 & 1 & 0 & 0 & 0 \\ 0 & 0 & 0 & 1 & 1 & 0 \\ 0 & 0 & 0 & 0 & 0 & 1 \\ 0 & 0 & 1 & 0 & -0.78 & 0 \\ 1 & -1 & 0 & -1 & 0 & -1 \\ 0 & 0 & 1 & 0 & 1 & 0 \end{pmatrix} \quad \mathbf{b} = \begin{pmatrix} 44.2 \\ 97.7 \\ 113.7 \\ 0 \\ 0 \\ 83.6 \end{pmatrix}$$

(Continues)

Example 3.19   Using Python to Solve Example 3.13 (*Continued*)

Now that **A** and **b** have been formulated, we can directly solve this problem using Python:

MATLAB Solution for Example 3.17

```
Ex3_19 Linear Eqns.py
##############################################################################
#                              NOMENCLATURE
#
#  A – the coefficient matrix for the set of linear equations
#  b – the constant vector for the set of linear equations
#  x – the vector of unknowns in the set of linear equations
#
#
##############################################################################

#                                PROGRAM
Import numpy as np
Import numpy.linalg
# Specify the coefficient matrix A
A=np.array([[0,1,1,0,0,0],[0,0,0,1,1,0],[0,0,0,0,0,1],
[0,0,1,0,-0.78,0],[1,-1,0,-1,0,-1],[0,0,1,0,1,0]])
k=numpy.linalg.rank(A)                              # Determine the
                                                   rank of A

b=np.array([44.2, 97.7, 113.7, 0, 0 , 83.6])       # Define the
                                                   constant vector b
#
# Apply function numpy.linalg.solve for the solution
#
xsol=numpy.linalg.solve(A,b)
#                              PROGRAM END
##############################################################################
```

IPython console:

```
In[1]: runfile(…
In[2]: xsol, k
Out[2]: array([255.60, 7.57, 36.63,50.73, 46.97, 197.30]), 6
```

Description of Program and Results. The elements of **A** and **b** are specified using function `np.array( )`. Then the `solve(A,b)` function is used to obtain the solution of the set of six linear equations. Developing the **A** matrix and the **b** vector from the system of linear equations in terms of the unknowns is a tedious process and one in which it is easy to make errors.

Self-Assessment Test

Questions

1. What function is used in Python to solve a set of linear equations.
2. How do you define a matrix in Python?

Answers

1. The function `numpy.linalg.solve`
2. Each row of a matrix is inside a square bracket, and the elements are separated by commas. The entire matrix is also inside a square matrix, and the entire expression is inside of the function `np.array( )`.

Summary

We began this chapter by demonstrating the connection between an industrial process and a schematic that is used to represent a full range of industrial processes. Moreover, the macroscopic material balances introduced and applied in this chapter can be used to quantify the overall behavior of these processes.

Steady-state material balances are based on the law of conservation of mass. Two forms of this balance were introduced: (1) based on specific amount of material, which can be applied to batch processes or continuous processes for a specified time interval, and (2) based on flow rates of material, which can be applied to continuous processes. By using a basis of a time interval, the latter form can be converted into the form of the former. Moreover, examples showed that these two forms of a steady-state material balance are equivalent except for the units of quantities balanced.

A systematic approach for solving material balances was introduced. This approach addresses all of the key issues associated with formulating and solving material balance problems in a consistent and organized fashion. The key issues associated with solving material balance problems are to make sure that you use only independent equations during the solution process and that you have a solvable set of equations. In addition, it is strongly recommended that you formulate only linear equations whenever possible. That is, if the flow rate and its composition are unknown, it is recommended to use component amounts or flow rates as unknowns instead of using both composition and flow rates as unknowns, which leads to the formation of nonlinear equations.

This chapter serves as a foundation for undertaking more complex material balance in later chapters, including material balances involving chemical reactions, processes with multiple units, and energy balances.

Glossary

accumulation   An increase or decrease in the material (e.g., mass or moles) in a system.

batch process   A process in which material is neither added to nor removed from a process during its operation.

batch reactor   Equipment in which a continuous process takes place.

closed system   A system that does not have material crossing the system boundary.

coefficient matrix   The coefficients of a set of linear equations stored in a matrix.

component balance   A material balance on a single chemical component in a system.

constant vector   The constant terms of a set of linear equations stored in a vector.

continuous process   A process in which material enters and/or exits continuously.

continuous reactor   Equipment in which a continuous process takes place.

degree-of-freedom analysis   Determination of the number of degrees of freedom in a problem.

degrees of freedom   The number of variables whose values are unknown minus the number of independent equations.

dependent equations   A set of equations that are not independent.

exactly specified   Describes a problem in which the degrees of freedom are zero.

explicit relationship   Equations that define the relationship between variables in a problem.

flow system   An open system.

implicit equation   An equation based on information not explicitly provided in a problem, such as the sum of mass fractions is 1.

independent equations   A set of equations for which the rank of the coefficient matrix formed from the equations is the same as the number of equations.

input   Material (e.g., mass, moles) that enters the system.

knowns   Variables whose values are known.

law of conservation of mass   States that mass cannot be created or destroyed.

macroscopic balance   A balance based on the overall behavior of a process.

material balance   The balance equation that corresponds to the conservation of mass.

matrix   A rectangular array of numbers, symbols, or functions.

open system   A system in which material crosses the system boundary.

output   Material (e.g., mass, moles) that leaves the system.

overspecified   Describes a set of equations (or a problem) that is composed of more equations than unknowns.

process   A system that processes material to produce products.

rank   The number of independent rows in a matrix.

specifications   Equations that define the relationship between variables in a problem.

steady-state system   A system in which all the conditions (e.g., temperature, pressure, amount of material) remain constant with time.

system   Any arbitrary portion of or whole process that is considered for analysis.

system boundary   The closed line that encloses the portion of the process that is to be analyzed.

tie component   A component that enters a process in only one stream and leaves the process in only one stream.

transient system   A system in which one or more of the conditions (e.g., temperature, pressure, amount of material) of the system vary with time; also known as an **unsteady-state system.**

underspecified   Describes a set of equations (or a problem) that is composed of fewer equations than unknowns.

unique solution   A single solution that exists for a set of equations (or a problem).

unknowns   Variables whose values are unknown.

unsteady-state system   A system in which one or more of the conditions (e.g., temperature, pressure, amount of material) of the system vary with time; also known as a **transient system.**

vector   A row or column of values, symbols, or functions.

Problems

Section 3.2   Introduction to Material Balances

**3.2.1   Examine Figure P3.2.1. What would be a good system to designate for this bioremediation process? Is your system open or closed? Is it steady state or unsteady state?

A system for treating soil above the water table (bioventing).

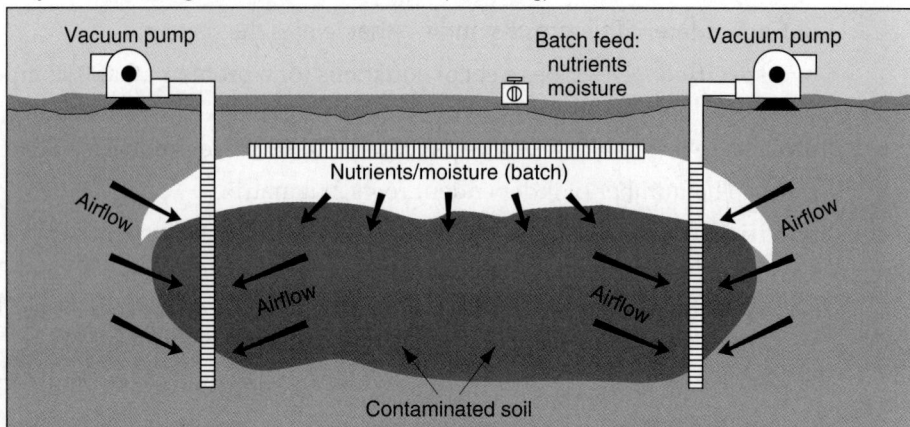

Figure P3.2.1

**3.2.2   Examine Figure P3.2.2, which shows a cylinder that is part of a Ford 2.9-liter V-6 engine. Pick a system and state it. Show by a crude sketch the system boundary. State whether your system is a flow system or a batch system and why (in one sentence).

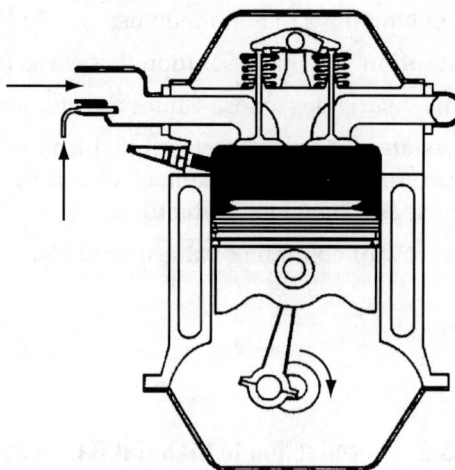

Figure P3.2.2

***3.2.3   State whether the following processes represent open or closed systems, and explain your answer very briefly.

a. Swimming pool (from the viewpoint of the water)

b. Home furnace

*3.2.4 Pick the correct answer(s): For a steady-state system
  a. The rate of input is zero.
  b. The rate of generation is zero.
  c. The rate of consumption is zero.
  d. The rate of accumulation is zero.

**3.2.5 State whether the following processes represent open or closed systems in making material balances:
  a. The global carbon cycle of the Earth
  b. The carbon cycle for a forest
  c. An outboard motor for a boat
  d. Your home air conditioner with respect to the coolant

**3.2.6 Read each one of the following scenarios. State what the system is. Draw the picture. Classify each as belonging to one or more of the following: open system, closed system, steady-state process, unsteady-state process.
  a. You fill your car radiator with coolant.
  b. You drain your car radiator.
  c. You overfill the car radiator and the coolant runs on the ground.
  d. The radiator is full and the water pump circulates water to and from the engine while the engine is running.

***3.2.7 State whether the process of a block of ice being melted by the sun (system: the ice) is an open or closed system, batch or flow, and steady state or unsteady state. List the choices vertically, and state beside each entry any assumptions you make.

**3.2.8 Examine the processes in Figure P3.2.8. Each box represents a system. For each, state whether
  a. The process is in the
    1. Steady state
    2. Unsteady state
    3. Unknown condition

  b. The system is
    1. Closed
    2. Open
    3. Neither
    4. Both

The wavy line represents the initial fluid level when the flows begin. In case (c), the tank stays full.

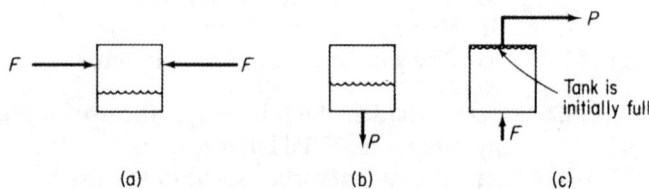

Figure P3.2.8

**3.2.9 In making a material balance, classify the following processes as (1) batch, (2) semi-batch, (3) continuous, (4) open or flow, (5) closed, (6) unsteady state, or (7) steady state. More than one classification may apply.

    a. A tower used to store water for a city distribution system
    b. A can of soda
    c. Heating up cold coffee
    d. A flush tank on a toilet
    e. An electric clothes dryer
    f. A waterfall
    g. Boiling water in an open pot

***3.2.10 Under what circumstances can a batch process that is carried out repeatedly be considered to be a continuous process?

**3.2.11 A manufacturer blends lubricating oil by mixing 300 kg/min of No. 10 oil with 100 kg/min of No. 40 oil in a tank. The oil is well mixed and is withdrawn at the rate of 380 kg/min. Assume the tank contains no oil at the start of the blending process. How much oil remains in the tank after 1 hr?

**3.2.12 One hundred kilograms of sugar are dissolved in 500 kg of water in a shallow open cylindrical vessel. After standing for 10 days, 300 kg of sugar solution are removed. Would you expect the remaining sugar solution to have a mass of 300 kg?

*3.2.13 A 1.0 g sample of solid iodine is placed in a tube, and the tube is sealed after all of the air is removed (Figure P3.2.13). The tube and the solid iodine together weigh 27.0 g.

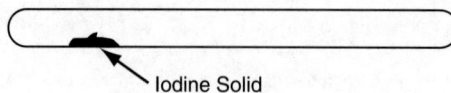

Iodine Solid

Figure P3.2.13

The tube is then heated until all of the iodine evaporates and the tube is filled with iodine gas. The weight after heating should be
a. Less than 26.0 g
b. 26.0 g
c. 27.0 g
d. 28.0 g
e. More than 28.0 g

***3.2.14 Heat exchangers are used to transfer heat from one fluid to another fluid, such as from a hotter fluid to a cooler fluid. Figure P3.2.14 shows a heat exchanger that transfers heat from condensing steam to a process stream. The steam condenses on the outside of the heat exchanger tubes while the process fluid absorbs heat as it passes through the inside of the heat exchanger tubes. The feed rate of the process stream is measured as 45,000 lb/h. The flow rate of steam is measured as 30,800 lb/h, and the exit flow rate of the process stream is measured as 50,000 lb/h. Perform a mass balance for the process stream. If the balance does not close adequately, what might be a reason for this discrepancy?

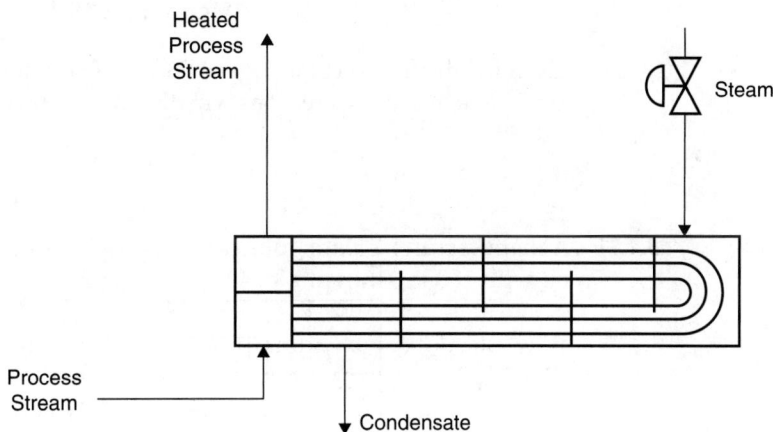

Figure P3.2.14

***3.2.15 Examine the flowsheet in Figure P3.2.15 [adapted from *Hydrocarbon Processing*, 159 (November 1974)] for the atmospheric distillation and pyrolysis of all atmospheric distillates for fuels and petrochemicals. Does the mass in equal the mass out? Give one or two reasons why the mass does or does not balance. Note: T/A is metric tons per year.

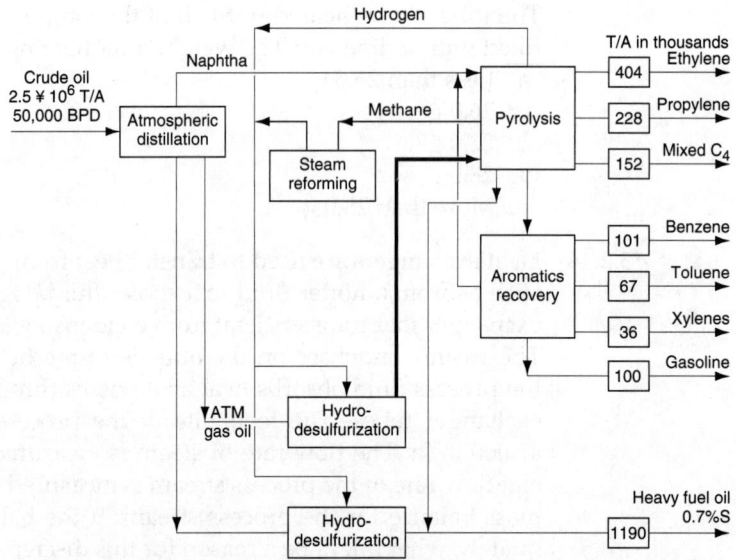

Figure P3.2.15

***3.2.16   Examine the flowsheet in Figure P3.2.16. Does the mass in equal the mass out? Give one or two reasons why the mass does or does not balance.

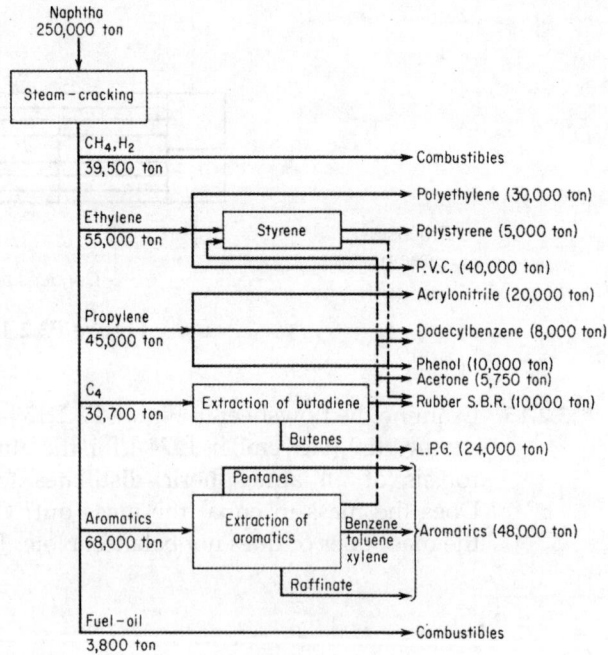

Figure P3.2.16

***3.2.17** Examine Figure P3.2.17. Is the material balance satisfactory? (t/wk means tons per week.)

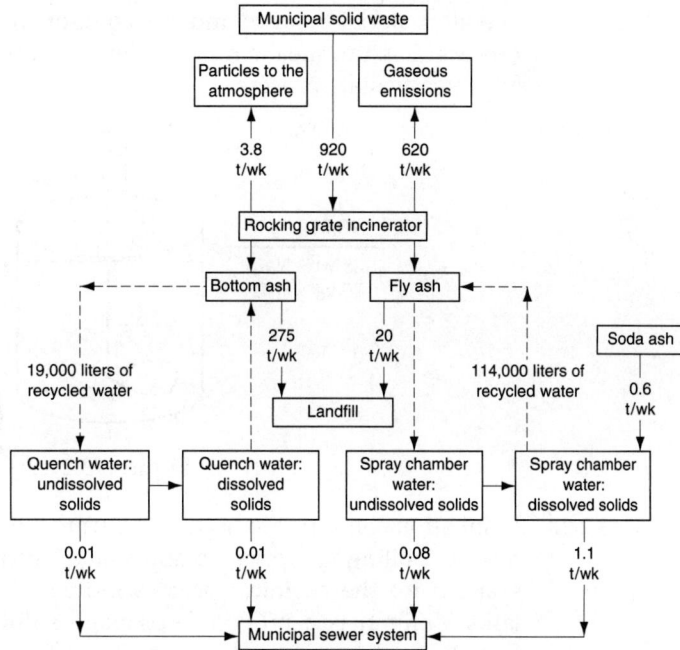

Figure P3.2.17

***3.2.18** Silicon rods used in the manufacture of chips can be prepared by the Czochralski (LEC) process in which a cylinder of rotating silicon is slowly drawn from a heated bath. Examine Figure P3.2.18. If the initial bath contains 62 kg of silicon, and a cylindrical ingot 17.5 cm in diameter is to be removed slowly from the melt at the rate of 3 mm/min, how long will it take to remove one half of the silicon? What is the accumulation of silicon in the melt? Assume that the silicon ingot has a specific gravity of 2.33.

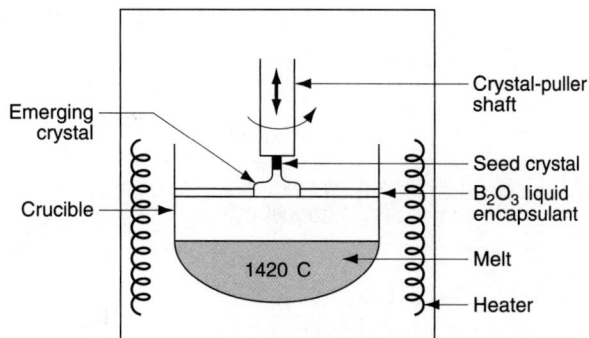

Figure P3.2.18

**3.2.19 Mixers can be used to mix streams with different compositions to produce a product stream with an intermediate composition. Figure P3.2.19 shows a diagram of such a mixing process. Evaluate the closure of the overall material balance and the component material balances for this process. Closure means how closely the inputs agree with the outputs for a steady-state process.

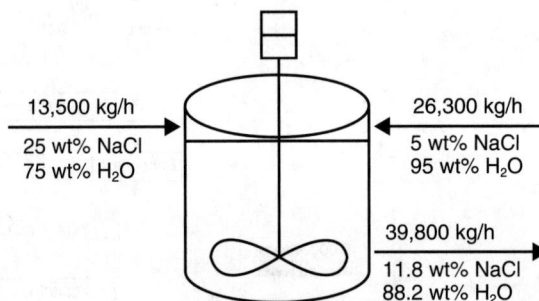

Figure P3.2.19

**3.2.20 Distillation columns are used to separate light boiling components from heavier boiling components and make up over 95% of the separation systems for the chemical process industries. A commonly used distillation column is a propylene-propane splitter. The overhead product from this column is used as a feedstock for the production of polypropylene, which is the largest quantity of plastic produced worldwide. Figure P3.2.20 shows a diagram of a propylene-propane splitter (C_3 refers to propane and $C_3=$ refers to propylene). The steam is used to provide energy and is not involved in the process material balance. Assume that the composition and flow rates listed on this diagram came from process measurements. Determine if the overall material balance is satisfied for this system. Evaluate the component material balances as well. What can you conclude?

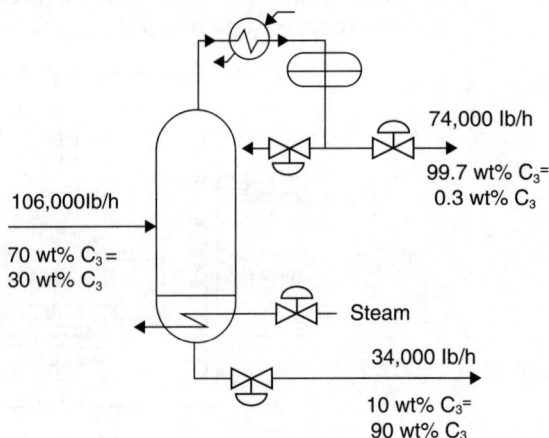

Figure P3.2.20

***3.2.21** A thickener in a waste disposal unit of a plant removes water from wet sewage sludge, as shown in Figure P3.2.21. How many kilograms of water leave the thickener per 100 kg of wet sludge that enter the thickener? The process is in the steady state.

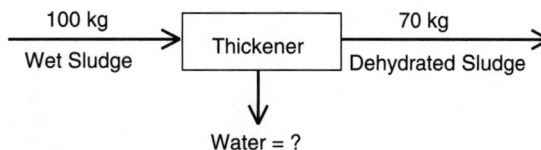

Figure P3.2.21

Section 3.3   A General Strategy for Solving Material Balance Problems

*3.3.1** Consider a water heater in a house. Assume that the metal shell of the tank is the system boundary.
   a. What is in the system?
   b. What is outside the system?
   c. Does the system exchange material with the outside of the system?
   d. Could you pick another system boundary?

3.3.2 For the process shown in Figure P3.3.2, how many material balance equations can be written? Write them. How many independent material balance equations are there in the set?

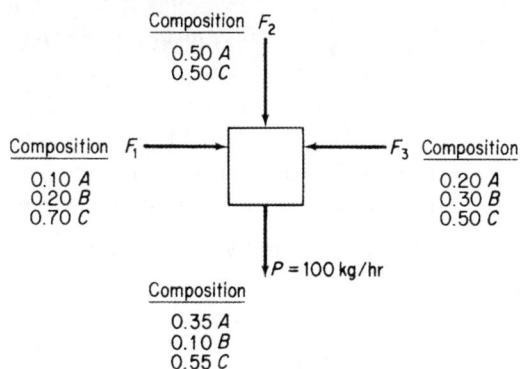

Figure P3.3.2

**3.3.3   Examine the process in Figure P3.3.3. No chemical reaction takes place, and x stands for mole fraction. How many variables are unknown? How many are concentrations? Can this problem be solved uniquely for the unknowns?

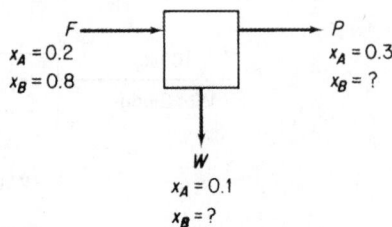

$$
\begin{array}{ccc}
F \longrightarrow & \boxed{} & \longrightarrow P \\
x_A = 0.2 & & x_A = 0.3 \\
x_B = 0.8 & & x_B = ? \\
& \downarrow & \\
& W & \\
& x_A = 0.1 & \\
& x_B = ? &
\end{array}
$$

Figure P3.3.3

3.3.4   Are the following equations independent? Do they have a unique solution? Explain your answers.

 *a. $x_1 + 2x_2 = 1$
 $x_1 + 2x_2 = 3$

 ****b. $(x_1 - 1)^2 + (x_2 - 1)^2 = 0$
 $x_1 + x_2 = 1$

**3.3.5   For one process, your assistant has prepared four valid material balances:

$$0.25\, m_{NaCl} + 0.25\, m_{KCl} + 0.55\, m_{H_2O} = 0.30$$

$$0.35\, m_{NaCl} + 0.20\, m_{KCl} + 0.40\, m_{H_2O} = 0.30$$

$$0.40\, m_{NaCl} + 0.45\, m_{KCl} + 0.05\, m_{H_2O} = 0.40$$

$$1.00\, m_{NaCl} + 1.00\, m_{KCl} + 1.00\, m_{H_2O} = 1.00$$

The assistant says that because the four equations exceed the number of unknowns, three, no solution exists. Is that statement correct? Explain briefly whether it is possible to achieve a unique solution.

***3.3.6   Do the following sets of equations have a unique solution?

 a. $u + v + w = 0$
 $u + 2v + 3w = 0$
 $3u + 5v + 7w = 1$

 b. $u + w = 0$
 $5u + 4v + 9w = 0$
 $2u + 4v + 6w = 0$

**3.3.7 Indicate whether the following statements are true or false:
   a. When the flow rate of one stream is given in a problem, you must choose it as the basis.
   b. If all of the stream compositions are given in a problem, but none of the flow rates are specified, you cannot choose one of the flow rates as the basis.
   c. The maximum number of material balance equations that can be written for a problem is equal to the number of species in the problem.

**3.3.8 In the steady-state process (with no reactions occurring) shown in Figure P3.3.8, you are asked to determine if a unique solution exists for the values of the variables. Does it? Show all calculations.

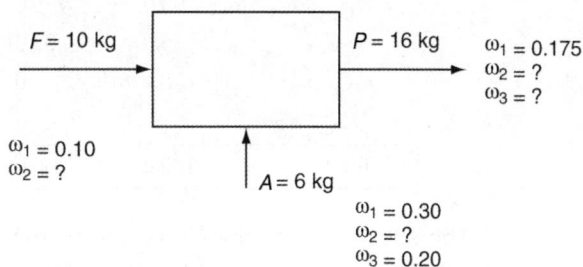

Figure P3.3.8

ω_i is the mass fraction of component i.

**3.3.9 Three gaseous mixtures, A, B, and C, with the compositions listed in the table, are blended into a single mixture.

Gas	A	B	C
CH_4	25	25	60
C_2H_6	35	30	25
C_3H_8	40	45	15
Total	100	100	100

A new analyst reports that the composition of the mixture is 25% CH_4, 25% C_2H_6, and 50% C_3H_8. Without making any detailed calculations, explain how you know the analysis is incorrect.

***3.3.10   A problem is posed as follows: It is desired to mix three LPG (liquefied petroleum gas) streams denoted by A, B, and C in certain proportions so that the final mixture will meet certain vapor-pressure specifications. These specifications will be met by a stream of composition D, as indicated in the following table. Calculate the proportions in which streams A, B, and C must be mixed to give a product with a composition of D. The values are liquid volume percent, but the volumes are additive for these compounds.

Component	Stream A	Stream B	Stream C	Stream D
C_2	5.0			1.4
C_3	90.0	10.0		31.2
iso-C_4	5.0	85.0	8.0	53.4
n-C_4		5.0	80.0	12.6
iso-C_5^+			12.0	1.4
Total	100.0	100.0	100.0	100.0

The subscripts on the Cs represent the number of carbons, and the + sign on C_5^+ indicates all compounds of higher molecular weight as well as iso-C_5. Does this problem have a unique solution?

***3.3.11   In preparing 2.50 moles of a mixture of three gases, SO_2, H_2S, and CS_2, gases from three tanks are combined into a fourth tank. The tanks have the following compositions (mole fractions):

Combined Tanks Mixture

		Tanks		Combined Mixture
Gas	1	2	3	4
SO_2	0.23	0.20	0.54	0.25
H_2S	0.36	0.33	0.27	0.23
CS_2	0.41	0.47	0.19	0.52

In the right-hand column is listed the supposed composition obtained by analysis of the mixture. Does the set of three mole balances for the three compounds have a solution for the number of moles taken from each of the three tanks and used to make up the mixture? If so, what does the solution mean?

3.3.12 You have been asked to check out the process shown in Figure P3.3.12. What will be the minimum number of measurements to make in order to compute the value of each of the stream flow rates and stream concentrations? Explain your answer.

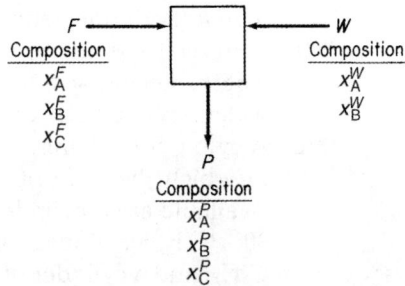

Figure P3.3.12

***3.3.13 Effluent from a fertilizer plant is processed by the system shown in Figure P3.3.13. How many additional concentration and stream flow measurements must be made to completely specify the problem (so that a unique solution exists)? Does only one unique set of specifications exist?

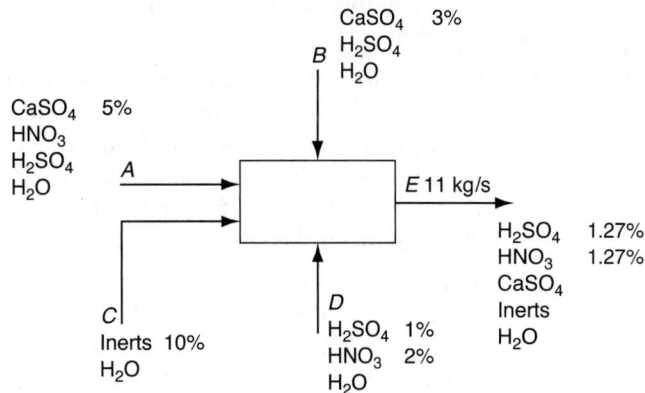

Figure P3.3.13

3.3.14 For each of the following three problems:
1. Draw a figure.
2. Put the data in the problem on the figure.
3. Pick a basis.
4. Determine the number of unknowns and independent equations.
5. Write the material balances needed to solve the problem.

6. Write down any other pertinent equations and specifications.
7. Solve the problem if possible.

**a. Tank A containing 90% nitrogen is mixed with Tank B containing 30% nitrogen to get Tank C containing 65% nitrogen. You are asked to determine the ratio of the gas used from Tank A to that used from Tank B.

**b. A dryer takes in wet timber (20.1% water) and reduces the water content to 8.6%. You want to determine the kilograms of water removed per kilogram of timber that enters the process.

***c. A cylinder containing CH_4, C_2H_6, and N_2 has to be prepared in which the ratio of the moles of CH_4 to C_2H_6 is 1.3 to 1. Available are a cylinder containing a mixture of 70% N_2 and 30% CH_4, a cylinder containing a mixture of 90% N_2 and 10% C_2H_6, and a cylinder of pure N_2. Determine the proportions in which the respective gases from each cylinder should be used.

**3.3.15 After you read a problem statement, what are some of the things you should think about to solve it? List them. This problem does not ask for the ten stages described in the chapter but for brainstorming.

3.4 Material Balances for Single Unit Systems

*3.4.1 You buy 100 kg of cucumbers that contain 99% water. A few days later, they are found to be 98% water. Is it true that the cucumbers now weigh only 50 kg?

*3.4.2 The fern *Pteris vittata* has been shown [*Nature*, 409, 579 (2001)] to effectively extract arsenic from soils. The study showed that in normal soil, which contains 6 ppm of arsenic, in two weeks the fern reduced the soil concentration to 5 ppm while accumulating 755 ppm of arsenic. In this experiment, what was the ratio of the soil mass to the plant mass? The initial arsenic in the fern was 5 ppm.

*3.4.3 Sludge is wet solids that result from the processing in municipal sewage systems. The sludge has to be dried before it can be composted or otherwise handled. If a sludge containing 70% water and 30% solids is passed through a dryer, and the resulting product contains 25% water, how much water is evaporated per ton of sludge sent to the dryer?

*3.4.4 Figure P3.4.4 is a sketch of an artificial kidney, a medical device used to remove waste metabolites from the blood in cases of kidney malfunction. The dialyzing fluid passes across a hollow membrane, and the waste products diffuse from the blood into the dialyzing fluid.

If the blood entering the unit flows at the rate of 220 mL/min, and the blood exiting the unit flows at the rate of 215 mL/min, how much water and urea (the main waste product) pass into the dialysate if the entering concentration of urea is 2.30 mg/mL and the exit concentration of urea is 1.70 mg/mL?

If the dialyzing fluid flows into the unit at the rate of 1500 mL/min, what is the concentration of the urea in the dialysate?

Hollow-Fiber Artificial Kidney

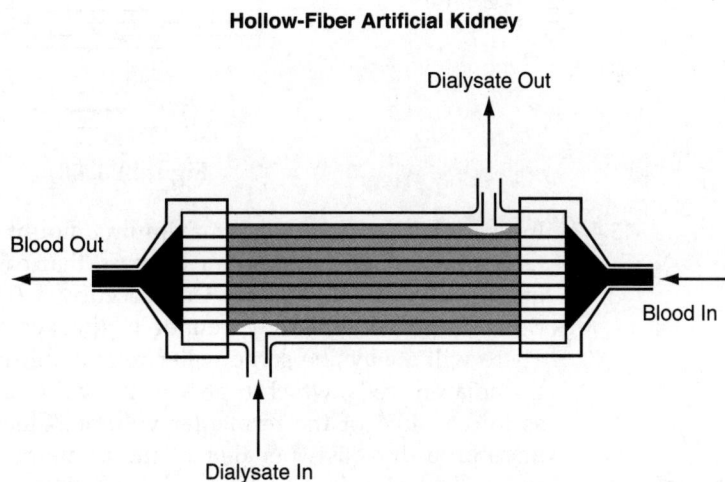

Figure P3.4.4

*3.4.5 A multiple-stage evaporator concentrates a weak NaOH solution from 3% to 18% and processes 2 tons of feed solution per day. How much product is made per day? How much water is evaporated per day?

*3.4.6 A liquid adhesive consists of a polymer dissolved in a solvent. The amount of polymer in the solution is important to the application. An adhesive dealer receives an order for 3000 lb of an adhesive solution containing 13% polymer by weight. On hand are 500 lb of 10% solution and very large quantities of 20% solution and pure solvent. Calculate the weight of each that must be blended together to fill this order. Use all of the 10% solution.

*3.4.7 A lacquer plant must deliver 1000 lb of an 8% nitrocellulose solution. The plant has in stock a 5.5% solution. How much dry nitrocellulose must be dissolved in the solution to fill the order?

**3.4.8   A gas containing 80% CH_4 and 20% He is sent through a quartz diffu-
sion tube (see Figure P3.4.8) to recover the helium. Twenty percent by
weight of the original gas is recovered, and its composition is 50% He.
Calculate the composition of the waste gas if 100 kg moles of gas are
processed per minute.

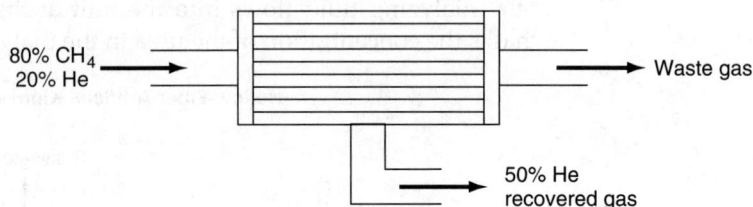

Figure P3.4.8

*3.4.9   In many fermentations, the maximum amount of cell mass must be obtained. However, the amount of mass that can be made is ultimately
limited by the cell volume. Cells occupy a finite volume and have a
rigid shape so that they cannot be packed beyond a certain limit.
There will always be some water remaining in the interstices between
the adjacent cells, which represent the void volume that at best can be
as low as 40% of the fermenter volume. Calculate the maximum cell
mass on a dry basis per liter of the fermenter that can be obtained if
the wet cell density is 1.1 g/cm^3. Note that cells themselves consist of
about 75% water and 25% solids, and cell mass is reported as dry
weight in the fermentation industry.

**3.4.10   A polymer blend is to be formed from the three compounds whose
compositions and approximate formulas are listed in the following
table. Determine the percentages of each compound A, B, and C to be
introduced into the mixture to achieve the desired composition.

	Compound (%)			
Composition	A	B	C	D (Desired Mixture)
$(CH_4)_x$	25	35	55	30
$(C_2H_6)_x$	35	20	40	30
$(C_3H_8)_x$	40	45	5	40
Total	100	100	100	100

How would you decide to blend compounds A, B, and C to achieve the
desired mixture D [$(CH_4)_x = 10\%$, $(C_2H_6)_x = 30\%$, $(C_3H_8)_x = 60\%$]?

**3.4.11 Your boss asks you to calculate the flow through a natural-gas pipeline. Since it is 26 in. in diameter, it is impossible to run the gas through any kind of meter or measuring device. You decide to add 100 lb of CO_2 per minute to the gas through a small 1/2 in. piece of pipe, collect samples of the gas downstream, and analyze them for CO_2. Several consecutive samples after 1 hr are given in the following table.

Time	% CO_2
1 hr, 0 min	2.0
10 min	2.2
20 min	1.9
30 min	2.1
40 min	2.0

    a. Calculate the flow of gas in pounds per minute at the point of injection.

    b. Unfortunately for you, the gas upstream of the point of injection of CO_2 already contained 1.0% CO_2. How much was your original flow estimate in error (in percent)?

Note: In part a., the natural gas is all methane, CH_4.

*3.4.12 Ammonia is a gas for which reliable analytical methods are available to determine its concentration in other gases. To measure flow in a natural-gas pipeline, pure ammonia gas is injected into the pipeline at a constant rate of 72.3 kg/min for 12 min. Five miles downstream from the injection point, the steady-state ammonia concentration is found to be 0.382 wt %. The gas upstream from the point of ammonia injection contains no measurable ammonia. How many kilograms of natural gas are flowing through the pipeline per hour?

*3.4.13 Water pollution in the Hudson River has claimed considerable attention, especially pollution from sewage outlets and industrial wastes. To determine accurately how much effluent enters the river is quite difficult because to catch and weigh the material is impossible, weirs are hard to construct, and so on. One suggestion that has been offered is to add a tracer of Br ion to a given sewage stream, let it mix well, and sample the sewage stream after it mixes. On one test of the proposal, you add 10 pounds of NaBr per hour for 24 hr to a sewage stream with essentially no Br in it. Somewhat downstream of the introduction point, a sampling of the sewage stream shows 0.012% NaBr. The sewage density is 60.3 lb/ft^3, and river water density is 62.4 lb/ft^3. What is the flow rate of the sewage in pounds per minute?

**3.4.14   A process for separating a mixture of incompatible polymers, such as polyethylene terephthalate (PET) and polyvinyl chloride (PVC), promises to expand the recycling and reuse of plastic waste. The first commercial plant, at Celanese's recycling facility in Spartanburg, South Carolina, has been operating at a PET capacity of 15 million lb/yr. Operating cost: 0.5¢/lb.

Targeted to replace the conventional sorting of individual PET bottles from PVC containers upstream of the recycling step, this process first chops the mixed waste with a rotary-blade cutter to 0.5 in. chips. The materials are then suspended in water, and air is forced through to create a bubblelike froth that preferentially entraps the PVC because of its different surface-tension characteristics. A food-grade surfactant is also added to enhance the separation. The froth is skimmed away along with the PVC, leaving behind the PET material. For a feed with 2% PVC, the process has recovered almost pure PET with an acceptable PVC contamination level of 10 ppm. How many pounds of PVC are recovered per year from this process?

***3.4.15   If 100 g of Na_2SO_4 are dissolved in 200 g of H_2O and the solution is cooled until 100 g of $Na_2SO_4 \cdot 10H_2O$ crystallize out, find (a) the composition of the remaining solution (*mother liquor*) and (b) the grams of crystals recovered per 100 g of initial solution.

Hint: Treat the hydrated crystals as a separate stream leaving the process.

****3.4.16   A chemist attempts to prepare some very pure crystals of borax (sodium tetraborate, $Na_2B_4O_7 \cdot 10H_2O$) by dissolving 100 g of $Na_2B_4O_7$ in 200 g of boiling water. He then carefully cools the solution slowly until some $Na_2B_4O_7 \cdot 10H_2O$ crystallizes out. Calculate the grams of $Na_2B_4O_7 \cdot 10H_2O$ recovered in the crystals per 100 g of total initial solution ($Na_2B_4O_7$ plus H_2O), if the residual solution at 55°C after the crystals are removed contains 12.4% $Na_2B_4O_7$.

Hint: Treat the hydrated crystals as a separate stream leaving the process.

***3.4.17   One thousand kilograms of $FeCl_3 \cdot 6H_2O$ are added to a mixture of crystals of $FeCl_3 \cdot H_2O$ to produce a mixture of $FeCl_3 \cdot 2.5H_2O$ crystals. How much $FeCl_3 \cdot H_2O$ must be added to produce the most $FeCl_3 \cdot 2.5H_2O$?

Hint: Treat the hydrated crystals as a separate stream leaving the process.

***3.4.18   The solubility of barium nitrate at 100°C is 34 g/100 g of H_2O and at 0°C is 5.0 g/100 g of H_2O. If you start with 100 g of $Ba(NO_3)_2$ and make a saturated solution in water at 100°C, how much water is required? If the saturated solution is cooled to 0°C, how much $Ba(NO_3)_2$ is precipitated out of solution? The precipitated crystals carry along with them on their surface 4 g of H_2O per 100 g of crystals.

Hint: Treat the hydrated crystals as a separate stream leaving the process.

****3.4.19 A water solution contains 60% $Na_2S_2O_2$ together with 1% soluble impurity. Upon cooling to 10°C, $Na_2S_2O_2 \cdot 5H_2O$/lb crystallizes out. The solubility of this hydrate is 1.4 lb $Na_2S_2O_2 \cdot 5H_2O$/lb free water. The crystals removed carry as adhering solution 0.06 lb solution/lb crystals. When dried to remove the remaining water (but not the water of hydration), the final dry $Na_2S_2O_2 \cdot 5H_2O$ crystals must not contain more than 0.1% impurity. To meet this specification, the original solution, before cooling, is further diluted with water. On the basis of 100 lb of the original solution, calculate (a) the amount of water added before cooling, and (b) the percentage recovery of the $Na_2S_2O_2$ in the dried hydrated crystals.

Hint: Treat the hydrated crystals as a separate stream leaving the process.

**3.4.20 Paper pulp is sold on the basis that it contains 12% moisture; if the moisture exceeds this value, the purchaser can deduct any charges for the excess moisture and also deduct for the freight costs of the excess moisture. A shipment of pulp became wet and was received with a moisture content of 22%. If the original price for the pulp was $40/ton of air-dry pulp and if the freight is $1.00/100 lb shipped, what price should be paid per ton of pulp delivered?

**3.4.21 A laundry can purchase soap containing 30% water for a price of $0.30/kg FOB the soap manufacturing plant (i.e., at the soap plant before shipping costs, which are owed by the purchaser of the soap). It can also purchase a different grade of soap that contains only 5% water. The freight rate between the soap plant and the laundry is $6.05/100 kg. What is the maximum price the laundry should pay for the 5% soap?

***3.4.22 A manufacturer of briquettes has a contract to make briquettes for barbecuing that are guaranteed to not contain over 10% moisture or 10% ash. The basic material used has this analysis: moisture 12.4%, volatile material 16.6%, carbon 57.5%, and ash 13.5%. To meet the specifications (at their limits) the manufacturer plans to mix with the base material a certain amount of petroleum coke that has this analysis: volatile material 8.2%, carbon 88.7%, and moisture 3.1%. How much petroleum coke must be added per 100 lb of the base material?

****3.4.23 In a gas-separation plant, the feed to the process has the following constituents:

Component	Mol %
C_3	1.9
$i\text{-}C_4$	51.5
$n\text{-}C_4$	46.0
C_5^+	0.6
Total	100.0

The flow rate is 5804 kg mol/day. If the overhead and bottoms streams leaving the process have the following compositions, what are the flow rates of the overhead and bottoms streams in kilogram moles per day?

	Mol %	
Component	**Overhead**	**Bottoms**
C_3	3.4	—
$i\text{-}C_4$	95.7	1.1
$n\text{-}C_4$	0.9	97.6
C_5^+	—	1.3
Total	100.0	100.0

****3.4.24 The organic fraction in the wastewater is measured in terms of the biological oxygen demand (BOD) material, namely, the amount of dissolved oxygen required to biodegrade the organic contents. If the dissolved oxygen (DO) concentration in a body of water drops too low, the fish in the stream or lake may die. The Environmental Protection Agency has set the minimum summer levels for lakes at 5 mg/L of DO.

a. If a stream is flowing at 0.3 m^3/s and has an initial BOD of 5 mg/L before reaching the discharge point of a sewage treatment plant, and the plant discharges 3.785 ML/day of wastewater, with a concentration of 0.15 g/L of BOD, what will be the BOD concentration immediately below the discharge point of the plant?

b. The plant reports a discharge of 15.8 ML/day having a BOD of 72.09 mg/L. If the EPA measures the flow of the stream before the discharge point at 530 ML/day with 3 mg/L of BOD, and measures the downstream concentration of 5 mg/L of BOD, is the report correct?

****3.4.25 Suppose that 100 L/min are drawn from a fermentation tank and passed through an extraction tank in which the fermentation product (in the aqueous phase) is mixed with an organic solvent, and then the aqueous phase is separated from the organic phase. The concentration of the desired enzyme (3-hydroxybutyrate dehydrogenase) in the aqueous feed to the extraction tank is 10.2 g/L. The pure organic extraction solvent runs into the extraction tank at the rate of 9.5 L/min. If the ratio of the enzyme in the exit product stream (the organic phase) from the extraction tank to the concentration of the enzyme in the exit waste stream (the aqueous phase) from the tank is $D = 18.5$ (g/L organic)/(g/L aqueous), what is the fraction recovery of the enzyme and the amount recovered per minute? Assume negligible miscibility between the aqueous and organic liquids in each other, and ignore any change in density on removal or addition of the enzyme to either stream.

****3.4.26   Consider the following process for recovering NH_3 from a gas stream composed of N_2 and NH_3 (see Figure P3.4.26). Flowing upward through the process is the gas stream, which can contain NH_3 and N_2 but *not* solvent S, and flowing downward through the device is a liquid stream which can contain NH_3 and liquid S but *not* N_2. The weight fraction of NH_3 in the gas stream A leaving the process is related to the weight fraction of NH_3 in the liquid stream B leaving the process by the following empirical relationship:

$$\omega_{NH_3}^{A} = 2\omega_{NH_3}^{B}$$

Given the data shown in Figure P3.4.26, calculate the flow rates and compositions of streams A and B.

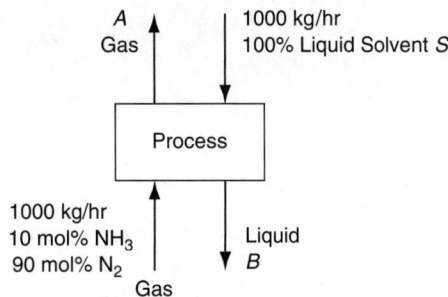

Figure P3.4.26

***3.4.27   MTBE (methyl tert-butyl ether) is added to gasoline to increase the oxygen content of the gasoline. MTBE is soluble in water to some extent and becomes a contaminant when the gasoline gets into surface or underground water. The gasoline used by boats has an MTBE content of 10%. The boats operate in a well-mixed flood control pond having the dimensions 3 km long, 1 km wide, and 3 m deep on the average. Suppose that each of the 25 boats on the pond spills 0.5 L of gasoline during 12 hr of daylight. The flow of water (that contains no MTBE) into the pond is 10 m^3/hr, but no water leaves because the water level is well below the spillway of the pond. By how much will the concentration of MTBE increase in the pond after the end of 12 hr of boating? Data: The specific gravity of gasoline is 0.72.

***3.4.28   Salt in crude oil must be removed before the oil undergoes processing in a refinery. The crude oil is fed to a washing unit where freshwater fed to the unit mixes with the oil and dissolves a portion of the salt contained in the oil. The oil (containing some salt but no water), being less dense than the water, can be removed at the top of the washer. If the "spent" wash water contains 15% salt and the crude oil contains 5% salt, determine the concentration of salt in the "washed" oil product if the ratio of crude oil (with salt) to water used is 4 to 1.

CHAPTER 4

Material Balances with Chemical Reaction

Chapter Objectives

- Write and balance chemical reaction equations.
- Identify the limiting and excess reactants in a reaction and calculate the fraction or percent excess reactant(s); the percent conversion, or completion; the yield; and the extent of reaction for a chemical reaction with the reactants given in non-stoichiometric proportions.
- Carry out a degree-of-freedom analysis for processes involving chemical reaction(s).
- Formulate and solve material balances using (a) species balances and (b) element balances.
- Decide when element balances can be used as material balances.
- Understand the meaning of stack gas, flue gas, Orsat analysis, dry basis, wet basis, theoretical air (oxygen), and excess air (oxygen), and employ these concepts in combustion problems.

Introduction

Consider a reactor in which ammonia is produced. Ammonia is used in the production of a wide range of chemical products. including textiles, dyes, plastics, and explosives, but more than 80% of the ammonia production is used to produce agricultural fertilizers as a source of nitrogen for plants. Ammonia is produced industrially using the catalytic Haber-Bosch process

(Figure 4.1). The nitrogen used in this process is produced by cryogenic separation of nitrogen from air, and hydrogen is produced from light hydrocarbon sources, such as natural gas.

Equation (3.1), Mass in = Mass out, and Equation (3.2), $(\dot{m}_i)_{in} = (\dot{m}_i)_{out}$, can be applied to the overall mass balance for this reactor, but not for the component balances or an overall mole balance. For example, the inlet flow rate of NH_3 is zero, but due to reaction in the reactor, the outlet flow of ammonia is not zero. Also, because four moles of reactants are required to produce two moles of the ammonia product, the total number of moles entering the reactor is less than the number of moles leaving the reactor. Therefore, to apply material balances for a reacting system, we must take into account any reaction taking place in the system. This chapter addresses how to take into account a single reaction or a number of reactions when solving material balance problems.

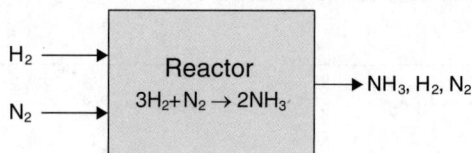

Figure 4.1   Schematic of an ammonia reactor

The heart of many plants is the reactor in which products and by-products are produced. Material balances considering reactions are used to design reactors. Moreover, these material balances can also be used to identify the most efficient operation of the reactors (i.e., process optimization). Of course, computer programs can make the calculations for you, but you have to formulate the necessary material balances correctly.

4.1   Stoichiometry

You are probably aware that chemical engineers differ from most other engineers because of their application of chemistry. When chemical reactions occur in contrast to the physical changes of material such as evaporation or dissolution, you want to be able to predict the mass or moles required for the reaction(s) and the mass or moles of each species remaining after the reaction has occurred. Reaction stoichiometry allows you to accomplish this. The word *stoichiometry* (stoi-ki-OM-e-tri) derives from two Greek words: *stoicheion* (meaning "element") and *metron* (meaning "measure"). **Stoichiometry** provides a quantitative means of relating the amount of products produced by a chemical reaction(s) to the amount of reactants, or vice versa.

As you already know, the chemical reaction equation provides both qualitative and quantitative information concerning chemical reactions that occur in a process. Specifically, the chemical reaction equation provides information of two types:

1. It tells you what substances are reactants (those being used up) and what substances are products (those being made).

2. The coefficients of a *balanced* chemical reaction equation tell you what the mole ratios are among the substances that react or are produced. (In 1803, John Dalton, an English chemist, was able to explain much of the experimental results on chemical reactions of the day by assuming that reactions occurred with fixed ratios of elements. This discovery led to the *law of constant proportionality*, which states that chemical reactions proceed with fixed ratios of the number of reactants and products involved in the reaction.)

You should take the following steps when solving problems involving stoichiometry:

1. Make sure the chemical equation is correctly balanced. How can you tell if the reaction equation is balanced? Make sure the total quantities of each of the elements on the left side equal those on the right side. For example,

$$CH_4 + O_2 \rightarrow CO_2 + H_2O$$

is not a balanced stoichiometric equation because there are four atoms of H on the reactant side (left side) of the equation but only two atoms of H on the product side (right side). In addition, the oxygen atoms do not balance. The balanced equation is given by

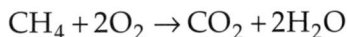

$$CH_4 + 2O_2 \rightarrow CO_2 + 2H_2O$$

Note that the sum of each of the elements present in the chemical reaction equation (C, H, and O) is the same for the reactants (left side of the chemical reaction equation) as for the products (right side of the chemical reaction equation). The coefficients in the balanced reaction equation have the units of moles of a species based on a particular reaction equation. For example, for the previous chemical reaction equation, for every mole of CH_4 that reacts, 2 moles of O_2 are consumed and 1 mole of CO_2 and 2 moles of H_2O are produced. If you multiply each term in a chemical reaction equation by the same constant, say, 2, the absolute stoichiometric coefficient in each term doubles, but the coefficients still occur in the same relative proportions, thus the reaction equation remains balanced. For example, for the previous chemical reaction equation, if 2 moles of CH_4 react, 4 moles of O_2 are consumed, and 2 moles of CO_2 and 4 moles of H_2O are produced.

2. Use the proper degree of completion for the reaction. If you do not know how much of the reaction has occurred, you may assume a reactant reacts completely in this text.

3. Use molecular weights to convert mass to moles for the reactants and moles to mass for the products.

4. Use the coefficients in the chemical equation to obtain the relative molar amounts of products produced and reactants consumed in the reaction.

Steps 3 and 4 can be applied in a fashion similar to that used in carrying out the conversion of units, as explained in Chapter 2. As an example, consider the combustion of heptane:

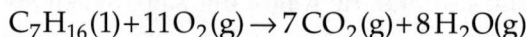

$$C_7H_{16}(1)+11\,O_2(g) \rightarrow 7\,CO_2(g)+8\,H_2O(g)$$

(Note that we have put the states of the compounds in parentheses after the species formula, information not needed for this chapter but that will be vital in subsequent sections of this book.)

What can you learn from this equation? The **stoichiometric coefficients** in the chemical reaction equation (1 for C_7H_{16}, 11 for O_2, and so on) tell you the relative amounts of moles of chemical species that react and are produced by the reaction. The units of a stoichiometric coefficient for species i are the change in the moles of species i divided by the moles reacting according to a specific chemical equation. In taking ratios of coefficients, the denominators cancel, and you are left with the ratio of the moles of one species divided by another. For example, for the combustion of heptanes:

$$\frac{1 \text{ mol } C_7H_{16}}{\text{moles reacting}} \div \frac{11 \text{ mol } O_2}{\text{moles reacting}} = \frac{1 \text{ mol } C_7H_{16}}{11 \text{ mol } O_2}$$

We will abbreviate the units of a coefficient for species i simply as mol i/moles reacting when appropriate, but frequently in practice, the units are ignored. You can conclude that 1 mole (*not* lb_m or kg) of heptane will react with 11 moles of oxygen to give 7 moles of carbon dioxide plus 8 moles of water. These may be pound moles, gram moles, kilogram moles, or any other type of mole. Another way to use the chemical reaction equation is to conclude that 1 mole of CO_2 is formed from each 1/7 mole of C_7H_{16} and 1 mole of H_2O is formed with each 7/8 mole of CO_2. The ratios indicate the **stoichiometric ratios** that can be used to determine the relative proportions of products and reactants.

Suppose you are asked how many kilograms of CO_2 will be produced as product if 10 kg of C_7H_{16} react completely with the **stoichiometric quantity** of O_2. On the basis of 10 kg of C_7H_{16}:

$$\frac{10 \text{ kg C}_7\text{H}_{16}}{} \left| \frac{1 \text{ kg mol C}_7\text{H}_{16}}{100.1 \text{ kg C}_7\text{H}_{16}} \right| \frac{7 \text{ kg mol CO}_2}{1 \text{ kg mol C}_7\text{H}_{16}} \left| \frac{44.0 \text{ kg CO}_2}{1 \text{ kg mol CO}_2} \right. = 30.8 \text{ kg CO}_2$$

$$\text{Conversion from mass to moles} \quad \text{Mole ratio} \quad \text{Conversion from moles to mass}$$

Let's now write a general chemical reaction equation as

$$cC + dD \leftrightharpoons aA + bB \tag{4.1}$$

where a, b, c, and d are the stoichiometric coefficients for the species A, B, C, and D, respectively. Equation (4.1) can be written in a general form:

$$v_A A + v_B B + v_C C + v_D D = \sum v_i S_i = 0 \tag{4.2}$$

where v_i is the stoichiometric coefficient for species S_i. The products are defined to have *positive values* for stoichiometric coefficients and the reactants to have *negative values* for stoichiometric coefficients. The ratios of stoichiometric coefficients are unique for a given reaction. Specifically for Equation (4.1) written in the form of Equation (4.2):

$$v_C = -c \qquad v_A = a \qquad v_D = -d \qquad v_B = b$$

If a species is not present in an equation, the value of its stoichiometric coefficient is deemed to be zero. As an example, in the reaction

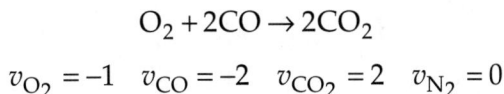

$$O_2 + 2CO \rightarrow 2CO_2$$

$$v_{O_2} = -1 \quad v_{CO} = -2 \quad v_{CO_2} = 2 \quad v_{N_2} = 0$$

Example 4.1   Balancing a Reaction Equation for a Biological Reaction

Problem Statement

The primary energy source for cells is the aerobic catabolism (oxidation) of glucose ($C_6H_{12}O_6$, aa sugar). The overall oxidation of glucose produces CO_2 and H_2O by the following reaction:

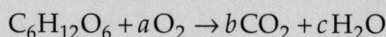

$$C_6H_{12}O_6 + aO_2 \rightarrow bCO_2 + cH_2O$$

Solution

Basis: The given chemical reaction equation

By inspection, the carbon balance gives $b = 6$, the hydrogen balance gives $c = 6$, and an oxygen balance yields the following equation:

$$6 + 2a = 6 \times 2 + 6$$

which gives $a = 6$. Therefore, the balanced reaction equation is

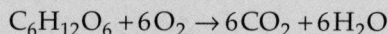

$$C_6H_{12}O_6 + 6O_2 \rightarrow 6CO_2 + 6H_2O$$

As a consistency check, verify that, for each element, the number of elements in the reactants is equal to the number of elements in the products.

Example 4.2   Use of the Chemical Reaction Equation to Calculate the Mass of Reactants Given the Mass of Products

Problem Statement

In the combustion of heptane with oxygen, CO_2 is produced. Assume that you want to produce 500 kg of dry ice per hour and that 50% of the CO_2 can be converted into dry ice, as shown in Figure E4.2. How many kilograms of heptane must be burned per hour?

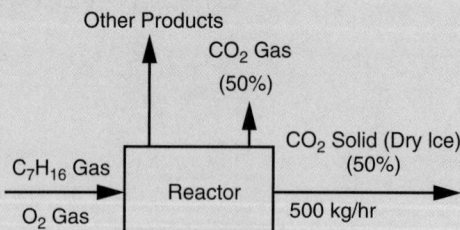

Figure E4.2

Solution

From the problem statement you can conclude that you want to use the product mass of CO_2 to calculate a reactant mass, the C_7H_{16}. The procedure is first to convert kilograms of CO_2 to moles, apply the chemical equation to get moles of C_7H_{16}, and finally calculate the kilograms of C_7H_{16}. We use Figure E4.2 in the analysis.

Look at the back inside cover to get the molecular weight of CO_2 (44.0) and C_7H_{16} (100.1). The chemical equation is

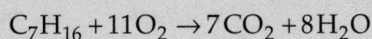

$$C_7H_{16} + 11O_2 \rightarrow 7CO_2 + 8H_2O$$

The next step is to select a basis.

Basis: 500 kg of dry ice (equivalent to 1 hr)

The calculation of the amount of C_7H_{16} can be made in one sequence of calculations:

$$\frac{500 \text{ kg dry ice}}{} \left| \frac{1 \text{ kg } CO_2 \text{ formed}}{0.5 \text{ kg dry ice}} \right| \frac{1 \text{ kg mol } CO_2}{44.0 \text{ kg } CO_2} \right|$$

$$\frac{1 \text{ kg mol } C_7H_{16}}{7 \text{ kg mol } CO_2} \left| \frac{100.1 \text{ kg } C_7H_{16}}{\text{kg mol } C_7H_{16}} \right. = 325 \text{ kg } C_7H_{16}$$

Therefore, the answer to this problem is 325 kg C_7H_{16}/hr. Finally, you could check your answer by reversing the sequence of calculations.

Example 4.3   Application of Stoichiometry When More than One Reaction Occurs

Problem Statement

A limestone analysis is:

$CaCO_3$	92.89%
$MgCO_3$	5.41%
Unreactive	1.7%

(*Continues*)

Example 4.3   Application of Stoichiometry When More than One Reaction Occurs (*Continued*)

By heating the limestone, you recover oxides that together are known as *lime*.

   a. How many pounds of calcium oxide can be made from 1 ton of this limestone?

   b. How many pounds of CO_2 can be recovered per pound of limestone?

   c. How many pounds of limestone are needed to make 1 ton of lime?

Solution

Steps 1 and 3

Read the problem carefully to fix in mind exactly what is required. The carbonates are decomposed to oxides. You should recognize that lime (oxides of Ca and Mg) will also include all of the impurities present in the limestone that remain after the CO_2 has been driven off.

Step 2

Next, draw a picture of what is going on in this process. See Figure E4.3.

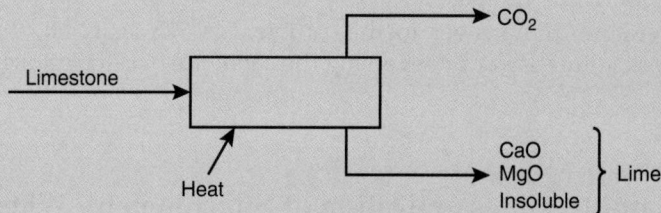

Figure E4.3

Step 4

To complete the preliminary analysis, you need the following chemical reaction equations:

$$CaCO_3 \rightarrow CaO + CO_2$$
$$MgCO_3 \rightarrow MgO + CO_2$$

Additional data that you need to look up (or calculate) are the molecular weights of the species:

	$CaCO_3$	$MgCO_3$	CaO	MgO	CO_2
Mol. wt.:	100.1	84.32	56.08	40.32	44.0

Step 5

The next step is to pick a basis:

Basis: 100 lb of limestone

This basis was selected because pounds of each component will be equal to its weight percent. You could also pick 1 lb of limestone if you wanted, or 1 ton.

Steps 6–9

Calculations of the percent composition and pound moles of the limestone and products in the form of a table will serve as an adjunct to Figure E4.3 and prove to be most helpful in answering the questions posed.

Limestone			Solid Products		
Component	lb = percent	lb mol	Compound	lb mol	lb
$CaCO_3$	92.89	0.9280	CaO	0.9280	52.04
$MgCO_3$	5.41	0.0642	MgO	0.0642	2.59
Unreactive	1.70		Unreactive		1.70
Total	100.00	0.9920	Total	0.9920	56.33

The quantities listed under Solid Products are calculated from the chemical equations. For example:

$$\frac{92.89 \text{ lb } CaCO_3}{} \left| \frac{1 \text{ lb mol } CaCO_3}{100.1 \text{ lb } CaCO_3} \right| \frac{1 \text{ lb mol } CaO}{1 \text{ lb mol } CaCO_3} \left| \frac{56.08 \text{ lb } CaO}{1 \text{ lb mol } CaO} \right.$$

$$= 52.04 \text{ lb } CaO$$

$$\frac{5.41 \text{ lb } MgCO_3}{} \left| \frac{1 \text{ lb mol } MgCO_3}{84.32 \text{ lb } MgCO_3} \right| \frac{1 \text{ lb mol } MgO}{1 \text{ lb mol } MgCO_3} \left| \frac{40.32 \text{ lb } MgO}{1 \text{ lb mol } MgO} \right.$$

$$= 2.59 \text{ lb } MgO$$

(Continues)

Example 4.3   Application of Stoichiometry When More than One Reaction Occurs (*Continued*)

The production of CO_2:

0.9280 lb mol CaO is equivalent to 0.9280 lb mol CO_2
0.0642 lb mol MgO is equivalent to <u>0.0642</u> lb mol CO_2
                    Total   0.9920 lb mol CO_2

$$\frac{0.9920 \text{ lb mol } CO_2}{} \left| \frac{44.0 \text{ lb } CO_2}{1 \text{ lb mol } CO_2} \right| = 44.67 \text{ lb } CO_2$$

Alternatively, you could have calculated the pounds of CO_2 from a total balance: $100 - 56.33 = 44.67$. Note that the total pounds of products equal the 100 lb of entering limestone. If they were not equal, what would you do? Check your molecular weight values and your calculations.

Now let's calculate the quantities originally asked for by converting the units of the previously calculated quantities:

a. CaO produced $\dfrac{52.04 \text{ lb CaO}}{100 \text{ lb stone}} \left| \dfrac{2000 \text{ lb}}{1 \text{ ton}} \right. = 1041$ lb CaO/ton limestone

b. CO_2 recovered $\dfrac{43.67 \text{ lb } CO_2}{100 \text{ lb limestone}} \bigg| = 0.437$ lb CO_2/lb limestone

c. Limestone required $\dfrac{100 \text{ lb stone}}{56.33 \text{ lb lime}} \left| \dfrac{2000 \text{ lb}}{1 \text{ ton}} \right. = 3550$ lb stone/ton lime

Self-Assessment Test

Question

For the following reaction

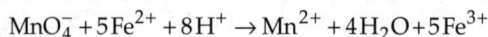

$$MnO_4^- + 5Fe^{2+} + 8H^+ \rightarrow Mn^{2+} + 4H_2O + 5Fe^{3+}$$

a student wrote the following to determine how many moles of MnO_4^- would react with 3 moles of Fe^{2+}:

1 mol $MnO_4^- = 5$ mol Fe^{2+}. Divide both sides by 5 mol Fe^{2+} to get 1 mol $MnO_4^- / 5$ mol $Fe^{2+} = 1$. The number of mol of MnO_4^- that reacts with 3 mol of Fe^{2+} is

$$\frac{1 \text{ mol MnO}_4^-}{5 \text{ mol Fe}^{2+}} \left| \frac{3 \text{ mol Fe}^{2+}}{} = 0.6 \text{ mol MnO}_4^-.\right.$$

Is the calculation correct?

Answer

When writing a chemical reaction, you can use the stoichiometric ratios of the reactants and products to determine changes due to a specific reaction, but 1 mol MnO_4^- is not equal to 5 mol of Fe^{2+}, and their ratio is not equal to a dimensionless one. That line in the solution is wrong, but the solution is correct.

Problems

1. Write balanced reaction equations for the following reactions:
   a. C_9H_{18} and oxygen to form carbon dioxide and water
   b. FeS_2 and oxygen to form Fe_2O_3 and sulfur dioxide
2. If 1 kg of benzene (C_6H_6) is oxidized with oxygen, how many kilograms of O_2 are needed to convert all of the benzene to CO_2 and H_2O?
3. Can you balance the following chemical reaction equation?

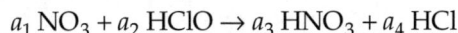

$$a_1 \text{ NO}_3 + a_2 \text{ HClO} \rightarrow a_3 \text{ HNO}_3 + a_4 \text{ HCl}$$

Answers

1. (a) $C_9H_{18} + 13.5O_2 \rightarrow 9CO_2 + 9H_2O$

   (b) $4FeS_2 + 11O_2 \rightarrow Fe_2O_3 + 8SO_2$

2. 7.5 mol of O_2 per mol C_6H_6. $\dfrac{1 \text{ kg Bz}}{} \left| \dfrac{\text{kg mol Bz}}{78 \text{ kg Bz}} \right| \dfrac{7.5 \text{ kg mol O}_2}{\text{kg mol Bz}} \left| \dfrac{32 \text{ kg O}_2}{\text{kg mol O}_2} = 3.08 \text{ kg O}_2 \right.$

3. No

4.2   **Terminology for Reaction Systems**

So far, we have discussed the stoichiometry of reactions in which the proper stoichiometric ratio of reactants is fed into a reactor, and the reaction goes to completion; no reactants remain in the reactor. What if (a) some other ratio of reactants is fed or (b) the reaction is incomplete? In such circumstances, you need to be familiar with a number of terms used to describe these types of cases.

4.2.1 Extent of Reaction, ξ

You will find the **extent of reaction** useful in solving material balances involving chemical reactions if you can ascertain the reaction equations. The extent of reaction applies to each species in the reaction. The extent of reaction, ξ is **based on a specified stoichiometric equation** and denotes how much reaction occurs. Its units are "moles reacting." The extent of reaction is calculated by dividing the change in the number of moles of a species that occurs in a reaction, for either a reactant or a product, by the associated stoichiometric coefficient (which has the units of the change in the moles of species i divided by the moles reacting). For example, consider the chemical reaction equation for the combustion of carbon monoxide:

$$2\,CO + O_2 \;\rightarrow\; 2\,CO_2$$

If 20 moles of CO are combined with 10 moles of O_2 to form 15 moles of CO_2, the extent of reaction can be calculated from the amount of CO_2 that is produced. The value of the change in the moles of CO_2 is $15 - 0 = 15$ mol. The value of the stoichiometric coefficient for the CO_2 is 2 mol CO_2/moles reacting. Then the extent of reaction is

$$\xi = \frac{(15-0)\,mol\,CO_2}{2\ mol\ CO_2/moles\ reacting} = 7.5\ moles\ reacting$$

Let's next consider a more formal definition of the extent of reaction, one that takes into account incomplete reaction and involves the initial amounts of reactants and products. The extent of reaction for a reaction is defined as follows for a single reaction involving component i:

$$\xi = \frac{n_i - n_{io}}{v_i} \tag{4.3}$$

where n_i = moles of species i present in the system after the reaction occurs
$\quad\;\; n_{io}$ = moles of species i present in the system when the reaction starts
$\quad\;\; v_i$ = stoichiometric coefficient for species i in the specified chemical reaction equation (moles of species per moles reacting)
$\quad\;\; \xi$ = extent of reaction (moles reacting according to the assumed reaction stoichiometry)

The stoichiometric coefficients of the products in a chemical reaction are assigned positive values, and the reactants are assigned negative values. Note that $(n_i - n_{i0})$ is equal to the generation by reaction of component i when the quantity is positive, and the consumption of component i by reaction when it is negative.

Equation (4.3) can be rearranged to calculate the final number of moles of component i from the value of the extent of reaction if known plus the value of the initial amount of component i:

$$n_i = n_{io} + \xi v_i \qquad (4.4)$$

As shown in the next example, the production or consumption of one species can be used to calculate the production or consumption of any of the other species involved in a reaction once you calculate, or are given, the value of the extent of reaction. Remember that Equations (4.3) and (4.4) assume that component i is involved in only one reaction.

Example 4.4   Calculation of the Extent of Reaction

Problem Statement

NADH (nicotinamide adenine dinucleotide) supplies hydrogen in living cells for biosynthesis reactions such as

$$CO_2 + 4H \rightarrow CH_2O + H_2O$$

If you saturate 1 L of deaerated water with CO_2 gas at 20°C (the solubility is 1.81 g CO_2/L) and add enough NADH to provide 0.057 g of H into a bioreactor used to imitate the reactions in cells, and obtain 0.7 g of CH_2O, what is the extent of reaction for this reaction? Use the extent of reaction to determine the number of grams of CO_2 left in solution.

Solution

Basis: 1 L water saturated with CO_2

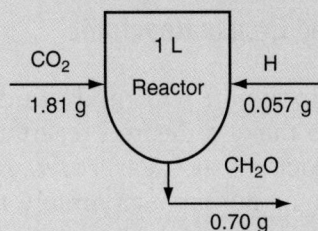

Figure E4.4

(Continues)

Example 4.4   Calculation of the Extent of Reaction (*Continued*)

The extent of reaction can be calculated by applying Equation (4.3) based on the value given for CH_2O:

$$n_{CH_2O}^{final} = \frac{0.70 \text{ g } CH_2O}{} \left| \frac{1 \text{ g mol } CH_2O}{30.02 \text{ g } CH_2O} = 0.0233 \text{ g mol } CH_2O\right.$$

$$n_{CH_2O}^{initial} = \frac{0 \text{ g } CH_2O}{} \left| \frac{1 \text{ g mol } CH_2O}{30.02 \text{ g } CH_2O} = 0 \text{ g mol } CH_2O\right.$$

$$\xi = \frac{n_i - n_{i0}}{v_i} = \frac{(0.0233 - 0) \text{ g mol } CH_2O)}{1 \text{ g mol } CH_2O/\text{moles reacting}} = 0.0233 \text{ moles reacting}$$

The number of moles of CO_2 left in solution can be obtained by using Equation (4.4) or Equation (4.3) for the CO_2:

$$n_{CO_2}^{initial} = \frac{1.81 \text{ g } CO_2}{} \left| \frac{1 \text{ g mol } CO_2}{40.00 \text{ g } CO_2} = 0.041 \text{ g mol } CO_2\right.$$

$$n_{CO_2}^{final} = 0.041 + (-1)(0.0233) = 0.0177 \text{ g mol } CO_2$$

$$m_{CO_2}^{final} = \frac{0.0177 \text{ g mol } CO_2}{} \left| \frac{44.00 \text{ g } CO_2}{1 \text{ g mol } CO_2} = 0.78 \text{ g } CO_2\right.$$

To sum up, the important characteristic of the extent of reaction, ξ, defined in Equation (4.3) is that it has the same value for each molecular species involved in a reaction. Thus, given the initial mole numbers of all species and a value for ξ (or the change in the number of moles of one species from which the value of ξ can be calculated, as is done in Example 4.4), you can easily compute the number of all other moles in the system using Equation (4.4).

4.2.2 Limiting and Excess Reactants

In industrial reactors, you will rarely find exact stoichiometric amounts of materials used. To make a desired reaction take place or to use up a costly reactant, excess reactants are nearly always used. The excess material comes out together with, or perhaps separately from, the product and sometimes can be used again. The **limiting reactant** is defined as the species in a chemical reaction that theoretically would be the first to be completely consumed if the reaction were to proceed to completion according to the chemical equation **even if the reaction does not proceed to completion!** All of the other reactants are called **excess reactants.** For example, using the chemical reaction equation in Example 4.2,

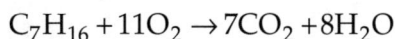

$$C_7H_{16} + 11O_2 \rightarrow 7CO_2 + 8H_2O$$

if 1 g mol of C_7H_{16} and 12 g mol of O_2 are mixed so as to react, C_7H_{16} would be the limiting reactant even if the reaction does not take place. The amount of the **excess reactant** O_2 would be calculated as 12 g mol of initial reactant less the 11 g mole needed to react with 1 g mol of C_7H_{16}, or 1 g mol of O_2. Therefore, if the reaction were to go to completion, the amount of product that would be produced is controlled by the amount of the limiting reactant, namely, C_7H_{16} in this example.

As a straightforward way of determining which species is the limiting reactant, you can calculate the **maximum extent of reaction,** a quantity that is based on **assuming the complete reaction** of each reactant. **The reactant with the smallest maximum extent of reaction is the limiting reactant.** For Example 4.2, for 1 g mol of C_7H_{16} and 12 g mol of O_2, you can calculate

$$\xi^{max}(\text{based on } O_2) = \frac{0 \text{ g mol } O_2 - 12 \text{ g mol } O_2}{-11 \text{ g mol } O_2/\text{moles reacting}} = 1.09 \text{ moles reacting}$$

$$\xi^{max}(\text{based on } C_7H_{16}) = \frac{0 \text{ g mol } C_7H_{16} - 1 \text{ g mol } C_7H_{16}}{-1 \text{ g mol } C_7H_{16}/\text{moles reacting}} = 1.00 \text{ moles reacting}$$

Therefore, heptane (C_7H_{16}) is the limiting reactant, and oxygen is the excess reactant.

Example 4.5   Calculation of the Limiting and Excess Reactants Given the Mass of Reactants

Problem Statement

In this example, let's use the same data as in Example 4.4. The basis is the same and the figure is the same.

    a. What is the maximum number of grams of CH_2O that can be produced?

    b. What is the limiting reactant?

    c. What is the excess reactant?

(Continues)

Example 4.5   Calculation of the Limiting and Excess Reactants Given the Mass of Reactants (*Continued*)

Solution

The first step is to determine the limiting reactant by calculating the maximum extent of reaction based on the complete reaction of both CO_2 and H.

$$\xi^{max}(\text{based on } CO_2) = \frac{0 - 0.041 \text{ g mol } CO_2}{-1 \text{ g mol } CO_2/\text{moles reacting}} = 0.041 \text{ moles reacting}$$

$$\xi^{max}(\text{based on } H) = \frac{0 - 0.057 \text{ g mol } H}{-4 \text{ g mol } H/\text{moles reacting}} = 0.014 \text{ moles reacting}$$

You can conclude that (b) H is the limiting reactant, and that (c) CO_2 is the excess reactant. The excess CO_2 is $(0.041 - 0.014) = 0.027$ g mol. To answer question (a), the maximum amount of CH_2O that can be produced is based on assuming complete reaction of the limiting reactant:

$$\frac{0.014 \text{ g mol } H}{} \left| \frac{1 \text{ g mol } CH_2O}{4 \text{ g mol } H} \right| \frac{30.02 \text{ g } CH_2O}{1 \text{ g mol } CH_2O} = 0.017 \text{ g } CH_2O$$

Finally, you should check your answer by working from the answer to the given reactant, or, alternatively, by summing up the mass of the C and the mass of excess H. What should the sums be?

4.2.3 Conversion and Degree of Completion

Conversion and **degree of completion** are terms not as precisely defined as are the extent of reaction and limiting and excess reactant. Rather than cite all of the possible usages of these terms, many of which conflict, we shall define them as follows: **Conversion (or the degree of completion) is the fraction of the limiting reactant in the feed that is converted into products.** Conversion is related to the **degree of completion** of a reaction. The numerator and denominator of the fraction contain the same units, so the fraction conversion is dimensionless. Thus, percent conversion is

$$\% \text{ conversion} = 100 \; \frac{\text{moles of the limiting reactant in the feed that reacts}}{\text{moles of the limiting reactant introduced in the feed}} \quad (4.5)$$

Note that Equation (4.5) can also be applied using the mass of the limiting reactant although the molar form in Equation (4.5) is more commonly used.

For complex reaction systems, it is important that if conversion or degree of completion is used that they are explicitly defined in terms of the relevant reactions.

For example, for the reaction equation used in Example 4.2, if 14.4 kg of CO_2 are formed in the reaction of 10 kg of C_7H_{16}, you can calculate the percent of the C_7H_{16} that is converted to CO_2 (reacts) as follows:

$$\left.\begin{array}{l} C_7H_{16} \text{ in feed equivalent to} \\ \text{the } CO_2 \text{ in the product} \end{array}\right\} \frac{14.4 \text{ kg } CO_2}{} \left| \frac{1 \text{ kg mol } CO_2}{44.0 \text{ kg } CO_2} \right| \frac{1 \text{ kg mol } C_7H_{16}}{7 \text{ kg mol } CO_2}$$

$$= 0.0468 \text{ kg mol } C_7H_{16}$$

$$\left.\begin{array}{l} \text{initial } C_7H_{16} \\ \text{in the reactants} \end{array}\right\} \frac{10 \text{ kg } C_7H_{16}}{} \left| \frac{1 \text{ kg mol } C_7H_{16}}{100.1 \text{ kg } C_7H_{16}} = 0.0999 \text{ kg mol } C_7H_{16}$$

$$\% \text{ conversion} = \frac{0.0468 \text{ kg mol reacted}}{0.0999 \text{ kg mol fed}} \, 100 = 46.8\% \text{ of the } C_7H_{16}$$

The conversion can also be calculated by using the extent of reaction as follows: Conversion is equal to the extent of reaction based on the formation of CO_2 (i.e., the actual extent of reaction) divided by the extent of reaction, assuming complete reaction of C_7H_{16} (i.e., the maximum possible extent of reaction):

$$\text{conversion} = \frac{\text{extent of reaction that actually occurs}}{\text{extent of reaction that would occur if complete reaction took place}}$$

$$= \frac{\xi}{\xi^{max}} \tag{4.6}$$

4.2.4 Selectivity

Selectivity is the ratio of the moles of a particular (usually the desired) product produced to the moles of another (usually undesired or by-product) product produced in a single reaction or group of reactions. For example, methanol (CH_3OH) can be converted into ethylene (C_2H_4) or propylene (C_3H_6) by the reactions

$$2 \, CH_3OH \rightarrow C_2H_4 + 2H_2O$$
$$3 \, CH_3OH \rightarrow C_3H_6 + 3H_2O$$

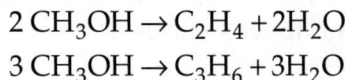

Of course, for the process to be economical, the value of the products has to be greater than the value of the reactants. Examine the data in Figure 4.2 for the concentrations of the products of the reactions. What is the selectivity of C_2H_4 relative to the C_3H_6 at 80% conversion of the CH_3OH. Proceed upward at 80% conversion to get for $C_2H_4 \cong 19$ and for $C_3H_6 \cong 8$ mol%. Because the basis for both values is the same, you can compute the selectivity $19/8 \cong 2.4 \, mol \, C_2H_4$ per mol C_3H_6.

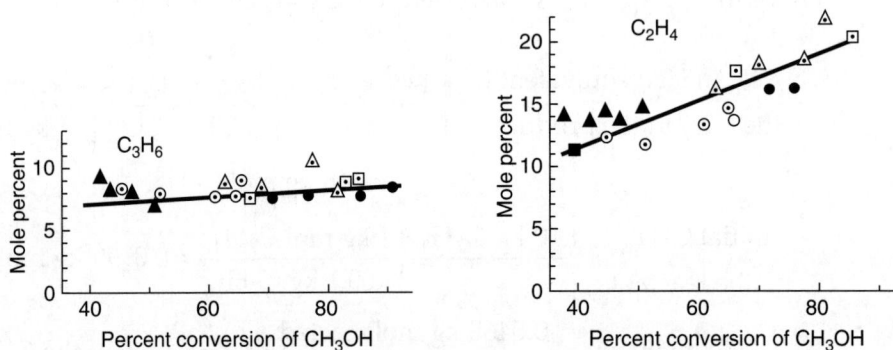

Figure 4.2   Products from the conversion of ethanol

4.2.5 Yield

No universally agreed-upon definitions exist for yield—in fact, quite the contrary is true. Here are three common ones:

- **Yield** (based on feed): The amount (mass or moles) of desired product obtained divided by the amount of the key (frequently the limiting) reactant fed.

- **Yield** (based on reactant consumed): The amount (mass or moles) of desired product obtained divided by the amount of the key (frequently the limiting) reactant consumed.

- **Yield** (based on 100% conversion): The amount (mass or moles) of a product obtained divided by the theoretical (expected) amount of the product that would be obtained based on the limiting reactant in the chemical reaction equation(s) if it were completely consumed. Note that this is a fractional (dimensionless) yield because the numerator and denominator have the same units, whereas the previous two definitions of yield are not dimensionless.

Why doesn't the actual yield in a reaction equal the theoretical yield predicted from the chemical reaction equation? Several reasons exist:

- Impurities among the reactants
- Leaks to the environment
- Side reactions
- Reversible reactions

As an illustration, suppose you have a reaction sequence as follows:

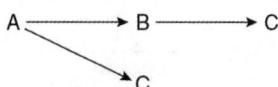

$$A \longrightarrow B \longrightarrow C$$
$$\searrow C$$

with B being the desired product and C the undesired one. The yield of B according to the first two definitions is the moles (or mass) of B produced divided by the moles (or mass) of A fed or consumed. The yield according to the third definition is the moles (or mass) of B actually produced divided by the maximum amount of B that could be produced in the reaction sequence (i.e., complete conversion of A to B). The selectivity of B is the moles of B divided by the moles of C produced.

Yield and *selectivity* are terms that measure the degree to which a desired reaction proceeds relative to competing alternative (undesirable) reactions. As a designer of equipment, you want to maximize production of the desired product and minimize production of the unwanted products. Do you want high or low selectivity? Yield?

The next example shows you how to calculate all of the terms discussed in this section.

Example 4.6   Calculation of Various Terms Pertaining to Reactions

Problem Statement

Semenov[1] described some of the chemistry of alkyl chlorides. The two reactions of interest for this example are

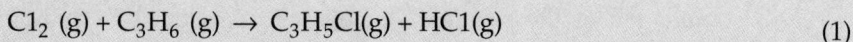

$$Cl_2 \, (g) + C_3H_6 \, (g) \rightarrow C_3H_5Cl(g) + HCl(g) \tag{1}$$

$$Cl_2 \, (g) + C_3H_6 \, (g) \rightarrow C_3H_6Cl_2(g) \tag{2}$$

---

[1] N. N. Semenov, *Some Problems in Chemical Kinetics and Reactivity*, Vol. II (Princeton, NJ: Princeton University Press, 1959), 39–42.

(Continues)

Example 4.6   Calculation of Various Terms Pertaining to Reactions (*Continued*)

C_3H_6 is propene (MW = 42.08)

C_3H_5Cl is alyl chloride (3-chloropropene) (MW = 76.53)

$C_3H_6Cl_2$ is propylene chloride (1,2-dichloropropane) (MW = 112.99)

The species recovered after the reaction takes place for some time are listed in Table E4.6.

Table E4.6

Species	g mol
Cl_2	141.0
C_3H_6	651.0
C_3H_5Cl	4.6
$C_3H_6Cl_2$	24.5
HCl	4.6

Based on the product distribution in Table E4.6, assuming that the feed consisted only of Cl_2 and C_3H_6, calculate the following:

a. How much Cl_2 and C_3H_6 were fed to the reactor in gram moles?

b. What was the limiting reactant?

c. What was the excess reactant?

d. What was the fractional conversion of C_3H_6 to C_3H_6Cl?

e. What was the selectivity of C_3H_5Cl relative to $C_3H_6Cl_2$?

f. What was the yield of C_3H_5Cl expressed in grams of C_3H_5Cl to the grams of C_3H_6 fed to the reactor?

g. What were the extents of reaction of Reactions (1) and (2)?

Steps 1–4

Examination of the problem statement reveals that the amount of feed is not given, and consequently you have to calculate the gram moles fed to the reactor even if the amounts were not asked for. The molecular weights were given. Figure E4.6 illustrates the process as an open-flow system.

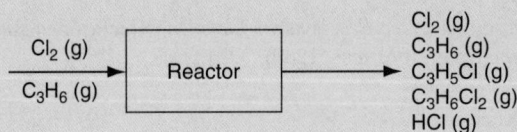

Figure E4.6

Step 5

A convenient basis is what is given in the product list in Table E4.6.

Steps 7–9

Use the chemical equations to calculate the moles of species in the feed. Start with the Cl_2.

Reaction (1):

$$\frac{4.6 \text{ g mol } C_3H_5Cl}{} \left| \frac{1 \text{ g mol } Cl_2}{1 \text{ g mol } C_3H_5Cl} = 4.6 \text{ g mol } Cl_2 \text{ reacts}\right.$$

Reaction (2):

$$\frac{24.5 \text{ g mol } C_3H_6Cl_2}{} \left| \frac{1 \text{ g mol } Cl_2}{1 \text{ g mol } C_3H_6Cl_2} = 24.5 \text{ g mol } Cl_2 \text{ reacts}\right.$$

Total	$\overline{29.1}$ g mol Cl_2 reacts
Unreacted Cl_2 in product	$\underline{141.0}$
a. Total Cl_2 fed	170.1

What about the amount of C_3H_6 in the feed? From the chemical equations, you can see that if 29.1 g mol of Cl_2 reacts in total by Reactions (1) and (2), 29.1 g mol of C_3H_6 must react. Since 651.0 g mol of C_3H_6 exist unreacted in the product, $651 + 29.1 = 680.1$ g mol of C_3H_6 were fed to the reactor.

You can check those answers by adding up the gram moles of Cl, C, and H in the product and comparing the value with that calculated in the feed:

      In product

Cl $2(141.0) + 1(4.6) + 2(24.5) + 1(4.6) = 340.2$

C $3(651) + 3(4.6) + 3(24.5) = 2040.3$

H $6(651) + 5(4.6) + 6(24.5) + 1(4.6) = 4080.6$

      In feed

Cl $2(170.1) = 340.2$ OK

C $3(680.1) = 2040.3$ OK

H $6(680.1) = 4080.6$ OK

(Continues)

Example 4.6 Calculation of Various Terms Pertaining to Reactions (*Continued*)

We will not go through detailed analysis for the remaining calculations but simply determine the desired quantities based on the data prepared for parts a, b, and c. In this particular problem, since both reactions involve the same reaction stoichiometric coefficients, both reactions will have the same limiting and excess reactants:

$$\xi^{max}(\text{based on }C_3H_6) = \frac{-680.1 \text{ g mol } C_3H_6}{-1 \text{ g mol } C_3H_6/\text{moles reacting}} = 680.1 \text{ mol reacting}$$

$$\xi^{max}(\text{based on }Cl_2) = \frac{-170.1 \text{ g mol } Cl_2}{-1 \text{ g mol } Cl_2/\text{mol reacting}} = 170.1 \text{ mol reacting}$$

Thus, C_3H_6 was the excess reactant and Cl_2 the limiting reactant.

    d. The fractional conversion of C_3H_6 to C_3H_6Cl was

$$\frac{29.1 \text{ g mol } C_3H_6 \text{ that reacted}}{680.1 \text{ g mol } C_3H_6 \text{ fed}} = 0.043$$

    e. The selectivity was

$$\frac{4.6 \text{ g mol } C_3H_5Cl}{24.5 \text{ g mol } C_3H_6Cl_2} = 0.19 \frac{\text{g mol } C_3H_5Cl}{\text{g mol } C_3H_6Cl_2}$$

    f. The yield was

$$\frac{(76.53)\,(4.6) \text{ g } C_3H_5Cl}{(42.08)\,(680.1) \text{ g } C_3H_6} = 0.012 \frac{\text{g } C_3H_5Cl}{\text{g } C_3H_6}$$

    g. Because C_3H_5Cl is produced only by the first reaction, the extent of reaction of the first reaction is

$$\xi_1 = \frac{n_i - n_{io}}{v_i} = \frac{4.6 - 0}{1} = 4.6$$

Because $C_3H_5Cl_2$ is produced only by the second reaction, the extent of reaction of the second reaction is

$$\xi_2 = \frac{n_i - n_{io}}{v_i} = \frac{24.5 - 0}{1} = 24.5$$

Self-Assessment Test

Questions

1. What is the extent of reaction based on?

2. How is the extent of reaction used to identify the limiting reactant?

Answers

1. It is based on a specific reaction and the changes in the amounts of reactants and/or products due to the reaction.

2. By calculating the maximum extent of reaction based on each reactant. Then the limiting reactant is the reactant with the smallest maximum extent of reaction.

Problem

Two well-known gas phase reactions take place in the dehydration of ethane:

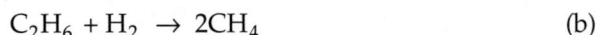

$$C_2H_6 \rightarrow C_2H_4 + H_2 \qquad (a)$$

$$C_2H_6 + H_2 \rightarrow 2CH_4 \qquad (b)$$

Given the product distribution measured in the gas phase reaction of C_2H_6 as follows and assuming that only C_2H_6 and H_2 were initially present:

$$
\begin{array}{ll}
C_2H_6 & 27\% \\
C_2H_4 & 33\% \\
H_2 & 13\% \\
CH_4 & 27\%
\end{array}
$$

a. What species was the limiting reactant?
b. What species was the excess reactant?
c. What was the conversion of C_2H_6 to CH_4?
d. What was the degree of completion of the reaction?
e. What was the selectivity of C_2H_4 relative to CH_4?
f. What was the yield of C_2H_4 expressed in kilogram moles of C_2H_4 produced per kilogram mole of C_2H_6?
g. What was the extent of reaction of C_2H_6?

Answer

a. C_2H_6 is the limiting reactant for the first reaction and H_2 is the limiting reactant for the second reaction. b. No excess reactant for the first reaction, while C_2H_6 is the excess reactant for the second reaction. c. Basis: 100 mol of product gas. Initial amount of C_2H_6 is 27 + 33 + 13.5 = 73.5 mol. Fractional conversion = 13.5/73.5 = 0.184. d. The degree of completion was 27/73.5 = 0.367 for the first reaction and 13.5/73.5 = 0.184 for the second reaction. e. The selectivity is equal to 33/(2*13.5) = 1.22. f. Yield = 33/73.5 = 0.449. g. The extent of reaction is 33 mol for the first reaction and 13.5 mol for the second reaction based on the basis selected.

4.3   Species Mole Balances

4.3.1   Processes Involving a Single Reaction

Do you recall from Section 3.1 that the material balance for a species without reaction was simply what comes in equals what goes out? When chemical reactions occur, the consumption and/or generation of the species by reaction must be considered. In terms of moles of species i, the material balance for a species for a steady-state system is

$$\left\{\begin{array}{l} \text{moles of } i \\ \text{leaving} \\ \text{the system} \end{array}\right\} = \left\{\begin{array}{l} \text{moles of } i \\ \text{entering} \\ \text{the system} \end{array}\right\} + \left\{\begin{array}{l} \text{moles of } i \\ \text{generated} \\ \text{by reaction} \end{array}\right\} - \left\{\begin{array}{l} \text{moles of } i \\ \text{consumed} \\ \text{by reaction} \end{array}\right\} \qquad (4.7)$$

Note that this equation becomes Equation (3.1) when no reaction occurs and that it is consistent with physical intuition if you consider that generation of a species by reaction will increase the amount of the species leaving the system and the consumption by reaction will decrease it. Also note that we have written Equation (4.7) in terms of moles rather than mass. The generation and consumption terms are more conveniently represented in terms of moles because reactions are usually written in terms of molar ratios.

Fortunately, you only have to add one additional variable to account for the generation or consumption of each species i present in the system *if you make use of the extent of reaction* that was presented in Section 4.2. To make the idea clear, let's examine the reaction of N_2 and H_2 to form NH_3 in the gas phase from the introductory example. Figure 4.3 presents the process as a steady-state, open system operating for a fixed interval. Figure 4.3 shows the measured values for the flows in gram moles.

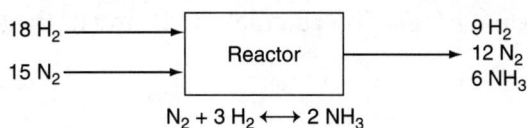

Figure 4.3   A reactor to produce NH_3

For this simple example, you can calculate by inspection or by Equation (4.7) a value in gram moles for the generation and/or consumption for each of the three species in the reaction:

$$NH_3: 6-0 = 6 \text{ g mol (product generated)}$$
$$H_2: 9-18 = -9 \text{ g mol (reactant consumed)}$$
$$N_2: 12-15 = -3 \text{ g mol (reactant consumed)}$$

Because of the stoichiometry of the chemical reaction equation

$$N_2 + 3\,H_2 \ \rightarrow \ 2\,NH_3$$

the three respective generation and consumption terms are related. For example, given the value for the generation of NH_3, you can calculate the values for the consumption of H_2 and N_2 using the reaction equation. The ratio of hydrogen to nitrogen consumed in the reactants and maintained in the product ammonia is always 3 to 1. Thus, you cannot specify more than one value of the N_2 and H_2 pair left over from the reaction without introducing a redundant or possibly an inconsistent specification. In general, if you specify the value for the generation or consumption of one species in a reaction, you are able to calculate the values of the other species from a solo chemical reaction equation.

Here is where the extent of reaction ξ becomes useful. Recall that Equation (4.3) relates the extent of reaction to the change in moles of a species i divided by the stoichiometric coefficient v_i of the species in the reaction equation, where S is the number of reaction components.

$$\xi = \frac{n_i^{out} - n_i^{in}}{v_i} \qquad\qquad i = 1, \ldots, S \qquad\qquad (4.3)$$

for the NH_3 reaction $v_{NH_3} = 2 \quad v_{H_2} = -3 \quad v_{N_2} = -1$

and the extent of reaction calculated via any of the species is

$$\xi = \frac{n_{NH_3}^{out} - n_{NH_3}^{in}}{v_{NH_3}} = \frac{6-0}{2} = 3$$

$$\xi = \frac{n_{H_2}^{out} - n_{H_2}^{in}}{v_{H_2}} = \frac{9-18}{-3} = 3$$

$$\xi = \frac{n_{N_2}^{out} - n_{N_2}^{in}}{v_{N_2}} = \frac{12-15}{-1} = 3$$

You can conclude for the case of a single chemical reaction that the specification of the extent of reaction provides one independent quantity that will determine all of the values of the generation and consumption terms for the various species in the respective implementations of Equation (4.7) because the molar ratios of the reactants and products are fixed by the solo independent chemical reaction equation. The three species balances corresponding to the process in Figure 4.3 are listed in the following table based on Equation (4.8), an equation directly derived from Equation (4.7) for an open, steady-state process:

$$n_i^{out} - n_i^{in} = v_i\xi \tag{4.8}$$

Component	Out	In	=	Generation or Consumption
I	n_i^{out}	n_i^{in}	=	$v_i\xi$
NH_3	6	0	=	$2(3) = 6$
H_2	9	18	=	$-3(3) = -9$
N_2	12	15	=	$-1(3) = -3$

The term $v_i\xi$ corresponds to the moles of species i generated or consumed in Equation 4.7. Can you determine by inspection that the three material balances are independent? Remember, in general you can write *one material balance* equation for each species present in the system. If Equation (4.8) is applied to each species that reacts in a steady-state system, the resulting set of material balances will contain an additional variable, namely, the extent of reaction, ξ. For a species that does not react, $v_i = 0$.

Frequently Asked Questions

1. Does it make any difference how the chemical reaction equation is written as long as the equation is balanced? No. For example, write the decomposition of ammonia as

$$NH_3 \rightarrow \frac{1}{2}N_2 + \frac{3}{2}H_2$$

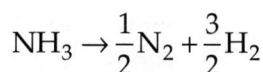

Let's calculate ξ given that zero moles of NH_3 are introduced into a reactor and that 6 moles exit, as in Figure 4.3. Then

$$\xi = \frac{6-0}{-1} = -6$$

The material balance for NH_3 is just

$$6 - 0 = (-1)(-6) = 6$$

If you calculate ξ for H_2, what result do you get? The key point here is that product $v_i \xi$, which is used in each species material balance, remains unchanged regardless of the size of the stoichiometric coefficients in a chemical reaction equation. Thus, if the chemical reaction equation is multiplied by a factor of 2, so that the stoichiometric coefficients are each twice their previous value, ξ will decrease by a factor of 2. As a result, the product $v_i \xi$ remains constant.

2. What does the negative sign in front of ξ mean? The negative sign signifies that the chemical reaction equation was written to represent a direction that is the reverse of the one in which the reaction actually proceeds.

3. Are species balances always independent? Almost always. An example of a nonindependent set of species balances occurs for the decomposition of NH_3, as shown in FAQ 1. Given one piece of information, namely, that 1 mole of NH_3 decomposes completely, the products of the reaction would be $3/2$ H_2 and $1/2$ N_2, and the H_2 and N_2 mole balances will not be independent because their ratio would always be 3 to 1. Would partial decomposition of the 1 mole of NH_3 lead to the same conclusion? Note that if H_2 and N_2 were present in the feed in nonstoichiometric amounts, the species balances for H_2 and N_2 would form independent equations.

You can calculate the total molar flow in, F^{in}, and the total molar flow out, F^{out}, by adding all of the species flows in and out respectively:

$$F^{in} = \sum_{i=1}^{S} n_i^{in} \tag{4.9a}$$

$$F^{out} = \sum_{i=1}^{S} n_i^{out} \tag{4.9b}$$

where S is the total number of species in the system.

Neither of the Equations (4.9) is an independent equation, since both simply represent the sum of all of the species flows but can be substituted for one of the species balances if convenient. Only S independent equations can be written for the system. Equations (4.9) apply only to open, steady-state systems and the net result of a batch process.

If the unknown, ξ, occurs in a set of S otherwise independent species equations, you will, of course, have to augment the existing information by one more independent bit of information in order to be able to solve a problem. For example, you might be told that complete conversion of the limiting reactant occurs, or be given the value of the fractional conversion f of the limiting reactant; ξ is related to f by

$$\xi = \frac{(-f)n^{in}_{\text{limiting reactant}}}{v_{\text{limiting reactant}}} \tag{4.10}$$

You can calculate the value of ξ from the value for the fraction conversion (or vice versa) plus information identifying the limiting reactant. In other cases, you are given sufficient information about the moles of a species entering and leaving the process so that ξ can be calculated directly from Equation (4.8).

Now let's look at some examples using the concepts discussed so far.

Example 4.7 Reaction in Which the Fractional Conversion Is Specified

Problem Statement

The chlorination of methane occurs by the following reaction:

$$CH_4 + Cl_2 \rightarrow CH_3Cl + HCl$$

You are asked to determine the product composition if the conversion of the limiting reactant is 67%, and the feed composition in mole percent is 40% CH_4, 50% Cl_2, and 10% N_2.

Solution

Steps 1–4

Assume the reactor is an open, steady-state process. Figure E4.7 is a sketch of the process with the known information placed on it.

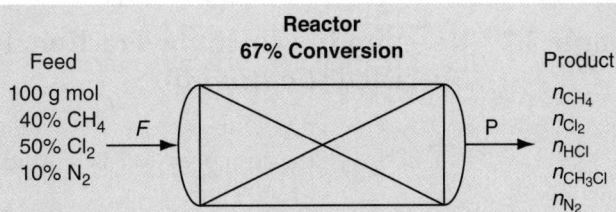

Figure E4.7

Step 5

$$\text{Basis: 100 g mol feed}$$

Step 4 Again

You have to determine the limiting reactant if you are to make use of the information about the 67% conversion. By comparing the maximum extent of reaction for each reactant, you can identify the limiting reactant as being the species that has the smallest ξ^{max}, as explained previously.

$$\xi^{max}(CH_4) = \frac{n_{i0}}{v_i} = \frac{40}{-(-1)} = 40$$

$$\xi^{max}(Cl_2) = \frac{n_{i0}}{-v_i} = \frac{50}{-(-1)} = 50$$

Therefore, CH_4 is the limiting reactant. You can now calculate the extent of reaction using the specified conversion and Equation (4.10):

$$\xi = \frac{-f\, n_{\text{lim reactant}}^{\text{in}}}{v_{\text{lim reactant}}} = \frac{(-0.67)(40)}{-1} = 26.8 \text{ g mol reacting}$$

One unknown can now be assigned a value, namely, ξ.

Steps 6 and 7

The next step is to carry out a degree-of-freedom analysis:
  Number of variables: 11

$$n_{CH_4}^{in},\ n_{Cl_2}^{in},\ n_{N_2}^{in},\ n_{CH_4}^{out},\ n_{Cl_2}^{out},\ n_{HCl}^{out},\ n_{CH_3Cl}^{out},\ n_{N_2}^{out},\ F,\ P,\ \xi$$

but you can assign values for the first three variables plus F, and have calculated ξ; hence, the *number of unknowns* has been reduced to just 6.

  Number of independent equations needed: 6
  Species material balances: 5 CH_4, Cl_2, HCl, CH_3Cl, N_2
  Specifications: 0 (*f* was used to calculate ξ in Step 4)
  Implicit equations: 1   (Why only 1 instead of 2?)

(*Continues*)

Example 4.7   Reaction in Which the Fractional Conversion Is Specified (*Continued*)

$$\sum n_i^{\text{out}} = P \text{ (the sum} \sum n_i^{\text{in}} = F \text{ is redundant)}$$

The degrees of freedom are zero.

Steps 8 and 9

The species material balances (in moles) using Equation (4.8) give a direct solution for each species in the product:

$$n_{\text{CH}_4}^{\text{out}} = 40 - 1(26.8) = 13.2$$

$$n_{\text{Cl}_2}^{\text{out}} = 50 - 1(26.8) = 23.2$$

$$n_{\text{CH}_3\text{Cl}}^{\text{out}} = 0 + 1(26.8) = 26.8$$

$$n_{\text{HCl}}^{\text{out}} = 0 + 1(26.8) = 26.8$$

$$n_{\text{N}_2}^{\text{out}} = 10 - 0(26.8) = \underline{10.0}$$
$$P = \overline{100.0}$$

Therefore, the composition of the product stream is 13.2% CH_4, 23.2% Cl_2, 26.8% CH_3Cl, 26.8% HCl, and 10% N_2 because the total number of product moles is conveniently 100 g mol. There are 100 g mol of products because there are 100 g mol of feed, and the chemical reaction equation results in the same number of moles for reactants as products. What would you have to do if the total moles in P did not amount to 100 g mol? Remember the camels!

Step 10

The fact that the redundant overall mole balance equation is satisfied can serve as a consistency check for this problem.

4.3.2 Processes Involving Multiple Reactions

In practice, reaction systems rarely involve just a single reaction. There may be a primary reaction (e.g., the desired reaction), but often there are additional or side reactions. To extend the concept of the extent of reaction to processes involving multiple reactions, the question is: Do you just include a ξ_i for every reaction? Usually the answer is yes, but more precisely, the answer is no! You should **include in the species material balances only the ξ_s**

associated with a (nonunique) set of independent chemical reactions called the **minimal set**[2] of reaction equations. What this latter term means is the smallest set of chemical reaction equations that can be assembled so as to include *all* of the species involved in the process. It is analogous to a set of independent linear algebraic equations, and you can form any other reaction equation by a linear combination of the reaction equations contained in the minimal set. Usually, the minimal set is equal to the full collection of reaction equations, but you should make sure that each set of reaction equations represents an independent set of reactions.

For example, look at the following set of reaction equations:

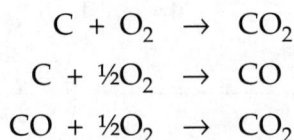

$$C + O_2 \rightarrow CO_2$$
$$C + \tfrac{1}{2}O_2 \rightarrow CO$$
$$CO + \tfrac{1}{2}O_2 \rightarrow CO_2$$

By inspection, you can see that if you subtract the second equation from the first one, you obtain the third equation. Only two of the three equations are independent; hence, the minimal set will be composed of any two of the three equations.

Example 4.8   Use MATLAB or Python to Determine the Minimal Set

Problem Statement

Determine the minimal set for the following reactions:

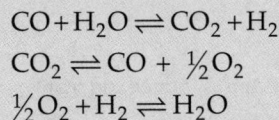

$$CO + H_2O \rightleftharpoons CO_2 + H_2$$
$$CO_2 \rightleftharpoons CO + \tfrac{1}{2}O_2$$
$$\tfrac{1}{2}O_2 + H_2 \rightleftharpoons H_2O$$

The first reaction is known as the water gas shift reaction, and it is used to convert CO into H_2.

(Continues)

---

[2] Sometimes called the maximal set.

Example 4.8   Use MATLAB or Python to Determine the Minimal Set (*Continued*)

Solution

First, let's convert this set of equations into a matrix of coefficients based on the following indices: 1 equals CO, 2 equal H_2O, 3 equal CO_2, 4 equal H_2, and 5 equal O_2.

$$\mathbf{A} = \begin{bmatrix} 1 & 1 & -1 & -1 & 0 \\ -1 & 0 & 1 & 0 & -0.5 \\ 0 & -1 & 0 & 1 & 0.5 \end{bmatrix}$$

MATLAB Solution:

MATLAB Solution for Example 4.8

```
%%%%%%%%%%%%%%%%%%%%%%%%%%%%%%%%%%%%%%%%%%%%%%
%          NOMENCLATURE
%
% A - the coefficient matrix for the set of chemical
%     reactions
%
%%%%%%%%%%%%%%%%%%%%%%%%%%%%%%%%%%%%%%%%%%%%%%%
%             PROGRAM
function Ex4_8_rank
clear; clc;
A=[1,1,-1,-1,0;-1,0,1,0,-0.5;0,-1,0,1,0.5];
k=rank(A);        % Apply function rank()
fprintf('Rank =%2.0d \n',k)
end
%          PROGRAM END
%%%%%%%%%%%%%%%%%%%%%%%%%%%%%%%%%%%%%%%%%%%%%%%
 Rank = 2
```

Python Solution:

Python Code for Example 4.8

```
Ex4_8 rank.py
###############################################
#                  NOMENCLATURE
#
# A - the coefficient matrix for the set of chemical reactions
#
###############################################
Import numpy as np
Import numpy.linalg
# Specify the coefficient matrix A
A=np.array([[1,1,-1,-1,0], [-1,0,1,0,-0.5], [0,-1,0,1,0.5]])
Rank=numpy.linalg.rank(A)
#                  PROGRAM END
###############################################
```

IPython console:

> In[1]: runfile(…
> In[2]: Rank
> Out[2]: 2

Therefore, only two of these reaction equations are independent. That is, two of these equations will form the minimal reaction set.

With these ideas in mind, we can state that for steady-state, open processes with multiple reactions, Equation (4.8) in moles becomes for component i

$$n_i^{\text{out}} = n_i^{\text{in}} + \sum_{j=1}^{R} v_{ij}\xi_j \tag{4.11}$$

where

v_{ij} is the stoichiometric coefficient of species i in reaction j in the minimal reaction set.

ξ_j is the extent of reaction for the jth reaction in which component i is present in the minimal set.

R is the number of independent chemical reaction equations (the size of the minimal set).

The total moles N exiting the reactor are

$$N = \sum_{i=1}^{S} n_i^{\text{out}} = \sum_{i=1}^{S} n_i^{\text{in}} + \sum_{i=1}^{S}\sum_{j=1}^{R} v_{ij}\,\xi_j \tag{4.12}$$

where S is the number of species in the system.

Example 4.9   Material Balances Involving Two Ongoing Reactions

Problem Statement

Formaldehyde (CHO_2) is produced industrially by the catalytic oxidation of methanol (CH_3OH) by the following reaction:

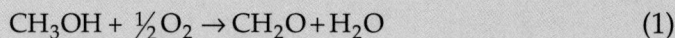

$$CH_3OH + \tfrac{1}{2}O_2 \rightarrow CH_2O + H_2O \tag{1}$$

Unfortunately, under the conditions used to produce formaldehyde at a profitable rate, a significant portion of the formaldehyde can react with oxygen to produce CO and H_2O:

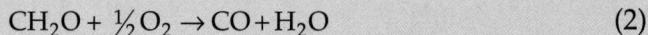

$$CH_2O + \tfrac{1}{2}O_2 \rightarrow CO + H_2O \tag{2}$$

Assume that methanol and twice the stoichiometric amount of air needed for complete oxidation of the CH_3OH are fed to the reactor, that 90% conversion of the methanol to formaldehyde results, and that a 75% yield of formaldehyde occurs [based on the theoretical production of CH_2O by Reaction (1)]. Determine the composition of the product gas leaving the reactor.

Solution

Steps 1–4

Figure E4.9 is a sketch of the process with y_i indicating the mole fraction of the respective components in P (a gas).

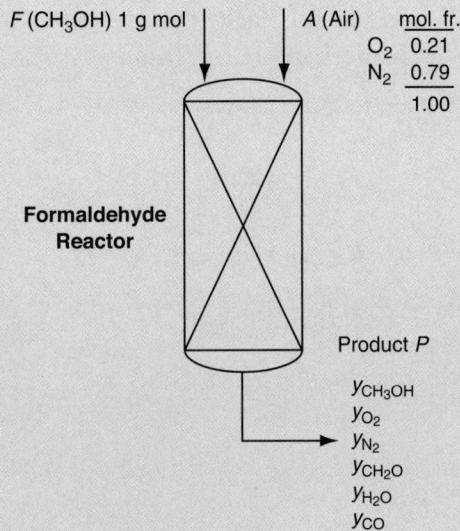

Figure E4.9

Step 5

$$\text{Basis: 1 g mol } F$$

Step 6

In this step the idea is to assign the values of variables that are specified directly or indirectly (using a related specification). Not all of the calculated values that follow are accompanied by their units (to reduce complexity), but the respective units can easily be inferred. The first calculation to make in this problem is to use the specified conversion of methanol and the yield of formaldehyde to determine the extents of reaction for the two reactions. Let ξ_1 represent the extent of reaction for Reaction (1) and ξ_2 represent the extent of reaction for Reaction (2). The limiting reactant is CH_3OH. Note that

$$n_{CH_2O}^{out,1} = n_{CH_2O}^{in,2}$$

because CH_2O is produced by reaction 1 and consumed by reaction 2. Based on the specified conversion, the extent of reaction for Reaction (1):

$$\xi_1 = \frac{-0.90}{-1}(1) = 0.9 \, \text{g mol reacting}$$

(*Continues*)

Example 4.9   Material Balances Involving Two Ongoing Reactions (*Continued*)

Yield is related to ξ_i as follows:

$$\text{yield of CH}_2\text{O} = \frac{n_{\text{CH}_2\text{O}}^{\text{out},2}}{F}$$

By Reaction (1): $n_{\text{CH}_2\text{O}}^{\text{out},1} = n_{\text{CH}_2\text{O}}^{\text{in},1} + 1(\xi_1) = 0 + \xi_1 = \xi_1$

By Reaction (2): $n_{\text{CH}_2\text{O}}^{\text{out},2} = n_{\text{CH}_2\text{O}}^{\text{in},2} - 1(\xi_2) = n_{\text{CH}_2\text{O}}^{\text{out},1} - \xi_2 = \xi_1 - \xi_2$

The specified yield: $\dfrac{n_{\text{CH}_2\text{O}}^{\text{out},2}}{F} = \dfrac{\xi_1 - \xi_2}{1} = 0.75$

$$\xi_2 = (0.90 - 0.75) = 0.15 \,\text{g mol reacting}$$

You should next calculate the amount of air (A) that enters the process. The entering oxygen is twice the required oxygen based on Reaction (1), namely,

$$n_{\text{O}_2}^{A} = 2\left(\frac{1}{2}F\right) = 2\left(\frac{1}{2}\right)(1.00) = 1.00\,\text{g mol}$$

$$A = \frac{n_{\text{O}_2}^{A}}{0.21} = \frac{100}{0.21} = 4.76\,\text{g mol}$$

$$n_{\text{N}_2}^{A} = 4.76 - 1.00 = 3.76\,\text{g mol}$$

Steps 7

The degree-of-freedom analysis is as follows:

Number of variables: 11

$$F, A, P, y_{\text{CH}_3\text{OH}}^{P}, y_{\text{O}_2}^{P}, y_{\text{N}_2}^{P}, y_{\text{CH}_2\text{O}}^{P}, y_{\text{H}_2\text{O}}^{P}, y_{\text{CO}}^{P}, \xi_1, \xi_2$$

You can assign the following values to certain of the variables:

Calculated values in Step 6 that can be assigned for F, A, ξ_1, ξ_2: total is 4

Number of unknowns: $11 - 4 = 7$

Number of independent equations needed: $11 - 4 = 7$; here are 7 to select:

Species material balances: 6

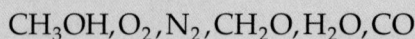

$$\text{CH}_3\text{OH}, \text{O}_2, \text{N}_2, \text{CH}_2\text{O}, \text{H}_2\text{O}, \text{CO}$$

Implicit equation: 1

$$\sum y_i^{P} = 1$$

Step 8

Because the variables in Figure E4.8 are y_i^P and not n_i^P, direct use of y_i^P in the material balances will involve the nonlinear terms $y_i^P \, P$. We could use the variable n_i^P, analogous to the material balances in previous examples, but for the purposes of illustration, let us write the equations in terms of y_i^P. Then we will calculate P using Equation (4.12).

$$P = \sum_{i=1}^{S} n_i^{in} + \sum_{i=1}^{S} \sum_{j=1}^{R} v_{ij} \xi_j$$

$$= 1 + 4.76 + \sum_{i=1}^{6} \sum_{j=1}^{2} v_{ij} \xi_j$$

$$= 5.76 + [(-1) + (-0.5) + (1) + (0) + (1) + 0]0.9$$

$$+ [0 + (-0.5) + (-1) + (0) + (1) + (1)]0.15 = 6.28 \text{ g mol}$$

The species material balances after entering their assigned values and $P = 6.28$ are

$$n_{CH_3OH}^{out} = y_{CH_3OH}(6.28) = 1 - (0.9) + 0 = 0.10$$

$$n_{O_2}^{out} = y_{O_2}(6.28) = 1.0 - (0.5)(0.9) - (0.5)(0.15) = 0.475$$

$$n_{CH_2O}^{out} = y_{CH_2O}(6.28) = 0 + 1(0.9) - 1(0.15) = 0.75$$

$$n_{H_2O}^{out} = y_{H_2O}(6.28) = 0 + 1(0.9) + 1(0.15) = 1.05$$

$$n_{CO}^{out} = y_{CO}(6.28) = 0 + 0 + 1(.15) = 0.15$$

$$n_{N_2}^{out} = y_{N_2}(6.28) = 3.76 - 0 - 0 = 3.76$$

Step 9

You can check the value of P by adding all of the n_i^{out} above.

Step 10

The six equations can be solved for y_i. Did you get the following answer (in mole percent)?

$$y_{CH_3OH} = 1.6\%, \quad y_{O_2} = 7.6\%, \quad y_{N_2} = 59.8\%, \quad y_{CH_2O} = 11.9\%,$$

$$y_{H_2O} = 16.7\%, \quad \text{and} \quad y_{CO} = 2.4\%$$

Example 4.10   Analysis of a Bioreactor

Problem Statement

A bioreactor is a vessel in which biological reactions are carried out involving enzymes, microorganisms, and/or animal and plant cells. In the anaerobic (in the absence of oxygen) fermentation of grain, the yeast *Saccharomyces cerevisiae* digests glucose ($C_6H_{12}O_6$) from plants to form the products ethanol (C_2H_5OH) and propenoic acid ($C_2H_3CO_2H$) by the following overall reactions:

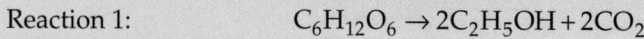

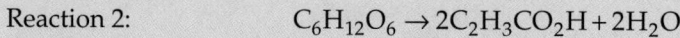

Reaction 1: $C_6H_{12}O_6 \rightarrow 2C_2H_5OH + 2CO_2$

Reaction 2: $C_6H_{12}O_6 \rightarrow 2C_2H_3CO_2H + 2H_2O$

In a process, a tank is initially charged with 4000 kg of a 12% solution of glucose in water. Then the glucose solution is inoculated with the yeast. After fermentation, 120 kg of CO_2 have been produced and 90 kg of unreacted glucose remain in the broth. What are the weight (mass) percents of ethanol and propenoic acid in the broth at the end of the fermentation process? Assume that none of the glucose is metabolized by the microorganisms.

Solution

You can treat this process as an unsteady-state process in a closed system. For component i, Equation (4.11) can be applied

$$n_i^{\text{final}} = n_i^{\text{initial}} + \sum_{j=1}^{R} v_{ij}\xi_j \tag{4.11}$$

The bioorganisms do not have to be included in the solution of the problem because they presumably exist in a small amount and are catalysts for the reaction, not reactants.

Steps 1–4

Figure E4.10 is a schematic of the process.

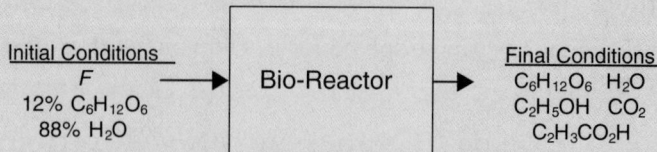

Initial Conditions	Bio-Reactor	Final Conditions
F		$C_6H_{12}O_6$ H_2O
12% $C_6H_{12}O_6$		C_2H_5OH CO_2
88% H_2O		$C_2H_3CO_2H$

Figure E4.10

Step 5

$$\text{Basis: 4000 kg } F$$

Step 4 Again

You should first convert the 4000 kg into moles of H_2O and $C_6H_{12}O_6$ because the reaction equations are based on moles:

$$n_{H_2O}^{initial} = \frac{4000(0.88)}{18.02} = 195.3 \, \text{g mol}$$

$$n_{C_6H_{10}O_6}^{initial} = \frac{4000(0.12)}{180.1} = 2.665 \, \text{g mol}$$

so that $F = 198.01$, or rounded to 198 g mol.

Steps 6 and 7

The degree-of-freedom analysis is as follows (note that units of gram moles have been suppressed):

Number of variables: 9

$$n_{H_2O}^{initial}, n_{C_6H_{12}O_6}^{initial}, n_{H_2O}^{Final}, n_{C_6H_{12}O_6}^{Final}, n_{C_2H_5OH}^{Final}, n_{C_2H_3CO_2H}^{Final}, n_{CO_2}^{Final}, \xi_1, \xi_2$$

Assign values to their respective variables:

From the specifications, you can assign the following values:

$$\left. \begin{array}{l} F = 198 \\ n_{H_2O}^{Initial} = 195.34 \\ n_{C_6H_{12}O_6}^{Initial} = 2.665 \end{array} \right\} \text{only 2 are independent (Why?)}$$

$$n_{C_6H_{12}O_6}^{Final} = \frac{90}{180.1} = 0.500$$

$$n_{CO_2}^{Final} = \frac{120}{44.0} = 2.727$$

Thus, a net of five unknowns exists for this problem.

You can make five species balances:

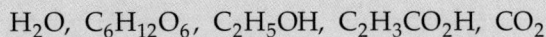

$$H_2O, \; C_6H_{12}O_6, \; C_2H_5OH, \; C_2H_3CO_2H, \; CO_2$$

Therefore, the degrees of freedom are zero.

(Continues)

Example 4.10   Analysis of a Bioreactor (*Continued*)

Step 8

The set of material balance equations, after introducing the known values for the variables, is

$$H_2O: \quad n_{H_2O}^{Final} = 195.3 + (0)\xi_1 + (2)\xi_2 \tag{1}$$

$$C_2H_{12}O_6: \quad 0.500 = 2.665 + (-1)\xi_1 + (-1)\xi_2 \tag{2}$$

$$C_2H_5OH: \quad n_{C_2H_5OH}^{Final} = 0 + 2\xi_1 + (0)\xi_2 \tag{3}$$

$$C_2H_3CO_2H: \quad n_{C_2H_3CO_2H}^{Final} = 0 + (0)\xi_1 + (2)\xi_2 \tag{4}$$

$$CO_2: \quad 2.727 = 0 + (2)\xi_1 + (0)\xi_2 \tag{5}$$

Step 9

Do any of these five equations involve only one unknown? If so, you could solve it in your head or with a calculator and reduce the number of simultaneous equations to be solved by one. In fact, Equation (5) contains only ξ_1 as an unknown. Therefore, you can determine ξ_1 using Equation (5). Then Equation (2) can be used to calculate the value for ξ_2. Finally, Equations (1), (3), and (4) can be applied to determine the remaining unknowns.

With this solution sequence, you get

$$\xi_1 = 1.364 \text{ kg mol reacting} \quad \xi_2 = 0.8015 \text{ kg mol reacting}$$

Species	Results kg mol %	MW	Conversion to Mass Percent kg	mass
H_2O	196.95	18.01	3547.1	89.2
C_2H_5OH	2.727	46.05	125.6	3.2
$C_2H_3CO_2H$	1.603	72.03	115.5	2.9
CO_2	2.727	44.0	120.0	2.5
$C_6H_{12}O_6$	0.500	180.1	90.1	2.3
			3998.3	1.00

Computer Solution:

Using the following definitions for unknowns (i.e., x_1–x_5): x_1 is equal to ξ_1; x_2 is equal to ξ_2; x_3 is equal to $n_{\text{H}_2\text{O}}^{\text{final}}$; x_4 is equal to $n_{\text{C}_2\text{H}_5\text{OH}}^{\text{final}}$; x_5 is equal to $n_{\text{C}_2\text{H}_3\text{CO}_2\text{H}}^{\text{final}}$; the coefficient matrix $\mathbf{A}$ and the constant vector $\mathbf{b}$ are

$$
\mathbf{A} = \begin{bmatrix} 0 & -2 & 1 & 0 & 0 \\ 1 & 1 & 0 & 0 & 0 \\ -2 & 0 & 0 & 1 & 0 \\ 0 & -2 & 0 & 0 & 1 \\ 2 & 0 & 0 & 0 & 0 \end{bmatrix} \quad \mathbf{b} = \begin{bmatrix} 195.3 \\ 2.165 \\ 0 \\ 0 \\ 2.727 \end{bmatrix}
$$

MATLAB:

MATLAB Solution for Example 4.10

```
%%%%%%%%%%%%%%%%%%%%%%%%%%%%%%%%%%%%%%%%%%%%%%%%%%%%%%%%%%%
%
%                    NOMENCLATURE
%
% A - the coefficient matrix for the set of linear equations
% b - the constant vector for the set of linear equations
% x - the unknowns in the set of linear equations
%
%
%%%%%%%%%%%%%%%%%%%%%%%%%%%%%%%%%%%%%%%%%%%%%%%%%%%%%%%%
%                      PROGRAM
function Ex4_10
clear; clc;
A=zeros(5); b(1:5)=0; b=b';      % Insert zeros for the
                                 %   elements of A and b
                                 % Insert elements for A
A=[0,-2,1,0,0;1,1,0,0,0;-2,0,0,0,1;0,-2,0,0,1;2,0,0,0,0];
k=rank(A)                        % Determine rank of A
b=[195.3;2.165;0;0;2.727];       % Insert non-zero values for b
x=A\b;                           % Apply function A\b
fprintf('x1 =%8.4f x2=%8.4f x3 =%8.4f\n',x(1),x(2),x(3))
fprintf('x4 =%8.4f x5=%8.4f k=%2d\n',x(4),x(5),k)
end
%                    PROGRAM END
%%%%%%%%%%%%%%%%%%%%%%%%%%%%%%%%%%%%%%%%%%%%%%%%%%%%%%%%

x1 = 1.3635    x2 =  0.8015    x3 =196.903
X4 = 2.727     x5 =  1.6030    k=5
```

(Continues)

Example 4.10   Analysis of a Bioreactor (*Continued*)

Python:

Python Code for Example 4.10

```
Ex4_10 Linear Eqns.py

####################################################################
                                #
                            NOMENCLATURE
#
#  A - the coefficient matrix for the set of linear equations
#  b - the constant vector for the set of linear equations
#  x - the vector of unknowns in the set of linear equations
#
#
####################################################################
                                #
                             PROGRAM
Import numpy as np
Import numpy.linalg
# Specify the coefficient matrix A
A=np.array([[0,-2 ,1,0,0],[1,1,0,0,0],[-2,0,0,1,0],[0,-2,0,0,1],
[2,0,0 ,0,0]])
k=numpy.linalg.rank(A)                  # Determine the rank of A
b=np.array([195.3, 2.165, 0, 0, 2.727])     # Define the constant vector b
#
# Apply function numpy.linalg.solve for the solution
#
xsol=numpy.linalg.solve(A,b)
                                #
                           PROGRAM END
####################################################################
```

IPython console:

> In[1]: runfile(...
>
> In[2]: xsol, k
>
> Out[2]: array([1.3635, 0.8015, 196.903, 2.727, 1.6030]), 5

Step 10

The total mass of 3998.3 kg is close enough to 4000 kg of feed to validate the results of the calculations.

Self-Assessment Test

Questions

1. Indicate whether the following statements are true or false:
   a. If a chemical reaction occurs, the total masses entering and leaving the system for an open, steady-state process are equal.
   b. In the combustion of carbon, all of the moles of carbon that enter a steady-state, open process exit from the process.
   c. The number of moles of a chemical compound entering a steady-state process in which a reaction occurs with that compound can never equal the number of moles of the same compound leaving the process.

2. List the circumstances for a steady-state process in which the number of moles entering the system equals the number of moles leaving the system.

3. Equation (4.3) can be applied to processes in which a reaction and also no reaction occurs. For what types of balances does the simple relation "the input equals the output" hold for *steady-state, open processes*? Fill in the blanks with yes or no.

Type of Balance	Without Chemical Reaction	With Chemical Reaction
Total balances		
Total mass	[ ]	[ ]
Total moles	[ ]	[ ]
Component balances		
Mass of a compound	[ ]	[ ]
Moles of a compound	[ ]	[ ]

4. Explain how the extent of reaction is related to the fractional conversion of the limiting reactant.

5. Explain why for a process you want to determine the rank of the component matrix.

Answers

1. (a) T; (b) T if you are thinking of C as an element, but F if you are thinking of C as a compound (CO_2); (c) F—several reactions may occur with the effect of no net change in the compound.

2. If the number of moles of the reactants is equal to the number of moles of the products, the total number of moles will not change due to reaction.

3. Total mass: yes, yes; Total moles: yes, no; Mass of a compound: yes, no; Moles of a compound: yes, no

4. The extent of reaction is directly proportional to the fractional conversion of the limiting reactant.

5. Because the rank of the component matrix will indicate the number of independent equations.

Problems

1. Corrosion of pipes in boilers by oxygen can be alleviated through the use of sodium sulfite. Sodium sulfite removes oxygen from boiler feedwater by the following reaction:

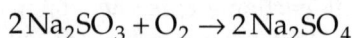

$$2\,Na_2SO_3 + O_2 \rightarrow 2\,Na_2SO_4$$

   How many pounds of sodium sulfite are theoretically required (for complete reaction) to remove the oxygen from 8,330,000 lb of water (10^6 gal) containing 10.0 ppm of dissolved oxygen?

2. Consider a continuous, steady-state process in which the following two reactions take place:

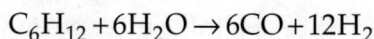

$$C_6H_{12} + 6H_2O \rightarrow 6CO + 12H_2$$
$$C_6H_{12} + H_2 \rightarrow C_6H_{14}$$

   In the process, 250 moles of C_6H_{12} and 800 moles of H_2O are fed into the reactor each hour. The yield of H_2 is 40.0%, and the selectivity of the first reaction compared to the second reaction is 12.0. Calculate the molar flow rates of all five components in the output stream.

Answers

1. 8.33×10^6 lb water $\left|\dfrac{10\ lb\ O_2}{10^6\ lb\ water}\right.\left|\dfrac{lb\ mol\ O_2}{32\ lb\ O_2}\right.\left|\dfrac{2\ lb\ mol\ Na_2SO_3}{1\ lb\ mol\ O_2}\right.$

   $\left|\dfrac{126.1\ lb\ Na_2SO_3}{lb\ mol\ Na_2SO_3}\right. = 656.6$ lb Na_2SO_3

2. $12\,\xi_1 - \xi_2 = 0.4(250) = 100$ mol/h; $\xi_1/\xi_2 = 12$; $\xi_1 = 8.3916$; $\xi_2 = 0.6993$; C_6H_{12}: $250 - 8.3916 - 0.6993 = 240.9$ mol/h; H_2: $12(8.3916) - 0.6993 = 100$ mol/h; H_2O: $800 - 6(8.3916) = 749.65$ mol/h; C_6H_{14}: 0.6993 mol/h; CO: $6(8.3916) = 50.35$ mol/h

4.4   Element Material Balances

In the previous section, you learned how to use species mole balances for reacting systems. Equation (4.7), which includes terms for the generation and consumption of each reacting species, was used for these problems to include the effects of species generation and consumption due to reaction. As you probably know, the elements in a process are conserved regardless of whether reactions are occurring, and consequently you can directly apply Equations (3.1) and (3.2) to the elements. Because elements are neither generated nor consumed for a steady-state process, Equation (3.1) applies to element balances:

$$\text{In} = \text{Out} \tag{4.13}$$

Why not use element balances to solve material balance problems rather than species balances? For most problems, it is easier to apply mole balances, but for some problems, such as problems with complex or unknown reaction equations, element balances are preferred.

You can use element balances, but just make sure that the element balances are independent. Here is an illustration of the issue. Let's use the decomposition of ammonia again as the illustration. In the gas phase, NH_3 can decompose as follows:

$$NH_3 \rightarrow N_2 + 3H_2$$

Figure 4.4 shows some data for the partial decomposition of NH_3.

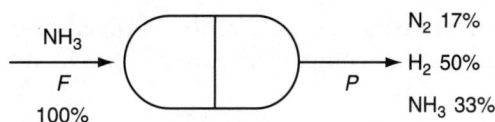

Figure 4.4   Schematic showing a possible case of the decomposition of ammonia. Note: Numbers are rounded.

Two unknowns exist, F and P, and the process involves two elements, N and H. It might appear that in Step 7 of the proposed solution strategy, you could use the two element balances to solve for the values of the unknowns, F and P, but you can't! Try it. The reason is that the two element balances are not independent. As explained in Section 4.3.2, only one of the element balances is independent. Look at the element material balances just below (in moles or mass?) for the decomposition of ammonia, and you will observe that the hydrogen balance is three times the nitrogen balance (if you ignore the rounding of the numbers):

$$\text{N balance:} (1)F = (2)[P(0.17)] + P(0.33) = 0.67\,P$$
$$\text{H balance:} (3)F = (2)[P(0.50)] + (3)[P(0.33)] = 2.0\,P$$

If you add a piece of information, say, by picking a basis of $P = 100$, the degrees of freedom become zero, and then you can solve for F. What happens if you apply species balances to solve the same problem of ammonia decomposition?

Example 4.11   Use of Element Balances to Solve a Hydrocracking Problem

Problem Statement

Hydrocracking is an important refinery process for converting low-valued heavy hydrocarbons to more valuable lower-molecular-weight hydrocarbons by exposing the feed to a zeolite catalyst at high temperature and pressure in the presence of hydrogen. Researchers study the hydrocracking of pure components, such as octane (C_8H_{18}) to understand the behavior of cracking reactions. In one such experiment, the cracked products had the following composition in mole percent: 19.5% C_3H_8, 59.4% C_4H_{10}, and 21.1% C_5H_{12}. You are asked to determine the molar ratio of hydrogen consumed to octane reacted for this experiment.

Solution

We use element balances to solve this problem because the reactions involved in the process are very complex and are not specified here.

Steps 1–4

Figure E4.11 is a sketch of the laboratory hydrocracker reactor together with the data for the streams. The process is assumed to be open, steady state.

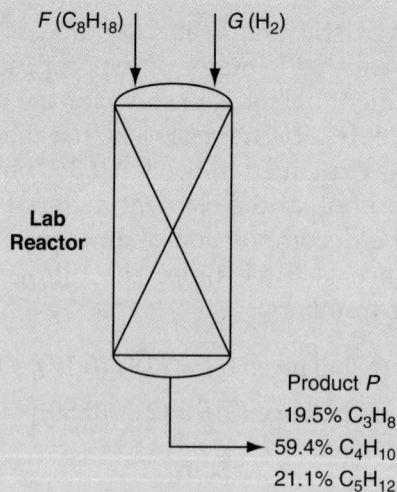

Figure E4.11

Step 5

$$\text{Basis: } P = 100 \text{ g mol}$$

Steps 6 and 7

The degree-of-freedom analysis is as follows:

Variables: 3 F, G, P

Because P is selected as the basis, you can assign $P = 100$ g mol; hence there are two unknowns: G and P.

Equations needed: 2

Element balances: 2   (H, C)

Therefore, this problem has zero degrees of freedom.

Steps 8 and 9

The element balances after introducing the specifications and the basis are (units are in gram moles):

$$\text{C: } F(8) + G(0) = 100[(0.195)(3) + (0.594)(4) + (0.211)(51)]$$

$$\text{H: } F(18) + G(2) = 100[(0.195)(8) + (0.594)(10) + (0.211)(12)]$$

and solving this set of linear equations simultaneously yields

$$F = 50.2 \text{ g mol} \quad G = 49.8 \text{ g mol}$$

The ratio

$$\frac{\text{H}_2 \text{consumed}}{\text{C}_8\text{H}_{18} \text{ reacted}} = \frac{49.8 \text{ g mol}}{50.2 \text{ g mol}} = 0.992$$

Step 10

$$\text{Check: } 50.2 + 49.8 = P = 100 \text{ g mol}$$

Example 4.12   Fusion of BaSO$_4$

Problem Statement

Barite (entirely composed of BaSO$_4$) is fused (reacted in the solid state) in a crucible with coke (composed of 94.0% C and 6.0% ash). The ash does not react. The final fusion mass of 193.7 g is removed from the crucible, analyzed, and found to have the following composition:

Component	%
BaSO$_4$	11.1
Bas	72.8
C	13.9
Ash	2.2
	100

The gas evolved from the crucible does not smell, indicating the absence of sulfur dioxide, and contains 1.13 mole O per 1 mole C. What was the mass ratio of BaSO$_4$ to C (excluding the ash) in the reactants put into the crucible?

Solution

Steps 1–4

The process is a batch process in which the solid components are loaded together, the reaction is carried out, and the fusion mass is removed. Figure E4.12 is a diagram of the process with the known data entered. You want to calculate the mass ratio of $m_{BaSO_4}^{in}$ to m_C^{in}.

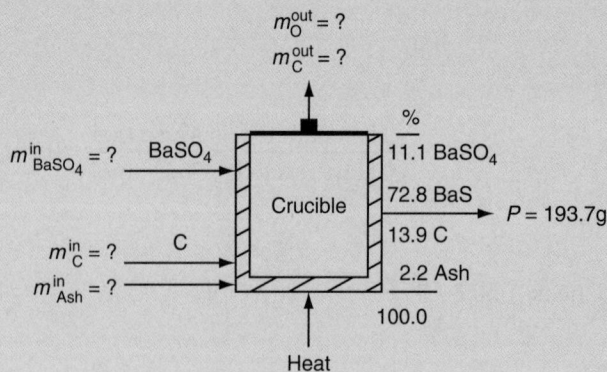

Figure E4.12

Step 5

You could take 193.7 g of P as the basis, but it is slightly easier to take 100 g P as the basis because the analysis of P is given in percent, and you only have to calculate the relative amounts of the inputs of C and BaSO, not the absolute amounts.

Basis: P=100 g

Step 4 Again

Some preliminary calculations of the composition of P will be helpful in the solution.

Compound	g	MW	g mol	Ba	S	O	C	Ash
				Composition in P in g mol				
$BaSO_4$	11.1	233.3	0.0477	0.0477	0.0477	0.19	–	–
BaS	72.8	169.3	0.431	0.430	0.430	–	–	–
C	13.9	12.0	1.16	–	–	–	1.16	–
Ash	2.2	–	–	–	–	–	–	*
Total	100.0			0.479	0.479	0.191	1.16	*
			MW:	137.34	32.064	16.0	12.0	*
			g:	65.8	15.36	3.06	13.9	2.2

Steps 6–8

From Figure E4.12, after assigning known values to variables, it appears that five unknowns exist, ignoring those variables whose values are zero (mass values are in grams below).

Assigned values:

$$P = 100 \quad m_O^P = 3.06 \quad m_{Ba}^P = 65.8 \quad m_C^P = 13.9 \quad m_S^P = 15.36 \quad m_{Ash}^P = 2.2$$

Unknowns:

$$m_{BaSO_4}^{in} \quad m_C^{in} \quad m_{Ash}^{in} \quad m_C^{out} \quad m_O^{out}$$

You can make Ba, S, O, C, and ash balances; hence, the problem seems to have zero degrees of freedom. The process is a batch process; hence, if the crucible

(Continues)

Example 4.12   Fusion of $BaSO_4$ (*Continued*)

is empty at the beginning and end of the fusion, the material balance equation to use is "What goes in must come out." Here are the element mass balances:

$$\text{Ash:} \quad m_{\text{Ash}}^{\text{in}} = 2.2 \text{ g}$$

In   Out

$$\text{Ba:} \frac{m_{\text{BaSO}_4\text{g}}^{\text{in}}}{} \left| \frac{1\,\text{g mol BaSO}_4}{233.3\,\text{g BaSO}_4} \right| \frac{1\,\text{g mol Ba}}{1\,\text{g mol BaSO}_4} \left| \frac{137.34\,\text{g Ba}}{1\,\text{g mol Ba}} \right. \quad = 65.8\,\text{g}$$

$$m_{\text{BaSO}_4}^{\text{in}} = 111.7\,\text{g}$$

In

$$\text{S:} \frac{m_{\text{BaSO}_4\text{g}}^{\text{in}}}{} \left| \frac{1\,\text{g mol BaSO}_4}{233.3\,\text{g BaSO}_4} \right| \frac{1\,\text{g mol S}}{1\,\text{g mol BaSO}_4} \left| \frac{32.064\,\text{g S}}{1\,\text{g mol S}} \right. \quad = 15.36$$

$$m_{\text{BaSO}_4}^{\text{in}} = 111.7\,\text{g, a redundant result}$$

In   Out

$$\text{O:} \frac{m_{\text{BaSO}_4\text{g}}^{\text{in}}}{} \left| \frac{1\,\text{g mol BaSO}_4}{233.3\,\text{g BaSO}_4} \right| \frac{4\,\text{g mol O}}{1\,\text{g mol BaSO}_4} \left| \frac{16.0\,\text{g 0}}{1\,\text{g mol 0}} \right. \quad = m_{\text{O}}^{\text{out}} + 3.06$$

Introduce $m_{\text{BaSO}_4}^{\text{in}} = 111.7$ into the O balance to get $m_{\text{O}}^{\text{out}} + 3.06$. The only balance left is the carbon balance.

Step 9

Because the Ba and S balances are not independent, we need one more piece of information, namely, that $m_{\text{C}}^{\text{out}} = m_{\text{O}}^{\text{out}} / 1.13 = 27.56 / 1.13 = 24.4$, which is needed in the carbon balance.

$$m_{\text{C}}^{\text{in}} = 24.4 + 13.92 = 38.3\,\text{g}$$

$$\frac{m_{\text{BaSO}_4}^{\text{in}}}{m_{\text{C}}^{\text{in}}} = \frac{111.7}{38.3} = 2.92$$

Step 10

Check using the redundant total material balance:

$$m_{BaSO_4}^{in} + m_C^{in} = m_O^{out} + m_C^{out} + P$$

$$111.7 + 38.3 \overset{?}{=} 27.56 + 24.4 + 100.0$$

$$150 \cong 152 \quad \text{close enough}$$

Element balances are especially useful when you do not know what reactions occur in a process and you only know information about the input and output stream components.

Self-Assessment Test

Questions

1. Do you have to write element material balances with the units of each term being moles rather than mass? Explain your answer.

2. Will the degrees of freedom be smaller or larger using element balances in place of species balances?

3. How can you determine whether a set of element balances are independent?

4. Can the number of independent element balances ever be larger than the number of species balances in a problem?

Answers

1. It makes no difference whether element balances are applied in terms of mass or moles.

2. The degree-of-freedom analysis should be the same if the problem is properly formulated.

3. (a) Get the rank of the coefficient matrix; (b) apply MATLAB or Python to solve the equations; (c) examine the equations for redundancy.

4. No

Problems

1. Consider a system used in the manufacture of electronic materials (all gases except Si):

$$SiH_4, Si_2, H_6, SiH_2, H_2, Si$$

How many independent element balances can you make for this system?

2. Methane burns with O_2 to produce a gaseous product that contains CH_4, O_2, CO_2, CO, H_2O, and H_2. How many independent element balances can you write for this system?

3. In the reaction of $KClO_3$ with HCl, the following products were measured: KCl, ClO_2, Cl_2, and H_2O. How many element material balances can you make for this system?

Answers

1. Two
2. Three
3. Four

4.5   Material Balances for Combustion Systems

In this section, we consider combustion as a special topic involving material balances that include chemical reactions. Combustion is, in general, the reaction of oxygen with materials containing hydrogen, carbon, and sulfur, and the associated release of heat and the generation of product gases such as H_2O, CO_2, CO, and SO_2. Typical examples of combustion are the combustion of coal, heating oil, and natural gas used to generate electricity in utility power stations, and engines that operate using the combustion of gasoline or diesel fuel. More complicated oxidation processes take place in the human body but are not called combustion. Most combustion processes use air as the source of oxygen. For our purposes, you can assume that air contains 79% N_2 and 21% O_2, neglecting the other components that amount to a total of less than 1.0%, and that air has an average molecular weight of 29. Although a small amount of N_2 oxidizes to NO and NO_2, gases called NO_x, a pollutant, the amount is so small that we treat N_2 as a nonreacting component of air and fuel. Figure 4.5 shows how the CO, unburned hydrocarbons, and NO_x vary with the air-fuel ratio in combustion.

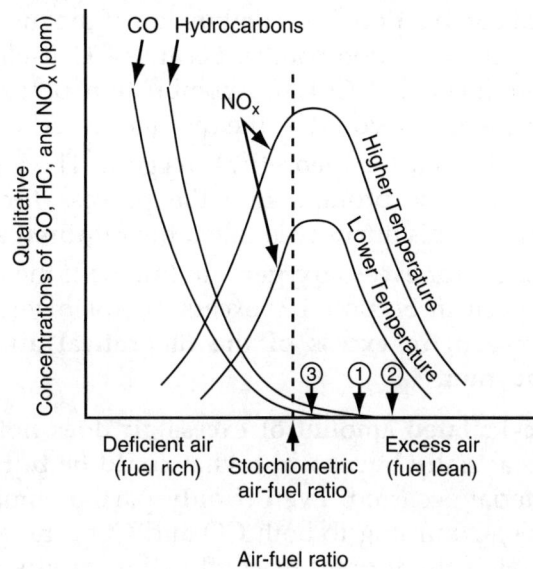

Figure 4.5   Pollutants resulting from combustion of natural gas vary with the air-fuel ratio and the temperature of combustion. The amounts of hydrocarbons and CO increases with deficient air. Efficiency goes down with too much excess air, but so does the NO_x. (1) Represents the optimal air-fuel ratio (no CO or unburned hydrocarbons). (2) Represents inefficient combustion (excess air has to be heated). (3) Represents unsatisfactory combustion (because not all of the fuel is burned and the unburned products are released to the atmosphere).

You should become acquainted with some of the special terms associated with combustion:

- **Flue or stack gas:** All of the gases resulting from a combustion process, including the water vapor, sometimes known as a **wet basis.**

- **Orsat analysis, or dry basis:** All of the gases resulting from a combustion process *not including the water vapor.* (Orsat analysis refers to a type of gas analysis apparatus in which the volumes of the respective gases are measured over and in equilibrium with water; hence, each component is saturated with water vapor. The net result of the analysis is to eliminate water as a component that is measured.) See Figure 4.6. To convert from one analysis to another, you have to adjust the percentages of the components to the desired basis, as explained in Chapter 2.

- **Complete combustion:** The complete reaction of the fuel producing CO_2 and H_2O.

- **Partial combustion:** The combustion of the fuel producing at least some CO from the carbon source. Because CO itself can react with oxygen, the production of CO in a combustion process does not produce as much energy as would be the case if only CO_2 were produced.
- **Theoretical air** (or **theoretical oxygen**): The amount of air (or oxygen) required to be brought into the process **for complete combustion.** Sometimes this quantity is called the **required air** (or **oxygen**).
- **Excess air** (or **excess oxygen**): In line with the definition of excess reactant given in Section 4.2, excess air (or oxygen) is the amount of air (or oxygen) **in excess of the theoretical air required for complete combustion.**

The calculated amount of excess air does not depend on how much material is actually burned but what could be burned if complete reaction of the material occurred. **Even if only partial combustion takes place, as,** for example, C burning to both CO and CO_2, **the excess air (or oxygen) is computed as if the process of combustion produced only CO_2.** A wooden chair sitting in a room will not burn at normal temperatures, but nevertheless, if the volume or mass of the chair is known, the excess air in the room for the combustion of the chair can be calculated.

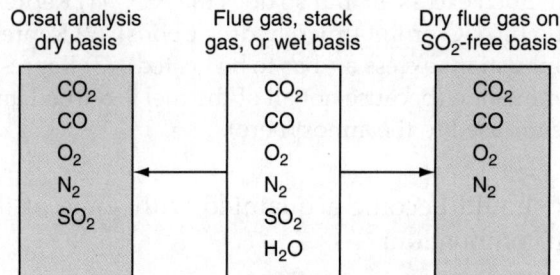

Orsat analysis dry basis	Flue gas, stack gas, or wet basis	Dry flue gas on SO_2-free basis
CO_2	CO_2	CO_2
CO	CO	CO
O_2	O_2	O_2
N_2	N_2	N_2
SO_2	SO_2	
	H_2O	

Figure 4.6   Comparison of gas analysis on different bases

The percent excess air is identical to the percent excess O_2 (a quantity often more convenient to use in calculations):

$$\% \text{ excess air} = \frac{\text{excess air}}{\text{required air}} = 100\frac{\text{excess } O_2/0.21}{\text{required } O_2/0.21} = 100\frac{\text{excess } O_2}{\text{required } O_2} \quad (4.14)$$

Note that the ratio $1/0.21$ of air to O_2 cancels out in Equation (4.14). Percent excess air may also be computed as

$$\% \text{ excess air} = 100\frac{O_2 \text{ entering process} - O_2 \text{ required}}{O_2 \text{ required}} \qquad (4.15)$$

or

$$\% \text{ excess air} = 100\frac{\text{excess } O_2}{O_2 \text{ entering} - \text{excess } O_2} \qquad (4.16)$$

In calculating the degrees of freedom in a problem, if the percent excess air is specified (and the chemical reaction equation for the process is known), you can calculate how much air enters with the fuel; hence, the specification of the amount of excess air can be used to assign a value to one of the variables like other specifications.

Now, let us apply these definitions with some examples.

Example 4.13   Calculation of Excess Air

Problem Statement

Fuels for motor vehicles other than gasoline are being evaluated because they generate lower levels of pollutants than does gasoline. Compressed propane has been suggested as a source of power for vehicles. Suppose that in a test, 20 kg of C_3H_8 is burned with 400 kg of air to produce 44 kg of CO_2 and 12 kg of CO. What was the percent excess air?

Solution

This is a problem involving the following reaction (is the reaction equation correctly balanced?):

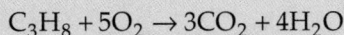

$$C_3H_8 + 5O_2 \rightarrow 3CO_2 + 4H_2O$$

$$\text{Basis: 20 kg of } C_3H_8$$

Since the percentage of excess air is based on the *complete combustion* of C_3H_8 to CO_2 and H_2O, the fact that combustion is not complete has no influence on the calculation of excess air. The required O_2 on the basis of 20 kg of C_3H_8 is

$$\frac{20 \text{ kg } C_3H_8}{} \left| \frac{1 \text{ kg mol } C_3H_8}{44.09 \text{ kg } C_3H_8} \right| \frac{5 \text{ kg mol } O_2}{1 \text{ kg mol } C_3H_8} = 2.27 \text{ kg mol } O_2$$

(*Continues*)

Example 4.13   Calculation of Excess Air (*Continued*)

The entering O_2 is

$$\frac{400 \text{ kg air}}{} \left| \frac{1 \text{ kg mol air}}{29 \text{ kg air}} \right| \frac{21 \text{ kg mol O}_2}{100 \text{ kg mol air}} = 2.90 \text{ kg mol O}_2$$

The percentage excess air is

$$100 \, \frac{\text{excess O}_2}{\text{required O}_2} = 100 \, \frac{\text{entering O}_2 - \text{required O}_2}{\text{required O}_2}$$

$$\% \text{ excess air} = \frac{2.90 \text{ kg mol O}_2 - 2.27 \text{ kg mol O}_2}{2.27 \text{ kg mol O}_2} \left| \frac{100}{} \right. = 28\%$$

In calculating the amount of excess air, remember that the excess is the amount of air that enters the combustion process over and above that required for complete combustion. Suppose there is some oxygen in the material being burned. For example, suppose that a gas containing 80% C_2H_6 and 20% O_2 is burned in an engine with 200% excess air. Eighty percent of the ethane goes to CO_2, 10% goes to CO, and 10% remains unburned. What is the amount of the excess air per 100 moles of the C_2H_6? First, you can ignore the information about the CO and the unburned ethane because the basis of the calculation of excess air is *complete combustion* of the C_2H_6. Specifically, the products of reaction are assumed to be the highest oxidation state. For example, C goes to CO_2, S to SO_2, H to H_2O, CO to CO_2, and so on.

Second, the oxygen in the fuel cannot be ignored. Based on the reaction

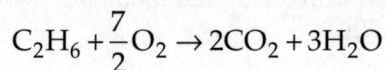

$$C_2H_6 + \frac{7}{2}O_2 \rightarrow 2CO_2 + 3H_2O$$

80 moles of C_2H_6 require C_2H_6 of 3.5(80) = 280 moles of O_2 for complete combustion. However, the gas contains 20 moles of O_2, so only 280 − 20 = 260 moles of O_2 are needed in the entering air for complete combustion. Thus, 260 moles of O_2 is the required O_2, and the calculation of the 200% excess O_2 (air) is based on 260, not 280, moles of O_2:

Entering with air	Moles O_2
Required O_2	260
Excess O_2 (2.00)(260)	520
Total O_2	780

Example 4.14   A Fuel Cell to Generate Electricity from Methane

Problem Statement

"A Fuel Cell in Every Car" is the headline of an article in *Chemical and Engineering News* (March 5, 2001, p. 19). In essence, a fuel cell is an open system into which fuel and air are fed, and out of which come electricity and waste products. Figure E4.14 is a sketch of a fuel cell in which a continuous flow of methane (CH_4) and air (O_2 plus N_2) produces electricity plus CO_2 and H_2O. Special membranes and catalysts are needed to promote the oxidation of the CH_4.

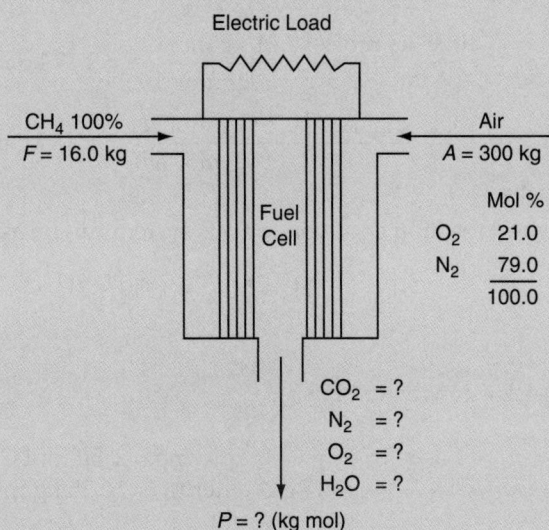

Figure E4.14

Based on the data given in Figure E4.14, you are asked to calculate the composition of the products in P.

Solution

Steps 1–4

This is a steady-state process with reaction. Can you assume that a complete reaction occurs? Yes. How? No CH_4 or CO appears in P. The system is the fuel cell (open, steady state). Because the process output is a gas, the composition will be in mole fractions (or moles); hence, it is more convenient to use kilogram moles rather than mass in this problem even though the quantities of CH_4 and air are

(Continues)

Example 4.14   A Fuel Cell to Generate Electricity from Methane (*Continued*)

stated in kilograms. You can carry out the necessary conversions from kilograms to kilogram moles as follows:

$$\frac{300 \text{ kg } A}{} \left| \frac{1 \text{ kg mol } A}{29.0 \text{ kg } A} \right. = 10.35 \text{ kg mol } A \text{ in}$$

$$\frac{16.0 \text{ kg CH}_4}{} \left| \frac{1 \text{ kg mol CH}_4}{16.0 \text{ kg CH}_4} \right. = 1.00 \text{ kg mol CH}_4 \text{ in}$$

$$\frac{10.35 \text{ kg mol } A}{} \left| \frac{0.21 \text{ kg mol O}_2}{1 \text{ kg mol } A} \right. = 2.17 \text{ kg mol O}_2 \text{ in } A$$

$$\frac{10.35 \text{ kg mol } A}{} \left| \frac{0.79 \text{ kg mol N}_2}{1 \text{ kg mol } A} \right. = 8.18 \text{ kg mol N}_2 \text{ in } A$$

The chemical reaction equation for this system can be assumed to be

$$CH_4 + 2O_2 \rightarrow CO_2 + 2H_2O$$

Step 5

We will pick a convenient basis.

Basis: 16.0 kg CH$_4$ entering = 1 kg mol CH$_4$ plus
300 kg A entering = 10.35 kg mol of air

Steps 6 and 7

The degree-of-freedom analysis is as follows:

Variables: 10 F, P, A, $n_{CO_2}^{P}$, $n_{N_2}^{P}$, $n_{O_2}^{P}$, $n_{H_2O}^{P}$, $n_{O_2}^{A}$, $n_{N_2}^{A}$ plus ξ

Given the basis and the quantities calculated above, four of these variables $\left(F, A, n_{O_2}^{A}, n_{N_2}^{A} \right)$ can be assigned values. Therefore, there are six unknowns remaining.

Equations: 6

Five (5) independent species balances: $CH_4, O_2, N_2, CO_2, H_2O$

One (1) independent implicit equation for P: $\sum n_i^{P} = P$

Therefore, the degrees of freedom are zero.

Step 8

The species mole balances are as follows:

Compound	Out		In		$v_i\xi$		g mol
CH_4	$n^P_{CH_4}$	=	1.0	−	ξ	=	0
O_2	$n^P_{O_2}$	=	2.17	−	2ξ	=	0.17
N_2	$n^P_{N_2}$	=	8.18	−	$0(\xi)$	=	8.18
CO_2	$n^P_{CO_2}$	=	0	+	ξ	=	1.0
H_2O	$n^P_{H_2O}$	=	0	+	2ξ	=	2.0

Step 9

The solution of this set of equations gives

$$n^P_{CH_4} = 0, \ n^P_{O_2} = 0.17, \ n^P_{N_2} = 8.18, \ n^P_{CO_2} = 1.0,$$
$$n^P_{CO_2} = 1.0, \ n^P_{H_2O} = 2.0, \ P = 11.35$$

and the mole percentage composition of P is

$$y_{O_2} = 1.5\%, \qquad y_{N_2} = 72.1\%, \qquad y_{CO_2} = 8.8\%, \text{ and} \qquad y_{H_2O} = 17.6\%$$

You could also use element balances without knowing the reaction to get the same solution using four element balances and one implicit equation for P (ξ would no longer be a variable).

Step 10

You can check the answer by determining the total mass of the exit gas and comparing it to total mass in (316 kg), but we will omit this step here to save space.

Example 4.15   Combustion of Coal

Problem Statement

A local utility burns coal having the following composition on a dry basis. (Note that the coal analysis below is a convenient one for our calculations but is not necessarily the only type of analysis that is reported for coal. Some analyses contain much less information about each element.)

Component	Percent
C	83.05
H	4.45
O	3.36
N	1.08
S	0.70
Ash	7.36
Total	100.0

The average Orsat analysis of the gas from the stack during a 24 hr test was

Component	Percent
$CO_2 + SO_2$	15.4
CO	0.0
O_2	4.0
N_2	80.6
Total	100.0

Moisture (H_2O) in the fuel was 3.90%, and the air on the average contained 0.0048 lb H_2O/lb dry air. The refuse showed 14.0% unburned coal, with the remainder being ash. The unburned coal in the refuse can be assumed to be of the same composition as the coal that serves as fuel.

What is the percent excess air used for this process as shown in Figure E4.15?

Solution

Note that for this problem you are asked only for the percent excess air used. Therefore, Figure E4.15 contains more information than is required to solve this problem.

Steps 1–4

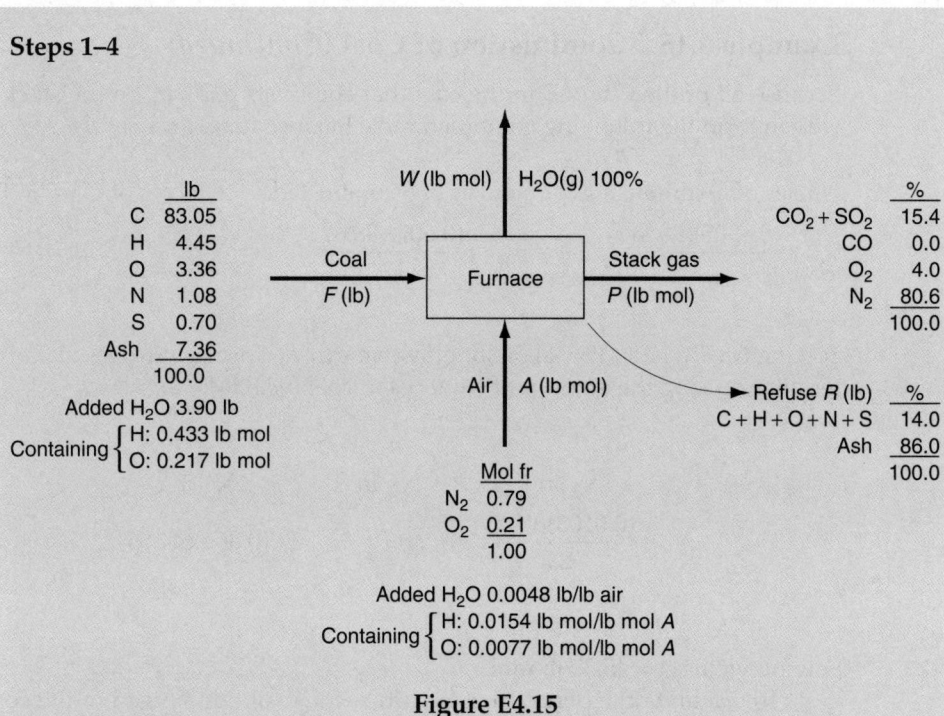

	lb
C	83.05
H	4.45
O	3.36
N	1.08
S	0.70
Ash	7.36
	100.0

Added H_2O 3.90 lb

Containing $\begin{cases} \text{H: 0.433 lb mol} \\ \text{O: 0.217 lb mol} \end{cases}$

Coal
F (lb)

W (lb mol) H_2O(g) 100%

Furnace

Stack gas
P (lb mol)

	%
$CO_2 + SO_2$	15.4
CO	0.0
O_2	4.0
N_2	80.6
	100.0

Air A (lb mol)

	Mol fr
N_2	0.79
O_2	0.21
	1.00

Refuse R (lb)

	%
C + H + O + N + S	14.0
Ash	86.0
	100.0

Added H_2O 0.0048 lb/lb air

Containing $\begin{cases} \text{H: 0.0154 lb mol/lb mol } A \\ \text{O: 0.0077 lb mol/lb mol } A \end{cases}$

Figure E4.15

Step 5

Pick a basis of $F = 100$ lb as convenient.

Steps 6–9

We need to solve for the feed rate of air (A), but to do this, we first need to solve for the unknown flow rates: R and P. Note that ash is a tie component. Applying a material balance for ash yields

$$7.36 = 0.86R \Rightarrow R = 8.56 \text{ lb}$$

Note that a portion of the coal does not react and leaves the process in R. We will assume here that the portion of R that is not ash has the same composition as F. Therefore, we simply subtract this from F to determine the amount of coal that is combusted:

$$\text{Amount of coal combusted} = 100. - 0.14R = 98.8 \text{ lb}$$

Continues)

Example 4.15   Combustion of Coal (*Continued*)

Because all of the C and S in the coal that combusts ends up in the stack gas (P), we can form the following combined mole balance for C and S:

moles of C combusted   moles of S combusted   moles of C + S in flue gas

$$\frac{(0.8305)(98.8)}{12} \quad + \quad \frac{(0.007)(98.8)}{32} \quad = \quad 0.154P$$

Solving for P yields $P = 44.54$ lb. Now we can perform a nitrogen balance to determine A using the amount of combusted coal (98.8 lb):

$\quad\quad$ N$_2$ in F $\quad\quad\quad$ N$_2$ in A $\quad\quad\quad$ N$_2$ in P

$$\frac{(0.0108)(98.8)}{28} \quad + \quad 0.79\,A \quad = \quad (0.806)(44.54)$$

Solving yields $A = 45.39$ lb mol.

$\quad$ To calculate the percent excess air, because of the oxygen in the coal available for combustion and the existence of unburned combustibles including O, we use the total oxygen in and the required oxygen as shown previously in Equation (4.15):

$$\% \text{ excess air} = 100\left(\frac{O_2 \text{ entering} - O_2 \text{ required}}{O_2 \text{ required}}\right)$$

The required O_2 is equal to the stoichiometric requirements for complete combustion of C, H, and S minus the O_2 present in the coal (not based on what actually combusted):

Component	Reaction	lb	lb mol	Required O_2 (lb mol)
C	$C + O_2 \rightarrow CO_2$	83.05	6.921	6.921
H	$H + \frac{1}{2}O_2 \rightarrow H_2O$	4.45	4.415	1.104
O	–	3.36	0.210	(0.105)
N	–	–	–	–
S	$S + O_2 \rightarrow SO_2$	0.70	0.022	0.022
Total				7.942

The oxygen in the entering air is $(45.39)(0.21) = 9.532$ lb mol. Therefore,

$$\% \text{ excess air} = 100\left(\frac{9.532 - 7.942}{7.942}\right) = 20.0\%$$

If you (incorrectly) calculated the percent excess air from the wet stack gas analysis alone, you would have ignored the oxygen in the coal.

From the viewpoint of the concern about the increase of the CO_2 concentration in the atmosphere, would the CH_4 in Example 4.14 or the coal in Example 4.15 contribute more CO_2 per kilogram of fuel? The H/C ratio in moles in CH_4 was 4/1, whereas in the coal it was $0.0537/0.0897 = 0.60$. Thus, the coal results in the larger emission of CO_2 per unit of energy produced. Figure 4.7 shows how the H/C ratio varies with various types of fuel.

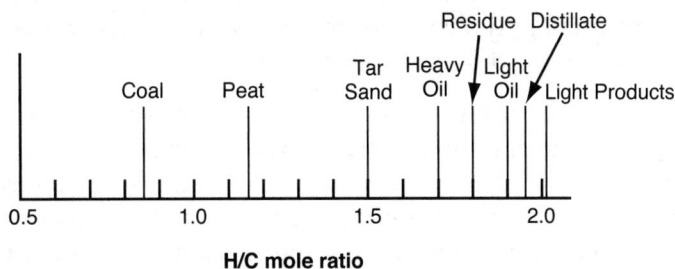

Figure 4.7   Variation of the H/C ratio in selected fuels

Oxygen use in the growth of biomass does not usually focus on the assumption of complete conversion of the reactants to CO_2 and H_2O. The chemical reaction equation includes the substrate (in solution or suspension)

as a reactant and a product along with the CO_2 and H_2O, and rarely is all of the reactant(s) used up by the dissolved O_2 (acronym DO). DO is normally considered in excess as long as it is greater than 20% of saturation and can be controlled by the agitation rate, air injection rate, or the injection rate of pure oxygen.

If you write the chemical reaction equation for the growth of biomass and possible by-products as follows, on the basis of 1 mole of C in the substrate, you would get Equation (4.17):

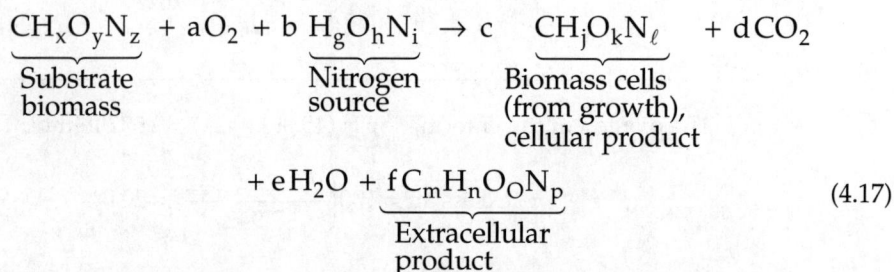

$$\underbrace{CH_xO_yN_z}_{\substack{\text{Substrate}\\\text{biomass}}} + aO_2 + b\underbrace{H_gO_hN_i}_{\substack{\text{Nitrogen}\\\text{source}}} \rightarrow c\underbrace{CH_jO_kN_\ell}_{\substack{\text{Biomass cells}\\\text{(from growth),}\\\text{cellular product}}} + dCO_2$$

$$+ eH_2O + \underbrace{fC_mH_nO_ON_p}_{\substack{\text{Extracellular}\\\text{product}}} \tag{4.17}$$

- **Substrate:** The primary carbon source nutrient (e.g., glucose) used by the cell mass as it grows and then produces product.
- **Nitrogen source:** The source of nitrogen, such as ammonia, ammonium, urea, nitrate, secondary or treatment wastes, for biomass growth.
- **Cellular product:** The biomass produced in the reaction(s). Growth takes place throughout the biomass, not just on the surface. A product may occur jointly with biomass growth that probably will have to be separated (at some cost) from the biomass.
- **Extracellular product:** Various metabolites, such as acetate, citrate, formulate, glycerol, pyramate, and so on, as well as the CO_2 and H_2O. Another product produced jointly.

The initial phase of a bioreactor, which usually takes about one-third of the total reaction time, normally involves exponential growth. After the initial phase, the reaction rate becomes limited usually by the availability of the substrate but can be limited by the availability of other nutrients.

For the case of a hydrocarbon fuel with components of C and H reacting with O_2 to produce only CO_2 and H_2O (i.e., $z = 0$; e.g., glucose), the coefficients b, c, and f are zero in Equation (4.17), leaving only a, d, and e, given the assumption of complete combustion. You can make three element balances, and the degrees of freedom are zero. But if all six terms and associated coefficients are involved in the reaction, you can make only four element balances (C, H, O, N), leaving two degrees of freedom to be specified. Presumably you know the chemical composition of each species in the reaction or additional specifications or assumptions have to be made. Thus, the molecular weights (MW) of the substrate and biomass product are usually known, and if the composition of the extracellular product is not known, you can use as an estimate $CH_{1.8}O_{0.5}N_{0.2}$ [J. A. Roels, *Biotechnology Bioengineering*, 22, 2457(1980)]. A cell contains elements other than C, H, O, and N, but the minor components such as P, S, K, Ca, Mg, Cl, and Fe require so little O_2 that they are often treated as ash, a nonreacting component.

With these assumptions, the values of c and f still have to be obtained by measurements:

$$c = \frac{\dfrac{\text{g biomass produced}}{\text{MW biomass}}}{\dfrac{\text{g substrate consumed}}{\text{MW substrate}}}$$

$$f = \frac{\dfrac{\text{g extracellular product produced}}{\text{MW of extracellular product}}}{\dfrac{\text{g substrate consumed}}{\text{MW substrate}}}$$

It is also possible to use the respiratory quotient (RQ):

$$RQ = \frac{\text{mol } CO_2 \text{ produced}}{\text{mol } O_2 \text{ consumed}} = \frac{d}{a}$$

although the RQ can be noisy in certain situations due to intermittent feeding of O_2 or other feedstocks.

As a result of the positive degrees of freedom, a number of *empirical* measures of O_2 usage in biosystems are employed to measure what is called **oxygen demand.** The definitions of these measures vary with what you read in books or on the Internet, but they are precisely defined by standard tests that are carried out or with instruments that serve as substitutes for the standard tests.

- **Total oxygen demand (TOD):** The quantity of oxygen required to completely oxidize all of the organic and inorganic compounds present in a sample (in water) as determined by a COD test (using a strong oxidant such as sulfuric acid or potassium dichromate) or combustion. The amount of O_2 is reported in milligrams O_2 per liter containing the sample or as grams O_2 per gram sample.
- **Chemical oxygen demand (COD):** The same as TOD except only the organic components are considered. Sometimes the COD is defined to

be the same as TOD. The COD value is higher than the BOD value (defined next) because the COD includes slowly biodegradable and recalcitrant organic compounds not degraded by microorganisms as in a BOD test. The value of the COD is reported in units of grams O_2 per gram sample.

- **Biochemical oxygen demand (BOD):** The quantity of oxygen required by microorganisms to oxidize the *organic compounds* in a sample (in water) as determined by a BOD test. The test is carried out for a number of days at 20°C. Five days is the most common duration. The quantity of O_2 used is determined by the difference in the dissolved O_2 (DO) in the water at the beginning of the test and at the end of five days. BOD is used mainly in evaluating water and wastewater quality. The value of the BOD5 (BOD_5) is reported in grams O_2 per gram sample.

- **Theoretical oxygen demand (ThOD):** The quantity of O_2 required according to a valid balanced chemical reaction equation (Equation 4.17) to oxidize the reactant(s) to CO_2, H_2O, and the highest oxidation state of other products, ignoring any extracellular products from a sample (f = 0). The ThOD is the same as the theoretical O_2 discussed previously. The highest stage of nitrogen compounds is nitrate. For example, in the oxidation of glycine, $CH_2(NH_3)COOH$, the overall reaction equation is

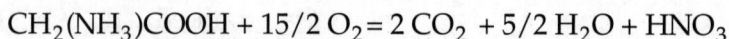

$$CH_2(NH_3)COOH + 15/2\, O_2 = 2\, CO_2 + 5/2\, H_2O + HNO_3$$

and the theoretical O_2 is 15/2 mol.

Self-Assessment Test

Questions

1. Explain the difference for a gas between a flue gas analysis and an Orsat analysis; wet basis and dry basis.

2. What does an SO_2-free basis mean?

3. Write down the equation relating percent excess air to the required air and entering air.

4. Will the percent excess air always be the same as the percent excess oxygen in combustion (by oxygen)?

5. In a combustion process in which a specified percentage of excess air is used, and in which CO is one of the products of combustion, will the analysis of the resulting exit gases contain more or less oxygen than if all the carbon had burned to CO_2?

6. Indicate whether the following statements are true or false:
   a. Excess air for combustion is calculated using the assumption of complete reaction whether or not a reaction takes place.
   b. For the typical combustion process, the products are CO_2 gas and H_2O vapor.
   c. In combustion processes, since any oxygen in the coal or fuel oil is inert, it can be ignored in the combustion calculations.
   d. The concentration of N_2 in a flue gas is usually obtained by direct measurement.

Answers

1. The Orsat analysis (dry analysis) does not include water vapor in the analysis while the flue gas does.

2. SO_2 is not included in the analysis.

3. % excess air $= \dfrac{\text{entering air - required air}}{\text{required air}}$

4. Yes

5. More

6. (a) T; (b) T; (c) F; (d) F

Problems

1. Pure carbon is burned in oxygen. The flue gas analysis is

CO_2	75 mol %
CO	14 mol %
O_2	11 mol %

   What was the percent excess oxygen used?

2. Toluene, C_7H_8, is burned with 30% excess air. A bad burner causes 15% of the carbon to form soot (pure C) deposited on the walls of the furnace. What is the Orsat analysis of the gases leaving the furnace?

3. A synthesis gas analyzing CO_2 6.4%, O_2 0.2%, CO 40.0%, and H_2 50.8% (the balance is N_2) is burned with excess dry air. The problem is to determine the composition of the flue gas. How many degrees of freedom exist in this problem; that is, how many additional variables have to have their values specified?

4. A hydrocarbon fuel is burned with excess air. The Orsat analysis of the flue gas shows 10.2% CO_2, 1.0% CO, 8.4% O_2, and 80.4% N_2. What is the atomic ratio of H to C in the fuel?

Answers

1. Basis: 100 mol. Required $O_2 = 75 + 14 = 89$ mol; Excess $O_2 = 11 - 14/2 = 4$ mol; % excess $= 4/89 = 4.49\%$.

2. Basis: 1 mol C_7H_8. $C_7H_8 + 9O_2 \rightarrow 7CO_2 + 4H_2O$; $\xi = 0.85$ mol; $(O_2)_{in} = 9(1.3) = 11.7$; $N_2 = 11.7(0.79)/0.21 = 44.01$; O_2 balance: $11.7 - 9(.85) = 4.05$; CO_2 balance: $7(0.85) = 5.95$; mols minus water $= 54.01$; 7.50% O_2; 11.02% CO_2; 81.48% N_2

3. One (e.g., the percent excess air)

4. Basis: 100 mol; Inlet $O_2 = 80.4(0.21)/0.79 = 21.37$; O_2 that reacted to form H_2O: $21.37 - 10.2 - 0.5 - 8.4 = 2.27$. For every mole of O_2, 4 H's. Moles of H $= 9.08$; mole of C $= 10.2 + 1 = 11.2$; therefore, $H/C = 9.08/11.2 = 0.812$.

Summary

In this chapter, we explained how the chemical reaction equations can be used to calculate quantitative relations among reactants and products. We also defined a number of terms used by engineers in making calculations involving chemical reactions. We applied Equation (4.11) and its analogs to processes involving reactions. If you make element balances, the generation and consumption terms are zero so that you can use Equations (3.1) and (3.2). If you make species balances, the accumulation and consumption terms are not zero, and you have to use the extents of reactions. You simply apply the general material balance with reaction to these systems, recognizing their characteristics.

Glossary

conversion   The fraction of the feed or some *key* material in the feed that is converted into products.

degree of completion   The percent or fraction of the limiting reactant converted into products.

excess reactant   All reactants other than limiting reactants.

extent of reaction   The mole of reactions that occur according to the chemical reaction equation.

limiting reactant   The species in a chemical reaction that would theoretically run out first (would be completely consumed) if the reaction were to proceed to completion according to the chemical equation, even if the reaction did not take place.

required air   The amount of air (or oxygen) required to be brought into the process for complete combustion. Sometimes this quantity is called theoretical air (or theoretical oxygen).

selectivity The ratio of the moles of a particular (usually the desired) product produced to the moles of another (usually undesired or by-product) product produced in a set of reactions.

stoichiometry Concerns calculations about the moles and masses of reactants and products involved in a chemical reaction(s).

stoichiometric coefficient Indicates the relative amounts of moles of chemical species that react and are produced in a chemical reaction.

stoichiometric quantity Quantity of material based on the stoichiometric ratio.

stoichiometric ratio Mole ratio obtained by using the coefficients of the species in the chemical equation, including both reactants and products.

theoretical air The amount of air (or oxygen) required to be brought into the process for complete combustion. Sometimes this quantity is called required air (or required oxygen).

yield Based on feed: The amount (mass or moles) of desired product obtained divided by amount of the key (frequently the limiting) reactant fed.

Based on reactant consumed: The amount (mass or moles) of desired product obtained divided by the amount of the key (frequently the limiting) reactant consumed.

Based on theory: The amount (mass or moles) of a product obtained divided by the theoretical (expected) amount of the product that would be obtained based on the limiting reactant in the chemical reaction equation(s) being completely consumed.

Problems

Section 4.1   Stoichiometry

*4.1.1 $BaCl_2 + Na_2SO_4 \rightarrow BaSO_4 + 2NaCl$

    a. How many grams of barium chloride will be required to react with 5.00 g of sodium sulfate?

    b. How many grams of barium chloride are required for the precipitation of 5.00 g of barium sulfate?

    c. How many grams of barium chloride are needed to produce 5.00 g of sodium chloride?

    d. How many grams of sodium sulfate are necessary to react with 5.00 g of barium chloride?

    e. How many grams of sodium sulfate have been added to barium chloride if 5.00 g of barium sulfate is precipitated?

    f. How many pounds of sodium sulfate are equivalent to 5.00 lb of sodium chloride?

    g. How many pounds of barium sulfate are precipitated by 5.00 lb of barium chloride?

    h. How many pounds of barium sulfate are precipitated by 5.00 lb of sodium sulfate?

    i. How many pounds of barium sulfate are equivalent to 5.00 lb of sodium chloride?

*4.1.2 $AgNO_3 + NaCl \rightarrow AgCl + NaNO_3$

    a. How many grams of silver nitrate are required to react with 5.00 g of sodium chloride?

    b. How many grams of silver nitrate are required for the precipitation of 5.00 g of silver chloride?

    c. How many grams of silver nitrate are equivalent to 5.00 g of sodium nitrate?

    d. How many grams of sodium chloride are necessary for the precipitation of 5.00 g of silver nitrate?

    e. How many grams of sodium chloride have been added to silver nitrate if 5.00 g of silver chloride are precipitated?

    f. How many pounds of sodium chloride are equivalent to 5.00 lb of sodium nitrate?

    g. How many pounds of silver chloride are precipitated by 5.00 lb of silver nitrate?

    h. How many pounds of silver chloride are precipitated by 5.00 lb of sodium chloride?

    i. How many pounds of silver chloride are equivalent to 5.00 lb of silver nitrate?

*4.1.3 Balance the following reactions (find the values of $a_1 - a_6$):

    a. $a_1As_2S_3 + a_2H_2O + a_3HNO_3 \rightarrow a_4NO + a_5H_3AsO_4 + a_6H_2SO_4$

    b. $a_1KClO_3 + a_2HCl \rightarrow a_3HCl + a_4ClO_2 + a_5Cl_2 + a_6H_2O$

*4.1.4 The formula for vitamin C is as follows:

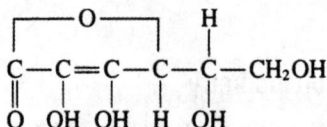

Figure P4.1.4

How many pounds of this compound are contained in 2 g mol?

*4.1.5 Acidic residue in paper from the manufacturing process causes paper based on wood pulp to age and deteriorate. To neutralize the paper, a vapor-phase treatment must employ a compound that would be volatile enough to permeate the fibrous structure of paper within a mass of books but that would have a chemistry that could be manipulated to yield a mildly basic and essentially nonvolatile compound. George Kelly and John Williams successfully attained this objective in 1976 by designing a mass deacidification process employing gaseous diethyl zinc (DEZ).

    At room temperature, DEZ is a colorless liquid. It boils at 117°C. When it is combined with oxygen, a highly exothermic reaction takes place:

$$(C_2H_5)_2Zn + 7O_2 \rightarrow ZnO + 4CO_2 + 5H_2O$$

Because liquid DEZ ignites spontaneously when exposed to air, a primary consideration in its use is the exclusion of air. In one case, a fire caused by DEZ ruined the neutralization center. Is the equation shown balanced? If not, balance it. How many kilograms of DEZ must react to form 1.5 kg of ZnO? If 20 cm^3 of water are formed on reaction, and the reaction was complete, how many grams of DEZ reacted?

**4.1.6 The following reaction was carried out:

$$Fe_2O_3 + 2X \rightarrow 2Fe + X_2O_3$$

It was found that 79.847 g of Fe_2O_3 reacted with X to form 55.847 g of Fe and 50.982 g of X_2O_3. Identify the element X.

**4.1.7 A combustion device was used to determine the empirical formula of a compound containing only carbon, hydrogen, and oxygen. A 0.6349 g sample of the unknown produced 1.603 g of CO_2 and 0.2810 g of H_2O. Determine the empirical formula of the compound.

**4.1.8 A hydrate is a crystalline compound in which the ions are attached to one or more water molecules. We can dry these compounds by heating them to get rid of the water. You have a 10.407 g sample of hydrated barium iodide. The sample is heated to drive off the water. The dry sample has a mass of 9.520 g. What is the mole ratio between barium iodide, BaI_2, and water, H_2O? What is the formula of the hydrate?

**4.1.9 Sulfuric acid can be manufactured by the contact process according to the following reactions:

1. $S + O_2 \rightarrow SO_2$
2. $2SO_2 + O_2 \rightarrow 2SO_3$
3. $SO_3 + H_2O \rightarrow H_2SO_4$

You are asked as part of the preliminary design of a sulfuric acid plant with a production capacity of 2000 tons/day of 66° Be (Baumé) (93.2% H_2SO_4 by weight) to calculate the following:
a. How many tons of pure sulfur are required per day to run this plant?
b. How many tons of oxygen are required per day?
c. How many tons of water are required per day for reaction 3?

**4.1.10 Seawater contains 65 ppm of bromine in the form of bromides. In the Ethyl-Dow recovery process, 0.27 lb of 98% sulfuric acid is added per ton of water, together with the theoretical Cl_2 for oxidation; finally, ethylene (C_2H_4) is united with the bromine to form $C_2H_4Br_2$. Assuming complete recovery and using a basis of 1 lb of bromine, find the weights of the 98% sulfuric acid, chlorine, seawater, and ethane dibromide involved.

$$2Br^- + Cl_2 \rightarrow 2Cl^- + Br_2$$
$$Br_2 + C_2H_4 \rightarrow C_2H_4Br_2$$

**4.1.11

BID EVALUATION

TO: *J. Coadwell* DEPT: *Water Waste Water* DATE: *9-29*

BID INVITATION: 0374-AV

REQUISITION: 135949 COMMODITY: *Ferrous Sulfate*

DEPARTMENT EVALUATION COMMENTS

It is recommended that the bid from VWR of $83,766.25 for 475 tons of Ferrous Sulfate Heptahydrate be accepted as they were the low bidder for this product as delivered. It is further recommended that we maintain the option of having this product delivered either by rail in a standard carload of 50 tons or by the alternate method by rail in piggy-back truck trailers.

What would another company have to bid to match the VWR bid if the bid they submitted was for ferrous sulfate ($FeSO_4 \cdot H_2O$)? For ($FeSO_4 \cdot 4H_2O$)?

**4.1.12 Three criteria must be met if a fire is to occur: (1) There must be fuel present; (2) there must be an oxidizer present; and (3) there must be an ignition source. For most fuels, combustion takes place only in the gas phase. For example, gasoline does not burn as a liquid. However, when gasoline is vaporized, it burns readily.

A minimum concentration of fuel in air exists that can be ignited. If the fuel concentration is less than this lower flammable limit (LFL) concentration, ignition will not occur. The LFL can be expressed as a volume percent, which is equal to the mole percent under conditions at which the LFL is measured (atmospheric pressure and 25°C). There is also a minimum oxygen concentration required for ignition of any fuel. It is closely related to the LFL and can be calculated from the LFL. The minimum oxygen concentration required for ignition can be estimated by multiplying the LFL concentration by the ratio of the number of moles of oxygen required for complete combustion to the number of moles of fuel being burned.

Above the LFL, the amount of energy required for ignition is quite small. For example, a spark can easily ignite most flammable mixtures. There is also a fuel concentration called the upper flammable limit (UFL) above which the fuel-air mixture cannot be ignited. Fuel-air mixtures in the flammable concentration region between the LFL and the UFL can be ignited. Both the LFL and the UFL have been measured for most of the common flammable gases and volatile liquids. The LFL is usually the more important of the flammability concentrations because if a fuel is present in the atmosphere in concentrations above the UFL, it will certainly be present within the flammable concentration region at some location. LFL concentrations for many materials can be found in the NFPA Standard 325M, *Properties of Flammable Liquids*, published by the National Fire Protection Association.

Estimate the minimum permissible oxygen concentration for *n*-butane. The LFL concentration for *n*-butane is 1.9 mol %. This problem was originally based on a problem in the text *Chemical Process Safety: Fundamentals with Applications*, by D. A. Crowl and J. F. Louvar, published by Prentice Hall, Englewood Cliffs, NJ, and has been adapted from problem 10 of the AIChE publication *Safety, Health, and Loss Prevention in Chemical Processes* by J. R. Welker and C. Springer, New York (1990).

**4.1.13 In a paper mill, soda ash (Na_2CO_3) can be added directly in the causticizing process to form, on reaction with calcium hydroxide, caustic soda (NaOH) for pulping. The overall reaction is $Na_2CO_3 + Ca(OH)_2 \rightarrow 2NaOH + CaCO_3$. Soda ash also may have potential in the on-site production of precipitated calcium carbonate, which is used as a paper filler. The chloride in soda ash (which causes corrosion of equipment) is 40 times less than in regular-grade caustic soda (NaOH), which can also be used; hence the quality of soda ash is better for pulp mills. However, a major impediment to switching to soda ash is the need for excess causticization capacity, generally not available at older mills.

Severe competition exists between soda ash and caustic soda produced by electrolysis. Average caustic soda prices are about $265 per metric ton FOB (free on board, i.e., without charges for delivery to or loading on carrier), while soda ash prices are about $130/metric ton FOB.

To what value would caustic soda prices have to drop in order to meet the price of $130/metric ton based on an equivalent amount of NaOH?

***4.1.14 A plant makes liquid CO_2 by treating dolomitic limestone with commercial sulfuric acid. The dolomite analyzes 68.0% $CaCO_3$, 30.0% $MgCO_3$, and 2.0% SiO_2, the acid is 94% H_2SO_4 and 6% H_2O. Calculate (a) pounds of CO_2 produced per ton of dolomite treated (b) pounds of acid used per ton of dolomite treated.

***4.1.15 A hazardous waste incinerator has been burning a certain mass of dichlorobenzene ($C_6H_4Cl_2$) per hour, and the HCl produced was neutralized with soda ash (Na_2CO_3). If the incinerator switches to burning an equal mass of mixed tetrachlorobiphenyls ($C_{12}H_6Cl_4$), by what factor will the consumption of soda ash be increased?

Section 4.2   Terminology for Reaction Systems

*4.2.1 Odors in wastewater are caused chiefly by the products of the anaerobic reduction of organic nitrogen- and sulfur-containing compounds. Hydrogen sulfide is a major component of wastewater odors; however, this chemical is by no means the only odor producer since serious odors can also result in its absence. Air oxidation can be used to remove odors, but chlorine is the preferred treatment because it not only destroys H_2S and other odorous compounds, but it also retards the growth of bacteria that cause the compounds in the first place. As a specific example, HOCl reacts with H_2S as follows in low-pH solutions:

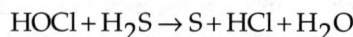

$$HOCl + H_2S \rightarrow S + HCl + H_2O$$

If the actual plant practice calls for 100% excess HOCl (to make sure of the destruction of the H_2S because of the reaction of HOCl with other substances), how much HOCl (5% solution) must be added to 1 L of a solution containing 50 ppm H_2S?

*4.2.2 Phosgene gas is probably most famous for being the first toxic gas used offensively in World War I, but it is also used extensively in the chemical processing of a wide variety of materials. Phosgene can be made by the catalytic reaction between CO and chlorine gas in the presence of a carbon catalyst. The chemical reaction is

$$CO + Cl_2 \rightarrow COCl_2$$

Suppose that you have measured the reaction products from a given reactor and found that they contained 3.00 lb mol of chlorine, 10.00 lb mol of phosgene, and 7.00 lb mol of CO. Calculate the extent of reaction, and using the value calculated, determine the initial amounts of CO and Cl_2 that were used in the reaction.

*4.2.3 In the reaction in which 135 moles of methane and 45.0 moles of oxygen are fed into a reactor, if the reaction goes to completion, calculate the extent of the reaction.

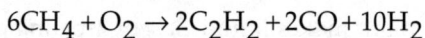

$$6CH_4 + O_2 \rightarrow 2C_2H_2 + 2CO + 10H_2$$

*4.2.4 FeS can be roasted in O_2 to form FeO:

$$2FeS + 3O_2 \rightarrow 2 FeO + 2SO_2$$

If the slag (solid product) contains 80% FeO and 20% FeS, and the exit gas is 100% SO_2, determine the extent of reaction and the initial number of moles of FeS. Use 100 g or 100 lb as the basis.

**4.2.5 Aluminum sulfate is used in water treatment and in many chemical processes. It can be made by reacting crushed bauxite (aluminum ore) with 77.7 weight percent sulfuric acid. The bauxite ore contains 55.4 weight percent aluminum oxide, the remainder being impurities. To produce crude aluminum sulfate containing 2000 lb of pure aluminum sulfate, 1080 lb of bauxite and 2510 lb of sulfuric acid solution (77.7% acid) are used.
   a. Identify the excess reactant.
   b. What percentage of the excess reactant was used?
   c. What was the degree of completion of the reaction?

4.2.6 A barite composed of 100% $BaSO_4$ is fused with carbon in the form of coke containing 6% ash (which is infusible). The composition of the fusion mass is

$BaSO_4$	11.1%
BaS	72.8
C	13.9
Ash	2.2
	100.0%

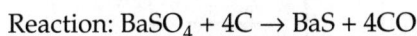

Reaction: $BaSO_4 + 4C \rightarrow BaS + 4CO$

Find the excess reactant, the percentage of the excess reactant, and the degree of completion of the reaction.

4.2.7 Read problem 4.2.2 again. Suppose that you have measured the reaction products from a given reactor and found that they contained 3.00 kg of chlorine, 10.00 kg of phosgene, and 7.00 kg of CO. Calculate the following:
a. The percent excess reactant used
b. The percentage conversion of the limiting reactant
c. The kilogram moles of phosgene formed per kilogram mole of total reactants fed to the reactor

4.2.8 The specific activity of an enzyme is defined in terms of the amount of solution catalyzed under a given set of conditions divided by the product of the time interval for the reaction times the amount of protein in the sample:

$$\text{specific activity} = \frac{\mu \text{ mol of solution converted}}{(\text{time interval in minutes}) (\text{mg protein in the sample})}$$

A 0.10 mL sample of pure β-galactosidase (β-g) solution that contains 1.00 mg of protein per liter hydrolyzed 0.10 m mol of o-nitrophenyl galactoside (o-n) in 5 min. Calculate the specific activity of the β-g.

4.2.9 One method of synthesizing the aspirin substitute acetaminophen involves a three-step procedure as outlined in Figure P4.2.9. First, p-nitrophenol is catalytically hydrogenated in the presence of aqueous hydrochloric acid to the acid chloride salt of p-aminophenol with an 86.9% degree of completion. Next, the salt is neutralized to obtain p-aminophenol with a 0.95 fractional conversion.

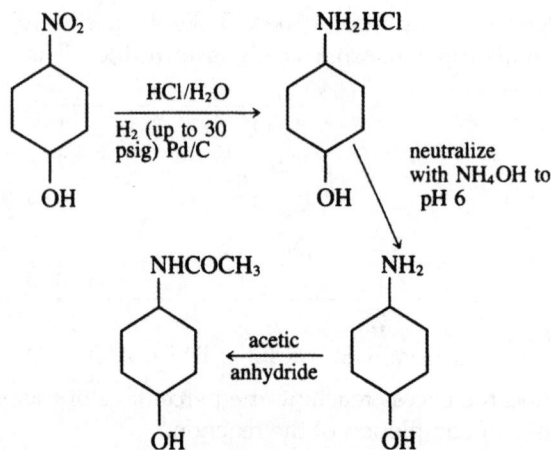

Figure P4.2.9

Finally, the *p*-aminophenol is acetylated by reacting with acetic anhydride, resulting in a yield of 3 kg mol of acetaminophen per 4 kg mol. What is the overall conversion fraction of *p*-nitrophenol to acetaminophen?

**4.2.10   The most economic method of sewage wastewater treatment is bacterial digestion. As an intermediate step in the conversion of organic nitrogen to nitrates, it is reported that the *Nitrosomonas* bacteria cells metabolize ammonium compounds into cell tissue and expel nitrite as a by-product by the following overall reaction:

$$5CO_2 + 55NH_4^+ + 76O_2 \rightarrow C_5H_7O_2N(\text{tissue}) + 54NO_2^- + 52H_2O + 109H^+$$

If 20,000 kg of wastewater containing 5% ammonium ions by weight flows through a septic tank inoculated with the bacteria, how many kilograms of cell tissue are produced, provided that 95% of the NH_4^+ is consumed?

**4.2.11   The overall yield of a product on a substrate in some bioreactions is the absolute value of the production rate divided by the rate of consumption of the feed in the substrate (the liquid containing the cells, nutrients, etc.). The overall chemical reaction for the oxidation of ethylene (C_2H_4) to epoxide (C_2H_4O) is

$$2C_2H_4 + O_2 \rightarrow 2C_2H_4O \tag{a}$$

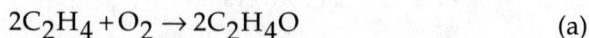

Calculate the theoretical yield (100% conversion C_2H_4) of C_2H_4O in moles per mole for Reaction (a).

The biochemical pathway for the production of epoxide is quite complex. Cofactor regeneration is required, which is assumed to originate by partial further oxidation of the formed epoxide. Thus, the amount of ethylene consumed to produce 1 mole of epoxide is larger than that required by Reaction (a). The following two reactions, (b1) and (b2), when summed approximate the overall pathway:

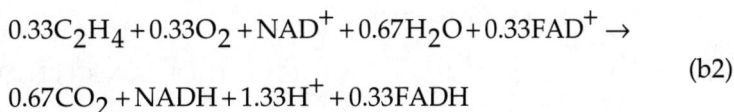

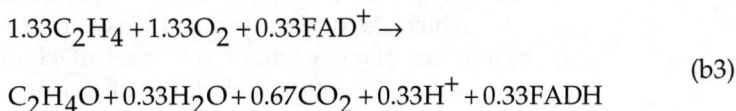

$$C_2H_4 + O_2 + NADH + H^+ \rightarrow C_2H_4O + H_2O + NAD^+ \qquad \text{(b1)}$$

$$0.33C_2H_4 + 0.33O_2 + NAD^+ + 0.67H_2O + 0.33FAD^+ \rightarrow$$
$$0.67CO_2 + NADH + 1.33H^+ + 0.33FADH \qquad \text{(b2)}$$

$$1.33C_2H_4 + 1.33O_2 + 0.33FAD^+ \rightarrow$$
$$C_2H_4O + 0.33H_2O + 0.67CO_2 + 0.33H^+ + 0.33FADH \qquad \text{(b3)}$$

Calculate the theoretical yield for Reaction (b3) of the epoxide.

***4.2.12 Antimony is obtained by heating pulverized stibnite (Sb_2S_3) with scrap iron and drawing off the molten antimony from the bottom of the reaction vessel:

$$Sb_2S_3 + 3Fe \rightarrow 2Sb + 3FeS$$

Suppose that 0.600 kg of stibnite and 0.250 kg of iron turnings are heated together to give 0.200 kg of Sb metal. Determine
a. The limiting reactant
b. The percentage of excess reactant
c. The degree of completion (fraction)
d. The percent conversion based on the limiting reactant
e. The yield in kilograms of Sb produced per kilogram of Sb_2S_3 fed to the reactor

***4.2.13 One can view the blast furnace from a simple viewpoint as a process in which the principal reaction is

$$Fe_2O_3 + 3C \rightarrow 2Fe + 3CO$$

but some other undesired side reactions occur, mainly

$$Fe_2O_3 + C \rightarrow 2FeO + CO$$

After 600.0 lb of carbon (coke) are mixed with 1.00 ton of pure iron oxide, Fe_2O_3, the process produces 1200.0 lb of pure iron, 183 lb of FeO, and 85.0 lb of Fe_2O_3. Calculate the following items:

a. The percentage of excess carbon furnished, based on the principal reaction
b. The percentage conversion of Fe_2O_3 to Fe
c. The pounds of carbon used up and the pounds of CO produced per ton of Fe_2O_3 charged
d. The selectivity in this process (of Fe with respect to FeO)

***4.2.14   A common method used in manufacturing sodium hypochlorite bleach is by the reaction

$$Cl_2 + 2NaOH \rightarrow NaCl + NaOCl + H_2O$$

Chlorine gas is bubbled through an aqueous solution of sodium hydroxide, after which the desired product is separated from the sodium chloride (a by-product of the reaction). A water-NaOH solution that contains 1145 lb of pure NaOH is reacted with 851 lb of gaseous chlorine. The NaOCl formed weighs 618 lb.

a. What was the limiting reactant?
b. What was the percentage excess of the excess reactant used?
c. What is the degree of completion of the reaction, expressed as the moles of NaOCl formed to the moles of NaOCl that would have formed if the reaction had gone to completion?
d. What is the yield of NaOCl per amount of chlorine used (on a weight basis)?
e. What was the extent of reaction?

***4.2.15   In a process for the manufacture of chlorine by direct oxidation of HCl with air over a catalyst to form Cl_2 and H_2O (only), the exit product is composed of HCl (4.4%), Cl_2 (19.8%), H_2O (19.8%), O_2 (4.0%), and N_2 (52.0%). What were (a) the limiting reactant; (b) the percent excess reactant; (c) the degree of completion of the reaction; and (d) the extent of reaction?

***4.2.16   A well-known reaction to generate hydrogen from steam is the so-called water gas shift reaction:

If the gaseous feed to a reactor consists of 30 moles of CO per hour, 12 moles of CO_2* per hour, and 35 moles of steam per hour at 800°C, and 18 moles of H_2 are produced per hour, calculate:

a. The limiting reactant
b. The excess reactant
c. The fraction conversion of steam to H_2
d. The degree of completion of the reaction
e. The kilograms of H_2 yielded per kilogram of steam fed
f. The moles of CO_2 produced by the reaction per mole of CO fed
g. The extent of reaction

***4.2.17** In the production of m-xylene (C_8H_{10}) from mesitylene (C_9H_{12}) over a catalyst, some of the xylene reacts to form toluene (C_7H_8):

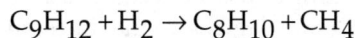

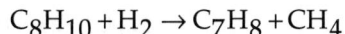

$$C_9H_{12} + H_2 \rightarrow C_8H_{10} + CH_4$$

$$C_8H_{10} + H_2 \rightarrow C_7H_8 + CH_4$$

The second reaction is undesirable because m-xylene sells for $0.65/lb, whereas toluene sells for $0.22/lb.

The CH_4 is recycled in the plant. One pound of catalyst is degraded per 500 lb of C_7H_8 produced, and the spent catalyst has to be disposed of in a landfill that handles low-level toxic waste at a cost of $25/lb. If the overall selectivity of C_8H_{10} to C_7H_8 is changed from 0.7 mole of xylene produced per mole of toluene produced to 0.8 by changing the residence time in the reactor, what is the gain or loss in dollars per 100 lb of mesitylene reacted?

Section 4.3   Species Mole Balances

*4.3.1 Pure A in gas phase enters a reactor. Fifty percent (50%) of this A is converted to B through the reaction A → 3B. What is the mole fraction of A in the exit stream? What is the extent of reaction?

**4.3.2 A low-grade pyrite containing 32% S is mixed with 10 lb of pure sulfur per 100 lb of pyrites so the mixture will burn readily with air, forming a burner gas that analyzes 13.4% SO_2, 2.7% O_2, and 83.9% N_2. No sulfur is left in the cinder. Calculate the percentage of the sulfur fired that burned to SO_3. (The SO_3 is not detected by the analysis.)

**4.3.3 Examine the reactor in Figure P4.3.3. Your boss says something has gone wrong with the yield of CH_2O, and it is up to you to find out what the problem is. You start by making material balances (naturally!). Show all calculations. Is there some problem?

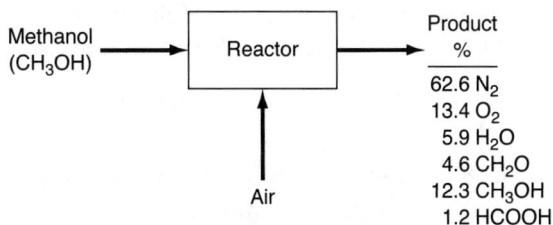

Methanol (CH_3OH) →	Reactor	→	Product %

Product
%
_____
62.6 N_2
13.4 O_2
5.9 H_2O
4.6 CH_2O
12.3 CH_3OH
1.2 HCOOH

Air

Figure P4.3.3

**4.3.4 A problem statement was the following: A dry sample of limestone is completely soluble in HCl and contains no Fe or Al. When a 1.000 g sample is ignited, the loss in weight is found to be 0.450 g. Calculate the percent $CaCO_3$ and $MgCO_3$ in the limestone. The solution was

$$\frac{x}{84.3} + \frac{(1.000-x)}{100} = \frac{0.450}{44.}$$

$$100x + 84.3 - 84.3\,x = (0.450)(84.3)(100)\,/\,44$$

$$x = 0.121 \; MgCO_3 = 12.1\%$$

$$CaCO_3 = 87.9\%$$

Answer the following questions:
a. What information in addition to that in the problem statement had to be obtained?
b. What would a diagram for the process look like?
c. What was the basis for the problem solution?
d. What were the known variables in the problem statement, their values, and their units?
e. What were the unknown variables in the problem statement and their units?
f. What are the types of material balances that could be made for this problem?
g. What type(s) of material balance was made for this problem?
h. What was the degree of freedom for this problem?
i. Was the solution correct?

***4.3.5 One of the most common commercial methods for the production of pure silicon that is to be used for the manufacture of semiconductors is the Siemens process (see Figure P4.3.5) of chemical vapor deposition (CVD). A chamber contains a heated silicon rod and a mixture of high-purity trichlorosilane mixed with high-purity hydrogen that is passed over the rod. Pure silicon (EGS - electronic grade silicon) deposits on the rod as a polycrystalline solid. (Single crystals of Si are later made by subsequently melting the EGS and drawing a single crystal from the melt.) The reaction is

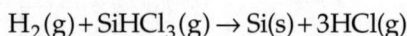

$$H_2(g) + SiHCl_3(g) \rightarrow Si(s) + 3HCl(g)$$

The rod initially has a mass of 1460 g, and the mole fraction of H_2 in the exit gas is 0.223. The mole fraction of H_2 in the feed to the reactor is 0.580, and the feed enters at the rate of 6.22 kg mol/hr. What will be the mass of the rod at the end of 20 min?

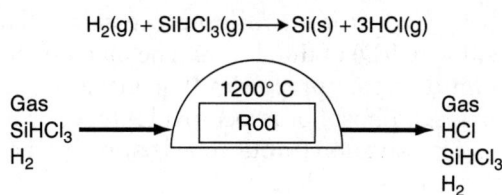

$$H_2(g) + SiHCl_3(g) \longrightarrow Si(s) + 3HCl(g)$$

Gas
SiHCl$_3$ $\longrightarrow$
H$_2$

1200° C
Rod

$\longrightarrow$ Gas
HCl
SiHCl$_3$
H$_2$

Figure P4.3.5

***4.3.6** Copper as CuO can be obtained from an ore called Covellite which is composed of CuS and gange (inert solids). Only part of the CuS is oxidized with air to CuO. The gases leaving the roasting process analyze SO_2 (7.2%), O_2 (8.1%), and N_2 (84.7%). Unfortunately, the method of gas analysis could not detect SO_3 in the exit gas, but SO_3 is known to exist. Calculate the percent of the sulfur in the part of the CuS that reacts that forms SO_3. Hint: You can consider the unreacted CuS as a compound that comes in and out of the process untouched and thus is isolated from the process and can be ignored.

***4.3.7** A reactor is used to remove SiO_2 from a wafer in semiconductor manufacturing by contacting the SiO_2 surface with HF. The reactions are

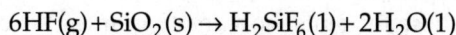

$$6HF(g) + SiO_2(s) \rightarrow H_2SiF_6(l) + 2H_2O(l)$$
$$H_2SiF_6(l) \rightarrow SiF_4(g) + 2HF(g)$$

Assume the reactor is loaded with wafers having a silicon oxide surface, a flow of 50% HF and 50% nitrogen is started, and all the H_2SiF_6 reacts. In the reaction 10% of the HF is consumed. What is the composition of the exhaust stream?

***4.3.8** In the anaerobic fermentation of grain, the yeast *Saccharomyces cerevisiae* digests glucose from plants to form the products ethanol and propenoic acid by the following overall reactions:

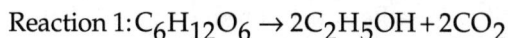

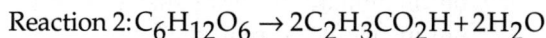

$$\text{Reaction 1:} C_6H_{12}O_6 \rightarrow 2C_2H_5OH + 2CO_2$$
$$\text{Reaction 2:} C_6H_{12}O_6 \rightarrow 2C_2H_3CO_2H + 2H_2O$$

In an open flow reactor 3500 kg of a 12% glucose-water solution flow in. During fermentation, 120 kg of carbon dioxide are produced together with 90 kg of unreacted glucose. What are the weight percents of ethyl alcohol and propenoic acid that exit in the broth? Assume that none of the glucose is assimilated into the bacteria.

***4.3.9   Semiconductor microchip processing often involves chemical vapor depo-
sition (CVD) of thin layers. The material being deposited needs to have certain desirable properties. For instance, to overlay on aluminum or other
bases, a phosphorus-pentoxide-doped silicon dioxide coating is deposited
as a passivation (protective) coating by the simultaneous reactions

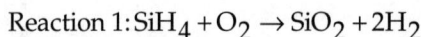

$$\text{Reaction 1:} SiH_4 + O_2 \rightarrow SiO_2 + 2H_2$$
$$\text{Reaction 2:} 4PH_3 + 5O_2 \rightarrow 2P_2O_5 + 6H_2$$

Determine the relative masses of SiH_4 and PH_3 required to deposit a film of
5% by weight of phosphorus oxide (P_2O_5) in the protective coating.

***4.3.10   Printed circuit boards are used in the electronics industry to both connect
and hold components in place. In production, 0.03 in. of copper foil is lami-
nated to an insulating plastic board. A circuit pattern made of a chemically
resistant polymer is then printed on the board. Next, the unwanted copper
is chemically etched away by using selected reagents. If copper is treated
with $Cu(NH_3)_4Cl_2$ (cupric ammonium chloride) and NH_4OH (ammonium
hydroxide), the products are water and $Cu(NH_3)_4Cl$ (cuprous ammonium
chloride). Once the copper is dissolved, the polymer is removed by solvents, leaving the printed circuit ready for further processing. If a single-
sided board 4 in. by 8 in. is to have 75% of the copper layer removed using
these reagents, how many grams of each reagent will be consumed? Data:
The density of copper is $8.96\ g/cm^3$.

***4.3.11   The thermal destruction of hazardous wastes involves the controlled
exposure of waste to high temperatures (usually 900°C or greater) in an oxidizing environment. Types of thermal destruction equipment include
high-temperature boilers, cement kilns, and industrial furnaces in which
hazardous waste is burned as fuel. In a properly designed system, primary
fuel (100% combustible material) is mixed with waste to produce a feed for
the boiler.

a. Sand containing 30% by weight of 4,4'-dichlorobiphenyl [an example of
a polychlorinated biphenyl (PCB)] is to be cleaned by combustion with
excess hexane to produce a feed that is 60% combustible by weight. To
decontaminate 8 tons of such contaminated sand, how many pounds of
hexane would be required?

b. Write the two reactions that would take place under ideal conditions if
the mixture of hexane and the contaminated sand were fed to the thermal oxidation process to produce the most environmentally satisfactory
products. How would you suggest treating the exhaust from the
burner? Explain.

c. The incinerator is supplied with an oxygen-enriched airstream containing 40% O_2 and 60% N_2 to promote high-temperature operation. The exit
gas is found to have a composition of $x_{CO_2} = 0.1654$ and $x_{O_2} = 0.1220$.

Use this information and the data about the feed composition to find (a) the complete exit gas concentrations and (b) the percent excess O_2 used in the reaction.

****4.3.12   In order to neutralize the acid in a waste stream (composed of H_2SO_4 and H_2O), dry ground limestone (composition 95% $CaCO_3$ and 5% inerts) is mixed in. The dried sludge collected from the process is only partly analyzed by firing it in a furnace, which results in only CO_2 being driven off. By weight, the CO_2 represents 10% of the dry sludge. What percent of the pure $CaCO_3$ in the limestone did not react in the neutralization? Solve this problem using mole balances.

Section 4.4   Element Material Balances

****4.4.1   In order to neutralize the acid in a waste stream (composed of H_2SO_4 and H_2O), dry ground limestone (composition 95% $CaCO_3$ and 5% inerts) is mixed in. The dried sludge collected from the process is only partly analyzed by firing it in a furnace, which results in only CO_2 being driven off. By weight, the CO_2 represents 10% of the dry sludge. What percent of the pure $CaCO_3$ in the limestone did not react in the neutralization? Solve this problem using element balances.

Section 4.5   Material Balances Combustion Systems

**4.5.1   A synthesis gas analyzing 6.4% CO_2, 0.2% O_2, 40.0% CO, and 50.8% H_2 (the balance is N_2) is burned with 40% dry excess air. What is the composition of the flue gas?

**4.5.2   Thirty pounds of coal (analysis 80% C and 20% H, ignoring the ash) are burned with 600 lb of air, yielding a gas having an Orsat analysis in which the ratio of CO_2 to CO is 3 to 2. What is the percent excess air?

**4.5.3   A gas containing only CH_4 and N_2 is burned with air yielding a flue gas that has an Orsat analysis of CO_2 8.7%, CO 1.0%, O_2 3.8%, and N_2 86.5%. Calculate the percent excess air used in combustion and the composition of the CH_4-N_2 mixture.

**4.5.4   A natural gas consisting entirely of methane (CH_4) is burned with an oxygen-enriched air of composition 40% O_2 and 60% N_2. The Orsat analysis of the product gas as reported by the laboratory is CO_2 20.2%, O_2 4.1%, and N_2 75.7%. Can the reported analysis be correct? Show all calculations.

**4.5.5   Dry coke composed of 4% inert solids (ash), 90% carbon, and 6% hydrogen is burned in a furnace with dry air. The solid refuse left after combustion contains 10% carbon and 90% inert ash (and no hydrogen). The inert ash content does not enter into the reaction.

The Orsat analysis of the flue gas gives 13.9% CO_2, 0.8% CO, 4.3% O_2, and 81.0% N_2. Calculate the percent of excess air based on complete combustion of the coke.

**4.5.6 A gas with the following composition is burned with 50% excess air in a furnace. What is the composition of the flue gas?

$$CH_4\ 60\%,\ C_2H_6\ 20\%,\ CO\ 5\%,\ O_2\ 5\%,\ N_2\ 10\%$$

**4.5.7 In underground coal combustion in the gas phase several reactions take place, including

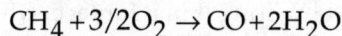

$$CO + 1/2O_2 \rightarrow CO_2$$

$$H_2 + 1/2O_2 \rightarrow H_2O$$

$$CH_4 + 3/2O_2 \rightarrow CO + 2H_2O$$

where the CO, H, and CH_4 come from coal pyrolysis.

If a gas phase composed of CO 13.54%, CO_2 15.22%, H_2 15.01%, CH_4 3.20%, and the balance N_2 is burned with 40% excess air, (a) how much air is needed per 100 moles of gas, and (b) what will be the analysis of the product gas on a wet basis?

**4.5.8 Solvents emitted from industrial operations can become significant pollutants if not disposed of properly. A chromatographic study of the waste exhaust gas from a synthetic fiber plant has the following analysis in mole percent:

CS_2	40%
SO_2	10%
H_2O	50%

It has been suggested that the gas be disposed of by burning with an excess of air. The gaseous combustion products are then emitted to the air through a smokestack. The local air pollution regulations say that no stack gas is to analyze more than 2% SO_2 by an Orsat analysis averaged over a 24-hour period. Calculate the minimum percent excess air that must be used to stay within this regulation.

**4.5.9 The products and by-products from coal combustion can create environmental problems if the combustion process is not carried out properly. Your boss asks you to carry out an analysis of the combustion in boiler

No. 6. You carry out the work assignment using existing instrumentation and obtain the following data:

Fuel analysis (coal): 74% C, 14% H, and 12% ash

Flue gas analysis on a dry basis: 12.4% CO_2, 1.2% CO, 5.7% O_2, and 80.7% N_2

What are you going to report to your boss?

**4.5.10 The Clean Air Act requires automobile manufacturers to warrant their control systems as satisfying the emission standards for 50,000 mi. It requires owners to have their engine control systems serviced exactly according to manufacturers' specifications and to always use the correct gasoline. In testing, an engine exhaust having a known Orsat analysis of 16.2% CO_2, 4.8% O_2, and 79% N_2 at the outlet, you find to your surprise that at the end of the muffler the Orsat analysis is 13.1% CO_2. Can this discrepancy be caused by an air leak into the muffler? (Assume that the analyses are satisfactory.) If so, compute the moles of air leaking in per mole of exhaust gas leaving the engine.

**4.5.11 One of the products of sewage treatment is sludge. After microorganisms grow in the activated sludge process to remove nutrients and organic material, a substantial amount of wet sludge is produced. This sludge must be dewatered, one of the most expensive parts of most treatment plant operations.

How to dispose of the dewatered sludge is a major problem. Some organizations sell dried sludge for fertilizer, some spread the sludge on farmland, and in some places it is burned. To burn a dried sludge, fuel oil is mixed with it, and the mixture is burned in a furnace with air. If you collect the following analysis for the sludge and for the product gas:

Sludge (%)		Product Gas (%)	
S	32	SO_2	1.52
C	40	CO_2	10.14
H_2	4	O_2	4.65
O_2	24	N_2	81.67
		CO	2.02

a. Determine the weight percent of carbon and hydrogen in the fuel oil.
b. Determine the ratio of pounds of dry sludge to pounds of fuel oil in the mixture fed to the furnace.

**4.5.12 Many industrial processes use acids to promote chemical reactions or produce acids from the chemical reactions occurring in the process. As a result, these acids many times end up in the wastewater stream from the process and must be neutralized as part of the wastewater treatment process before the water can be discharged from the process. Lime (CaO) is a cost-effective neutralization agent for acid wastewater. Lime is dissolved in water by the following reaction:

$$CaO + 1/2O_2 \rightarrow Ca(OH)_2$$

which reacts directly with acid; for example, for H_2SO_4,

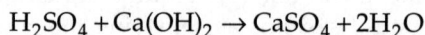

$$H_2SO_4 + Ca(OH)_2 \rightarrow CaSO_4 + 2H_2O$$

Consider an acidic wastewater stream with a flow rate of 1000 gal/min with an acid concentration of 2% H_2SO_4. Determine the flow rate of lime in pounds per minute necessary to neutralize the acid in this stream if 20% excess lime is used. Calculate the production rate of $CaSO_4$ from this process in tons per year. Assume that the specific gravity of the acidic wastewater stream is 1.05.

4.5.13 Nitric acid (HNO_3) that is used industrially for a variety of reactions can be produced by the reaction of ammonia (NH_3) with air by the following overall reaction:

$$NH_3 + 2O_2 \rightarrow HNO_3 + H_2O$$

The product gas from such a reactor has the following composition (on a water-free basis):

$$0.8\%\,NH_3$$
$$9.5\%\,HNO_3$$
$$3.8\%\,O_2$$
$$85.9\%\,N_2$$

Determine the percent conversion of NH_3 and the percent excess air used.

4.5.14 Ethylene oxide (C_2H_4O) is a high-volume chemical intermediate that is used to produce glycol and polyethylene glycol. Ethylene oxide is produced by the partial oxidation of ethylene C_2H_4 using a solid catalyst in a fixed-bed reactor:

$$C_2H_4 + \tfrac{1}{2}O_2 \rightarrow C_2H_4O$$

In addition, a portion of the ethylene reacts completely to form CO_2 and H_2O:

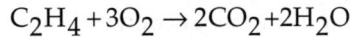

$$C_2H_4 + 3O_2 \rightarrow 2CO_2 + 2H_2O$$

The product gas leaving a fixed-bed ethylene oxide reactor has the following water-free composition: 20.5% C_2H_4O, 72.7, N_2, 2.3 O_2, and 4.5% CO_2. Determine the percent excess air based in the desired reaction, and the pounds per hour of ethylene feed required to produce 100,000 ton/yr of ethylene oxide.

***4.5.15 A flare is used to convert unburned gases to innocuous products such as CO_2 and H_2O. If a gas of the following composition (in percent) is burned in the flare—70%, C_3H_8 5%, CO 15%, O_2 5%, N_2 5%—and the flue gas contains 7.73% CO_2, 12.35% H_2O, and the balance is O_2 and N_2, what was the percent excess air used?

***4.5.16 Hydrogen-free carbon in the form of coke is burned (a) with complete combustion using theoretical air, (b) with complete combustion using 50% excess air, or (c) using 50% excess air but with 10% of the carbon burning to CO only. In each case calculate the gas analysis that will be found by testing the flue gases on a dry basis.

***4.5.17 Ethanol (CH_3CH_2OH) is dehydrogenated in the presence of air over a catalyst, and the following reactions take place:

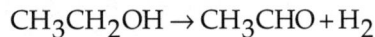

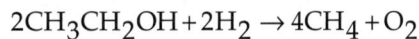

$$CH_3CH_2OH \rightarrow CH_3CHO + H_2$$

$$2CH_3CH_2OH + 3O_2 \rightarrow 4CO_2 + 6H_2$$

$$2CH_3CH_2OH + 2H_2 \rightarrow 4CH_4 + O_2$$

Separation of the product, CH_3CHO (acetaldehyde), as a liquid leaves an output gas with the following Orsat analysis: CO_2 0.7%, O_2 2.1%, CO 2.3%, H_2 7.1%, CH_4 2.6%, and N_2 85.2%. How many kilograms of acetaldehyde are produced per kilogram of ethanol fed into the process?

***4.5.18 Refer to Example 4.14. Suppose that during combustion, a very small amount (0.24%) of the entering nitrogen reacts with oxygen to form nitrogen oxides (NO_x). Also, suppose that the CO produced is 0.18% and the SO_2 is 1.4% of the $CO_2 + SO_2$ in the flue gas. The emissions listed by the EPA in the load units (ELU) per kilogram of gas are

NO_x	0.22
CO	0.27
CO_2	0.09
SO_2	0.10

What is the total ELU for the stack gas? Note: The ELU are additive.

****4.5.19 Glucose ($C_6H_{12}O_6$) and ammonia form a sterile solution (no live cells) fed continuously into a vessel. Assume that glucose and ammonia are fed in stoichiometric proportions and that they react completely. One product formed from the reaction contains ethanol, cells ($CH_{1.8}O_{0.5}N_{0.2}$), and water. The gas produced is CO_2. If the reaction occurs anaerobically (without the presence of oxygen), what is the minimum amount in kilograms of feed (ammonia and glucose) required to produce 4.6 kg of ethanol? Only 60% of the moles of glucose are converted to ethanol. The remainder is converted to cell mass, carbon dioxide, and water.

CHAPTER 5

Material Balances for Multiunit Processes

Chapter Objectives

- Write a set of independent material balances for a process involving more than one unit.
- Apply the 10-step strategy to solve multiunit steady-state problems (with and without chemical reactions) involving sequential, recycle, and/or bypass, and/or purge streams.
- Solve by hand problems in which a modest number of interconnected units are involved by making appropriate balances, or apply MATLAB or Python to solve the resulting set of linear equations.
- Use the overall conversion and single-pass (once-through) conversion concepts in solving recycle problems involving reactors.
- Explain the purpose of a recycle stream, a bypass stream, and a purge stream.
- Understand in a general sense how material balances are used in industry.

Introduction

In this chapter, we consider material balances applied to systems with multiple units, including a sequential arrangement of units, systems with recycle (i.e., instances in which material is returned to a process from downstream), and systems with purge and bypass. In addition, we comment on the application of material balances in industrial systems.

5.1   Preliminary Concepts

A **process flowsheet (flowchart)** is a graphical representation of a process, as pointed out in Chapter 3. A flowsheet describes the actual process in sufficient detail that you can use it to formulate material (and energy) balances. Flowsheets are also used for troubleshooting, control of operating conditions, and optimization of process performance. You will find that flowsheets are also prepared for proposed processes that involve new techniques or modifications of existing processes.

Figure 5.1 is a picture of a section of an ammonia plant. Figure 5.2a is a flowsheet of the process indicating the equipment sequence and the flow of materials. Figure 5.2b is a **block diagram** corresponding to Figure 5.2a. The units appear as simple boxes called **subsystems** rather than as the more elaborate portrayal in Figure 5.2a. You should note that the operations of mixing and splitting are clearly denoted by boxes in Figure 5.2b, whereas the same functions appear only as intersecting lines in Figure 5.2a. A **mixer** (e.g., Mix 1 in Figure 5.2b) combines two or more streams that have different compositions, yielding a product stream with a different composition. On the other hand, a **splitter** (Split in Figure 5.2b) has one feed stream and produces two or more product streams, all with the same composition as the feed stream. A **separator** can have more than one stream entering, each of a different composition, and two or more streams exiting, each of a different composition (not shown in Figure 5.2).

Figure 5.1   Section of a large ammonia plant showing the equipment in place (Image courtesy of nepper77/Shutterstock)

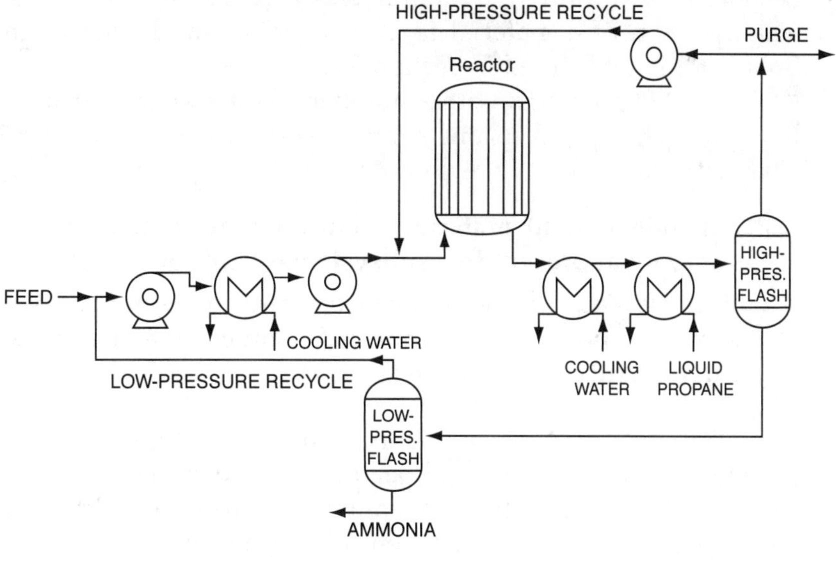

a.

Figure 5.2   (a) Flowsheet of the ammonia plant that includes major pieces of equipment and shows the materials flow

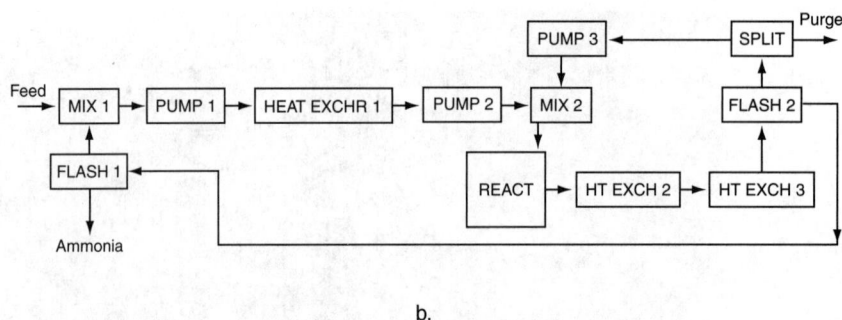

b.

Figure 5.2   (b) block diagram corresponding to Figure 5.2a

Note that Figure 5.2b contains elements of a sequential series of units, a topic treated in Section 5.2. An example of such a sequence is the flow from Pump 1 to Heat Exchanger 1 and then on to Pump 2. The figure also shows an example of **recycle,** as in the flow of material from Mix 1 to the end of the diagram (follow the arrows) and back through Flash 1 to Mix 1 again. A second example of recycle is from Pump 2 through the Reactor, Heat Exchangers 2 and 3, Flash 2, Split, and then back to Pump 2. Recycle is covered in Section 5.3. In addition, the figure shows an example of **purge** (removing a small portion of a material to prevent its accumulation in the process) at Split, a topic considered in Section 5.4.

To solve problems with a number of subsystems, such as those shown in Figure 5.2, apply the same strategy as discussed in Chapter 3 and used in Chapter 4. Note that

1. **An independent material balance equation can be written for each component present in a subsystem except for a splitter.**

2. **Because each of the streams entering or leaving a splitter has the same composition, only one independent material balance equation can be written for a splitter.**

Warning! For most problems involving multiple units, the number of pertinent *independent* material balance equations that can be written is considerably lower than the total number of material balance equations that could be written. Therefore, when solving multiunit problems, be careful to choose appropriate equations to use to ensure that the resulting set of equations is independent. Some of the specifications when assigned to variables and/or some of the implicit equations may cause one or more material balance equations to become redundant. Moreover, if you are going to solve multiunit problems by hand, you have to carefully choose the order in which you solve the various equations to avoid having to solve too many equations simultaneously.

5.2   **Sequential Multiunit Systems**

Let's first examine a multiunit system composed of a **sequential combination of units**. Figure 5.3a illustrates a sequential combination of mixing and splitting stages. Streams 1 and 2 combine to form the first mixing point, streams 3 and 4 also combine in the box for the second mixing point, and stream 5 splits (at presumably a pipe junction) into streams 6 and 7. Note that the first mixing point occurs at the combination of streams 1 and 2 (presumably a junction of pipes), and the second mixing occurs where streams 3 and 4 enter the box (presumably representing a process such as a mixing tank). You will encounter both types of mixing in the problems and examples in this book and in flowsheets in professional practice.

Examine Figure 5.3a. Which streams must have the same composition? Do streams 5, 6, and 7 have the same composition? Yes, because streams 6 and 7 flow from a splitter. Do streams 3, 4, and 5 have the same composition? It's quite unlikely. Stream 5 is some type of average of the compositions of streams 3 and 4 (in the absence of reaction). What is the composition inside the system (the box)? It will have the same composition as stream 5 only if streams 3 and 4 are really **well mixed** in the subsystem represented by the box.

How many material balances can you formulate for the system and subsystems shown in Figure 5.3a? First, let's examine how many total material balances you can write. You can write an **overall total material balance,** namely, a total balance on the system that includes all of the three subsystems within the overall system boundary, which is denoted by the dashed line labeled I in Figure 5.3b.

In addition, you can make a total balance on each of the three subsystems that make up the overall system, as denoted by the boundaries indicated by the dashed lines II, III, and IV in Figure 5.3c. Finally, you can make a balance about each of the combinations of two subsystems, as indicated by the dashed-line boundaries V and VI in Figures 5.3d and 5.3e, respectively.

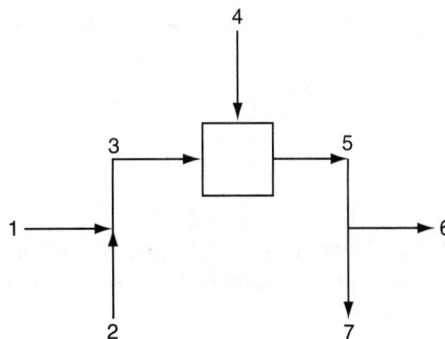

Figure 5.3a   Serial mixing and splitting in a process without reaction

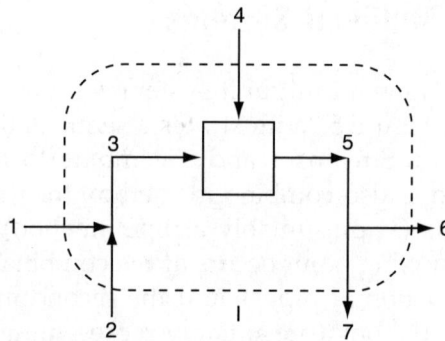

Figure 5.3b   Dashed line I designates the boundary for an overall total material balance made on the system in Figure 5.3a

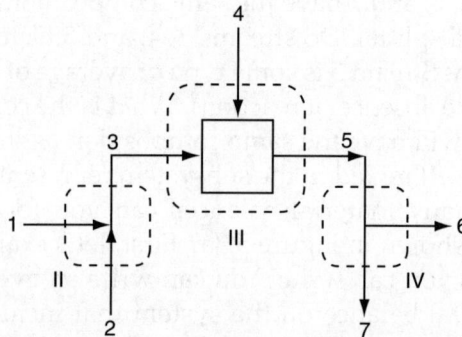

Figure 5.3c   Dashed lines II, III, and IV denote the boundaries for material balances around each of the individual units constituting the overall process

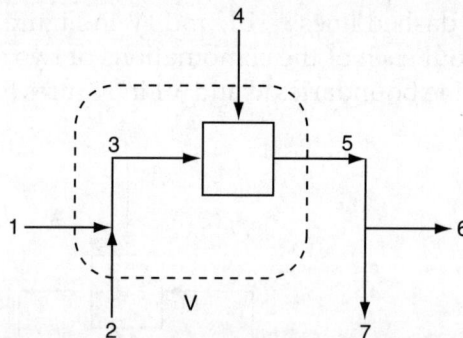

Figure 5.3d   Dashed line V denotes the boundary for material balances around a subsystem composed of the first mixing point plus the subsystem portrayed by the box

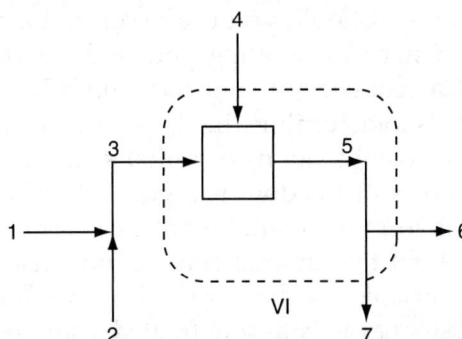

Figure 5.3e   Dashed line VI denotes the boundary for material balances about a subsystem composed of the process portrayed by the box plus the splitter

You can conclude for the system shown in Figure 5.3a that you can make total material balances on six different combinations of subsystems.

The important question is: How many independent material balance equations can be written for the process illustrated in Figure 5.3a if more than one component exists? In Section 5.1, we stated that you can write one independent equation for each component in each subsystem except for the splitter, for which you can write only one independent material balance equation. For Figure 5.3a, assume that three components are present in each of the separate subsystems shown in Figure 5.3c. You can write three independent material balance equations about the first pipe junction, three for the box, plus one independent equation for the splitter, for a total of seven (7) independent material balance equations. How many material balances are possible if you include *all of the redundant equations,* total as well as component balances?

Figure 5.3b: 3 components plus 1 total

Figure 5.3c: $3 \times 3 = 9$ components plus 3 total

Figure 5.3d: 3 components plus 1 total
Figure 5.3e: 3 components plus 1 total

The total is 24. Which 7 of the 24 equations would be the most appropriate to choose to retain independence and solve easily?

Be careful to **select an independent set of equations.** As an example of what not to do, do not select three component balances for (a) the first pipe junction shown in Figure 5.3c, (b) the box, and (c) one balance for the splitter plus an overall total balance (Figure 5.3b). This set of equations would not be independent because, as you know, the overall balance is just the sum of the respective species balances for the individual unit. However, the total balance could be substituted for one of the component balances.

What strategy should you use to select the particular unit or subsystem with which to start formulating your independent equations for a process composed of a sequence of connected units? A good, but time-consuming, way to decide is to determine the degrees of freedom for various subsystems (single units or combinations of units) selected by inspection. A subsystem with zero degrees of freedom is a good starting point. Frequently, the best way to start is to make material balances for the **overall process,** ignoring information about the internal **connections.** If you ignore all of the internal streams and variables within a set of connected subsystems, you can treat the overall system exactly as you treated a single system in Chapters 3 and 4.

Example 5.1   Determination of the Number of Independent Material Balances in a Process with Multiple Units

Problem Statement

Lactic acid ($C_3H_6O_3$) produced by fermentation is used in the food, chemical, and pharmaceutical industries. Figure E5.1 illustrates the mixing of components to form a suitable fermentation broth. The whole system is steady state and open. The arrows designate the direction of the flows. No reaction occurs in any of the subsystems.

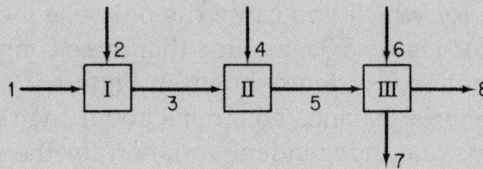

Figure E5.1

The mass compositions of each stream are as follows:

1. Water (W): 100%

2. Glucose (G): 100%

3. W and G, concentrations known: $\omega_W = 0.800$ and $\omega_G = 0.200$

4. *Lactobacillus* (L): 100%

5. W, G, and L, concentrations known: $\omega_W = 0.769, \omega_G = 0.192, \omega_L = 0.0385$

6. Vitamin G with amino acids and phosphate (V): 100%

7. $\omega_W = 0.962, \omega_V = 0.0385$

8. $\omega_G = 0.833, \omega_L = 0.167$

What is the maximum number of independent mass balances that can be generated for this system?

Solution

From taking into account each of the three units as subsystems, you certainly can make nine component equations as follows: Let's bypass any total balance for any of the three units as well as any overall component or total balance, or any of the possible balances for combinations of units.

	Total Number of Component Balances
At unit I, two components are involved	2
At unit II, three components are involved	3
At unit III, four components are involved	4
Total	9

However, not all of the balances are independent. In the following list of the component balances, all of the known component concentrations have been inserted. F_i represents the stream flow designated by the subscript.

Subsystem I:

Component Balances $\begin{cases} A: F_1(1.00) + F_2(0) = F_3(0.800) \\ B: F_1(0) + F_2(1.00) = F_3(0.20) \end{cases}$

$\qquad$ (a)

$\qquad$ (b)

Subsystem II:

Component Balances $\begin{cases} A: F_3(0.800) + F_4(0) = F_5(0.769) \\ B: F_3(0.200) + F_4(0) = F_5(0.192) \\ C: F_3(0) + F_4(1.00) = F_5(0.0385) \end{cases}$

$\qquad$ (c)

$\qquad$ (d)

$\qquad$ (e)

Subsystem III:

Component Balances $\begin{cases} A: F_5(0.769) + F_6(0) = F_7(0.962) + F_8(0) \\ B: F_5(0.192) + F_6(0) = F_7(0) + F_8(0.833) \\ C: F_5(0.0385) + F_6(0) = F_7(0) + F_8(0.167) \\ D: F_5(0) + F_6(1.00) = F_7(0.086) + F_8(0) \end{cases}$

$\qquad$ (f)

$\qquad$ (g)

$\qquad$ (h)

$\qquad$ (i)

If you take as an arbitrary basis $F_1 = 100$, seven values of F_i are unknown; hence, only seven independent equations need to be written. Can you recognize by inspection that among the entire set of nine equations, two are indeed redundant,

(Continues)

Example 5.1   Determination of the Number of Independent Material Balances in a Process with Multiple Units (*Continued*)

and hence a unique solution can be obtained using the seven independent equations?

If you solved the nine equations by hand *sequentially*, starting with Equation (a) and ending with Equation (i), along the way you would notice that Equation (d) is redundant with Equation (c), and Equation (h) is redundant with Equation (g). The redundancy of Equations (c) and (d) becomes apparent if you recall that the sum of the mass fractions in a stream is unity, hence an implicit relation exists between Equations (c) and (d) so they are not independent. Why are Equations (g) and (h) not independent?

As you inspect the set of Equations (a) through (i) with the viewpoint of solving them sequentially, you will note that each one can be solved for one variable. Look at the following list:

Equation	Determines	Equation	Determines
(a)	F_3	(f)	F_7
(b)	F_2	(g)	F_8
(c)	F_5	(h)	F_8
(d)	F_5	(i)	F_6
(e)	F_4		

If you entered Equations (a) through (i) into a software program (e.g., MATLAB or Python) that solves linear equations, you would receive an error notice of some type because the set of equations includes redundant equations. Also, the rank of the set of linear equations would be 7 and not 9.

If you make one or more component mass balances around the combination of subsystems I plus II, or II plus III, or I plus III in Example 5.1, or around the entire set of three units, no additional *independent* mass balances will be generated. Can you substitute one of the indicated alternative mass balances for an independent species mass balance? In general, yes.

In calculating the degree-of-freedom analysis for problems involving multiple units, you must be careful to involve only independent material balances and not miss any essential unknowns. All the same principles that were discussed in Chapters 3 and 4 apply to processes with multiple units. The following is a simplified checklist to help you keep in mind the possible unknowns to include in an analysis, and the possible specifications and equations that should be taken into account.

Make sure that you have accounted for all of the variables involved in the problem and not included irrelevant ones. Check for

- Flow variables entering and leaving each subsystem
- Species or components entering and leaving for each subsystem
- Reaction variables and extents of reaction

Make sure that you have not missed any of the information in the problem that should be included as an equation. Check for

- Selection of a basis for a subsystem or the overall system
- Material balances for each species, element, or component in each subsystem and the overall system as needed
- Specifications for a subsystem and the overall system that can be used to assign values to variables or add equations to the solution set (remember the assignments correspond to possible equations)
- Implicit equations (sum of mole or mass fractions) used explicitly or implicitly

---

Formally reviewing the checklist will help if you are perplexed or uncertain as to how to identify which variables and equations to employ in solving for the unknowns, and, of course, to ensure that the degrees of freedom are zero before starting to solve the set of equations. If you have doubts that you have formulated a set of equations that are independent, simply determine the rank of the coefficient matrix using MATLAB (Section 3.6) or Python (Section 3.7) and compare the rank with the number of unknowns.

Frequently Asked Questions

1. In carrying out a degree-of-freedom analysis, do you have to include at the start of the analysis every one of the variables and equations that are involved in the entire process? No. You pick a system for analysis, and you only have to take into account the unknowns and equations pertaining to the streams crossing the system boundary. For example, note in Example 5.1 that each subsystem could be treated independently. If you picked as the system the overall system (all three of the units) and the system was in the steady state, only the variables and equations pertaining to streams 1, 2, 4, 6, 7, and 8 would be involved in the analysis.

2. Should you use element material balances or species material balances in solving problems that involve multiple units? For processes that do not involve reaction, use species balances; element balances are quite inefficient. For processes that do involve reaction, if you are given the reactions and information that enable you to calculate the extent of reaction, use species balances. Even if you are not specifically given the reactions, you can often formulate them based on your experience, such as C burning with O_2 to yield CO_2. But if such data are missing, element balances are easier to use. Just make sure that they are independent!

We next look at some examples of making and solving material balances for systems composed of multiple units.

Example 5.2   Material Balances for Multiple Units in Which No Reaction Occurs

Problem Statement

Acetone is used in the manufacture of many chemicals and also as a solvent. In its latter role, many restrictions are placed on the release of acetone vapor to the environment. You are asked to design an acetone recovery system having the flowsheet illustrated in Figure E5.2. All of the concentrations shown in Figure E5.2 of both the gases and liquids are specified in *weight percent* in this special case to make the calculations simpler. Calculate A, F, W, B, and D in kilograms per hour. Assume that $G = 1400$ kg/hr.

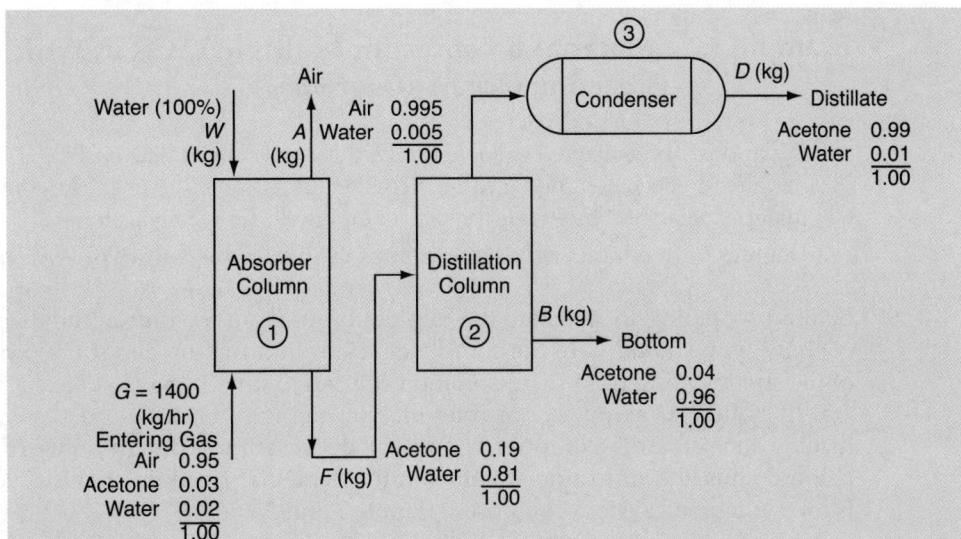

Figure E5.2

Solution

This is an open, steady-state process without reaction. Three subsystems exist, as labeled in Figure E5.2.

Steps 1–4

All of the stream compositions are given. All of the unknown stream flows are designated by letter symbols in the figure.

Step 5

Pick 1 hr as a basis so that $G = 1400$ kg.

Steps 6 and 7

We could start the analysis of the degrees of freedom with overall balances, but since the subsystems are connected serially, we start the analysis with unit 1 (absorber column), then proceed to unit 2 (distillation column), and then to unit 3 (condenser). In each step of the solution of the problem, as is usual, the variables whose values are not specifically listed in Figure E5.2 will be assumed to be equal to zero and not included in the count of variables.

Unit 1 (absorber)
   Variables: Eight component mass fractions whose values are known and are assigned (see Figure E5.2) plus three variables whose values are not known (the unknowns): W, F, and A (the three unknown flow streams).

(Continues)

Example 5.2　Material Balances for Multiple Units in Which No Reaction Occurs (*Continued*)

Equations: Three needed to solve for the three unknowns. The basis, $G = 1400$ kg, has already been assigned a value. What relations are left to use? The component material balances: three (one for each component: air, water, acetone).

Degrees of freedom: Zero. Thus, you can solve for the values of W, F, and A.

Before proceeding to calculate the degrees of freedom for unit 2 (the distillation column), you should note the complete lack of information about the properties of the stream going from the distillation column to unit 3 (the condenser). In general, it is best to avoid, if possible, making material balances on systems that include such streams as inputs or outputs, because they contain no useful information. Thus, the substitute system and the degree-of-freedom analysis we select is for a combined system composed of units 2 and 3.

Units 2 and 3 (distillation column plus condenser)

Variables: D and B (two streams) are the only *unknowns* because F is known from the absorber analysis; the component mass fractions in each stream, $2 \times 3 = 6$, are known values and have already been assigned values.

Equations: Two needed to achieve zero degrees of freedom. You can make two component balances, one for acetone and the other for water.

Degrees of freedom: Zero.

What would happen if a correct analysis of the degrees of freedom for a subsystem gave the result of +1? Then you would hope that the value for one of the unknowns in the subsystem could be determined from another subsystem in the overall system. In fact, for this example, if you started the analysis of the degrees of freedom with the combined units 2 plus 3, you would obtain a value of +1 because the value of F would not be known prior to solving the equations for unit 1.

Step 8 (Unit 1)

The mass balances for unit 1 are as follows:

	In	Out	
Air	400 (0.95)	$= A$ (0.995)	(a)
Acetone	1400 (0.03)	$= F$ (0.19)	(b)
Water	1400 (0.02) + W(1.00)	$= F$ (0.81) + A (0.005)	(c)

Check to make sure that the equations are independent.

Step 9

Solve Equation (a) for A, solve (b) for F, and then with these results solve (c) for W to get

$$A = 1336.7 \text{ kg/hr}$$
$$F = 221.05 \text{ kg/hr}$$
$$W = 157.7 \text{ kg/hr}$$

Step 10

(Check) Use the total mass balance equation:

$$G + W = A + F$$

$$
\begin{array}{rcl}
1400 & = & 1336 \\
157.7 & = & 221.05 \\
\hline
1557.7 & & 1557.1 \qquad \text{Close enough}
\end{array}
$$

Step 8 (Units 2 and 3)

The mass balances for units 2 plus 3 are

Acetone:	221.05 (0.19)	=	$D(0.99)+$	$B(0.04)$	(d)
Water:	221.05 (0.81)	=	$D(0.01)+$	$B(0.96)$	(e)

Step 9

Solve Equations (d) and (e) simultaneously to get

$$D = 34.91 \text{ kg/hr}$$
$$B = 186.1 \text{ kg/hr}$$

Step 10

(Check) Use the total balance:

$$F = D + B, \text{ or } 221.05; \quad 34.91 + 186.1 = 221.01 \quad \text{Close enough}$$

As a matter of interest, what other mass balances could be written for the system and substituted for any one of the Equations (a) through (e)? Typical balances would be the overall balances:

(Continues)

Example 5.2   Material Balances for Multiple Units in Which No Reaction Occurs (*Continued*)

	In	Out	
Air	G (0.95)	$= A$ (0.995)	(f)
Acetone	G (0.03)	$= D$ (0.99) + B (0.04)	(g)
Water	G (0.02)	$= A$ (0.005) + D (0.01) + B (0.96)	(h)
Total	$G + W$	$= A + D + B$	(i)

Equations (f) through (i) do not add any extra information to the problem; the degrees of freedom are still zero. But any of the equations can be substituted for one of Equations (a) through (e) as long as you make sure that the resulting set of equations is composed of independent equations.

Example 5. 3   Material Balances for Multiple Units in Which a Reaction Occurs

Problem Statement

In the face of higher fuel costs and the uncertainty of the supply of a particular fuel, many companies operate two furnaces, one fired with natural gas and the other with fuel oil. In the RAMAD Corp., each furnace has its own supply of oxygen; the oil furnace uses as a source of oxygen a stream that has the following composition: O_2 20%, N_2 76%, and CO_2 4%. The stack gases from both furnaces exit using a common stack. See Figure E5.3.

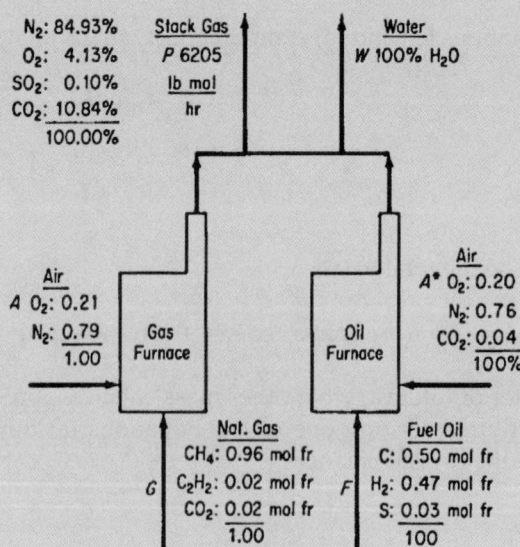

Figure E5.3

Note that two outputs are shown in the common stack to point out that the stack gas analysis is on a dry basis but *water vapor also exists*. The fuel oil composition is given in mole fractions to save you the bother of converting mass fractions to mole fractions.

During one blizzard, all transportation to the RAMAD Corporation was cut off, and officials were worried about the dwindling reserves of fuel oil because the natural-gas supply was being used at its maximum rate possible. The reserve of fuel oil was only 560 bbl. How many hours could the company operate before shutting down if no additional fuel oil was attainable? How many pound moles per hour of natural gas were being consumed? The minimum heating load for the company when translated into the stack gas output was 6205 lb mol/hr of dry stack gas. Analysis of the fuels and stack gas at this time is shown in Figure E5.3. The molecular weight of the fuel oil was 7.91 lb/lb mol, and its density was 7.578 lb/gal.

Solution

This is a steady-state open process with reaction. Two subsystems exist. We want to calculate F and G in pound moles per hour and then F in barrels per hour.

Steps 1–4

Even though species material balances could be used in solving this problem, we use element material balances because element balances are easier to apply for this problem. The units of all of the variables whose values are unknown will be pound moles. Rather than making balances for each furnace, since we do not have any information about the individual outlet streams of each furnace, we choose to make overall balances and thus draw the system boundary about both furnaces.

Step 5

$$\text{Basis: 1 hr, so that } P = 6205 \text{ lb mol}$$

Steps 6 and 7

The simplified degree-of-freedom analysis is as follows: You have five elements in the problem and five streams whose values are unknown: A, G, F, A^*, and W; hence, if the elemental mole balances are independent, you can obtain a unique solution for the problem.

(Continues)

Example 5.3   Material Balances for Multiple Units in Which a Reaction Occurs (*Continued*)

Step 8

The overall balances for the elements are (in pound moles):

	In	Out
H	$G[(0.96)(4) + (.02)(2)] + F(0.47)(2)$	$= 2W$
N	$A(0.79)(2) + (0.76)(2)A^*$	$= 6205(0.8493)(2)$
O	$A (0.21)(2) + A^*(0.20 + 0.04)(2) + G(0.02)(2)$	$= 6205(0.0413 + 0.001 + 0.1084)(2) + W$
S	$F(0.03)$	$= 6205(0.0010)$
C	$G(0.96 + (2)(0.02) + 0.02) + F(0.50) + 0.04A^*$	$= 6205(0.1084)$

The balances can be shown to be independent.

Step 9

Solve the set of this set of 5 equations and 5 unknowns. Although F can be calculated directly, a sequential solution by hand of the remaining equations is not convenient. Therefore, we employ a computer solution of these equations.

Computer Solution

Using the following definitions for unknowns (i.e., x_1–x_5): x_1 is equal to G; x_2 is equal to F; x_3 is equal to W; x_4 is equal to A; x_5 is equal to A^*, the coefficient matrix **A** and the constant vector **b** are

$$\mathbf{A} = \begin{bmatrix} 3.88 & 0.94 & -2 & 0 & 0 \\ 0 & 0 & 0 & 1.58 & 1.52 \\ 0.04 & 0 & -1 & 0.42 & 0.48 \\ 0 & 0.03 & 0 & 0 & 0 \\ 1.02 & 0.5 & 0 & 0 & 0.40 \end{bmatrix} \qquad \mathbf{b} = \begin{bmatrix} 0 \\ 10,540 \\ 1870.2 \\ 6.05 \\ 672.6 \end{bmatrix}$$

MATLAB:

MATLAB Solution for Example 5.3

``
%   PROGRAM
function Ex5_3
clear; clc;
A=[3.88,0.94,-2,0,0;0,0,6.25,1.58,1.52;0.04,0,
-1,0.42,0.48;0,0.03,0,0,0;1.02,0.5,0,0.04,0];
k=rank(A);
b=[0,10540,1870.2,6.205,672.7]';
x=A\b;
fprintf('x1 =%8.2f x2=%8.2f\n',x(1),x(2))
fprintf('x3 =%8.2f x4=%8.2f\n',x(3),x(4))
fprintf('x5 =%8.2f rank=%2d\n',x(5),k)
end
%   PROGRAM END
   x1 = 501.06 x2 = 201.67
   x3 =1066.67 x4 =5211.46
   x5 =1517.13 rank = 5
``

Python:

Python Code for Example 5.11

``
Ex5_3.py
 	
# PROGRAM 	
 	
Import numpy as np
Import numpy.linalg
# Specify the coefficient matrix A 	
A=np.array([[3.88,0.94,-2,0,0],[0,0,6.25,1.58,1.52],[0.04,0,-1,
0.42,0.48],[0,0.03,0,0,0],[1.02,0.5,0,0.04,0]])
k=numpy.linalg.rank(A)
b=np.array([0,10540,1870.2 ,6.205,672.7])
xsol=numpy.linalg.solve(A,b)
 	
# PROGRAM END 	
``

(*Continues*)

Example 5.3   Material Balances for Multiple Units in Which a Reaction Occurs (*Continued*)

IPython console:

In[1]: runfile(…
In[2]: xsol, k
Out[2]: array([501.06, 201.67, 1066.84, 5211.36, 1517.13]),5

Even though the precision of the SO_2 concentration in the fuel oil is only one significant figure, the calculated results are

$$F = 202 \text{ lb mol/hr}$$
$$G = 501 \text{ lb mol/hr}$$
$$W = 1067 \text{ lb mol/hr}$$
$$A = 5211 \text{ lb mol/hr}$$
$$A^* = 1517 \text{ lb mol/hr}$$

Finally, the fuel oil consumption is

$$\frac{202 \text{ lb mol}}{\text{hr}} \left| \frac{7.91 \text{ lb}}{\text{lb mol}} \right| \frac{\text{gal}}{7.578 \text{ lb}} \left| \frac{\text{bbl}}{42 \text{ gal}} \right. = 5.02 \text{ bbl/hr}$$

If the fuel oil reserves were only 560 bbl, they could last at the most

$$\frac{560 \text{ bbl}}{5.02 \dfrac{\text{bbl}}{\text{hr}}} = 109 \text{ hr}$$

Example 5.4   Analysis of a Sugar Recovery Process Involving Multiple Serial Units

Problem Statement

Figure E5.4 shows the process and the known data. You are asked to calculate the compositions of every flow stream, and the fraction of the sugar in the cane (*F*) that is recovered in *M*.

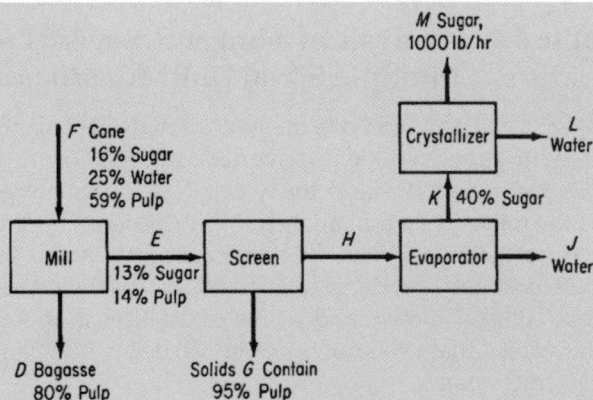

Figure E5.4

Solution

Steps 1–4

All of the known data have been placed on Figure E5.4. The process is an open, steady-state system without reaction. If you examine the figure, two questions naturally arise: What basis should you pick, and what system should you pick to start the analysis? The selection of certain bases and certain systems can lead to more equations that have to be solved simultaneously than others. You could pick as a basis $F = 100$ lb, or $M = 1000$ lb (the same as 1 hr), or the value of any of the intermediate flow streams. You could pick an overall process as the system to start with, or any of the individual units, or any consecutive combinations of units. We pick M because its value is specified.

Step 5

$$\text{Basis: 1 hr } (M = 1000 \text{ lb})$$

Steps 6 and 7

Another important determination you must make is what are the compositions of streams D, E, G, and H. Stream F has three components, and presumably stream K contains only sugar and water. Does stream H contain pulp? Presumably not, because if you inspect the process flowsheet, you will not find any pulp exiting anywhere downstream of the evaporator. Presumably streams D and G contain sugar and water because the problem implies that not all of the sugar in stream F is recovered. What happens if you assume streams D and G contain neither water

(Continues)

Example 5.4   Analysis of a Sugar Recovery Process Involving Multiple Serial Units (*Continued*)

nor sugar? Then you would write a set of material balances that are not independent and/or are inconsistent (have no solution). Try it. Let S stand for sugar, P stand for pulp, and W stand for water. Their units are pounds. Pick the crystallizer as the initial system to analyze. Why? Because (a) if you check the degrees of freedom for the crystallizer, the values of only a small set of unknowns must be determined to get zero degrees of freedom; (b) the crystallizer involves the feed M whose value is known; and (c) the crystallizer is at one end of the process. Assign the values that are known to the variables (the components in the streams are mass fractions):

$$M = 1000 \qquad L = ? \qquad\qquad\qquad K = ?$$

$$\omega^{M}_{sugar} = 1.00 \quad \omega^{L}_{water} = 1.00 \quad \omega^{K}_{sugar} = 0.40 \ \text{ and } \ \omega^{K}_{water} = 0.60$$

The unknowns are K and L. You can make two species balance, sugar and water. What about the implicit equations? We have used all three to assign the compositions in M, K, and L. Consequently, the degrees of freedom are zero, and the crystallizer seems to be a good subsystem with which to start.

If you had picked another basis, say, $F = 100$ lb, and another subsystem, say, the mill, to start with, you would have four unknowns: D, E, ω^{D}_{sugar}, and ω^{D}_{water} (assuming you had assigned $\omega^{E}_{water} = 0.73$ initially). You could make three species balance and employ one implicit equation, $\sum \omega^{D}_{i} = 1$; hence, the degrees of freedom would be zero. But you would have to solve four simultaneous equations. Therefore, **as a general rule, the basis you choose and the unit with which you start the analysis in a multiunit process affect the degree of complexity of your calculations.**

Steps 8 and 9

For the crystallizer the equations are

$$\text{Sugar: } K(0.40) = L(0) + 1000$$
$$\text{Water: } K(0.60) = L + 0$$

from which you get $K = 2500$ lb and $L = 1500$ lb.

Step 10

Check using the total flows:

$$2500 = 1500 + 1000 = 2500$$

Next, we need to choose another system for applying material balances. Choosing the evaporator will be problematic because the composition of stream H is unknown. Instead, let's lump the evaporator and the screen together so that stream H is inside the system that we are analyzing. But the composition of sugar and water in G is unknown and amounts to 5% of stream G. Let's look at the screening process physically. The screening process separates most of the liquid from the pulp. So, let's assume that the sugar solution in stream G has the same composition as the liquid in stream E. Therefore, therefore, the composition of sugar in G is equal to $5\% \times 13\%/86\% = 0.756\%$. This assumption in effect assumes that there is no undissolved sugar in the pulp in stream G.

Now that all the compositions for the screen and evaporator are known, this system has three unknowns (E, G, and J), but we can write three independent mass balance equations so that there are zero degrees of freedom. Consider the pulp and sugar balance because these equations do not involve stream J because stream J is pure water. Using the pulp balance, we can eliminate G from the sugar balance so that E can be calculated. Then the pulp balance can be applied to calculate G. Finally, the water balance is used to determine J. The results are

$$E = 7759 \text{ lb} \quad G = 1143 \text{ lb} \quad J = 4115 \text{ lb}$$

Using these values, it is direct to calculate H and its composition:

$$H = 6615 \text{ lb and } \omega_S^H = 15.1\% \text{ and } \omega_W^H = 84.9\%$$

At this point, only F, D, and D's composition remain to be determined. If we use a pulp balance and an overall balance on the mill, we can calculate D and F by solving a set of two equations and two unknowns, which yields

$$F = 24{,}385 \text{ lb} \quad \text{and} \quad D = 16{,}626 \text{ lb}$$

Finally, a sugar balance is used to determine the concentration of sugar in stream D. Then the implicit equation determines the concentration of water in stream D.

$$\omega_S^D = 17.34\% \text{ and } \omega_W^D = 2.65\%$$

Then the percent sugar recovered is $1000/[(24{,}385)(0.16)] \times 100\% = 25.6\%$.

Example 5.5   Acetylene Process

Problem Statement

In the process for the production of pure acetylene, C_2H_2 (see Figure E5.5), pure methane (CH_4) and pure oxygen are combined in the burner, where the following reactions occur:

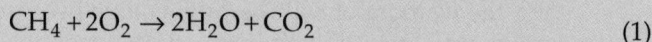

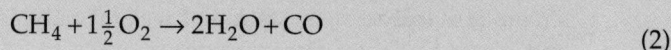

$$CH_4 + 2O_2 \rightarrow 2H_2O + CO_2 \tag{1}$$

$$CH_4 + 1\tfrac{1}{2}O_2 \rightarrow 2H_2O + CO \tag{2}$$

$$2CH_4 \rightarrow C_2H_2 + 3H_2 \tag{3}$$

a. Calculate the ratio of the moles of O_2 to the moles of CH_4 fed to the burner.

b. On the basis of 100 lb mol of gases leaving the condenser, calculate how many pounds of water are removed by the condenser.

c. What is the overall percentage yield of product (pure) C_2H_2, based on the methane feed entering the burner?

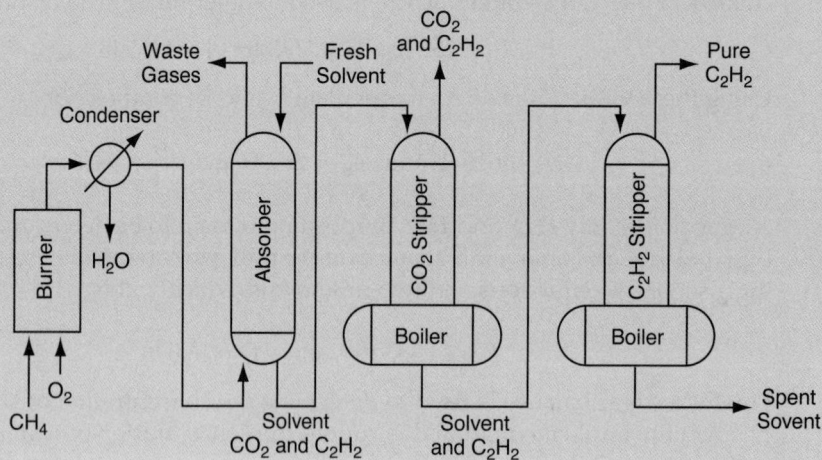

Figure E5.5

The gases from the burner are cooled in the condenser that removes all of the water. The analysis of the gases leaving the condenser is as follows:

Component	mol %
C_2H_2	8.5
H_2	25.5
CO	58.3
CO_2	3.7
CH_4	4.0
Total	100.0

These gases are sent to an absorber where 97% of the C_2H_2 and essentially all of the CO_2 are removed with the solvent. The solvent from the absorber is sent to the CO_2 stripper, where all of the CO_2 is removed. The analysis of the gas stream leaving the top of the CO_2 stripper is 7.5 mole % C_2H_2 and 92.5 mol % CO_2. The solvent from the CO_2 stripper is pumped to the C_2H_2 stripper, which removes all of the C_2H_2 as a pure product.

Solution

Steps 1–4

The problem statement contains the diagram and the specified information to solve this problem.

Step 5

Basis: 100 lb mol of gas leaving the condenser

Steps 6–9

Instead of applying each step individually, we solve this problem directly without a degree-of-freedom analysis. From the basis and the known reactions, we can directly determine the extent of reaction for each of the three reactions:

$\xi_1 = 3.7$ lb mol reacting   (based on CO_2 generation)

$\xi_2 = 58.3$ lb mol reacting   (based on CO generation)

$\xi_3 = 25.5 / 3 = 8.5$ lb mol reacting   (based on H_2 generation)

(Continues)

Example 5.5 Acetylene Process (*Continued*)

From the condenser gas analysis, you can see that O_2 was the limiting reactant and CH_4 was the excess reactant. Because O_2 is completely consumed by reactions 1 and 2, the corresponding feed rate of O_2 is equal to $2\xi_1 + 1.5\xi_2 = 94.84$ lb mol. The feed rate of CH_4 is equal to $\xi_1 + \xi_2 + 2\xi_3 + 4 = 83$ lb mol. Therefore,

(a) The ratio of the feed rate of O_2 to CH_4 is 1.14.

From the process diagram, note that all the water produced by reaction is removed by the condenser. Therefore,

(b) Water removed by condenser $= 2\xi_1 + 2\xi_2 = 124$ lb mol water $= 2234$ lb_m.

From Figure E5.5, you can see that all the CO_2 produced in the burner leaves the process in the overhead of the CO_2 stripper. And all the CO_2 is produced by the first reaction. Therefore, the amount of CO_2 leaving the overhead of the CO_2 stripper is ξ_1. In addition, the composition of the overhead of the CO_2 stripper is provided in the problem statement. The amount of C_2H_2 lost in the overhead of the CO_2 stripper is equal to $0.075\ \xi_1/0.925 = 0.300$ lb mol. Then the corresponding production rate of C_2H_2 is $2\xi_3 - 0.300 = 16.7$ lb mol. Therefore,

(c) The percent yield of C_2H_2 based on CH_4 feed $= 2 \times 16.7/83 = 40.2\%$. Note that the factor of 2 appears in this calculation because it takes 2 moles of CH_4 to produce 1 mole of C_2H_2.

Self-Assessment Test

Questions

1. Can a system be composed of more than one unit or piece of equipment?
2. Can one piece of equipment be treated as a set of several subsystems?
3. Does a flowsheet for a process have to show one subsystem for each process unit that is connected to one or more other process units?
4. If you count the degrees of freedom for each individual unit (subsystem) and sum them up, can their total be different from the degrees of freedom for the overall system?

Answers

1. Yes
2. Yes, even if the process is continuous.

3. It does not have to, but it is usually helpful if it does.

4. No, if the degrees of freedom for each subsystem are determined properly.

Problems

1. A two-stage separation unit is shown in Figure SAT5.2P1. Given that the input stream $F1$ is 1000 lb/hr, calculate the value of $F2$ and the composition of $F2$.

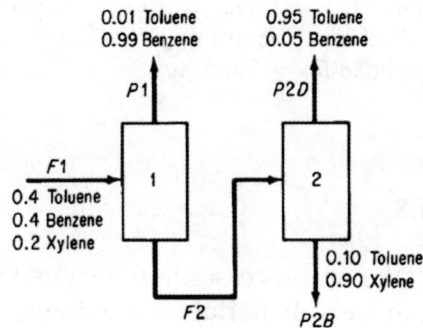

Figure SAT5.2P1

2. A simplified process for the production of SO_3 to be used in the manufacture of sulfuric acid is illustrated in Figure SAT5.2P2. Sulfur is burned with 100% excess air in the burner, but for the reaction $S + O_2 \rightarrow SO_2$, only 90% conversion of the S to SO_2 is achieved in the burner. In the converter, the conversion of SO_2 to SO_3 is 95% complete. Calculate the kilograms of air required per 100 kg of sulfur burned, and the concentrations of the components in the exit gas from the burner and from the converter in mole fractions.

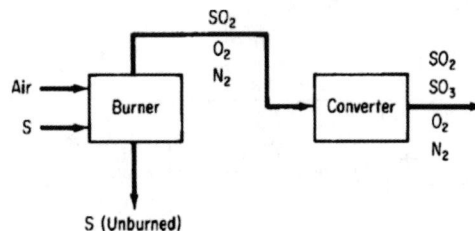

Figure SAT5.2P2

Answers

1. Xylene is a tie component: $P2_{Xyl} = 200$. $P2B = 200/.9 = 222.22$; overall toluene (Tol) balance: $0.01P1 + 0.95P2D = 377.77$; overall benzene (Bz) balance: $0.99P1 + 0.05P2D = 400$. Solving these two equations simultaneously yields $P2D = 393.40$ and $P1 = 384.17$. Tol balance on first unit: $F2_{tol} = 396.16$; Bz balance on first unit: $F2_{Bz} = 19.67$; $F2 = 615.83$, $x_{F2,Tol} = 64.3\%$, $x_{F2,Bz} = 3.2\%$, $x_{F2,Tol} = 32.5\%$.

2. Basis: 100 kg S = 3.1188 kg mol S; O_2 feed = 6.2375; air feed = 29.7025 kg mol. Burner: N_2: 23.465; SO_2: 2.807; O_2: $6.2375 - .9(3.1188) = 3.431$; $y_{SO2} = 9.45\%$; $y_{O2} = 11.55\%$; $y_{N2} = 79.00\%$. Converter: N_2: 23.465; SO_2: 0.1404; SO_3: 2.6667; O_2: $3.431 - .95(0.5)(2.807) = 2.098$; $y_{SO3} = 9.40\%$; $y_{SO2} = 0.50\%$; $y_{O2} = 7.40\%$; $y_{N2} = 82.71\%$.

5.3   Recycle Systems

In this section we take up processes in which material is **recycled,** that is, a stream containing the feed material from a downstream unit is routed back to an upstream unit, as shown in Figure 5.4c. The stream containing the re-cycled material is known as a **recycle stream**. What is a **recycle system**? A recycle system is a system that includes one or more recycle streams.

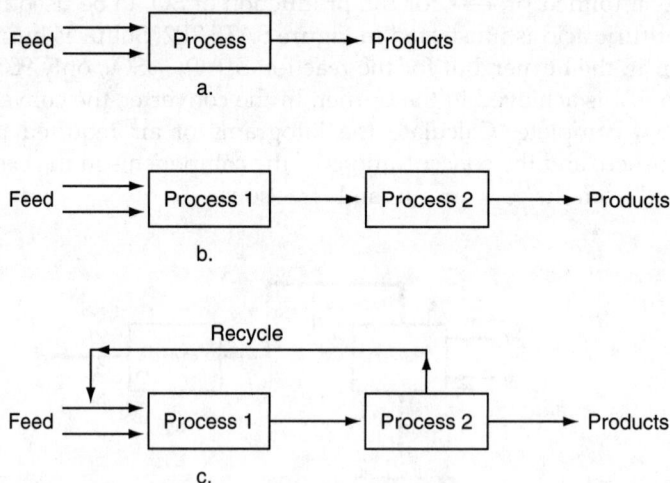

Figure 5.4   (a) A single unit with serial flows; (b) multiple units but still with serial flows; (c) multiple units with the addition of recycle

You can see in Figure 5.4c that the recycle stream is mixed with the feed stream, and the combination is fed to Process 1. In Process 2 (Figure 5.4c), the outputs from Process 1 are separated into (a) the products and (b) the recycle stream. The recycle stream is returned to Process 1 for further processing.

Recycle systems can be found in everyday life. Used newspaper is collected from households, processed to remove the ink, and used to print new newspapers. Clearly, the more newspapers recycled, the fewer trees are consumed to produce newspapers. Recycling of glass, aluminum cans, plastics, copper, and iron are also common and directly affect sustainability (Section 1.4).

Recycle systems also occur in nature. For example, consider the water cycle shown in Figure 5.5. If a region of Earth is denoted as the system, the recycle stream consists of evaporated water that condenses, falls to Earth as precipitation, and subsequently becomes the flow of water in creeks and rivers into the body of water. Evaporation from a body of water returns (recycles) water into the clouds.

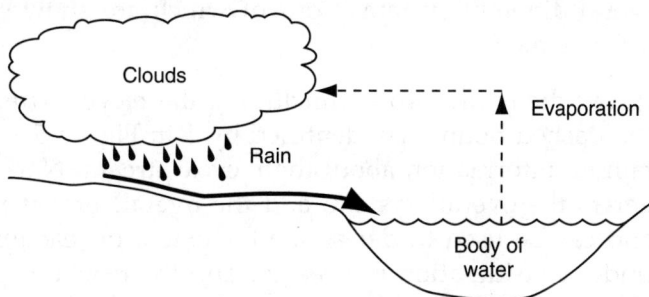

Figure 5.5   A portion of the water cycle

Because of the relatively high cost of industrial feedstocks, when chemical reactions are involved in a process, recycle of unused reactants to the reactor can offer significant economic savings for high-volume processing systems. Heat recovery within a processing unit (comprising energy recycle) reduces the overall energy consumption of a process.

You can formulate material balances for recycle systems without reaction using the same strategy explained in Chapter 3 and applied in subsequent chapters. Examine Figure 5.6.

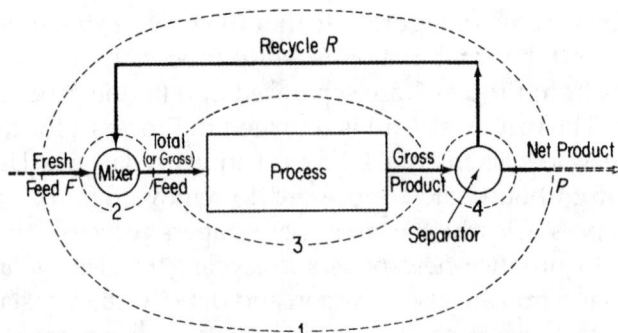

Figure 5.6   Process with recycle. The numbers and dashed lines designate possible system boundaries for the material balances; see the text.

The main new aspect of solving problems involving recycle is picking the proper sequence of systems to analyze. You can write material balances for several different systems, four of which are shown by dashed lines in Figure 5.6, namely:

1. About the entire process including the recycle stream, as indicated by the dashed boundary identified by 1 in Figure 5.6. Such balances contain no information about the recycle stream. Note that the **fresh feed** enters the overall system, and the **overall** or **net product** is removed and can be used to determine the extent of reaction when the process under consideration is a reactor and the reactions are known. You can also employ element balances, as explained in Chapter 4.

2. About the junction point at which the fresh feed is combined with the recycle stream (identified by 2 in Figure 5.6).

3. About the process itself (identified by 3 in Figure 5.6). Note that the **process feed** enters the process and the **gross product** is removed.

4. About the separator at which the gross product is separated into recycle and overall (net) product (identified by 4 in Figure 5.6).

In addition, you can make balances about combinations of subsystems, such as the process plus the separator. Only three of the four balances you make for the systems denoted by 1 to 4 are independent, whether made for the total mass or a particular component mass. However, balance 1 will not include the recycle stream, so the balance will not be directly useful in calculating a value for the recycle R. Balances 2 and 4 do include R. You could write a material balance for the combination of subsystems 2 and 3 or 3 and 4 and include the recycle stream.

Note that in Figure 5.6, the recycle stream is associated both with the mixer, which is located at the beginning of the process, and with the separator, which is located at the end of the process. As a result, recycle problems lead to coupled equations that have to be solved simultaneously. You will find that overall material balances (1 in Figure 5.6), particularly with a tie component, are usually a good place to start when solving recycle problems. If you solve an overall material balance(s) successfully, that is, you are able to calculate all or some of the unknowns by using the overall balance, the rest of the problem can usually be solved by sequentially applying single-unit material balances traversing sequentially through the process. Otherwise, you must write a number of material balance equations and solve them simultaneously.

Frequently Asked Question

If you feed material to a stream continuously as in Figure 5.4c, why does the value of the material in the recycle stream not increase and continue to build up? What is done in this chapter and subsequent chapters is to assume, often without so stating, that the entire process including all units *is in the steady state*. As the process starts up or shuts down, the flows in many of the streams change, but once the steady state is reached, "what goes in must come out" applies to the recycle stream as well as the other streams in the process.

5.3.1   Recycle without Reaction

Recycle of material occurs in a variety of processes that do not involve chemical reaction, including distillation, crystallization, and heating and refrigeration systems. As an example of a recycle system used to dilute a process stream, look at the process of drying lumber shown in Figure 5.7. If dry air is used to dry the wood, the lumber will warp and crack. By recycling the moist air that exits from the dryer and mixing it with outdoor dry air, the inlet air can be maintained at a safe water content to prevent warping and cracking of the lumber while slowly drying it.

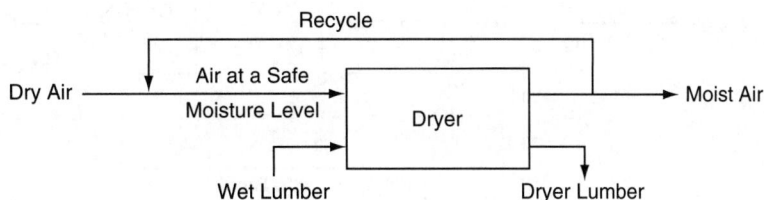

Figure 5.7   Lumber-drying process

Another example, which is shown in Figure 5.8, is a two-product distillation column. Note that a portion of the exit flow from the accumulator is recycled into the column as liquid **reflux** while the reboiler vaporizes part of the liquid in the bottom of the column to create the vapor flow up the column. The recycle of vapor from the reboiler and return of the liquid from the accumulator into the column maintain good vapor and liquid contact on the trays inside the column. This contact makes it possible to concentrate the more volatile components into the overhead vapor stream, leading to the condenser and then to the accumulator, and concentrating the less volatile components into the liquid collected in the bottom of the column.

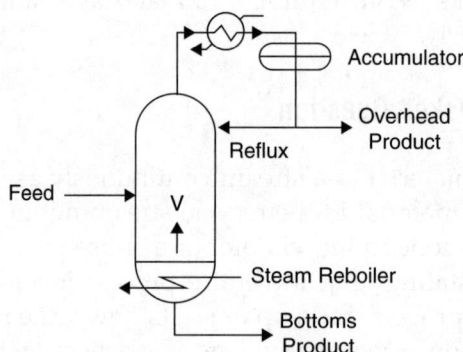

Figure 5.8   Schematic of a two-product distillation column

Example 5.6   Continuous Filtration Involving a Recycle Stream

Problem Statement

Figure E5.6 is a schematic of a process for the production of biomass (denoted as Bio) that is to be used in the production of drugs.

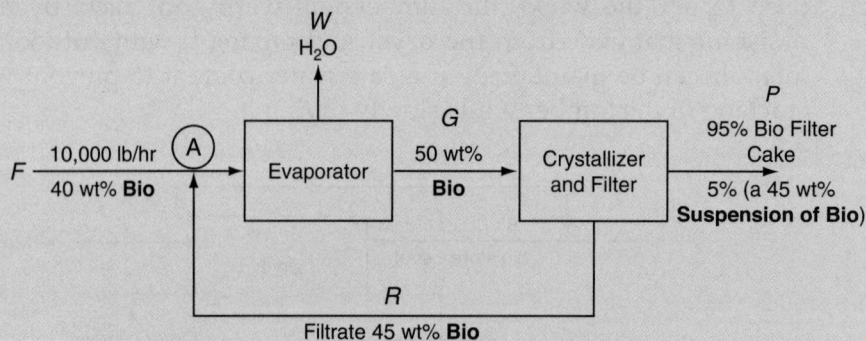

Figure E5.6   Schematic of a process to produce biomass

The fresh feed F to the process is 10,000 lb/hr of a 40 wt % aqueous biomass in suspension. The fresh feed is combined with the recycled filtrate from the filter and fed to the evaporator, where water is removed to produce a 50 wt % Bio solution, which is fed to the filter. The filter produces a filter cake that is composed of 95 wt % dry biomass wet by a 5 wt % solution that in the lab proves to be composed of 55 wt % water with the rest dry biomass. The filtrate contains 45 wt % biomass.

a. Determine the flow rate of water removed by the evaporator and the recycle rate for this process.

b. Assume that the same production rate of filter cake occurs, but that the filtrate is not recycled. What would be the process feed rate of 40 wt % biomass then? Assume that the product solution from the evaporator still contains 50 wt % biomass in water.

Solution

Solution a

Steps 1–4

Figure E5.6 contains the information needed to solve the problem.

Step 5

Basis: 10,000 lb fresh feed (equivalent to 1 hr)

Steps 6 and 7

The unknowns are W, G, P, and R, since you can assign values to F and all of the mass fractions in the respective streams using appropriate implicit equations. You can make two component balances about three subsystems: the mixing point A, the evaporator, and the filter (as well as a total balance for each). You can also make similar balances for various combinations of subsystems and the overall system. What balances should you choose to solve the problem? If you put the equations in an equation solver, it does not make any difference as long as the equations are independent. But if you solve the problem by hand and are going to make two component balances for each of the three subsystems just mentioned and the overall system, you can count the number of unknowns involved for each of the three subsystems and the overall system:

Mixing point: P plus feed (and compositions) to evaporator (not labeled in Figure E5.6)
Evaporator: W, G, and feed to evaporator
Filter: G, P, and R
Overall: W and P

(Continues)

Example 5.6   Continuous Filtration Involving a Recycle Stream (*Continued*)

Can you conclude that by using just two overall component balances you can determine the values of at least W and P? Consequently, let's start with overall balances.

Steps 8 and 9

Overall Bio balance: $(0.4)(10,000) = [0.95 + (0.45)(0.05)]P$;

$$P = 4113 \text{ lb}$$

Overall H_2O balance: $(0.6)(10,000) = W + [(0.55)(0.05)](4113)$;

$$W = 5887 \text{ lb}$$

The total amount of Bio exiting with P is $[(0.95) + (0.45)(0.05)](4113)$

$$= 4000 \text{ lb}$$

Are you surprised at this result? The amount of water in P is 113 lb. As a check, $113 + 5887 = 6000$ lb as expected.

Steps 6 and 7 (repeated)

Now that you know W and P, the next step is to make balances on a subsystem that involves the stream R. Choose either the mixing point A or the filter. Which one would you pick? The filter involves three variables, and you now know the value of P, so only two unknowns would be involved as opposed to more if you chose mixing point A as the system.

Bio balance on the filter: $0.5G = 4000 + 0.45R$
H_2O balance on the filter: $0.5G = 113 + 0.55R$; $R = 870$ lb in 1 hr
As a check:
$$10,000 + 870 - 5887 = G = 4983 = 4113 + 870 \quad \text{OK}$$

Solution b

Now, suppose that recycle from the filter does not occur, but the production rate of P remains the same. Note that for this case, R would be discharged instead of recycled and mixed with the fresh feed. How should you proceed? Do you recognize that the problem is one of the class that you read about in Section 5.2?

Step 5 (repeated)

The basis is now $P = 4113$ lb (the same as 1 hr).

Steps 6 and 7 (repeated)

The unknowns are now *F*, *W*, *G*, and *R*. You can make two component balances on the evaporator and two on the filter, plus two overall component balances. Only four are independent. The evaporator balances would involve *F*, *W*, and *G*. The crystallizer balances would involve *G* and *R*, while the overall balances would involve *F*, *W*, and *R*. Which balances are best to start with? If you put the equations in an equation solver, it makes no difference which four equations you use as long as they are independent. The filter balances are best because you have to solve just two pertinent equations for *G* and *R*.

Steps 8 and 9

> Bio balance on the filter: $0.5G = [(0.95) + (0.05)(0.45)](4113) + 0.45R$
> H_2O balance on the filter: $0.5G = [(0.05)(0.55)](4113) + 0.55R$
> Solve simultaneously: R = 38,870 lb in 1 hr

Note that without recycle, the feed rate must be 5.37 times larger to produce the same amount of product, not to mention the fact that you would have to dispose of a large volume of filtrate, which is 45 wt % Bio.

5.3.2   Recycle with Chemical Reaction

The most common application of recycle for systems involving chemical reaction is the recycle of reactants, an application that is used to increase the overall conversion in a reactor. Figure 5.9 shows a simple example for the reaction (A → 2B).

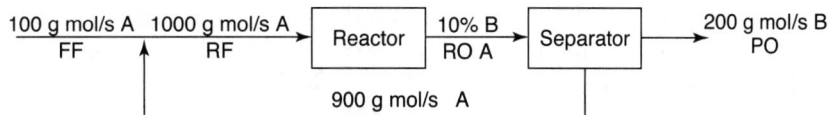

Figure 5.9   A simple recycle system with chemical reaction (A → 2B): FF (fresh feed); RF (reactor feed); RO (reactor outlet); PO (process outlet)

You will encounter two different types of conversions that can be applied to processes in which reaction occurs:

Overall conversion is based on what enters and leaves the overall process:

Overall conversion = f_{OA} =

$$\frac{\text{(moles of limiting reactant in the fresh feed)} - \text{(moles of limiting reactant in the output of the overall process)}}{\text{(moles of limiting reactant in the fresh feed)}} \quad (5.1)$$

The single-pass conversion is based on what enters and leaves the reactor:

Single - pass conversion = f_{SP} =

$$\frac{\text{(moles of limiting reactant fed into the reactor)} - \text{(moles of limiting reactant exiting the reactor [gross process])}}{\text{(moles of limiting reactant fed into the reactor)}} \quad (5.2)$$

For the data in Figure 5.9, what is the overall fraction conversion?

$$f_{OA} = \frac{100 \text{ A} - 0}{100 \text{ A}} = 1.00$$

$$f_{SP} = \frac{1000 \text{ A} - 900}{1000 \text{ A}} = 0.10$$

Therefore, by using recycle, a reactor with a 10% conversion per pass can be used to produce an overall 100% conversion when the separation of the feed from the product is highly effective.

When the fresh feed (FF) consists of more than one reactant, the conversion can be expressed for a defined single component, usually the limiting reactant, or the most important (expensive) reactant. Recall from Chapter 4 that conversion in a reactor (the single-pass conversion) can be limited by chemical equilibrium and/or chemical kinetics. On the other hand, the overall conversion is strongly dependent upon the efficiency of the separator in separating compounds to be recycled from the other compounds.

How are the overall conversion and single-pass conversion related to the extent of reaction, ξ, a term that was discussed in Chapter 4? Do you recall the relation given by Equation (4.10)?

$$\xi = \frac{(-f)n_{\text{limiting reactant}}^{\text{in}}}{v_{\text{limiting reactant}}}$$

To be more specific in the notation, let $n_{\text{limiting reactant}}^{\text{in}}$ be denoted by $n_{\text{LR}}^{\text{FF}}$, $v_{\text{limiting reactant}}$ be denoted by v_{LR}, and f be denoted by f_{OA} or f_{SP}, respectively. Then

$$\text{Overall conversion for the limiting reactant} = f_{\text{OA}} = \frac{-v_{\text{LR}}\xi}{n_{\text{LR}}^{\text{FF}}} \tag{5.3}$$

and for the single-pass conversion where n_{LR}^{RF} is the total feed rate of the limiting reactant to the reactor (i.e., limiting reactant in fresh feed plus limiting reactant in recycle stream).

$$\text{Single-pass conversion} = f_{\text{SP}} = \frac{-v_{\text{LR}}\xi}{n_{\text{LR}}^{\text{RF}}} \tag{5.4}$$

If you solve Equations (5.3) and (5.4) for the extent of reaction, equate the extents of reaction, and use a material balance at the junction of the fresh feed and the recycle stream (a mixing point), $n_{\text{LR}}^{\text{RF}} = n_{\text{LR}}^{\text{FF}} + n_{\text{LR}}^{\text{recycle}}$, you can obtain the following relationship between overall and single-pass conversion:

$$\frac{f_{\text{SP}}}{f_{\text{OP}}} = \frac{n_{\text{LR}}^{\text{FF}}}{n_{\text{LR}}^{\text{FF}} + n_{\text{LR}}^{\text{recycle}}} = \frac{n_{\text{LR}}^{\text{FF}}}{n_{\text{LR}}^{\text{RF}}} \tag{5.5}$$

If you now apply Equation (5.3) to the simple recycle example in Figure 5.9, what value do you get for the ratio of the single-pass to overall conversion? Do you get 0.1, which agrees with the result stated previously? The same relationship can be demonstrated mathematically rather than numerically. In general, **the extent of reaction is the same regardless of whether an overall material balance is used or a material balance for the reactor is used.** This fact is quite useful when solving material balances for recycle systems with reactions.

Example 5.7   Recycle in a Process in Which a Reaction Occurs

Problem Statement

Cyclohexane (C_6H_{12}) can be made by the reaction of benzene (Bz, short for C_6H_6) with hydrogen according to the following reaction:

$$C_6H_6 + 3H_2 \rightarrow C_6H_{12}$$

(*Continues*)

Example 5.7   Recycle in a Process in Which a Reaction Occurs (*Continued*)

For the process shown in Figure E5.7, determine the ratio of the recycle stream to the fresh feed stream if the overall conversion of benzene is 95% and the single-pass conversion through the reactor is 20%. Assume that 20% excess hydrogen is used in the fresh feed, and that the composition of the recycle stream is 22.74 mol % benzene and 78.26 mol % hydrogen.

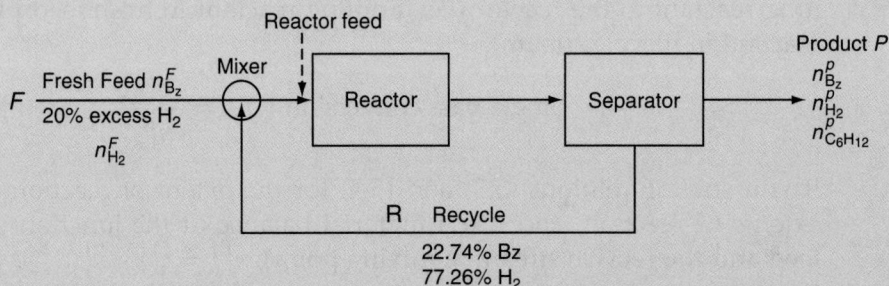

Figure E5.7   Schematic of a recycle reactor

Note that in this example, there is a relatively low conversion per pass (20%) and the overall conversion is relatively high (95%). A low conversion per pass can be desirable in certain cases, for example, a case in which the yield decreases as the conversion increases. By using recycle, you can obtain a high yield and at the same time a high overall conversion (i.e., high utilization of the limiting reactant).

Solution

The process is open and steady-state.

Steps 1–4

Figure E5.7 contains all of the information needed to solve the problem.

Step 5–7

$$\text{Basis: 100 moles of fresh benzene feed } (n_{B_z}^F)$$

The amount of H_2 feed, which is in 20% excess (for complete reaction, remember), is

$$n_{H_2}^F = 100(3)(1+0.20) = 360\,\text{mol}$$

and the total fresh feed is 460 mol.

From Equation (5.3) for benzene ($v_{Bz} = -1$),

$$0.95 = \frac{-(-1)\xi}{100}$$

you can calculate that $\xi = 95$ reacting moles.

The unknowns are R, n_{Bz}^P, $n_{H_2}^P$, and $n_{C_6H_{12}}^P$ You can write three species balances for each of the three systems—the mixing point, the reactor and the separator—plus overall balances (not all of which are independent, of course). Which systems should you choose to start with? The overall process, because then you can use the value calculated for the extent of reaction.

Steps 8 and 9

The species overall balances are

$$n_i^{out} = n_i^{in} + v_i\xi$$

$$\text{Bz: } n_{Bz}^P = 100 + (-1)(95) = 5\,\text{mol}$$

$$\text{H}_2\text{: } n_{H_2}^P = 360 + (-3)(95) = 75\,\text{mol}$$

$$\text{C}_6\text{H}_{12}\text{: } n_{C_6H_{12}}^P = 0 + (1)(95) = 95\,\text{mol}$$

$$\text{Total:} \qquad\qquad\qquad P = 175\,\text{mol}$$

The next step is to use the final piece of information, the information about the single-pass conversion plus Equation (5.4), to get R. The amount of the Bz feed to the reactor is $100 + 0.2274R$, and $\xi = 95$ (the same as was calculated from the overall conversion). Thus,

$$0.20 = \frac{95}{100 + 0.2274R}$$

and solving for R yields

$$R = 1649\ \text{mol}$$

Finally, the ratio of recycle to fresh feed is

$$\frac{R}{F} = \frac{1659\ \text{mol}}{460\ \text{mol}} = 3.59$$

Example 5.8   Recycle in a Process with a Reaction Occurring

Problem Statement

Immobilized glucose isomerase is used as a catalyst in producing fructose from glucose in a fixed-bed reactor (water is the solvent). For the system shown in Figure E5.8a, what percent conversion of glucose results on one pass through the reactor when the ratio of the exit stream to the recycle stream in mass units is equal to 8.33? The reaction is

$$C_{12}H_{22}O_{11} \rightarrow C_{12}H_{22}O_{11}$$

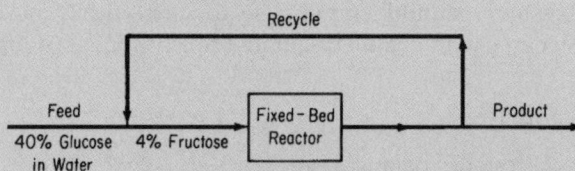

Figure E5.8a

Solution

Steps 1–4

The process is a steady-state process with a reaction occurring and recycle. Figure E5.8b includes all of the known and unknown values of the variables using appropriate notation (W stands for water, G for glucose, and F for fructose). Note that the recycle stream and product stream have the same composition, and consequently, the same mass symbols are used in the diagram for each stream.

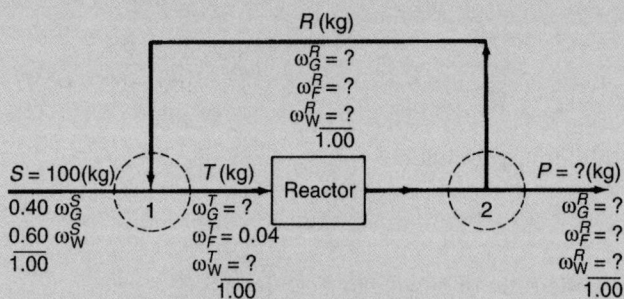

Figure E5.8b

Step 5

Pick as a basis $S = 100$ kg given the data shown in Figure E5.8b.

Step 6

We have not provided notation for the reactor exit stream and composition as we will not be using these values in our balances. Let f be the fractional conversion for one pass through the reactor. The unknowns are R, P, T, ω_G^R, ω_F^R, ω_W^R, ω_G^T, ω_W^T, and f, for a total of nine.

Step 7

The balances are $\sum \omega_i^R = 1$, $\sum \omega_i^T = 1$, $R = P/8.33$, plus three species balances each on the mixing point 1, the separator 2, and the reactor, as well as overall balances. We will assume we can find nine independent balances among the lot and proceed. We do not have to solve all of the equations simultaneously. The units are mass (kilograms).

Steps 8 and 9

Overall balances:

$$\text{Total: } P = 100 \text{ kg (How simple!)}$$

Consequently,

$$R = \frac{100}{8.33} = 12.00 \text{ kg}$$

Overall, no water is generated or consumed (i.e., water is not involved in the reaction); hence,

$$100(0.60) = P\omega_W^R = 100\omega_W^R$$

Water

$$\omega_W^R = 0.60$$

We now have six unknowns left for which to solve. We start somewhat arbitrarily with mixing point 1 to calculate some of the unknowns.

Mixing Point 1:

No reaction occurs at the mixing point, so species balances can be used without involving the extent of reaction:

$$\text{Total: } 100 + 12 = T = 112$$

(Continues)

Example 5.8 Recycle in a Process with a Reaction Occurring (*Continued*)

$$\text{Glucose:} \quad 100(0.40)+12\omega_G^P = 112\omega_G^T$$

$$\text{Fructose:} \quad 0+12\omega_F^R = 112(0.04)$$

Solving for ω_F^R from the last equation yields $\omega_F^R = 0.373$.

Also, because $\omega_F^R + \omega_G^R + \omega_W^R = 1$,

$$\omega_G^R = 1-0.373-0.600 = 0.027$$

Next, from the glucose balance, rather than make separate balances on the reactor and separator, we combine the two into one system (and thus avoid having to calculate values associated with the reactor exit stream).

Reactor Plus Separator 2:

$$\text{In} \quad - \quad \text{Out} \quad = \text{Consumed}$$
$$\text{Glucose:} \ (0.360)(112)-(112)(0.027)= f(0.360)(112)$$
$$f = 0.93$$

Step 10

Check by using Equation (5.5), including the extent of reaction.

Self-Assessment Test

Questions

1. Explain the purpose of using recycle in a process.
2. For what systems can you make material balances in processes that involve recycle?
3. Can you formulate sets of equations that are not independent if recycle occurs in a system?
4. If the components in the feed to a process appear in stoichiometric quantities and the subsequent separation process is complete so that all of the unreacted reactants are recycled, what is the ratio for reactants in the recycle stream?
5. Indicate whether the following statements are true or false:
   a. The general material balance applies for processes that involve recycle with reaction as it does for other processes.

b. The key extra piece of information in material balances on processes with recycle in which a reaction takes place is the specification of the fractional conversion or extent of reaction.

c. The degrees of freedom for a process with recycle that involves chemical reaction are the same as for a process without recycle.

Answers

1. A recycle stream is used to improve the feedstock utilization and thus the economics of a process.

2. Essentially all systems; see text for some examples.

3. Yes

4. The stoichiometric ratio

5. All true

Problems

1. How many recycle streams occur in Figure SAT5.3P1?

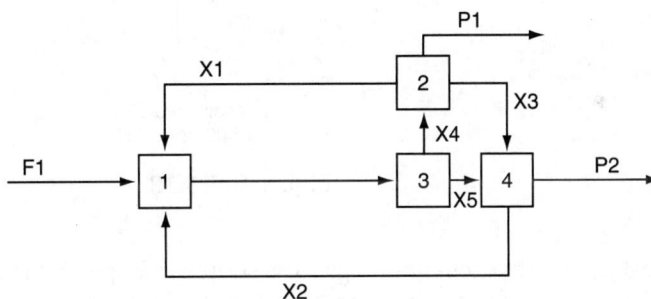

Figure SAT5.3P1

2. A ball mill grinds plastic to make a very fine powder. Look at Figure SAT5.3P2.

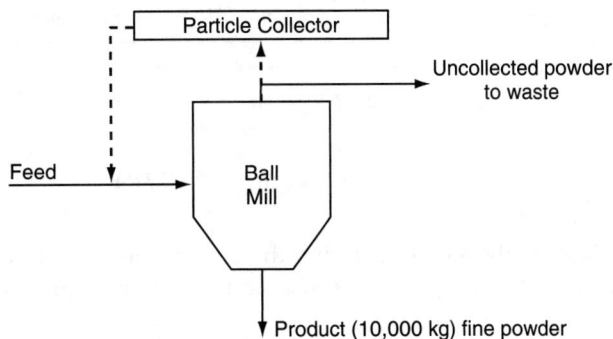

Figure SAT5.3P2

At the present time, 10,000 kg of powder are produced per day. You observe that the process (shown by the solid lines) is inefficient because 20% of the feed is not recovered as powder—i.e., it goes to waste.

You make a proposal (designated by the dashed lines) to recycle the uncollected material back to the feed so that it can be remilled. You plan to recycle 75% of the 2500 kg of uncollected material back to the feed stream. If the feed costs $1.20/ kg, how much money would you save per day while producing 10,000 kg of fine powder?

3. Seawater is to be desalinized by reverse osmosis using the scheme indicated in Figure SAT5.3P3. Use the data given in the figure to determine (a) the rate of waste brine removal (B), (b) the rate of desalinized water (called potable water) production (P), and (c) the fraction of the brine leaving the reverse osmosis cell (which acts in essence as a separator) that is recycled.

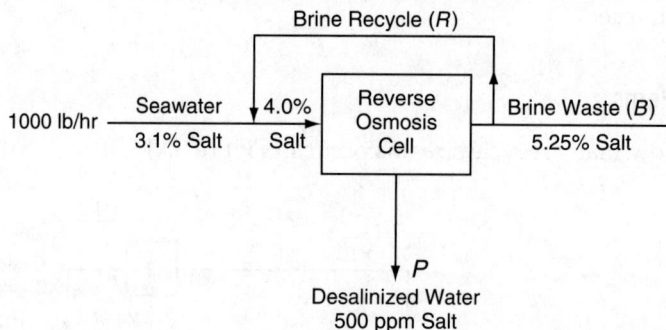

Figure SAT5.3P3

4. A catalytic dehydrogenation process shown in Figure SAT5.3P4 produces 1, 2-bu-tadiene (C_4H_6) from pure normal butane (C_4H_{10}). The product stream contains 75 mol/hr of H_2 and 13 mol/hr of C_4H_{10} as well as C_4H_6. The recycle stream is 30% (mol) C_4H_{10} and 70% (mol) C_4H_6, and the flow is 24 mol/hr.

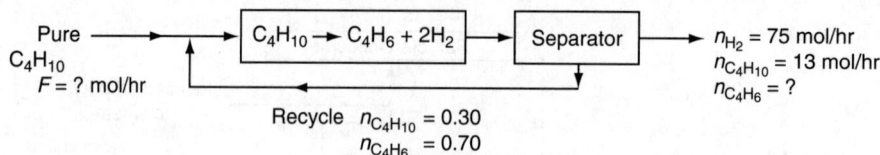

Figure SAT5.3P4

a. What are the feed rate, F, and the product flow rate of C_4H_6 leaving the process?
b. What is the single-pass conversion of butane in the process?

5. Pure propane (C_3H_8) from El Paso is dehydrogenated catalytically in a continuous process to obtain propylene (C_3H_6). All of the hydrogen formed is separated from the reactor exit gas with no loss of hydrocarbon. The hydrocarbon mixture is then fractionated to give a product stream containing 88 mol % propylene and 12 mol % propane. The other stream, which is 70 mol % propane and 30 mol % propylene, is recycled. The one-pass conversion in the reactor is 25%, and 1000 kg of fresh propane are fed per hour. Find (a) the kilograms of product stream per hour, and (b) the kilograms of recycle stream per hour.

Answers

1. Two (X1 and X2)

2. Basis: 10,000 kg product; feed = 12,500 kg; recycle 75% of 2,500 kg. Savings = 0.75 × 2500 × $1.20 = $2250 assuming that all the recycled particles end up in the product.

3. Basis: Feed = 1000 lb/h; salt balance: 31 = 0.0525B + 0.005P; overall balance: 1000 = B + P. Solving (a), B = 586.5 lb/hr; from overall balance: (b) P = 413.5 lb/hr; salt balance on mixer: 31 + 0.0525R = 0.04(1000 + R); R = 720; (c) R/(R + B) = 0.551

4. From H_2 production, $\xi = 37.5$; (a) $C_4H_6 = 37.5$ mol/hr; C_4H_{10} bal: $F - \xi = 13$; and $F = 50.5$ mol/hr; (b) $f_{SP} = \xi/(50.5 + 24 \times 0.3)100\% = 65.0\%$

5. Basis: 1000 kg/h propane feed; 22.73 kg mol propane; $\xi = 0.88(22.73) = 20.0$ kg mol; propane bal: 22.73 − 20 = 2.727 kg mol; (a) mass of product = 960 kg/h; $f_{SP} = \xi/(0.7R + 22.73)$; R = 81.814 kg mol; R = 3550 kg/h

5.4   Bypass and Purge

Two additional commonly encountered types of process streams are shown in Figures 5.10 and 5.11:

1. A **bypass** stream is one that skips one or more stages of the process and goes directly to another downstream stage (Figure 5.10).

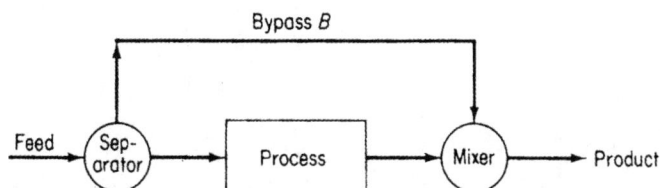

Figure 5.10   A process with a bypass stream

A bypass stream can be used to control the composition of a final exit stream from a unit by mixing the bypass stream and the unit exit stream in suitable proportions to obtain the desired final composition.

2. A **purge** stream is a stream bled off from the process to remove an accumulation of inerts or unwanted material that might otherwise build up in the recycle stream with time of operation (Figure 5.11).

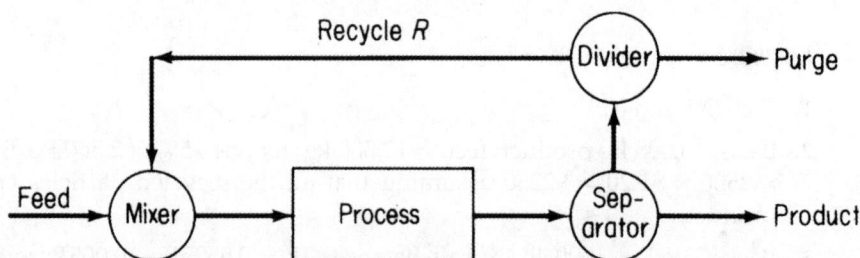

Figure 5.11   A process with a recycle stream with purge

Many companies have had the unfortunate experience that on start-up of a new process, trace components not considered in the material balances used in the design of the process (because the amounts were so small) build up in one or more recycle loops. That is, if a trace species that enters a process is not removed from the process, it will accumulate in the process until the process no longer functions properly. For example, distillation columns without an overhead purge and with light trace impurity in the feed will experience a steady increase in column pressure due to the accumulation of trace impurity, undermining the condensation in the overhead condenser, leading to a steady reduction in separation performance of the column. Calculations for processes involving bypass and purge streams introduce no new principles or techniques beyond those presented so far. Some examples will make that clear.

Example 5.9   Bypass Calculations

Problem Statement

In the feedstock preparation section of a plant manufacturing natural gasoline, isopentane is removed from butane-free gasoline. Assume for purposes of simplification that the process and components are as shown in Figure E5.9. What fraction of the butane-free gasoline is passed through the isopentane tower? Detailed steps will not be listed in the analysis and solution of this problem. The process is in the steady state and no reaction occurs.

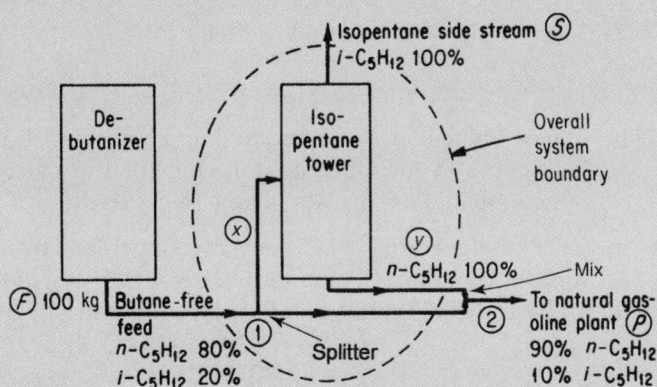

Figure E5.9

Solution

By examining the flow diagram, you can see that part of the butane-free gasoline bypasses the isopentane tower and proceeds to the next stage in the natural gasoline plant. All of the compositions (the streams are liquid) are known. The units are all kilograms. Select a basis:

Basis: 100 kg of feed

a. With what system or subsystem should you start the solution? Let's start with the overall system, usually the easiest to work with, especially in view of the information given. What are the unknowns? If you assign $F = 100$ kg, the unknowns are S and P. What balances can you write for the overall system? You can write a total balance and two component balances, but only two of the three are independent.

Total overall material balance:

$$\frac{\text{In}}{100} = \frac{\text{Out}}{S + P} \tag{a}$$

(Continues)

Example 5.9   Bypass Calculations (*Continued*)

Component material balance for n-C_5 (a tie component):

$$\frac{\text{In}}{100(0.80)} = \frac{\text{Out}}{S(0) + P(0.90)} \tag{b}$$

Consequently,

$$P = 100\left(\frac{0.80}{0.90}\right) = 88.9 \,\text{kg and}$$
$$S = 100 - 88.9 = 11.1 \,\text{kg}$$

b. The overall balances will not tell you anything about the fraction of the feed going to the isopentane tower; for this calculation you need another system. Pick the isopentane tower itself.

   Total material balance around isopentane tower: From Figure E5.9, X is the kilograms of butane-free gasoline going to the isopentane tower, and Y is the kilograms of n-C_5H_{12} leaving the isopentane tower.

$$\text{In} = \text{Out}$$
$$X = 11.1 + Y \tag{c}$$

Component balance (n-C_5, a tie component):

$$X(0.80) = Y \tag{d}$$

   Combine (c) and (d) to get $X = 55.5$ kg, so the desired fraction is 0.55.

c. Another approach to this problem would be to make material balances for the subsystem defined by mixing points 1 and 2.

   Total material balance around mixing point 2:

$$(100 - X) + Y = 88.9 \tag{e}$$

Component balance

$$(100 - X)(0.20) + 0 = 88.9(0.10) \tag{f}$$

Because Equation (f) does not contain Y, it can be solved directly for X: $X = 55.5$ kg as before. Would you expect the value to be different?

Example 5.10   Purge for an Ammonia Process

Problem Statement

In the famous Haber process (Figure E5.10) used to manufacture ammonia, the reaction is carried out at pressures of 800 to 1000 atm and at 500°C to 600°C using a suitable catalyst. Only a small fraction of the material entering the reactor reacts on one pass, so recycle is needed. Also, because the nitrogen is obtained from the air, it contains almost 1% rare gases (chiefly argon) that do not react. The rare gases would continue to build up in the recycle until their effect on the reaction equilibrium became adverse, so a small purge stream is used.

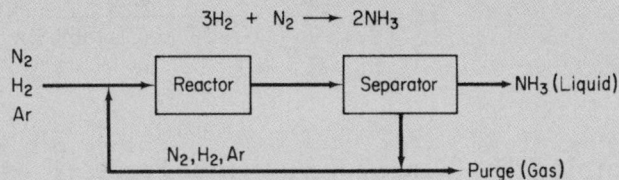

Figure E5.10

The fresh feed of gas composed of 75.16% H_2, 24.57% N_2, and 0.27% Ar is mixed with the recycled gas and enters the reactor with a composition of 79.52% H_2. The gas stream leaving the ammonia separator contains 80.01% H_2 and no ammonia. The product ammonia contains no dissolved gases. Per 100 moles of fresh feed:

  a. How many moles are recycled and purged?

  b. What is the percent conversion of the limiting reactant per pass?

Solution

After using 100 moles of fresh feed for the basis and skipping the degree-of-freedom analysis, we can perform overall balances for N_2, H_2, and Ar:

$$N_2:\qquad 24.57 - \xi = P_{N_2}$$
$$H_2:\qquad 75.16 - 3\xi = P_{H_2}$$
$$Ar:\qquad 0.27 = P_{A_r}$$

Adding these three balances together yields $P = 100 - 4\xi$. Then, applying an overall H_2 balance using the specified concentration of H_2 in the recycle:

$$75.16 - 3\xi = 0.8001P = 0.8001(100 - 4\xi) \quad \Rightarrow \quad \xi = 24.20 \text{ mol}$$

(Continues)

Example 5.10   Purge for an Ammonia Process (*Continued*)

Then $P = 100 - 4\xi = 3.19$ mol. Applying the overall N_2 balance because N_2 is the limiting reactant, the mole fraction of N_2 in the purge is equal to 11.6 mol %.

Now, we can apply a H_2 balance on the mixer for the feed and recycle:

$$75.16 + 0.8001R = .7952(R + 100) \quad \Rightarrow \quad R = 889.8 \text{ mol}$$

Finally, applying the definition of the single pass conversion [Equation (5.4)]:

$$f_{SP} = \frac{-v_{LR}\xi}{F_{LR} + R_{LR}} = \frac{24.20}{24.57 + 0.116(889.8)} 100\% = 18.9\%$$

Example 5.11   Purge for Methanol Process

Problem Statement

Considerable interest exists in the conversion of coal into more convenient liquid products for subsequent production of chemicals. Two of the main gases that can be generated under suitable conditions from in situ (in the ground) coal combustion in the presence of steam (as occurs naturally in the presence of groundwater) are H_2 and CO. After cleanup, these two gases can be combined to yield methanol according to the following equation:

$$CO + 2H_2 \rightarrow CH_3OH$$

Figure E5.11a illustrates a steady-state, open process for the production of methanol. All of the compositions are in mole fractions or percent. The stream flows will be in moles.

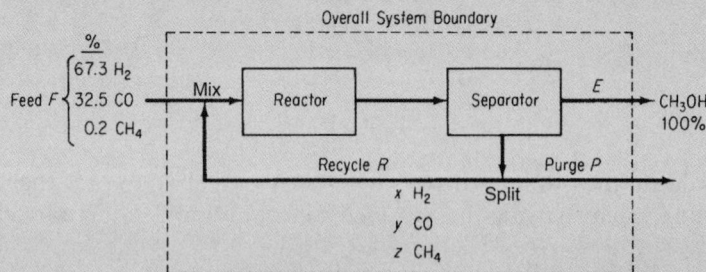

Figure E5.11

You will note in Figure E5.11 that some CH_4 enters the process. However, the CH_4 does not participate in the reaction. A purge stream is used to maintain the CH_4 concentration in the exit stream from the separator going to R and P at no more than 3.2 mol %, and to prevent H_2 from accumulating in the system. The once-through conversion of the CO in the reactor is 18%.

Compute the moles of recycle, R, the moles of CH_3OH, E, and the moles of purge, P, per 100 moles of feed, and also compute the purge gas composition.

Solution

Steps 1–4

Most of the known information is shown in Figure E5.11. The process is in the steady state with reaction. The purge and recycle streams have the same composition as implied by the split of one of the outputs of the separator into $P + R$ in the figure. The mole fractions of the components in the purge stream have been designated as x, y, and z for H_2, CO, and CH_4, respectively.

Step 5

Select a convenient basis: Let $F = 100$ mol

Step 6

After assigning known values to variables, the variables whose values are unknown are x, y, z, E, P, R, and the extent of reaction, ξ, the latter being required if you plan to use species material balances rather than element balances. You can ignore the stream between the reactor and separator as no questions are asked about it.

Steps 7–9

We can write the set of equations for this problem:

Overall H_2 balance:	$67.3 - 2\xi = xP$
Overall CO balance:	$32.5 - \xi = yP$
Overall CH_4 balance:	$0.2 = zP$
Implicit equation:	$x + y + z = 1$
Single pass conversion:	$0.18 = \dfrac{\xi}{32.5 + yR}$
Specification:	$z = 0.032$
Overall CH_3OH balance:	$E = \xi$

(Continues)

Example 5.11　Purge for Methanol Process (*Continued*)

Therefore, the degrees of freedom is zero, but notice that several of these equations are nonlinear so that if we solved all seven equations simultaneously, we would have to use software that solves a system of nonlinear equations although this set of equations is not highly nonlinear. Even if we applied the material balances using component flows, we will still have nonlinear equations. Let's see if we can find a way to avoid having to solve a system of nonlinear equations.

Note that with the specification of the value of z, we can calculate the value of P using the overall CH_4 balance, that is, $P = 0.2/0.032 = 6.25$ mol. Thus, all the nonlinear equations have been eliminated except for the equation for the single-pass conversion. Also notice that if we apply the first two equations plus the fourth equation using these results, we will have a set of three linear equations containing three unknowns (ξ, x, and y). Once the values of ξ, x, and y are calculated, the remainder of the unknowns (i.e., R and E) can be determined directly.

Computer Solution

Using the following definitions for unknowns (i.e., x_1–x_3): x_1 is equal to ξ; x_2 is equal to x; x_3 is equal to y; the coefficient matrix **A** and the constant vector **b** are

$$\mathbf{A} = \begin{bmatrix} 2 & 6.25 & 0 \\ 1 & 0 & 6.25 \\ 0 & 1 & 1 \end{bmatrix} \quad \mathbf{b} = \begin{bmatrix} 67.3 \\ 32.5 \\ 0.968 \end{bmatrix}$$

MATLAB:

MATLAB Solution for Example 5.11

```
%                          PROGRAM
function Ex5_11_LinEqnSolver
clear; clc;
A=[2,6.25,0;1,0,6.25;0,1,1];
k=rank(A);
b=[67.3;32.5;0.968];
x=A\b;
fprintf('x1 =%8.4f x2=%8.4f\n',x(1),x(2))
fprintf('x3 =%8.4f rank=%8.4f\n',x(3),k)
end
%                          PROGRAM END
    x1 = 31.25 x2 = 0.768
    x3 =0.200  rank = 3
```

Python Code for Example 5.11

Ex5_11 Linear Eqns.py

```
#                             PROGRAM
Import numpy as np
Import numpy.linalg
# Specify the coefficient matrix A
A=np.array([[2 ,1,0],[1,0,1],[0,1,1]])
k=numpy.linalg.rank(A)
b=np.array([67.3, 32.5, 0.968])
xsol=numpy.linalg.solve(A,b)
#                             PROGRAM END
```

IPython console:

```
In[1]: runfile(…
In[2]: xsol, k
Out[2]: array([31.25, 0.768, 0.200]), 3
```

Rearranging the single-pass conversion equation yields

$$R = \frac{\xi/.18 - 32.5}{y} = 705.6 \text{ mol}$$

E	CH_3OH	31.25
P	purge	6.25
R	recycle	705.6
x	H_2	0.768
y	CO	0.200
z	CH_4	0.032

Step 10

Check each equation to determine if they all balanced

Self-Assessment Test

Questions

1. Explain what bypassing means in words.

2. Indicate whether the following statements are true or false:
   a. Purge is used to maintain a concentration of a minor component of a process stream below some set point so that it does not accumulate in the process.
   b. Bypassing means that a process stream enters the process in advance of the feed to the process.
   c. A trace component in a stream or produced in a reactor will have negligible effect on the overall material balance when recycle occurs.

3. Is the waste stream the same as a purge stream in a process?

Answers

1. Bypassing is a stream that skips some intermediate processing unit or system and joins the process downstream.

2. a. T; b. F; c. F

3. Not necessarily. A purge is a stream that removes a small amount of a component to prevent it from building up in the process. A waste stream usually contains a relatively large amount of an undesirable component.

Problem

1. Figure SAT5.4P1 shows a schematic for making freshwater from seawater by freezing. The prechilled seawater is sprayed into a vacuum at a low pressure. The cooling required to freeze some of the feed seawater comes from evaporation of a fraction of the water entering the chamber. The concentration of the brine stream, B, is 4.8% salt. The pure salt-free water vapor is compressed and fed to a melter at a higher pressure, where the heat of condensation of the vapor is removed through the heat of fusion of the ice, which contains no salt. As a result, pure cold water and concentrated brine (6.9%) leave the process as products.
   a. Determine the flow rates of streams W and D if the feed is 1000 kg/hr.
   b. Determine the flow rates of streams C, B, and A per hour.

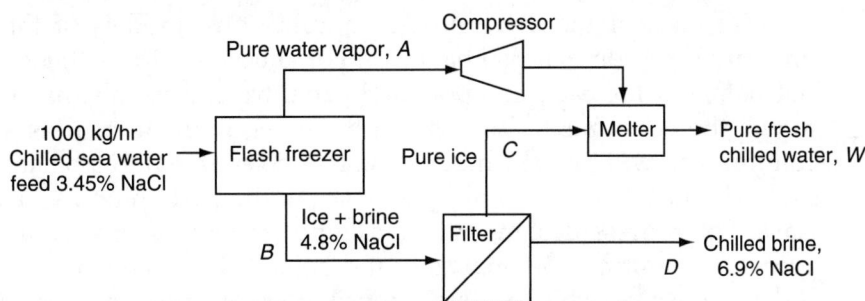

Figure SAT5.4P1

Answer

1. For overall balance, salt is tie component: 34.5 kg = 0.069D; D = 500 kg. From overall mass balance: W = 500 kg; from flash freezer balance: 34.5 = 0.048B; B = 718.75 kg; A = 281.25 kg; from overall balance on melter: C = 500 − 281.25 = 218.75 kg.

5.5   The Industrial Application of Material Balances

Process engineers use **process simulators** for a number of important activities, including process design, process analysis, and process optimization. **Process design** involves selecting suitable processing units (e.g., reactors, mixers, and distillation columns) and sizing them so that the feed to the process can be efficiently converted into the desired products. **Process analysis** involves comparing predictions of process variables using models of the process units with the measurements made in the operating process. By comparing corresponding values of variables, you can determine if a particular process unit is functioning properly. If discrepancies exist, the predictions from the model can provide insight into the root causes of these discrepancies. In addition, process models can be used to carry out studies that evaluate alternate processing approaches and studies of debottlenecking, that is, methods designed to increase the production rate of the overall process. **Process optimization** is directed at determining the most profitable way to operate the process. For process optimization, models of the major processing units in the process are used to determine the operating conditions, such as product compositions and reactor temperatures, that yield the maximum profit for the process (subject to appropriate constraints).

For each of the three process applications, models of the processing units are based on material balances. For simple equipment, just a few material balances for each component in the system are all that is needed to model the equipment. For more complex equipment such as a distillation column, you will find the model involves a set of material balance equations for each tray in the column, and some industrial columns have over 200 trays. For process design and many applications of process analysis, each processing unit can be analyzed and solved separately. Modern computer codes make it possible to solve extensive sets of simultaneous equations. For example, the optimization model for an ethylene plant usually has over 150,000 equations with material balances constituting over 90% of the equations and requiring many solutions to determine the optimum set of operating conditions for the entire plant.

Now consider material balance closure for industrial processes. One important way in which individual material balances are applied industrially is to check that "in = out"—that is, to determine how well the material balances match the process operation using process measurements in the material balance equations. You look for what is called *closure,* namely, that the error between "in" and "out" is acceptable. The measured flow rates and measured compositions for all of the streams entering and exiting a process unit are substituted into the appropriate material balance equations. Ideally, the amount (mass) of each component entering the system should equal the amount of that component leaving the system. Unfortunately, the amount of a component entering a process rarely equals the amount leaving the process when you make such calculations. The lack of closure for material balances on an industrial process occurs for several reasons:

1. The process is rarely operating in the steady state. Industrial processes are almost always in a state of flux and rarely approach true steady-state behavior.

2. The flow and composition measurements have a variety of errors associated with them. First, sensor readings have noise (variations in the measurement due to more or less random variations in the readings that do not correspond to changes in the process). The sensor readings can also be inaccurate for a wide variety of other reasons. For example, a sensor may require recalibration because it degrades, or it may be used for a measurement for which it was not designed.

3. A component of interest may be generated or consumed inside the process by reactions that the process engineer has not considered.

As a result, material balance closure during a relatively steady-state period of operation to within ±5% for material balances for most industrial processes is considered reasonable. (Here closure is defined as the calculated difference between the amount of a particular material entering and exiting the process divided by the amount entering multiplied by 100%). If special attention is paid to calibrating sensors, material balance closure to ±2% to ±3% can be attained. If special high-accuracy sensors are used, smaller closure of the material balances can be attained, but if faulty sensor readings are used, much greater errors in material balances can be observed. In fact, material balances can be used to determine when faulty sensor readings exist, which is known as **data reconciliation.**

Summary

In this chapter, you have seen how systems composed of more than one subsystem can be treated by the same principles that you used to treat single unit systems. Whether you use combinations of material balances from each of several subsystems or lump all of the units into one system, all you have to do is check to see that the number of independent equations you prepared was adequate to solve for the variables whose values are unknown. Flowsheets can help in the preparation of the equation set. The one new factor brought out in this chapter is that recycle for a reactor usually involves information about the fractional conversion of a reactant for the overall process and for the reactor only.

Glossary

block diagram   A sequence of boxes, circles, and other shapes used to represent operational features of a process flowsheet.

bypass stream   A stream that skips one or more units of a process and goes directly to a downstream unit.

connections   Streams flowing between subsystems (units).

data reconciliation   The process of evaluating the validity of process measurements using material balances.

flowchart   A graphical representation of a process layout.

fresh feed   The overall feed to a system.

gross product   The product stream that leaves a reactor.

mixer   Apparatus to combine two or more flow streams.

once-through fractional conversion   The conversion of a reactant based on the amount of material that enters and leaves a reactor.

overall fractional conversion   The conversion of a reactant in a process with recycle based on the fresh feed of the reactant and the overall products.

overall process   The entire system composed of subsystems (units).

overall products   The streams that exit a process.

process feed   The feed stream that enters a reactor, usually used in a process with a reactor and recycle.

process flowsheet   A graphical representation of a process. See **flowchart.**

purge   A stream bled off from a process to remove the accumulation of inerts or unwanted material that might otherwise build up in the recycle streams.

recycle   Material (or energy) that leaves a process unit that is downstream and is returned to the same unit or an upstream unit for processing again.

recycle stream   The stream that recycles material (or energy).

recycle system   A system that includes one or more recycle streams.

separator   Apparatus that produces two or more streams of different composition from the fluid(s) entering the apparatus.

sequential combination of units   A set of units arranged in a row.

single-pass fractional conversion   The conversion based on what enters and leaves a reactor. See **once-through fractional conversion.**

splitter   Apparatus that divides a flow into two or more streams.

subsystem   A designated part of a complete system.

well-mixed system   Material within a system (equipment) is of uniform composition, and the exit stream(s) is of the same composition as the material inside the system.

Problems

5.1   Preliminary Concepts

*5.1.1   For Figure P5.1.1, how many independent equations are obtained from the overall balance around the entire system plus the overall balances on units A and B? Assume that only one component exists in each stream.

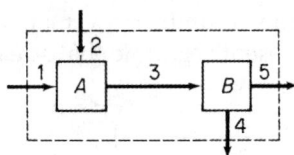

Figure P5.1.1

**5.1.2 What is the maximum number of independent material balances that can be written for the process in Figure P5.1.2?

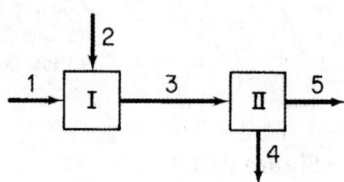

Stream composition (known)

1   Pure A
2   Pure B
3
4 } Mixture of A and B
5 } with different compositions
       in each stream

Stream flows (unknown)

Figure P5.1.2

**5.1.3 What is the maximum number of independent material balances that can be written for the process in Figure P5.1.3? The stream flows are unknown. Suppose you find out that A and B are always combined in each of the streams in the same ratio. How many independent equations could you write?

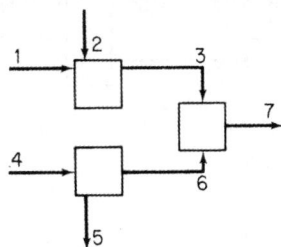

Stream composition (known)

1 A and B
2 Pure C
3 $A,B,$ and C
4 D and E
5 Pure E
6 Pure D
7 $A,B,C,$ and D

Figure P5.1.3

**5.1.4 The diagram in Figure P5.1.4 represents a typical but simplified distillation column. Streams 3 and 6 consist of water and steam, respectively, and do not come in contact with the fluids in the column that contains two components. Write the total and component material balances for the three sections of the column. How many independent equations would these balances represent? Assume that stream 1 contains n components.

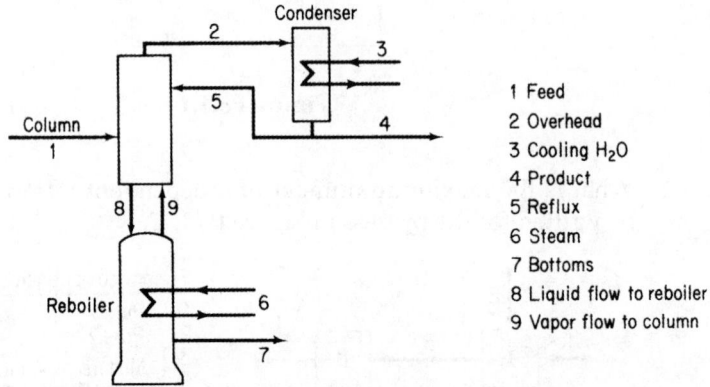

1 Feed
2 Overhead
3 Cooling H_2O
4 Product
5 Reflux
6 Steam
7 Bottoms
8 Liquid flow to reboiler
9 Vapor flow to column

Figure P5.1.4

**5.1.5   A distillation process is shown in Figure P5.1.5. You are asked to solve for all of the values of the stream flows and compositions. How many variables and unknowns are there in the system? How many independent material balance equations can you write? Explain each answer and show all details of how you reached your decision. For each stream (except F), the only components that occur are labeled below the stream

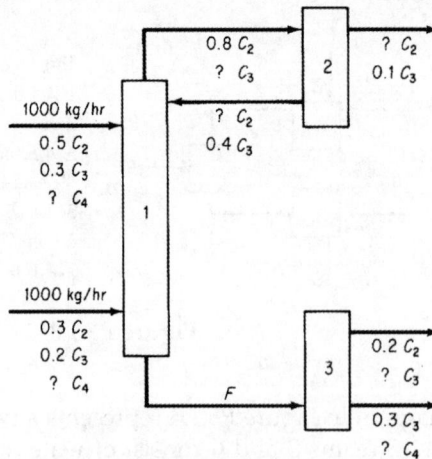

Figure P5.1.5

***5.1.6** In Figure P5.1.6 you see two successive liquid separation columns operating in tandem in the steady state (and with no reaction taking place). The compositions of the feed and products are as shown in the figure. The amount of W_2 is 20% of the feed. Focus only on the material balances. Write down the names of the material balances for each column treated as separate units (one set for each column), along with the balances themselves placed next to the names of the balances. Also, place an asterisk in front of the names of the balances that will constitute a set of independent equations for each column. Write down in symbols the unknowns for each column. Calculate the degrees of freedom for each column separately.

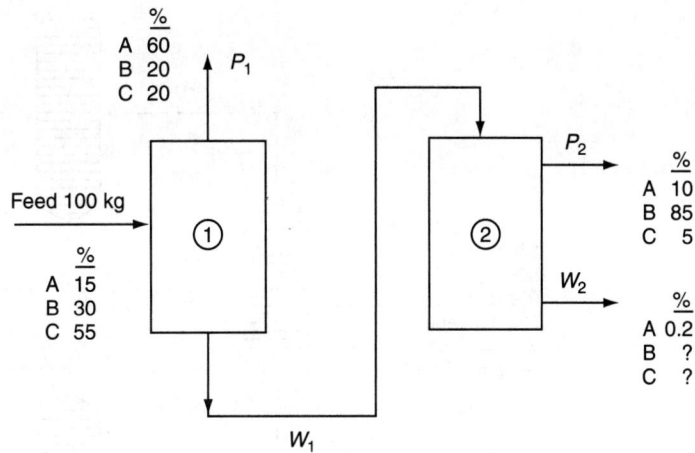

Figure P5.1.6

Then determine the number of independent equations for the overall system composed of the two columns together by listing the duplicate variables (by symbols) and unused specifications as well as the redundant equations, and appropriately adding or subtracting them from the values found in the preceding paragraph. Determine the degrees of freedom for the overall system.

Check the results obtained in the second paragraph by repeating the entire analysis of independent equations, unknowns, and degrees of freedom for the overall system. Do they agree? They should. Do not solve any of the equations in this problem.

5.2   Sequential Multiunit Systems

**5.2.1   Examine Figure P5.2.1, provided through the courtesy of Professor Mike
Cutlip.

   a.   Calculate the molar flow rate of $D1$, $D2$, $B1$, and $B2$.

   b.   Reduce the feed flow rate for each one of the compounds by 1% in turn.
Calculate the flow rates of $D1$, $D2$, $B1$, and $B2$ again. Do you notice
something unusual?

   Explain your results.

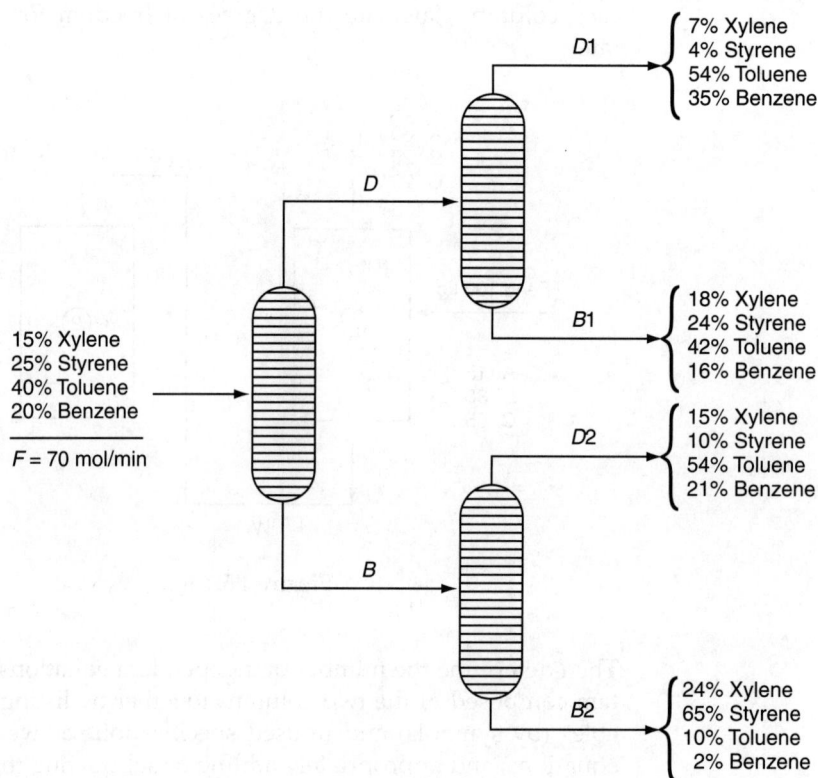

Figure P5.2.1

**5.2.2   Figure P5.2.2 shows a simplified process to make ethylene dichloride
($C_2H_4Cl_2$). The feed data have been placed on the figure. Ninety percent
conversion of the C_2H_4 occurs on each pass through the reactor. The
overhead stream from the separator contains 98% of the Cl_2 entering the
separator, 92% of the entering C_2H_4, and 0.1% of the entering $C_2H_4Cl_2$. Five
percent of the overhead from the separator is purged. Calculate (a) the flow
rate and (b) the composition of the purge stream.

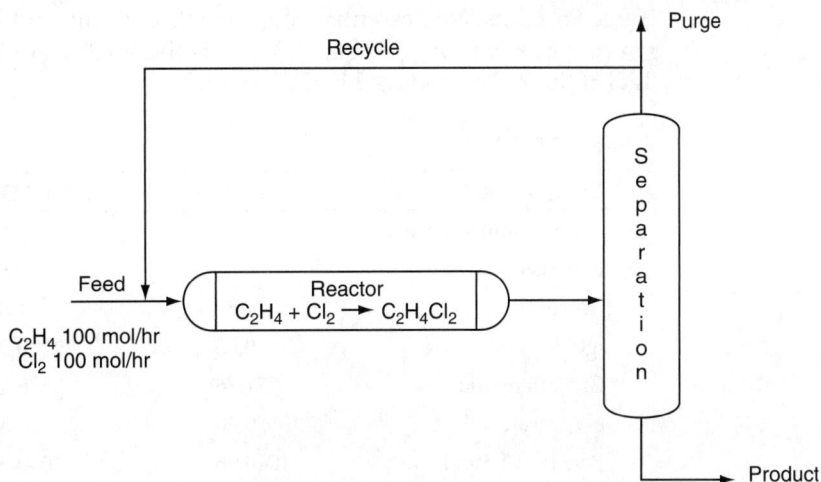

Figure P5.2.2

**5.2.3 Monoclonal antibodies are used to treat various diseases as well as in diagnostic tests. Figure P5.2.3 shows a typical process used to produce monoclonal antibodies. A stirred tank bioreactor grows the cells of the antibody of interest, namely, immunoglobulin G (IgG). After fermentation in the reactor, a batch of 2200 L contains 220 g of the product IgG. The batch is processed through a number of stages as shown in Figure P5.2.3 before the purified product is obtained. In the diafiltration stage, 95% of the IgG entering the filter is recovered; in the ultrafiltration stage, 95% of the entering IgG is recovered; and in the chromatography, 90% is recovered.

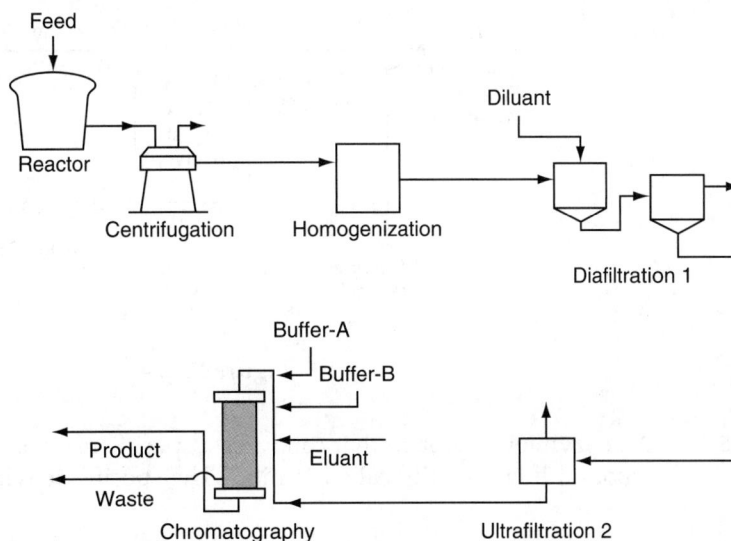

Figure P5.2.3

Table P5.2.3 lists the essential components entering and leaving the overall process in kilograms per batch. What is the fractional yield of the product IgG of the 220 g produced in the reactor?

Table P5.2.3

Component	Total Inlet	Total Outlet	Product
Ammonium sulfate	64.69	64.69	
Biomass	0.00	0.87	
Glycerol	1.85	1.85	
IgG	0.00	0.22	0.14
Growth media	21.76	8.41	
Na$_3$ citrate	0.80	0.80	
Phosphoric acid	1040.96	1040.96	
Sodium hydrophosphate	6.83	6.81	
Sodium chloride	55.18	55.19	
Tris-HCl	0.69	0.69	
Water	11,459.59	11,458.80	
Injection water	18,269.54	18,269.54	
Total	**30,928.72**	**30,928.72**	**0.14**

**5.2.4   In a tissue paper machine (Figure P5.2.4), stream N contains 85% fiber. Find the unknown fiber values (all values in the figure are in kilograms) in kilograms for each stream.

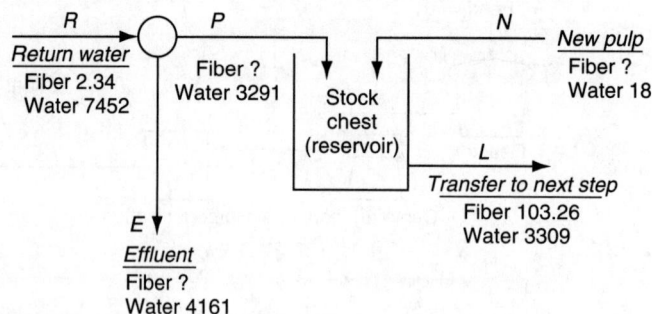

Figure P5.2.4

**5.2.5   An enzyme is a protein that catalyzes a specific reaction, and its activity is reported in a quantity called "units." The specific activity is a measure of the purity of an enzyme. The fractional recovery of an enzyme can be calculated from the ratio of the specific activity (units per milligram) after processing occurs to the initial specific activity. A three-stage process for the purification of an enzyme involves

1. Breakup of the cells in a biomass to release the intercellular products
2. Separation of the enzyme from the intercellular product
3. Further separation of the enzyme from the output of stage 2

Based on the following data for one batch of biomass, calculate the percent recovery of the enzyme after each of the three stages of the process. Also calculate the purification of the enzyme that is defined as the ratio of the specific activity to the initial specific activity. Fill in the blank columns of the following table.

Stage No.	Activity (units)	Protein Present (mg)	Specific Activity (units/mg)	Percent Recovery	Purification
1	6860	76,200			
2	6800	2200			
3	5300	267			

***5.2.6 Several streams are mixed as shown in Figure P5.2.6. Calculate the flows of each stream in kilograms per second.

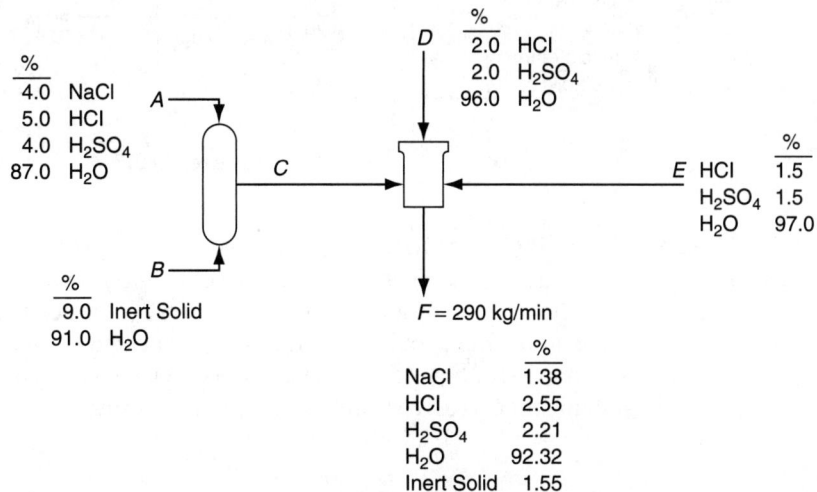

Figure P5.2.6

***5.2.7   In 1988, the U.S. Chemical Manufacturers Association (CMA) embarked upon an ambitious and comprehensive environmental improvement effort, the Responsible Care initiative. Responsible Care committed all of the 185 members of the CMA to ensure continual improvement in the areas of health, safety, and environmental quality, as well as in eliciting and responding to public concerns about their products and operations.

One of the best ways to reduce or eliminate hazardous waste is through source reduction. Generally, this means using different raw materials or redesigning the production process to eliminate the generation of hazardous by-products. As an example, consider the following countercurrent extraction process (Figure P5.2.7) to recover xylene from a stream that contains 10% xylene and 90% solids by weight.

The stream from which xylene is to be extracted enters Unit 2 at a flow rate of 2000 kg/hr. To provide a solvent for the extraction, pure benzene is fed to Unit 1 at a flow rate of 1000 kg/hr. The mass fractions of the xylene in the solids stream (F) and the clear liquid stream (S) have the following relations:

$$\omega_{Xylene}^{F^1} = \omega_{Xylene}^{S^2} \quad \text{and} \quad \omega_{Xylene}^{F^2} = \omega_{Xylene}^{S^1}.$$

Determine the benzene and xylene concentrations in all of the streams. What is the percent recovery of the xylene entering the process at Unit 2?

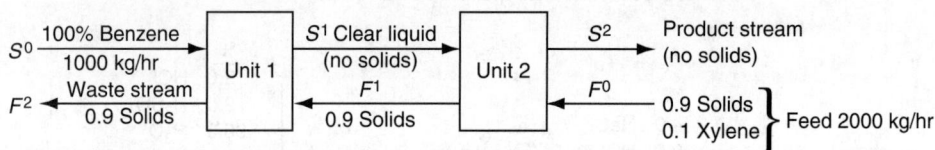

Figure P5.2.7

***5.2.8   Figure P5.2.8 shows a three-stage separation process. The ratio of P_3/D_3 is 3, the ratio of P_2/D_2 is 1, and the ratio of A to B in stream P_2 is 4 to 1. Calculate the composition and percent of each component in stream E. Hint: Although the problem comprises connected units, application of the standard strategy of problem solving will enable you to solve it without solving an excessive number of equations simultaneously.

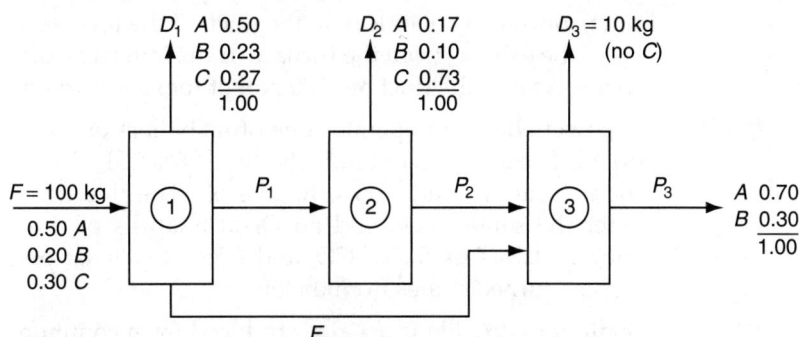

Figure P5.2.8

***5.2.9** Metallurgical-grade silicon is purified to electronic grade for use in the semiconductor industry by chemically separating it from its impurities. The Si metal reacts in varying degrees with hydrogen chloride gas at 300°C to form several polychlorinated silanes. Trichlorosilane is liquid at room temperature and is easily separated by fractional distillation from the other gases. If 100 kg of silicon is reacted as shown in Figure P5.2.9, how much trichlorosilane is produced?

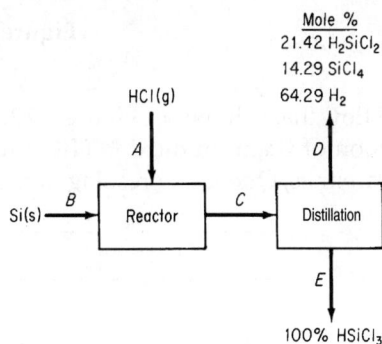

Figure P5.2.9

***5.2.10** A furnace burns fuel gas of the following composition 70% methane (CH_4), 20% hydrogen (H_2), and 10% ethane (C_2H_6 with excess air). An oxygen probe placed at the exit of the furnace reads 2% oxygen in the exit gases. The gases are then passed through a long duct to a heat exchanger. At the entrance to the heat exchanger the Orsat analysis of the gas reads 6% O_2. Is the discrepancy due to the fact that the first analysis is on a wet basis and the second analysis is on a dry basis (no water condenses in the duct), or due to an air leak in the duct? If the former, give the Orsat analysis of the exit gas from the furnace. If the latter, calculate the amount of air that leaks into the duct per 100 mol of fuel gas burned.

***5.2.11 A power company operates one of its boilers on natural gas and another on oil. The analyses of the fuels show 96% CH_4, 2% C_2H_2, and 2% CO_2 for the natural gas and $C_nH_{1.8n}$ for the oil. The flue gases from both groups enter the same stack, and an Orsat analysis of this combined flue gas shows 10.0% CO_2 0.63% CO, and 4.55% O_2. What percentage of the total carbon burned comes from the oil?

***5.2.12 Sodium hydroxide is usually produced from common salt by electrolysis. The essential elements of the system are shown in Figure P5.2.12.
  a. What is the percent conversion of salt to sodium hydroxide?
  b. How much chlorine gas is produced per pound of product?
  c. Per pound of product, how much water must be evaporated in the evaporator?

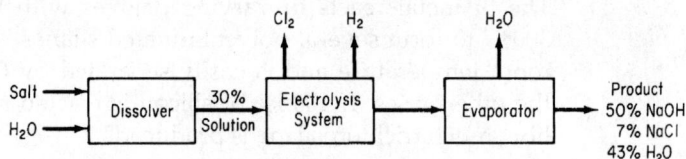

Figure P5.2.12

****5.2.13 The flowsheet shown in Figure P5.2.13 represents the process for the production of titanium dioxide (TiO_2) used by Canadian Titanium Pigments at Varennes, Quebec. Sorel slag of the following analysis

	wt %
TiO_2	70
Fe	8
Inert silicates	22

is fed to a digester and reacted with H_2SO_4, which enters as 67% by weight H_2SO_4 in a water solution. The reactions in the digester are as follows:

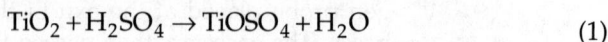

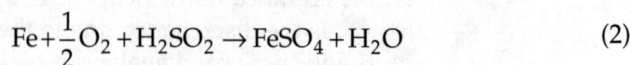

$$TiO_2 + H_2SO_4 \rightarrow TiOSO_4 + H_2O \tag{1}$$

$$Fe + \frac{1}{2}O_2 + H_2SO_2 \rightarrow FeSO_4 + H_2O \tag{2}$$

Both reactions are complete. The theoretically required amount of H_2SO_4 for the Sorel slag is fed. Pure oxygen is fed in the theoretical amount for all of the Fe in the Sorel slag. Scrap iron (pure Fe) is added to the digester to reduce the formation of ferric sulfate to negligible amounts. Thirty-six pounds of scrap iron are added per pound of Sorel slag.

The products of the digester are sent to the clarifier, where all of the inert silicates and unreacted Fe are removed. The solution of $TiOSO_4$ and $FeSO_4$ from the clarifier is cooled, crystallizing the $FeSO_4$, which is completely removed by a filter. The product $TiOSO_4$ solution from the filter is evaporated down to a slurry that is 82% by weight $TiOSO_4$.

The slurry is sent to a dryer from which a product of pure hydrate, $TiOSO_4 \cdot H_2O$, is obtained. The hydrate crystals are sent to a direct-fired rotary kiln, where the pure TiO_2 is produced according to the following reaction:

$$TiOSO_4 \cdot H_2O \rightarrow TiO_2 + H_2SO_4 \qquad (3)$$

Reaction (3) is complete. On the basis of 100 lb of Sorel slag feed, calculate (a) the pounds of water removed by the evaporator; (b) the exit pounds of H_2O per pound of dry air from the dryer if the air enters having 0.036 moles of H_2O per mole of dry air and the air rate is 18 lb mol of dry air per 100 lb of Sorel slag; (c) the pounds of product TiO_2 produced.

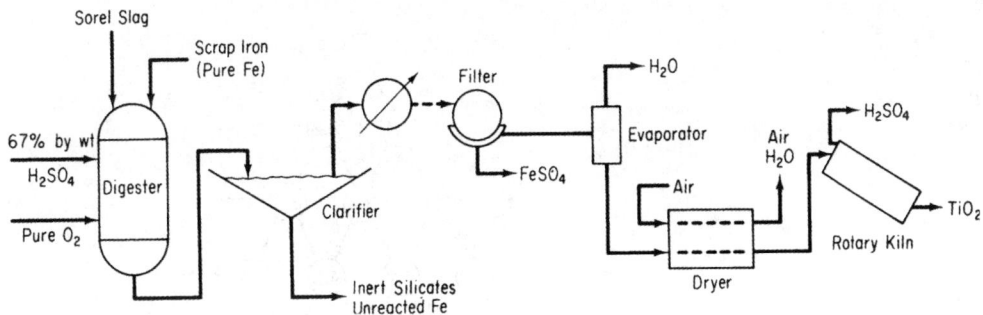

Figure P5.2.13

5.3 Recycle Systems

*5.3.1 **How many recycle streams exist in each of the following processes?**

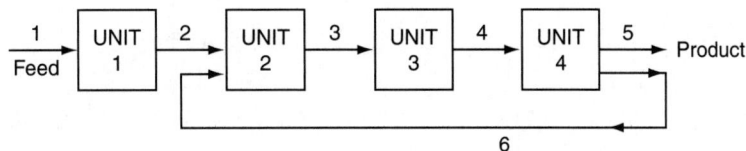

Figure P5.3.1a

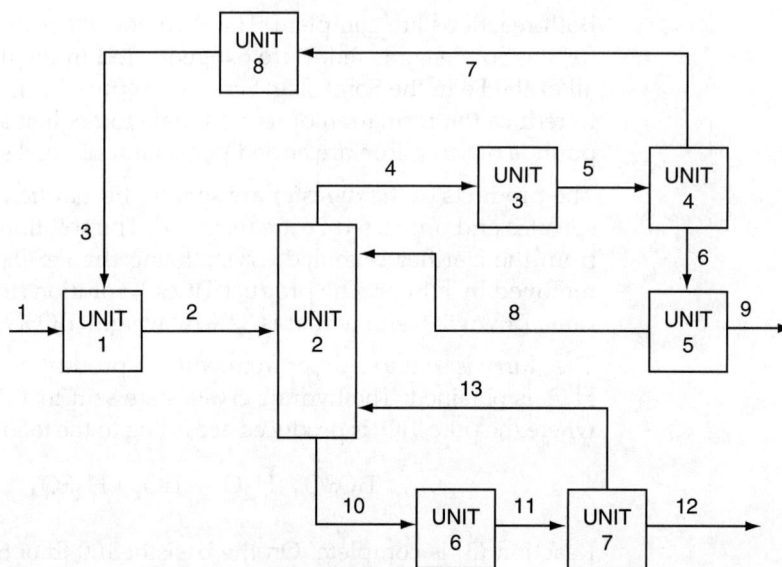

Figure P5.3.1b

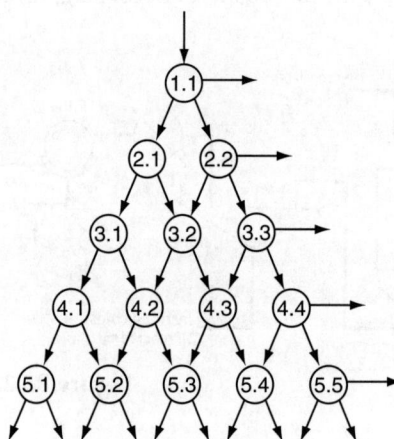

Figure P5.3.1c

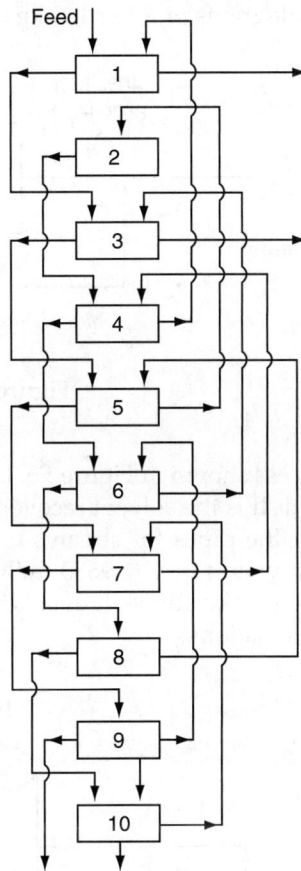

Figure P5.3.1d

*5.3.2   Based on Figure P5.3.2, find the kilograms of recycle per kilogram of feed if the amount of waste (W) is 60 kg of A.

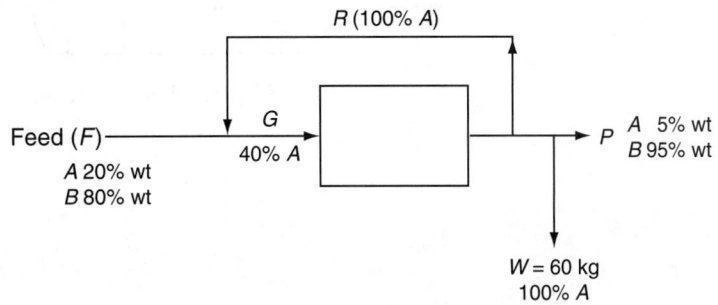

Figure P5.3.2

*5.3.3   Find the kilograms of R per 100 kg of fresh feed.

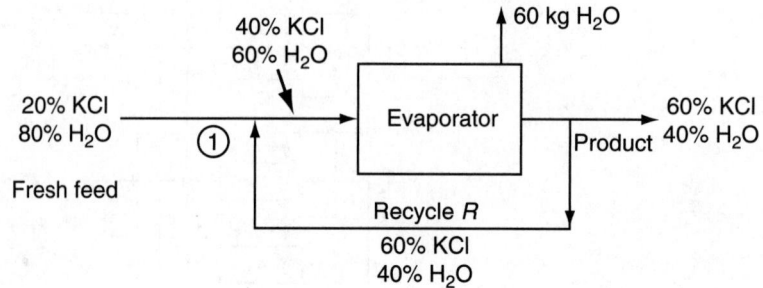

Figure P5.3.3

**5.3.4   In the process shown in Figure P5.3.4, unit I is a liquid-liquid solvent extrac-
tor and unit II is the solvent recovery system. For the purposes of designing
the size of the pipes for streams C and D, the designer obtained from the
given data values of $C = 9630$ lb/hr and $D = 1510$ lb/hr. Are these values
correct? Be sure to show all details of your calculations or explain if you do
not use calculations.

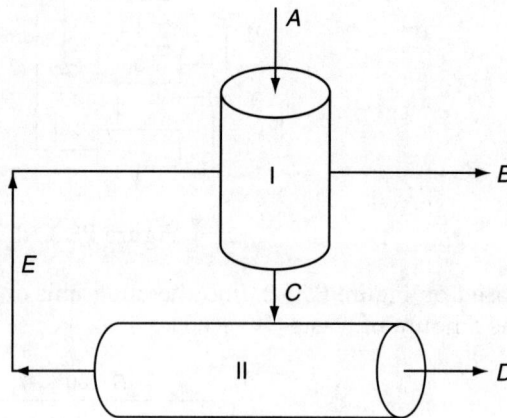

Figure P5.3.4

Known Data

	Flow Rate (lb/hr)	Butene	Butadiene	Solvent
			Composition	
A	5000	0.75	0.25	
B			1.00	
C				
D		0.05	0.95	
E	10,000		0.01	0.99

**5.3.5 The ability to produce proteins through genetic engineering of microbial and mammalian cells and the need for high-purity therapeutic proteins has established a need for efficient large-scale protein purification schemes.

The system of continuous affinity-recycle extraction (CARE) combines the advantages of well-accepted separation methods, such as affinity chromatography, liquid extraction, and membrane filtration, while avoiding the drawbacks inherent in batch and column operations.

The technical feasibility of the CARE system was studied using β-galactosidase affinity purification as a test system. Figure P5.3.5 shows the process. What is the recycle flow rate in milliliters per hour in each stream? Assume that the concentrations of U are equivalent to the concentrations of the β-galactosidase in solution and that steady state exists.

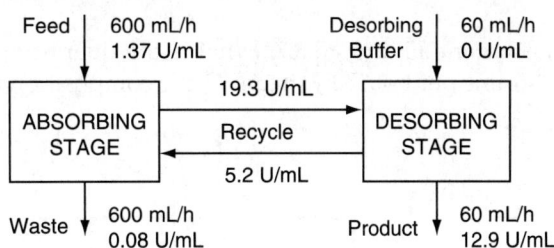

Figure P5.3.5

**5.3.6 Cereal is being dried in a vertical dryer by air flowing countercurrent to the cereal. To prevent breakage of the cereal flakes, exit air from the dryer is recycled. For each 1000 kg/hr of wet cereal fed to the dryer, calculate the input of moist fresh air in kilograms per hour and the recycle rate in kilograms per hour. Data on stream compositions (note that some are mass and others mole fractions):

	Fresh air	Wet cereal	Exit air	Dried cereal	Air entering drier
H_2O	0.0132	0.200	0.263	0.050	0.066

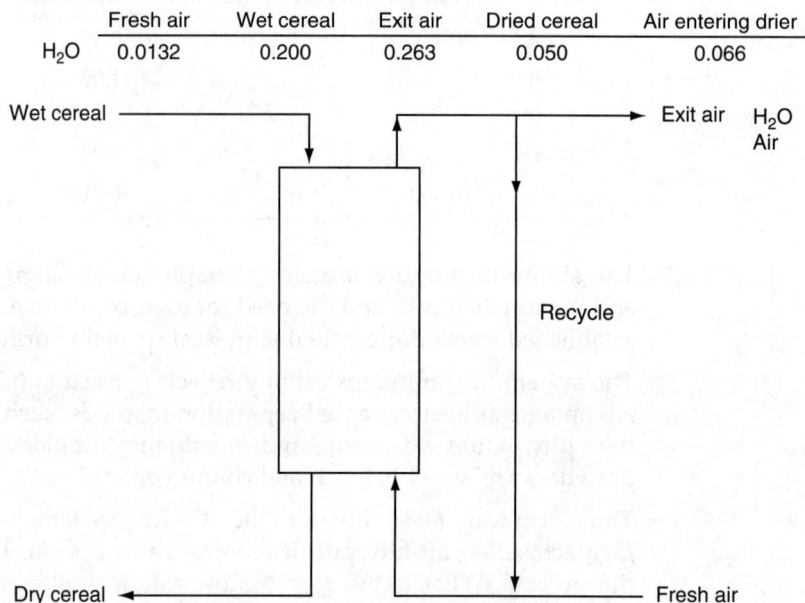

Figure P5.3.6

**5.3.7 Examine Figure P5.3.7. What is the quantity of the recycle stream in kilograms per hour? In stream C, the composition is 4% water and 96% KNO_3.

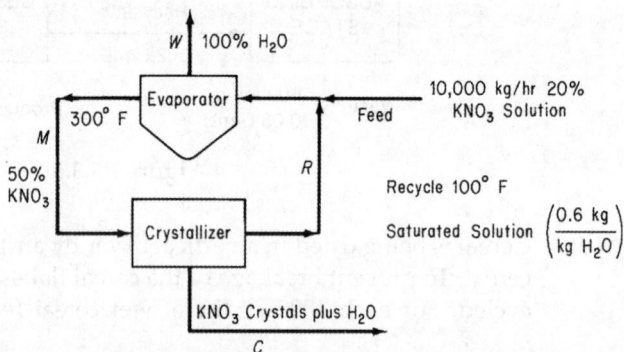

Figure P5.3.7

5.3.8 Examine Figure P5.3.8 (data for 1 hr).
  a. What is the single-pass conversion of H_2 in the reactor?
  b. What is the single-pass conversion of CO?
  c. What is the overall conversion of H_2?
  d. What is the overall conversion of CO?

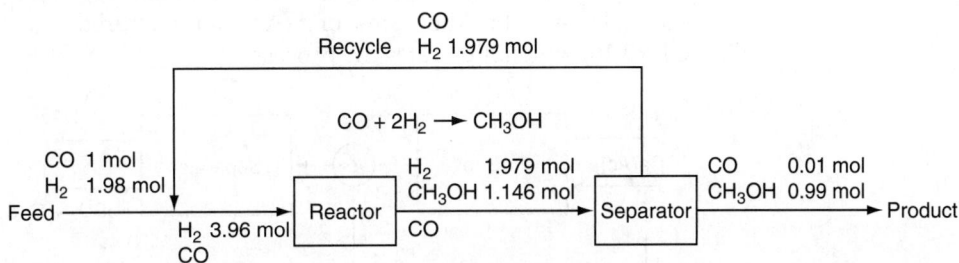

Figure P5.3.8

5.3.9 Hydrogen, important for numerous processes, can be produced by the shift reaction:

$$CO + H_2O \rightarrow CO_2 + H_2$$

In the reactor system shown in Figure P5.3.9, the conditions of conversion have been adjusted so that the H_2 content of the effluent from the reactor is 3 mol %. Based on the data in Figure P5.3.9:
  a. Calculate the composition of the fresh feed.
  b. Calculate the moles of recycle per mole of hydrogen produced.

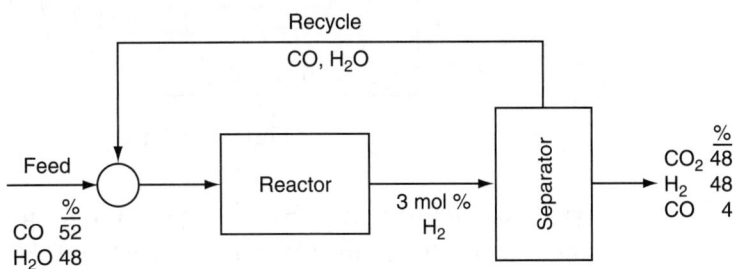

Figure P5.3.9

**5.3.10 Acetic acid (HAc) is to be generated by the addition of 10% excess sulfuric acid to calcium acetate (Ca(Ac)$_2$). The reaction Ca(Ac)$_2$ + H$_2$SO$_4$ → CaSO$_4$ + 2HAc goes to 90% completion based on a single pass through the reactor. The unused Ca(Ac)$_2$ is separated from the products of the reaction and recycled. The HAc is separated from the remaining products. Find the amount of recycle per hour based on 1000 kg of Ca(Ac)$_2$ feed per hour, and also calculate the kilograms of HAc manufactured per hour. See Figure P5.3.10, which illustrates the process.

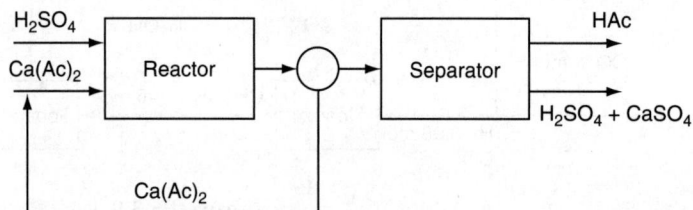

Figure P5.3.10

**5.3.11 The reaction of ethyl-tetrabromide with zinc dust proceeds as shown in Figure P5.3.11.

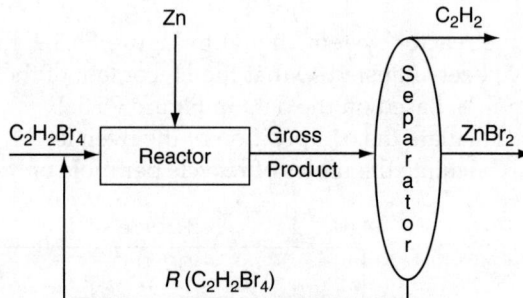

Figure P5.3.11

The reaction is C$_2$H$_2$Br$_4$ + 2Zn → C$_2$H$_2$ + 2ZnBr$_2$. Based on one pass through the reactor, 80% of the C$_2$H$_2$Br$_4$ is reacted and the remainder recycled. On the basis of 1000 kg of C$_2$H$_2$Br$_4$ fed to the reactor per hour, calculate (a) how much C$_2$H$_2$ is produced per hour (in kilograms), (b) the rate of recycle in kilograms per hour, (c) the feed rate necessary for Zn to be 20% in excess, and (d) the mole ratio of ZnBr$_2$ to C$_2$H$_2$ in the final products.

**5.3.12 Examine Figure P5.3.12. NaCl and the feed solution react to form $CaCl_2$. In the reactor, the conversion of $CaCO_3$ is 76% complete. Unreacted $CaCO_3$ is recycled. Calculate (a) the kilograms of Na_2CO_3 exiting the separator per 1000 kg of feed, and (b) the kilograms of $CaCO_3$ recycled per 1000 kg of feed.

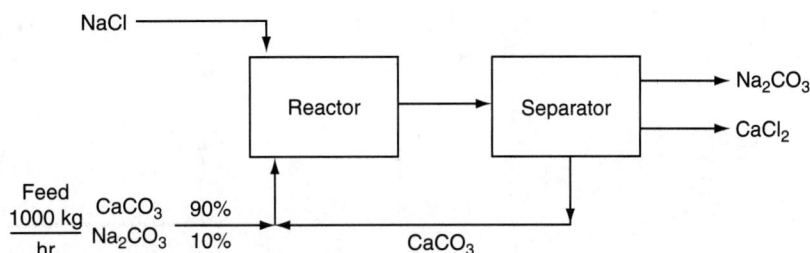

Figure P5.3.12

**5.3.13 Natural gas (CH_4) is burned in a furnace using 15% excess air based on the complete combustion of CH_4. One of the concerns is that the exit concentration of NO (from the combustion of N_2) is about 415 ppm. To lower the NO concentration in the stack gas to 50 ppm, it is suggested that the system be redesigned to recycle a portion of the stack gas back through the furnace. You are asked to calculate the amount of recycle required. Will the scheme work? Ignore the effect of temperature on the conversion of N_2 to NO; that is, assume the conversion factor is constant.

***5.3.14 A plating plant has a waste stream containing zinc and nickel in quantities in excess of that allowed to be discharged into the sewer. The proposed process to be used as a first step in reducing the concentration of Zn and Ni is shown in Figure P5.3.14. Each stream contains water. The concentrations of several of the streams are listed in the table. What is the flow (in liters per hour) of the recycle stream R_{32} if the feed is 1 L/hr?

	Concentration (g/L)	
Stream	Zn	Ni
F	100	10.0
P_0	190.1	17.02
P_2	3.50	2.19
R_{32}	4.35	2.36
W	0	0
D	0.10	1.00

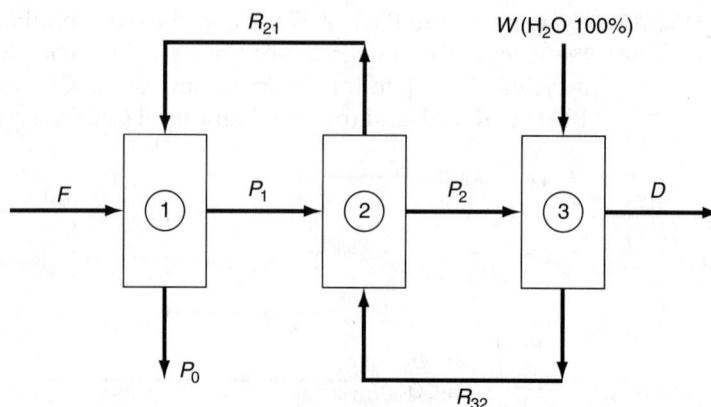

Figure P5.3.14

***5.3.15**   Ultrafiltration is a method for cleaning up input and output streams from a number of industrial processes. The lure of the technology is its simplicity, merely putting a membrane across a stream to sieve out physically undesirable oil, dirt, metal particles, polymers, and the like. The trick, of course, is coming up with the right membrane. The screening material has to meet a formidable set of conditions. It has to be very thin (less than 1 μm), highly porous, yet strong enough to hold up month after month under severe stresses of liquid flow, pH, particle abrasion, temperature, and other plant operating characteristics.

A commercial system consists of standard modules made up of bundles of porous carbon tubes coated on the inside with a series of proprietary inorganic compositions. A standard module is 6 in. in diameter and contains 151 tubes, each 4 ft long, with a total working area of 37.5 ft^2 and daily production of 2000 to 5000 gal of filtrate. Optimum tube diameter is about 0.25 in. A system probably will last at least two to three years before the tubes need replacing from too much residue buildup over the membrane. A periodic automatic chemical cleanout of the tube bundles is part of the system's normal operation. On passing through the filter, the exit stream concentration of oil plus dirt is increased by a factor of 20 over the entering stream.

Calculate the recycle rate in gallons per day (g.p.d.) for the setup shown in Figure P5.3.15, and calculate the concentration of oil plus dirt in the stream that enters the filtration module. The circled values in Figure P5.3.15 are the known concentrations of oil plus dirt.

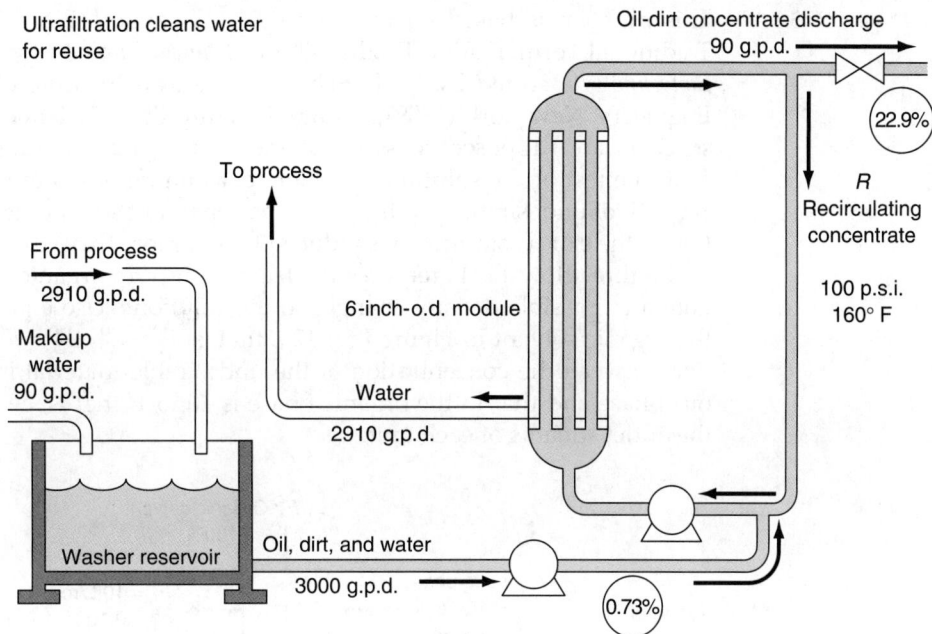

Figure P5.3.15

***5.3.16** To save energy, stack gas from a furnace is used to dry rice. The flowsheet and known data are shown in Figure P5.3.16. What is the amount of recycle gas (in pound moles) per 100 lb of P if the concentration of water in the gas stream entering the dryer is 5.20%?

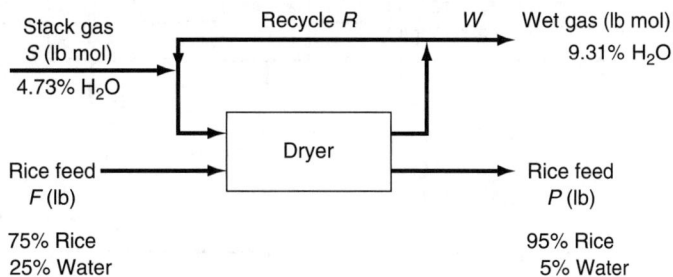

Figure P5.3.16

***5.3.17   This problem is based on the data of G. F. Payne ["Bioseparations of Traditional Fermentation Products," in *Chemical Engineering Problems in Biotechnology*, edited by M. L. Schuler, American Institute of Chemical Engineers, New York (1989)]. Examine Figure P5.3.17. Three separation schemes are proposed to separate the desired fermentation products from the rest of the solution. Ten liters per minute of a broth containing 100 g/L of undesirable product are to be separated so that the concentration in the exit waste stream is reduced to (not more than) 0.1 g/L. Which of the three flowsheets requires the least fresh pure organic solvent? Ignore any possible density changes in the solutions. Use equal values of the organic solvent in Figure P5.3.17b, that is, $F_1^O + F_2^O + F_3^O = F^O$. The relation between the concentration of the undesirable material in the aqueous phase and that in the organic phase is 10 to 1, that is, $c^A/c^O = 10$ in the outlet streams of each unit.

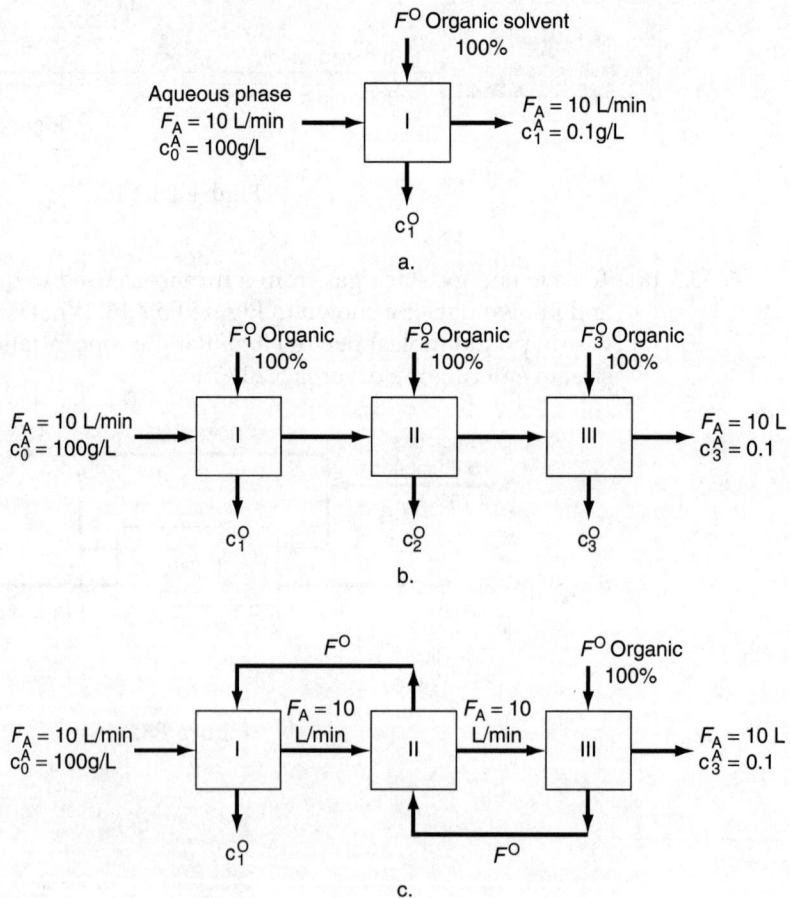

Figure P5.3.17

***5.3.18 In the process sketched in Figure P5.3.18, Na_2CO_3 is produced by the reaction $Na_2S + CaCO_3 \rightarrow Na_2CO_3 + CaS$. The reaction is 90% complete on one pass through the reactor, and the amount of $CaCO_3$ entering the reactor is 50% in excess of that needed. Calculate on the basis of 1000 lb/hr of fresh feed (a) the pounds of Na_2S recycled, and (b) the pounds of Na_2CO_3 solution formed per hour.

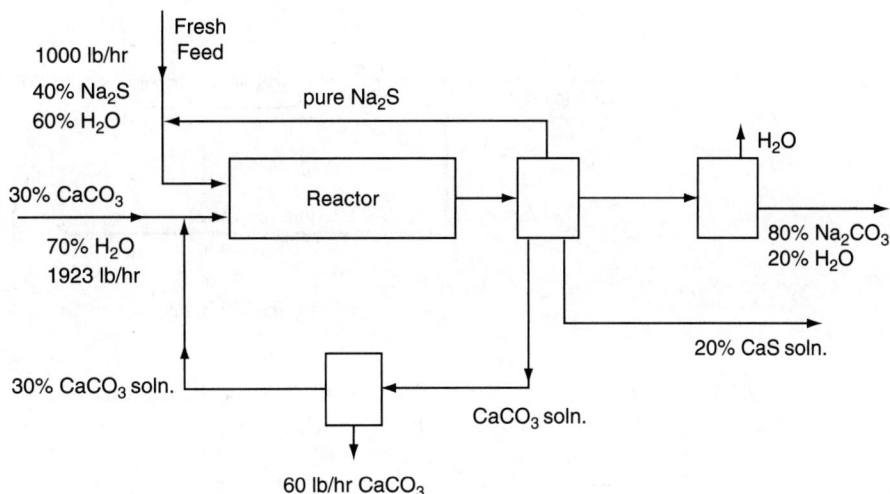

Figure P5.3.18

***5.3.19 Toluene reacts with H_2 to form benzene (B), but a side reaction occurs in which a by-product, diphenyl (D), is formed:

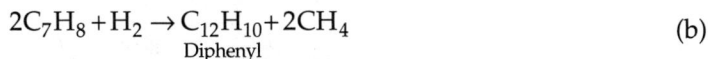

$$\underset{\text{Toluene}}{C_7H_8} + \underset{\text{Hydrogen}}{H_2} \rightarrow \underset{\text{Benzene}}{C_6H_6} + \underset{\text{Methane}}{CH_4} \qquad (a)$$

$$2C_7H_8 + H_2 \rightarrow \underset{\text{Diphenyl}}{C_{12}H_{10}} + 2CH_4 \qquad (b)$$

The process is shown in Figure P5.3.19. Hydrogen is added to the gas recycle stream to make the ratio of H_2 to CH_4 1 to 1 before the gas enters the mixer. The ratio of H_2 to toluene entering the reactor at G is $4H_2$ to 1 toluene. The conversion of toluene to benzene on one pass through the reactor is 80%, and the conversion of toluene to the by-product diphenyl is 8% on the same pass.

Calculate the moles of R_G and moles of R_L per hour.

Data:	Compound:	H_2	CH_4	C_2H_6	C_7H_8	$C_{12}H_{10}$
	MW:	2	16	78	92	154

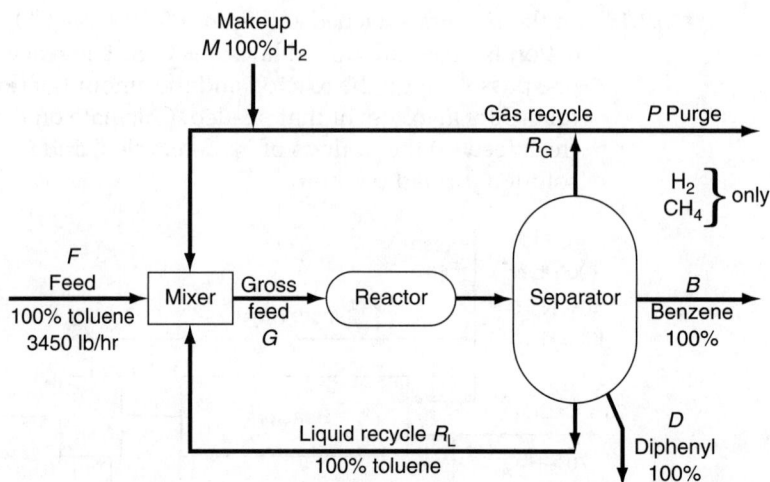

Figure P5.3.19

***5.3.20** The process shown in Figure P5.3.20 is the dehydrogenation of propane (C_3H_8) to propylene (C_3H_6) according to the reaction

$$C_3H_8 \rightarrow C_3H_6 + H_2$$

The conversion of propane to propylene based on the *total* propane feed into the reactor at F_2 is 40%. The product flow rate F_5 is 50 kg mol/hr.
a. Calculate all six flow rates F_1 to F_6 in kilogram moles per hour.
b. What is the percent conversion of propane in the reactor based on the fresh propane fed to the process (F_1)?

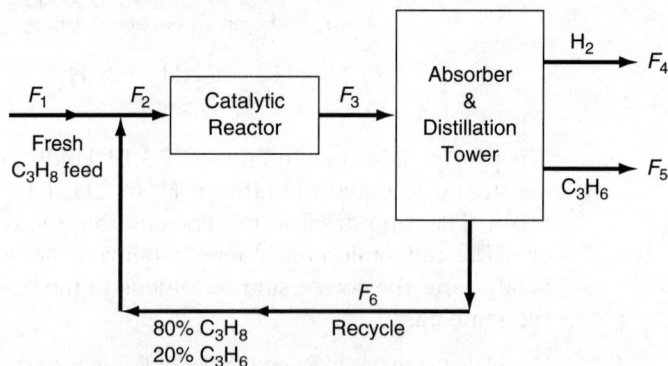

Figure P5.3.20

***5.3.21** Sulfur dioxide may be converted to SO_3, which has many uses, including the production of H_2SO_4 and sulfonation of detergent. A gas stream having the composition shown in Figure P5.3.21 is to be passed through a two-stage converter. The fractional conversion of the SO_2 to SO_3 (on one pass though) in the first stage is 0.75 and in the second stage, 0.65. To boost the overall conversion to 0.95, some of the exit gas from stage 2 is recycled back to the inlet of stage 2. How much must be recycled per 100 mol of inlet gas (stream F)? Ignore the effect of temperature on the conversion.

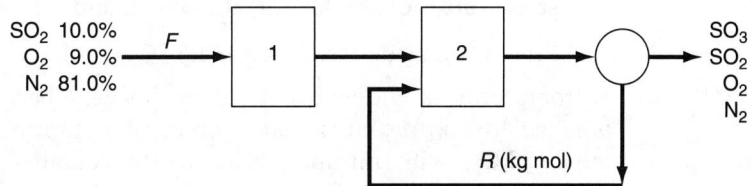

Figure P5.3.21

****5.3.22 Benzene, toluene, and other aromatic compounds can be recovered by solvent extraction with sulfur dioxide. As an example, a catalytic reformate stream containing 70% by weight benzene and 30% non-benzene material is passed through the countercurrent extractive recovery scheme shown in Figure P5.3.22. One thousand kilograms of the reformate stream and 3000 kg of sulfur dioxide are fed to the system per hour. The benzene product stream contains 0.15 kg of sulfur dioxide per kilogram of benzene. The waste stream contains all of the initially charged non-benzene material as well as 0.25 kg of benzene per kilogram of the non-benzene material. The remaining component in the waste stream is the sulfur dioxide.

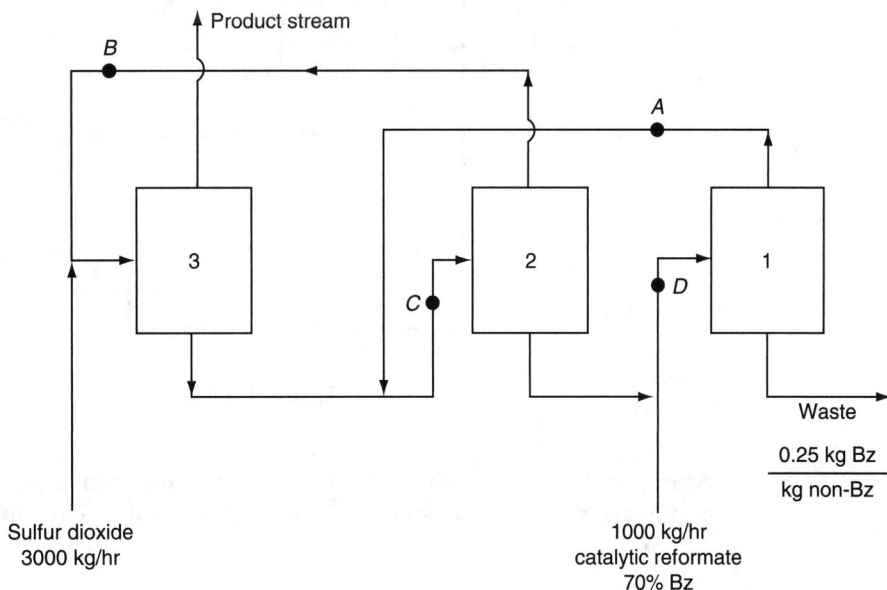

Figure P5.3.22

a. How many kilograms of benzene are extracted per hour (are in the product stream)?

b. If 800 kg of benzene containing 0.25 kg of the non-benzene material per kilogram of benzene are flowing per hour at point A and 700 kg of benzene containing 0.07 kg of the non-benzene material per kilogram of benzene are flowing at point B, how many kilograms (exclusive of the sulfur dioxide) are flowing at points C and D?

Product benzene includes 0.15 kg SO_2/kg Bz.

****5.3.23 Nitroglycerine, a widely used high explosive, when mixed with wood flour is called "dynamite." It is made by mixing high-purity (99.9+% pure) glycerine ($C_3H_8O_3$) with nitration acid, which contains 50.00% H_2SO_4, 43.00% HNO_3, and 7.00% water by weight. The reaction is

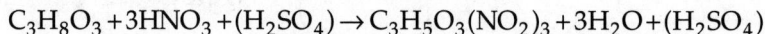

$$C_3H_8O_3 + 3HNO_3 + (H_2SO_4) \rightarrow C_3H_5O_3(NO_2)_3 + 3H_2O + (H_2SO_4)$$

The sulfuric acid does not take part in the reaction but is present to "catch" the water formed. Conversion of the glycerine in the nitrator is complete, and there are no side reactions, so all of the glycerine fed to the nitrator forms nitroglycerine. The mixed acid entering the nitrator (stream G) contains 20.00% excess HNO_3 to assure that all of the glycerine reacts. Figure P5.3.23 is a process flow diagram.

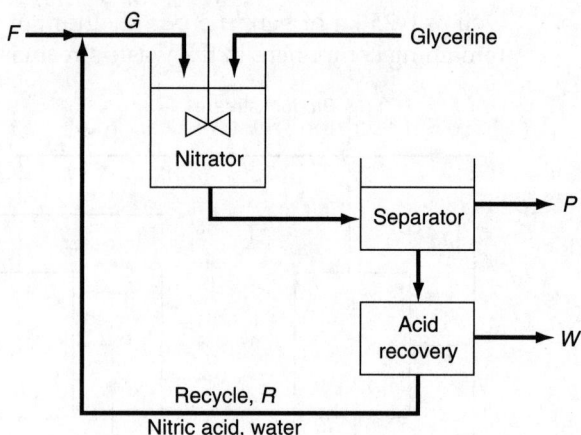

Figure P5.3.23

After nitration, the mixture of nitroglycerine and spent acid (HNO_3, H_2SO_4, and water) goes to a separator (a settling tank). The nitroglycerine is insoluble in the spent acid, and its density is less, so it rises to the top. It is carefully drawn off as product stream P and sent to wash tanks for purification. The spent acid is withdrawn from the bottom of the separator and sent to an acid recovery tank, where the HNO_3 and H_2SO_4 are separated. The H_2SO_4 – H_2O mixture is stream W and is concentrated and sold for industrial purposes. The recycle stream to the nitrator is a 70.00% by weight solution of HNO_3 in water. In the diagram, product stream P is 96.50% nitroglycerine and 3.50% water by weight.

To summarize:

Stream F = 50.00 wt % H_2SO_4, 43.00% HNO_3, 7.00% H_2O

Stream G contains 20.00% excess nitric acid

Stream P = 96.50 wt % nitroglycerine, 3.50 wt % water

Stream R = 70.00 wt % nitric acid, 30.00% water

a. If 1000×10^3 of glycerine per hour are fed to the nitrator, how many kilograms per hour of stream P result?
b. How many kilograms per hour are in the recycle stream?
c. How many kilograms of fresh feed, stream F, are fed per hour?
d. Stream W is how many kilograms per hour? What is its analysis in weight percent? Molecular weights: glycerine = 92.11, nitroglycerine = 227.09, nitric acid = 63.01, sulfuric acid = 98.08, and water = 18.02

Caution: Do not try this process at home.

****5.3.24 The following problem is condensed from the book writter by D. T. Allen and D. R. Shonnard, *Green Engineering* [Prentice Hall, Upper Saddle River, NJ (2002)]. Acrylonitrile (AN) can be produced by the reaction of propylene with ammonia in the gas phase:

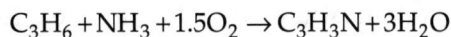

$$C_3H_6 + NH_3 + 1.5O_2 \rightarrow C_3H_3N + 3H_2O$$

Figure P5.3.24 is the flowsheet for the process with the data superimposed. The only contaminate of concern is the ammonia.

Answer the following questions:
a. Can any of the waste streams that are collected and sent to treatment be used to replace some of the boiler water feed?
b. What streams might be considered as candidates to replace some of the feed to the scrubber?
c. If the discharge stream from the condenser associated with the distillation column is recycled back to the scrubber to replace 0.7 kg/s of the water used in the scrubber, what changes in the flows and concentrations will occur in the process?

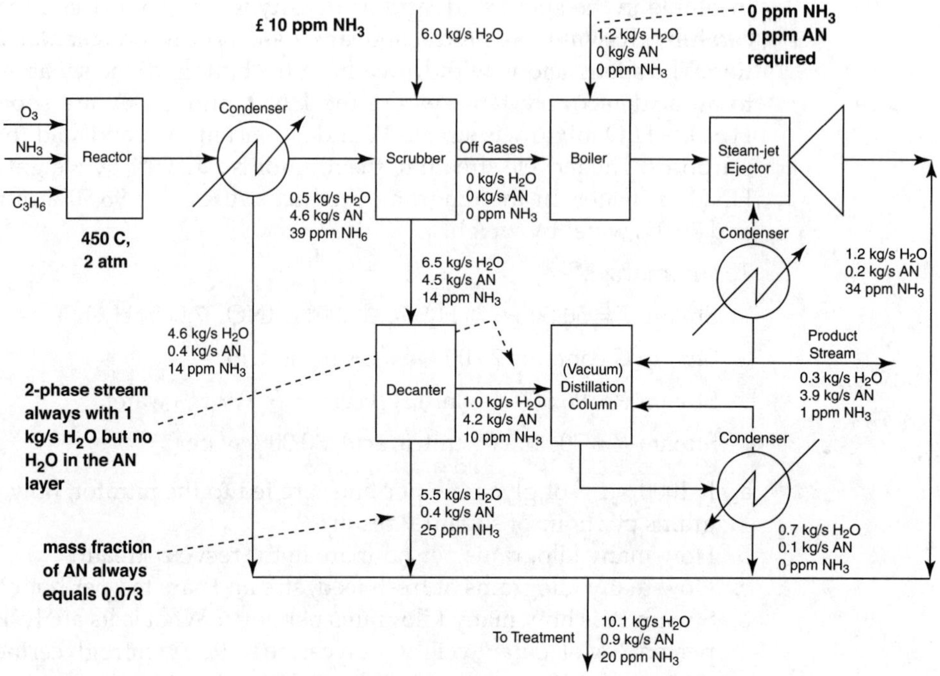

Figure P5.3.24

5.4   Bypass and Purge

**5.4.1 Many chemical processes generate emissions of volatile compounds that need to be controlled. In the process shown in Figure P5.4.1, the exhaust of CO is eliminated by its separation from the reactor effluent and recycling of 100% of the CO generated in the reactor together with some reactant back to the reactor feed.

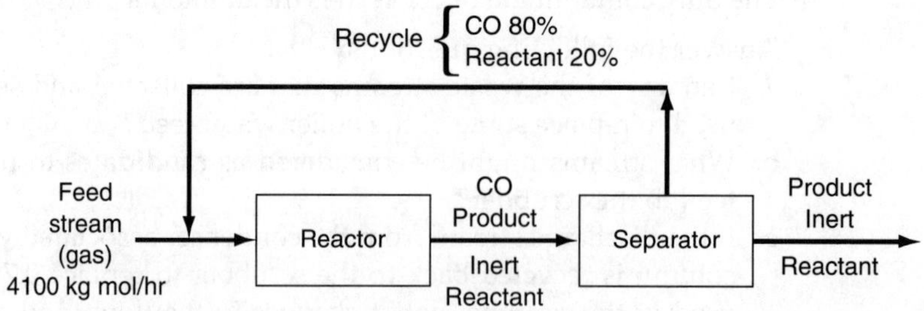

Figure P5.4.1

Although the product is proprietary, information is provided that the feed stream contains 40% reactant, 50% inert, and 10% CO, and that on reaction, 2 mol of reactant yield 2.5 mol of product. Conversion of reactant to product is only 73% on one pass through the reactor and 90% overall. You are asked to calculate the ratio of moles of recycle to moles of product. What do you discover is wrong with this problem?

**5.4.2 Alkyl halides are used as an alkylating agent in various chemical transformations. The alkyl halide ethyl chloride can be prepared by the following chemical reaction:

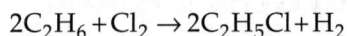

$$2C_2H_6 + Cl_2 \rightarrow 2C_2H_5Cl + H_2$$

In the reaction process shown in Figure P5.4.2, fresh ethane and chlorine gas and recycled ethane are combined and fed into the reactor. A test shows that if 100% excess chlorine is mixed with ethane, a single-pass optimal conversion of 60% results, and of the ethane that reacts, all is converted to products and none goes into undesired products. You are asked to calculate (a) the fresh feed concentrations required for operation, and (b) the moles of C_2H_5Cl produced in P per mole of C_2H_6 in the fresh feed F_1.

What difficulties will you discover in the calculations?

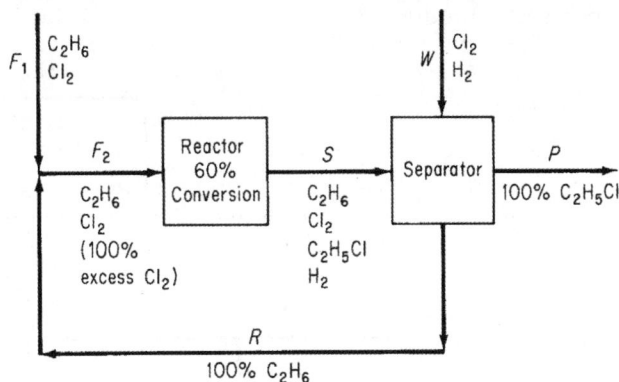

Figure P5.4.2

****5.4.3 A process for methanol synthesis is shown in Figure P5.4.3. The pertinent chemical reactions involved are

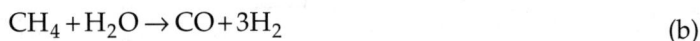

$$CH_4 + 2H_2O \rightarrow CO_2 + 4H_2 \tag{a}$$

$$CH_4 + H_2O \rightarrow CO + 3H_2 \tag{b}$$

$$2CO + O_2 \rightarrow 2CO_2 \quad \text{(CO converter reaction)} \tag{c}$$

$$CO_2 + 3H_2 \rightarrow CH_3OH + H_2O \quad \text{(methanol synthesis reaction)} \tag{d}$$

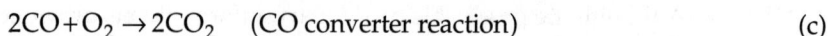

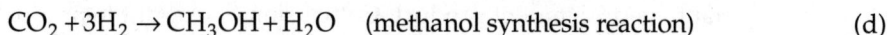

Ten percent excess steam, based on Reaction (a), is fed to the reformer, and conversion of methane is 100%, with a 90% yield of CO_2. Conversion in the methanol reactor is 55% on one pass through the reactor.

A stoichiometric quantity of oxygen is fed to the CO converter, and the CO is completely converted to CO_2. Additional makeup CO_2 is then introduced to establish a 3-to-1 ratio of H_2 to CO_2 in the feed stream to the methanol reactor.

The methanol reactor effluent is cooled to condense all of the methanol and water, with the noncondensable gases recycled to the methanol reactor feed. The H_2/CO_2 ratio in the recycle stream is also 3 to 1.

Because the methane feed contains 1% nitrogen as an impurity, a portion of the recycle stream must be purged as shown in Figure P5.4.3 to prevent the accumulation of nitrogen in the system. The purge stream analyzes 5% nitrogen.

On the basis of 100 mol of methane feed (including the impurity):
a. How many moles of H_2 are lost in the purge?
b. How many moles of makeup CO_2 are required?
c. What is the recycle-to-purge ratio in moles per mole?
d. How much methanol solution (in kilograms) of what strength (weight percent) is produced?

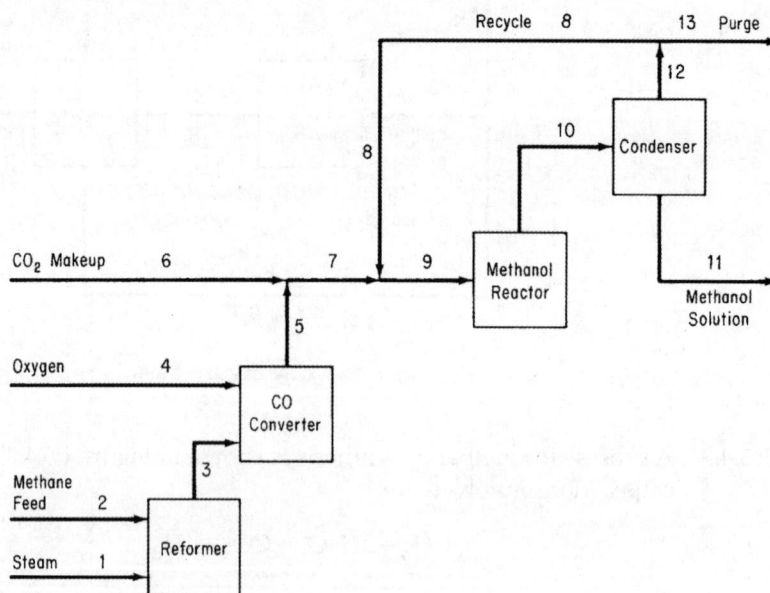

Figure P5.4.3

PART III
GASES, VAPORS, AND LIQUIDS

CHAPTER 6

Ideal and Real Gases

Chapter Objectives

- Understand the conditions under which the ideal gas law applies, and the conditions for which real gas relations should be used.
- Remember that the values of p and T used in relations to determine gas properties are absolute, not relative, values.
- Use partial pressure in calculations involving multicomponent ideal gases.
- Solve material balances involving ideal or real gases.

Introduction

Consider the furnace and associated stack for discharging the flue gas to the atmosphere, shown in Figure 6.1. Assume that you want to design the stack (e.g., determine its diameter) based on the average velocity of the flue gas in the stack. From material and energy balances on the furnace, you are able to calculate the molar flow rates of each component in the flue gas, but you need the molar density and other properties of the flue gas in order to calculate the average velocity in the stack. This chapter introduces a variety of ways to calculate the properties of gases, including the molar density, temperature, and pressure.

Figure 6.1   Schematic of a furnace and flue gas stack

In this chapter, we begin to consider the physical properties of pure components and mixtures. By **property**, we mean any measurable characteristic of a substance, such as pressure, volume, or temperature, or a characteristic that can be calculated or deduced, such as internal energy, which is discussed in Chapter 8. The **state** of a system gives the condition of a system as specified by its properties. You can find values for properties of compounds and mixtures in many formats, including

1. Experimental data
2. Tables (developed from experimental data or theory)
3. Graphs
4. Equations

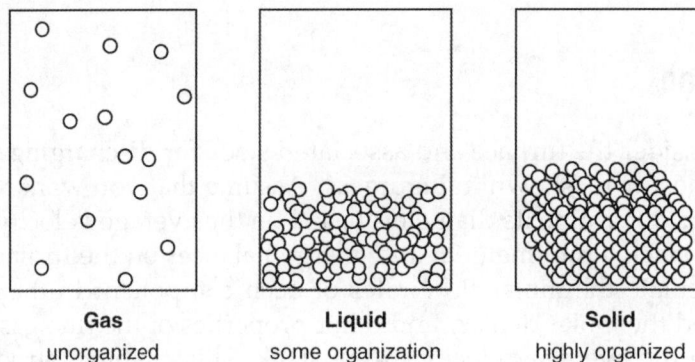

| **Gas** | **Liquid** | **Solid** |
| unorganized | some organization | highly organized |

Figure 6.2   Three phases of a compound showing the classification by degree of organization

Clearly, you cannot realistically expect to have reliable, detailed experimental data at hand or in a database for the properties of all of the useful pure compounds and mixtures with which you will be involved. Consequently, in the absence of experimental information, you have to estimate (predict)

properties based on empirical correlations or graphs so that you can introduce appropriate parameters in material and energy balances.

A compound (or a mixture of compounds) may consist of one or more
phases. A **phase** is defined as a completely homogeneous and uniform state
of matter. Look at Figure 6.2. Liquid water would be a phase, and ice would
be another phase. Two immiscible liquids in the same container, such as
mercury and water, would represent two different phases because each liquid is separate and homogeneous and has different properties.

6.1   Ideal Gases

You have no doubt been exposed to the concept of the ideal gas in chemistry
and physics. Why go over ideal gases again? At least two reasons exist. First,
the experimental and theoretical properties of ideal gases are far simpler
than the corresponding properties of liquids and solids. Second, use of the
ideal gas concept is of considerable industrial importance.

In this section, we explain how the ideal gas law can be used to calculate the pressure, temperature, volume, or number of moles in a quantity of
gas, and we define the partial pressure of a gas in a mixture of gases. We also
discuss how to calculate the specific gravity and density of a gas. Then we
apply these concepts to solving material balances.

6.1.1   Ideal Gas

The most famous and widely used equation that relates p, V, n, and T for a
gas is the **ideal gas law:**

$$pV = nRT \qquad (6.1)$$

where p is the **absolute pressure** of the gas
  V is the total volume occupied by the gas
  n is the number of moles of the gas
  R is the ideal (universal) gas constant in appropriate units
  T is the **absolute temperature** of the gas

You can find values of R in various units inside the back cover of this
book. Sometimes, the ideal gas law is written as

$$p\hat{V} = RT \qquad (6.1a)$$

Note that in Equation (6.1a), $\hat{V}$ is the *specific molar* volume (volume per mole, V/n) of the gas. When gas volumes are involved in a problem, $\hat{V}$ will be the volume per mole and not the volume per mass. The inverse of $\hat{V}$ is the molar density, moles per volume. Figure 6.3 illustrates the surface generated by Equation (6.1a) in terms of the three properties p, $\hat{V}$, and T. Look at the projections of the surface in Figure 6.3 onto the two-parameter planes. The interpretation is as follows:

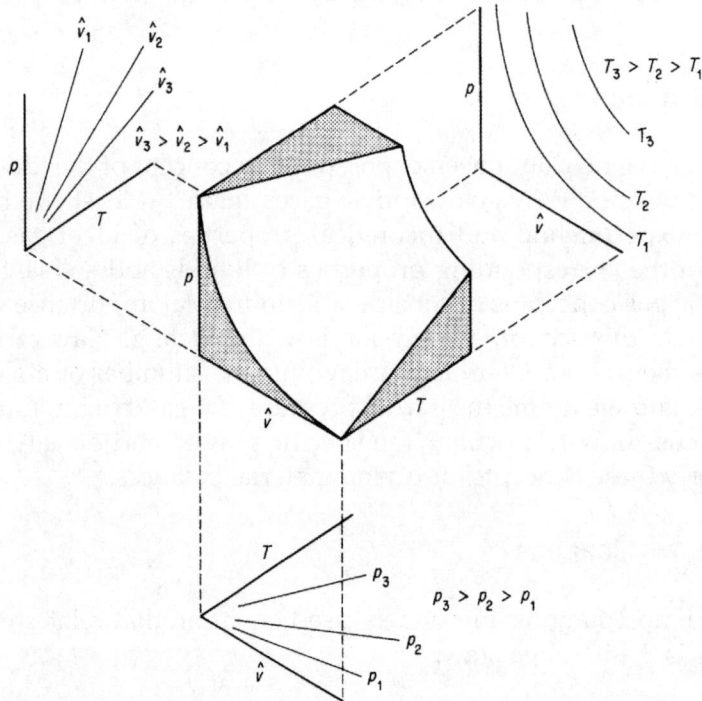

Figure 6.3   Representation of the ideal gas law in three dimensions as a surface

1. The projection to the upper left onto the p-T plane shows straight lines for constant values of $\hat{V}$. Why? Equation (6.1a) for constant specific volume reduces to $p = KT$, where K is a constant that yields the equation of a straight line that passes through the origin.

2. The projection to the upper right onto the p-$\hat{V}$ plane shows curves for values of constant T. What kinds of curves are they? For constant T, Equation (6.1a) becomes $p\hat{V}$ = constant, namely, a hyperbola.

3. The projection downward onto the T-$\hat{V}$ plane again shows straight lines. Why? Equation (6.1a) for constant p is $\hat{V}$ = (constant)(T).

For an ideal gas, Equation (6.1) can be applied to a pure component or to a mixture. What are the conditions for a gas to behave as predicted by the ideal gas law? The major ones for a gas to be ideal are as follows:

1. The molecules do not occupy any space; they are infinitesimally small.

2. No attractive forces exist between the molecules, so the molecules move completely independently of each other.

3. The gas molecules move in random, straight-line motion and the collisions between the molecules, and between the molecules and the walls of the container, are perfectly elastic.

Gases at low pressure and/or high temperature meet these conditions. Solids, liquids, and gases at high density—that is, high pressure and/or low temperature—do not. From a practical viewpoint, within reasonable error, you can treat air, oxygen, nitrogen, hydrogen, carbon dioxide, methane, and even water vapor, under most of the ordinary conditions you encounter, as ideal gases.

Several equivalent standard states known as **standard conditions** (S.C., or STP, for "standard temperature and pressure") of temperature and pressure have been specified for gases by custom. Refer to Table 6.1. Note that the standard conditions for the International System of Units (SI), Universal Scientific, Universal Scientific B, and American Engineering (AE) systems are exactly the same conditions, but in different units. The natural gas industry, however, uses a different reference temperature (15°F) but the same reference pressure (1 atm). Also, note that the values of R are listed for each set of units.

Table 6.1   Common Standard Conditions for the Ideal Gas

System	T	p	$\hat{V}$	R
SI	273.15 K	101.325 kPa	22.415 m^3/kg mol	8.315 kPa m^3 kg mol^{-1} K^{-1}
Universal Scientific	0.0°C	760 mm Hg	22.415 L/g mol	62.325 mm Hg L g mol^{-1} K^{-1}
Universal Scientific B	273.15 K	1 atm	22.415 L/g mol	0.08206 atm L g mol^{-1} K^{-1}
American Engineering	491.67°R (32°F)	1 atm	359.05 ft^3/lb mol	0.7303 atm ft^3 lb mol^{-1} °R^{-1}
Natural gas industry	59.0°F (15.0°C)	14.696 psia	379.4 ft^3/lb mol	10.73 psia ft^3 lb mol^{-1} °R^{-1}

You can insert the values at S.C. into the ideal gas equation to calculate R in any set of units you want. For example, what is R for 1 g mol of ideal gas with a volume in cubic centimeters, pressure in atmospheres, and temperature in kelvin?

$$R = \frac{p\hat{V}}{T} = \frac{1\,\text{atm}}{273.15\,\text{K}} \left| \frac{22,415\,\text{cm}^3}{1\,\text{g mol}} = 82.06 \frac{(\text{cm}^3)(\text{atm})}{(\text{K})(\text{g mol})} \right.$$

The fact that a substance may not exist as a gas at 0°C and 1 atm is immaterial. Thus, as we shall see later, water vapor at 0°C cannot exist at a pressure greater than its vapor pressure of 0.61 kPa (0.18 in. Hg) without condensation occurring. However, you can calculate the imaginary volume at standard conditions, and it is just as useful a quantity in the calculation of volume-mole relationships as though it could exist. In what follows, the symbol V stands for total volume and the symbol $\hat{V}$ for volume per mole.

Because the SI, Universal Scientific, Universal Scientific B, and AE standard conditions all refer to the same point in the p, $\hat{V}$, and T space, you can use the values in Table 6.1 with their units to change from one system of units to another. If you memorize the standard conditions, you will find it easy to work with mixtures of units from different systems.

The following example illustrates how you can use the standard conditions to convert mass or moles to volume.

Example 6.1   Use of Standard Conditions to Calculate Volume from Mass

Problem Statement

Calculate the volume, in cubic meters, occupied by 40 kg of CO_2 at standard conditions, assuming CO_2 acts as an ideal gas.

Solution

$$\text{Basis: 40 kg of } CO_2$$

$$\frac{40\,\text{kg } CO_2}{} \left| \frac{1\,\text{kg mol } CO_2}{44\,\text{kg } CO_2} \right| \frac{22.42\,\text{m}^3\,CO_2}{1\,\text{kg mol } CO_2} = 20.4\,\text{m}^3\,CO_2 \text{ at S.C.}$$

Notice in Example 6.1 that the information that 22.42 m³ of gas at S.C. = 1 kg mol of gas is applied to transform a known number of moles into an equivalent number of cubic meters. An alternative way to calculate the volume at standard conditions is to use Equation (6.1) directly. Incidentally, whenever you report a volumetric value, **you must establish the conditions of temperature and pressure at which the volumetric measure exists** because the term m³ or ft³, standing alone, is really not any particular *quantity* of a gaseous material.

In many processes going from an initial state to a final state, you use the ratio of the ideal gas laws in the respective states and thus eliminate R as follows (the subscript 1 designates the initial state, and the subscript 2 designates the final state):

$$\frac{p_1 V_1}{p_2 V_2} = \frac{n_1 R T_1}{n_2 R T_2}$$

or

$$\left(\frac{p_1}{p_2}\right)\left(\frac{V_1}{V_2}\right) = \left(\frac{n_1}{n_2}\right)\left(\frac{T_1}{T_2}\right)$$

(6.2)

Note that Equation (6.2) involves ratios of the same variable. This result has the convenient feature that the pressures may be expressed in any system of units you choose, such as kilopascals, inches of Hg, millimeters of Hg, atmospheres, and so on, as long as the same units are used for both conditions of pressure (do not forget that the pressure must be *absolute* pressure in both cases). Similarly, the ratio of the *absolute* temperatures and ratio of the volumes result in dimensionless ratios. Also, note how the ideal gas constant R is eliminated in taking the ratios.

Let's see how you can apply the ideal gas law to problems in the form of both Equation (6.2) and Equation (6.1).

Example 6.2   Application of the Ideal Gas Law to Calculate a Volume

Problem Statement

Calculate the volume occupied by 88 lb of CO_2 at 15°C and a pressure of 32.2 ft of water.

Solution

Examine Figure E6.2. To use Equation (6.2), the initial volume has to first be calculated as shown in Example 6.1.

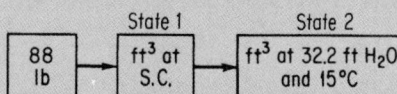

Figure E6.2

(*Continues*)

Example 6.2   Application of the Ideal Gas Law to Calculate a Volume (*Continued*)

Then the final volume can be calculated via Equation (6.2) in which both R and (n_1/n_2) cancel. Table 6.1 does not list the pressure in feet of H_2O at S.C. Where do you get the value? Look in Chapter 2 at the discussion of pressure, or calculate it ($\rho g h = p$).

$$V_2 = V_1 \left(\frac{P_1}{P_2} \right) \left(\frac{T_2}{T_1} \right)$$

Assume that the given pressure is absolute pressure.

At S.C. (state 1)	At p and T (state 2)
$p = 33.91$ ft H_2O	$p = 32.2$ ft H_2O
$T = 273$ K	$T = 273 + 15 = 288$ K

Basis: 88 lb of CO_2

$$\frac{88 \text{ lb } CO_2}{} \left| \frac{1 \text{ lb mol } CO_2}{44 \text{ lb } CO_2} \right| \frac{359 \text{ ft}^3}{1 \text{ lb mol}} \left| \frac{288}{273} \right| \frac{33.91}{32.2} = \begin{array}{c} 798 \text{ ft}^3 \; CO_2 \\ \text{at } 32.2 \text{ ft } H_2O \text{ and } 15°C \end{array}$$

You can mentally check your calculations by saying to yourself: The temperature goes up from 0°C at S.C. to 15°C at the final state; hence, the volume must increase from S.C., and the temperature ratio must be greater than unity. Similarly, you can say: The pressure goes down from S.C. to the final state, so the volume must increase from S.C.; hence, the pressure ratio must be greater than unity.

The same result can be obtained by using Equation (6.1). First obtain the value of R in the same units as the variables p, $\hat{V}$, and T. Look it up or calculate the value from p, $\hat{V}$, and T at S.C.:

$$R = \frac{p\hat{V}}{T}$$

At S.C.: $p = 33.91$ ft H_2O $\hat{V} = 359$ ft^3/lb mol $T = 273$ K

$$R = \frac{33.91}{} \left| \frac{359}{273} \right| = 44.59 \frac{(\text{ft } H_2O)(\text{ft}^3)}{(\text{lb mol})(K)}$$

Now, using Equation (6.1), insert the given values, and perform the necessary calculations.

Basis: 88 lb of CO_2

$$V = \frac{nRT}{p} = \frac{88 \text{ lb } CO_2}{44 \text{ lb } CO_2} \left| \frac{44.59 \,(\text{ft } H_2O)\,(\text{ft}^3)}{(\text{lb mol})\,(K)} \right| \frac{288 \text{ K}}{32.2 \text{ ft } H_2O}$$

$$= 798 \,\text{ft}^3 \, CO_2 \text{ at } 32.2 \,\text{ft } H_2O \text{ and } 15°C$$

If you inspect the two solutions, you will observe that the same numbers appear in both and that the results are identical.

To calculate the **volumetric flow rate** of a gas, $\dot{V}$, such as in cubic meters or cubic feet per second, through a pipe, you divide the volume of the gas passing through the pipe in a time interval by the value of the time interval. To get the **velocity**, $\dot{v}$, of the flow, you divide the volumetric flow rate by the area, A, of the pipe:

$$\dot{V} = A\dot{v} \text{ hence } \dot{v} = \dot{V}/A \tag{6.3}$$

The (mass) **density of a gas** is defined as the mass per unit volume and can be expressed in various units, including kilograms per cubic meter, pounds per cubic foot, grams per liter, and so on. Inasmuch as the mass contained in a unit volume varies with the temperature and pressure, as we have previously mentioned, you should always be careful to specify these two conditions in calculating density. If not otherwise specified, the densities are presumed to be at S.C. As an example, what is the density of N_2 at 27°C and 100 kPa in SI units?

Basis: 1 m^3 of N_2 at 27°C and 100 kPa

$$\frac{1 m^3}{} \left| \frac{273 \text{ K}}{300 \text{ K}} \right| \frac{100 \text{ kPa}}{101.3 \text{ kPa}} \left| \frac{1 \text{ kg mol}}{22.4 \text{ m}^3} \right| \frac{28 \text{ kg}}{1 \text{ kg mol}} = 1.123 \,\text{kg/m}^3$$

of N_2 at 27°C (300 K) and 100 kPa

In addition to the mass density, sometimes the "density" of a gas refers to the molar density, namely, moles per unit volume. How can you tell the difference if the same symbol is used for the density?

The **specific gravity** of a gas is usually defined as the ratio of the density of the gas at a desired temperature and pressure to that of air (or any specified reference gas) at a certain temperature and pressure. The use of specific gravity occasionally may be confusing because of the imprecise manner in which the values of specific gravity are reported without citing T and p, such as "What is the specific gravity of methane?" The answer to the question is not clear; hence, assume S.C. for both the gas and the reference gas:

$$\text{sp. gr.} = \frac{\text{density of methane at S.C.}}{\text{density of air at S.C.}}$$

6.1.2   Ideal Gas Mixtures

Frequently, as an engineer, you will want to make calculations for *mixtures of gases* instead of individual gases. You can use the ideal gas law, under the proper assumptions, of course, for a mixture of gases by interpreting p as the total absolute pressure of the mixture, V as the volume occupied by the mixture, n as the total number of moles of all components in the mixture, and T as the absolute temperature of the mixture. As the most obvious example, air is composed of N_2, O_2, Ar, CO_2, Ne, He, and other trace gases, but you can treat air as a single compound in applying the ideal gas law.

Engineers use a fictitious but useful quantity called the **partial pressure** in many of their calculations involving gases. The partial pressure of Dalton, p_i, namely, the pressure that would be exerted by a single component in a gaseous mixture if it existed alone in the *same volume* as that occupied by the mixture and at the *same temperature* as the mixture, is defined by

$$p_i V_{total} = n_i R T_{total} \tag{6.4}$$

where p_i is the partial pressure of component i in the mixture. If you divide Equation (6.4) by Equation (6.1), you find that

$$\frac{p_i V_{total}}{p_{total} V_{total}} = \frac{n_i R T_{total}}{n_{total} R T_{total}}$$

and

$$p_i = p_{total} \frac{n_i}{n_{total}} = p_{total} y_i \tag{6.5}$$

where y_i is the mole fraction of component i. In air, the percent oxygen is 20.95; hence, at the standard condition of 1 atm, the partial pressure of oxygen is $p_{O_2} = 0.2095(1) = 0.2095$ atm. Can you show that Dalton's law of the summation of partial pressures is true using Equation (6.5)?

$$p_1 + p_2 + \ldots + p_n = p_{total} \tag{6.6}$$

Although you cannot easily measure the partial pressure of a gaseous component directly with commercial instruments, you can calculate the value from Equation (6.5) and/or Equation (6.6). To illustrate the significance of Equation (6.5) and the meaning of partial pressure, suppose that you carried out the following experiment with two nonreacting ideal gases. Examine Figure 6.4. Two tanks of 1.50 m³ volume, one containing gas A at 300 kPa and the other gas B at 400 kPa (both gases being at the same temperature of 20°C), are connected to an empty third tank (C) of the same volume. All of the gas in tanks A and B is forced into tank C isothermally. Now you have a 1.50 m³ tank of A + B at 700 kPa and 20°C for this mixture. According to Equation (6.5), you could say that gas A exerts a partial pressure of 300 kPa and gas B exerts a partial pressure of 400 kPa in tank C. Of course, you cannot put a pressure gauge on the tank and check this conclusion because the pressure gauge will read only the total pressure. These partial pressures are hypothetical pressures in tank C that the individual gases would exert if each was put into separate but identical volumes at the same temperature.

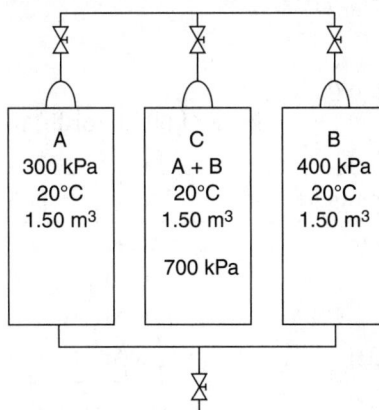

Figure 6.4   Illustration of the meaning of partial pressure of the components of an ideal gas mixture

When the $150 million Biosphere project in Arizona began in September 1991, it was billed as a sealed utopian planet in a bottle, where everything would be recycled. Its eight inhabitants lived for two years in the first large self-contained habitat for humans. But slowly the oxygen disappeared from the air—four women and four men in the 3.15 acres of glass domes eventually were breathing air with an oxygen content similar to that found at an altitude of about 13,400 ft. The "thin" air left the group members so fatigued and aching that they sometimes gasped for breath. Finally, the leaders of Biosphere 2 had to pump 21,000 lb of oxygen into the domes to raise the oxygen level from 14.5% to 19.0%. Subsequent investigation found the cause of the decrease in oxygen to be microorganisms in the soil that took up oxygen, a factor not accounted for in the design of the biosphere.

Example 6.3   Calculation of the Partial Pressures of the Components in a Gas from a Gas Analysis

Problem Statement

Few organisms are able to grow in solution using organic compounds that contain just one carbon atom such as methane or methanol. However, the bacterium *Methylococcus capsulates* can grow under aerobic conditions (in the presence of air) on C-1 carbon compounds. The resulting biomass is a good protein source that can be used directly as feed for domestic animals or fish.

In one process, the off-gas analyzes 14.0% CO_2, 6.0% O_2, and 80.0% N_2. It is at 300°F and 765.0 mm Hg pressure. Calculate the partial pressure of each component.

Solution

Use Equation (6.5):

Basis: 1.00 kg (or lb) mol of off-gas

Component	kg (or lb) mol	p_i (mm Hg)
CO_2	0.140	107.1
O_2	0.060	45.9
N_2	0.800	612.0
Total	1.000	765.0

On the basis of 1.00 mol of off-gas, the mole fraction y_i of each component, when multiplied by the total pressure, gives the partial pressure of that component. If you find that the temperature measurement of the flue gas was actually 337°F but the total pressure measurement was correct, would the partial pressures change? Hint: Is the temperature involved in Equation (6.5)?

6.1.3   Material Balances Involving Ideal Gases

Now that you have had a chance to review the ideal gas law applied to simple problems, let's apply it in material balances. The only difference between the subject matter of Chapters 3 through 5 and this chapter is that here the amount of material can be specified in terms of p, V, and T rather than solely as mass or moles. For example, the basis for a problem, or the quantity to be solved for, might be a volume of gas at a given temperature and pressure rather than a mass of gas. The next two examples illustrate balances for problems similar to those you have encountered before, but now involving gases.

Example 6.4   Material Balances for a Process Involving Combustion

Problem Statement

To evaluate the use of renewable resources, an experiment was carried out with rice hulls. After pyrolysis, the product gas analyzed 6.4% CO_2, 0.1% O_2, 39% CO, 51.8% H_2, 0.6% CH_4, and 2.1% N_2. It entered a combustion chamber at 90°F and a pressure of 35.0 in. Hg and was burned with 40% excess air (dry) at 70°F and an atmospheric pressure of 29.4 in. Hg; 10% of the CO remains. How many cubic feet of air were supplied per cubic foot of entering gas? How many cubic feet of product gas were produced per cubic foot of entering gas if the exit gas was at 29.4 in. Hg and 400°F?

Solution

This is an open, steady-state system with reaction. The system is the combustion chamber.

Steps 1–4

Figure E6.4 illustrates the process and notation. With 40% excess air, certainly all of the CO, H_2, and CH_4 should burn to CO_2 and H_2O; apparently, for some unknown reason, not all of the CO burns to CO_2. No CH_4 or H_2 appears in the product gas. The components of the product gas are shown in the figure.

(Continues)

Example 6.4   Material Balances for a Process Involving Combustion (*Continued*)

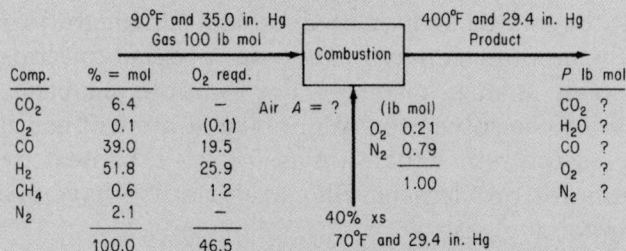

Figure E6.4

Step 5

You could take 1 ft³ of the feed material at 90°F and 35.0 in. Hg as the basis and convert the volume to moles, but it is just as easy to take 100 lb (or kg) mol as a basis because then % = lb (or kg) mol. Because only ratios of volumes are asked for, not absolute amounts, at the end of the problem, you can convert pound (or kilogram) moles to cubic feet.

$$\text{Basis: 100 lb mol of pyrolysis gas}$$

Step 4 (continued)

The entering air can be calculated from the specified 40% excess air; the reactions for complete combustion are

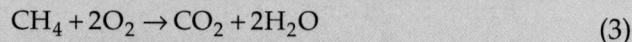

$$CO + \frac{1}{2}O_2 \rightarrow CO_2 \tag{1}$$

$$H_2 + \frac{1}{2}O_2 \rightarrow H_2O \tag{2}$$

$$CH_4 + 2O_2 \rightarrow CO_2 + 2H_2O \tag{3}$$

The moles of oxygen required are listed in Figure E6.4 (i.e., 46.3 lb mol). (We omit the units pound moles in what follows.) The excess oxygen is

$$\text{Excess } O_2: 0.4(46.3) = 18.6$$

$$\text{Total } O_2: 46.5 + 18.6 = 65.1$$

$$N_2 \text{ in is } 65.1\left(\frac{79}{21}\right) = 244.9$$

Steps 6 and 7

Degree-of-freedom analysis:

Unknowns (5): $n_{CO_2}, n_{O_2}, n_{N_2}, n_{H_2O}, P$

Equations (5):

Element balances (4): C, H, O, N

Implicit equations (1): $P = \sum n_i$

Steps 8 and 9

Make the element balances in moles to calculate the unknown quantities, and substitute the value of 3.9 for the number of moles of CO exiting.

In	Out
N: (2)(2.1) + (2)(244.9)	$= 2n_{N_2}$
C: 6.4 + 39.0 + 0.6	$= n_{CO_2} + 3.9$
H: (2)(51.8) + (0.6)(2)	$= 2n_{H_2O}$
O: (2)(6.4) + (2)(0.1) + 39 + (2)(65.1)	$= 2n_{O_2} + 2n_{CO_2} + n_{H_2O} + n_{CO}$

The solutions of these equations are

$$n_{N_2} = 247 \quad n_{CO_2} = 42.1 \quad n_{H_2O} = 53.0 \quad n_{O_2} = 20.55$$

The total moles exiting calculated from the implicit equation sum to 366.6 mol.

Finally, you can convert the pound moles of air and products that were calculated on the basis of 100 lb mol of pyrolysis gas to the volumes of gases at the states requested using the ideal gas law:

$$T_{gas} = 90 + 460 = 550°R \rightarrow 306\,K$$

$$T_{air} = 70 + 460 = 530°R \rightarrow 294\,K$$

$$T_{product} = 400 + 460 = 860°R \rightarrow 478\,K$$

$$\text{ft}^3 \text{ of gas:} \frac{100 \text{ lb mol entering gas}}{} \left| \frac{359 \text{ ft}^3 \text{ at S.C.}}{1 \text{ lb mol}} \right| \frac{550°R}{492°R} \left| \frac{29.92 \text{ in. Hg}}{35.0 \text{ in. Hg}} \right. = 343 \times 10^2$$

(*Continues*)

Example 6.4   Material Balances for a Process Involving Combustion (*Continued*)

$$\text{ft}^3 \text{ of air:} \frac{310 \text{ lb mol air}}{} \left| \frac{359 \text{ ft}^3 \text{ at S.C.}}{1 \text{ lb mol}} \right| \frac{530°R}{492°R} \left| \frac{29.92 \text{ in. Hg}}{29.4 \text{ in. Hg}} = 1220 \times 10^2$$

$$\text{ft}^3 \text{ of product:} \frac{366.6 \text{ lb mol } P}{} \left| \frac{359 \text{ ft}^3 \text{ at S.C.}}{1 \text{ lb mol}} \right| \frac{860°R}{429°R} \left| \frac{29.92 \text{ in. Hg}}{35.0 \text{ in. Hg}} = 2331 \times 10^2$$

The answers to the questions are

$$\frac{1220 \times 10^2}{343 \times 10^2} = 3.56 \frac{\text{ft}^3 \text{ air at 530°R and 29.4 in. Hg}}{\text{ft}^3 \text{ gas at 550°R and 35.0 in. Hg}}$$

$$\frac{2255 \times 10^2}{343 \times 10^2} = 6.57 \frac{\text{ft}^3 \text{ product at 860°R and 29.4 in. Hg}}{\text{ft}^3 \text{ gas at 550°R and 35.0 in. Hg}}$$

Example 6.5   Material Balance without Reaction

Problem Statement

Gas at 15°C and 105 kPa is flowing through an irregular duct. To determine the rate of flow of the gas, CO_2 from a tank is steadily passed into the gas stream. The flowing gas, just before mixing with the CO_2, analyzes 1.2% CO_2 by volume. Downstream, after mixing, the flowing gas analyzes 3.4% CO_2 by volume. As the CO_2 that was injected exited the tank, it was passed through a rotameter and found to flow at the rate of 0.0917 m³/min at 7°C and 131 kPa. What was the rate of flow of the entering gas in the duct in cubic meters per minute?

Solution

This is an open, steady-state system without reaction. The system is the duct. Figure E6.5 is a sketch of the process.

Steps 1–4

The data are presented in Figure E6.5.

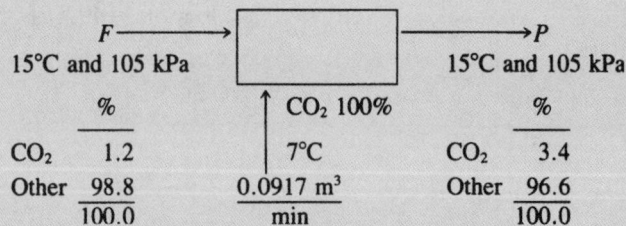

	F			P
	15°C and 105 kPa			15°C and 105 kPa
	%	CO_2 100%		%
CO_2	1.2	7°C	CO_2	3.4
Other	98.8	0.0917 m³	Other	96.6
	100.0	min		100.0

Figure E6.5

Both F and P are at the same temperature and pressure.

Step 5

Should you take as a basis 1 min → 0.0917 m³ of CO_2 at 7°C and 131 kPa? The gas analysis is in volume percent, which is the same as mole percent. We could convert all of the gas volumes to moles and solve the problem in terms of moles, but there is no need to do so because we can just as easily convert the known flow rate of the addition of CO_2 to 15°C and 105 kPa—that is, $0.0917[(273.15 + 7)/(273.15 + 15)]$ $(131/105) = 0.1112$ m³—and solve the problem using cubic meters for each stream since all of the streams will be at the same conditions. Note that in the calculation of the volume of CO_2 at 15°C and 105 kPa, changing to a lower temperature reduced the volume, but moving to a lower pressure increased the volume.

Steps 6 and 7

The unknowns are F and P, and you can make two independent component balances, CO_2 and "other"; hence, the problem has zero degrees of freedom.

Steps 7–9

The "other" balance (in cubic meters at 15°C and 105 kPa) is

$$F(0.988) = P(0.966) \tag{a}$$

The CO_2 balance (in cubic meters at 15°C and 105 kPa) is

$$F(0.012) + 0.1112 = P(0.034) \tag{b}$$

The total balance (in cubic meters at 15°C and 105 kPa) is

$$F + 0.1112 = P \tag{c}$$

Note that the "other" is a tie component. Select Equations (a) and (c) to solve. The solution of Equations (a) and (c) gives

$$F = 4.884 \text{ m}^3/\text{min at 15°C and 105 kPa}$$

Step 10 (check)

Use the redundant Equation (b):

$$4.884(0.012) + 0.1112 = 0.1698 \overset{?}{=} \left(4.884 \frac{0.988}{0.966} \right)(0.034) = 0.1698$$

(*Continues*)

Example 6.5   Material Balance without Reaction (*Continued*)

The equation checks out to a satisfactory degree of precision. Note that if you incorrectly calculated the volume of CO_2 at 15°C and 105 kPa, but correctly solved the set of equations, the redundant equation would be satisfied, but the answer would be incorrect. This results because the redundant equation only verifies that the set of equation was properly solved and not whether the set of equation was, in fact, correct.

Self-Assessment Test

Questions

1. What are the dimensions of T, P, V, n, and R?

2. List the standard conditions for a gas in the SI and AE systems of units.

3. How do you calculate the density of an ideal gas at S.C.?

4. Can you use the respective specific molar densities (mole/volume) of the gas and the reference gas to calculate the specific gravity of a gas?

5. A partial pressure of oxygen in the lungs of 100 mm Hg is adequate to maintain oxygen saturation of the blood in a human. Is this value higher or lower than the partial pressure of oxygen in the air at sea level?

6. An exposure to a partial pressure of N_2 of 1200 mm Hg in air has been found by experience not to cause the symptoms of N_2 intoxication to appear. Will a diver at 60 m be affected by the N_2 in the air being breathed?

Answers

1. T in absolute degrees (e.g., kelvin or Rankine degrees), p in force per unit area absolute, V in volume, n in moles, and R has units that depend on the units of T, p, V, and n.

2. SC for SI: 273.15 K; 101.325 kPa; 22.415 m^3/kg mol; SC for AE: 32°F; 1 atm; 359.05 ft^3/ lb mol

3. $\rho = Mn/V = Mp/RT$

4. Yes

5. Lower because at sea level $0.21 \times 760 = 159.6$ mm Hg

6. Yes, because $0.79 \times 760 \times 30$ m/10 m per atm = 1801.2 mm Hg partial pressure of N_2 is greater than 1400 mm Hg

Problems

1. Calculate the volume in cubic feet of 10 lb mol of an ideal gas at 68°F and 30 psia.

2. A steel cylinder of volume 2 m³ contains methane gas (CH_4) at 50°C and 250 kPa absolute. How many kilograms of methane are in the cylinder?

3. What is the value of the ideal gas constant R to use if the pressure is to be expressed in atmospheres, the temperature in kelvin, the volume in cubic feet, and the quantity of material in pound moles?

4. Twenty-two kilograms per hour of CH_4 are flowing in a gas pipeline at 30°C and 920 mm Hg. What is the volumetric flow rate of the CH_4 in cubic meters per hour?

5. A gas has the following composition at 120°F and 13.8 psia:

Component	mol %
N_2	2
CH_4	79
C_2H_6	19

   a. What is the partial pressure of each component?
   b. What is the volume fraction of each component?

6. A furnace is fired with 1000 ft³/hr at 60°F and 1 atm of a natural gas having the following volumetric analysis: CH_4 80%, C_2H_6 16%, O_2 2%, CO_2 1%, and N_2 1%. The exit flue gas temperature is 800°F, and the pressure is 760 mm Hg absolute; 15% excess air is used, and combustion is complete. Calculate (a) the volume of CO_2 produced per hour; (b) the volume of H_2O vapor produced per hour; (c) the volume of N_2 produced per hour; (d) the total volume of flue gas produced per hour.

Answers

1. (359.05 ft³/lb mol)(10 lb mol)(14.69/30.)(459 + 68 − 32)/459 = 1896 ft³

2. (250 kPa)(2 m³)(16 kg/kg mol)/323.15 K/8.315 = 2.98 kg

3. $R = pV/nT$ = (1 atm)(359.05 ft³/kg mol)/(273.15 K) = 1.3144 atm ft³ kg mol⁻¹ K⁻¹

4. (22 kg)/(16 kg/kg mol)(22.414 m³/kg mol)(303.15/273.15)(760/920) = 28.26 m³/h

5. (a) P_{N2} = 0.276 psia; P_{CH4} = 10.9 psia; P_{C2H6} = 2.62 psia; (b) same as mol percent

6. First calculate the moles of feed = (1000 ft³/h)/(359.05 ft/lb mol)(491.16/519.67) = 2.635 lb mol. Multiply by mole % to get inlet feed. Based on this, ξ_1 = 2.108 and ξ_2 = 0.4216. $(O_2)_{required}$ = $2\xi_1$ +3.5ξ_2 − 0.0527 = 5.639; air = 5.639(1.15)/0.21 = 30.88; from component mole balance, flue gas composition is 8.83% CO_2, 16.25% H_2O, 2.50% O_2, 72.40% N_2. Volume of flue gas = (33.73 lb mol)(359.05 ft³/lb mol) (459.67 + 800)/491.67 = **31,208 ft3/hr.** Volume per hour of each component is equal to mole fraction times total flue gas volume flow rate.

6.2   Real Gases: Equations of State

Gases whose properties cannot be represented by the ideal gas law are called **nonideal** gases or **real gases**. Real gas properties can be predicted by equations called **equations of state**.

6.2.1   Equations of State

The simplest example of what is called an equation of state is the ideal gas law itself. Equations of state for nonideal gases can be just empirical relations selected to fit a data set, or they can be based on theory or a combination of the two. Figure 6.5 shows the measurements by Thomas Andrews in 1863 of the pressure versus the specific volume for CO_2 at various constant temperatures. Note that point C at 31°C is the highest temperature at which liquid and gaseous CO_2 can coexist in equilibrium. Above 31°C, only **critical** fluid exists, so that what is called the **critical temperature** for CO_2 is 31°C (304 K)—that is, point C in Figure 6.5. The corresponding gas (fluid) **pressure** is 72.9 atm (7385 kPa). Also note that at higher temperatures, such as 50°C, the data can be represented by the ideal gas law because pV is constant, a hyperbola. You can find experimental values of the critical temperature (T_c) and the critical pressure (p_c) for various compounds by consulting. Reid et al. (refer to the references at the end of this chapter), who describe and evaluate methods of estimating critical constants for various compounds.

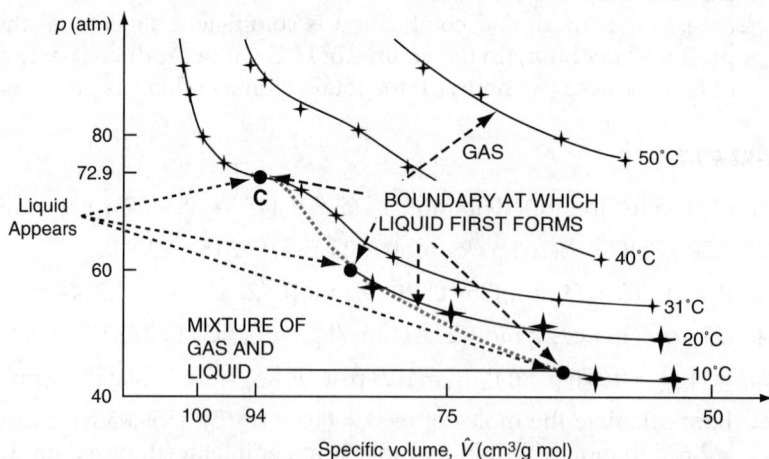

Figure 6.5   Experimental measurements of carbon dioxide by Andrews.[1] The solid lines represent smoothed data. C is the highest temperature at which any liquid exists. At the big solid dots, liquid and vapor start to coexist. Note the nonlinear scale on the horizontal axis.

---

[1]Andrews, Thomas. "The Bakerian lecture: On the Continuity of the Gaseous and Liquid States Of Matter," *Philosophical Transactions of the Royal Society* (London), **159** : 575-590 (1869).

How can you predict the p, $\hat{V}$, and T properties of a gas between the region in which the ideal gas law is valid and the region in which the gas condenses into liquid? One way is to use one or more equations of state. Where substantial changes in curvature occur, perhaps several different equations must be used to cover a region accurately. Table 6.2 lists some of the well-known single equations of state.

Table 6.2   Examples of Equations of State in Terms of $\hat{V}$*

van der Waals:	Redlich-Kwong (RK equation):
$\left(p + \dfrac{a}{\hat{V}^2}\right)(\hat{V} - b) = RT$	$p = \dfrac{RT}{(\hat{V} - b)} - \dfrac{a}{T^{1/2}\hat{V}(\hat{V} + b)}$
$a = \left(\dfrac{27}{64}\right)\dfrac{R^2 T_c^2}{p_c}$	$a = 0.42748\dfrac{R^2 T_c^{2.5}}{p_c}$
$b = \left(\dfrac{1}{8}\right)\dfrac{RT_c}{p_c}$	$b = 0.08664\dfrac{RT_c}{p_c}$

Soave-Redlich-Kwong (SRK equation):	Kammerlingh-Onnes (a virial equation):
$p = \dfrac{RT}{\hat{V} - b} - \dfrac{a'\lambda}{\hat{V}(\hat{V} + b)}$	$p\hat{V} = RT\left(1 + \dfrac{B}{\hat{V}} + \dfrac{C}{\hat{V}^2} + ...\right)$
$a' = \dfrac{0.42748 R^2 T_c^2}{p_c}$	**Holborn (a virial equation):**
$b = \dfrac{0.08664\, RT_c}{p_c}$	$p\hat{V} = RT(1 + B'p + C'p^2 + ...)$
$\lambda = [1 + \kappa(1 - T_r^{1/2})^2]$	
$\kappa = (0.480 + 1.574\omega - 0.176\omega^2)$	

* $\hat{V}$ is the specific volume, T_c and p_c are explained in the text, and ω is the acentric factor, also explained in the text.

The units used in calculating the coefficients in the equations in Table 6.2 are determined by the units selected for R.

Some of the classical equations of state are formulated as a power series (called the **virial** form) with p being a function of $1/\hat{V}$ or $\hat{V}$ being a function of p with three to six terms. You should note that the coefficients in the van der Waals, Redlich-Kwong (RK), and Soave-Redlich-Kwong (SRK) equations can be calculated from certain physical properties (discussed shortly), whereas in the virial equations, the coefficients are strictly determined from experimental measurements. Because of its accuracy, the databases in many commercial process simulators make extensive use of the SRK equation of state, particularly for hydrocarbons. These equations in general will not predict p-$\hat{V}$-T values across a phase change from gas to liquid very well. Keep in mind that under conditions such that the gas starts to liquefy, the *gas laws apply only to the vapor phase portion* of the system for this book.

How accurate are equations of state? Cubic equations of state such as van der Waals, RK, and SRK equations listed in Table 6.2 can exhibit an accuracy of 1% to 2% over a large range of conditions for many compounds. Equations of state in databases may have as many as 30 or 40 coefficients to achieve high accuracy (e.g., see the AIChE DIPPR reports that can be located on the AIChE website). Keep in mind that you must know the region of validity of any equation of state and not extrapolate outside that region, particularly not into the liquid region, by ignoring the possibility of condensation for gases such as CO_2, NH_3, and low-molecular-weight organic compounds such as acetone, ethyl alcohol, and so on. If you plan to use a specific equation of state such as one of those listed in Table 6.2, you have numerous choices, none of which will consistently give the best results based on a full range of components.

Other than the use of equations of state to make predictions of values of p, $\hat{V}$, and T, what good are they?

1. They permit a concise summary of a large mass of experimental data and also permit accurate interpolation between experimental data points.
2. They provide a continuous function to facilitate calculation of physical properties based on differentiation and integration of p-$\hat{V}$-T relationships.
3. They provide a point of departure for the treatment of the properties of mixtures.

In addition, some of the advantages and disadvantages of using equations of state versus other methods to make predictions are the following:

Advantages:

1. Values of p-$\hat{V}$-T can be predicted with reasonable error in regions where no data exist.
2. Only a few values of coefficients are needed in the equation to be able to predict gas properties versus collecting large amounts of data by experiment for tables and graphs. The coefficients are based on physical properties that are readily available, that is, the critical temperature and pressure.
3. The equations can be manipulated on a computer, whereas graphics methods of prediction cannot.

Disadvantages:

1. The form of an equation is hard to change to fit new or better data.
2. Inconsistencies may exist between equations for p-$\hat{V}$-T and equations for other physical properties.
3. Usually the equation is quite complicated and may not be easy to solve for $\hat{V}$ because of its nonlinearity.

Disadvantage 3 prevented the widespread use of equations of state until computers and computer programs for solving nonlinear algebraic equations came into the picture. You can see that it is easy to solve the SRK equation in Table 6.2 for p given values for T and $\hat{V}$, or for T given values of p and $\hat{V}$, but requires the solution of a cubic equation to solve for $\hat{V}$ given values for T and p. Similar remarks apply to specific virial equations.

For example, look at the RK equation. Given p and T, is the RK equation cubic in $\hat{V}$? Yes. Given p, T, and V, is the RK equation cubic in n? Yes. Given p and $\hat{V}$, is it cubic in T? No.

Example 6.6   Use of the RK Equation to Calculate p or T

Problem Statement

Determine the pressure (in atmospheres) of 1 g mol of C_2H_4 at 300 K with $V = 0.647$ L using the RK equation for C_2H_4: $T_c = 282.8$ K and $p_c = 50.44$ atm.

Solution

The RK equation is (from Table 6.2)

$$p = \frac{RT}{(\hat{V}-b)} - \frac{a}{T^{1/2}\hat{V}(\hat{V}+b)}$$

What should you do first? Take a basis of the given values of V, n, and T. Then calculate a, b, and $\hat{V}$:

$$a = 0.42748\frac{R^2 T_c^{2.5}}{p_c} = \frac{0.42748}{}\left|\frac{(0.08206)^2 (L)^2 (atm)^2}{(g\,mol)^2 (K)^2}\right|\frac{(282.8)^{2.5}(K)^{2.5}}{50.44\,atm}$$

$$= 76.75\frac{(L)^2 (atm)(K)^{0.5}}{(g\,mol)^2}$$

$$b = 0.08664\frac{RT_c}{p_c} = \frac{0.08664}{}\left|\frac{(0.08206)(L)(atm)}{(g\,mol)(K)}\right|\frac{282.8\,K}{50.44\,atm} = 0.03986\frac{L}{g\,mol}$$

$$\hat{V} = \frac{0.674\,L}{1\,g\,mol} = 0.674\,L/g\,mol$$

(Continues)

Example 6.6   Use of the RK Equation to Calculate p or T (*Continued*)

Next, insert the known values of the variables and coefficients:

$$p = \frac{(0.08206)(L)^2(atm)}{1(g\,mol)(K)} \Bigg| \frac{300\,K}{} \Bigg| \frac{1\,g\,mol}{(0.674 - 0.03986)L} -$$

$$\frac{76.75(L)^2(atm)(K)^{0.5}}{1(g\,mol)^2} \Bigg| \frac{1}{(300\,K)^2} \Bigg| \frac{1\,g\,mol}{0.674\,L} \Bigg| \frac{1\,g\,mol}{(0.674 + 0.03986)L} = 29.6\,atm$$

If you know p and $\hat{V}$ instead of p and T, you can solve explicitly for T by rearranging the equation or by using an equation solver. Figure E6.6 compares the prediction of p by the ideal gas law and the RK equation.

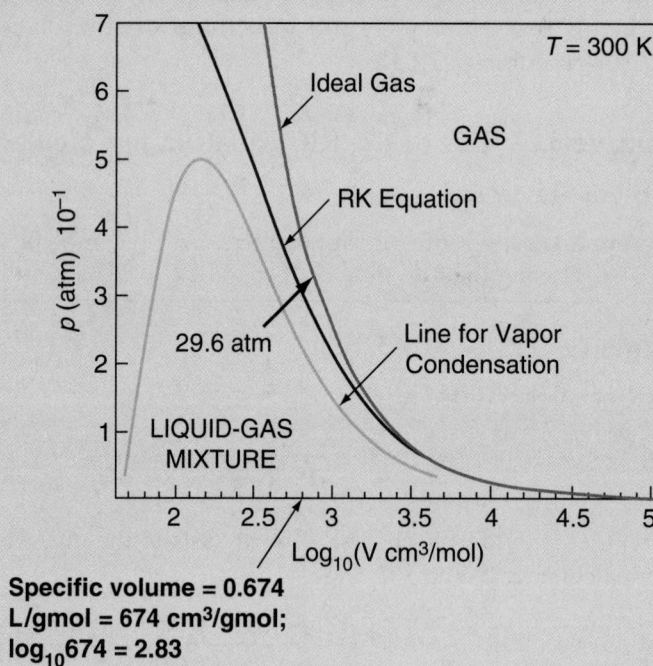

Specific volume = 0.674 L/gmol = 674 cm³/gmol; $\log_{10} 674 = 2.83$

Figure E6.6

6.2.2   Nonlinear Equation Solvers

MATLAB and Python offer built-in nonlinear equation solvers that can be used to solve equations of state that are nonlinear (e.g., solving for $\hat{V}$ for the RK equation). In addition, these built-in nonlinear equation solvers are used later in this text to solve other nonlinear equation problems.

MATLAB

Single nonlinear equations can be solved using MATLAB's *fzero* function. The call statement for *fzero* is

```
x=fzero(fx,x0)
```

where x is the solution to the nonlinear equation, fx is the function handle for the function that is to be zeroed, and x0 is the initial estimate of the solution.

Example 6.7   Use of the RK Equation to Calculate $\hat{V}$ Using MATLAB

Problem Statement

Using the physical property data given in Example 6.6 for ethylene, determine the specific volume of C_2H_4 at 300 K and 30 atm using the RK equation. Because with the RK equation you cannot explicitly solve for $\hat{V}$, we use function *fzero* to solve this problem.

Solution

From Example 6.6, the values of a and b are

$$a = 0.42748\frac{R^2 T_c^{2.5}}{p_c} = 76.75\frac{(L)^2(atm)(K)^{0.5}}{(g\,mol)^2} \qquad b = 0.08664\frac{RT_c}{p_c} = 0.03986\frac{L}{g\,mol}$$

because the critical properties of ethylene remain constant. Converting the RK equation into a single nonlinear equation yields

$$\text{RK equation: } p = \frac{RT}{(\hat{V}-b)} - \frac{a}{T^{1/2}\hat{V}(\hat{V}+b)}$$

$$\text{Nonlinear equation: } f(x) = p - \frac{RT}{(x-b)} + \frac{a}{T^{1/2}x(x+b)}$$

(Continues)

Example 6.7   Use of the RK Equation to Calculate $\hat{V}$ Using MATLAB (*Continued*)

where x is the unknown molar volume, $\hat{V}$. Following is the MATLAB code that solves this nonlinear equation using function fzero.

```
                    MATLAB Results for Example 6.7
%%%%%%%%%%%%%%%%%%%%%%%%%%%%%%%%%%%%%%%%%%%%%%%%%%%%%%
%                   NOMENCLATURE
%  a - parameter in RK equation
%  b - parameter in RK equation
%  p - pressure (30 atm)
%  R - gas constant (0.08206 atm L/gmol-K)
%  T - temperature (300 K)
%  xsoln - the root of f(x) determined by function fzero
%%%%%%%%%%%%%%%%%%%%%%%%%%%%%%%%%%%%%%%%%%%%%%%%%%%%%%
%                   PROGRAM
function Ex6_7
clear; clc;
a=76.75; b=0.03986; p=30;      % Input the problem data
T=300; R=0.08206;              % Input the problem data
x0=R*T/p;                      % Set initial guess for x
                               % Define anonymous function
fx=@(x)p-R*T/(x-b)+a/(T^0.5*x*(x+b));
xsoln=fzero(fx,x0)             % Call built-in function fzero
end
%                 PROGRAM END
%%%%%%%%%%%%%%%%%%%%%%%%%%%%%%%%%%%%%%%%%%%%%%%%%%%%%%%%%%
x =
    0.6629
```

Description of Results. Note that the ideal gas law is used to make the initial guess for the numerical solution for the specific volume. Because the RK method is cubic in $\hat{V}$, its solution can have up to three different numerical values. Therefore, if you use a very poor initial guess for $\hat{V}$, you may obtain a solution to the equation that does not correspond to the solution for the vapor (e.g., a solution corresponding to the liquid). Also note that this problem is almost exactly the same as the case shown in Example 6.6 where the pressure was calculated explicitly. Therefore, the solution of this problem (0.6629 L/g mol) should be relatively close to the specific volume used in Example 6.6 (0.647 L/g mol), which it is, thus validating the numerical solution.

Python

Single nonlinear equations can be solved by using function scipy.optimize. newton:

$$\texttt{xsoln=scipy.optimize.newton (f, x0)}$$

where f is the user-defined function for $f(x)$, x0 is the initial guess for a solution of $f(x)$, and xsoln is the calculated value of x that renders $f(x)$ equal to zero.

Example 6.8   Use of the RK Equation to Calculate $\hat{V}$ Using Python

Problem Statement

Using the physical property data given in Example 6.6 for ethylene, determine the specific volume of C_2H_4 at 300 K and 30 atm using the RK equation. Because with the RK equation you cannot explicitly solve for $\hat{V}$, we use function scipy.optimize. newton to solve this problem.

Solution

From Example 6.6, the values of a and b are

$$a = 0.42748 \frac{R^2 T_c^{2.5}}{p_c} = 76.75 \frac{(L)^2 (atm)(K)^{0.5}}{(g\,mol)^2} \qquad b = 0.08664 \frac{RT_c}{p_c} = 0.03986 \frac{L}{g\,mol}$$

because the critical properties of ethylene remain constant. Converting the RK equation into a single nonlinear equation yields

$$\text{RK equation: } p = \frac{RT}{(\hat{V} - b)} - \frac{a}{T^{1/2}\hat{V}(\hat{V} + b)}$$

$$\Rightarrow \text{ Nonlinear equation: } f(x) = p - \frac{RT}{(x-b)} + \frac{a}{T^{1/2}x(x+b)}$$

where x is the unknown molar volume, $\hat{V}$. Following is the Python code that solves this nonlinear equation using function scipy.optimize.newton.

(*Continues*)

Example 6.8   Use of the RK Equation to Calculate $\hat{V}$ Using Python (*Continued*)

Python Code for Example 6.8

```
Ex6_8 Soln RK method.py
#########################################################################
#                              NOMENCLATURE
#  a - parameter in RK equation
#  b - parameter in RK equation
#  f - the UD function (f(x))
#  fx - the value of the function f(x)
#  p - pressure (30 atm)
#  R - gas constant (0.08206)
#  T - temperature (300 K)
#  x0 - the initial guess for the root of function f(x) using ideal gas law.
#  xsol - the root of f(x) determined by function scipy.optimize.newton
#########################################################################
#                               PROGRAM
import numpy as np
import scipy.optimize
# Define the UD function for the function value
def f(x):
    fx=p-R*T/(x-b)+a/(T**0.5*x*(x+b))
    return fx
a, b, p, R, T = 76.75, 0.03986, 30, 0.08206, 300     # Input data
x0=R*T/p                                                          #
Initial guess based on ideal gas law
# Apply function scipy.optimize.newton to determine a root of the
nonlinear equation
xsol=scipy.optimize.newton(f, x0)
#                             PROGRAM END
#########################################################################
```

```
IPython console:
In[1]: runfile(…
In[2]: %precision 4
In[3]: xsol
Out[3]: 0.6629
```

Description of Results. Note that the ideal gas law is used to make the initial guess for the numerical solution for the specific volume. Because the RK method is cubic in $\hat{V}$, its solution can have up to three different numerical values. Therefore, if you use a very poor initial guess for $\hat{V}$, you may obtain a solution to the equation that does not correspond to the solution for the vapor (e.g., a solution corresponding to the liquid). Also note that this problem is almost exactly the same as the case shown in Example 6.6 where the pressure was calculated explicitly. Therefore, the solution of this problem (0.6629 L/g mol) should be relatively close the specific volume used in Example 6.6 (0.647 L/g mol), which it is, thus validating the numerical solution.

6.2.3   The Critical State and Compressibility

We mentioned the critical pressure p_c and critical temperature T_c in connection with Figure 6.4. The **critical state (point)** for the gas-liquid transition is the set of physical conditions at which the density and other properties of the liquid and vapor become identical. In Figure 6.6 the points on the constant temperature lines at which the gas (vapor) starts to condense have been connected by a dashed line (---). On the opposite side of the figure, the long-short-long dashed curve (— - —) shows the locus of points at the respective temperatures at which completion of the condensation occurs; that is, the vapor becomes all liquid. Between the two bounds, a mixture of vapor and liquid exists.

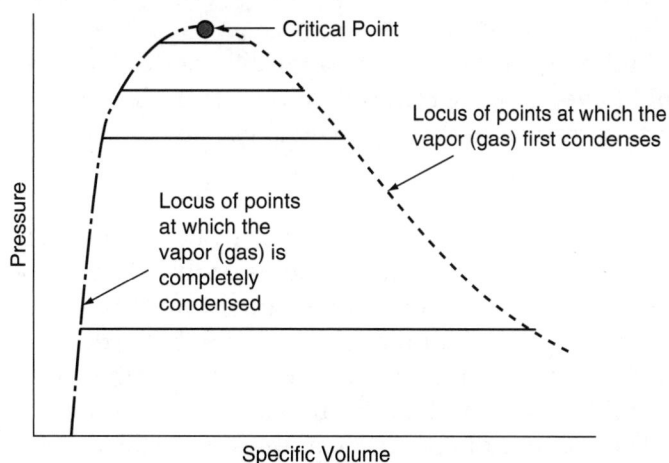

Figure 6.6   The critical point is located where the lengths of the (solid) lines are zero. The solid lines connect the points at which condensation starts at various temperatures to the corresponding points for that temperature at which condensation is complete.

The intersection of the two bounds is denoted as the **critical point,** and it occurs at the highest temperature and pressure possible ($T_r = 1$, $p_r = 1$) at which gas and liquid can coexist.

A **supercritical fluid** is a compound in a state above its critical point. Supercritical fluids are used to replace solvents such as trichloroethylene and methylene chloride, the emissions from which, and the contact with which, have been severely limited. For example, coffee decaffeination, the removal of cholesterol from egg yolk with CO_2, the production of vanilla extract, and the destruction of undesirable organic compounds all can take place using supercritical water. Supercritical water has been shown to destroy 99.99999% of all of the major types of toxins in these organic compounds.

Other terms with which you should become familiar are the **reduced variables.** These are conditions of temperature, pressure, or specific volume *normalized* (divided) by their respective critical conditions, as follows:

$$\text{Reduced temperature:} \quad T_r = \frac{T}{T_c}$$

$$\text{Reduced pressure:} \quad p_r = \frac{p}{p_c}$$

$$\text{Reduced specific volume:} \quad \tilde{V}_r = \frac{\hat{V}}{\hat{V}_c}$$

In theory, the **law of corresponding states** indicates that any compound should have the same reduced volume at the same reduced temperature and reduced pressure so that a universal gas law might be

$$P_r \tilde{V} = kT_r \tag{6.7}$$

Unfortunately, Equation (6.7) does not universally make accurate predictions. You can check this conclusion by selecting a compound such as water, applying Equation (6.7) at some low temperature and high pressure to calculate $\hat{V}$, and comparing your results with the value obtained for $\hat{V}$ with the corresponding conditions from the tables for water vapor that are in the folder in the back of this book.

The concept of reduced variables nevertheless has been applied to the prediction of real gas properties. One common way is to modify the ideal gas law by inserting an adjustable coefficient z, the **compressibility factor,** a factor that compensates for the nonideality of the gas and can be looked at as a measure of nonideality. Thus, the ideal gas law is turned into a real gas law called a **generalized equation of state:**

$$pV = znRT \qquad (6.8)$$

or

$$p\hat{V} = zRT \qquad (6.8a)$$

One way to look at z is to consider it to be a factor that makes Equation (6.8) an equality. Note that $z = 1$ is for an ideal gas. Although we treat only gases in this chapter, Equation (6.8) has been applied to liquids.

If you plan on using Equation (6.8), where can you find the values of z to use in it? Equations exist in the literature for specific compounds and classes of compounds, such as those found in petroleum refining. Theoretical calculations based on molecular structure sometimes prove to be useful. Usually, you will find graphs or tables of z to be quite convenient sources for engineering purposes. If the compressibility factor derived from experiment is plotted for a given temperature against the pressure for different gases, figures such as Figure 6.7a result. However, if the compressibility factor is plotted against the reduced pressure as a function of the reduced temperature, then for like gases, the compressibility values at the same reduced temperature and reduced pressure fall at about the same point, as illustrated in Figure 6.7b.

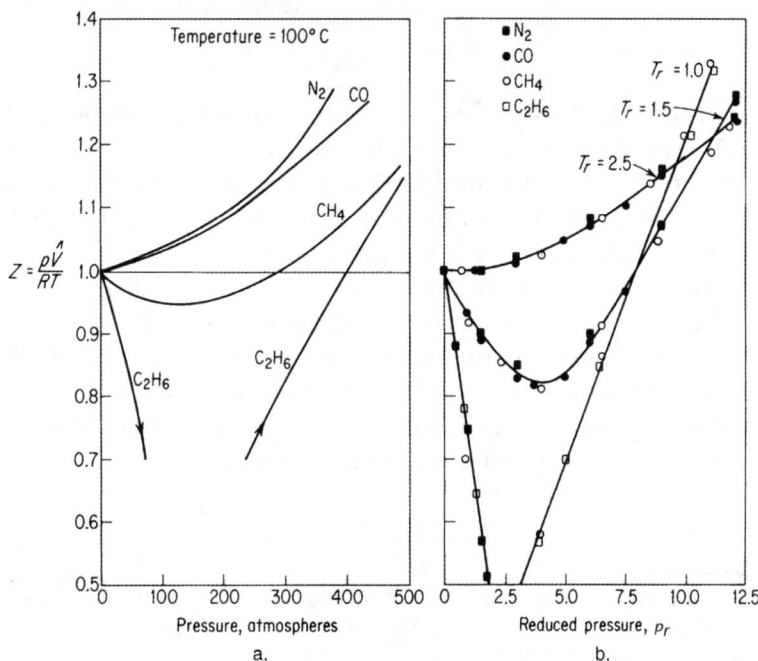

Figure 6.7   (a) Compressibility at 100°C for several gases as a function of pressure; (b) compressibility factor for several gases as a function of reduced temperature and reduced pressure

The **acentric factor** ω indicates the degree of acentricity or nonsphericity of a molecule. For helium and argon, ω is equal to zero. For higher-molecular-weight hydrocarbons and for molecules with increased polarity, the value of ω increases. Values of the acentric factor for a few selected substances are listed in Table 6.3.

Table 6.3   Selected Values of the Pitzer* Acentric Factor

Compound	Acentric Factor	Compound	Acentric Factor
Acetone	0.309	Hydrogen sulfide	0.100
Benzene	0.212	Methane	0.008
Ammonia	0.250	Methanol	0.559
Argon	0.000	n-Butane	0.193
Carbon dioxide	0.225	n-Pentane	0.251
Carbon monoxide	0.049	Nitric oxide	0.607
Chlorine	0.073	Nitrogen	0.040
Ethane	0.098	Oxygen	0.021
Ethanol	0.635	Propane	0.152
Ethylene	0.089	Propylene	0.148
Freon-12	0.176	Sulfur dioxide	0.251
Hydrogen		Water vapor	0.344

* K. S. Pitzer, *J. Am. Chem. Soc.*, **77**, 3427 (1955).

A different way of predicting p, $\hat{V}$, and T properties is the **group contribution method,** which has been successful in estimating properties of pure components. This method is based on combining the contribution of each functional group of a compound. The key assumption is that a group such as $-CH_3$ or $-OH$ behaves identically irrespective of the molecule in which it appears. This assumption is not quite true, so any group contribution method yields approximate values for gas properties. Probably the most widely used group contribution method is **UNIFAC,**[2] which forms a part of many computer databases. **UNIQUAC** is a variant of UNIFAC and is widely used in the chemical industry in the modeling of nonideal systems (systems with strong interaction between the molecules).

---

[2] A. Fredenslund, J. Gmehling, and P. Rasmussen, *Vapor-Liquid Equilibria Using UNIFAC* (Amsterdam: Elsevier, 1977); D. Tiegs, J. Gmehling, P. Rasmussen, and A. Fredenslund, *Ind. Eng. Chem. Res.*, **26**, 159 (1987).

Self-Assessment Test

Questions

1. Explain why the van der Waals and Redlich-Kwong equations (of state) are easy to solve for p and hard to solve for $\hat{V}$.

2. Under what conditions will an equation of state be the most accurate?

3. What are the units of a and b in the SI system for the Redlich-Kwong equation?

Answers

1. For these equations of state, you can solve for p or T explicitly, but $\hat{V}$ appears as a cubic equation.

2. At low pressures and high temperatures

3. b is in m^3/g mol; a is $(K)^{0.5}(m)^6(kPa)/g$ mol^2

Problems

1. Convert the virial (power series) equations of Kammerlingh-Onnes and Holborn (in Table 6.2) to a form that yields an expression for z.

2. Calculate the temperature of 2 g mol of a gas using van der Waals' equation with $a = 1.35 \times 10^{-6}$ m^6 (atm) (g mol^{-2}), $b = 0.0322 \times 10^{-3}$ (m^3) (g mol^{-1}) if the pressure is 100 kPa and the volume is 0.0515 m^3.

3. Calculate the pressure of 10 kg mol of ethane in a 4.86 m^3 vessel at 300 K using two equations of state: (a) ideal gas and (b) Soave-Redlich-Kwong. Compare your answer with the experimentally observed value of 34.0 atm.

Answers

1. $z = \left(1 + \dfrac{B}{\hat{V}} + \dfrac{C}{\hat{V}^2} + \ldots\right)$ and $z = (1 + B'p + C'p^2 + \ldots)$

2. $T = \left(p + \dfrac{a}{\hat{V}^2}\right)(\hat{V} - b))/R = \left(100/101.3 + a/0.02575^2\right)$

   $(0.02575 - b/0.02575)/8.206 \; 10^{-5} = 295\,K$

3. (a) 50.7 atm; (b) 34.0 atm

6.3   Real Gases: Compressibility Charts

Calculation of any of the variables p, T, $\hat{V}$, and z using the generalized equation $p\hat{V} = zRT$ can be assisted by using graphs called **generalized compressibility charts**, or z factor charts.

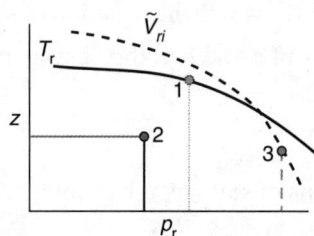

Figure 6.8   A compressibility chart involves four parameters: z, p_r, T_r, and $\hat{V}_{ri}$.

Four parameters are displayed in Figure 6.8. Any two values will fix a point from which you can determine the other two. For example, if p_r and T_r are known (point 1), the value of $\hat{V}_{ri}$ can be determined by interpolating between the two closest curves of $\hat{V}_{ri}$, and z can be determined by drawing a horizontal line from point 1 to the z axis.

Figures 6.9a and 6.9b show two examples of the **generalized compressibility factor** charts prepared by Nelson and Obert.[3] These charts are based on data for 30 nonpolar gases, including light hydrocarbons, CO_2 and N_2. Figure 6.9a represents z for 26 gases (excluding H_2, He, NH_3, and H_2O) with a maximum deviation of 1% for z values greater than 0.6 and H_2 and H_2O within a deviation of 1.5%. Figure 6.9b is for nine gases, and errors can be as high as 5%. Note that the vertical axis in Figure 6.9b is not z but zT_r. To use the charts for H_2 and He (only), make corrections to the actual constants to get **pseudocritical** constants as follows:

$$T_c' = T_c + 8\,\mathrm{K}$$

$$p_c' = p_c + 8\,\mathrm{atm}$$

[3] L. C. Nelson and E. F. Obert, *Chem. Eng.*, **61**, No. 7, 203–8 (1954). Figures 6.8a and 6.8b include data reported by P. E. Liley, *Chem. Eng.*, 123 (July 20, 1987).

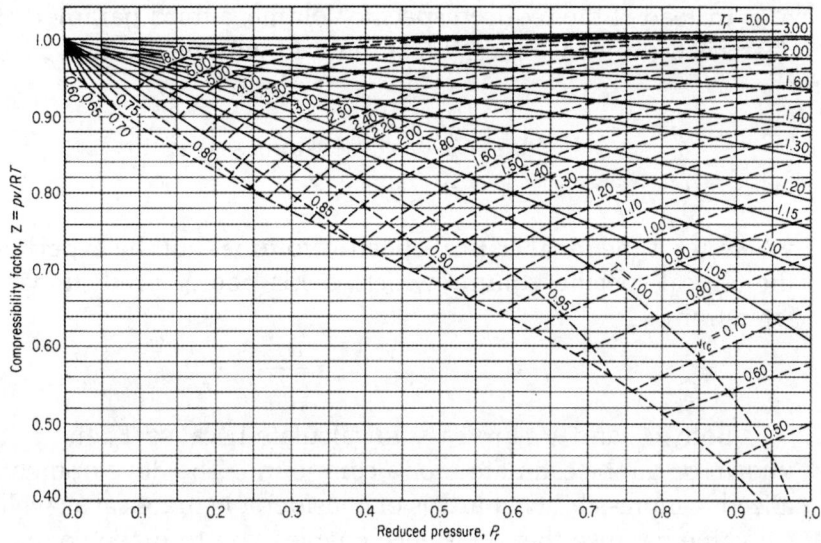

Figure 6.9a   Generalized compressibility chart for lower pressures showing z as a function of p_r, T_r, and $\hat{V}_{ri}$.

Then you can use Figures 6.9a and 6.9b for these two gases using the pseudocritical constants as replacements for their true values.

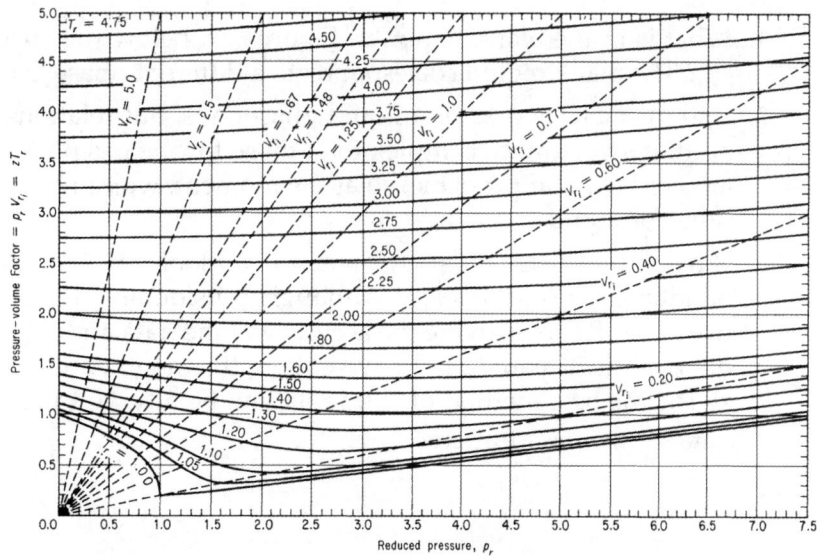

Figure 6.9b   Generalized compressibility chart for higher values of p_r

Instead of the reduced specific volume, a third parameter shown on the charts is the dimensionless **ideal reduced volume defined** by

$$\tilde{V}_{ri} = \frac{\tilde{V}}{\tilde{V}_{ci}} \text{(a dimensionless quantity)} \tag{6.9}$$

where $\hat{V}_{ci}$ is the **ideal critical specific volume** (not the experimental value of the critical specific volume which yields poorer predictions) and is calculated from

$$\tilde{V}_{ci} = \frac{RT_c}{p_c} \tag{6.10}$$

Both $\hat{V}_{ci}$ and $\hat{V}_{ri}$ are easy to calculate because T_c and p_c are presumed known or can be estimated for a compound. The development of the generalized compressibility charts is of considerable practical as well as pedagogical value because their existence enables you to make engineering calculations with considerable ease, and it also permits the development of thermodynamic functions for gases for which no experimental data are available.

Frequently Asked Questions

1. What is in the blank region in Figure 6.9a below the curves for T_r and $\hat{V}_{ri}$? The blank region corresponds to a different phase: a liquid.

2. Will $p\hat{V} = zRT$ work for a liquid phase? Yes, but relations to calculate z accurately are more complex than those for the gas phase. Also, liquids are not very compressible, so at the moment, we can bypass p-$\hat{V}$-T relations for liquids.

3. Why should I use $p\hat{V} = zRT$ when I can look up the data needed in a handbook or on the Web? Although considerable data exist, you can use $p\hat{V} = zRT$ to evaluate the accuracy of the data and interpolate within data points. If you do not have data in the range you want, use of $p\hat{V} = zRT$ is the best method of extrapolation. Finally, you may not have any data for the gas of interest.

Example 6.9   Use of the Compressibility Factor in Calculating a Specific Volume

Problem Statement

In spreading liquid ammonia fertilizer, the charges for the amount of NH_3 used are based on the time involved plus the pounds of NH_3 injected into the soil. After the liquid has been spread, there is still some ammonia left in the source tank (volume = 120 ft^3), but in the form of a gas. Suppose that your weight tally, which is obtained by difference, shows a net weight of 125 lb of NH_3 left in the tank at 292 psig. Because the tank is sitting in the sun, the temperature in the tank is 125°F.

Your boss complains that calculations show the specific volume of the NH_3 gas is 1.20 ft^3/lb, and hence there are only 100 lb of NH_3 in the tank. Could your boss be correct? See Figure E6.9.

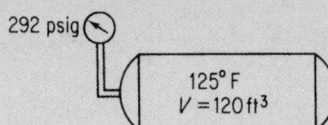

Figure E6.9

Solution

The simplest calculation to make to get the specific volume of the ammonia in the tank is to select a pound or pound mole as a basis:

Basis: 1 lb of NH_3

Apparently, your boss used the ideal gas law ($z = 1$) in getting the figure of 1.20 ft^3/lb of NH_3 gas:

$$R = 10.73 \frac{(psia)(ft^3)}{(lb\,mol)(°R)} \qquad p = 292 + 14.7 = 306.7\,psia$$

$$T = 125°F + 460 = 585°R \qquad n = \frac{1\,lb}{17\,lb/lb\,mol}$$

$$\hat{V} = \frac{nRT}{p} = \frac{\frac{1}{17}(10.73)(585)}{306.7} = 1.20\,ft^3/lb$$

(Continues)

Example 6.9   Use of the Compressibility Factor in Calculating a Specific Volume (*Continued*)

What should you do? Ammonia probably does not behave like an ideal gas under the observed conditions of temperature and pressure. You can apply $pV = znRT$ to calculate n and determine the real amount of NH_3 in the tank if you include the correct compressibility factor in the real gas law. Let's compute z; z is a function of T_r and p_r. You can look up all of the values of the necessary parameters in Appendix E (available online).

$$T_c = 405.5\,\text{K} \Rightarrow 729.9°\text{R} \quad p_c = 111.3\,\text{atm} \Rightarrow 1636\,\text{psia}$$

Then, since

$$T_r = \frac{T}{T_c} = \frac{585°\text{R}}{729.9°\text{R}} = 0.801 \quad p_r = \frac{p}{p_c} = \frac{306.7\,\text{psia}}{1636\,\text{psia}} = 0.187$$

From the Nelson and Obert (N&O) chart, Figure 6.9a, you can read $z \cong 0.855$. The value may be somewhat in error because ammonia was not one of the gases included in the preparation of the figure. Rather than calculating the specific volume directly, let's calculate it from the ratio of $pV_{real} = z_{real}nRT$ to $pV_{ideal} = z_{ideal}nRT$, the net result of which is

$$\frac{\hat{V}_{real}}{\hat{V}_{ideal}} = \frac{z_{real}}{z_{ideal}}$$

On the basis of 1 lb NH_3,

$$\hat{V}_{real} = \frac{1.20\,\text{ft}^3\ \text{ideal}}{\text{lb}} \left|\frac{0.855}{1}\right. = 1.03\,\text{ft}^3/\text{lb}\ NH_3$$

On the basis of 120 ft³ in the tank,

$$\frac{1\,\text{lb}\ NH_3}{1.03\,\text{ft}^3} \left|\frac{120\,\text{ft}^3}{}\right. = 117\,\text{lb}\ NH_3$$

Certainly, 117 lb is a more realistic figure than 100 lb, but it still could be in error, considering that the residual weight of the NH_3 in the tank is determined by difference.

As a matter of interest, as an alternative to making these calculations, you could look up the specific volume of NH_3 at the conditions in the tank in a handbook. You would find that $\hat{V} = 0.973$ ft^3/lb, equivalent to 123 lb of NH_3, the correct value. Would you tell your boss to use the right compressibility factor, or state that you used the handbook value of $\hat{V}$?

Example 6.10   Use of the Compressibility Factor in Calculating a Pressure

Problem Statement

Liquid oxygen is used in the steel industry, in the chemical industry, in hospitals, as rocket fuel oxidant, and for wastewater treatment, as well as in many other applications. A tank sold to hospitals contains 0.0284 m^3 of volume filled with 3.500 kg of liquid O_2 that will vaporize at –25°C. After all of the O_2 in the tank vaporizes, will the pressure in the tank exceed the safety limit for the tank specified as 10^4 kPa?

Solution

$$\text{Basis: 3.500 kg of } O_2$$

In Appendix E on the website, you can find that for oxygen

$$T_c = 154.4\,\text{K} \quad p_c = 49.7\,\text{atm} \Rightarrow 5035\,\text{kPa}$$

However, you cannot proceed to solve this problem in exactly the same way as the preceding problem because you do not know the pressure of the O_2 in the tank to begin with. But you can use the pseudoparameter, $\hat{V}_{ri}$, which is available as a parameter on the N&O charts, as a second parameter to fix a point on the compressibility charts.

First calculate

$$\hat{V}\,(\text{specific molar volume}) = \frac{0.0284\ \text{m}^3}{3.500\ \text{kg}} \left| \frac{32\ \text{kg}}{1\ \text{kg mol}} = 0.260\ \text{m}^3/\text{kg mol} \right.$$

Note that the *specific molar volume* must be used in calculating $\hat{V}_{ri}$ because $\hat{V}_{ci}$ is the volume per mole.

$$\hat{V}_{ci} = \frac{RT_c}{p_c} = \frac{8.313(\text{m}^3)(\text{kPa})}{(\text{kg mol})(\text{K})} \left| \frac{154.4\ \text{K}}{5,035\ \text{kPa}} = 0.255\ \frac{\text{m}^3}{\text{kg mol}} \right.$$

(Continues)

Example 6.10   Use of the Compressibility Factor in Calculating a Pressure (*Continued*)

Then

$$\hat{V}_{ri} = \frac{\hat{V}}{\hat{V}_{ci}} = \frac{0.260}{0.255} = 1.02$$

Now you know the values of two parameters, $\hat{V}_{ri}$ and

$$T_r = \frac{248\,\text{K}}{154.4\,\text{K}} = 1.61$$

From the N&O chart (Figure 6.9b) you can read
$$p_r = 1.43$$
Then

$$p = p_r p_c = 1.43(5035) = 7200\,\text{kPa}$$

The pressure of 10^4 kPa will not be exceeded. Even at room temperature the pressure will be less than 10^4 kPa.

To get one snapshot of the difference between estimates of z by two of the methods discussed in this chapter, Table 6.4 compares the experimental values of z for ethylene with predictions by two methods: N&O charts and the ideal gas law for three different set of conditions.

Table 6.4   A Comparison of Values of the Compressibility Factor z for Ethylene* Determined via Two Different Methods with the Associated Experimental Values

	At 350 K and 500 kPa		At 300 K and 3000 kPa		At 274 K and 3600 kPa	
	z	% Deviation	z	% Deviation	z	% Deviation
Experimental value	0.983	-	0.812	-	0.563	-
Ideal gas law	1	1.8	1	23.1	1	78
N&O chart	0.982	0.0	0.815	0.0	0.57	1

* $T_c = 282.8$ K; $p_c = 50.5$ atm.

Self-Assessment Test

Questions

1. What is the pseudocritical volume? What is the advantage of using $\hat{V}_{ci}$?

2. Indicate whether the following statements are true or false:
   a. Two fluids, which have the same values of reduced temperature and pressure and the same reduced volume, are said to be in corresponding states.
   b. It is expected that all gases will have the same z at a specified T_r and p_r. Thus, a correlation of z in terms of T_r and p_r would apply to all gases.
   c. The law of corresponding states implies that at the critical state (T_c, p_c), all substances should behave alike.
   d. The critical state of a substance is the set of physical conditions at which the density and other properties of the liquid and vapor become identical.
   e. Any substance (in theory), by the law of corresponding states, should have the same reduced volume at the same reduced T_r and P_r.
   f. The equation $pV = znRT$ cannot be used for ideal gases.
   g. By definition, a fluid becomes supercritical when its temperature and pressure exceed the critical point.
   h. Phase boundaries do not exist under supercritical conditions.
   i. For some gases under normal conditions, and for most gases under conditions of high pressure, values of the gas properties that might be obtained using the ideal gas law would be at wide variance with the experimental evidence.

3. Explain the meaning of the following equation for the compressibility factor:
$$z = f(T_r, p_r)$$

4. What is the value of z at $p_r = 0$?

Answers

1. $\hat{V}_{ci} = RTc/p_c$; it can be used as a parameter on the compressibility charts when the value of p or T is unknown.

2. All are true except f.

3. It means that z is a function of the reduced temperature T_r and the reduced pressure p_r.

4. $z = 1.00$

Problems

1. Calculate the compressibility factor z and determine whether the following gases can be treated as ideal at the listed temperature and pressure:
   a. Water at 1000°C and 2000 kPa
   b. Oxygen at 35°C and 1500 kPa
   c. Methane at 10°C and 1000 kPa

2. A carbon dioxide fire extinguisher has a volume of 40 L and is to be charged to a pressure of 20 atm at a storage temperature of 20°C. Determine the mass in kilograms of CO_2 in the fire extinguisher.

3. Calculate the pressure of 4.00 g mol of CO_2 contained in a 6.25×10^{-3} m^3 fire extinguisher at 25°C.

Answers

1. (a) Yes; (b) Yes; (c) No ($z = 0.98$)

2. $T_r = 0.964$; $p_r = 0.275$; from chart, $z = 0.88$; $n = PV/(zRT) = (20)(40)/(0.88(293.15)$ $(0.08206)) = 37.8$ g mol $= 1.66$ kg

3. $\hat{V}_{ci} = 0.3428$ L/g mol; $V_r = 4.92$; $T_r = 0.98$; from chart, $z = 0.92$; $p = znRT/V = (0.92)$ $(0.08206)(298.15)/1.69 = 13.3$ atm

6.4   Real Gas Mixtures

To this point, we have discussed predicting *p-V-T* properties for *pure* components of real gases. How should you treat mixtures of real gases? The actual critical points of binary mixtures are not linear combinations of the properties of the two components as shown in Figure 6.10 for combinations of CO_2 and SO_2. Too many dimensions are involved to draw pictures for three or more components.

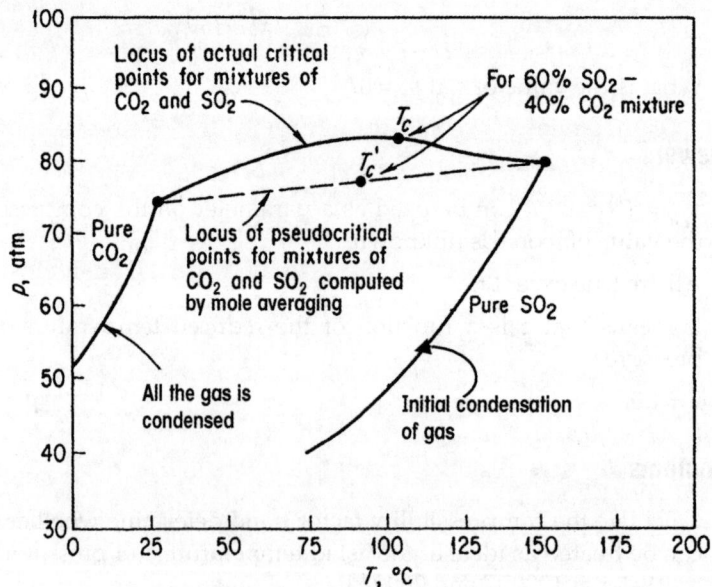

Figure 6.10   Critical and pseudocritical points for mixtures of CO_2 and SO_2

One way you can make reasonable predictions for z and $\hat{V}_{ri}$ for engineering purposes is to use Kay's method[4] and the compressibility charts. In **Kay's method, pseudocritical** values for mixtures of gases are calculated on the assumption that each component in the mixture contributes to the pseudocritical value in the same proportion as the mole fraction of that component in the gas. Thus, the pseudocritical values are computed as mole averages as follows:

$$p_c' = p_{c_A} y_A + p_{c_B} y_B + \cdots \tag{6.11}$$

$$T_c' = T_{c_A} y_A + T_{c_B} y_B + \cdots \tag{6.12}$$

where y_i is the mole fraction, p'_c is the pseudocritical pressure, and T'_c is the pseudocritical temperature. You can see that these are linearly weighted mole average pseudocritical properties. Look at Figure 6.10, which compares the true critical values of a gaseous mixture of CO_2 and SO_2 with the respective pseudocritical values. The respective **pseudoreduced** variables are

$$p_r' = \frac{p}{p_c'} \qquad\qquad T_r' = \frac{T}{T_c'}$$

Kay's method is known as a two-parameter rule since only p_c and T_c for each component are involved in the calculation of z. If a third parameter such as z_c, or the Pitzer acentric factor, or $\hat{V}_{ci}$, is included in the determination of the compressibility factor, you would have a three-parameter rule. Other pseudocritical methods with additional parameters provide better accuracy in predicting p-$\hat{V}$-T properties than Kay's method, but Kay's method can suffice for our work, and it is easy to use.

In instances in which the temperature or pressure of a gas mixture is unknown, to avoid a trial-and-error solution using the generalized compressibility charts, you can compute the **pseudocritical ideal volume** and a **pseudoreduced ideal volume** $\hat{V}_{ri}$, thus

$$\tilde{V}'_{ci} = \frac{RT'_c}{p'_c} \text{ and } \tilde{V}'_{ri} = \frac{\hat{V}}{\tilde{V}_{ci}}$$

$\hat{V}_{ri}$ can be used in lieu of p'_r or T'_r in the compressibility charts.

An enormous literature exists describing proposals for *mixing rules* for equations of state, that is, rules to weight the coefficients or the predictions of each pure component so that the weighted values can be used with the same equations of state as are used for a pure component. Refer to the references at the end of this chapter, or look on the Internet for examples.

---

[4] W. B. Kay, "Density of Hydrocarbon Gases and Vapors at High Temperature and Pressure," *Ind. Eng. Chem.*, **28**, 1014–19 (1936).

Example 6.11   Calculation of *p-V-T* Properties for a Real Gas Mixture

Problem Statement

A gaseous mixture has the following composition (in mole percent):

Methane, CH_4	20
Ethylene, C_2H_4	30
Nitrogen, N_2	50

at 90 atm pressure and 100°C. Compare the volume per mole as computed by the methods of (a) the ideal gas law and (b) the pseudoreduced technique (Kay's method). What other types of averaging might you use?

Solution

Basis: 1 g mol of gas mixture

Additional data needed are

Component	T_c (K)	p_c (atm)
CH_4	191	45.8
C_2H_4	283	50.5
N_2	126	33.5

The units used are fixed by the units of R. Let R be $R = 82.06 \dfrac{(cm^3)(atm)}{(g\,mol)(K)}$.

a. Ideal gas law:

$$\hat{V} = \frac{RT}{p} = \frac{(82.06)(373)}{90} = 340\,cm^3/g \text{ mol at 90 atm and 373 K}$$

b. According to Kay's method, you first calculate the pseudocritical values for the mixture:

$$p_c' = p_{c_A} y_A + p_{c_B} y_B + p_{c_C} y_C = (45.8)(0.2) + (50.5)(0.3) + (33.5)(0.5)$$
$$= 41.1\,atm$$

$$T_c' = T_{c_A} y_A + T_{c_B} y_B + T_{c_C} y_C = (191)(0.2) + (283)(0.3) + (126)(0.5) = 186\,K$$

Then you calculate the pseudoreduced values for the mixture:

$$p_r' = \frac{p}{p_c'} = \frac{90}{41.2} = 2.19, \quad T_r' = \frac{T}{T_c'} = \frac{373}{186} = 2.01$$

With the aid of these two parameters, you can find from Figure 6.9b that $zT_r' = 1.91$, and thus $z = 0.95$. Then

$$\hat{V} = \frac{zRT}{p} \approx \frac{0.95(1)(82.06)(373)}{90} = 323 \text{ cm}^3/\text{g mol at 90 atm and 373 K}$$

Two of the many possible ways of averaging are to use an equation of state with mole-averaged coefficients or use the mole-averaged predictions of $\hat{V}$ obtained from the individual equation of state.

Self-Assessment Test

Question

1. How is Kay's method used to predict the p-$\hat{V}$-T behavior of mixtures of gases?

Answer

1. Kay's method uses the molar average of the critical pressure and critical temperature to estimate the critical pressure and temperature of the mixture. Then these estimates of the reduced properties are used to calculate the compressibility factor for the mixture.

Problem

1. A mixture containing 40 mol % of N_2 and 60 mol % of C_2H_4 at 50°C is at a pressure of 30 atm. What is the specific volume of the gas in the vessel in $\text{ft}^3/\text{lb mol}$? Compute your answer by Kay's method.

Answer

1. $T_c' = 220.2$ K; $p_c' = 43.7$ atm; $T_r = 1.47$; $p_r = 0.68$; from Figure 6.9a, $z = 0.94$

$$\hat{V} = \frac{zRT}{p} = \frac{0.94}{} \left| \frac{0.7303 \text{ atm ft}^3}{\text{lb mol °R}} \right| \frac{581.67 \text{ °R}}{30 \text{ atm}} = 13.3. \text{ ft}^3/\text{lb mol}$$

Summary

We reviewed the ideal gas law and showed how to use it in conjunction with material balances. The law of corresponding states was introduced, and it was shown how to correct the ideal gas law by calculating compressibility factors from tables based on reduced conditions. Several commonly used equations of state were presented, and it was shown how to use them to calculate unknown properties of nonideal gases. In addition, it was demonstrated how to use MATLAB and Python to solve equations of state that are nonlinear.

Glossary

acentric factor   A parameter that indicates the degree of nonsphericity of a molecule.

compressibility charts   Graphs of the compressibility factor as a function of reduced temperature, pressure, and ideal reduced volume.

compressibility factor   A factor that is introduced into the ideal gas law to compensate for the nonideality of a gas.

corrected   Normalized.

corresponding states   Any gas should have the same reduced volume at the same reduced temperature and reduced pressure.

critical state   The set of physical conditions at which the density and other properties of liquid and vapor become identical.

Dalton's law   The summation of each of the partial pressures of the components in a system equals the total pressure. The other related law (of partial pressures) is that the total pressure times the mole fraction of a component in a system is the partial pressure of the component.

density of gas   Mass per unit volume expressed in kilograms per cubic meter, pounds per cubic foot, grams per liter, or equivalent units.

generalized compressibility   See **compressibility charts**.

generalized equation of state   The ideal gas law converted to a real gas law by inserting a compressibility factor.

group contribution method   A technique of estimating physical properties of compounds by using properties of molecular groups of elements in the compound.

Holborn   A multiple-parameter equation of state expanded in p.

ideal critical volume $\hat{V}_{ci} = RT_c/p_c$.

ideal gas constant   The constant in the ideal gas law (and other equations) denoted by the symbol R.

ideal gas law   Equation relating p, V, n, and T that applies to many gases at low density (high temperature and/or low pressure).

ideal reduced volume $V_{ri} = \hat{V}/\hat{V}_{ci}$.

Kammerlingh-Onnes   A multiple-parameter equation of state expanded in $\hat{V}^{-1}$.

Kay's method   Rule for calculating the compressibility factor for a mixture of gases.

law of corresponding states   See **corresponding states**.

partial pressure   The pressure that would be exerted by a single component in a gaseous mixture if it existed alone in the same volume as occupied by the mixture and at the same temperature as the mixture.

Pitzer acentric factor   See **acentric factor**.

pseudocritical   Temperatures, pressures, and/or specific volumes adjusted to be used with charts or equations used to calculate the compressibility factor.

real gases   Gases whose behavior does not conform to the assumptions underlying ideality.

reduced variables   Corrected or normalized conditions of temperature, pressure, and volume, normalized by their respective critical conditions.

Soave-Redlich-Kwong (SRK)   A three-parameter equation of state.

specific gravity   Ratio of the density of a gas at a temperature and pressure to the density of a reference gas at some temperature and pressure.

standard conditions (S.C.)   Arbitrarily specified standard states of temperature and pressure established for gases by custom.

supercritical fluid   Material in a state above its critical point.

UNIFAC   A group contribution method of estimating physical properties.

UNIQUAC   An extension of the UNIFAC method of estimating physical properties.

van der Waals   A two-parameter equation of state.

virial equation of state   An equation of state expanded in successive terms of one of the physical properties.

Supplementary References

Ben-Amotz, D., A. Gift, and R. D. Levine. "Updated Principle of Corresponding States," *J. Chem. Edu.*, **81**, No. 1 (2004).

Castillo, C. A. "An Alternative Method for the Estimation of Critical Temperatures of Mixtures," *AIChE J.*, **33**, 1025 (1987).

Chao, K. C., and R. L. Robinson. *Equations of State in Engineering and Research*, American Chemical Society, Washington, DC (1979).

Copeman, T. W., and P. M. Mathias. "Recent Mixing Rules for Equations of State," *ACS Symposium Series*, **300**, 352–69, American Chemical Society, Washington, DC (1986).

Eliezer, S., A. K. Ghatak, and H. Hora. *An Introduction to Equations of State: Theory and Applications*, Cambridge University Press, Cambridge, UK (1986).

Elliott, J. R., and T. E. Daubert. "Evaluation of an Equation of State Method for Calculating the Critical Properties of Mixtures," *Ind. Eng. Chem. Res.*, **26**, 1689 (1987).

Gibbons, R. M. "Industrial Use of Equations of State," in *Chemical Thermodynamics in Industry*," edited by T. I. Barry, Blackwell Scientific, Oxford, UK (1985).

Lawal, A. S. "A Consistent Rule for Selecting Roots in Cubic Equations of State," *Ind. Eng. Chem. Res.*, **26**, 857–59 (1987).

Manavis, T., M. Volotopoulos, and M. Stamatoudis. "Comparison of Fifteen Generalized Equations of State to Predict Gas Enthalpy," *Chem. Eng. Commun.*, **130**, 1–9 (1994).

Masavetas, K. A. "The Mere Concept of an Ideal Gas," *Math. Comput. Modelling*, **12**, 651–57 (1989).

Mathias, P. M., and M. S. Benson. "Computational Aspects of Equations of State," *AIChE J.*, **32**, 2087 (1986).

Mathias, P. M., and H. C. Klotz. "Take a Closer Look at Thermodynamic Property Models," *Chem. Eng. Progress*, 67–75 (June, 1994).

Orbey, H., S. I. Sander, and D. S. Wong. "Accurate Equation of State Predictions at High Temperatures and Pressures Using the Existing UNIFAC Model," *Fluid Phase Equil.*, **85,** 41–54 (1993).

Reid, R. C., J. M. Prausnitz, and B. E. Poling. *The Properties of Gases and Liquids*, 4th ed., McGraw-Hill, New York (1987).

Sandler, S. I., H. Orbey, and B. I. Lee. "Equations of State," in *Modeling for Thermodynamic and Phase Equilibrium Calculations*, Chapter 2, edited by S. I. Sander, Marcel Dekker, New York (1994).

Span, R. *Multiparameter Equations of State*, Springer, New York (2000).

Sterbacek, Z., B. Biskup, and P. Tausk. *Calculation of Properties Using Corresponding State Methods*, Elsevier Scientific, New York (1979).

Yaws, C. L., D. Chen, H. C. Yang, L. Tan, and D. Nico. "Critical Properties of Chemicals," *Hydrocarbon Processing*, **68**, 61 (July, 1989).

Problems

6.1 Ideal Gases

*6.1.1 How many pounds of H_2O are in 100 ft^3 of vapor at 15.5 mm Hg and 23°C?

*6.1.2 One liter of a gas is under a pressure of 780 mm Hg. What will be its volume at standard pressure, the temperature remaining constant?

*6.1.3 A gas occupying a volume of 1 m^3 under standard pressure is expanded to 1.200 m^3, the temperature remaining constant. What is the new pressure?

*6.1.4 Determine the mass specific volume and molar specific volume for air at 78°F and 14.7 psia.

*6.1.5 Divers work as far as 500 ft below the water surface. Assume that the water temperature is 45°F. What is the molar specific volume (in cubic feet per pound mole) for an ideal gas under these conditions?

*6.1.6 A 25 L glass vessel is to contain 1.1 g mol of nitrogen. The vessel can withstand a pressure of only 20 kPa above atmospheric pressure (taking into account a suitable safety factor). What is the maximum temperature to which the N_2 can be raised in the vessel?

**6.1.7 An oxygen cylinder used as a standby source of oxygen contains O_2 at 70°F. To calibrate the gauge on the O_2 cylinder, which has a volume of 1.01 ft^3, all of the oxygen, initially at 70°F, is released into an evacuated tank of known volume (15.0 ft^3). At equilibrium, the gas pressure was measured as 4 in. H_2O gauge and the gas temperature in both cylinders was 75°F. See Figure P6.1.7. The barometer read 29.99 in. Hg.

What did the pressure gauge on the oxygen tank initially read in psig if it was a Bourdon gauge?

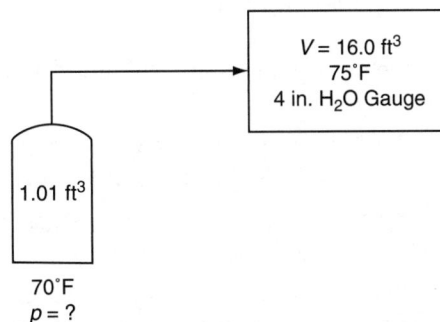

Figure P6.1.7

*6.1.8   An average person's lungs contain about 5 L of gas under normal conditions. If a diver makes a free dive (no breathing apparatus), the volume of the lungs is compressed when the pressure equalizes throughout the body. If compression occurs below 1 L, irreversible lung damage will occur. Calculate the maximum safe depth for a free dive in seawater (assume the density is the same as freshwater).

*6.1.9   An automobile tire when cold (at 75°F) reads 30 psig on a tire gauge. After driving on the freeway, the temperature in the tire becomes 140°F. Will the pressure in the tire exceed the pressure limit of 35 psi the manufacturer stamps on the tire?

*6.1.10   You are making measurements on an air conditioning duct to test its load capacity. The warm air flowing through the circular duct has a density of 0.0796 lb/ft^3. Careful measurements of the velocity of the air in the duct disclose that the average air velocity is 11.3 ft/s. The inside radius of the duct is 18.0 in. What are (a) the volumetric flow rate of the air in cubic feet per hour and (b) the mass flow rate of the air in pounds per day?

*6.1.11   One pound mole of flue gas has the following composition. Treat it as an ideal gas.

CO_2(11.2%), CO(1.2%), SO_2(1.2%), O_2(5.3%), N_2(81.0%), H_2O(0.1%)

How many cubic feet will the gas occupy at 100°F and 1.54 atm?

*6.1.12   From the known standard conditions, calculate the value of the gas law constant R in the following sets of units:
a. cal/(g mol)(K)
b. Btu/(lb mol)(°R)
c. (psia)(ft^3)/(lb mol)(°R)
d. J/(g mol)(K)
e. (cm^3)(atm)/(g mol)(K)
f. (ft^3)(atm)/(lb mol)(°R)

*6.1.13   What is the density of O_2 at 100°F and 740 mm Hg in (a) pounds per cubic foot and (b) grams per liter?

*6.1.14   What is the density of propane gas (C_3H_8) in kilograms per cubic meter at 200 kPa and 40°C? What is the specific gravity of propane?

*6.1.15   What is the specific gravity of propane gas (C_3H_8) at 100°F and 800 mm Hg relative to air at 60°F and 760 mm Hg?

*6.1.16   What is the mass of 1 m^3 of H_2 at 5°C and 110 kPa? What is the specific gravity of this H_2 compared to air at 5°C and 110 kPa?

*6.1.17 A gas used to extinguish fires is composed of 80% CO_2 and 20% N_2. It is stored in a 2 m^3 tank at 200 kPa and 25°C. What is the partial pressure of the CO_2 in the tank in kilopascals?

*6.1.18 A natural gas has the following composition by volume:

CH_4	94.1%
N_2	3.0
H_2	1.9
O_2	1.0
	100.0%

This gas is piped from the well at a temperature of 20°C and a pressure of 30 psig. It may be assumed that the ideal gas law is applicable. Calculate the partial pressure of the oxygen.

*6.1.19 A liter of oxygen at 760 mm Hg is forced into a vessel containing a liter of nitrogen at 760 mm Hg. What will be the resulting pressure? What assumptions are necessary for your answer?

*6.1.20 Indicate whether the following statements are true or false:
   a. The volume of an ideal gas mixture is equal to the sum of the volumes of each individual gas in the mixture.
   b. The temperature of an ideal gas mixture is equal to the sum of the temperatures of each individual gas in the mixture.
   c. The pressure of an ideal gas mixture is equal to the sum of the partial pressures of each individual gas in the mixture.

**6.1.21 An oxygen cylinder used as a standby source of oxygen contains 1.000 ft^3 of O_2 at 70°F and 200 psig. What will be the volume of this O_2 in a dry-gas holder at 90°F and 4.00 in. H_2O above atmospheric? The barometer reads 29.92 in. Hg.

**6.1.22 You have 10 lb of CO_2 in a 20 ft^3 fire extinguisher tank at 30°C. Assuming that the ideal gas law holds, what will the pressure gauge on the tank read in a test to see if the extinguisher is full?

**6.1.23 The U-tube manometer depicted in Figure P6.1.23 has a left leg 20 in. high and a right leg 40 in. high. The manometer initially contains mercury to a depth of 12 in. in each leg. Then the left leg is closed with a cork, and mercury is poured in the right leg until the mercury in the left (closed) leg reaches a height of 14 in. How deep is the mercury in the right leg from the bottom of the manometer?

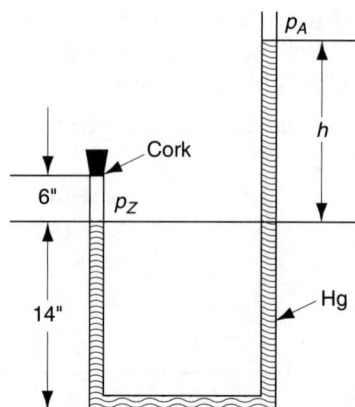

Figure P6.1.23

**6.1.24 One of the experiments in the fuel-testing laboratory has been giving some trouble because a particular barometer gives erroneous readings owing to the presence of a small amount of air above the mercury column. At a true atmospheric pressure of 755 mm Hg the barometer reads 748 mm Hg, and at a true 740 mm Hg the reading is 736 mm Hg. What will the barometer read when the actual pressure is 760 mm Hg?

**6.1.25 Flue gas at a temperature of 1800°F is introduced to a scrubber through a pipe that has an inside diameter of 4.0 ft. The inlet velocity to and the outlet velocity from the scrubber are 25 ft/s and 20 ft/s, respectively. The scrubber cools the flue gas to 550°F. Determine the duct size required at the outlet of the unit.

**6.1.26 Calculate the number of cubic meters of hydrogen sulfide, measured at a temperature of 30°C and a pressure of 15.71 cm Hg, which may be produced from 10 kg of iron sulfide (FeS).

**6.1.27 Monitoring of hexachlorobenzene (HCB) in a flue gas from an incinerator burning 500 lb/hr of hazardous wastes is to be conducted. Assume that all of the HCB is removed from a sample of the flue gas and concentrated in 25 mL of solvent. The analytical detection limit for HCB is 10 μg/ml in the solvent. Determine the minimum volume of flue gas that has to be sampled to detect the existence of HCB in the flue gas. Also, calculate the time needed to collect a gas sample if you can collect 1.0 L/min. The flue gas flow rate is 427,000 ft^3/hr measured at standard conditions.

**6.1.28 Ventilation is an extremely important method of reducing the level of toxic airborne contaminants in the workplace. Since it is impossible to eliminate absolutely all leakage from a process into the workplace, some method is always needed to remove toxic materials from the air in closed rooms when such materials are present in the process streams. The Occupational Safety and Health Administration (OSHA) has set the permissible exposure limit (PEL) of vinyl chloride (VC, MW = 78) at 1.0 ppm as a maximum time-weighted average (TWA) for an 8-hr workday, because VC is believed to be a human carcinogen. (A carcinogen is an agent that, when exposed to humans, might increase the likelihood of the subject getting cancer at some time in the future.) If VC escapes into the air, its concentration must be maintained at or below the PEL. If dilution ventilation were to be used, you can estimate the required airflow rate by assuming complete mixing in the workplace air, and then assuming that the volume of airflow through the room will carry VC out with it at the concentration of 1.0 ppm.

If a process loses 10 g/min of VC into the room air, what volumetric flow rate of air will be necessary to maintain the PEL of 1.0 ppm by dilution ventilation? (In practice we must also correct for the fact that complete mixing will not be realized in a room, so you must multiply the calculated airflow rate by a safety factor, say, a factor of 10.)

If the safety analysis or economics of ventilation do not demonstrate that a safe concentration of VC exists, the process might have to be moved into a hood so that no VC enters the room. If the process is carried out in a hood with an opening of 30 in. wide by 25 in. high, and the "face velocity" (average air velocity through the hood opening) is 100 ft/s, what is the volumetric airflow rate at S.C.? Which method of treating the pollution problem seems to be better to you? Explain why dilution ventilation is not recommended for maintaining air quality. What might be a problem with the use of a hood? The problem is adapted with permission from the publication *Safety, Health, and Loss Prevention in Chemical Processes* published by the American Institute of Chemical Engineers, New York (1990).

**6.1.29 Ventilation is an extremely important method of reducing the level of toxic airborne contaminants in the workplace. Trichloroethylene (TCE) is an excellent solvent for a number of applications and is especially useful in degreasing. Unfortunately, TCE can lead to a number of harmful health effects, and ventilation is essential. TCE has been shown to be carcinogenic in animal tests. It is also an irritant to the eyes and respiratory tract. Acute exposure causes depression of the central nervous system, producing symptoms of dizziness, tremors, and irregular heartbeat, plus others.

Since the molecular weight of TCE is approximately 131.5, it is much denser than air. As a first thought, you would not expect to find a high concentration of this material above an open tank because you might assume that the vapor would sink to the floor. If this were so, we would place the inlet of a local exhaust hood for such a tank near the floor. However, toxic concentrations of many materials are not much denser than the air itself, so where there can be mixing with the air we may not assume that all the vapors will go to the floor. For the case of trichloroethylene

OSHA has established a time-weighted average 8 hr PEL of 100 ppm. What is the fraction increase in the density of a mixture of TCE in air over that of air if the TCE is at a concentration of 100 ppm and at 25°C? This problem has been adapted from *Safety, Health, and Loss Prevention in Chemical Processes*, Vol. 3, American Institute of Chemical Engineers, New York (1990).

**6.1.30 Benzene can cause chronic adverse blood effects such as anemia and possibly leukemia with chronic exposure. Benzene has a PEL for an 8-hr exposure of 1.0 ppm. If liquid benzene is evaporating into the air at a rate of 2.5 cm^3 of liquid/min, what must the ventilation rate be in volume per minute to keep the concentration below the PEL? The ambient temperature is 68°F and the pressure is 740 mm Hg. This problem has been adapted from *Safety, Health, and Loss Prevention in Chemical Processes*, Vol. 6, American Institute of Chemical Engineers, New York (1990).

**6.1.31 A recent newspaper report states:

> Home meters for fuel gas measure the volume of gas usage based on a standard temperature, usually 60 degrees. But gas contracts when it's cold and expands when warm. East Ohio Gas Co. figures that in chilly Cleveland, the homeowner with an outdoor meter gets more gas than the meter says he does, so that's built into the company's gas rates. The guy who loses is the one with an indoor meter: If his home stays at 60 degrees or over, he'll pay for more gas than he gets. (Several companies make temperature-compensating meters, but they cost more and aren't widely used. Not surprisingly, they are sold mainly to utilities in the North.)

Suppose that the outside temperature drops from 60°F to 10°F. What is the percentage increase in the mass of the gas passed by a noncompensated outdoor meter that operates at constant pressure? Assume that the gas is CH_4.

**6.1.32 Soft ice cream is a commercial ice cream mixture whipped usually with CO_2 (as the O_2 in the air causes deterioration). You are working in the Pig-in-a-Poke Drive-In and need to make a machine full (4 gal) of soft ice cream. The unwhipped mix has a specific gravity of 0.95 and the local ordinance forbids you to make ice cream of less than a specific gravity of 0.85. Your CO_2 tank is a No. 1 cylinder (9 in. diameter by 52 in. high) of commercial-grade (99.5% min. CO_2) carbon dioxide. In examining the pressure gauge, you note it reads 68 psig. Do you have to order another cylinder of CO_2? Be sure to specifically state all the assumptions you make for this problem.

**6.1.33 A natural gas has the following composition:

CH_4	(methane)	87%
C_2H_6	(ethane)	12%
C_3H_8	(propane)	1%

a. What is the composition in weight percent?
b. What is the composition in volume percent?
c. How many cubic meters will be occupied by 80.0 kg of the gas at 9°C and 600 kPa?
d. What is the density of the gas in kilograms per cubic meter at S.C.?
e. What is the specific gravity of this gas at 9°C and 600 kPa referred to air at S.C.?

**6.1.34 A mixture of bromine vapor in air contains 1% bromine by volume.
a. What weight percent bromine is present?
b. What is the average molecular weight of the mixture?
c. What is its specific gravity?
d. What is its specific gravity compared to bromine?
e. What is its specific gravity at 100°F and 100 psig compared to air at 60°F and 30 in. Hg?

**6.1.35 The contents of a gas cylinder are found to contain 20% CO_2, 60% O_2, and 20% N_2 at a pressure of 740 mm Hg and at 20°C. What are the partial pressures of each of the components? If the temperature is raised to 40°C, will the partial pressures change? If so, what will they be?

**6.1.36 Methane is completely burned with 20% excess air, with 30% of the carbon going to CO. What is the partial pressure of the CO in the stack gas if the barometer reads 740 mm Hg, the temperature of the stack gas is 300°F, and the gas leaves the stack at 250 ft above the ground level?

**6.1.37 A 0.5 m³ rigid tank containing hydrogen at 20°C and 600 kPa is connected by a valve to another 0.5 m³ rigid tank that holds hydrogen at 30°C and 150 kPa. Now the valve is opened and the system is allowed to reach thermal equilibrium with the surroundings, which are at 15°C. Determine the final pressure in the tank.

**6.1.38 A 400 ft³ tank of compressed H_2 is at a pressure of 55 psig. It is connected to a smaller tank with a valve and short line. The small tank has a volume of 50 ft³ and contains H_2 at 1 atm absolute and the same temperature. If the interconnecting valve is opened and no temperature change occurs, what is the final pressure in the system?

**6.1.39 A tank of N_2 has a volume of 100 ft³ and an initial temperature of 80°F. One pound of N_2 is removed from the tank, and the pressure drops to 100 psig while the temperature of the gas in the tank drops to 60°F. Assuming N_2 acts as an ideal gas, calculate the initial pressure reading on the pressure gauge.

**6.1.40 Measurement of flue gas flow rates is difficult by traditional techniques for various reasons. Tracer gas flow measurements using sulfur hexafluoride (SF_6) have proved to be more accurate. Figure P6.1.40 shows the stack arrangement and the injection and sampling points for the SF_6. Here are the data for one experiment:

Volume of SF_6 injected (converted to S.C.):   28.2 m³/min
Concentration of SF_6 at the flue gas sample point: 4.15 ppm
Relative humidity correction: none

Calculate the volume of the exit flue gas per minute.

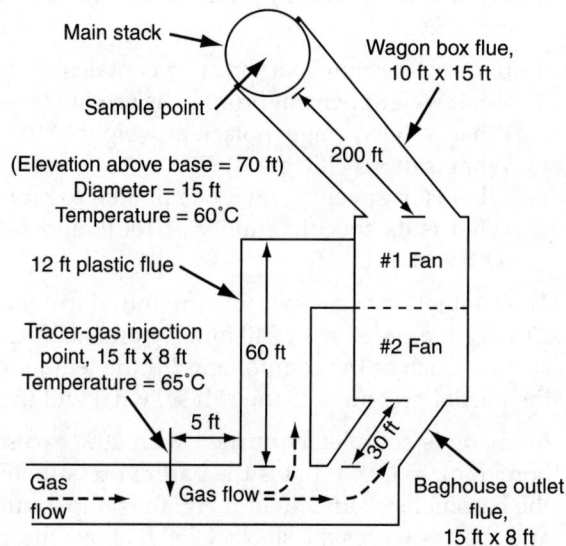

Figure P6.1.40

**6.1.41 An ideal gas at 60°F and 31.2 in. Hg (absolute) is flowing through an irregular duct. To determine the flow rate of the gas, CO_2 is passed into the gas stream. The gas analyzes 1.2 mol % CO_2 before and 3.4 mol % after addition. The CO_2 tank is placed on a scale and found to lose 15 lb in 30 min. What is the flow rate of the entering gas in cubic feet per minute?

*6.1.42 In the manufacture of dry ice, a fuel is burned to a flue gas which contains 16.2% CO_2, 4.8% O_2, and the remainder N_2. This flue gas passes through a heat exchanger and then goes to an absorber. The data show that the analysis of the flue gas entering the absorber is 13.1% CO_2 with the remainder O_2 and N_2. Apparently, something has happened. To check your initial assumption that an air leak has developed in the heat exchanger, you collect the following data on a dry basis on the heat exchanger:

Entering flue gas in a 2-min period: 47,800 ft³ at 600°F and 740 mm of Hg
Exiting flue gas in a 2-min period: 30,000 ft³ at 60°F and 720 mm of Hg

Was your assumption about an air leak a good one, or was perhaps the analysis of the gas in error? Or both?

**6.1.43 Three thousand cubic meters per day of a gas mixture containing methane and n-butane at 21°C enters an absorber tower. The partial pressures at these conditions are 103 kPa for methane and 586 kPa for n-butane. In the absorber, 80% of the butane is removed and the remaining gas leaves the tower at 38°C and a total pressure of 550 kPa. What is the volumetric flow rate of gas at the exit? How many moles per day of butane are removed from the gas in this process? Assume ideal behavior.

**6.1.44 A heater burns normal butane (n-C_4H_{10}) using 40.0% excess air. Combustion is complete. The flue gas leaves the stack at a pressure of 100 kPa and a temperature of 260°C.
   a. Calculate the complete flue gas analysis.
   b. What is the volume of the flue gas in cubic meters per kilogram mole of n-butane?

**6.1.45 The majority of semiconductor chips used in the microelectronics industry are made of silicon doped with trace amounts of materials to enhance conductivity. The silicon initially must contain less than 20 ppm of impurities. Silicon rods are grown by the following chemical deposition reaction of trichlorosilane with hydrogen:

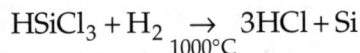

$$HSiCl_3 + H_2 \underset{1000°C}{\rightarrow} 3HCl + Si$$

Assuming that the ideal gas law applies, what volume of hydrogen at 1000°C and 1 atm must be reacted to increase the diameter of a rod 1 m long from 1 cm to 10 cm? The density of solid silicon is 2.33 g/cm^3.

**6.1.46 The oxygen and carbon dioxide concentrations in the gas phase of a 10 L bioreactor operating in the steady state control the dissolved oxygen and pH in the liquid phase where the biomass exists.
   a. If the rate of oxygen uptake by the liquid is 2.5×10^{-7} g mol/(1000 cells)(hr), and if the culture in the liquid phase contains 2.9×10^6 cells/mL, what is the rate of oxygen uptake in millimoles per hour?
   b. If the gas supplied to the gas phase is 45 L/hr containing 40% oxygen at 110 kPa and 25°C, what is the rate of oxygen supplied to the bioreactor in millimoles per hour?
   c. Will the oxygen concentration in the gas phase increase or decrease by the end of 1 hr compared to the initial oxygen concentration?

**6.1.47 When natural gas (mainly CH_4) is burned with 10% excess air, in addition to the main gaseous products of CO_2 and H_2O, other gaseous products result in minor quantities. The Environmental Protection Agency (EPA) lists the following data:

Emission Factors (kg/10^6 m^3 at S.C.)

	SO_2	NO_2	CO	CO_2
Large utility boiler, uncontrolled	9.6	3040	1344	1.9×10^6
Large utility boiler, controlled gas recirculation	9.6	1600	1344	1.9×10^6
Residential furnace	9.6	1500	640	1.9×10^6

The data are based on 10^6 m^3 measured at S.C. of methane burned. What is the approximate mole fraction of SO_2, NO_2, and CO (on a dry basis) for each class of combustion equipment?

**6.1.48 Estimate the emissions of each compound produced in cubic meters measured at S.C. per metric ton (1000 kg) of No. 6 fuel oil burned in an oil-fired burner with no emission controls given the following data:

Pollution Emission Factors from the EPA (kg/10^3 L oil)

SO_2	SO_3	NO_2	CO	CO_2	Particulate Matter
19S*	0.69S*	8	0.6	3025	1.5

* S = weight percent sulfur

The No. 6 fuel oil contains 0.84% sulfur and has a specific gravity of 0.86 at 15°C.

**6.1.49 The composition from Perry of No. 6 fuel oil with a specific gravity of 0.86 is in mass percent:

C	87.26
H	10.49
O	0.64
N	0.28
S	0.84
Ash	0.04

Compute the kilograms of each component per 10^3 L of oil, and compare the resulting emissions with those listed in the EPA analysis in problem P6.1.48

***6.1.50 One important source of emissions from gasoline-powered automobile engines that causes smog is the nitrogen oxides NO and NO_2. They are formed whether combustion is complete or not as follows: At the high temperatures that occur in an internal combustion engine during the burning process, oxygen and nitrogen combine to form nitric oxide (NO). The higher the peak temperatures and the more oxygen available, the more NO is formed. There is insufficient time for the NO to decompose back to O_2 and N_2 because the burned gases cool too rapidly during the expansion and exhaust cycles in the engine. Although both NO and nitrogen dioxide (NO_2) are significant air pollutants (together termed NO_x), the NO_2 is formed in the atmosphere as NO is oxidized.

Suppose that you collect a sample of a NO-NO_2 mixture (after having removed the other combustion gas products including N_2, O_2, and H_2O by various separation procedures) in a 100 cm^3 standard cell at 30°C. Certainly some of the NO will have been oxidized to NO_2

$$2NO + O_2 \rightarrow 2NO_2$$

during the collection, storage, and processing of the combustion gases, so measurement of NO alone will be misleading. If the standard cell contains 0.291 g of NO_2 plus NO and the pressure measured in the cell is 170 kPa, what percent of the NO + NO_2 is in the form of NO?

***6.1.51 Ammonia at 100°C and 150 kPa is burned with 20% excess O_2:

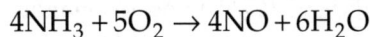

$$4NH_3 + 5O_2 \rightarrow 4NO + 6H_2O$$

The reaction is 80% complete. The NO is separated from the NH_3 and water, and the NH_3 is recycled as shown in Figure P6.1.51.

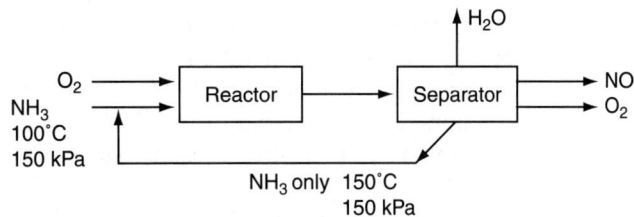

Figure P6.1.51

Calculate the cubic meters of NH_3 recycled at 150°C and 150 kPa per cubic meter of NH_3 fed at 100°C and 150 kPa.

***6.1.52   Benzene (C_6H_6) is converted to cyclohexane (C_6H_{12}) by direct reaction with H_2. The fresh feed to the process is 260 L/min of C_6H_6 plus 950 L/min of H_2 at 100°C and 150 kPa. The single-pass conversion of H_2 in the reactor is 48% while the overall conversion of H_2 in the process is 75%. The recycle stream contains 90% H_2 and the remainder benzene (no cyclohexane). See Figure P6.1.52.

    a. Determine the molar flow rates of H_2, C_6H_6, and C_6H_{12} in the exiting product.

    b. Determine the volumetric flow rates of the components in the product stream if it exits at 100 kPa and 200°C.

    c. Determine the molar flow rate of the recycle stream and the volumetric flow rate if the recycle stream is at 100°C and 100 kPa.

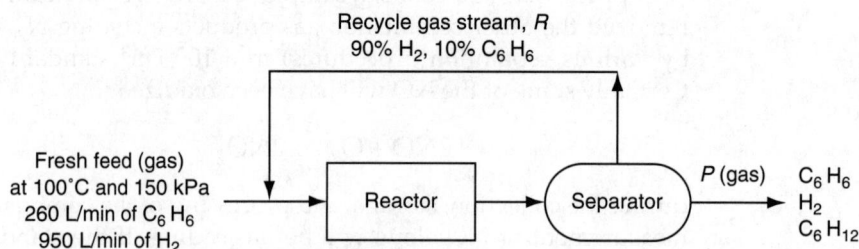

Figure P6.1.52

***6.1.53   Pure ethylene (C_2H_4) and oxygen are fed to a process for the manufacture of ethylene oxide (C_2H_4O):

$$C_2H_4 + \tfrac{1}{2}O_2 \rightarrow C_2H_4O$$

    Figure P6.1.53 is the flow diagram for the process. The catalytic reactor operates at 300°C and 1.2 atm. At these conditions, single-pass measurements on the reactor show that 50% of the ethylene entering the reactor is consumed per pass, and of this, 70% is converted to ethylene oxide. The remainder of the ethylene reacts to form CO_2 and water.

$$C_2H_4 + 3O_2 \rightarrow 2CO_2 + 2H_2O$$

For a daily production of 10,000 kg of ethylene oxide:

    a. Calculate the cubic meters per hour of total gas entering the reactor at S.C. if the ratio of the $O_2(g)$ fed to fresh $C_2H_4(g)$ is 3 to 2.

    b. Calculate the recycle ratio, cubic meters at 10°C and 100 kPa of C_2H_4 recycled per cubic meter at S.C. of fresh C_2H_4 fed.

    c. Calculate the cubic meters of the mixture of O_2, CO_2, and H_2O leaving the separator per day at 80°C and 100 kPa.

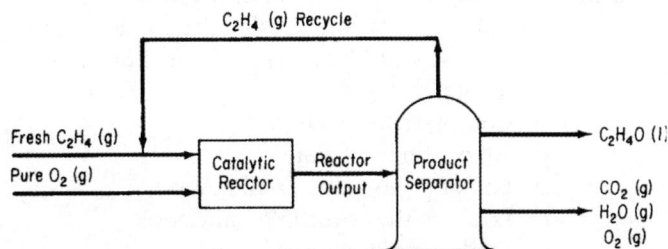

Figure 6.1.53

***6.1.54   An incinerator produces a dry exit gas of the following Orsat composition measured at 60°F and 30 in. Hg absolute: 4.0% CO_2, 26.0% CO, 2.0% CH_4, 16.0% H_2, and 52.0% N_2. A dry natural gas of the following (Orsat) composition—80.5% CH_4, 17.8% C_2H_6, and 1.7% N_2—is used at the rate of 1200 ft³/min at 60°F and 30 in. Hg absolute to burn the incineration off-gas with air. The final products of combustion analyze on a dry basis 12.2% CO_2, 0.7% CO, 2.4% O_2, and 84.7% N_2.

    Calculate (a) the rate of flow in cubic feet per minute of the incinerator exit gas at 60°F and 30 in. Hg absolute on a dry basis, and (b) the rate of airflow in cubic feet per minute, dry, at 80°F and 29.6 in. Hg absolute.

***6.1.55   A gaseous mixture consisting of 50 mol % hydrogen and 50 mol % acetaldehyde (C_2H_4O) is initially contained in a rigid vessel at a total pressure of 760 mm Hg absolute. The formation of ethanol (C_2H_6O) occurs according to

$$C_2H_4 + H_2 \rightarrow C_2H_6O$$

    After a time, it was noted that the total pressure in the rigid vessel had dropped to 700 mm Hg absolute. Calculate the degree of completion of the reaction at that time using the following assumptions: (a) All reactants and products are in the gaseous state, and (b) the vessel and its contents were at the same temperature when the two pressures were measured.

****6.1.56   Biomass ($CH_{1.8}O_{0.5}N_{0.5}$) can be converted to glycerol by anaerobic (in the absence of air) reaction with ammonia and glucose. In one batch of reactants, 52.4 L of CO_2 measured at 300 K and 95 kPa were obtained per mole of glucose in the reactor. The molar stoichiometric ratio of nitrogen produced to ammonia reacted in the reaction equation is 1 to 1, and the mol CO_2/ mol $C_6H_{12}O_6$ = 2.

In gram moles, (a) how much glycerol was produced, and (b) how much biomass reacted to produce the 52.4 L of CO_2?

6.2   Real Gases: Equations of State

6.2.1 You want to obtain an answer immediately as to the specific volume of ethane at 700 kPa and 25°C. List in descending order the techniques you would use with the most preferable one at the top of the list:

a. Ideal gas law
b. Compressibility charts
c. An equation of state
d. Look up the value on the Web
e. Look up the value in a handbook
   Explain your choices.

6.2.2 Which procedure would you recommend to calculate the density of carbon dioxide at 120°F and 1500 psia? Explain your choice.

a. Ideal gas law
b. Redlich-Kwong equation of state
c. Compressibility charts
d. Look up the value on the Web
e. Look up the value in a handbook

6.2.3 Finish the following sentence:
Equations of state are preferred in P-V-T calculations because _____.

6.2.4 Use the Kammerlingh-Onnes virial equation with four terms to answer the following questions for CH_4 at 273 K:

a. Up to what pressure is one term (the ideal gas law) a good approximation?
b. Up to what pressure is the equation truncated to two terms a good approximation?
c. What is the error in using a and using b for CH_4?
   Data: At 273 K the values of the virial coefficients are
$$B = -53.4 \text{ cm}^3/\text{mol}$$
$$C = 2620 \text{ cm}^6/\text{mol}^{-2}$$
$$D = 5000 \text{ cm}^9/\text{mol}^{-3}$$

6.2.5 The Peng-Robinson equation is listed in Table 6.2. What are the units of a, b, and α in the equation if p is in atmospheres, $\hat{V}$ is in liters per gram mole, and T is in kelvin?

6.2.6 The pressure gauge on an O_2 cylinder stored outside at 0°F in the winter reads 1375 psia. By weighing the cylinder (whose volume is 6.70 ft³) you find that the net weight, that is, the O_2, is 63.9 lb. Is the reading on the pressure gauge correct? Use an equation of state to make your calculations.

6.2.7 First commercialized in the 1970s as extractants in "natural" decaffeination processes, SCFs (supercritical fluids)—particularly carbon dioxide and water—are finding new applications, as better, less expensive equipment lowers processing costs, and regulations drive the chemical process industries away from organic solvents.

SCFs' extraction capabilities are now being exploited in a range of new pharmaceutical and environmental applications, while supercritical extraction, oxidation, and precipitation are being applied to waste cleanup challenges.

A compressor for carbon dioxide compresses 2000 m^3/min at 20°C and 500 kPa to 110°C and 4800 kPa. How many cubic meters per minute are produced at the high pressure? Use van der Waals' equation.

***6.2.8   You are asked to design a steel tank in which CO_2 will be stored at 290 K. The tank is 10.4 m^3 in volume, and you want to store 460 kg of CO_2 in it. What pressure will the CO_2 exert? Use the Redlich-Kwong equation to calculate the pressure in the tank. Repeat using the SRK equation. Is there a significant difference in the predictions of pressure between the equations?

***6.2.9   What pressure would be developed if 100 ft^3 of ammonia at 20 atm and 400°F were compressed into a volume of 5.0 ft^3 at 350°F? Use the Peng-Robinson equation to get your answer.

***6.2.10   An interesting patent (U.S. 3,718,236) explains how to use CO_2 as the driving gas for aerosol sprays in a can. A plastic pouch is filled with small compartments containing sodium bicarbonate tablets. Citric acid solution is placed in the bottom of the pouch, and a small amount of carbon dioxide is charged under pressure into the pouch as a starter propellant. As the CO_2 is charged into the pouch, it ruptures the lowest compartment membrane, thus dropping bicarb tablets into the citric acid. That generates more carbon dioxide, giving more pressure in the pouch, which expands and helps push out more product. (The CO_2 does not escape from the can, just the product.)

How many grams of $NaHCO_3$ are needed to generate a residual pressure of 81.0 psig in the can to deliver the very last cubic centimeter of product if the cylindrical can is 8.10 cm in diameter and 17.0 cm high? Assume the temperature is 25°C. Use the Peng-Robinson equation.

***6.2.11   Find the molar volume (in cubic centimeters per gram mole) of propane at 375 K and 21 atm. Use the Redlich-Kwong and Peng-Robinson equations, and solve for the molar volume; the acentric factor for propane to use in the Peng-Robinson equation is 0.1487.

***6.2.12   The tank cited in problem 6.2.8 is constructed and tested, and your boss informs you that you forgot to add a safety factor in the design of the tank. It tests out satisfactorily to 3500 kPa, but you should have added a safety factor of 3 to the design; that is, the tank pressure should not exceed (3500/3) = 1167 kPa, say, 1200 kPa. How many kilograms of CO_2 can be stored in the tank if the safety factor is applied? Use the Redlich-Kwong equation. Hint: Polymath will solve the equation for you.

***6.2.13   A graduate student wants to use van der Waals' equation to express the pressure-volume-temperature relations for a gas. Her project requires a reasonable degree of precision in the p-V-T calculations. Therefore, she made the following experimental measurements with her setup to get an idea of how easy the experiment would be:

Temperature, K	Pressure, atm	Volume, ft³/lb mol
273.1	200	1.860
273.1	1000	0.741

Determine values of constants a and b to be used in van der Waals' equation that best fit the experimental data.

***6.2.14   An 80 lb block of ice is put into a 10 ft³ container and heated to 900 K. What is the final pressure in the container? Do this problem two ways: (a) Use the compressibility factor method and (b) use the Redlich-Kwong equation. Compare your results.

***6.2.15   What weight of ethane is contained in a gas cylinder that is 1.0 ft³ in volume if the gas is at 100°F and 2000 psig? Do this problem two ways: (a) Use van der Waals' equation, and (b) use the compressibility factor method. The experimental value is 21.4 lb.

***6.2.16   Answer the following questions:
   a. Will the constant a in van der Waals' equation be higher or lower for methane than for propane? Repeat for the other van der Waals constant b.
   b. Will the constant a' in the SRK equation be higher or lower for methane than for propane? Repeat for the other SRK constant b.

****6.2.17   A 5 L tank of H_2 is left out overnight in Antarctica. You are asked to determine how many gram moles of H_2 are in the tank. The pressure gauge reads 39 atm gauge and the temperature is −50°C How many gram moles of H_2 are in the tank?

   Use the van der Waals and Redlich-Kwong equations of state to solve this problem. (Hint: A nonlinear-equation-solving program, such as the ones described in section 6.2.2, makes the execution of the calculations quite easy.)

****6.2.18   A 6250 cm³ vessel contains 4.00 g mol of CO_2 at 298.15 K and 14.5 atm. Use the nonlinear-equation-solving program on the CD in the back of the book to solve the Redlich-Kwong equation for the molar volume. Compare the calculated molar volume of the CO_2 in the vessel with the experimental value.

6.3   Real Gases: Compressibility Charts

**6.3.1 Seven pounds of N_2 are stored in a cylinder 0.75 ft^3 volume at 120°F. Calculate the pressure in the cylinder in atmospheres (a) assuming N_2 to be an ideal gas and (b) assuming N_2 is a real gas and using compressibility factors.

**6.3.2 Two gram moles of ethylene (C_2H_4) occupy 418 cm^3 at 95°C. Calculate the pressure. (Under these conditions ethylene is a nonideal gas.) Data: $T_c = 283.1$ K, $p_c = 50.5$ atm.

**6.3.3 The critical temperature of a real gas is known to be 500 K, but its critical pressure is unknown. Given that 3 lb mol of the gas at 252°C occupy 50 ft^3 at a pressure of 463 psia, estimate the critical pressure.

**6.3.4 The volume occupied by 1 lb of n-octane at 27 atm is 0.20 ft^3. Calculate the temperature of the n-octane.

**6.3.5 A block of dry ice weighing 50 lb was dropped into an empty steel tank, the volume of which was 5.0 ft^3. The tank was heated until the pressure gauge read 1600 psi. What was the temperature of the gas? Assume all of the CO_2 became gas.

**6.3.6 A cylinder containing 10 kg of CH_4 exploded. It had a bursting pressure of 14,000 kPa gauge and a safe operating pressure of 7000 kPa gauge. The cylinder had an internal volume of 0.0250 m^3. Calculate the temperature when the cylinder exploded.

**6.3.7 A cylinder has a volume of 1.0 ft^3 and contains dry methane at 80°F and 200 psig. What weight of methane (CH_4) is in the cylinder? The barometric pressure is 29.0 mm Hg.

**6.3.8 How many kilograms of CO_2 can be put into a 25 L cylinder at room temperature (25°C) and 200 kPa absolute pressure?

**6.3.9 A natural gas composed of 100% methane is to be stored in an underground reservoir at 1000 psia and 120°F. What volume of reservoir is required for 1,000,000 ft^3 of gas measured at 60°F and 14.7 psia?

**6.3.10 Calculate the specific volume of propane at a pressure of 6000 kPa and a temperature of 230°C.

**6.3.11 State whether the following gases can be treated as ideal gases in calculations:
  a. Nitrogen at 100 kPa and 25°C
  b. Nitrogen at 10,000 kPa and 25°C
  c. Propane at 200 kPa and 25°C
  d. Propane at 2000 kPa and 25°C
  e. Water at 100 kPa and 25°C
  f. Water at 1000 kPa and 25°C
  g. Carbon dioxide at 1000 kPa and 0°C
  h. Propane at 400 kPa and 0°C

**6.3.12 One gram mole of chlorobenzene (C_6H_5Cl) just fills a tank at 230 kPa and 380 K. What is the volume of the tank?

**6.3.13 You have been asked to settle an argument. The argument concerns the maximum allowable working pressure (MAWP) permitted in an A1 gas cylinder. One of your coworkers says that calculating the pressure in a tank via the ideal gas law is best because it gives a conservative (higher) value of the pressure than can actually occur in the tank. The other coworker says that everyone knows the ideal gas law should not be used to calculate real gas pressures as it gives a lower value than the true pressure. Which coworker is correct?

***6.3.14 A size A1 cylinder of ethylene (T_c = 9.7°C) costs $45.92 FOB New Jersey. The outside cylinder dimensions are 9 in. diameter, 52 in. high. The gas is 99.5% (minimum) C_2H_4, and the cylinder charge is $44.00. Cylinder pressure is 1500 psig, and the invoice says it contains "165 cu.ft." of gas. An identical cylinder of CP-grade methane at a pressure of 2000 psig is 99.0% (minimum) CH_4 and costs $96.00 FOB Illinois. The CH_4 cylinder contains "240 cu.ft." of gas. The ethylene cylinder is supposed to have a gross weight (including cylinder) of 163 lb, while the CH_4 cylinder has a gross weight of 145 lb. Answer the following questions:
   a. What do the "165 cu.ft." and "240 cu.ft." of gas probably mean? Explain with calculations.
   b. Why does the CH_4 cylinder have a gross weight less than the C_2H_4 cylinder when it seems to contain more gas? Assume the cylinders are at 80°F.
   c. How many pounds of gas are actually in each cylinder?

***6.3.15 Safe practices in modern laboratories call for placing gas cylinders in hoods or in utility corridors. In case of leaks, a toxic gas can be properly taken care of. A cylinder of CO that has a volume of 175 ft³ at 1 atm and 25°C is received from the distributor of gases on Friday with a gauge reading of 2000 psig and is placed in the utility corridor. On Monday when you are ready to use the gas, you find the gauge reads 1910 psig. The temperature has remained constant at 76°F, as the corridor is air-conditioned, so you conclude that the tank has leaked CO (which does not smell).
   a. What has been the leak rate from the tank?
   b. If the tank was placed in a utility corridor whose volume is 1600 ft³, what would be the minimum time that it would take for the CO concentration in the hallway to reach the threshold limit value ceiling (TLV-C) of 100 ppm set by the state Air Pollution Control Commission if the air conditioning did not operate on the weekend?
   c. In the worst case, what would be the concentration of CO in the corridor if the leak continued from Friday, 3 PM, to Monday, 9 AM?
   d. Would either case b or c occur in practice? Why or why not?

***6.3.16** Levitating solid materials during processing is the best way known to ensure their purity. High-purity materials, which are in great demand in electronics, optics, and other areas, usually are produced by melting a solid. Unfortunately, the containers used to hold the material also tend to contaminate it. And heterogeneous nucleation occurs at the container walls when molten material is cooled. Levitation avoids these problems because the material being processed is not in contact with the container.

Electromagnetic levitation requires that the sample be electrically conductive, but with a levitation method based on buoyancy, the density of the material is the only limiting factor.

Suppose that a gas such as argon is to be compressed at room temperature so that silicon (sp. gr. 2.0) just floats in the gas. What must the pressure of the argon be? If you wanted to use a lower pressure, what different gas might be selected? Is there a limit to the processing temperature for this manufacturing strategy?

***6.3.17** While determining the temperature that occurred in a fire in a warehouse, the arson investigator noticed that the relief valve on a methane storage tank had popped open at 3000 psig, the rated value. Before the fire started, the tank was presumably at ambient conditions, about 80°F, and the gauge read 1950 psig. If the volume of the tank was 240 ft³, estimate the temperature during the fire. List any assumptions you make.

6.4 Real Gas Mixtures

6.4.1 A gas has the following composition:

CO_2	10%
CH_4	40%
C_2H_4	50%

It is desired to distribute 33.6 lb of this gas per cylinder. Cylinders are to be designed so that the maximum pressure will not exceed 2400 psig when the temperature is 180°F. Calculate the volume of the cylinder required by Kay's method.

6.4.2 A gas composed of 20% ethanol and 80% carbon dioxide is at 500 K. What is its pressure if the volume per gram mole is 180 cm³/g mol?

**6.4.3   A sample of natural gas taken at 3500 kPa absolute and 120°C is separated by chromatography at standard conditions. It was found by calculation that the grams of each component in the gas were as follows:

Component	g
Methane (CH_4)	100
Ethane (C_2H_6)	240
Propane (C_3H_8)	150
Nitrogen (N_2)	50
Total	540

What was the density of the original gas sample?

**6.4.4   A gaseous mixture has the following composition (in mole percent):

C_2H_4	57
Ar	40
He	3

at 120 atm pressure and 25°C. Compare the experimental volume of 0.14 L/g mol with that computed by Kay's method.

***6.4.5   You are in charge of a pilot plant using an inert atmosphere composed of 60% ethylene (C_2H_4) and 40% argon (Ar). How big a cylinder (or how many) must be purchased if you are to use 300 ft^3 of gas measured at the pilot plant conditions of 100 atm and 300°F? Buy the cheapest array.

Cylinder Type	Cost	Pressure (psig)	lb Gas
1A	$52.30	2000	62
2	42.40	1500	47
3	33.20	1500	35

State any additional assumptions. You can buy only one type of cylinder.

***6.4.6   A feed for a reactor has to be prepared composed of 50% ethylene and 50% nitrogen. One source of gas is a cylinder containing a large amount of gas with the composition 20% ethylene and 80% nitrogen. Another cylinder that contains pure ethylene at 1450 psig and 70°F has an internal volume of 2640 in^3. If all the ethylene in the latter cylinder is used up in making the mixture, how much reactor feed was prepared and how much of the 20% ethylene mixture was used?

***6.4.7   A gas is flowing at a rate of 100,000 scfh (standard cubic feet per hour). What is the actual volumetric gas flow rate if the pressure is 50 atm and the temperature is 600°R? The critical temperature is 40.0°F, and the critical pressure is 14.3 atm.

***6.4.8   A steel cylinder contains ethylene (C_2H_4) at 200 psig. The cylinder and gas weigh 222 lb. The supplier refills the cylinder with ethylene until the pressure reaches 1000 psig, at which time the cylinder and gas weigh 250 lb. The temperature is constant at 25°C. Calculate the charge to be made for the ethylene if the ethylene is sold at $0.41 per pound, and what the weight of the cylinder is for use in billing the freight charges. Also find the volume of the empty cylinder in cubic feet.

****6.4.9   In a high-pressure separation process, a gas having a *mass* composition of 50% benzene, 30% toluene, and 20% xylene is fed into the process at the rate of 483 m^3/hr at 607 K and 26.8 atm. One exit stream is a vapor containing 91.2% benzene, 7.2% toluene, and 1.6% xylene. A second exit stream is a liquid containing 6.0% benzene, 9.0% toluene, and 85.0% xylene.

What is the composition of the third exit stream if it is liquid flowing at the rate of 9800 kg/hr, and the ratio of the benzene to the xylene in the stream is 3 kg benzene to 2 kg xylene.

CHAPTER 7

Multiphase Equilibrium

Chapter Objectives

- Recognize the connection between multiphase equilibrium and separation technology.
- Understand phase diagrams and the associated terminology as well as the phase rule.
- Determine the vapor pressure of a pure component and to use it to determine the degree of vaporization into or condensation from a condensable gas.
- Understand vapor-liquid equilibrium for a binary system.

In this chapter, we introduce separation technology, which is used extensively in the process industries. Phase diagrams and the phase rule are first introduced, then the characteristics of a variety of systems of single-component two-phase systems, concluding with a discussion of the equilibrium of two-phase systems.

7.1   Introduction

The most common pieces of equipment in the process industries are separation devices, which remove one or more components from a stream and concentrate them in another stream. Mixing of components occurs regularly in nature (e.g., minerals dissolve in rainwater as the water flows down a creek bed), but to separate components requires separation equipment that uses energy and materials to accomplish the separation. It is well known that the value of products can greatly increase when the key component in a product is taken from a dilute solution by a separation device and transformed into a highly concentrated form (e.g., pharmaceutical products). Therefore, separation technologies can provide significant economic advantages for processing companies. This chapter deals with the description of multiphase systems that are used in the development and design of various types of separation systems. Examples of the application of separation systems include

- **Drinking water from seawater:** One way drinking water can be produced from seawater is by boiling the seawater to produce water vapor, which is then condensed, yielding drinking water. This process of boiling and condensing is a simple example of **distillation.**

- **Gasoline from crude oil:** Part of the gasoline produced by an oil refinery comes directly from the distillation of crude oil. The crude is distilled into a number of products, each with a different boiling point range; one of these products is gasoline.

- **Removal of pollutants from effluent streams:** Plants that discharge water and gas streams into the environment are required to reduce the concentration of pollutants to specified levels, a step that usually requires the application of a separation system(s). As an example, when coal is burned in a power plant, SO_2 is produced from the sulfur in the coal. SO_2 in the atmosphere is converted to sulfite, which forms acid rain. Therefore, coal-fired power plants are required to remove SO_2 from their flue gas (i.e., the combustion gases after most of the thermal energy has been removed) before discharging it to the atmosphere. Many power plants pass their flue gas through a scrubbing process, which exposes the flue gas to a lime-water mixture to absorb the SO_2. The column that accomplishes this contacting for SO_2 removal is known as an **absorber** because the lime-water mixture absorbs the SO_2 from the flue gas.

- **Pharmaceuticals:** Certain prescription drugs are produced by concentrating a dilute solution of the desired product, which was produced in a bioreactor, using an **extraction process.** An extraction process uses a liquid that has a much greater affinity for the desired product than the components in the reactor effluent. Thus, an extraction process is able to produce a very nearly pure product.

- **Typical chemical plant:** In a typical chemical plant, a reactor produces a mixture of products and unconverted feed that is fed to a separation train (i.e., a series of separation equipment) that concentrates the products into salable form and returns the unreacted feed to the reactor. For most chemical plants, the separation train is primarily composed of distillation columns with some absorbers and extractors.

7.2   Phase Diagrams and the Phase Rule

You can conveniently display the properties of compounds via phase diagrams. A pure substance can exist in many phases simultaneously of which, as you know, solid, liquid, and gas are the most common. Phase diagrams enable you to view the properties of two or more phases as functions of temperature, pressure, specific volume, concentration, and other variables.

We discuss phase diagrams in terms of water because presumably you are familiar with the three phases of water, namely, ice, water, and water vapor (steam), but the discussion applies to all other pure substances. The terms *vapor* and *gas* are used very loosely in practice. A gas that exists below its critical temperature is usually called a vapor because it can condense. We reserve the word **vapor** to describe a gas below its critical point in a process in which the phase change is of primary interest, while we use **gas** and **noncondensable gas** to describe, respectively, a gas above the critical point and a gas in a process at conditions under which it cannot condense.

Phase diagrams are based on equilibrium conditions. That is, for phase equilibrium, it is assumed that each phase remains invariant (i.e., constant quantity under constant conditions). On a molecular level, when two or more phases are present, there will always be molecules that move from one phase to another, but under phase equilibrium, the net flux is zero. For example, for a liquid and a vapor in phase equilibrium, the flux of molecules from the liquid into the vapor must be equal to the flux from the vapor to the liquid. In fact, when multiple phases exist, continuous exchange between phases occurs, even at equilibrium.

Suppose you carry out some experiments with the apparatus shown in Figure 7.1. Place a lump of ice in the chamber below the piston, and evacuate the chamber to remove all air (you want to retain only pure water in the chamber). Fix the volume of the chamber by fixing the position of the piston, and start slowly (so that the phases of water that result will be in equilibrium) heating the ice. If you plot the measured pressures as a function of temperature, you will get Figure 7.2, a phase diagram in which all of the measurements made have been fitted by a continuous smooth curve for clarity.

The initial conditions of p and T in the chamber are at point O in Figure 7.2 with the solid in equilibrium with the vapor.

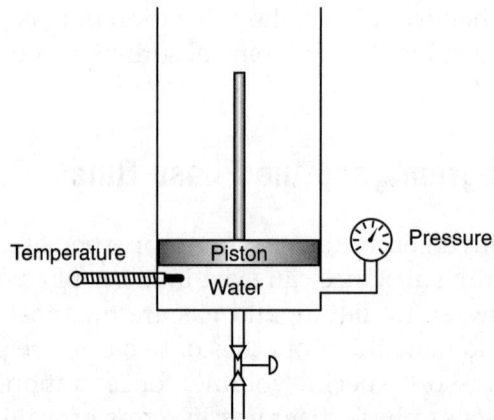

Figure 7.1   Apparatus used to explore the p, $\hat{V}$, and T properties of water

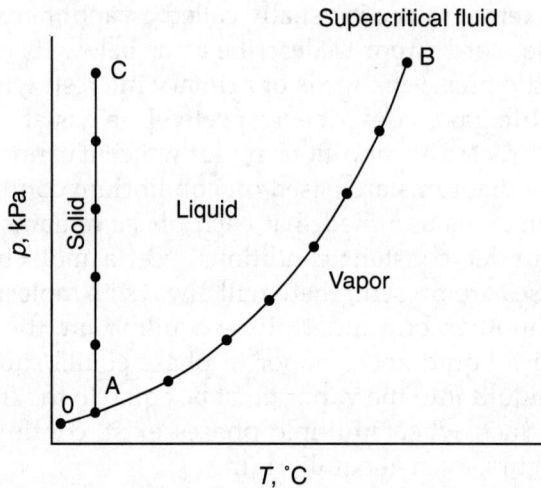

Figure 7.2   Results of the experiment of heating at constant volume shown on a phase diagram (p versus T at constant $\hat{V}$)

As you raise the temperature, the ice would start to melt at point A, the **triple point**, the one p-$\hat{V}$-T combination at which solid, liquid, and vapor can be in equilibrium. Further increase in the temperature causes the ice to abruptly melt before forming water vapor and the pressure to rise, which is indicated by the curve AB. B is the critical point at which vapor and liquid properties become the same.

If you had kept the temperature almost constant and raised the pressure on the ice, ice would still exist and be in equilibrium with liquid water along the line AC. The line AC is so vertical that **you can use the saturated liquid properties for the properties of the compressed liquid**. Ice skating is possible because the high pressure exerted by the thin blade on ice forms a liquid layer with low friction on the blade.

If the vapor and liquid of a pure component are in **equilibrium,** the **equilibrium pressure is called the vapor pressure**, which we will denote by p^*. **At a given temperature, there is only one pressure at which the liquid and vapor phases of a pure substance may exist in equilibrium.** Either phase alone may exist, of course, over a wide range of conditions.

We next take up some terminology associated with processes that are conveniently represented on a p^*-T phase chart such as Figure 7.3 (in the definitions of terms that follow, the letters in parentheses refer to the corresponding process denoted in Figure 7.3 by the same sequence of letters):

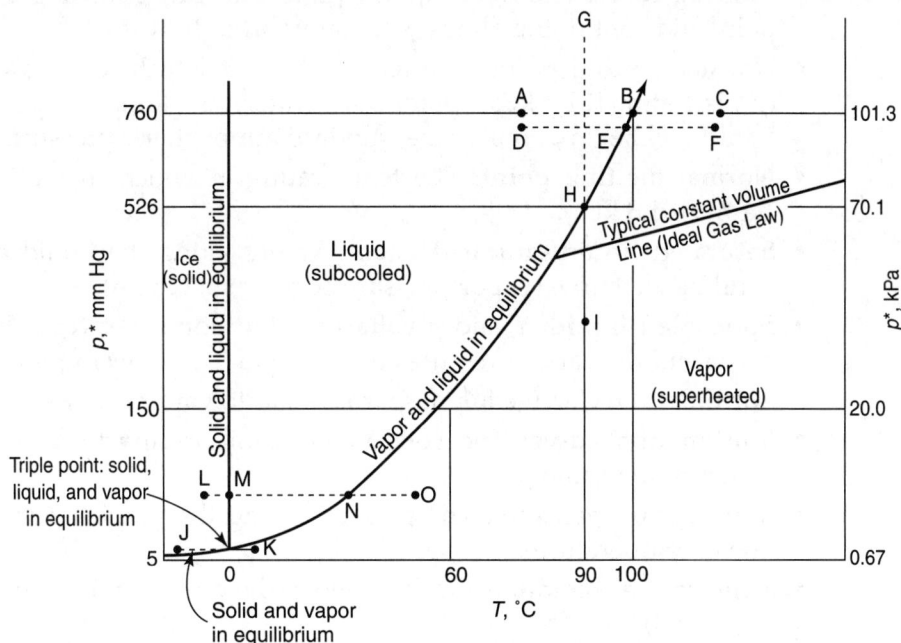

Figure 7.3   Various common processes as represented on a p^*-T diagram

- **Boiling:** The change of phase from liquid to vapor (e.g., B, E, N; note that because boiling occurs at a constant temperature and pressure, the process of boiling appears as a point in a p^*-T diagram).

- **Bubble point:** The temperature at which a liquid just starts to vaporize (N, H, and E are examples).
- **Condensation:** The change of phase from vapor to liquid (e.g., N, E, B; note that because condensation occurs at a constant temperature and pressure, the process of condensation appears as a point in a p-T diagram).
- **Dew point:** The temperature at which the vapor just begins to condense at a specified pressure, namely, temperature values on the horizontal axis read from the vapor pressure curve (N, H, and E are examples).
- **Evaporation:** The change of phase from liquid to vapor (e.g., D to F, A to C, or M to O).
- **Freezing (solidifying):** The change of phase from liquid to solid (N to L).
- **Melting (fusion):** The change in phase from solid to liquid (L to M; similarly to boiling, the process of melting or fusion appears as a single point in a p-T diagram).
- **Melting curve:** The solid-liquid equilibrium curve starting at the triple point and continuing almost vertically through M.
- **Normal boiling point:** The temperature at which the vapor pressure (p^*) is 1 atm (101.3 kPa) (point B for water); the temperature at which a liquid will begin to boil at the standard atmospheric pressure.
- **Normal melting point:** The temperature at which the solid melts at 1 atm (101.3 kPa).
- **Saturated liquid/saturated vapor:** Values along the liquid and vapor equilibrium curve (vapor-pressure curve, e.g., N to B).
- **Subcooled liquid:** T and p values for the liquid between the melting curve and the vapor-pressure curve (liquid D is an example).
- **Sublimation:** Change in phase from solid to vapor (J to K).
- **Sublimation curve:** The solid-vapor equilibrium curve from J (and lower) to the triple point.
- **Sublimation pressure:** The pressure along the melting curve (a function of temperature).
- **Supercritical region:** p-T values above the critical point (not shown in Figure 7.3).
- **Superheated vapor:** Values of vapor at temperatures and pressure exceeding those at saturation; I is an example. The **degrees of superheat** are the differences in temperature between the actual T and the saturated T at the given pressure. For example, steam at 500°F and 100 psia (the saturation temperature for 100 psia is 327.8°F) has $(500 - 327.8) = 172.2$°F of superheat.
- **Vaporization:** The change of phase from liquid to vapor (for example, D to F).

In Figure 7.3, the process of evaporation (A to C) and condensation (C to A) of water at 1 atm is represented by the line ABC with the phase transformation occurring at 100°C. Suppose that you went to the top of Pikes Peak and repeated the process of evaporation and condensation in the open air. What would happen then? The process would be the same (points DEF) with the exception of the temperature and pressure at which the water would begin to boil, or condense. Since the pressure of the atmosphere at the top of Pikes Peak is lower than 101.3 kPa, the water would start to boil at a lower temperature. Some unfortunate consequences might result if you expected to kill certain types of disease-causing bacteria by boiling the water! In addition, it will take longer to cook rice at that elevation due to the lower boiling point for water at higher elevations.

To conclude, *at equilibrium* you can see that (a) at any given temperature water exerts its unique vapor pressure; (b) as the temperature goes up, the vapor pressure goes up, and vice versa; and (c) it makes no difference whether water vaporizes into air, into a cylinder closed by a piston, into an evacuated cylinder, or into the atmosphere; at any temperature it still exerts the same vapor pressure as long as the liquid water is in equilibrium with its vapor.

A pure compound can change phase at constant volume from a liquid to a vapor, or the reverse, via a constant temperature process as well as a constant pressure process. A process of **vaporization,** or **condensation, at constant temperature** is illustrated by the lines GHI or IHG, respectively, in Figure 7.3. Water would vaporize or condense at constant temperature as the pressure reached point H on the vapor-pressure curve. The change that occurs at H is the increase or decrease in the fraction of vapor, or liquid, respectively, at the fixed temperature. The pressure does not change until all of the vapor, or liquid, has completed the phase transition.

Now let's go back to the experimental apparatus and collect data to prepare a p-$\hat{V}$ phase chart. This time you want to hold the temperature in the chamber constant and adjust the volume while measuring the pressure. Start with compressed liquid water (subcooled water) rather than ice, and raise the piston so that water eventually vaporizes. Figure 7.4 illustrates by dashed lines the measurements for two different temperatures, T_1 and T_2. As the pressure is reduced at constant T_1, $\hat{V}$ increases very slightly (liquids are not very compressible) until the liquid pressure reaches p^*, the vapor pressure, at point A.

Then, as the piston still rises (i.e., as $\hat{V}$ increases), both the pressure and temperature remain constant until all of the liquid is vaporized by point B on the saturated vapor line. Subsequently, starting from point B, as the pressure reduces, the value of $\hat{V}$ can be calculated via an ideal or real gas equation. Compression at constant T_2 is just a reversal of the process at T_1. The dots in Figure 7.4 represent just the measurements made when saturation of liquid and vapor coexist and are deemed to form the envelope for the **two-phase** region that from a different angle appears in Figures 7.2 and 7.3 as the vapor-pressure curve. The two-phase region (e.g., A to B or D to C) represents the conditions under which liquid and vapor can exist at equilibrium. Note from Figure 7.4 the discontinuous change in the specific volume in going from a liquid to a solid at the triple point. In other words, water expands when it freezes, and this is why ships trapped in the polar ice can be crushed by the force of the expanding ice. By comparing Figures 7.3 and 7.4, you can see that lines AB and CD in the p-$\hat{V}$ phase diagram (Figure 7.4) correspond to a single point each in the p-T diagram (Figure 7.3).

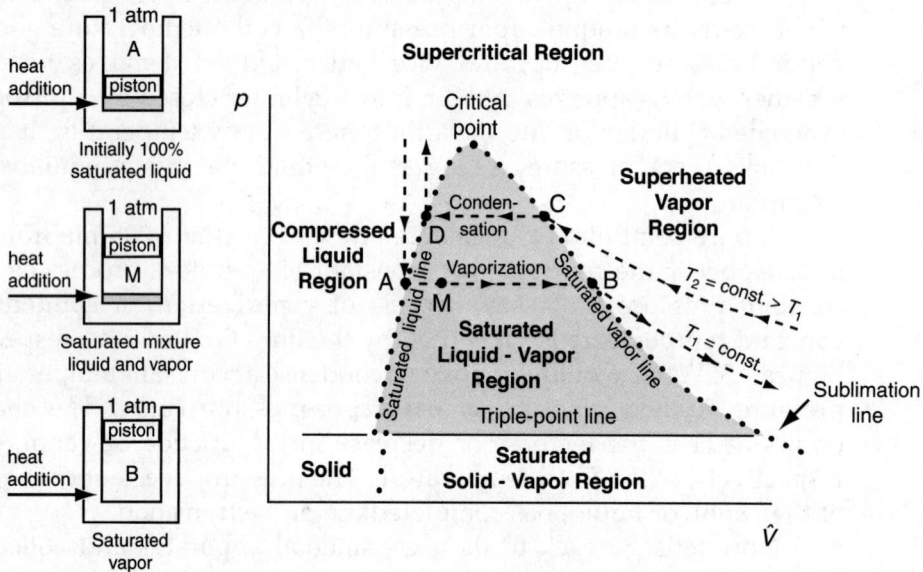

Figure 7.4   Experiments to obtain a p-$\hat{V}$ phase diagram. The dashed lines are measurements made at constant temperatures T_1 and T_2. The dots represent the points at which vaporization, or condensation, respectively, of the saturated liquid, or vapor, occurs; they form an envelope about the two-phase region.

Figure 7.4 involves a new term, **quality**, the fraction or percent of the total vapor and liquid mixture that is vapor (wet vapor). Examine Figure 7.5. You can calculate the volume of the liquid-vapor mixture at B in Figure 7.5 by adding a volume fraction of material that is saturated liquid to the volume fraction that is saturated vapor:

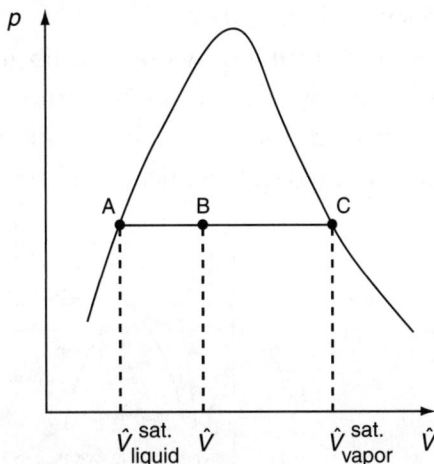

Figure 7.5   Representation of quality on a p-$\hat{V}$ phase diagram. A is saturated liquid, and C is saturated vapor. The compound at B is part liquid and part vapor, and the fraction vapor is called the quality.

$$\hat{V} = (1-x)\hat{V}_{\text{liquid}}^{\text{sat.}} + x\hat{V}_{\text{vapor}}^{\text{sat.}} \tag{7.1}$$

where x is the fractional quality. Solving for x yields

$$x = \frac{\hat{V} - \hat{V}_{\text{liquid}}^{\text{sat.}}}{\hat{V}_{\text{vapor}}^{\text{sat.}} - \hat{V}_{\text{liquid}}^{\text{sat.}}}$$

That is, by examining the location of $\hat{V}$ in relation to $\hat{V}_{\text{vapor}}^{\text{sat.}}$ and $\hat{V}_{\text{liquid}}^{\text{sat.}}$, you can determine the quality.

Figures 7.3 and 7.4 can be reconciled by looking at the three-dimensional surface that illustrates the p-$\hat{V}$-T (see Figure 7.6).

You can see that vapor pressure is the two-dimensional projection, yielding a curve, of a three-dimensional surface into the p-T plane. Note that the vapor-pressure curve in a p-T plane is actually a surface in the three-dimensional representation because a vapor and liquid at equilibrium are at the same temperature (see Figure 7.6). Figure 7.3 thus proves to be a portion of the complete region shown in Figure 7.6.

Now we consider the phase rule, which defines key relationships between the phases in a phase diagram. The phase rule pertains only to systems at equilibrium. Equilibrium means

- A state of absolute rest

- No tendency to change state
- No processes operating (physical equilibrium)
- No fluxes of energy, mass, or momentum
- No temperature, pressure, or concentration gradients
- No reactions occurring (chemical equilibrium)

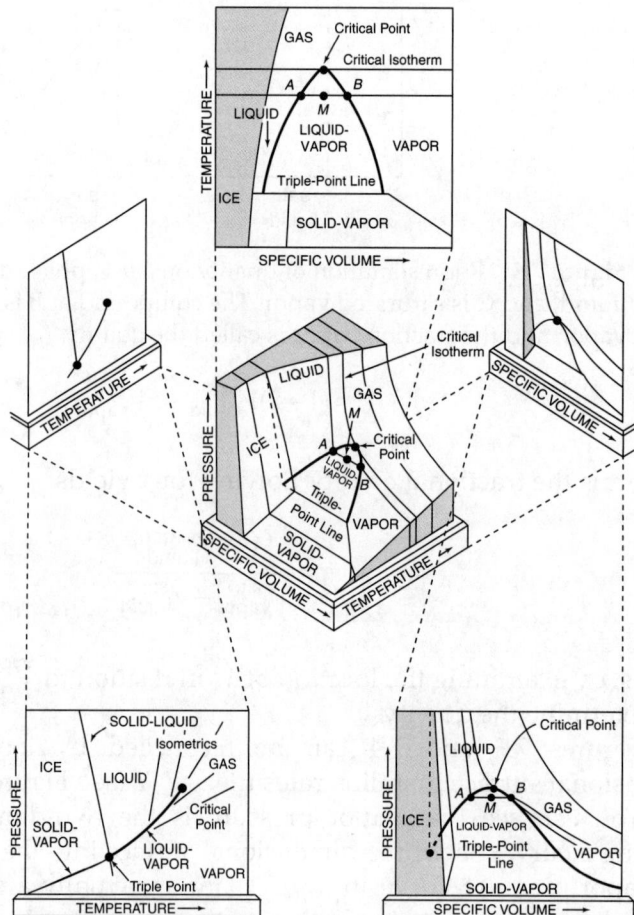

Figure 7.6   The p-$\hat{V}$-T surface for water (a compound that expands on freezing) in three dimensions showing also two-dimensional projections for sequential pairs of the three variables

Thus, **phase equilibrium** means that the phases present in a system are invariant as are the phase properties. By **phase**, we mean a part of a system that is *chemically and physically* uniform throughout. This definition does not necessarily imply that a phase is continuous. For example, ice cubes in water represent a system that consists of two phases. The important concept of phase for you to retain is that a gas and liquid at equilibrium can each be treated as having a uniform domain. Each ice cube is chemically and physically the same; hence, all the cubes are considered to make up one phase. The decision about whether a solid is one or more phases is not always clear.

If you mechanically mix table salt and sugar, you have a solid system, but it consists of two distinct solid phases. Small particles of one phase are intermingled with small particles of the other. Particles of sugar are not the same chemically as those of salt, even though they may appear to be the same physically. On the other hand, it should be emphasized here that most gases and liquids at equilibrium can be assumed to be uniform.

The phase rule is concerned only with the **intensive** properties of the system. By intensive, we mean **properties that do not depend on the quantity of material present**. If you think about the properties we have employed so far in this book, do you get the feeling that pressure and temperature are independent of the amount of material present? Concentration is an intensive variable, but what about volume? The total volume of a system is called an **extensive variable** because it does depend on how much material you have; on the other hand, the **specific volume** (the volume per mass) or the **density** (mass per volume) are **intensive properties** because they are not independent of the amount of material present. You should remember that the specific (per unit mass) values are intensive properties; the total quantities are extensive properties. Furthermore, the state of a system is specified by the intensive variables, not the extensive ones.

You will find **Gibbs phase rule** to be a useful guide in establishing how many intensive properties, such as pressure and temperature, have to be specified to definitely fix all of the remaining intensive properties and number of phases that can coexist for any physical system. **The rule can be applied only to systems in equilibrium** and is given by Equation (7.2), assuming that **no chemical reaction occurs**:

$$F = 2 - P + C \qquad\qquad (7.2)$$

where

$F =$ number of degrees of freedom (i.e., the number of independent properties that have to be specified to determine all of the intensive properties of each phase of the system of interest)—*not to be confused* with the degrees of freedom calculated in solving material balances that can involve both intensive *and* extensive variables.

P = number of phases that exist in the system; a phase is a homogeneous quantity of material such as a gas, a pure liquid, a solution, or a homogeneous solid.

C = number of independent components (chemical species) in the system.

Let's look at Figure 7.7, which shows the surface of part of Figure 7.6. Consider the vapor phase.

You will remember for a pure gas that we had to specify three of the four variables in the ideal gas equation $pV = nRT$ in order to be able to determine the remaining one unknown. You might conclude that $F = 3$. If we apply the phase rule, for a single phase $P = 1$ and for a pure gas $C = 1$.

$$F = 2 - P + C = 2 - 1 + 1 = 2 \text{ variable to be specified}$$

How can we reconcile this apparent paradox with our previous statement? Easily! Since the phase rule is concerned with *intensive properties* only, the following are the phase rule variables to be included in the ideal gas law:

$$\left. \begin{array}{l} p \\ \hat{V}\text{(specific molar volume)} \\ T \end{array} \right\} 3 \text{ intensive properties}$$

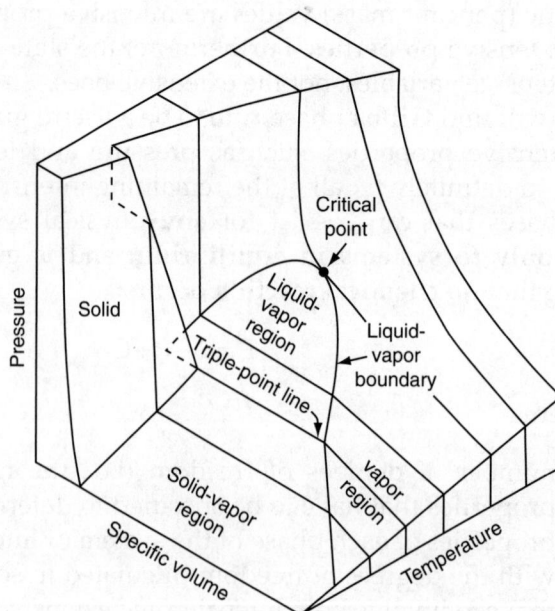

Figure 7.7   The surface of the solid-liquid-vapor phases of water with the coordinates of p, $\hat{V}$, and T

Thus, the ideal gas law would be written

$$p\hat{V} = RT \qquad (7.3)$$

and in this form, you can see that when two intensive variables are specified ($F = 2$), the third can be calculated. Thus, in the superheated region in the steam tables, you can fix all of the properties of the water vapor by specifying two intensive variables.

An **invariant** system is one in which no variation of conditions is possible without one phase disappearing. In Figure 7.7, a system that is composed of ice, water, and water vapor exists at only one temperature (0.01°C) and pressure (0.611 kPa), namely, along the triple-point line (a point in a p-T diagram), and represents one of the invariant states in the water system:

$$F = 2 - P + C = 2 - 3 + 1 = 0$$

With all three phases present, none of the physical conditions of p, T, or $\hat{V}$ can be varied without one phase disappearing. As a corollary, if the three phases are present, the temperature, the specific volume, and so on, must always be fixed at the same values. This phenomenon is useful in calibrating thermometers and other instruments. Now let's look at some examples of the application of the phase rule.

Example 7.1   Application of the Phase Rule

Problem Statement

Calculate the number of degrees of freedom (how many additional intensive variables must be specified to fix the system) from the phase rule for the following materials at equilibrium:

  a. Pure liquid benzene

  b. A mixture of ice and water only

  c. A mixture of liquid benzene, benzene vapor, and helium gas

  d. A mixture of salt and water designed to achieve a specific vapor pressure

What variables might be specified in each case?

(Continues)

Example 7.1   Application of the Phase Rule (*Continued*)
Solution

a. $P = 1$, and $C = 1$; hence, $F = 2 - 1 + 1 = 2$. The temperature and pressure might be specified in the range in which benzene remains a liquid.

b. $P = 2$, and $C = 1$; hence, $F = 2 - 2 + 1 = 1$. Once either the temperature or the pressure is specified, the other intensive variables are fixed.

c. $P = 2$, and $C = 2$; hence, $F = 2 - 2 + 2 = 2$. A pair from temperature, pressure, or mole fraction can be specified.

d. $P = 2$, and $C = 2$; hence, $F = 2 - 2 + 2 = 2$. Since a particular pressure is to be achieved, you would adjust the salt concentration and the temperature of the solution.

Note that in a and b it is likely that a vapor phase would exist in practice, increasing P by 1 and reducing F by 1.

Self-Assessment Test

Questions

1. Why does dry ice sublime at room temperature and pressure?

2. List two intensive and two extensive properties.

3. Indicate whether the following statements are true or false:
   a. A phase is an agglomeration of matter having distinctly identifiable properties such as a distinct refractive index, viscosity, density, X-ray pattern, and so on.
   b. A solution containing two or more compounds constitutes a single phase.
   c. A mixture of real gases constitutes a single phase.

4. Fill in the following table for water:

Number of Phases P	Example	Degrees of Freedom F	Number of Variables That Can Be Adjusted at Equilibrium
1	Steam		
2	Steam and water		
3	Steam, water, and ice		

Answers

1. The point representing ambient temperature and pressure falls below the liquid region, where only solid and vapor exist in equilibrium.

2. Intensive: any p, T, c, ρ, etc.; extensive: any of V, m, n, etc.

3. All true.

4. Number of phases = 1: 2 DOF; number of phases = 2: 1 DOF; number of phases = 3: 0 DOF. Note that the number of DOF is equal to the number of variables that can be adjusted at equilibrium.

Problems

1. Determine the number of degrees of freedom from the phase rule for the following systems at equilibrium:
   a. Liquid water, water vapor, and nitrogen
   b. Liquid water with dissolved acetone in equilibrium with their vapors
   c. $O_2(g)$, $CO(g)$, $CO_2(g)$, and $C(s)$ at high temperature

2. A tank contains 1000 kg of acetone (C_3H_6O), half of which is liquid and the other half of which is in the vapor phase. Acetone vapor is withdrawn slowly from the tank, and a heater in each phase maintains the temperature of each of the two phases at 50°C. Determine the pressure in the tank after 100 kg of vapor have been withdrawn.

3. Draw a p-T phase diagram for water. Label the following clearly: vapor-pressure curve, dew point curve, saturated region, superheated region, subcooled region, and triple point. Show the processes of evaporation, condensation, and sublimation by arrows.

Answers

1. a. $F = 2 - 2 + 2 = 2$; **b.** $F = 2 - 2 + 2 = 2$; **c.** $F = 2 - 2 + 4 = 4$

2. If equilibrium is maintained, the pressure is the vapor pressure of acetone at 50°c (i.e., 610 mm hg).

3. See Figures 7.2 and 7.3.

7.3   Single-Component Two-Phase Systems (Vapor Pressure)

You can understand the behavior of single-component two-phase systems by examining the phase diagram of the component of interest. For example, consider the p^*-versus-T diagram (at constant $\hat{V}$) for water shown in Figure 7.3. The relationship between temperature and pressure for steam and liquid water phases in equilibrium is represented by the line from the triple point up to point B. In the remainder of this section, we explain how to determine values for the vapor pressure given the temperature, or the temperature given the vapor pressure.

7.3.1   Prediction via Equations

You can see from Figure 7.2 (line AB) that the function of p^* versus T is not a linear function (except as an approximation over a very small temperature range). Many functional forms have been proposed to predict p^* from T, but we use the **Antoine equation** in this text because it has sufficient accuracy for our needs, and coefficients for the equation exist in the literature for over 5000 compounds:

$$\ln(p^*) = A - \frac{B}{C+T} \qquad (7.4)$$

where A, B, C = constants for each substance, and T = temperature, kelvin.

Refer to Appendix F (available online) for the values of A, B, and C for various compounds.

You can estimate the values of A, B, and C in Equation (7.4) from experimental data by using a regression program such as found in MATLAB and Python. With just three experimental values for the vapor pressure versus temperature, you can fit Equation (7.4), but more values are better!

Example 7.2   Vaporization of Metals for Thin Film Deposition

Problem Statement

Three methods of providing vaporized metals for thin film deposition are evaporation from a boat, evaporation from a filament, and transfer via an electronic beam. Figure E7.2 illustrates evaporation from a boat placed in a vacuum chamber.

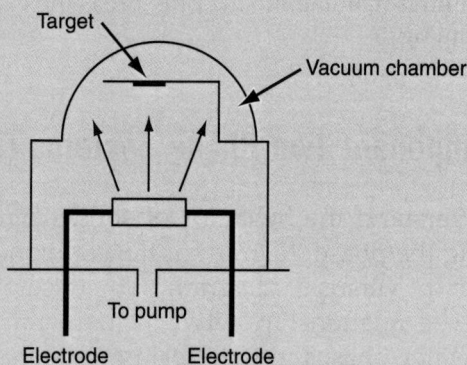

Figure E7.2

The boat made of tungsten has a negligible vapor pressure at 972°C, the operating temperature for the vaporization of aluminum (which melts at 660°C and fills the boat). The approximate rate of evaporation m is given in g/(cm²)(s) by

$$m = 0.437 \frac{p*(MW)^{1/2}}{T^{1/2}}$$

where $p*$ is the vapor pressure in kilopascals, MW is the molecular weight and T is the temperature in kelvin. What is the vaporization rate for Al at 972°C in g/(cm²)(s)?

Solution

You have to calculate $p*$ for Al at 972°C. The Antoine equation is suitable if data are known for the vapor pressure of Al. Considerable variation exists in the data for Al at high temperatures, but we will use $A = 8.779$, $B = 1.615 \times 104$, and $C = 0$ with $p*$ in millimeters of Hg and T in kelvin.

$$\ln p*_{972°C} = 8.799 - \frac{1.615 \times 10^4}{972 + 273} = 0.0154 \text{ mm Hg} (0.00201 \text{ kPa})$$

$$m = 0.437 \frac{(0.00201)(26.98)^{1/2}}{(972 + 273)^{1/2}} = 1.3 \times 10^{-4} \text{ g/(cm}^2\text{)(s)}$$

7.3.2   Retrieving Vapor Pressures from the Tables

You can find the vapor pressures of substances listed in tables in handbooks, physical property books, and websites. We use water as an example. Tabulations of the properties of water and steam (water vapor) are commonly called the **steam tables,** although the tables are as much about water as they are about steam. The properties of water and steam from one source may not agree precisely with other sources because the values available from different sources are likely generated by using equations having different accuracy and precision.

Four classes of tables exist in the foldout:

1. A table of p^* versus T (saturated water and vapor) listing other properties such as $\hat{V}$

		Properties of Saturated Water	
		Volume, m³/kg	
Press. kPa	T (K)	$\hat{V}_l$	V_g
0.80	276.92	0.001000	159.7
1.0	280.13	0.001000	129.2
1.2	282.81	0.001000	108.7
1.4	285.13	0.001001	93.92
1.6	287.17	0.001001	82.76
1.8	288.99	0.001001	74.03
2.0	290.65	0.001002	67.00
2.5	294.23	0.001002	54.25
3.0	297.23	0.001003	45.67
4.0	302.12	0.001004	34.80

2. A table of T versus p^* (saturated water and vapor) containing other properties such as $\hat{V}$

		Properties of Saturated Water	
		Volume (m³/kg)	
T (K)	Press. (kPa)	$\hat{V}_l$	V_g
273.16	0.6113	0.001000	206.1
275	0.6980	0.001000	181.7
280	0.9912	0.001000	130.3
285	1.388	0.001001	94.67
290	1.919	0.001001	69.67
295	2.620	0.001002	51.90
300	3.536	0.001004	39.10
305	4.718	0.001005	29.78
310	6.230	0.001007	22.91
315	8.143	0.001009	17.80

3. A table listing superheated vapor (steam) properties as a function of T and p

		Superheated Steam*				
Abs. Press. lb/in.² (Sat. Temp. °F)		**Sat. Water**	**Sat. Steam**	**400°F**	**420°F**	**440°F**
	Sh			29.23	49.23	69.23
175	*v*	0.0182	2.601	2.730	2.814	2.897
(370.77)	*h*	343.61	1196.7	1215.6	1227.6	1239.9
	Sh			26.92	46.92	66.92
180	*v*	0.0183	2.532	2.648	2.731	2.812
(373.08)	*h*	346.07	1197.2	1214.6	1226.8	1239.2
	Sh			24.66	44.66	64.66
185	*v*	0.0183	2.466	2.570	2.651	2.731
(375.34)	*h*	348.47	1197.6	1213.7	1226.0	1238.4

*Sh = degrees of superheat, °F; v = specific volume, ft^3/lb_m; h = specific enthalpy, Btu/lb_m

4. A table of subcooled water (liquid) properties as a function of p and T (h and u in this table are the specific enthalpy and the specific internal energy, respectively, which are introduced in Chapter 8)

		Properties of Liquid Water		
p, kPa		**400**	**425**	**450**
Sat.	ρ, kg/m³	937.35	915.08	890.25
	h, kJ/kg	532.69	639.71	748.98
	u, kJ/kg	532.43	639.17	747.93
500	ρ, kg/m³	937.51	915.08	
	h, kJ/kg	532.82	639.71	
	u, kJ/kg	532.29	639.17	
700	ρ, kg/m³	937.62	915.22	
	h, kJ/kg	532.94	639.84	
	u, kJ/kg	532.19	639.07	

Locate each table in the foldout, and use the tables with the following explanations. How can you tell which table to use to get the properties you want? One way is to look at one of the phase diagrams for water. For example, do the conditions of 25°C and 4 atm refer to liquid water, a saturated liquid-vapor mixture, or water vapor? You can use the values in the steam tables plus what you know about phases to reach a decision about the state of the water. In the SI steam tables of T versus p^* for saturated water, T is just less than 300 K at which $p^* = 3.536$ kPa. Because the given pressure was about 400 kPa, much higher than the saturation pressure at 298 K, clearly the water is subcooled (compressed liquid).

Can you locate the point $p^* = 250$ kPa and $\hat{V} = 1.00$ m³/kg using Figure 7.8? Do you find the water is in the superheated steam region? The specified volume is larger than the saturated volume of 0.7187 m³/kg at 250 kPa. What about the point $T = 300$ K and $\hat{V} = 0.505$ m³/kg? Water at that state is a mixture of saturated liquid and vapor. You can calculate the quality of the water-water vapor mixture using Equation (7.1) as follows: From the steam tables, the specific volumes of the saturated liquid and vapor are

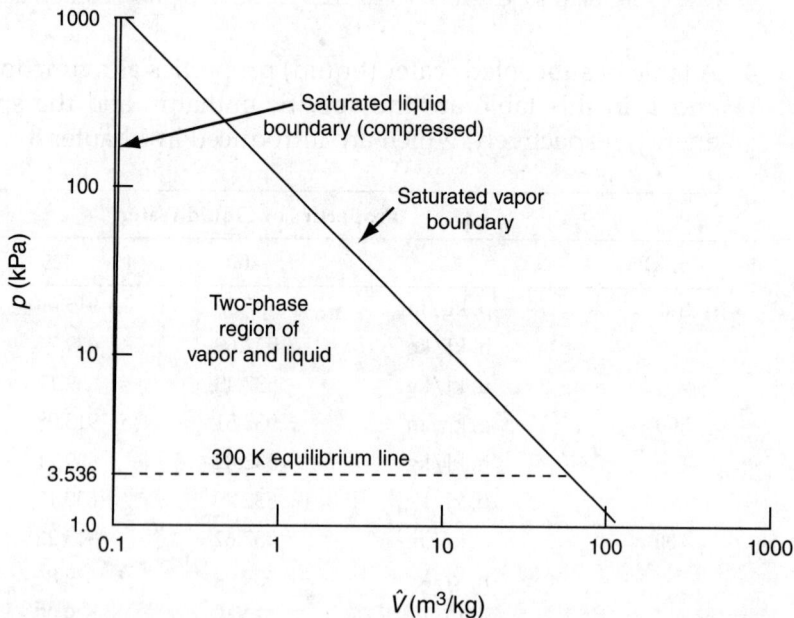

Figure 7.8   Portion of the p-$\hat{V}$ phase diagram for water (note that the axes are logarithmic scales)

$$\hat{V}_l = 0.001004 \text{ m}^3/\text{kg} \ \hat{V}_g = 39.10 \text{ m}^3/\text{kg}$$

Basis: 1 kg of wet steam mixture

Let x = mass fraction vapor. Then

$$\frac{0.001004 \text{ m}^3}{1 \text{ kg liquid}}\left|\frac{(1-x)\text{kg liquid}}{} + \frac{39.10 \text{ m}^3}{1 \text{ kg vapor}}\right|\frac{x \text{ kg vapor}}{} = 0.505 \text{ m}^3$$

$$x = 0.0129 \text{ (the fractional quality)}$$

If you are given a specific mass of saturated water plus steam at a specified temperature or pressure so that you know the state of the water is in the two-phase region, you can use the steam tables for various calculations. For example, suppose a 10.0 m³ vessel contains 2000 kg of water plus steam at 10 atm, and you are asked to calculate the volume of each phase. Let the volume of water be V_l and the volume of steam be Vg; then the masses of each phase are $V_l/\hat{V}_l$ and $Vg/\hat{V}g$ respectively. From your knowledge of the total volume and total mass:

$$V_l + V_g = 10$$

and

$$V_l/\hat{V}_l + V_g/\hat{V}_g = 2000$$

From the steam tables, $\hat{V}_l = 0.0011274$ m³/kg and $\hat{V}_g = 0.19430$ m³/kg. Solving these simultaneous equations for V_l and V_g gives the volume of the liquid as 2.21 m³ and the volume of the steam as 7.79 m³. The mass of the liquid is 1960 kg, and the mass of the steam is 40 kg.

Because the values in the steam tables are tabulated in discrete increments, for intermediate values, you will have to interpolate to retrieve values between the discrete values. (If interpolation does not appeal to you, use the physical property software on the website that accompanies this book.) The next example shows how to carry out interpolations in tables.

Example 7.3   Interpolating in the Steam Tables

Problem Statement

What is the saturation pressure of water at 312 K?

Solution

To solve this problem, you have to carry out a single interpolation. Look in the steam tables under the properties of saturated water to get p^* so as to bracket 312 K:

T (K)	p^* (kPa)
310	6.230
315	8.143

Figure E7.3 shows the concept of a linear interpolation between 310 K and 315 K. Find the change of p^* per unit change in T.

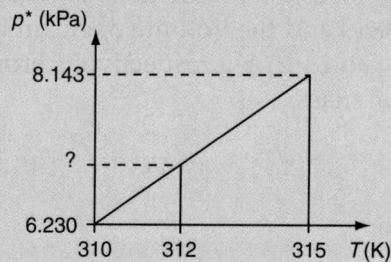

Figure E7.3

$$\frac{\Delta p^*}{\Delta T} = \frac{8.143 - 6.230}{315 - 310} = \frac{1.91}{5} = 0.383$$

Multiply the fractional change times the number of degrees increase from 310 K to get change in p^*, and add the result to the value of p^* at 310 K of 6.230 kPa:

$$p^*_{312\,\text{K}} = p^*_{310\,\text{K}} + \frac{\Delta p^*}{\Delta T}(T_{312} - T_{310}) = 6.230 + 0.383(2) = 7.000\,\text{kPa}$$

7.3.3   Predicting Vapor Pressures from Reference Substance Plots

Because of the curvature of the vapor-pressure data versus temperature (see Figure 7.2), no simple equation with two or three coefficients will fit the data accurately from the triple point to the critical point. Othmer proposed in numerous articles[1] that **reference substance plots** (the name will become clear in a moment) could convert the vapor-pressure-versus-temperature curve into a straight line. One well-known example is the Cox chart.[2] You can use the Cox chart to retrieve vapor-pressure values as well as to test the reliability of experimental data, to interpolate, and to extrapolate. Figure 7.9 is a Cox chart.

Here is how you can make a Cox chart:

1. Mark on the horizontal scale values of log p^* so as to cover the desired range of p^*.

2. Next, draw a straight line on the plot at a suitable angle, say, 45°, that covers the range of T that is to be marked on the vertical axis.

3. To calibrate the vertical axis in common integers such as 25, 50, 100, 200 degrees, and so on, you use a **reference substance**, usually water. For the first integer, say, $T = 100°F$, you look up the vapor pressure of water in the steam tables, or calculate it from the Antoine equation, to get 0.9487 psia. Locate this value on the horizontal axis, and proceed vertically until you hit the 45° straight line. Then proceed horizontally left until you hit the vertical axis. Mark the scale there as 100°F.

4. Pick the next temperature, say, 200°F, and get 11.525 psia. Proceed vertically from $p^* = 11.525$ to the 45° straight line, and then horizontally to the vertical axis. Mark the scale as 200°F.

5. Continue as in steps 3 and 4 until the vertical scale is established over the desired range for the temperature.

Other compounds will give straight lines for p^* versus T as shown in Figure 7.9. What proves useful about the Cox chart is that the vapor pressures of other substances plotted on this specially prepared set of coordinates will yield straight lines over extensive temperature ranges and thus facilitate the extrapolation and interpolation of vapor-pressure data. It has been found that lines so constructed for closely related compounds, such as hydrocarbons, all meet at a common point. Since straight lines can be obtained in a Cox chart, only *two points* of vapor-pressure data are needed to provide adequate information about the vapor pressure of a substance over a considerable temperature range.

---

[1] See, for example, D. F. Othmer, *Ind. Eng. Chem.*, **32**, 841 (1940); and J. H. Perry and E. R. Smith, *Ind. Eng. Chem.*, **25**, 195 (1933).

[2] E. R. Cox, *Ind. Eng. Chem.*, **15**, 592 (1923).

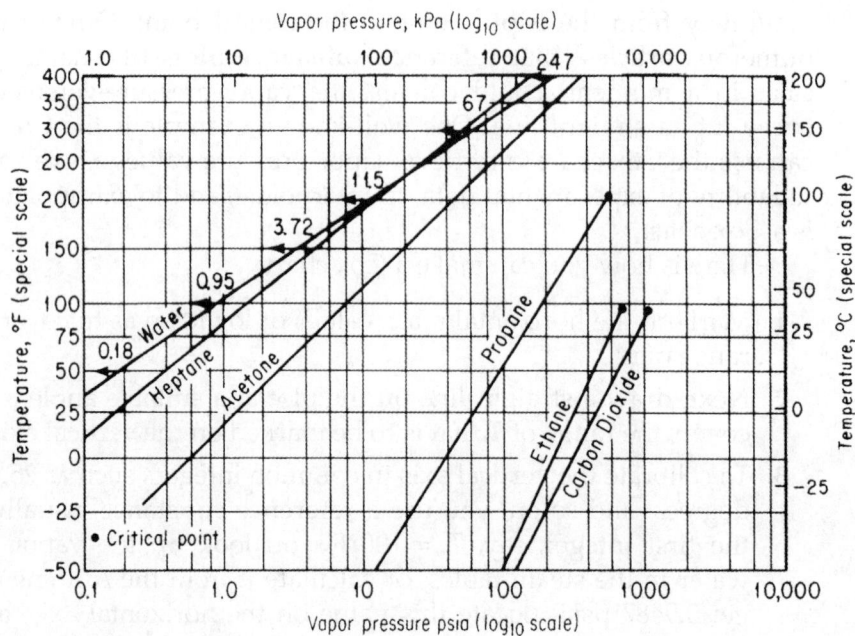

Figure 7.9 Cox chart. The vapor pressure of compounds other than water can be observed to fall on straight lines.

Let's look at an example of using a Cox chart.

Example 7.4   Extrapolation of Vapor-Pressure Data

Problem Statement

The control of solvents was first described in the *Federal Register* [**36**, No. 158 (August 14, 1971)] under Title 42, Chapter 4, Appendix 4.0, "Control of Organic Compound Emissions." Chlorinated solvents and many other solvents used in industrial finishing and processing, dry-cleaning plants, metal degreasing, printing operations, and so forth can be recycled and reused by the introduction of carbon adsorption equipment. To predict the size of the adsorber, you first need to know the vapor pressure of the compound being adsorbed at the process conditions.

The vapor pressure of chlorobenzene is 400 mm Hg absolute at 110°C and 5 atm at 205°C. Estimate the vapor pressure at 245°C and also at the critical point (359°C).

Solution

The vapor pressures will be estimated by use of a Cox chart. You construct the temperature scale (vertical) and vapor-pressure scale (horizontal) as described in connection with Figure 7.9. On the horizontal axis with p^* given in a $\log_{10}$ scale, mark the vapor pressures of water from 3.72 to 3094 psia corresponding to 150°F to 700°F, and mark the respective temperatures on the vertical scale as shown in Figure E7.4.

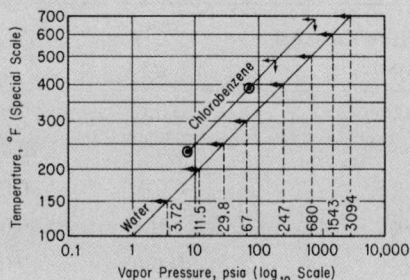

Figure E7.4

Next, convert the two given vapor pressures of chlorobenzene into psia:

$$\frac{400 \text{ mm Hg}}{} \left| \frac{14.7 \text{ psia}}{760 \text{ mm Hg}} = 7.74 \text{ psia } 110°C = 230°F \right.$$

$$\frac{5 \text{ atm}}{} \left| \frac{14.7 \text{ psia}}{1 \text{ atm}} = 73.5 \text{ psia } 205°C = 401°F \right.$$

and plot these two points on the graph paper. Examine the encircled dots. Finally, draw a straight line between the encircled points and extrapolate to 471°F (245°C) and 678°F (359°C). At these two temperatures, you can read off the estimated vapor pressures.

	471°F (245°C)	678° (359°C)
Estimated:	150 psia	700 psia
Experimental:	147 psia	666 psia

Experimental values are given for comparison.

7.3.4   Using MATLAB and Python to Interpolate Nonlinear Data

Linear interpolation can be highly accurate when applied to certain data sets and not so accurate for other. How can you tell if a set of data lends itself to linear interpolation or requires nonlinear interpolation? The following two data sets can be used to address this issue:

	Vapor Pressure and Specific Volume of Liquid Saturated Steam			
T (K)	Vapor Pressure (kpa)	$\dfrac{dp_{vp}}{dT}$	$\hat{V}_{liq}$ (m^3/kg)$\times 10^4$	$\dfrac{d\hat{V}_{liq}}{dT}\times 10^4$
275	0.6980	0.059	9.9959	0.0013
280	0.9912	0.059	10.0022	0.0012
285	1.388	0.079	10.0084	0.0012
290	1.919	0.106	10.0147	0.0012
295	2.620	0.140	10.0209	0.0019
300	3.536	0.183	10.0334	0.0025
305	4.718	0.236	10.0459	0.0037
310	6.230	0.302	10.0709	0.0044
315	8.143	0.383	10.0896	0.0037

The data sets come from the saturated steam data for the vapor pressure and the specific volume of liquid water as functions of temperature. For nonlinear data, the slope of the data changes significantly, while for relatively linear data, the slope changes gradually. Also, when you plot the data on a linear scale, the plot will have significant curvature for a nonlinear set of data. Note that the derivatives of the vapor pressure with respect to temperature changes by a factor of 6 for the range of temperatures shown, while the specific volume of the liquid changed by a factor of 3. Therefore, both sets of data contain significant nonlinearity.

Now note that the total change in the specific volume for this set of data is less than 1% while the total change for the vapor pressure is about 250%. Therefore, the relative change in the specific volume between data points is quite small. As a result, linear interpolation applied to the data for the specific volume will provide highly accurate approximations of the data. On the other hand, the relative change between data for the vapor pressure is significant. Therefore, the combination of the nonlinearity of the data and the magnitude of relative change between adjacent data values for the vapor pressure will produce significant error if linear interpolation is used. This point is demonstrated by the following two examples.

Both MATLAB and Python offer functions that apply interpolation based on a cubic spline approximating equation. A *spline* is a nonlinear function that can be viewed as a mathematical representation of a flexible curve because the function will pass through each data point but can have slopes that vary through the data set. The following examples demonstrate how to apply these cubic spline functions for interpolation of data that cannot be accurately interpolated using linear interpolation.

MATLAB

MATLAB offers a built-in function for cubic spline interpolation: function spline. The call statement for function `spline` is given by

$$fv=spline(\mathbf{xd}, \mathbf{fd}, xv)$$

where fv is the interpolated value based on xv, xd is a vector containing the values of x for the data used for the interpolation, fd is a vector containing the values of $f(x)$ for the data used for the interpolation, and xv is the value of x used for the interpolation. Function `spline` uses linear, quadratic, or cubic interpolation depending on the nonlinearity of the data.

Example 7.5   Cubic Spline Interpolation Using MATLAB

Problem Statement

Using the saturated steam data for vapor pressure in the last table, estimate the vapor pressure at 307.5 K.

Solution

MATLAB Solution for Example 7.5
%%%
%   <u>NOMENCLATURE</u>
%
% fv - the interpolated value
% xd - a vector containing the x values of the data
% fd - a vector containing the y values of the data
% xv- the value of x that is to be interpolated
%
%%

(Continues)

Example 7.5   Cubic Spline Interpolation Using MATLAB (*Continued*)

```
%                              PROGRAM

function Ex7_5_Spline
clear; clc;
                              % Insert data
xd=[275,280,285,290,295,300,305,310,315];
fd=[0.6968,0.9912,1.388,1.919,2.62,3.536,4.718,6.23,8.143];
xv=[307.5];                  % Set value to be interpolated
fv=spline(xd,fd,xv)          % Call built-in function spline
end

%                              PROGRAM END

%%%%%%%%%%%%%%%%%%%%%%%%%%%%%%%%%%%%%%%%%%%%%%%%%%%%%%%%%%%%%%%

fv =
    5.4284
```

Analysis of Results. If linear interpolation were used for this example, the result would be 5.474 kPa, and this amounts to a difference of 0.046 kPa between linear and cubic spline interpolation, which is a relatively small difference but nevertheless significant based on the number of digits reported in the vapor pressure data.

Python

Python offers a built-in function for interpolation: scipy.interpolate. interp1d. The call statement for interp1d is given by

$$f=scipy.interpolate.interp1d(\mathbf{xd}, \mathbf{yd}, kind='linear')$$

$$yi=\mathbf{f}(xv)$$

where **f** is a function that can be used for interpolation, **xd** is a vector containing the values of x for the data used for the interpolation, **yd** is a vector containing the values of $y(x)$ for the data used for the interpolation, xv is the value of x for which interpolation evaluation is to be made, and yi is the interpolated value for xv. scipy.interpolate.interp1d offers linear, quadratic, or cubic spline interpolation depending on the value of kind for which the default is linear (i.e., 'linear', 'quadratic', or 'cubic'). Note that function **f** determined by this function can be used for a single interpolation or for a number of interpolations.

Example 7.6   Cubic Spline Interpolation Using Python

Problem Statement

Using the saturated steam data for vapor pressure in the last table, estimate the vapor pressure at 307.5 K.

Solution

Python Code for Example 7.6

```
Ex7_6 Interpolation.py
##########################################################################
#                            NOMENCLATURE
#
#  xd - a vector containing the x values for the data
#  xv- a vector containing the values of x that are to be used for
     interpolation
#  yd - a vector containing the y values for the data
#  yi - the interpolated value
##########################################################################
#                              PROGRAM
import scipy.interpolate
import numpy as np
#  Input the x,y data
xd=np.array([275, 280, 285, 290, 295, 300, 305, 310, 315])
yd=np.array([0.6968, 0.9912, 1.388, 1.919, 2.62, 3.536, 4.718, 6.23,
    8.143])
#
#  Apply function scipy.interpolate.interp1d to implement cubic spine
    interpolation
#
f=scipy.interpolate.interp1d(xd, yd, kind='cubic')
#  Specify the value that will be used to perform the interpolation
xv=307.5
#  Use the function determined by scipy.interpolate.interp1d to perform
    the interpolation
```

(Continues)

Example 7.6   Cubic Spline Interpolation Using Python
(*Continued*)

```
yi=f(xv)
print("The interpolated value for x={0:5.2f} is {1:6.4f}".format(xv, yi))

#                              PROGRAM END

############################################################################
```

IPython Console:

```
In[1]: runfile(…
Out[1]:
The interpolated value for x=307.5 is 5.4284 kPa
```

Analysis of Results. If linear interpolation were used for this example, the result would be 5.474 kPa, and this amounts to a difference of 0.046 kPa between linear and cubic spline interpolation, which is a relatively small difference but nevertheless significant based on the number of digits reported in the vapor pressure data.

Although this chapter uses the vapor pressure of a pure component, we should mention that the term *vapor pressure* has been applied to solutions of multiple components as well. For example, to meet emission standards, refiners formulate gasoline and diesel fuel differently in the summer than in the winter. The rules on emissions are related to the vapor pressure of a fuel, which is specified in terms of the *Reid vapor pressure* (RVP), a value that is determined at 100°F in a bomb that permits partial vaporization. For a pure component, the RVP is the true vapor pressure, but for a mixture (as are most fuels), the RVP is lower than the true vapor pressure of the mixture (by roughly 10% for gasoline).[3]

Self-Assessment Test

Questions

1. As the temperature is increased, what happens to the vapor pressure of a compound?

---

[3] Refer to J. J. Vazquez-Esparragoza, G. A. Iglesias-Silva, M. W. Hlavinka, and J. Bulin, "How to Estimate RVP of Blends," *Hydrocarbon Processing*, **135** (August, 1992), for specific details about estimating the RVP.

2. Do the steam tables, and similar tables for other compounds, provide more accurate values of the vapor pressure than use of the Antoine equation or modifications of it?

3. When you need a vapor pressure outside the range of known values, what is the best way to estimate it?

4. Is it possible to prepare a Cox chart for water?

5. How do you determine if you should use linear interpolation or nonlinear interpolation for a set of data?

Answers

1. The vapor pressure of a compound increases as the temperature increases.

2. Yes

3. Using a Cox chart

4. Yes, if you use another substance as the reference substance.

5. If the data are nonlinear and have significant relative changes between data points, use nonlinear interpolation. Otherwise, use linear interpolation.

Problems

1. Describe the state and values of the pressure of water initially at 20°F as the temperature is increased to 250°F in a fixed volume.

2. Use the Antoine equation to calculate the vapor pressure of ethanol at 50°C, and compare with the experimental value.

3. Determine the normal boiling point of benzene from the Antoine equation.

4. Prepare a Cox chart from which the vapor pressure of toluene can be predicted over the temperature range –20°C to 140°C.

Answers

1. Initially, it is ice. As it is heated, the ice melts to liquid water at 32°F. As it is further heated, liquid and vapor exist at equilibrium up to 250°F.

2. The experimental value is 219.9 mm Hg; the calculated value is 220.9 mm Hg.

3. 80.1°C

4. Look at Figure 7.9.

7.4   Two-Component Gas/Single-Component Liquid Systems

From a single-component two-phase system, let's extend the discussion to a more complicated system, namely, a system with two components in the gas phase together with a single-component liquid system. An example of such a system is water and a noncondensable gas, such as air. The equilibrium relationships for the water and air help explain how rain is formed and lead to a number of meteorological terms, such as the *dew point* and the *humidity* of the air. Moreover, the equilibrium relationship for two-component gas/single-component liquid systems is used industrially to describe and design many systems, including cooling towers, in which water is cooled by evaporation, and stripping systems, in which a volatile component is removed from a liquid by contacting the liquid with a noncondensable gas.

7.4.1   Saturation

When any noncondensable gas (or a gaseous mixture) comes in contact with a liquid, the gas will acquire molecules from the liquid. If contact is maintained for a sufficient period of time, vaporization continues until equilibrium is attained, at which time the *partial pressure of the vapor in the gas will equal the vapor pressure* of the liquid at the temperature of the system. Regardless of the duration of contact between the liquid and gas, after equilibrium is reached, no more net liquid will vaporize into the gas phase. The gas is then said to be **saturated** with the particular vapor at the given temperature. We also say that the gas mixture is at its **dew point**. **The dew point for the mixture of pure vapor and noncondensable gas means the temperature at which the vapor would just start to condense** if the temperature were very slightly reduced. **At the dew point, the partial pressure of the vapor is equal to the vapor pressure of the volatile liquid**.

Consider a gas partially saturated with water vapor at p and T. If the partial pressure of the water vapor is increased by increasing the total pressure on the system, eventually the partial pressure of the water vapor will equal p^* at T of the system. Because the partial pressure of water cannot exceed p^* at that temperature, a further attempt to increase the pressure will result in water vapor condensing at constant T and p. Thus, p^* represents the maximum partial pressure that water can attain at that temperature, T.

Do you have to have liquid present for saturation to occur? Really, no; only a minute drop of liquid at equilibrium with its vapor will suffice.

What use can you make of the information or specification that a non-condensable gas is saturated? Once you know that a gas is saturated, you can determine the composition of the vapor-gas mixture from knowledge of the vapor pressure of the vapor (or the temperature of the saturated mixture) to use in material balances. From Chapter 6 you should recall that the ideal gas law applies to both air and water vapor at atmospheric pressure with excellent precision. Thus, we can say that the following relations hold *at saturation:*

$$\frac{p_{H_2O}V}{p_{air}V} = \frac{n_{H_2O}RT}{n_{air}RT} \tag{7.5}$$

or

$$\frac{p_{H_2O}}{p_{air}} = \frac{p_{H_2O}^*}{p_{air}} = \frac{n_{H_2O}}{n_{air}} = \frac{p_{total} - p_{air}}{p_{air}} \tag{7.6}$$

because V and T are the same for the air and water vapor.

Also,

$$y_{H_2O} = \frac{p_{H_2O}}{p_{total}} = \frac{p_{H_2O}}{p_{air} + p_{H_2O}} = 1 - y_{air} \tag{7.7}$$

As a numerical example, suppose you have a saturated gas, say, water in air at 51°C, and the pressure on the system is 750 mm Hg absolute. What is the partial pressure of the air? If the air is saturated, you know that the partial pressure of the water vapor is p^* at 51°C. You can use the physical property software on the website that accompanies this text, or use the steam tables, and find that $p^* = 98$ mm Hg. Then

$$P_{air} = 750 - 98 = 652 \text{ mm Hg}$$

Furthermore, the vapor-air mixture has the following composition:

$$y_{H_2O} = \frac{p_{H_2O}}{p_{total}} = \frac{98}{750} = 0.13$$

$$y_{air} = \frac{p_{air}}{p_{total}} = \frac{652}{750} = 0.87$$

Example 7.7   Calculation of the Dew Point of the Products of Combustion

Problem Statement

Oxalic acid ($H_2C_2O_4$) is burned at atmospheric pressure with 4% excess air so that 65% of the carbon burns to CO. Calculate the dew point of the product gas.

Solution

The solution of the problem involves the following steps:

1. Calculate the combustion products via material balances.

2. Calculate the mole fraction of the water vapor in the combustion products as indicated just prior to this example.

3. Assume a total pressure, say, 1 atm, and calculate $y_{H_2O} p_{total} = p_{H_2O}$ in the combustion products. At equilibrium p_{H_2O} will be the vapor pressure $p_{H_2O}^*$.

4. Condensation (at constant total pressure) would be possible when $p_{H_2O}^*$ equals the calculated p_{H_2O}. This value is the dew point.

5. Look up the temperature corresponding to p_{H_2O} in the saturated steam tables.

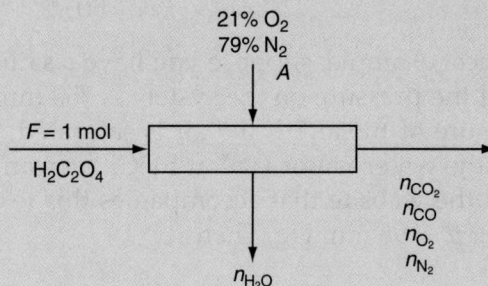

Figure E7.7

Steps 1–5

$$\text{Basis: 1 mol of } H_2C_2O_4$$

The figure and data are given. The chemical reaction equation for the combustion of oxalic acid is given as

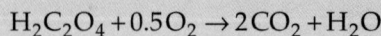

$$H_2C_2O_4 + 0.5O_2 \rightarrow 2CO_2 + H_2O$$
$$H_2C_2O_4 \rightarrow 2CO + H_2O + 0.5O_2$$

Step 4

O_2 required:

$$\frac{1\,\text{mol}\,H_2C_2O_4}{1}\frac{0.5\,\text{mol}\,O_2}{1\,\text{mol}\,H_2C_2O_4} = 0.5\,\text{mol}\,O_2\,(\text{note oxygen in oxalic acid})$$

Moles of O_2 in with air including excess:

$$(1 + 0.04)\,(0.5\,\text{mol}\,O_2) = 0.52\,\text{mol}\,O_2$$

Therefore, $0.52/0.21 = 2.48$ mol air enters containing 1.96 mol n_{N_2}.

Specifications: 65% of the carbon burns to CO: $(0.65)\,(2) = 1.30$.

The results:

Element Material Balances		
Element	**In (mol)**	**Out (mol)**
C	2	$n_{CO_2} + n_{CO}$ or $0.70 + 1.30$
H	2	$2n_{H_2O}$
N	1.96×2	1.96×2
O	$0.52 \times 2 + 4$	$2n_{CO_2} + n_{CO} + n_{O_2}$ or $2(0.70) + 1.30 + 1.0 + 2n_{O_2}$

$$n_{H_2O} = 1.0;\ n_{CO_2} = 0.7;\ n_{CO} = 1.3;\ n_{O_2} = 0.67;\ n_{N_2} = 1.96;\ \text{total mol} = 5.63$$
$$y_{H_2O} = 1\ \text{mol}\ H_2O/6.78\ \text{mol} = 0.178$$

The partial pressure of the water in the product gas (at an assumed atmospheric pressure) determines the dew point of the stack gas; that is, the temperature of saturated steam that equals the partial pressure of the water is equal to the dew point of the product gas:

$$p_{H_2O}^{*} = y_{H_2O}(p_{\text{total}}) = 0.178\,(101.3\,\text{kPa}) = 18.0\,\text{kPa}\,(\text{i.e., } 2.61\,\text{psia})$$

From the steam tables, $T = 136°F$.

Self-Assessment Test

Questions

1. What does the term *saturated gas* mean?

2. If a gas is saturated with water vapor, describe the state of the water vapor and the air if it is (a) heated at constant pressure, (b) cooled at constant pressure, (c) expanded at constant temperature, and (d) compressed at constant temperature.

3. How can you lower the dew point of a pollutant gas before analysis?

4. In a gas-vapor mixture, when is the vapor pressure the same as the partial pressure of the vapor in the mixture?

Answers

1. A saturated gas means that the gas is in equilibrium with a liquid; that is, the partial pressure of the condensable component in the gas is equal to the vapor pressure of the liquid.

2. (a) Both are in the gas phase; (b) some liquid water will condense from the gas; (c) both are in the gas phase; (d) some liquid water will condense from the gas.

3. Remove water from the gas or dilute it with an inert dry gas.

4. At saturation, that is, at equilibrium.

Problems

1. The dew point of water in atmospheric air is 82°F. What is the mole fraction of water vapor in the air if the barometric pressure is 750 mm Hg?

2. Calculate the composition in mole fractions of air that is saturated with water vapor at a total pressure of 100 kPa and 21°C.

3. An 8.00 L cylinder contains a gas saturated with water vapor at 25.0°C and a pressure of 770 mm Hg. What is the volume of the gas when dry at standard conditions?

Answers

1. Vapor pressure of water at 82°F = 0.5409 psia;

$$y = \frac{0.5409 \text{ psia}}{} \left| \frac{760 \text{ mm Hg}}{14.69 \text{ psia}} \right| \frac{}{750 \text{ mm Hg}} = 0.0373$$

2. The mole fraction water vapor is $2.501/100 = 0.0250$ and air is 0.9750.

3. Mole fraction water = 3.197(760)/101.3/770 = 0.0312; total moles = PV/RT = 0.3315 g mol; moles air = (1 − 0.0312)0.3315 = 0.3212 g mol; volume of dry gas at STP = 0.3212(22.415) = 7.20 L

7.4.2 Condensation

Modern automobiles are engineering marvels, no less than other flashier technology products. Properly maintained, these machines can last for a long time. To ensure longevity and high performance of the vehicle, any problems that may arise must be promptly addressed. Machinery in an automobile uses several types of fluids—for lubrication, cooling, and so on—and a puddle of a fluid under the vehicle is potentially a telltale sign of impending trouble. Should the owner/operator of a vehicle be worried about a watery, non-greasy puddle a few inches in diameter observed occasionally under the exhaust pipe? While there is some black carbon substance on the top of the puddle, it does not appear to be oil. The formation of the puddle is noticeable for shorter trips but rarely for the longer ones. On the contrary, the owner/operator's neighbor never observes any such puddle under their car, irrespective of the length of the trip.

While a puddle of fluid under the car is generally an unwelcome sign, this particular phenomenon is actually benign and attributable to the condensation of water formed during the combustion of the fuel, as the system never gets hot enough during the shorter trips to keep the products in the vapor state. The black carbon or soot is simply the result of incomplete combustion. This is the normal operating mode of the vehicle. And of course, the neighbor drives an electric vehicle.

Examine the setup for combustion gas analysis shown in Figure 7.10. What error has been made in the setup? If you do not heat the sample of gas collected by the probe and/or put an intermediate condenser before the pump, the analyzer will fill with liquid as the gas sample cools and will not function.

Condensation is the change of vapor in a noncondensable gas to liquid. Some typical ways of condensing a vapor that is in a gas are to

1. Cool it at constant total system pressure (the volume changes)
2. Cool it at constant total system volume (the pressure changes)
3. Compress it isothermally (the volume changes)

Combinations of the three as well as other processes are possible, of course.

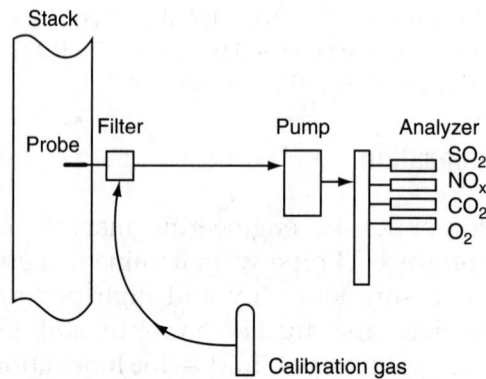

Figure 7.10   Instrumentation for stack gas analysis

As an example of condensation, let's look at cooling at constant total system pressure for a mixture of air and 10% water vapor. Pick the air-water vapor mixture as the system. If the mixture is cooled at constant total pressure from 51°C and 750 mm Hg absolute (point A in Figures 7.11a and b for the water), how low can the temperature go before condensation starts (at point B)? You can cool the mixture until the temperature reaches the dew point associated with the partial pressure of water of

$$p_{H_2O}^* \equiv p_{H_2O} = 0.10(750) = 75 \, \text{mm Hg}$$

From the steam tables, you can find that the corresponding temperature is $T = 46°C$ (point B on the vapor curve). After reaching $p^* = 75$ mm Hg at point B, if the condensation process continues, it continues at constant pressure (75 mm Hg) and constant temperature (46°C) until all of the water vapor has been condensed to liquid (point C). Further cooling will reduce the temperature of the liquid water below 46°C.

     If the air-water mixture with 10% water vapor starts at 60°C and 750 mm Hg and is cooled at constant pressure, at what temperature will condensation occur for the same process? Has the dew point changed? It is the same because $p_{H_2O} = 0.10(750) = 75 \, \text{mm Hg}$ still. The volume of both the air and the water vapor can be calculated from $pV = nRT$ until condensation starts, at which point the ideal gas law applies only to the residual water vapor, not the liquid. The number of moles of H_2O in the gas phase does not change from the initial number of moles until condensation occurs, at which point the number of moles of water in the gas phase starts to decrease. The number of moles of air in the system remains constant throughout the process.

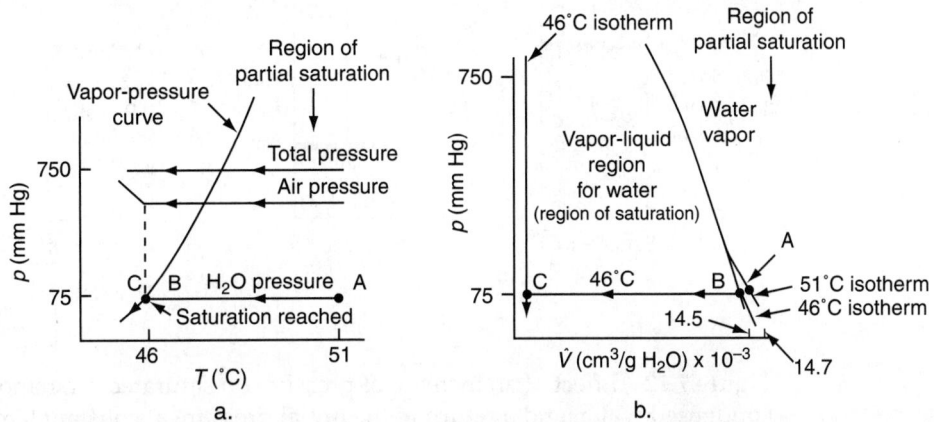

Figure 7.11   Cooling of an air-water mixture at constant total pressure. The lines and curves for the water are distorted for the purpose of illustration. The scales are not linear.

Condensation can also occur when the pressure on a vapor-gas mixture is increased. If a pound of saturated air at 75°F is isothermally compressed (with a reduction in volume, of course), liquid water will be condensed out of the air (see Figure 7.12).

For example, if a pound of saturated air at 75°F and 1 atm (the vapor pressure of water is 0.43 psia at 75°F) is compressed isothermally to 4 atm (58.8 psia), almost three-fourths of the original content of water vapor now will be in the form of liquid, and the air still has a dew point of 75°F. Remove the liquid water, expand the air isothermally back to 1 atm, and you will find that the dew point has been lowered to about 36°F. Mathematically (1 = state at 1 atm, 4 = state at 4 atm), with $z = 1.00$ for both components:

For saturated air at 75°F and 4 atm:

$$\left(\frac{n_{H_2O}}{n_{air}}\right)_4 = \left(\frac{p_{H_2O}^*}{p_{air}}\right)_4 = \frac{0.43}{58.4}$$

For the same air saturated at 75°F and 1 atm:

$$\left(\frac{n_{H_2O}}{n_{air}}\right)_1 = \left(\frac{p_{H_2O}^*}{p_{air}}\right)_1 = \frac{0.43}{14.3}$$

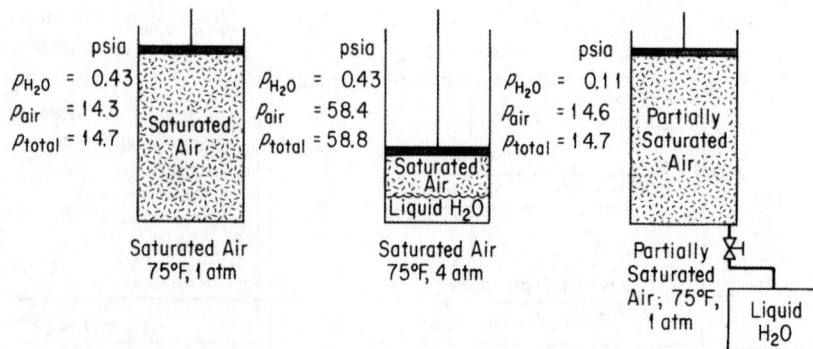

Figure 7.12 Effect of an increase of pressure on saturated air, removal of condensed water, and a return to the initial pressure at constant temperature

The material balance gives for the H_2O:

$$\left(\frac{n_4}{n_1}\right)_{H_2O} = \frac{\dfrac{0.43}{58.4}}{\dfrac{0.43}{14.3}} = \frac{14.3}{58.4} = 0.245$$

In other words, 24.5% of the original water will remain as vapor after compression.

After the air-water vapor mixture is returned to a total pressure of 1 atm, the following two equations now apply at 75°F:

$$p_{H_2O} + p_{air} = 14.7$$

$$\frac{p_{H_2O}}{p_{air}} = \frac{n_{H_2O}}{n_{air}} = \frac{0.43(0.245)}{14.3} = 0.00737$$

From these two relations you can find that

$$p_{H_2O} = 0.108 \text{ psia}$$

$$p_{air} = 14.6 \text{ psia}$$

$$p_{air} = 14.7 \text{ psia}$$

This pressure of the water vapor represents a dew point of about 36°F.

Now let's look at some examples of condensation from a gas-vapor mixture.

Example 7.8   Condensation of Benzene from a Vapor Recovery Unit

Problem Statement

Emission of volatile organic compounds from processes is closely regulated. Both the Environmental Protection Agency (EPA) and the Occupational Safety and Health Administration (OSHA) have established regulations and standards covering emissions and frequency of exposure. This problem concerns the first step of the removal of benzene vapor from an exhaust stream, shown in Figure E7.8a, designed to recover 95% of the benzene from air by compression. What is the exit pressure from the compressor?

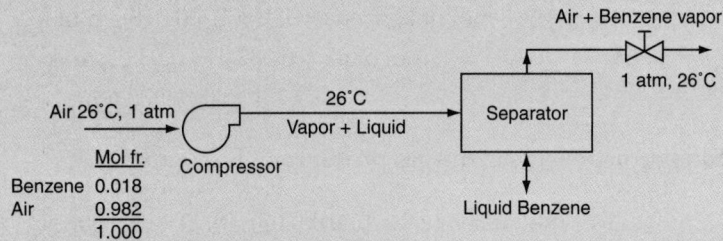

Figure E7.8a

Solution

Figure E7.8b illustrates on a $\hat{V}$ chart for benzene what occurs to the benzene vapor during the process. The process is an isothermal compression.

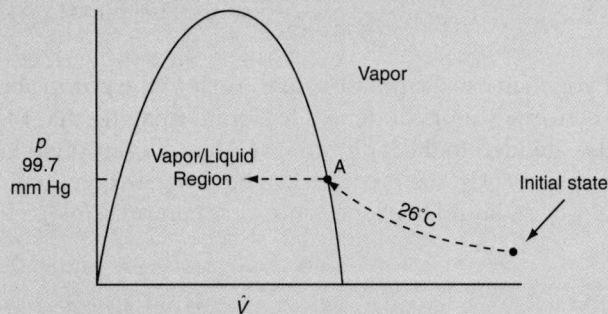

Figure E7.8b

(Continues)

Example 7.8   Condensation of Benzene from a Vapor Recovery Unit (*Continued*)

If you pick the compressor as the system, the compression is isothermal and yields a saturated gas. You can look up the vapor pressure of benzene at 26°C in a handbook or get it from the website that accompanies this book. It is $p^* = 99.7$ mm Hg.

Next you have to carry out a short material balance to determine the outlet concentrations from the compressor:

Basis: 1 g mol of entering gas at 26°C and 1 atm

Entering components to the compressor:

$$\text{mol of benzene} = 0.018(1) = 0.018 \text{ g mol}$$
$$\text{mol of air} = 0.982(1) = \underline{0.982} \text{ g mol}$$
$$\text{total gas} = 1.000 \text{ g mol}$$

Exiting components in the gas phase from the compressor:

$$\text{mol of benzene} = 0.018(0.05) = 0.90 \times 10^{-3} \text{ g mol}$$
$$\text{mol of air} = \underline{0.982} \text{ g mol}$$
$$\text{total gas} = 0.983 \text{ g mol}$$

$$y_{\text{Benzene exiting}} = \frac{0.90 \times 10^{-3}}{0.983} = 0.916 \times 10^{-3} = \frac{p_{\text{Benzene}}}{p_{\text{total}}}$$

Now the partial pressure of the benzene is 99.7 mm Hg, so

$$p_{\text{total}} = \frac{99.7 \text{ mm Hg}}{0.916 \times 10^{-3}} = 108 \times 10^3 \text{ mm Hg} \, (143 \, \text{atm})$$

Could you increase the pressure at the exit of the pump above 143 atm? Only if all of the benzene vapor condenses to liquid. Imagine that the dashed line in Figure E7.8b is extended to the left until it reaches the saturated liquid line (bubble point line). Subsequently, the pressure can be increased on the liquid (it would follow a vertical line as liquid benzene is not very compressible).

Example 7.9   Smokestack Emissions and Pollution

Problem Statement

A local pollution-solutions group has reported the Simtron Co. boiler plant as being an air polluter and has provided as proof photographs of heavy smokestack emissions on 20 different days. As the chief engineer for the Simtron Co. you know that your plant is not a source of pollution because you burn natural gas (essentially methane) and your boiler plant is operating correctly. Your boss believes the pollution-solutions group has made an error in identifying the stack: it must belong to the company next door that burns coal. Is he correct? Is the pollution-solutions group correct? See Figure E7.9a.

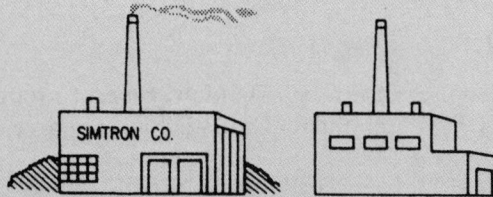

Figure E7.9a

Solution

Methane (CH_4) contains 2 kg mol of H_2/kg mol of C; you can look in Chapter 2 and see that coal contains 71 kg of C/5.6 kg of H_2 in 100 kg of coal. The coal analysis is equivalent to

$$\frac{71 \text{ kg C}}{} \left| \frac{1 \text{ kg mol C}}{12 \text{ kg C}} = 5.92 \text{ kg mol C} \qquad \frac{5.6 \text{ kg H}_2}{} \right| \frac{1 \text{ kg mol H}_2}{2.016 \text{ kg H}_2} = 2.78 \text{ kg mol H}_2$$

or a ratio of 2.78/5.92 = 0.47 kg mol of H_2/kg mol of C. Suppose that each fuel burns with 40% excess air and that combustion is complete. We can compute the mole fraction of water vapor in each stack gas and thus get the respective partial pressures.

Steps 1–4

The process is shown in Figure E7.9b.

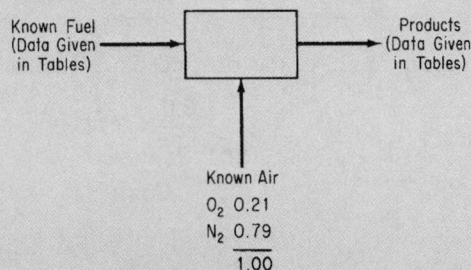

Known Fuel (Data Given in Tables)

Products (Data Given in Tables)

Known Air
O_2 0.21
N_2 0.79
1.00

Figure E7.9b

(Continues)

Example 7.9   Smokestack Emissions and Pollution (*Continued*)

Step 5

$$\text{Basis: 1 kg mol of C}$$

Steps 6 and 7

The combustion problem is a standard type of problem having zero degrees of freedom in which both the fuel and airflows are given, and the product flows are calculated directly.

Steps 7–9

Tables will make the analysis and calculations compact.

Natural gas:

$$CH_4 + 2O_2 \rightarrow CO_2 + 2H_2O$$
$$\text{Required } O_2 : 2$$
$$\text{Excess } O_2 : 2(0.40) = 0.80$$
$$N_2 : (2.80)(79/21) = 10.5$$

Composition of combustion gases (kilogram moles):

	Combustion Product Gases (kg mol)				
Components	**kg mol in**	**CO$_2$**	**H$_2$O**	**Excess O$_2$**	**N$_2$**
C	1.0	1.0			
H	4.0		2.0		
Air				0.80	10.5
Total		1.0	2.0	0.80	10.5

The total kilogram moles of gas produced are 14.3, and the mole fraction H$_2$O is

$$\frac{2.0}{14.3} = 0.14$$

Coal:

$$C + O_2 \rightarrow CO_2 \quad H_2 + \frac{1}{2}O_2 \rightarrow H_2O$$
$$\text{Required } O_2 : 1 + 0.47(1/2) = 1.24$$
$$\text{Excess } O_2 : (1.24)(0.40) = 0.49$$
$$N_2 : 1.40(79/21)[1 + 0.47(1/2)] = 6.50$$

Composition of Combustion Gases (kg mol)					
Components	**kg mol in**	CO_2	H_2O	**Excess O_2**	N_2
C	1	1			
H	0.94		0.47		
Air				0.49	6.5
Total		1	0.47	0.49	6.5

The total kilogram moles of gas produced are 8.46 and the mole fraction H_2O is

$$\frac{0.47}{8.46} = 0.056$$

If the barometric pressure is, say, 100 kPa, and if the stack gas became saturated so that water vapor would start to condense at $p^*_{H_2O}$, condensed vapor could be photographed:

	Natural Gas	**Coal**
Partial pressure (p^*)	100(0.14) = 14 kPa	100(0.056) = 5.6 kPa
Equivalent temperature	52.5°C	35°C

Thus, the stack will emit condensed water vapor at higher ambient temperatures for a boiler burning natural gas than for one burning coal. The public, unfortunately, sometimes concludes that all the emissions they perceive are pollution. Natural gas could appear to the public to be a greater pollutant than either oil or coal, whereas, in fact, the emissions are just water vapor. The sulfur content of coal and oil can be released as sulfur dioxide to the atmosphere, and the polluting capacities of mercury and heavy metals in coal and oil are much greater than those of natural gas when all three are being burned properly. The sulfur contents as delivered to consumers are as follows: natural gas, 4×10^{-4} mol % (as added mercaptans to provide smell for safety); No. 6 fuel oil, up to 2.6%; and coal, from 0.5% to 5%. In addition, coal may release particulate matter into the stack plume. By mixing the stack gas with fresh air, and by convective mixing above the stack, the mole fraction water vapor can be reduced, and hence the condensation temperature can be reduced. However, for equivalent dilution, the coal-burning plant will always have a lower condensation temperature.

    With the information calculated above, how would you resolve the questions that were originally posed?

Self-Assessment Test

Questions

1. Is the dew point of a vapor-gas mixture the same variable as the vapor pressure?
2. What variables can be changed, and how, in a vapor-gas mixture to cause the vapor to condense?
3. Can a gas containing a superheated vapor be made to condense?

Answers

1. No, the dew point is a temperature, not a pressure.
2. Reduce the temperature, increase the pressure, reduce the volume.
3. Yes; see the answers to the previous question.

Problems

1. A mixture of air and benzene contains 10 mol % benzene at 43°C and 105 kPa pressure. At what temperature can the first liquid form? What is the liquid?
2. Two hundred pounds of water out of 1000 lb is electrolytically decomposed into hydrogen and oxygen at 25°C and 740 mm Hg absolute. The hydrogen and oxygen are separated at 740 mm Hg and stored in two different cylinders, each at 25°C and a pressure of 5.0 atm absolute. How much water condenses from the gas in each cylinder?

Answers

1. P_{Bz} = 10.5 kPa = 0.105 bar; VP Bz: $\log_{10}(0.105) = 0.14591 - 39.165/(T - 261.236)$; solve for T = 296.56K = 23.4°C. The liquid is pure benzene.
2. At 25°C and 740 mm Hg, y_{H_2O} = 0.03215; at 25°C and 5 atm, y_{H_2O} = 0.00626; amount of H_2O initially in H_2 = 0.36905 lb mol; amount H_2O at 5 atm = 0.06774 lb mol; difference is 0.3013 lb mol = 5.42 lb H_2O condensed for H_2; for O_2, 0.5(5.42) = 2.71 lb H_2O condensed for O_2.

7.4.3 Vaporization

At equilibrium, you can vaporize a liquid into a noncondensable gas and raise the partial pressure of the vapor in the gas until the saturation pressure (vapor pressure) is reached. Figure 7.13 shows how the partial pressures of water and air change with time as water evaporates into dry air. On a *p-T* diagram such as Figure 7.2, the liquid would vaporize at the saturation temperature on line AB in the figure (the bubble point temperature which is equal to the dew point temperature) until the air was saturated.

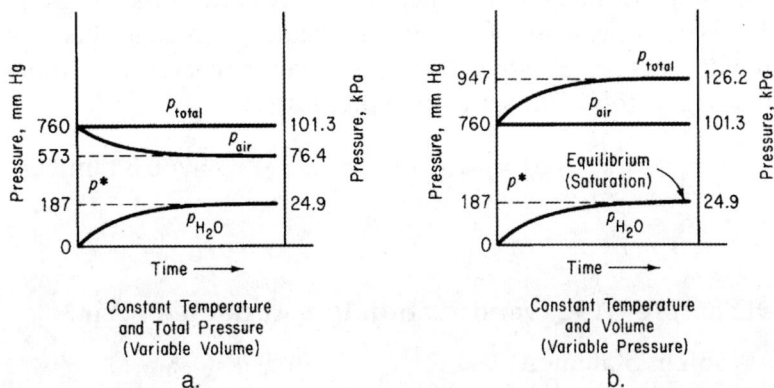

Figure 7.13   Change of partial and total pressure during the vaporization of water into initially dry air (a) at constant temperature and total pressure (variable volume); (b) at constant temperature and volume (variable pressure)

On the *p-$\hat{V}$* diagram (Figure 7.4), evaporation would occur from A to B at constant temperature and pressure until the air was saturated. At constant total pressure, as shown in Figure 7.14, the volume of the air would remain constant, but the volume of water vapor would increase, so the total volume of the mixture would increase.

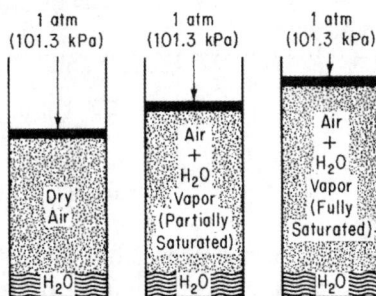

Figure 7.14   Evaporation of water at constant pressure and temperature of 65°C

You might ask: Is it possible to have the water evaporate into air and saturate the air, and yet maintain a constant temperature, pressure, and volume in the cylinder? (Hint: What would happen if you let some of the gas-vapor mixture escape from the cylinder?)

You can use Equations (7.5), (7.6), and (7.7) to solve vaporization problems. For example, if sufficient liquid water is placed in a volume of dry gas that is at 15°C and 754 mm Hg, and if the temperature and volume remain constant during the vaporization, what is the final pressure in the system? The partial pressure of the dry gas remains constant because n, V, and T for the dry gas are constant. The water vapor reaches its vapor pressure of 12.8 mm Hg at 15°C. Thus, the total pressure is

$$p_{tot} = p_{H_2O} + p_{air} = 12.8 + 754 = 766.8 \text{ mm Hg}$$

Example 7.10   Vaporization to Saturate Dry Air

Problem Statement

What is the minimum number of cubic meters of dry air at 20°C and 100 kPa that are necessary to evaporate 6.0 kg of ethyl alcohol if the total pressure remains constant at 100 kPa and the temperature remains 20°C? Assume that the air is blown through the alcohol to evaporate it in such a way that the exit pressure of the air-alcohol mixture is at 100 kPa.

Solution

Look at Figure E7.10. The process is isothermal. The additional data needed are

$$p^*_{alcohol} \text{ at } 20°C = 5.93 \text{ kPa}$$
$$\text{Mol. wt. ethyl alcohol} = 46.07$$

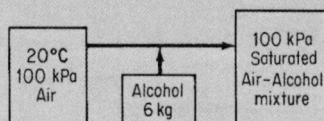

Figure E7.10

The minimum volume of air means that the resulting mixture is saturated; any condition less than saturated would require more air.

Basis: 6.0 kg of alcohol

The ratio of moles of ethyl alcohol to moles of air in the final gaseous mixture is the same as the ratio of the partial pressures of these two substances. Since we know the moles of alcohol, we can find the number of moles of air needed for the vaporization.

$$\frac{p_{alcohol}^{*}}{p_{air}} = \frac{n_{alcohol}}{n_{air}}$$

Once you calculate the number of moles of air, you can apply the ideal gas law. Since $p_{alcohol}^{*} = 5.93\,kPa$,

$$p_{air} = p_{total} - p_{alcohol}^{*} = (100 - 5.93)\,kPa = 94.07\,kPa$$

$$\frac{6.0\ kg\ alcohol}{} \left| \frac{1\ kg\ mol\ alcohol}{46.07\ kg\ alcohol} \right| \frac{94.07\ kg\ mol\ air}{5.93\ kg\ mol\ alcohol} = 2.07\ kg\ mol\ air$$

$$V_{air} = \frac{2.07\ kg\ mol\ air}{} \left| \frac{8.314\,(kPa)(m^3)}{(kg\ mol)\,(K)} \right| \frac{293\ K}{100\ kP} = 50.3\ m^3 \text{ at } 20°C \text{ and } 100\ kPa$$

Sublimation into a noncondensable gas can occur as well as vaporization.

The Chinook: A Wind That Eats Snow

Each year the area around the Bow River Valley in Southwestern Canada experiences temperatures that go down to −40°F. And almost every year, when the wind called the Chinook blows, the temperature climbs as high as 60°F. In just a matter of a few hours, this Canadian area experiences a temperature increase of about 100°F. How does this happen?

Air over the Pacific Ocean is always moist due to the continual evaporation from the ocean. This moist air travels from the Pacific Ocean to the foot of the Rocky Mountains because air masses tend to move from west to east.

As this moist air mass climbs up the western slopes of the Rocky Mountains, it encounters cooler temperatures, and the water vapor condenses out of the air. Rain falls, and the air mass becomes drier as it loses water in the form of rain. As the air becomes drier, it becomes heavier. The heavier dry air falls down the eastern slope of the Rockies, and the atmospheric pressure on the falling gas increases as the gas approaches the ground. . . . the air gets warmer as you increase the pressure. As the air falls down the side of the mountain, its temperature goes up 5.5°F for every thousand-foot drop. So we have warm, heavy, dry air descending upon the Bow River Valley at the base of the Rockies. The Chinook is this warm air (wind) moving down the Rocky Mountains at 50 mph.

Remember, it is now winter in the Bow River Valley, the ground is covered with snow, and it is quite cold (–40°F). Since the wind is warm, the temperature of the Bow River Valley rises very rapidly. Because the wind is very dry, it absorbs water from the melting snow. The word *Chinook* is an Indian word meaning "snow eater". The Chinook can eat a foot of snow off the ground overnight. The weather becomes warmer and the snow is cleared away. It's the sort of thing they have fantasies about in Buffalo, New York.

Reprinted with permission. Adapted from an article originally appearing in *Problem Solving in General Chemistry*, 2nd ed., by Ronald DeLorenzo, Wm. C. Brown Publ. (1993), pp. 130–32.

Self-Assessment Test

Questions

1. If a dry gas is isothermally mixed with a liquid in a fixed volume, will the pressure remain constant with time?

2. If dry gas is placed in contact with a liquid phase under conditions of constant pressure and allowed to come to equilibrium:
   a. Will the total pressure increase with time?
   b. Will the volume of the gas plus liquid plus vapor increase with time?
   c. Will the temperature increase with time?

Answers

1. No, it will increase slightly.

2. (a) No; (b) Yes; (c) No

Problems

1. Carbon disulfide (CS_2) at 20°C has a vapor pressure of 352 mm Hg. Dry air is bubbled through the CS_2 at 20°C until 4.45 lb of CS_2 are evaporated. What was the volume of the dry air required to evaporate this CS_2 (assuming that the air becomes saturated) if the air was initially at 20°C and 10 atm and the final pressure on the air-CS_2 vapor mixture is 750 mm Hg?

2. In an acetone recovery system, the acetone is evaporated into dry N_2. The mixture of acetone vapor and nitrogen flows through a 2-ft-diameter duct at 10 ft/s. At a sampling point, the pressure is 850 mm Hg and the temperature is 100°F.

The dew point is 80°F. Calculate the pounds of acetone per hour passing through the duct.

3. Toluene is used as a diluent in lacquer formulas. Its vapor pressure at 30°C is 36.7 mm Hg absolute. If the barometer falls from 780 mm Hg to 740 mm Hg, will there be any change in the volume of dry air required to evaporate 10 kg of toluene?

4. What is the minimum number of cubic meters of dry air at 21°C and 101 kPa required to evaporate 10 kg of water at 21°C?

Answers

1. $y_{CS_2} = 352/750 = 0.4693$; $y_{air} = 0.5307$; basis: 4.45 lb CS_2 = 0.05855 lb mol; lb mol air = 0.05855(0.5307)/0.4693 = 0.06621 lb mol air; $V = nRT/p = (0.06621)(0.7303)$ (527.67)/10 = 2.55 ft^3.

2. Vapor pressure of acetone at 80°F = 2.465 mm Hg; $y_A = 2.465/850 = 0.00290$; volumetric flow rate = 20π ft^3/s; $n = pV/(RT) = 20\pi(1.333)/((0.7303)(560)) = 0.1536$ lb mol/s; $n_A = (0.00290)(0.1536)(3600)(58.08) = 93.16$ lb acetone/hr.

3. Initially, $y_{Tol} = 36.7/780 = 0.04705$; $y_{air} = 0.9530$; afterwards, $y_{Tol} = 36.7/740 = 0.04959$, $y_{air} = 0.9504$; basis:1 g mol total evaporated. Initially, air flow = 0.9530/0.04705 = 20.26 g mol; afterwards, air flow = 0.9504/0.04959 = 19.17 g mol. Therefore, the required air flow decreases by 5.4%.

4. Vapor pressure of water at 21°C = 2.493 kPa; $y_{H_2O} = 2.493/101 = 0.02468$; $y_{air} = 0.9753$; basis: 10 kg water; $n_{air} = (10/18)(0.9753/0.02458) = 21.95$ kg mol; $V = 21.95(22.315)(294.15/273.15)(101/101.325) = 518.3$ m^3.

7.5   Two-Component Gas/Two-Component Liquid Systems

In Section 7.3, we discussed vapor-liquid equilibria of a pure component. In Section 7.4, we covered equilibria of a pure component in the presence of a noncondensable gas. In this section, we consider certain aspects of a more general set of circumstances, namely, cases in which both the liquid and vapor have two components; that is, the vapor and liquid phases each contain both components. Distilling moonshine from a fermented grain mixture is an example of binary vapor-liquid equilibrium in which water and ethanol are the primary components in the system and are present in both the vapor and the liquid.

The primary result of vapor-liquid equilibrium is that the more volatile component (i.e., the component with the larger vapor pressure at a given temperature) tends to accumulate in the vapor phase while the less volatile component tends to accumulate in the liquid phase. Distillation columns, which are used to separate a mixture into its components, are based on this principle. A distillation column is composed of a number of trays that provide contacting between liquid and vapor streams inside the column. At each tray, the concentration of the more volatile component is increased in the vapor stream leaving the tray, and the concentration of the less volatile component is increased in the liquid leaving the tray. In this manner, applying a number of trays in series, the more volatile component is concentrated in the overhead stream from the column while the less volatile component is concentrated in the bottom product. In order to design and analyze distillation, you must be able to quantitatively describe vapor-liquid equilibrium for these systems.

7.5.1 Ideal Solution Relations

An **ideal solution** is a mixture whose properties such as vapor pressure, specific volume, and so on can be calculated from the knowledge of only the corresponding properties of the pure components and the composition of the solution. For a solution to behave as an ideal solution:

- All of the molecules of all types should have the same size.
- All of the molecules should have the same intermolecular interactions.

Most solutions are not ideal, but some real solutions are nearly ideal.

Raoult's law
The best-known relation for ideal solutions is

$$p_i = x_i p_i^*(T) \tag{7.8}$$

where p_i = partial pressure of component i in the vapor phase
          x_i = mole fraction of component i in the liquid phase
     $p_i^*(T)$ = vapor pressure of component i at T

Figure 7.15 shows how the vapor pressure of the two components in an ideal binary solution sum to the total pressure at 80°C. Compare Figure 7.15 with Figure 7.16, which displays the pressures for a nonideal solution.

Raoult's law is used primarily for a component whose mole fraction spans the full range from 0 to 1 for solutions of components quite similar in chemical nature, such as straight-chain hydrocarbons.

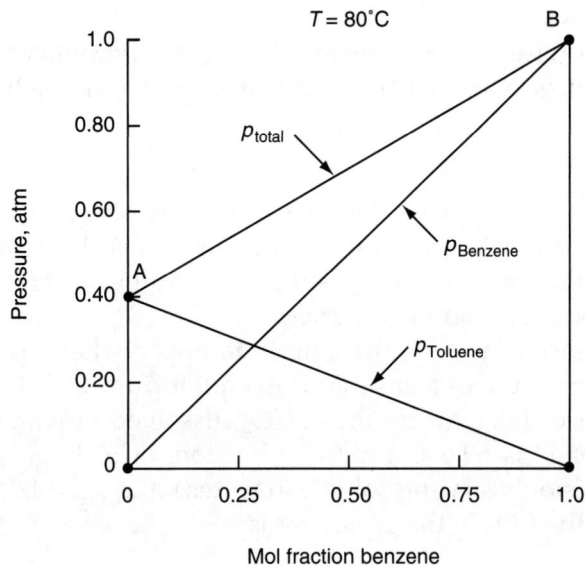

Figure 7.15   Application of Raoult's law to an ideal solution of benzene and toluene (like species) to get the total pressure as a function of composition. The respective vapor pressures are shown by points A and B at 0 and 1.0 mole fraction benzene.

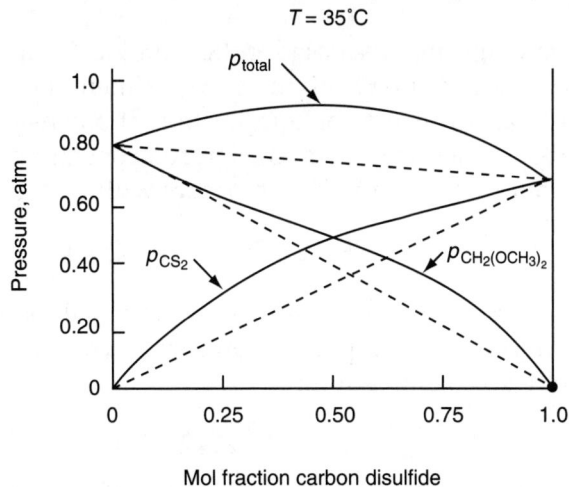

Figure 7.16   Plot of the partial pressures and total pressure (solid lines) exerted by a solution of carbon disulfide (CS_2) in methylal ($CH_2(OCH_3)_2$) as a function of composition. The dashed lines represent the pressures that would exist if the solution were ideal.

Henry's law

Henry's law is used primarily for a component whose mole fraction approaches zero, such as a dilute gas dissolved in a liquid:

$$p_i = H_i x_i \tag{7.9}$$

where p_i is the partial pressure in the gas phase of the dilute component at equilibrium at some temperature, and H_i is the *Henry's law constant*. Note that in the limit where $x_i \rightarrow 0, p_i \rightarrow 0$. Values of H_i can be found in several handbooks and on the Internet.

Henry's law is quite simple to apply when you want to calculate the partial pressure of a gas that is in equilibrium with the gas dissolved the liquid phase. Take, for example, CO_2 dissolved in water at 40°C for which the value of H is 69,600 atm/mol fraction. (The large value of H shows that $CO_2(g)$ is only sparing soluble in water.) If $x_{CO_2} = 4.2 \times 10^{-6}$, the partial pressure of the CO_2 in the gas phase is

$$p_{CO_2} = 69,600(4.2 \times 10^{-6}) = 0.29 \, \text{atm}$$

That is, the gas phase contains almost 30% CO_2 for 1 atm pressure, but the liquid phase would contain only 0.00042% CO_2.

7.5.2 Vapor-Liquid Equilibria Phase Diagrams

The phase diagrams discussed in Section 7.2 for a pure component can be extended to cover binary mixtures. Experimental data usually are presented as pressure as a function of composition at a constant temperature, or temperature as a function of composition at a constant pressure. For a pure component, vapor-liquid equilibrium occurs with only one degree of freedom:

$$F = 2 - P + C = 2 - 2 + 1 = 1$$

At 1 atm pressure, vapor-liquid equilibrium will occur at only one temperature: the normal boiling point. However, if you have a binary solution, you have two degrees of freedom:

$$F = 2 - 2 + 2 = 2$$

For a system at a fixed pressure, both the phase compositions and the temperature can be varied over a finite range. Figures 7.17 and 7.18 show the vapor-liquid envelope for a binary mixture of benzene and toluene, which is essentially ideal.

You can interpret the information on the phase diagrams as follows: Suppose you start in Figure 7.17 at a 50-50 mixture of benzene-toluene at 80°C and 0.30 atm in the vapor phase. Then you increase the pressure on the system until you reach the dew point at about 0.47 atm, at which point the vapor starts to condense. At 0.62 atm the mole fraction in the vapor phase will be about 0.75, and the mole fraction in the liquid phase will be about 0.38 as indicated by the tie line. As you increase the pressure from 0.70 atm, all of the vapor will have condensed to liquid. What will the composition of the liquid be? 0.50 benzene, of course! Can you carry out an analogous conversion of vapor to liquid on Figure 7.18, the temperature-composition diagram?

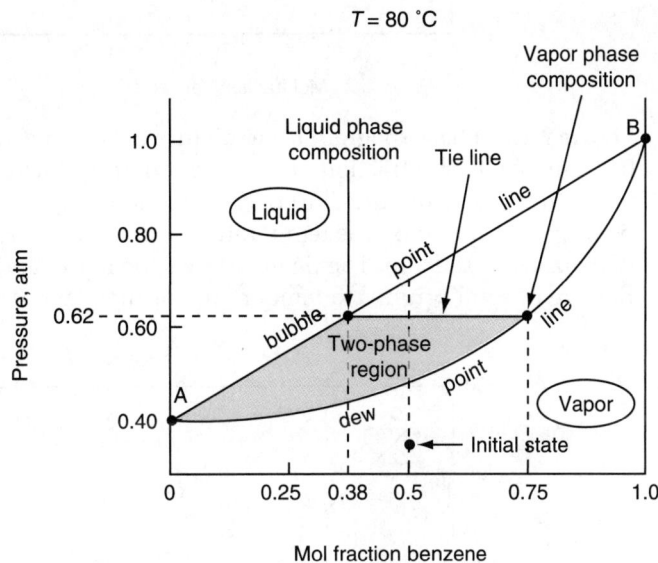

Figure 7.17   Phase diagram for a mixture of benzene and toluene at 80°C. At 0 mol fraction of benzene (point A), the pressure is the vapor pressure of toluene at 80°C. At a mole fraction of benzene of 1 (point B), the pressure is the vapor pressure of benzene at 80°C. The tie line shows the liquid and vapor compositions that are in equilibrium at a pressure of 0.62 atm (and 80°C).

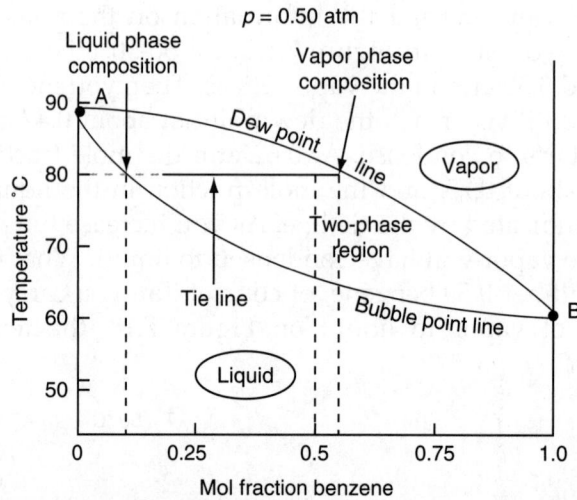

Figure 7.18 Phase diagram for a mixture of benzene and toluene at 0.50 atm. At 0 mole fraction of benzene (point A), the temperature is that when the vapor pressure of toluene is 0.50 atm. At a mole fraction of benzene of 1 (point B), the temperature is that when the vapor pressure of benzene is 0.50 atm. The tie line shows the liquid and vapor compositions that are in equilibrium at a temperature of 80°C (and 0.50 atm).

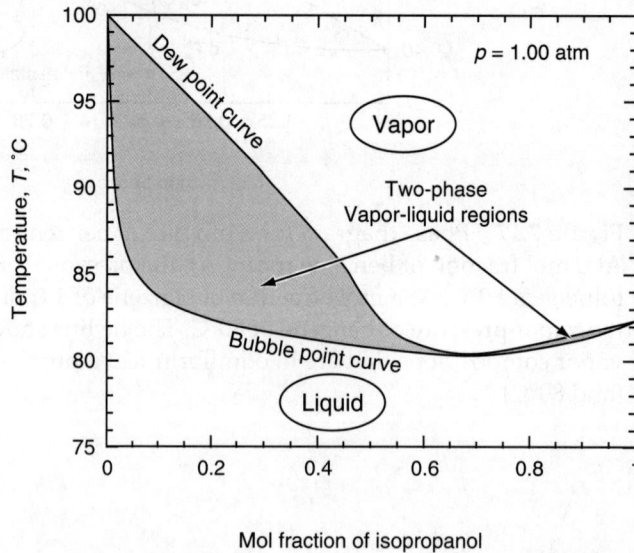

Figure 7.19 Phase diagram for a nonideal mixture of isopropanol and water at 1 atm

Phase diagrams for nonideal solutions abound. Figure 7.19 shows the temperature-composition diagram for isopropanol in water at 1 atm. Note the minimum boiling point at a mole fraction of isopropanol of about 0.68, a point called an **azeotrope** (a point at which on a y_i-versus-x_i plot the function of (y_i/x_i) crosses the function $y_i = x_i$, a straight line). An azeotrope makes separation by simple distillation difficult because it creates a pinch point because when the dew point and the bubble point coincide so that separation between the more volatile and the less volatile components does not occur.

7.5.3   *K*-value (Vapor-Liquid Equilibrium Ratio)

For nonideal as well as ideal mixtures that comprise two (or more) phases, it proves to be convenient to express the ratio of the mole fraction in one phase to the mole fraction of the same component in another phase in terms of a **distribution coefficient** or **equilibrium ratio** *K*, usually called a *K*-value. For example:

$$\text{Vapor-liquid ratio of component } i = \frac{y_i}{x_i} = K_i \tag{7.10}$$

and so on. If the ideal gas law $p_i = y_i\, p_{total}$ applies to the gas phase and the ideal Raoult's law $p_i = x_i p_i^*(T)$ applies to the liquid phase, then for an *ideal* system

$$K_i = \frac{y_i}{x_i} = \frac{p_i^*(T)}{p_{total}} \tag{7.10a}$$

Equation (7.10a) gives reasonable estimates of K_i values at low pressures for components well below their critical temperatures but yields values too large for components above their critical temperatures, at high pressures, and/or for polar compounds. For nonideal mixtures, Equation (7.10) can be employed if K_i is made a function of temperature, pressure, and composition so that relations for K_i can be fit by equations to experimental data and used directly, or in the form of charts, for design calculations, as explained in some of the references at the end of this chapter. Figure 7.20 shows how *K* varies for the nonideal mixture of acetone and water at 1 atm. *K* can be greater or less than 1 but never negative.

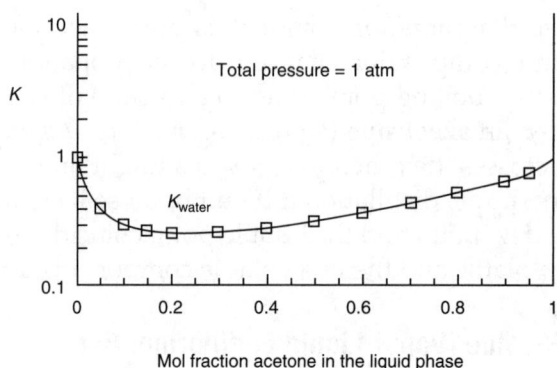

Figure 7.20   Change of K of water with composition at $p = 1$ atm

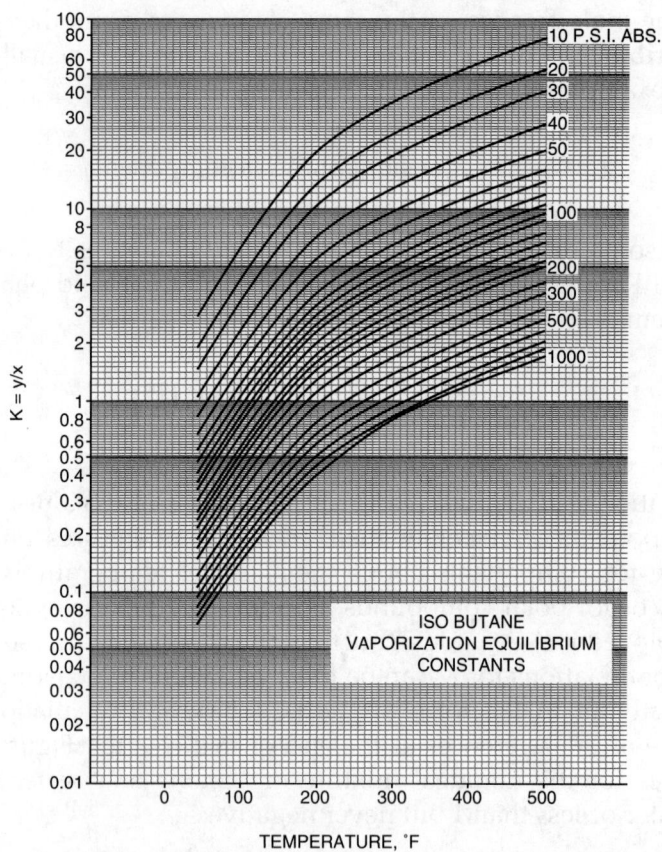

Figure 7.21 K-values for isobutane as a function of temperature and pressure. From *Natural Gasoline Association of America Technical Manual*, 4th ed. (1941) (based on data provided by George Granger Brown).

For ideal solutions, you can calculate values of K using Equation (7.10a). For nonideal solutions you can get approximate K-values from

1. Empirical equations such as[4]

$$\text{If } T_{c,i}/T > 1.2: K_i = \frac{(p_{c,i})\exp[7.224 - 7.534/T_{r,i} - 2.598\ln T_{r,i}]}{p_{\text{total}}}$$

2. Databases refer to the supplementary references (at the end of the chapter).
3. Charts such as Figure 7.21
4. Thermodynamic relations refer to the references (at the end of the chapter).

7.5.4   Bubble Point and Dew Point Calculations

Here are some typical problems you should be able to solve that involve the use of the equilibrium coefficient K_i and material balances:

1. **Calculate the bubble point temperature of a liquid mixture given the total pressure and liquid composition.**

   To calculate the **bubble point temperature** (given the total pressure and liquid composition), you can write Equation (7.10) as $y_i = K_i x_i$. Also, you know that $\sum y_i = 1$ in the vapor phase. Thus, for a binary,

   $$1 = K_1 x_1 + K_2 x_2 \tag{7.11}$$

   in which the K_i are functions of solely the temperature. Because each of the K_i increases with temperature, Equation (7.11) has only one positive root. A trial-and-error procedure is required to determine the bubble point temperature, which can be performed on the computer or by hand. If you choose to solve for the bubble point by hand, you have to assume varying temperatures so that you can look up or calculate K_i, and then calculate each term in Equation (7.11). After the sum $(K_1 x_1 + K_2 x_2)$ brackets 1, you can interpolate to get a T that satisfies Equation (7.11).

   For an ideal solution, Equation (7.11) becomes

   $$p_{\text{total}} = p_1^* x_1 + p_2^* x_2 \tag{7.12}$$

---

[4] S. I. Sandler, in *Foundations of Computer Aided Design*, Vol. 2, edited by R. H. S. Mah and W. D. Seider, American Institute of Chemical Engineers, New York (1981), p. 83.

and you might use Antoine's equation for p^*_i. Once the bubble point temperature is determined, the vapor composition can be calculated from

$$y_i = \frac{p^*_i x_i}{p_{total}}$$

A degree-of-freedom analysis for the bubble point temperature for a binary mixture shows that the degrees of freedom are zero:

Total variables = $2 \times 2 + 2 = 6$ variables: x_1, x_2; y_1, y_2; p_{total}; T

Prespecified values of = $2 + 1 = 3$ variables: x_1, x_2; p_{Total}

Independent equations = $2 + 1 = 3$ equations:
$$y_1 = K_1 x_1, \quad y_2 = K_2 x_2; \quad y_1 + y_2 = 1$$

Therefore, there are three unknowns and three equations with which to determine their values.

2. **Calculate the dew point temperature of a vapor mixture given the total pressure and vapor composition.**

   To calculate the **dew point temperature** (given the total pressure and vapor composition), you can write Equation (7.10) as $x_i = y_i / K_i$, and you know $\sum x_i = 1$ in the liquid phase. Consequently, you want to solve the equation

$$1 = \frac{y_1}{K_1} + \frac{y_2}{K_2} \tag{7.13}$$

   in which the Ks are functions of temperature as explained for the bubble point temperature calculation. For an ideal solution,

$$1 = p_{total} \left[\frac{y_1}{p^*_1} + \frac{y_2}{p^*_2} \right] \tag{7.13a}$$

   The degree-of-freedom analysis is similar to that for the bubble point temperature calculation.

In selecting a particular form of the equation to be used for your equilibrium calculations, you must select a method of solving the equation that has desirable convergence characteristics. Convergency to the solution should

- Lead to the desired root if the equation has multiple roots

- Be stable, that is, approach the desired root asymptotically rather than by oscillating
- Be rapid, and not become slower as the solution is approached

You can use MATLAB's **fzero** function or Python's **scipy.optimize.newton**, which were introduced in Chapter 6, Section 6.2.2, to solve bubble point and dew point equations because in both cases the equation to be solved is a single nonlinear equation in which the temperature is unknown. These equations can be more efficiently and reliably solved by first transforming them into a more linear form.[5]

Table 7.1 summarizes the usual phase equilibrium calculations.

The next example illustrates the details of vapor-liquid equilibrium calculations.

Table 7.1   Summary of the Information Associated with Typical Phase Equilibrium Calculations

Type	Known* Information	Variables to Be Calculated	Equation(s) to Use	Convergence Characteristics
Bubble point temperature	p_{total}, x_i	T, y_i	7.10	Good
Dew point temperature	p_{total}, y_i	T, x_i	7.12	Good
Bubble point pressure	T, x_i	p_{total}, y_i	7.10	Fair
Dew point pressure	T, y_i	p_{total}, x_i	7.12	Fair

* K_i is assumed to be a known function of T, p, and composition

Example 7.11   Bubble Point Calculation

Problem Statement

Suppose that a liquid mixture of 4.0% *n*-hexane in *n*-octane is vaporized. What is the composition of the first vapor formed if the total pressure is 1.00 atm?

Solution

Refer back to Figure 7.18 to view a relation of T versus x at constant p such as is involved in this problem. The mixture can be treated as an ideal mixture because the components are quite similar. As an intermediate step, you must calculate the

(Continues)

[5] J. B. Riggs, *Computational Methods for Chemical Engineers* (Austin, TX: Ferret Publishing, 2020), 95–97.

Example 7.11   Bubble Point Calculation (*Continued*)

bubble point temperature using Equation (7.12). You have to look up the coefficients of the Antoine equation to obtain the vapor pressures of the two components:

$$\ln(p^*) = A - \frac{B}{C+T}$$

where p^* is in millimeters of Hg and T is in kelvin:

	A	B	C
n-hexane	15.8737	2697.55	−48.784
n-octane	15.9798	3127.60	−63.633

Basis: 1 kg mol of liquid

You have to solve the following formidable equation to get the bubble point temperature. Use a nonlinear equation solver such as provided by MATLAB or Python:

$$760 = \exp\left(15.8737 - \frac{2697.55}{-48.784+T}\right)0.040 + \exp\left(15.9787 - \frac{3127.60}{-63.633+T}\right)0.960$$

MATLAB Results for Example 7.11

```
%%%%%%%%%%%%%%%%%%%%%%%%%%%%%%%%%%%%%%%%%%%%%%%%%%%%%%%%%%%%%%%%%%%%%

%                        NOMENCLATURE

%

%  x0 – initial guess for the root of the equation

%  xsoln - the root of f(x) determined by function fzero

%%%%%%%%%%%%%%%%%%%%%%%%%%%%%%%%%%%%%%%%%%%%%%%%%%%%%%%%%%%%%%%%%%%%%

%                          PROGRAM

function Ex7_11_Bubble Point
clear; clc;
x0=350.

                        % Define anonymous function
fx=@(x)760-exp(15.8737-2697.55/(-48.784+x))*0.04 …
-exp(15.9787-3127.6/(-63.633+x))*0.96;
xsoln=fzero(fx,x0)            % Call built-in function fzero
end

%                        PROGRAM END

%%%%%%%%%%%%%%%%%%%%%%%%%%%%%%%%%%%%%%%%%%%%%%%%%%%%%%%%%%%%%%%%%%%%%
xsoln =
   393.3
```

Python Code for Example 7.11

`Ex7_11 Bubble Point.py`

```
#############################################################################
#                          NOMENCLATURE
#
#  x0 - the initial guess for the root of function f(x)
#  xsol - the root of f(x) determined by function scipy.optimize.newton
#############################################################################
#                            PROGRAM
import numpy as np
import scipy.optimize
# Define the UD function for the function value
def f(x):
    fx=760-exp(15.8737-2697.55/(-48.784+x))*
    0.04-exp(15.9787-3127.6/(-63.633+x))*0.96
    return fx
x0=350.                                   Initial guess for x
# Apply function scipy.optimize.newton to determine a root of the
  nonlinear equation
xsol=scipy.optimize.newton(f, x0)

#                          PROGRAM END
#############################################################################
```

IPython console:

```
In[1]: runfile(...
In[2]: %precision 1
In[3]: xsol
Out[3]: 393.3
```

The solution is $T = 393.3$ K, for which the vapor pressure of a hexane is 3114 mm Hg and the vapor pressure of octane is 661 mm Hg. The mole fractions in the vapor phase are

$$y_{C_6} = \frac{p_{C_6}^*}{p_{tot}} x_{C_6} = \frac{3114}{760} = 0.164$$

$$y_{C_8} = 1 - 0.164 = 0.836$$

Self-Assessment Test

Questions

1. When should you use Henry's law, and when should you use Raoult's law?
2. Can you make a plot of the partial pressures and total pressure of a mixture of heptane and octane, given solely that the vapor pressure of heptane is 92 mm Hg and that of octane is 31 mm Hg at a temperature?

Answers

1. Henry's law is for a gas dissolved in a solvent. Raoult's law is for a solvent in an ideal solution.
2. Yes, because this system acts as an ideal solution

Problems

1. Calculate the boiling point temperature of 1 kg of a solution of 70% ethylene glycol (antifreeze, $C_2H_6O_2$) in water at 1 atm. Assume the solution is ideal.

Answer

1. This problem requires a trial-and-error solution. Recognizing that almost all of the total pressure will come from the water, we set the initial guess for temperature equal to the saturated water vapor pressure for $101.325/0.33 = 307.0$ kPa; that is, the initial guess: 134.5°C; by trial and error, the bubble point temperature = 135.1°C. At this temperature, vapor pressure of water = 312.095 kPa; vapor pressure of ethylene glycol = 10.819 kPa; therefore, $0.7(10.819) + 0.3(312.095) = 101.2$ kPa (close enough).

7.6   Multicomponent Vapor-Liquid Equilibrium

So-called white oil is pressurized and cooled gas well vapors. In Texas awhile back, independent producers operated refrigeration units (to as low as −20°F) at their wells, exploiting a 1977 letter from the legal counsel of the Texas Railroad Commission (the agency that governed oil and gas production) that said "white oil" could be deemed oil rather than gas. There were advantages to having a well classified as an oil well; namely, one oil well could be drilled on 10 acres, but a gas well required 640 acres, and at that time, white oil could be sold for six times the price of natural gas (which was under price control). In 1984, after years of contention, Judge Clark overruled the Railroad Commission. As a result, it issued a new order that said for a liquid to be counted as crude oil, it must be liquid in the reservoir, liquid in the well bore, and liquid at the surface. Over \$27 billion of gas reserves were involved in this controversy. You can see that determining the true state and composition of petroleum products can be a serious matter.

Multicomponent vapor-liquid equilibrium pertains to systems that contain three or more components in the vapor or liquid phases. Vapor-liquid equilibrium calculations for multicomponent systems are performed in a manner analogous to the ones for binary systems. For bubble point calculations, $\sum_i y_i = 1$; therefore, the bubble point equation can be written in terms of the known liquid compositions (x_i) and the component K-values (K_i):

$$\sum_i^n x_i K_i = 1$$

For dew point calculations, $\sum_i x_i = 1$; therefore, the dew point equation can be written in terms of the known vapor composition (y_i) and the component K-values (K_i):

$$\sum_i^n \frac{y_i}{K_i} = 1$$

Note that both **the bubble point and dew point equations are, in general, nonlinear equations with temperature as the only unknown.**

To calculate the bubble point or the dew point using these equations, you will need the K-values for each component in the system. Bubble point and dew point calculations are used by commercial process simulators to model distillation columns and other separation processes. Equations of state are used by commercial process simulators to calculate component K-values for bubble and dew point calculations. For example, the SRK method is routinely used to calculate the K-values for nonpolar systems (e.g., hydrocarbons), and the UNIQUAC equation of state is used to calculate the K-values for a wide range of polar systems (i.e., systems with strong intermolecular interactions, such as hydrogen bonding).

Summary

In this chapter, we introduced phase diagrams and the phase rule and showed the connection to pure component properties, such as vapor pressure. Pure component vapor pressures are used to represent the behavior of vapor-liquid systems with a noncondensable gas, allowing one to describe saturation conditions, condensation, and vaporization. Finally, binary vapor-liquid equilibrium relations were considered.

Glossary

absolute pressure   Pressure relative to a complete vacuum.

Antoine equation   Equation that relates vapor pressure to absolute temperature.

azeotrope   A constant boiling mixture with identical compositions in the liquid and vapor phases.

azeotrope point   Boiling point of the azeotrope at the specified pressure.

boiling   Change from liquid to vapor.

bubble point   The temperature at which liquid changes to vapor (at some pressure).

condensation   The change of phase from vapor to liquid.

degrees of superheat   The difference in temperature between the actual T and the saturated T at a given pressure.

dew point   The temperature at which the vapor just begins to condense at a specified pressure, that is, the value of the temperature along the vapor-pressure curve.

equilibrium   A state of the system in which there is no tendency to spontaneous change.

evaporation   The change of phase of a substance from liquid to vapor.

freezing   The change of phase of a substance from liquid to solid.

fusion   See **melting**.

Gibbs phase rule   A relation that gives the degrees of freedom for intensive variables in a system in terms of the number of phases and number of components.

Henry's law   A relation between the partial pressure of a gas in the gas phase and the mole fraction of the gas in the liquid phase at equilibrium.

ideal solution   A system whose properties, such as vapor pressure, specific volume, and so on, can be calculated from the knowledge only of the corresponding properties of the components and the composition of the solution.

invariant   A system in which no variation of conditions is possible without one phase disappearing.

K-value   A parameter (distribution coefficient) used to express the ratio of the mole fraction in one phase to the mole fraction of the same component in another phase.

melting   The change of phase from solid to liquid of a substance.

noncondensable gas   A gas at conditions under which it cannot condense to a liquid or solid.

normal boiling point   The temperature at which the vapor pressure of a substance (p^*) is 1 atm (101.3 kPa).

normal melting point   The temperature at which a solid melts at 1 atm (101.3 kPa).

phase diagram   Representation of the different phases of a compound on a two- (or three-) dimensional graph.

quality   Fraction or percent of the liquid-vapor mixture that is vapor (wet vapor).

Raoult's law   A relation that relates the partial pressure of one component in the vapor phase to the mole fraction of the same component in the liquid phase.

reference substance   The substance used as the reference in a reference substance plot.

reference substance plot   A plot of a property of one substance versus the same property of another (reference) substance that results in an approximate straight line.

saturated liquid   Liquid that is in equilibrium with its vapor.

saturated vapor   Vapor that is in equilibrium with its liquid.

steam tables   Tabulations of the properties of water and steam (water vapor).

subcooled liquid   Liquid at values of temperature and pressure less than those that exist at saturation.

sublimation   Change of phase of a solid directly to a vapor.

sublimation pressure   The pressure given by the melting curve (a function of temperature).

supercritical region   A portion of a physical properties plot in which the substance is at combined p-T values above the critical point.

superheated vapor   Vapor at values of temperature and pressure exceeding those that exist at saturation.

triple point   The one p-V-T combination at which solid, liquid, and vapor are in equilibrium.

two-phase (region)   Region on a plot of physical properties where both the liquid and vapor exist simultaneously.

vapor A gas below its critical point in a system in which the vapor can condense.

vaporization The change of a substance from liquid to vapor.

vapor-liquid equilibria Graphs showing the concentration of a component in a vapor-liquid system as a function of temperature and/or pressure.

Supplementary References

American National Standards, Inc. *ASTM D323-79 Vapor Pressure of Petroleum Products (Reid Method)*, Philadelphia (1979).

Bhatt, B. I., and S. M. Vora. *Stoichiometry (SI Units)*, Tata McGraw-Hill, New Delhi (1998).

Henley, E. J., and J. D. Seader. *Equilibrium—Stage Separation Operations in Chemical Engineering*, Wiley, New York (1981).

Horvath, A. L. *Conversion Tables in Science and Engineering*, Elsevier, New York (1986).

Jensen, W. B. "Generalizing the Phase Rule," *J. Chem. Educ.*, **78**, 1369–70 (2001).

Perry, R. H., et al. *Perry's Chemical Engineers' Handbook*, McGraw-Hill, New York (2000).

Rao, Y. K. "Extended Form of the Gibbs Phase Rule," *Chem. Engr. Educ.*, 40–49 (Winter, 1985).

Yaws, C. L. *Handbook of Vapor Pressure* (4 volumes), Gulf Publishing Co., Houston, TX (1993–1995).

Problems

7.2 Phase Diagrams and the Phase Rule

*7.2.1 Select the correct answer(s) in the following statements:

a. In a container of 1.00 L of toluene, the vapor pressure of the toluene is 103 mm Hg. The same vapor pressure will be observed in a container of (1) 2.00 L of toluene at the same temperature; (2) 1.00 L of toluene at one-half the absolute temperature; (3) 1.00 L of alcohol at the same temperature; (4) 2.00 L of alcohol at the same temperature.

b. The temperature at which a compound melts is the same temperature at which it (1) sublimes; (2) freezes; (3) condenses; (4) evaporates.

    c. At what pressure would a liquid boil first? (1) 1 atm; (2) 2 atm; (3) 200 mm Hg; (4) 101.3 kPa.

    d. When the vapor pressure of a liquid reaches the pressure of the atmosphere surrounding it, it will (1) freeze; (2) condense; (3) melt; (4) boil.

    e. Sublimation is the phase change from (1) the solid phase to the liquid phase; (2) the liquid phase to the solid phase; (3) the solid phase to the gas phase; (4) the gas phase to the solid phase.

    f. A liquid that evaporates rapidly at ambient conditions is more likely than not to have a (1) high vapor pressure; (2) low vapor pressure; (3) high boiling point; (4) strong attraction among the molecules.

*7.2.2 Draw a p-T diagram for a pure component. Label the curves and points that are listed in Figure 7.3 on it.

*7.2.3 A vessel contains liquid ethanol, ethanol vapor, and N_2 gas at equilibrium. How many phases, components, and degrees of freedom are there according to the phase rule?

*7.2.4 What is the number of degrees of freedom according to the phase rule for each of the following systems: (a) solid iodine in equilibrium with its vapor; (b) a mixture of liquid water and liquid octane (which is immiscible in water), both in equilibrium with their vapors?

*7.2.5 Liquid water in equilibrium with water vapor is a system with how many degrees of freedom?

*7.2.6 Liquid water in equilibrium with moist air is a system with how many degrees of freedom?

*7.2.7 You have a closed vessel that contains $NH_4Cl(s)$, $NH_3(g)$, and $HCl(g)$ in equilibrium. How many degrees of freedom exist in the system?

*7.2.8 In the decomposition of $CaCO_3$ in a sealed container from which the air was initially pumped out, you generate CO_2 and CaO. If not all of the $CaCO_3$ decomposes at equilibrium, how many degrees of freedom exist for the system according to the Gibbs phase rule?

**7.2.9 Based on the phase diagrams in Figure P7.2.9, answer the questions and explain your answers.

    a. What is the approximate normal melting point for compound A?

    b. What is the approximate normal boiling point for compound A?

    c. What is the approximate triple point temperature for compound B?

    d. Which compounds sublime at atmospheric pressure?

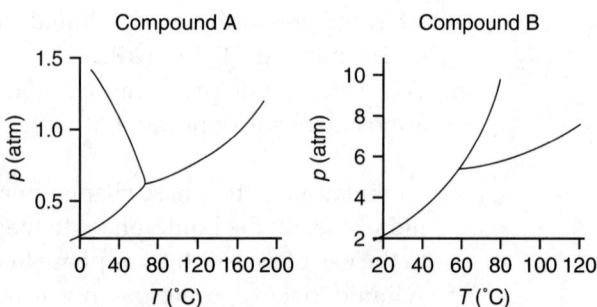

Figure P7.2.9

**7.2.10 One form of cooking is to place the food in a pressure cooker (a sealed pot). Pressure cookers decrease the time require to cook the food. Some explanations of how a pressure cooker works are as follows. Which of the explanations are correct?

a. We know $p_1 T_1 = p_2 T_2$. So if the pressure is doubled, the temperature should be doubled and result in quicker cooking.

b. Pressure cookers are based on the principle that $p \propto T$, that is, pressure is directly proportional to temperature. With the volume kept constant, if you increase the pressure, the temperature also increases, and it takes less time to cook.

c. Food cooks faster because the pressure is high. This means that there are more impacts of molecules per surface area, which in turn increases the temperature of the food.

d. If we increase the pressure under which food is cooked, we have more collisions of hot vapor with the food, cooking it faster. On an open stove, vapor escapes into the surroundings, without affecting the food more than once.

e. As the pressure inside the sealed cooker builds, as a result of the vaporization of water, the boiling point of water is increased, thereby increasing the temperature at which the food cooks—hotter temperature, less time.

**7.2.11 A mixture of water, acetic acid, and ethyl alcohol is placed in a sealed container at 40°C at equilibrium. How many degrees of freedom exist according to the phase rule for this system? List a specific variable for each degree of freedom.

**7.2.12 a. A system contains two components at equilibrium. What is the maximum number of phases possible with this system? Give reasons for your answer.

b. A two-phase system is specified by fixing the temperature, the pressure, and the amount of one component. How many components are there in the system at equilibrium? Explain.

7.3 Single-Component Two-Phase Systems (Vapor Pressure)

*7.3.1 Indicate whether the following statements are true or false:
a. The vapor-pressure curve separates the liquid phase from the vapor phase in a p-T diagram.
b. The vapor-pressure curve separates the liquid phase from the vapor phase in a p-V diagram.
c. The freezing curve separates the liquid phase from the solid phase in a p-T diagram.
d. The freezing curve separates the liquid phase from the solid phase in a p-V diagram.
e. At equilibrium at the triple point, liquid and solid coexist.
f. At equilibrium at the triple point, solid and vapor coexist.

*7.3.2 Explain how the pressure for a pure component changes (higher, lower, no change) for the following scenarios:
a. A system containing saturated liquid is compressed at constant temperature.
b. A system containing saturated liquid is expanded at constant temperature.
c. A system containing saturated liquid is heated at constant volume.
d. A system containing saturated liquid is cooled at constant volume.
e. A system containing saturated vapor is compressed at constant temperature.
f. A system containing saturated vapor is expanded at constant temperature.
g. A system containing saturated vapor is heated at constant volume.
h. A system containing saturated vapor is cooled at constant volume.
i. A system containing vapor and liquid in equilibrium is heated at constant volume.
j. A system containing vapor and liquid in equilibrium is cooled at constant volume.
k. A system containing a superheated gas is expanded at constant temperature.
l. A system containing superheated gas is compressed at constant temperature.

*7.3.3 Ice skates function because a lubricating film of liquid forms immediately below the small contact area of the skate blade. Explain by means of diagrams and words why this liquid film appears on ice at 25°F.

*7.3.4 Methanol has been proposed as an alternate fuel for automobile engines. Proponents point out that methanol can be made from many feedstocks such as natural gas, coal, biomass, and garbage, and that it emits 45% less ozone precursor gases than does gasoline. Critics say that methanol combustion emits toxic formaldehyde and that methanol rapidly corrodes automotive parts. Moreover, engines using methanol are hard to start at temperatures below 40°F. Why are engines hard to start? What would you recommend to ameliorate the situation?

*7.3.5   Calculate the vapor pressure of each compound listed below at the designated temperature using the Antoine equation and the coefficients in Appendix G. Compare your results with the corresponding values of the vapor pressures obtained from the Antoine equation found in the physical properties package on this book's website.
   a. Acetone at 0°C
   b. Benzene at 80°F
   c. Carbon tetrachloride at 300 K

**7.3.6   Estimate the vapor pressure of ethyl ether at 40°C using the Antoine equation based on the experimental values as follows:

p^*(kPa)	2.53	15.0	58.9
T (°C)	−40.0	−10.0	20.0

**7.3.7   At the triple point, the vapor pressures of liquid and solid ammonia are respectively given by $\ln p^* = 15.16 - 3063/T$ and $\ln p^* = 18.70 - 3754/T$, where p^* is in atmospheres and T is in kelvin. What is the temperature at the triple point?

**7.3.8   In a handbook, the vapor pressure of solid decaborane ($B_{10}H_{14}$) is given as

$$\log_{10} p^* = 8.3647 - \frac{2642}{T}$$

and of liquid $B_{10}H_{14}$ as

$$\log_{10} p^* = 10.3822 - \frac{3392}{T}$$

The handbook also shows that the melting point of $B_{10}H_{14}$ is 89.8°C. Can this be correct?

**7.3.9   Calculate the normal boiling point of benzene and of toluene using the Antoine equation. Compare your results with listed data in a handbook or database.

**7.3.10 Numerous methods are employed to evaporate metals in thin film deposition. The rate of evaporation is

$$W = 5.83 \times 10^{-2} \frac{p_v M^{1/2}}{T^{1/2}} \, g\,/\,(cm^2)(s)$$

(p_v in torr, T in K, M = molecular weight)

Since p_v is also temperature-dependent, it is necessary to define further the vapor pressure-temperature relationship for this rate equation. The vapor-pressure model is

$$\log_{10}p_v = A - \frac{B}{T}$$

where T is in kelvin. Calculate the temperature needed for an aluminum evaporation rate of 10^{-4} g/(cm²)(s). Data: $A = 8.79$, $B = 1.594 \times 10^4$.

**7.3.11 Calculate the specific volume for water (in cubic meters per kilogram for a and b, and in cubic feet per pound for c and d) that exists at the following conditions:
   a. $T = 100°C$, $p = 101.4$ kPa, $x = 0.5$
   b. $T = 406.70$ K, $p = 300.0$ kPa, $x = 0.5$
   c. $T = 100°F$, $p = 0.9487$ psia, $x = 0.3$
   d. $T = 860.97°R$, $p = 250$ psia, $x = 0.7$

**7.3.12 Indicate whether the following statements are true or false:
   a. A pot full of boiling water is tightly closed by a heavy lid. The water will stop boiling.
   b. Steam quality is the same thing as steam purity.
   c. Liquid water that is in equilibrium with its vapor is saturated.
   d. Water can exist in more than three different phases.
   e. Superheated steam at 300°C means steam at 300 degrees above the boiling point.
   f. Water can be made to boil without heating it.

**7.3.13 A vessel that has a volume of 0.35 m³ contains 2 kg of a mixture of liquid water and water vapor at equilibrium with a pressure of 450 kPa. What is the quality of the water vapor?

**7.3.14 A vessel with an unknown volume is filled with 10 kg of water at 90°C. Inspection of the vessel at equilibrium shows that 8 kg of the water is in the liquid state. What is the pressure in the vessel, and what is the volume of the vessel?

**7.3.15 What is the velocity in feet per second when 25,000 lb/hr of superheated steam at 800 psia and 900°F flow through a pipe of inner diameter 2.9 in.?

**7.3.16 Maintenance of a heater was carried out to remove water that had condensed in the bottom of the heater. By accident, hot oil at 150°C was released into the heater when the maintenance man opened the wrong valve. The resulting explosion caused serious damage both to the maintenance man and to the equipment he was working on. Explain what happened during the incident when you write up the accident report.

***7.3.17 In a vessel with a volume of 3.00 m³, you put 0.030 m³ of liquid water and 2.97 m³ of water vapor so that the pressure is 101.33 kPa. Then you heat the system until all of the liquid water just evaporates. What are the temperature and pressure in the vessel at that time?

***7.3.18 In a vessel with a volume of 10.0 ft³, you put a mixture of 2.01 lb of liquid water and water vapor. When equilibrium is reached, the pressure in the vessel is measured as 80 psia. Calculate the quality of the water vapor in the vessel, and the respective masses and volumes of liquid and vapor at 80 psia.

***7.3.19 Take 10 data points from the steam tables for the vapor pressure of water as a function of temperature from the freezing point to 500 K, and fit the following function:

$$p^* = \exp[a + b \ln T + c(\ln T)^2 + d(\ln T)^3]$$

where p is in kilopascals and T is in kelvin.

***7.3.20 For each of the conditions of temperature and pressure listed below for water, state whether the water is a solid phase, liquid phase, superheated, or a saturated mixture, and if the latter, indicate how you would calculate the quality. Use the steam tables (inside the back cover) to assist in the calculations.

State	p (kPa)	T (K)	$\hat{V}$(m³/kg)
1	2000	475	—
2	1000	500	0.2206
3	101.3	200	—
4	245.6	400	0.7308
5	1000	453.06	0.001127
6	200	393.38	0.8857

***7.3.22 Prepare a Cox chart for (a) acetone vapor, (b) heptane, (c) ammonia, (d) ethane from 0°C to the critical point (for each substance). Compare the estimated vapor pressure at the critical point with the critical pressure.

***7.3.23 Estimate the vapor pressure of benzene at 125°C from the following vapor-pressure data:

T (°F):	102.6	212
p^* (psia):	3.36	25.5

by preparing a Cox chart.

***7.3.24 Estimate the vapor pressure of aniline at 350°C based on the following vapor-pressure data:

T (°C):	184.4	212.8	254.8	292.7
p^* (atm):	1.00	2.00	5.00	10.00

***7.3.25 Exposure in the industrial workplace to a chemical can come about by inhalation and skin adsorption. Because skin is a protective barrier for many chemicals, exposure by inhalation is of primary concern. The vapor pressure of a compound is one commonly used measure of exposure in the workplace. Compare the relative vapor pressures of three compounds added to gasoline—methanol, ethanol, and MTBE (methyl tert-butyl ether)—with their respective OSHA permissible exposure limits (PEL) that are specified in parts per million (by volume):

Methanol	200
Ethanol	1000
MTBE	100

7.4 Two-Component Gas/Single-Component Liquid Systems

*7.4.1 Suppose that you place in a volume of dry gas that is in a flexible container a quantity of liquid and allow the system to come to equilibrium at constant temperature and total pressure. Will the volume of the container increase, decrease, or stay the same from the initial conditions? Suppose that the container is of a fixed instead of flexible volume, and the temperature is held constant as the liquid vaporizes. Will the pressure increase, decrease, or remain the same in the container?

**7.4.2 A large chamber contains dry N_2 at 27°C and 101.3 kPa. Water is injected into the chamber. After saturation of the N_2 with water vapor, the temperature in the chamber is 27°C.
   a. What is the pressure inside the chamber after saturation?
   b. How many moles of H_2O per mole of N_2 are present in the saturated mixture?

**7.4.3 The vapor pressure of hexane (C_6H_{14}) at –20°C is 14.1 mm Hg absolute. Dry air at this temperature is saturated with the vapor under a total pressure of 760 mm Hg. What is the percent excess air for combustion?

**7.4.4 In a search for new fumigants, chloropicrin (CCl_3NO_2) has been proposed. To be effective, the concentration of chloropicrin vapor must be 2.0% in air. The easiest way to get this concentration is to saturate air with chloropicrin from a container of liquid.

Assume that the pressure on the container is 100 kPa. What temperature should be used to achieve the 2.0% concentration? From a handbook, the vapor-pressure data are (T,°C; vapor pressure, mm Hg): 0, 5.7; 10, 10.4; 15, 13.8; 20, 18.3; 25, 23.8; 30, 31.1.

At this temperature and pressure, how many kilograms of chloropicrin are needed to saturate 100 m^3 of air?

**7.4.5   What is the dew point of a mixture of air and water vapor at 60°C and 1 atm in which the mole fraction of the air is 12%? The total pressure on the mixture is constant.

**7.4.6   Hazards can arise if you do not calculate the pressure in a vessel correctly. One gallon of a hazardous liquid that has a vapor pressure of 13 psia at 80°F is transferred to a tank containing 10 ft^3 of air at 10 psig and 80°F. The pressure seal on the tank containing air will rupture at 30 psia. When the transfer takes place, will you have to worry about the seal rupturing?

**7.4.7   A room contains 12,000 ft^3 of air at 75°F and 29.7 in. Hg absolute. The air has a dew point of 60°F. How many pounds of water vapor are in the air?

**7.4.8   One gallon of benzene (C_6H_6) vaporizes in a room that is 20 ft by 20 ft by 9 ft in size at a constant barometric pressure of 750 mm Hg absolute and 70°F. The lower explosive limit for benzene in air is 1.4%. Has this value been exceeded?

**7.4.9   A mixture of acetylene (C_2H_2) with an excess of oxygen measured 350 ft^3 at 25°C and 745 mm Hg absolute pressure. After explosion, the volume of the dry gaseous product was 300 ft^3 at 60°C and the same pressure. Calculate the volume of acetylene and of oxygen in the original mixture. The final gas was saturated. Assume that all of the water resulting from the reaction was in the gas phase after the reaction.

**7.4.10   In a science question-and-answer column, the following question was posed: On a trip to see the elephant seals in California, we noticed that when the male elephant seals were bellowing, you could see their breath. But we couldn't see our own breath. How come?

**7.4.11   One way that safety enters into specifications is to specify the composition of a vapor in air that could burn if ignited. If the range of concentration of benzene in air in which ignition could take place is 1.4% to 8.0%, what would be the corresponding temperatures for air saturated with benzene in the vapor space of a storage tank? The total pressure in the vapor space is 100 kPa.

**7.4.12   When you fill your gas tank or any closed vessel, the air in the tank rapidly becomes saturated with the vapor of the liquid entering the tank. Consequently, as air leaves the tank and is replaced by liquid, you can often smell the fumes of the liquid around the filling vent such as with gasoline.

Suppose that you are filling a closed 5-gal can with benzene at 75°F. After the air is saturated, what will be the moles of benzene per mole of air expelled from the can? Will this value exceed the OSHA limit for benzene in air (currently 0.1 mg/cm^3)? Should you fill a can in your garage with the door shut in the winter?

**7.4.13 All of the water is to be removed from moist air (a process called dehydration) by passing it through silica gel. If 50 ft^3/min of air at 29.92 in. Hg absolute with a dew point of 50°F are dehydrated, calculate the pounds of water removed per hour.

***7.4.14 In a dry-cleaning establishment, warm dry air is blown through a revolving drum in which clothes are tumbled until all of the Stoddard solvent is removed. The solvent may be assumed to be *n*-octane (C_8H_{18}) and have a vapor pressure of 2.36 in. Hg at 120°F. If the air at 120°F becomes saturated with octane, calculate (a) the pounds of air required to evaporate 1 lb of octane, (b) the percent octane by volume in the gases leaving the drum, and (c) the cubic feet of inlet air required per pound of octane. The barometer reads 29.66 in. Hg.

***7.4.15 When people are exposed to certain chemicals at relatively low but toxic concentrations, the toxic effects are experienced only after prolonged exposures. Mercury is such a chemical. Chronic exposure to low concentrations of mercury can cause permanent mental deterioration, anorexia, instability, insomnia, pain and numbness in the hands and feet, and several other symptoms. The level of mercury that can cause these symptoms can be present in the atmosphere without a worker being aware of it because such low concentrations of mercury in the air cannot be seen or smelled.

Federal standards based on the toxicity of various chemicals have been set by OSHA for PEL, the maximum level of exposure permitted in the workplace based on a time-weighted average (TWA) exposure. The TWA exposure is the average concentration permitted for exposure day after day without causing adverse effects. It is based on exposure for 8 hr/day for the worker's lifetime.

The present federal standard (OSHA/PEL) for exposure to mercury in air is 0.1 mg/m^3 as a ceiling value. Workers must be protected from concentrations greater than 0.1 mg/m^3 if they are working in areas where mercury is being used.

Mercury manometers are filled and calibrated in a small storeroom that has no ventilation. Mercury has been spilled in the storeroom and is not completely cleaned up because the mercury runs into cracks in the floor covering. What is the maximum mercury concentration that can be reached in the storeroom if the temperature is 20°C? You may assume that the room has no ventilation and that the equilibrium concentration will be reached. Is this level acceptable for worker exposure?

Data: $p_{Hg}^{*} = 1.729 \times 10^{-4}$ kPa; the barometer reads 99.5 kPa. This problem has been adapted from the problems in the publication *Safety, Health, and Loss Prevention in Chemical Processes* published by the American Institute of Chemical Engineers, New York (1990) with permission.

***7.4.16   Figure P7.4.16 shows a typical *n*-butane loading facility. To prevent explosions, either (a) additional butane must be added to the intake lines (a case not shown) to raise the concentration of butane above the upper explosive limit (UEL) of 8.5% butane in air, or (b) air must be added (as shown in the figure) to keep the butane concentration below the lower explosive limit (LEL) of 1.9%. The *n*-butane gas leaving the water seal is at a concentration of 1.5%, and the exit gas is saturated with water (at 20°C). The pressure of the gas leaving the water seal is 120.0 kPa. How many cubic meters of air per minute at 20.0°C and 100.0 kPa must be drawn through the system by the burner if the joint leakage from a single tank car and two trucks is 300 cm³/min at 20.0°C and 100.0 kPa?

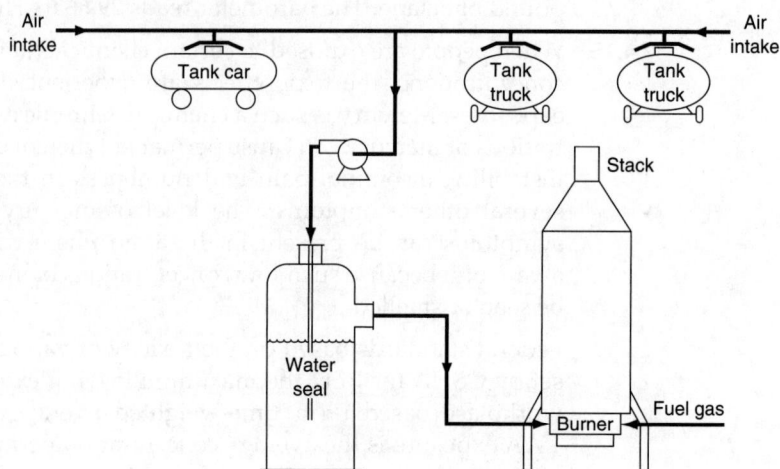

Figure P7.4.16

***7.4.17   Sludge containing mercury is burned in an incinerator. The mercury concentration in the sludge is 0.023%. The resulting gas (MW = 32) is 40,000 lb/hr, at 500°F and is quenched with water to bring it to a temperature of 150°F. The resulting stream is filtered to remove all particulates. What happens to the mercury? Assume the process pressure is 14.7 psia. (The vapor pressure of Hg at 150°F is 0.005 psia.)

***7.4.18   To prevent excessive ice formation on the cooling coils in a refrigerator room, moist air is partially dehydrated and cooled before passing it through the room (see Figure P7.4.18). The moist air from the cooler is passed into the refrigerator room at the rate of 20,000 ft^3/24 hr measured at the entrance temperature and pressure. At the end of 30 days, the refrigerator room must be allowed to warm in order to remove the ice from the coils. How many pounds of water are removed from the refrigerator room when the ice on the coils in it melts?

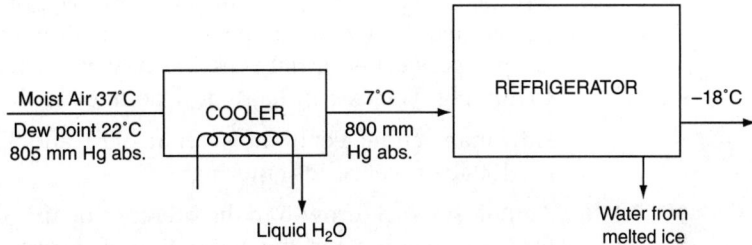

Figure P7.4.18

***7.4.19   Air at 25°C and 100 kPa has a dew point of 16°C. If you want to remove 50% of the initial moisture in the air (at a constant pressure of 100 kPa), to what temperature should you cool the air?

***7.4.20   One thousand cubic meters of air saturated with water vapor at 30°C and 99.0 kPa is cooled to 14°C and compressed to 133 kPa. How many kilograms of H_2O condense out?

***7.4.21   Ethane (C_2H_6) is burned with 20% excess air in a furnace operating at a pressure of 100 kPa. Assume complete combustion occurs. Determine the dew point temperature of the flue gas.

***7.4.22   A synthesis gas of the following composition—4.5% CO_2, 26.0% CO, 13.0% H_2, 0.5% CH_4, and 56.0% N_2—is burned with 10% excess air. The barometer reads 98 kPa. Calculate the dew point of the stack gas. To prevent condensation and consequent corrosion, stack gases must be kept well above their dew point.

***7.4.23 CH_4 is completely burned with air. The outlet gases from the burner, which contain no oxygen, are passed through an absorber where some of the water is removed by condensation. The gases leaving the absorber have a nitrogen mole fraction of 0.8335. If the exit gases from the absorber are at 130°F and 20 psia:
  a. To what temperature must this gas be cooled at constant pressure in order to start condensing more water?
  b. To what pressure must this gas be compressed at constant temperature before more condensation will occur?

***7.4.24   3M removes benzene from synthetic resin base sandpaper by passing it through a dryer where the benzene is evaporated into hot air. The air comes out saturated with benzene at 40°C (104°F). p^* of benzene at 40°C = 181 mm Hg; the barometer = 742 mm Hg. They recover the benzene by cooling to 10°C (p) and compressing to 25 psig. What fraction of the benzene do they recover? The pressure is then reduced to 2 psig, and the air is recycled in the dryer. What is the partial pressure of the benzene in the recycled air?

***7.4.25   Wet solids containing 40% moisture by weight are dried to 10% moisture content by weight by passing moist air over them at 200°F, 800 mm Hg pressure. The partial pressure of water vapor in the entering air is 10 mm Hg. The exit air has a dew point of 140°F.

How many cubic feet of moist air at 200°F and 800 mm Hg must be used per 100 lb of wet solids entering?

****7.4.26   Aerobic growth (growth in the presence of air) of a biomass involves the uptake of oxygen and the generation of carbon dioxide. The ratio of the moles of carbon dioxide produced per mole of oxygen consumed is called the respiratory quotient (RQ). Calculate the RQ for yeast cells suspended in the liquid in a well-mixed steady-state bioreactor based on the following data:

    a. Volume occupied by the liquid: 600 m^3
    b. Air (dry) flow rate into the gas (head) space: 600 m^3/hr at 120 kPa and 300 K
    c. Composition of the entering air: 21.0% O_2 and 0.055% CO_2
    d. Pressure inside the bioreactor: 120 kPa
    e. Temperature inside the bioreactor: 300 K
    f. Exit gas: saturated with water vapor and contains 8.04% O_2 and 12.5% CO_2
    g. Exit gas pressure: 110 kPa
    h. Exit gas temperature: 300 K

****7.4.27   Coal as fired contains 2.5% moisture. On a dry basis, the coal analysis is C 80%, H 6%, O 8%, ash 6%. The flue gas analyzes CO_2 14.0%, CO 0.4%, O_2 5.6%, N_2 80.0%. The air used has a dew point of 50°F. The barometer is 29.90 in. Hg. Calculate the dew point of the stack gas.

7.5   Two-Component Gas/Two-Component Liquid Systems

*7.5.1   Indicate whether the following statements are true or false:

    a. The critical temperature and pressure are the highest temperature and pressure at which a binary mixture of vapor and liquid can exist at equilibrium.
    b. Raoult's law is best used for a solute in dilute solutions.
    c. Henry's law is best used for a solute in concentrated solutions.
    d. A mixture of liquid butane and pentane can be treated as an ideal solution.

      e. The liquid phase region is found above the vapor phase region on a *p-x-y* chart.

      f. The liquid phase region is found below the vapor phase region on a *T-x-y* chart.

*7.5.2 Examine the following statements:

      a. "The vapor pressure of gasoline is about 14 psia at 130°F."

      b. "The vapor pressure of the system, water-furfural diacetate, is 760 mm Hg at 99.96°C."

Are the statements correct? If not, correct them. Assume the numerical values are correct.

**7.5.3 Determine if Henry's law applies to H_2S in H_2O based on the following measurements at 30°C:

Liquid Mole Fraction $\times 10^2$	Pressure (kPa)
0.0003599	20
0.0004498	30
0.0005397	40
0.0008273	50
0.0008992	60
0.001348	90
0.001528	100
0.003194	200
0.004712	300
0.007858	500
0.01095	700
0.01376	900
0.01507	1000

**7.5.4 Determine the equilibrium concentration in milligrams per liter of chloroform in water at 20°C and 1 atm, assuming that gas and liquid phases are ideal and the mole fraction of the chloroform in the gas phase is 0.024. The Henry constant for chloroform is $H = 170$ atm/mol fraction.

**7.5.5 Water in an enclosed vessel at 17°C contains a concentration of dissolved oxygen of 6.0 mg/L. At equilibrium, determine the concentration of oxygen in the space above the water in mole fraction and the total pressure in kilopascals. Henry's law constant is 4.07×10^6 kPa/mol fraction.

**7.5.6 A tank contains a liquid composed of 60 mol % toluene and 40 mol % benzene in equilibrium with the vapor phase and air at 1 atm and 60°F.

      a. What is the concentration of hydrocarbons in the vapor phase?

      b. If the lower flammability limit for toluene in air is 1.27% and benzene is 1.4%, is the vapor phase flammable?

**7.5.7 Fuel tanks for barbecues contain propane and *n*-butane. At 120°F, in an essentially full tank of liquid that contains liquid and vapor in equilibrium and exhibits a pressure of 100 psia, what is the overall (vapor plus liquid) mole fraction of butane in the tank?

**7.5.8 Based on the following vapor-pressure data, construct the temperature-composition diagram at 1 atm for the system benzene-toluene, assuming ideal solution behavior.

	Vapor Pressure (mm Hg)	
Temperature (°C)	Benzene	Toluene
80	760	300
92	1078	432
100	1344	559
110.4	1748	760

**7.5.9 Sketch a *T-x-y* diagram that shows an azeotrope, and locate and label the bubble and dew lines and the azeotrope point.

**7.5.10 Methanol has a flash point at 12°C at which temperature its vapor pressure is 62 mm Hg. What is the flash point (temperature) of a mixture of 75% methanol and 25% water? Hint: The water does not burn.

**7.5.11 You are asked to determine the maximum pressure at which steam distillation of naphtha can be carried out at 180°F (the maximum allowable temperature). Steam is injected into the liquid naphtha to vaporize it. If the distillation is carried out at 160°F, the liquid naphtha contains 7.8% (by weight) nonvolatile impurities, and the initial charge to the distillation equipment is 1000 lb of water and 5000 lb of impure naphtha, how much water will be left in the still when the last drop of naphtha is vaporized? Data: For naphtha, the MW is about 107, and $p^*(180°F)$ = 460 mm Hg, $p^*(160°F)$ = 318 mm Hg.

***7.5.12 You are asked to remove 90% of the sulfur dioxide in a gas stream of air and sulfur dioxide that flows at the rate of 85 m^3/min and contains 3% sulfur dioxide. The sulfur dioxide is to be removed by a stream of water. The entering water contains no sulfur dioxide. The temperature is 290 K, and the pressure on the process is 1 atm. Find (a) the kilograms of water per minute needed to remove the sulfur dioxide, assuming that the exit water is in equilibrium with the entering gas; and (b) the ratio of the water stream to the gas stream. The Henry's law constant for sulfur dioxide at 290 K is 43 atm/mol fraction.

***7.5.13** What are (a) the pressure in the vapor phase and (b) the composition of the vapor phase in equilibrium with a liquid mixture of 20% pentane and 80% heptane at 50°F? Assume the mixture is an ideal one at equilibrium.

***7.5.14** Two kilograms of a mixture of 50-50 benzene and toluene is at 60°C. As the total pressure on the system is reduced, at what pressure will boiling commence? What will be the composition of the first bubble of liquid?

***7.5.15** The normal boiling point of propane is –42.1°C and the normal boiling point of *n*-butane is –0.5°C.
   a. Calculate the mole fraction of the propane in a liquid mixture that boils at –31.2°C and 1 atm.
   b. Calculate the corresponding mole fraction of the propane in the vapor at –31.2°C.
   c. Plot the temperature versus propane mole fraction for the system of propane and butane.

***7.5.16** In the system *n*-heptane–*n*-octane at 200°F, determine the partial pressure of each component in the vapor phase at liquid mole fractions of *n*-heptane of 0, 0.2, 0.4, 0.6, 0.8, and 1.0. Also calculate the total pressure above each solution.

Plot your results on a *P-x* diagram with mole fractions of C_7 increasing to the right and mole fractions of C_8 increasing to the left. The ordinate should be pressure in psia.

Read from the plotted graph the total pressure and the partial pressure of each component at a mole fraction of $C_7 = 0.47$ in the liquid.

***7.5.17** Calculate the bubble point of a liquid mixture of 80 mol % *n*-hexane and 20 mol % *n*-pentane at 200 psia.

***7.5.18** Calculate the dew point of a vapor mixture of 80 mol % *n*-hexane and 20 mol % *n*-pentane at 100 psia.

***7.5.19** A mixture of 50 mol % benzene and 50 mol % toluene is contained in a cylinder at 39.36 in. Hg absolute. Calculate the temperature range in which a two-phase system can exist.

***7.5.20** A liquid mixture of *n*-pentane and *n*-hexane containing 40 mol % *n*-pentane is fed continuously to a flash separator operating at 250°F and 80 psia. Determine (a) the quantity of vapor and liquid obtained from the separator per mole of feed, and (b) the composition of both the vapor and the liquid leaving the separator.

***7.5.21   Most combustion reactions occur in the gas phase. For any flammable material to burn, both fuel and oxidizer must be present, and a minimum concentration of the flammable gas or vapor in the gas phase must also exist. The minimum concentration at which ignition will occur is called the lower flammable limit (LFL). The liquid temperature at which the vapor concentration reaches the LFL can be found experimentally.

It is usually measured using a standard method called a "closed cup flash point" test. The "flash point" of a liquid fuel is thus the liquid temperature at which the concentration of fuel vapor in air is large enough for a flame to flash across the surface of the fuel if an ignition source is present.

The flash point and the LFL concentration are closely related through the vapor pressure of the liquid. Thus, if the flash point is known, the LFL concentration can be estimated, and if the LFL concentration is known, the flash point can be estimated.

Estimate the flash point (the temperature) of liquid *n*-decane that contains 5.0 mol % pentane. The LFL for pentane is 1.8% and that for *n*-decane is 0.8%. Assume the propane–*n*-decane mixture is an ideal liquid. Assume the ambient pressure is 100 kPa. This problem has been adapted from *Safety, Health, and Loss Prevention in Chemical Processes*, edited by J. R. Welker and C. Springer, American Institute of Chemical Engineers, New York (1990), with permission.

***7.5.22   Late in the evening of August 21, 1986, a large volume of toxic gas was released from beneath and within Lake Nyos in the Northwest Province of Cameroon. An aerosol of water mixed with toxic gases swept down the valleys to the north of Lake Nyos, leaving more than 1700 dead and dying people in its wake. The lake had a surface area of 1.48 km^2 and a depth of 200 m to 250 m. It took 4 days to refill the lake; hence it was estimated to have lost about 200,000 tons of water during the gas emission. To the south of the lake and in the small cove immediately to the east of the spillway, a wave rose to a height of about 25 m.

The conclusion of investigators studying this incident was that the waters of Lake Nyos were saturated with CO_2 of volcanic origin. Late in the evening of August 21, a pulse of volcanic gas—mainly CO_2 but containing some H_2S—was released above a volcanic vent in the northeast corner of the lake. The stream of bubbles rising to the surface brought up more bottom waters highly charged with CO_2 that gushed out, increasing the gas flow and hence the flow of water to the surface much as a warm soda bottle overflows on release of pressure. At the surface, the release of gas transformed the accompanying water into a fine mist and sent a wave of water crashing across the lake. The aerosol of water and CO_2 mixed with a trace of H_2S swept down the valleys to the north of the lake, leaving a terrible toll of injury and death in its wake.

If the solution at the bottom of the lake obeyed Henry's law, how much CO_2 was released with the 200,000 metric tons of water, and what would be the volume of the CO_2 at S.C. in cubic meters? At 25°C the Henry's law constant is 1.7×10^3 atm/mol fraction.

***7.5.23** If the pressure in the head space (gas space) in a bioreactor is 110 kPa and 25°C, and the oxygen concentration in the head space is enriched to 39.7%, what is the mole fraction of the dissolved oxygen in the liquid phase? What is the percent excess oxygen dissolved in the liquid phase compared with the saturation value that could be obtained from air alone dissolved in the liquid?

****7.5.24** One hundred moles per minute of a binary mixture of 50% A and 50% B are separated in a two-stage (serial) process. In the first stage, the liquid and vapor flow rates exiting from the stage are each 50 mol/min. The liquid stream is then passed into a second separator that operates at the same temperature as the first stage, and the respective exit streams of liquid and vapor from the second stage are each 25 mol/min. The temperature is the same for each stage, and at that temperature, the vapor pressure of A is 10 kPa and the vapor pressure of B is 100 kPa. Treat the liquids and vapors as ideal.

Calculate the compositions of all of the streams in the process, and calculate the pressure in each stage.

7.6 Multicomponent Vapor-Liquid Equilibrium

7.6.1 Three separate waste discharge streams from a plant into a river contain the following respective chemicals in the water:

	Concentration (g/100 g water)	K
Glycerol	5.5	1.20×10^{-7}
Methyl ethyl ketone (MEK)	1.1	3.065
Phenol	2.1	0.00485

The K-values are from the Aspen tech process simulator at 20°C.

Estimate the concentration of the respective compounds in the gas phase above each discharge stream at 20°C. Will volatilization from the discharge stream be significant?

PART IV
ENERGY BALANCES

CHAPTER 8

Energy Balances without Reaction

Chapter Objectives

- Understand the terms associated with energy balances.
- Appreciate the development of the general energy balance.
- Know how to apply energy balances to unsteady-state, closed systems; steady-state, open systems; and fluid flow problems.
- Quickly locate the source of property values from tables, charts, equations, and computer databases.
- For special case systems, be able to simplify the general energy balance for the specifics of that particular problem.

Introduction

Consider the schematic of a shell and tube heat exchanger, shown in Figure 8.1. The cold fluid enters a header and flows through a number of parallel tubes while the hot fluid flows through the heat exchanger on the outside of the tubes. As a result, thermal energy is transferred from the hot fluid to the cold fluid.

561

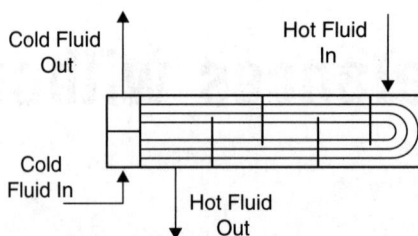

Figure 8.1   Schematic of a heat exchanger

Heat exchangers, which are one of the most common unit operations used in the chemical process industries, transfer thermal energy from one stream to another and are used to heat streams to a desired temperature, vaporize a liquid, condense a vapor, and recover thermal energy. For example, two heat exchangers are used on most distillation columns (Figure 8.2): the one at the bottom of the column, which supplies heat to create vapor flow up the column (the reboiler), and the one at the top of the column, which condenses the overhead vapor into a liquid that can be passed down the column (the condenser). The reboiler often uses condensing steam to create the vapor stream at the bottom of the column, while cooling water is used to condense the overhead vapor. The remainder of this chapter shows how to apply energy balances to describe the performance of heat exchangers and many other processes in terms of the various forms of energy.

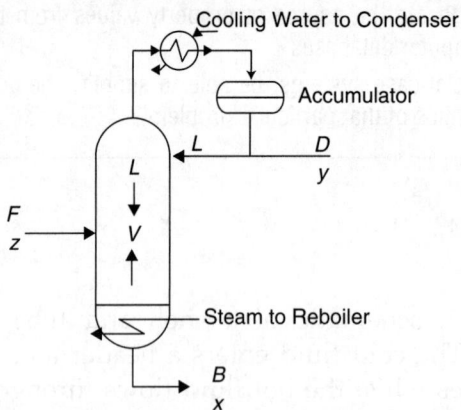

Figure 8.2   Schematic of a distillation column

Energy itself is often defined as the capacity to do work or transfer heat, a fuzzy concept. It is easier to understand specific types of energy. Two things energy is *not* are (a) some sort of invisible fluid and (b) something that can be measured directly.

Energy is a dominant element of any economy. When you consider the total energy consumption from cradle to grave (i.e., see Chapter 1, Section 1.4) for any product, energy is always a significant element, especially for hardware, electronics, and transportation. Our modern life is built around the use of energy from heating and cooling our homes to providing fuel for our automobiles to lighting our streets to powering our computers and the Internet. Energy is an essential element for useful products and services from collecting raw materials to manufacturing products to delivering the products to consumers.

No question exists as to the increase in the long-run use of energy. Figure 8.3 shows the history and forecast of U.S. energy demand and supply to the year 2050. Note the sharp decrease in the contribution from coal and its projected continued decline and the continued projected increase for renewable energy sources. Figure 8.4 shows the history and forecast for U.S. electric energy generation from selected fuels in billions of kilowatt-hours. The projections are that the nuclear and coal percentage contribution to the total will decrease by a factor of 2 by 2050 while renewables are expected to increase their contribution by a factor of 2. With respect to renewables, most of the growth is expected to come from solar, which is projected to increase by a factor of 8 from 2021 to 2050 while wind energy is expected to increase by a factor of 2.

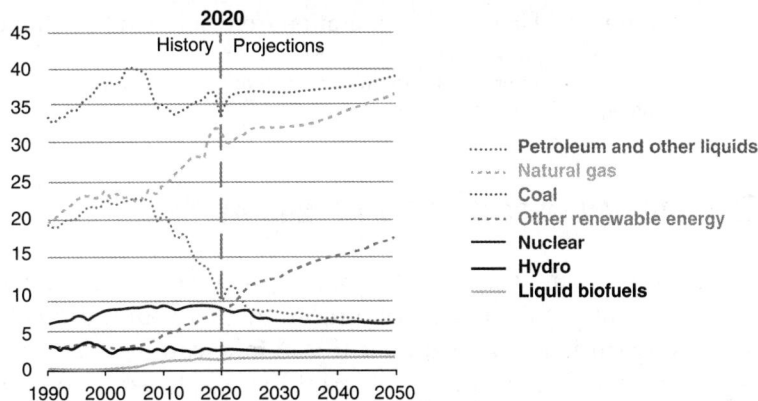

Source: U.S. Energy Information Administration, *Annual Energy Outlook 2021 (AEO2021)*

Figure 8.3   Past and predicted energy consumption by fuel (in quadrillion British thermal units)
Source: Energy Information Administration, *Annual Energy Outlook 2021*, AEO21(2021), Washington, DC (February, 2021).

In this chapter, we discuss energy balances together with the accessory background information needed to understand and apply them correctly. Our main attention is devoted to heat, work, enthalpy, internal energy, and carrying out energy balances in the absence of chemical reaction. Chapter 9 addresses energy balances with chemical reaction.

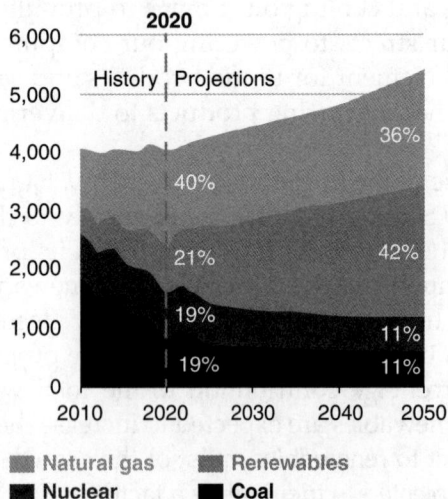

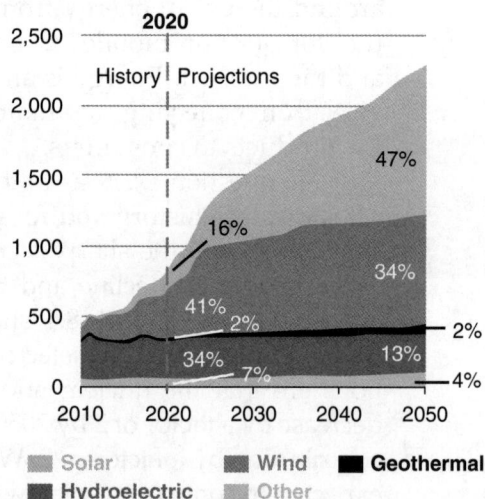

Source: U.S. Energy Information Administration, Annual Energy Outlook 2021 (AEO2021)

Figure 8.4   US Electric Generation from Selected Fuels (in quadrillion British thermal units)
Source: Energy Information Administration, *Annual Energy Outlook 2021*, AEO21(2021), Washington, DC (February, 2021).

8.1   Terminology Associated with Energy Balances

Some of the difficulty in analyzing processes involving energy balances occurs because of the failure of our language to communicate an exact meaning. Many difficulties disappear if you take care to learn the meaning of the terms listed in Tables 8.1 and 8.2. Table 8.1 lists terminology covered in previous chapters, and Table 8.2 lists new terminology that arises in connection with energy balances.

Table 8.1　Previously Defined Terminology That Pertains to Energy Balances

Term	Definition or Explanation
Boundary	The surface that separates a system from the surroundings. It may be a real or imaginary surface, either rigid or movable.
Closed system (nonflow system)	A system that does not interchange mass with the surroundings. However, heat and work can be exchanged.
Equilibrium (state)	The properties of a system are invariant despite flows of material or energy in and out; an implied state of balance. Types are thermal, mechanical, phase, and chemical equilibrium.
Extensive property	A property whose value depends on the amount of material present in a system, such as mass or volume.
Intensive property	A property whose value is independent of the amount of material present in a system, such as temperature or density (inverse of specific volume).
Macroscopic balance	A balance equation involving extensive variable, such as an amount of mass or energy.
Open system (flow system)	A system that is open to interchange of mass with the surroundings. Heat and work can also be exchanged.
Phase	A part (or whole) of a system that is physically distinct and macroscopically homogeneous of fixed or variable composition, such as gas, liquid, or solid.
Property	Observable (or calculable) characteristic of a system such as pressure, temperature, and volume.
State	Conditions of a system (specified by the values of temperature, pressure, composition, etc.).
Steady state	For this book, the accumulation in a system is zero. More generally, the flows in and out are constant, and the properties of the system are invariant.
Surroundings	Everything outside the system boundary.
System	The quantity of matter or region of space chosen for study enclosed by a boundary.
Unsteady state (transient state)	The system is not in the steady state.

Now for some comments about the new terminology listed in Table 8.2. The terms **adiabatic, isothermal, isobaric**, and **isochoric** are useful to specify conditions that do not change in a process. Isothermal, isobaric, and isochoric are short ways of saying no temperature change, no pressure change, and no volume change, respectively, occurs in the system; that is, the properties are invariant. Adiabatic means no heat transfer occurs between the system and its surroundings across the system boundary. Under what circumstances might a process be adiabatic?

- The system is insulated.
- Q is very small in relation to the other terms in the energy equation and may be neglected.
- The process takes place so fast that there is no time for heat to be transferred.

Table 8.2 Additional Terminology That Pertains to Energy Balances

Term	Definition or Explanation
Adiabatic system	A system that does not exchange heat with the surroundings during a process (i.e., perfectly insulated).
Isobaric system	A system in which the pressure is constant during a process.
Isochoric system	A system in which the volume is invariant during a process.
Isothermal system	A system in which the temperature is invariant during a process.
Path variable (function)	Any variable (function) whose value depends on how the process takes place and can differ for different histories (e.g., heat and work).
State variable (point function) (state function)	Any variable (function) whose value depends only on the state of the system and not on its previous history (e.g., internal energy).

The concept of a **state (or point) function** or **variable** is an important concept to understand. Temperature, pressure, and all of the other intensive variables are known as state variables because, between two different states, any change in value is the same no matter what path is taken between the two states. Look at Figure 8.5, which illustrates two processes, A and B. Both start at state 1 and terminate at state 2. The change in the value of a state variable is the same by both processes.

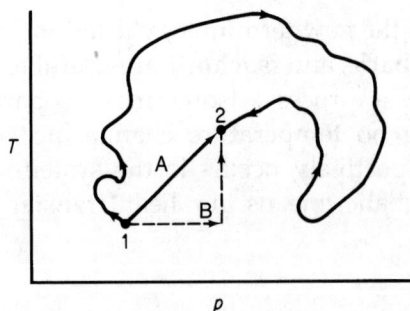

Figure 8.5 The value for the change in a state variable is the same for path A as for path B or for any other route between 1 and 2.

If two systems are in the same state, all of their state variables, such as temperature or pressure, must be identical. If the state of a system is changed, say, by heating, so that energy flows in, the values of its state variables change, and if the system is returned to its original state, say, by cooling, so that energy flows out, the values of its state variables return to their original values.

A process that proceeds first at constant pressure and then at constant temperature from state 1 to state 2 will yield exactly the same final value for a state variable as one that takes place first at constant temperature and then at constant pressure as long as the end point is the same. The concept of the state or point function is the same as that of an airplane passenger who plans to go straight to New York from Chicago but is detoured by way of Cincinnati because of bad weather. When the passenger arrives in New York, they are the same distance (the state variable) from Chicago no matter which route the plane traveled; hence, the value of the state variable depends only on the initial and final states. However, the fuel consumption of the plane may vary considerably; in analogous fashion, heat and work are two **path functions** or **variables** that are involved in an energy balance and may vary depending on the specific path chosen. If the passenger were to return to Chicago from New York, the distance from Chicago would be zero at the end of the return trip. Thus, the change in a state variable is zero for a cyclical process, which goes from state 1 to state 2 and back to state 1 again.

Let us now mention the units associated with energy. As you know, in the SI system, the unit of energy is the joule (J). In the American Engineering (AE) system, we use Btu, (ft)(lb_f), and (kW)(hr), among others. You can find conversion factors among the energy units listed on the inside of the front cover of this book.

What about the calorie? Is the unit archaic? It seems not. Most people are concerned about calories, not Btus or joules. A food package gives you information about calories (look at Figure 8.6). A calorie is only enough energy to raise the temperature of 1 mL of water 1 degree Celsius. The word *calorie* is misused in the context of food. The accurate term is *kilocalories* (kcal), generally expressed in writing as *Calorie* with a capital C. One kilocalorie equals 1000 real (thermochemical) calories.

Nutrition Facts

Serving Size 2 cookies (26 g)
Servings Per Container 8

Amount Per Serving		
Calories 110	Calories from Fat 35	

	% Daily Value*
Total Fat 4g	6%
Saturated Fat 3g	15%
Cholesterol 0mg	0%
Sodium 50mg	2%
Total Carbohydrate 18g	6%
Dietary Fiber 0g	0%
Sugars 12g	
Protein 1g	

Vitamin A 0%	•	Vitamin C 0%
Calcium 0%	•	Iron 2%

*Percent Daily Values are based on 2,000 calorie diet.
Your daily values may be higher or lower depending on
your calorie needs:

		Calories:	2,000	2,500
Total Fat	Less than		65g	80g
Sat Fat	Less than		20g	25g
Cholesterol	Less than		300mg	300mg
Sodium	Less than		2400mg	2400mg
Total Carbohydrate			300g	375g
Dietary Fiber			25g	30g

Figure 8.6   Information about the nutritional characteristics of a cookie

Thus, if your diet consists of 2000 Calories per day, you can calculate the number of joules involved per hour:

$$\frac{2000\ \text{Calories}}{\text{day}} \left| \frac{1\ \text{kcal}}{\text{Calorie}} \right| \frac{1000\ \text{cal}}{\text{kcal}} \left| \frac{4.184\ \text{J}}{\text{cal}} \right| \frac{1\ \text{day}}{24\ \text{hr}} = 350{,}000\ \text{J/hr}$$

Your body converts the food you eat into this amount of heat or work every hour, neglecting any energy stored as fat.

Self-Assessment Test

Questions

1. What is the essential difference between the system and the surroundings? Between an open and a closed system? Between a property and a phase?

2. Can a variable be both intensive and extensive variable at the same time?

3. Describe the difference between a state variable and a path variable.

Answers

1. The system is the region or a part of the universe selected for analysis. The surroundings are everything outside the system. An open system is a system in which mass enters and/or leaves the system. A closed system has no such mass

exchange. A phase is a physically distinct part or whole of a system. A property is an observable and calculable characteristic of a system.

2. No

3. The property change for a state variable depends only on the initial and final states, not on the path. The value for a path variable depends on how the variable moved from the initial to the final state.

Problems

1. What is the value of the change in the specific volume of a gas in a closed container that is first compressed to 100 atm, then heated to increase the temperature by 20%, and finally returned to its original state?

2. If you eat food containing 1800 Calories per day, according to an advertisement, you will lose weight. A handbook says that a person uses 20,000 kJ per day, given normal waking and sleeping activities. Will the person lose weight if he or she eats food as suggested by the advertisement?

Answers

1. No change

2. 20,000 kJ/day is about 4800 Calories per day: far too large to be realistic. Most diets call for about 1200 Calories per day; therefore, most individuals would likely not lose weight on a diet of 1800 Calories per day.

8.2   Overview of Types of Energy and Energy Balances

The principle of the **conservation of energy** states that the total energy of a system *plus the surroundings* can be neither created nor destroyed.[1] Julius Mayer (1814–1878) gave the first precise quantitative formulation of the principle of the conservation of energy. A journal refused to publish his ideas originally, and many of his contemporaries laughed at him when he

---

[1] Can mass be converted into energy according to $E = mc^2$? It is incorrect to say that $E = mc^2$ means mass is converted into energy. The equal sign can mean that two quantities have the same value as in measurements of two masses in an experiment, or it may mean (as in general relativity) that the two variables are the same or are equivalent things. It is in the latter sense that $E = mc^2$ applies, and not in terms of converting a *rest mass* into energy. You might write $\Delta E = c^2 \Delta m$. If Δm is negative, ΔE is also negative. What this means is that a loss of part of a body's mass is equivalent to the loss of an equivalent amount of energy as governed by the above equation. Treating the body as a system, this loss of energy from the system (negative ΔE) appears as a gain of energy by the surroundings.

explained his ideas. German physicist Johann Philipp Gustav Von Jolly said that if what Mayer proposed was true, it should be possible to heat water by shaking it—which, of course, you can do if your experiment is performed properly.

The principle is well founded based on experimental measurements. During an interaction between a system and its surroundings, the amount of energy gained by the system must be exactly equal to the amount of energy lost by the surroundings. Rather than the words *law of the conservation of energy*, in this book, we use *energy balance* to avoid confusion with the colloquial use of *energy conservation*, that is, reduction of energy waste or increased efficiency of energy utilization. Keep in mind three important points as you read what follows:

1. The energy balance is developed and applied from the macroscopic viewpoint (overall about the system) rather than from a microscopic viewpoint (i.e., an elemental volume within the system).

2. The energy balance is presented as a difference equation that incorporates net quantities over a time interval (i.e., extensive variables) analogous to the mass balances in Chapters 3, 4, and 5. Chapter 11 develops macroscopic energy balances that result in differential equations describing the dynamic behavior of these systems.

3. We examine only systems that are homogeneous, not charged, and without surface effects, in order to make the energy balance as simple as possible.

Many different forms of energy exist, but we limit our discussion to the following forms of energy that generally apply to energy balances for process systems:

- Q—Heat transfer across a system boundary into the system
- H—The energy content of a stream entering or leaving a system (i.e., enthalpy, which is defined in Section 8.4)
- ΔU—The change in thermal energy content of the material inside a system (i.e., internal energy, which is defined in Section 8.3)
- W—Work done on the system by the surroundings
- ΔPE—The change in potential energy (i.e., energy changes due to elevation changes)
- ΔKE—The change in kinetic energy (i.e., energy associated with changes in the velocity of material)

The **first law of thermodynamics**, which applies the principle of conservation of energy in a quantitative fashion, states that for a closed system, the change in total internal energy of the system (U_{total}) is equal to the sum of the heat added to the system (Q) and the work done on the system by the surroundings (W):

$$\Delta U_{total} = Q + W \tag{8.1}$$

The total internal energy for the material inside the system is made up of several components: potential energy (ΔPE), kinetic energy (ΔKE), and thermal energy (ΔU). Therefore, this equation states that the change in the total internal energy of the system for a period of time is equal to the total energy added to the system (i.e., Q and W) during that time period. Remember that this a macroscopic energy balance because it represents an overall representation of a system and involves extensive properties.

If we extend Equation (8.1) to the case for an open process, we must consider the addition of energy to the system provided by streams entering the system and the removal of energy by streams leaving the system. Here the energy entering and leaving the system with material is referred to as *convective energy* (ΔE_{conv}) and is made up of thermal energy (H), kinetic energy, and potential energy. Then the general energy balance equation for open systems becomes

$$\Delta U_{total} = Q + W - \Delta E_{conv} \tag{8.2}$$

where ΔE_{conv} is defined as the net energy content of the streams leaving minus entering the system (i.e., out minus in). Therefore, when ΔE_{conv} adds energy to the system, it has a negative sign (i.e., when the amount of energy entering is larger than the amount leaving). In simple terms, **Equation (8.2) states that the total change of energy of the material in a system for a period of time is equal to the total energy added** (heat transfer, work, and convective energy) **to the system during the time period.** The details of how to explicitly calculate ΔU_{total} and ΔE_{conv} are not presented here because the current emphasis is on developing an overall understanding of energy balances and, in general terms, how Equation (8.2) can be applied to the three major types of systems: closed, unsteady state (i.e., batch processes); open, steady state (i.e., steady-state continuous processes); and mechanical energy balances, which can be used to describe certain fluid flow systems. Sections 8.3 through 8.5 address in detail how to calculate each of the elements of ΔU_{total} and ΔE_{conv} and how to apply energy balances to a wide variety of process systems.

8.2.1   Closed, Unsteady-State Systems

This category pertains to batch processes for which the kinetic and potential energy changes are neglected. In addition, ΔE_{conv} is equal to zero because these are closed systems. In Section 8.3, we use certain gas/piston systems to demonstrate the application of this energy balance equation. Therefore, the general energy balance equation reduces to

$$\Delta U = Q + W \tag{8.3}$$

This equation states that the change in internal energy of a system for a period of time is equal to the heat transferred into the system plus the work done on the system by the surroundings for that time period.

8.2.2   Open, Steady-State Systems

This category is the most common type of energy balance employed for the units in large-scale continuous processes (e.g., units in a refinery or a plant that produces chemical intermediates such as ethylene). For these cases, we only need to consider heat transfer and the thermal energy content of the streams entering and leaving the system because these two forms of energy are so much larger than work, potential energy changes, and kinetic energy changes in most cases. Because these systems are considered to operate at steady state, the energy content of the material inside the system does not change with time, and therefore, ΔU_{total} is zero.

The energy balance for these systems is simply

$$\Delta H = Q \tag{8.4}$$

This equation states that the change in thermal energy content (i.e., enthalpy) of streams leaving and entering the process is equal to the heat transferred to the system. This equation assumes that no chemical reactions occur in the system. The effect of chemical reactions on energy balances is addressed in Chapter 9.

8.2.3   Mechanical Energy Balances

Mechanical energy balances are for open, steady-state systems that consider work done on the system, potential energy changes, and kinetic energy changes and can be used to describe certain fluid flow systems. For example, mechanical energy balances can be applied for pumping fluids through a piping system such as sizing pumps or calculating the flow rate through the system for continuous steady-state systems. In addition, the mechanical energy balance can be used to estimate the power generation for a hydro-electric process. Because this is a steady-state analysis and heat transfer is not considered, the mechanical energy balance reduces to

$$W = \Delta E_{conv} \tag{8.5}$$

This equation states that the work done on the system by the surroundings for a period of time is equal to the change in ΔE_{conv} for the same time period. Note that ΔE_{conv} contains terms for enthalpy, kinetic energy, and potential energy.

8.2.4 Special Cases

Closed, unsteady-state energy balances; open, steady-state energy balances; and mechanical energy balances address the vast majority of cases that a process engineer will encounter. Nevertheless, special cases can arise, such as an unsteady-state, open process. For such cases, you should return to the general energy balance equation, Equation (8.2), and simply retain each of the relevant terms. For example, for an unsteady-state continuous process, the general energy balance becomes

$$\Delta U = Q - \Delta H$$

assuming that kinetic and potential energy and work can be neglected. Likewise, for an open, steady-state process in which the work added is significant, the energy balance becomes

$$\Delta H = Q + W$$

Self-Assessment Test

Questions

1. The general energy balance states, in general terms, that the change in energy of a system is equal to the heat transferred to the system plus the work done on the system plus the net energy carried into the system by mass. True or false?

2. Why is $\Delta H = Q$ the most commonly used energy balance for steady-state, large-scale, continuous processing systems.

3. How do you apply an energy balance to a system that is neither a closed, unsteady-state process; an open, steady-state process; or a fluid flow problem?

4. What type of energy balance would you use for an open, steady-state process for which potential and kinetic energy changes are significant?

Answers

1. True

2. Because the system is at steady state, $\Delta E_{\text{total}} = 0$, and because work, kinetic energy, and potential energy are small compared to the heat transferred

3. Apply the general energy balance equation.

4. The mechanical energy balance equation

8.3   Energy Balances for Closed, Unsteady-State Systems

In this section, we address energy balances for batch systems. The energy balance for this class of systems, as given earlier in Equation (8.3), is

$$\Delta U = Q + W$$

because these systems do not involve kinetic and potential energy changes and are closed systems; thus, ΔE_{conv} is zero. Therefore, we first define heat transfer, work, and internal energy in some detail and present several example of how to apply this type of energy balance.

8.3.1   Heat Transfer (Q)

Heat transfer, Q, when used in the general energy balance, Equation (8.2), as a single term, is *the net amount of heat **transferred** to or from the system over a fixed time interval*. A process may involve more than one specified form of heat transfer, of course, the sum of which is Q. The rate of transfer will be designated by an overlay dot on Q ($\dot{Q}$), with the units of heat transfer per unit time, and the net heat transfer per unit mass would be designated by an overlay caret ($\hat{Q}$).

   In a discussion of *heat transfer*, we enter an area in which our everyday use of the term *heat* may cause some confusion because we use heat in a very restricted sense when we apply the laws governing energy changes. **Heat transfer** (Q) is commonly defined as that part of the total energy flow across a system boundary that is caused by a temperature difference (potential) between the system and the surroundings (or between two systems). See Figure 8.7. Engineers say "heat" when meaning "heat transfer" or "heat flow." Because heat is based on the transfer of energy, heat cannot be stored. **Heat is positive when transferred to the system and negative when removed from the system.** Heat is a **path variable.**

Keep in mind that a process in which no heat transfer occurs is an **adiabatic** process ($Q = 0$).

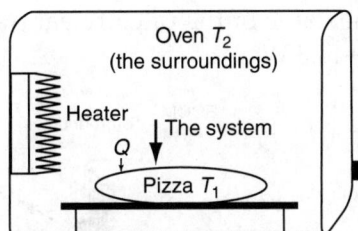

Figure 8.7   Heat transfer is energy that crosses the system boundary because of a temperature difference.

Here are some misconceptions about heat that you should avoid (saying *heat transfer* helps):

- Heat is a substance.
- Heat is proportional to temperature.
- A cold body contains no heat.
- Heating always results in an increase in temperature.
- Heat only travels upward.

Heat transfer is usually classified in three categories: conduction, convection, and radiation. To evaluate heat transfer *quantitatively*, you can apply various empirical formulas to estimate the heat transfer rate. One example of such a formula is the rate of heat transfer by convection that can be calculated from

$$\dot{Q} = UA(T_2 - T_1) \tag{8.6}$$

where $\dot{Q}$ is the rate of heat transfer (such as joules per second), A is the area for heat transfer (such as square meters), $(T_2 - T_1)$ is the temperature difference between the surroundings at T_2 and the system at T_1 (such as in degrees Celsius), and U is an empirical coefficient usually determined from experimental data for the equipment involved; it might have the units of $J/(s)(m^2)$ (°C). For example, ignoring conduction and radiation, the convective heat transfer from a person (the system) to a room (the surroundings) can be calculated using $U = 7\ W/(m^2)(°C)$; the data is described in Figure 8.8.

$$\dot{Q} = \frac{7W}{(m^2)(°C)}\bigg|\frac{1.6\ m^2}{}\bigg|\frac{(25-29)°C}{} = -44.8W\ or -44.8J/s$$

Note that $\dot{Q}$ is negative because heat is transferred from the system to the surroundings. Multiply $\dot{Q}$ by the time period in hours to get Q in watt-hours. What is the value of the rate of heat transferred into the air if the air is the system? It is +44.8 W.

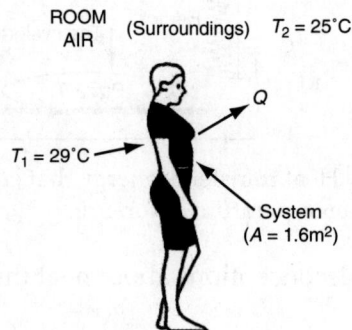

Figure 8.8 Heat transfer from a person

A device (system) that involves a high-temperature fluid and a low-temperature fluid to do work is known as a "heat engine." Examine Figure 8.9. Examples are power plants, steam engines, heat pumps, and so on.

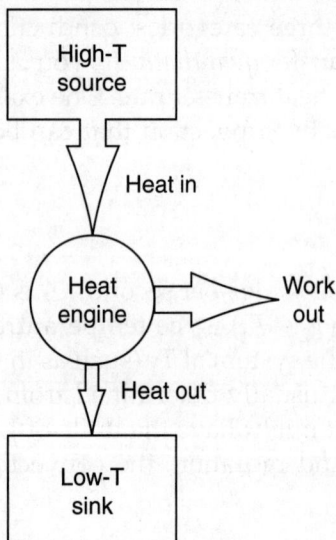

Figure 8.9 A heat engine produces work by operating between a high-temperature fluid and a low-temperature fluid.

Example 8.1   Energy Conservation

Problem Statement

Energy conservation is important for houses, commercial buildings, and so on. You have decided to replace an old single-pane window with an energy-efficient double-pane window with argon between the two panes. Because the double-pane window is so much more efficient than your current single-pane window, you decided to put in a bigger window going from a 2.5 ft by 5 ft window to a 5 ft by 5 ft window. For this example, assume

$$U_{\text{single pane}} = 5.5\,\text{Btu}/(\text{hr})\left(\text{ft}^2\right)(°\text{F}) \quad U_{\text{double pane}} = 0.36\,\text{Btu}/(\text{hr})\left(\text{ft}^2\right)(°\text{F}).$$

If the cost of energy is $8.50/10^6$ Btu, how many dollars per 30-day month are saved by changing the window if the given temperatures are constant assuming that the inside temperature is 75°F and the outside temperature is 25°F?

Solution

You could calculate $\dot{Q}$ for each case using Equation (8.1), but it is quicker to take a ratio:

$$\frac{\dot{Q}_2}{\dot{Q}_1} = \frac{U_{\text{new}}}{U_{\text{original}}}\frac{A_2}{A_1}\frac{\Delta T_2}{\Delta T_1} = \frac{(0.36)(25)}{(5.5)(12.5)} = 0.131$$

$$\text{Savings} = \frac{5.5\,\text{Btu}}{(\text{hr})\left(\text{ft}^2\right)(°\text{F})}\left|\frac{(2.5)(5)\,\text{ft}^2}{}\right|\left|\frac{(75-25)°\text{F}}{}\right|\frac{\$9.50}{10^6\,\text{Btu}}\left|\frac{24\,\text{hr}}{1\,\text{day}}\right|\frac{30\,\text{day}}{1\,\text{month}}\left|(1-0.131)\right.$$

$$= \$20.43/\text{month}$$

Note that if you went with the same sized window, you would save $40.84 per month.

Self-Assessment Test

Questions

1. There can be no heat transfer between two systems that are at the same temperature. True or false?

2. Which of the following are valid terms for heat transfer?

Heat addition	Heat generation
Heat rejection	Heat storage
Heat absorption	Electrical heating
Heat gain	Resistance heating
Heat loss	Frictional heating
Heat of reaction	Gas heating
Specific heat	Waste heat
Heat content	Body heat
Heat quality	Process heat
Heat sink	Heat source

3. Which of the following statements presents an incorrect view of heat?
   a. Los Angeles winters are mild because the ocean holds a lot of heat.
   b. Heat rises in the chimney of a fireplace.
   c. Your house won't lose much heat this winter because of the new insulation in the attic.
   d. A nuclear power plant dumps a lot of heat into the river.
   e. Close that door—don't let the heat out (in Minnesota) or in (in Texas).

Answers

1. True

2. Any of the terms that explicitly or implicitly denote the transfer of energy are okay; those that denote generation or storage are definitely wrong; those that give a mechanism of heat transfer and do not denote just heat are okay. Words such as *heat of reaction, specific heat, heat quality, heat sink, heat source,* and *body heat* are wrong.

3. (a), (b), (d), (e)

Problems

1. A calorimeter (a device to measure heat transfer) is being tested. It consists of a sealed cylindrical vessel containing water placed inside a sealed, well-insulated tank containing ice water at 0°C. The water in the cylinder is heated by an electric coil so that 1000 J of energy are introduced into the water. Then the water is allowed to cool until it reaches 0°C after 15 minutes and is in thermal equilibrium with the water in the ice bath.
   a. How much heat was transferred from the ice bath to the surrounding air during this test?
   b. How much heat was transferred from the cylinder to the ice bath during the test?
   c. If you pick two different systems composed of (1) the ice bath and (2) the cylinder, was the heat transfer to the ice bath *exactly* the same as the heat transfer from the ice bath to the cylinder?
   d. Is the interaction between the surroundings and the ice bath work, heat transfer, or both?

2. Classify the energy transfer in the following processes as work, heat transfer, both, or neither.
   a. A gas in a cylinder is compressed by a piston, and as a result, the temperature of the gas rises. The gas is the system.
   b. When an electric space heater is operating in a room, the temperature of the air goes up. The system is the room.
   c. The situation is the same as in b, but the system is the space heater.
   d. The temperature of the air in a room increases because of the sunshine passing through a window.

Answers

1. (a) 0 J because the ice bath is insulated; (b) 1000 J; (c) same amount but different sign; (d) heat transfer

2. (a) work; (b) work; (c) work and heat transfer; (d) heat transfer

8.3.2   Work (*W*)

The next type of energy we discuss is work (*W*). *Work* is a term that has wide usage in everyday life (e.g., "I am going to work") but has a specialized meaning in connection with energy balances. Work is a form of energy that represents a **transfer** of energy between the system and surroundings. Work cannot be stored. ***Work is a path variable. Work is positive when the*** *surroundings perform work on the system. Work is negative when the system performs work on the surroundings.* In some books, the sign is the opposite. In this text, the symbol W refers to the net work done over a period of time, *not* the *rate* of work. The latter is $\dot{W}$, the **power**, namely, the work per unit time. The work per unit mass will be designated by $\hat{W}$.

Many types of work can take place (which we will lump together under the notation W), among which are the following:

- **Mechanical work:** Work that occurs because of a mechanical force that **moves the boundary** of a system. You might calculate W on the system (or by the system with the appropriate sign) as

$$W = \int_{\text{state}1}^{\text{state}2} \vec{F} \cdot d\vec{s} \tag{8.7}$$

where F is an external force (a vector) in the direction of **s** (a vector) acting on the system boundary (or a system force acting at the boundary on the surroundings). However, the amount of mechanical work done by or on a system can be difficult to calculate because (a) the displacement $d\vec{s}$ may not be easy to define, and (b) the integration of $\vec{F} \cdot d\vec{s}$ does not necessarily give the amount of work actually being done on the system (or by the system) because some of the energy of W dissipates as Q into the contents of the system. For an example, look at Figure E8.2a in Example 8.2.

- **Electrical work:** Electrical work occurs when an electrical current passes through an electrical resistance in the circuit. If the system generates an electrical current (e.g., an electrical generator inside the system) and the current passes through an electrical resistance outside the system, the electrical work is negative because the electrical work is done on the surroundings. If the electrical work is done inside the system because of an applied voltage from outside the system, the electrical work is positive.

- **Shaft work:** Shaft work occurs when the system causes a shaft to turn against an external mechanical resistance. When a source of water outside the system circulates in the system and consequently causes a shaft to turn, the shaft work is positive. Look at Figure 8.10. When a shaft in the system turns a pump to pump water out of the system, the shaft work is negative. Does the shaft in Figure 8.10 rotate clockwise or counterclockwise? If water is pumped out of the system, which way does the shaft rotate?

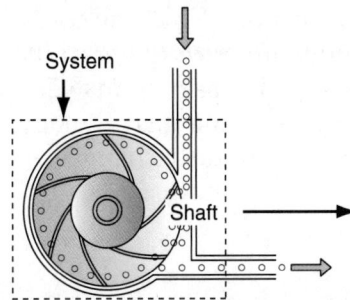

Figure 8.10 Shaft work. The force exerted on the impeller due to the water flow from the surroundings causes the shaft to rotate. Work is positive here, done by the surroundings on the system.

- **Flow work:** Flow work is performed on the system when a fluid is pushed into the system by the surroundings. Look at Figure 8.11. For example, when a fluid enters a pipe, some work is done on the system (the water that already is in the pipe) to force the new fluid into the pipe. Similarly, when fluid exits the pipe, the system does some work on the surroundings to push the exiting fluid into the surroundings. Flow work, which is part of the definition of enthalpy, is described in more detail when enthalpy is discussed in the next section.

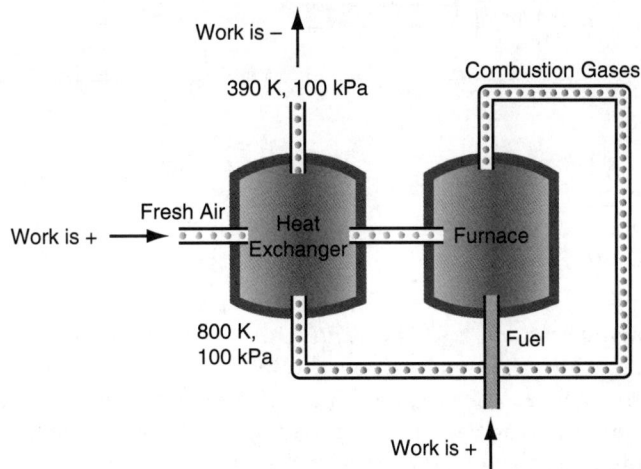

Figure 8.11 Flow work. Flow work occurs when the surroundings push an element of fluid into the system (sign is positive) or when the system pushes an element of fluid into the surroundings (sign is negative).

Suppose a gas in a fixed-volume container is heated so that its temperature is doubled. How much work was done on or by the gas during the process? Such a question is easy to answer: no work was done because the boundary of the system (the gas) remained fixed. Let us next look at an example in which the boundary changes.

Example 8.2   Calculation of Mechanical Work by a Gas on a Piston

Problem Statement

Suppose that an ideal gas at 300 K and 200 kPa is enclosed in a cylinder by a *frictionless (ideal) piston,* and the gas slowly forces the piston so that the volume of gas expands from 0.1 to 0.2 m^3. Examine Figure E8.2a. Calculate the work done by the gas on the piston (the only part of the system boundary that moves) if two different paths are used to go from the initial state to the final state.

Path *A*: The expansion occurs at constant pressure (**isobaric**) ($p = 200$ kPa).

Path *B*: The expansion occurs at constant temperature (**isothermal**) ($T = 300$ K).

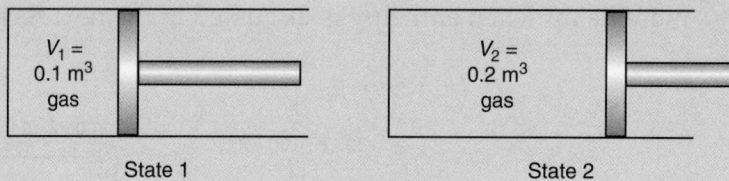

State 1   State 2

Figure E8.2a

Solution

The piston must be frictionless and the process ideal (occur very slowly) for the following calculations to be valid. Otherwise, some of the calculated work will be changed into a different form of unmeasured energy (i.e., thermal energy). The system is the gas. The piston is part of the surroundings. You use Equation (8.3) to calculate the work, but because you do not know the force exerted by the gas on the piston, you will have to use the pressure (force/area) as the driving force, which is okay because you do not know the area of the piston anyway and because p is exerted normally on the piston face. All of the data you need are provided in the problem statement. Let the basis be the amount of gas cited in the problem statement:

$$n = \frac{200\ \text{kPa}}{} \left| \frac{0.1\ \text{m}^3}{} \right| \frac{}{300\ \text{K}} \left| \frac{(\text{kg mol})(\text{K})}{8.314(\text{kPa})(\text{m}^3)} \right| = 0.00802\ \text{kg mol}$$

Figure E8.2b illustrates the two processes: an isobaric path and an isothermal path.

The mechanical work done by the system on the piston (in moving the system boundary) *per unit area* is

$$W = -\int_{\text{state 1}}^{\text{state 2}} \left(\frac{F}{A} \right)(A\,ds) = -\int_{V_1}^{V_2} p\,dV$$

Note that by definition, the work done by the system is *negative*. If the integral dV is positive (such as in expansion), the value of the integral will be positive and W negative (work done on the surroundings). If dV is negative, W will be positive (work done on the system).

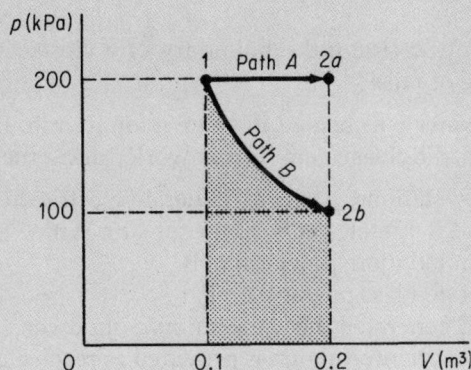

Figure E8.2b

Path *A* (the constant pressure process):

$$W = -p \int_{V_1}^{V_2} dV = -p(V_2 - V_1)$$

$$= -\frac{200 \times 10^3\ \text{Pa}}{} \left| \frac{1\ \text{N}}{1(\text{m}^2)(\text{Pa})} \right| \frac{0.1\,\text{m}^3}{} \left| \frac{1\ \text{J}}{1(\text{N})(\text{m})} \right| = -20\ \text{kJ}$$

(Continues)

Example 8.2   Calculation of Mechanical Work by a Gas on a Piston (*Continued*)

Path *B* (the constant temperature process):
   The gas is ideal. Then

$$W = -\int_{V_1}^{V_2} \frac{nRT}{V} dV = -nRT \ln\left(\frac{V_2}{V_1}\right)$$

$$= -\frac{0.00802 \,\text{kg mol}}{} \left|\frac{8.314 \,\text{kJ}}{(\text{kg mol})(\text{K})}\right| \frac{300\,\text{K}}{} \ln 2 = -13.86 \,\text{kJ}$$

In Figure E8.2b, the two integrals are areas under the respective curves in the *p-V* plot. By which path does the system do the most work?

Self-Assessment Test

Questions

1. If the energy crossing the boundary of a closed system is not heat, it must be work. True or false?

2. A pot of water was heated in a stove for 10 min. If the water was selected to be the system, did the system do any work during the 10 min?

3. Two power stations, A and B, generate electrical energy. A operates at 800 MW for 1 hr and B operates at 500 MW for 2 hr. Which is the correct statement?
   a. A generated more power than B.
   b. A generated less power than B.
   c. A and B generated the same amount of power.
   d. Not enough information is provided to reach a decision.

Answers

1. True based on the forms of energy considered here

2. Yes, in vaporization by pushing against the atmosphere

3. (a) because MW is power

Problems

1. A gas cylinder contains N_2 at 200 kPa and 80°C. As a result of cooling at night, the pressure in the cylinder drops to 190 kPa and the temperature to 30°C. How much work was done on the gas?

2. Nitrogen gas goes through four ideal process stages, as detailed in Figure SAT8.2.2P2. Calculate the work done by the gas on each stage in British thermal units.

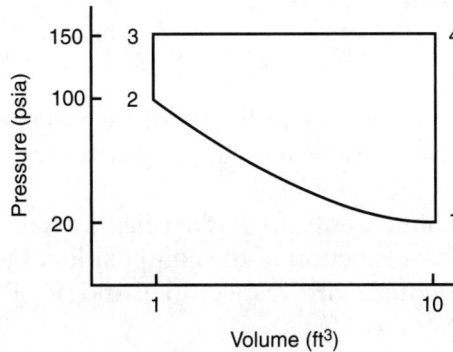

Figure SAT8.2.2P2

Answers

1. Heat was transferred, but no work was performed, ignoring the minute contraction of the cylinder itself.

2. The work from 3 to 4 is $W = p\Delta V = \dfrac{150 \text{ lb}_m}{\text{in}^2} \Big| \dfrac{9 \text{ ft}^3}{} \Big| \dfrac{144 \text{ in}^2}{\text{ft}^2} \Big| \dfrac{\text{Btu}}{778.2 \text{ lb}_m\text{-ft}}$

   $= 249.8 \text{ Btu}$; 1 to 2 and 4 to 1 are zero because volume is constant. For 1 to 2:

   $$W = \int_1^2 p\, dV \approx -\left\{ \left[\frac{100+60}{2} \right] 4.5 + \left[\frac{60+20}{2} \right] 4.5 \right\} \frac{144}{778.2} = -99.9 \text{ Btu based on trapezoi-}$$

   dal rule with $\Delta V = 4.5 \text{ ft}^3$.

8.3.3   Internal Energy

Internal energy (U) is a *macroscopic* concept that takes into account the molecular, atomic, and subatomic energies of entities, all of which follow definite microscopic conservation rules for dynamic systems. Specific internal energy is a state variable and can be stored. Because no instruments exist with which to measure internal energy directly on a macroscopic scale, internal energy must be calculated from certain other variables that can be measured macroscopically, such as pressure, volume, temperature, and composition.

To calculate the internal energy per unit mass ($\hat{U}$) from variables that can be measured, we make use of the phase rule. For a pure component in one phase, $\hat{U}$ can be expressed in terms of just two intensive variables according to the phase rule:

$$F = 2 - P + C = 2 - 1 + 1 = 2$$

Custom dictates the use of temperature and specific volume as the two variables. For a *single phase* and single component, we say that $\hat{U}$ is a function of only T and $\hat{V}$: $\hat{U} = \hat{U}(T, \hat{V})$.

If two components are in the phase, what is F? $C = 2$, hence, $F = 3$, and $\hat{U}$ would also be a function of the composition. Because $\hat{U}$ is a state function, $\hat{U}$ can be differential with respect to T and $\hat{V}$. By taking the total derivative, we find that

$$d\hat{U} = \left(\frac{\partial \hat{U}}{\partial T}\right)_{\hat{V}} dT + \left(\frac{\partial \hat{U}}{\partial \hat{V}}\right)_{T} d\hat{V} \tag{8.8}$$

By definition $(\partial \hat{U} / \partial T)_{\hat{V}}$ is the **heat capacity** (specific heat) at constant volume, given the special symbol C_v. C_v can also be defined to be the amount of heat necessary to raise the temperature of 1 kg of substance by 1 degree in a closed system and so has the SI units of J/(kg)(K), if the process is carried out at constant volume. For all practical purposes in this text, the term $(\partial \hat{U}/\partial \hat{V})_T$ is so small that the second term on the right side of Equation (8.5) can usually be neglected. (Note that in the steam tables, the second term on the right side of Equation (8.5) cannot be neglected.) Consequently, *changes in the internal energy* over a specified time interval can usually be computed by integrating Equation (8.5) as follows:

$$\Delta \hat{U} = \hat{U}_2 - \hat{U}_1 = \int_{\hat{U}_1}^{\hat{U}_2} d\hat{U} = \int_{T_2}^{T_2} C_v dT \tag{8.9}$$

For an ideal gas, $\hat{U}$ is a function of temperature only. Always keep in mind that **Equation (8.9) alone is *not* valid if a phase change occurs** during the interval.

Note that you can only calculate differences in internal energy or calculate the internal energy relative to a reference state *but cannot calculate absolute values* of internal energy. Look up the values of p and $\hat{V}$ for water for the reference state that has been assigned a zero value for $\hat{U}$. From the program for saturated steam, at 0°C, did you get $p = 0.612$ kPa corresponding to liquid water with $\hat{V} = 0.001000$ m²/kg? The reference internal energy cancels out when you calculate an internal energy difference as long as you use the same reference state for the variables:

$$\Delta \hat{U} = \left(\hat{U}_2 - \hat{U}_{ref} \right) - \left(\hat{U}_1 - \hat{U}_{ref} \right) = \hat{U}_2 - \hat{U}_1 \qquad (8.10)$$

If you do not use the same table, chart, or equation, you can't automatically cancel the $\hat{U}_{ref}$.

What would be the value of ΔU for a constant volume system if 1 kg of water at 100 kPa was heated from 0°C to 100°C, and then cooled back to 0°C and 100 kPa? Would $\Delta \hat{U} = 0$? Yes, because it is a state variable, and the integral in Equation (8.9) would be zero because $\hat{U}_2 = \hat{U}_1$.

The internal energy of a system containing more than one component is the sum of the internal energies of each component:

$$U_{tot} = m_1 \hat{U}_1 + m_2 \hat{U}_2 + \dots + m_n \hat{U}_n \qquad (8.11)$$

The heat of mixing, if any (discussed in Chapter 12), is neglected in Equation (8.11).

Equations, charts, or tables for C_v are rare; hence, you will usually have to calculate $\Delta \hat{U}$ by some other method than using Equation (8.9). But if you can find a relation for C_v, then getting ΔU is as simple as shown in the next example.

Example 8.3   Calculation of an Internal Energy Change Using the Heat Capacity

Problem Statement

What is the change in internal energy when 10 kg mol of air are cooled from 60°C to 30°C in a constant volume process?

Solution

Because you don't know the value of C_v, you have to look up the value. It is 2.1×10^4 J/(kg mol)(°C) over the temperature range. Use Equation (8.9) to carry out the calculation:

$$\Delta U = 10\,\text{kg} \int_{60°C}^{30°C} \left(2.1 \times 10^4 \frac{\text{J}}{(\text{kg mol})(°C)} \right) dT$$

$$= 2.1 \times 10^5 \, (30 - 60) = -6.3 \times 10^6 \, \text{J}$$

Self-Assessment Test

Questions

1. An entrance examination for graduate school asked the following two multiple-choice questions:
   a. The internal energy of a solid is equal to (1) the absolute temperature of the solid, (2) the total kinetic energy of its molecules, (3) the total potential energy of its molecules, or (4) the sum of the kinetic and potential energy of its molecules.
   b. The internal energy of an object depends on its (1) temperature only; (2) mass only; (3) phase only; (4) temperature, mass, and phase.

   Which answers would you choose?

2. If C_v is not constant over the temperature range in Equation (8.6), can you still integrate C_v to get ΔU?

Answers

1. Remember that only the internal energy change can be calculated, not the absolute value of the internal energy. Therefore, for (a) none of the answers would be acceptable, and for (b) also none, but (4) would be acceptable if the pressure were included.

2. Yes

Problems

1. A database lists an equation for $\hat{U}$ as

$$\hat{U} = 1.10 + 0.810T + 4.75 \times 10^{-4}T^2$$

   where $\hat{U}$ is in kilojoules per kilogram and T is in degrees Celsius. What are (a) the corresponding equation for C_v and (b) the reference temperature for $\hat{U}$?

2. Use the steam tables to calculate the change in internal energy between liquid water at 1000 kPa and 450 K, and steam at 3000 kPa and 800 K. How did you take into account the phase change in the water?

Answers

1. (a) $C_v = 9.5 \times 10^{-4}T + 0.810$; (b) Setting $\hat{U}$ equal to zero; the reference temperature is −1.36°C.

2. The saturated water internal energy for 450 K is equal to 749.0 kJ/kg. The internal energy for superheated steam at 3000 kPa and 450 K is 3154.7 kJ/kg. Therefore, $\Delta \hat{U}$ is equal to 2405.7 kJ/kg. The steam tables automatically incorporate phase changes into the data.

8.3.4   Closed, Unsteady-State Examples

For a closed, unsteady-state system, the energy balance is given by Equation (8.3):

$$\Delta U = Q + W$$

Because we have now discussed each of the terms in Equation (8.3), we are prepared to apply this equation to closed, unsteady-state systems.

If several components are involved in the process, then U_{inside} is the sum of the mass (or moles) of each component i times the respective specific internal energy of each component i, $\hat{U}_i$. Note that $(Q + W)$ represents the total flow of energy into the closed system. We do not put a Δ representing the symbol for a change in states before Q or W because they are not state variables. Remember that Q and W are both *positive* when the net transfer is into the system, and ΔU represents the change in the internal energy associated with *mass inside the system* itself. Be careful; in some books, W is defined as positive when done by the system. Also keep in mind that each term in Equation (8.3) represents the respective *net cumulative* amount of energy over the time interval from t_1 to t_2, not the respective energy per unit time, a rate that would be denoted by an overlay dot.

In closed systems, the values of ΔPE and ΔKE in ΔU_{total} are usually negligible or zero; hence, Equation (8.3) does not include these terms.

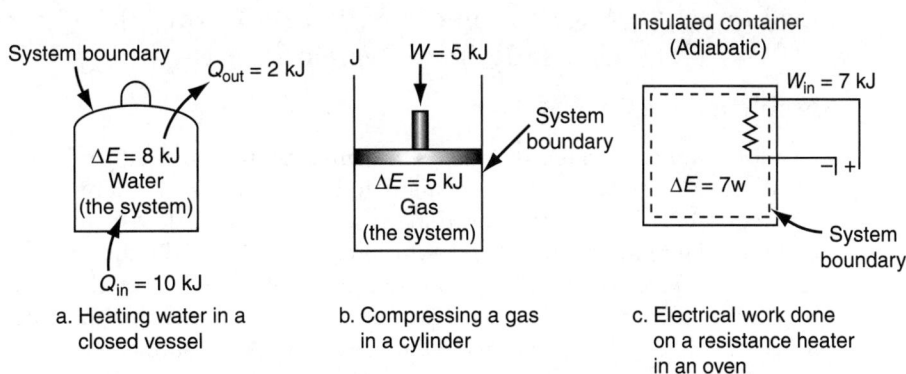

Figure 8.12   Examples of unsteady-state, closed systems that involve energy changes

If the sum of Q and W is positive, ΔU increases; if negative, ΔU decreases. Does $W_{system} = -W_{surrounds}$? Not necessarily, as you will learn subsequently. For example, in Figure 8.12c, the electrical work done by the surroundings on the system degrades into the internal energy (increase of temperature) of the system, not in expanding its boundaries.

Figure 8.12 illustrates three examples of applying Equation (8.3) to simple closed, unsteady-state systems. In Figure 8.12a, 10 kJ of heat are transferred through the fixed boundary (bottom) of a vessel with 2 kJ being transferred out at the top during the same time period. Thus, ΔU increases by 8 kJ. In Figure 8.12b, a piston does 5 kJ of work on a gas whose internal energy increases by 5 kJ. In Figure 8.12c, the voltage difference between the system and surroundings forces a current into a system in which no heat transfer occurs because of the insulation on the system.

Equation (8.3) involves several variables whose values have to be specified or solved for: ΔU, Q, and W. Furthermore, the specific internal energy, $\Delta \hat{U}$ is a function of T and $\hat{V}$ or, alternatively, T and p, inside the system. The net number of variables to be considered in a degree-of-freedom analysis involving Equation (8.3) is five using ΔU, or six substituting T, and p, for ΔU. With only one equation, Equation (8.3), the energy balance, the number of unknowns before specifications and material balances come into play will be two or three.

The next examples demonstrate how to apply energy balances to closed, unsteady-state systems.

Example 8.4   Application of an Energy Balance to a Closed, Unsteady-State System

Problem Statement

Alkaloids are chemical compounds containing nitrogen that can be produced by plant cells. In an experiment, an insulated closed vessel 1.673 m³ in volume was injected with a dilute water solution containing two alkaloids: ajmalicine and serpentine. The temperature of the solution was 10°C. To obtain an essentially dry residue of alkaloids, all of the water in the vessel was vaporized. Assume that the properties of water can be used in lieu of the properties of the solution. How much heat had to be transferred to the vessel if 1 kg of saturated liquid water initially at 10°C was completely vaporized to a final condition of 100°C and 1 atm? See Figure E8.4. Ignore any air present in the vessel (or assume an initial vacuum existed).

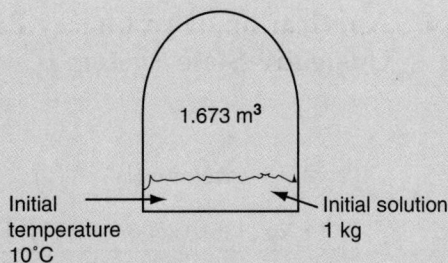

Figure E8.4

Solution

The system is the closed vessel; hence, from the viewpoint of the material balance, it is steady state, but from the viewpoint of the energy balance, it is unsteady state.

Sufficient data are given in the problem statement to fix the initial state and the final state of the water. You can look up the properties of water in steam tables. Note that the specific volume of steam at 100°C and 1 atm is 1.673 m^3/kg (!).

	Initial state (liquid)	Final state (gas)
p	vapor pressure	1 atm
T	10.0°C	100°C
$\hat{U}$	17.7 kJ/kg	2506.0 kJ/kg

You can look up additional properties of water such as $\hat{V}$ and $\hat{H}$, but they are not needed for the problem.

The system is closed, unsteady state, so Equation (8.3) applies:

$$\Delta U = Q + W$$

No work is involved (fixed tank boundary, no engine in the system). You can conclude, using

Basis: 1 kg of H$_2$O evaporated

(Continues)

Example 8.4   Application of an Energy Balance to a Closed, Unsteady-State System (*Continued*)

that

$$Q = \Delta U = m\Delta \hat{U} = m\left(\hat{U}_2 - \hat{U}_1\right)$$

$$Q = \frac{1 \text{ kg H}_2\text{O}}{} \left| \frac{(2506.0 - 17.7)\,\text{kJ}}{\text{kg}} = 2488 \text{ kJ}\right.$$

Data from steam tables are used in these calculations. **If you use two different sources of data, take care to account for differences in the reference values of each source**.

How would the solution differ if the container prior to the injection contained dry air at 1 atm?

Example 8.5   Application of the Energy Balance to a Closed System

Problem Statement

Ten pounds of water at room temperature (80°F) are stored in a sealed tank that has a volume of 4.0 ft³. How much heat must be transferred to the tank so that 40% of the water is vaporized? Also, what will be the final temperature and pressure?

Solution

This problem involves a closed, unsteady-state system (Figure E8.5) without reaction. The necessary property values can be obtained from the steam tables.

Steps 1–4

The specific volume of the solution is $4.0/10 = 0.40$ ft³/lb. Neglecting any air with the water in the tank, the initial internal energy is 48.02 Btu/lb.

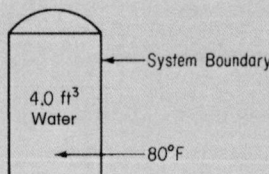

4.0 ft³
Water

System Boundary

80°F

Figure E8.5

Step 5

$$\text{Basis: 10 lb of } H_2O$$

Steps 6 and 7

The material balance is easy: the mass in the system is constant at 10 lb. As to the energy balance,

$$\Delta U = Q + W$$

W is zero because the boundary of the system is fixed, no electrical work is done, and so forth;

$$Q = \Delta U$$

Because the volume of the final state is fixed at 4 ft^3 and the quality of the final state is set at 40%, we can use these conditions to determine the temperature and pressure that satisfy these conditions. The necessary property values can be obtained from the steam tables. For 40% quality, there will be 4 lb of saturated steam and 6 lb of liquid. By trial and error, we can select temperatures and calculate the total volume based on the specific volume of the liquid and the specific volume of the vapor. For example, at 400°F, the specific volume of the vapor is 1.863 ft^3/lb and the specific volume of the liquid is 0.019 ft^3/lb. Therefore, the total volume for this case is

$$V(400°F) = 0.019(6) + 1.863(4) = 7.566 \text{ ft}^3$$

Therefore, the assumed temperature was too high. By continuing this trial-and-error procedure, a temperature of 375°F yields a total volume of 4 ft^3. And the corresponding pressure is 184.4 psia. In addition, the internal energy of the vapor and the liquid are provided by the software so that the total internal energy for the final state is 6538.5 Btu. Therefore,

$$Q = \Delta U = 6538.4 - 480.3 = 6058.3 \text{ Btu}$$

Example 8.6   Application of the Energy Balance to Plasma Etching

Problem Statement

Argon gas in an insulated plasma deposition chamber with a volume of 2 L is to be heated by an electric resistance heater. Initially, the gas, which can be treated as an ideal gas, is at 1.5 Pa and 300 K. The 1000-ohm heater draws current at 40 V for 5 min (i.e., 480 J of work is done on the system by the surroundings). What are the final gas temperature and pressure in the chamber? The mass of the heater is 12 g, and its heat capacity is 0.35 J/(g)(K). Assume that the heat transfer through the walls of the chamber from the gas at this low pressure and in the short time period involved is negligible.

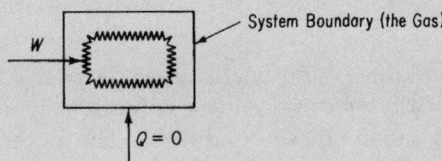

Figure E8.6

Solution

Pick the system shown in Figure E8.12. No reaction occurs. The fact that the electric coil is used to "heat" (raise the temperature of) the argon inside the system does not mean that heat transfer takes place to the selected system from the surroundings. Although the net result is an increase in temperature, only work is done, but the electrical work is all converted into thermal energy. The system does not exchange mass with the surroundings; hence it is steady state with regard to mass but is unsteady state with respect to energy.

Steps 1–4

Because of the assumption about the heat transfer from the chamber wall, $Q = 0$. W is given as +480 J (work done on the system) in 5 min.

Step 5

Basis: 5 min

Steps 6 and 7

The application of Equation 8.3 yields $\Delta U = 480$ J. One way to solve the problem is to find the T and p associated with a value of ΔU equal to 480 J. A table, equation, or chart for argon would make this procedure easy. In the absence of such a data source, we fall back on the assumption that the argon gas is an ideal gas (as is true), so that $pV = nRT$. Initially we know p, V, and T and thus can calculate the amount of the gas:

$$n = \frac{pV}{RT} = \frac{1.5 \text{ Pa}}{} \left| \frac{2 \text{ L}}{} \right| \frac{10^{-3} \text{ m}^3}{1 \text{ L}} \left| \frac{1 \text{ (g mol) (K)}}{8.314 \text{ (Pa) } \left(\text{m}^3\right)} \right| \frac{}{300 \text{ K}} = 1.203 \times 10^{-6} \text{g mol}$$

You are given the heater mass and its heat capacity of $C_v = 0.35 \text{ J}/(\text{g})(\text{K})$.

The C_v of the gas can be calculated. Because $C_p = \frac{5}{2}R$ (see Table 8.3),

$$C_v = C_p - R = \frac{5}{2}R - R = \frac{3}{2}R$$

You have to pick a reference temperature for the calculations. The most convenient reference state is 300 K. Then ΔU for the gas and the heater can be calculated assuming both the heater and the gas end up at the same temperature:

Gas: $$\Delta U_g = 1.203 \times 10^{-6} \int_{300}^{T} C_v \, dT = 1.203 \times 10^{-6} \left(\frac{3}{2}R\right)(T - 300)$$

Heater: $$\Delta U_h = 12 \text{ g} \left(\frac{0.35 \text{ J}}{(\text{g})(\text{K})}\right)(T - 300)$$

The mass balance is trivial: the mass in the chamber does not change. The unknown is T, and one equation is involved, the energy balance, so the degrees of freedom are zero.

Steps 8 and 9

Because $\Delta U = 480 \text{ J}$, you can calculate T from the energy balance:

$$\Delta U = 480 \text{ J} = (12)(0.35)(T - 300) + \left(2.302 \times 10^{-6}\right)\left(\frac{3}{2}\right)(8.314)(T - 300)$$

$$T = 414 \text{ K}$$

The final pressure is obtained from

$$\frac{p_2 V_2}{p_1 V_1} = \frac{n_2 R T_2}{n_1 R T_1} \text{ or}$$

$$p_2 = p_1 \left(\frac{T_2}{T_1}\right) = 1.5 \left(\frac{414}{300}\right) = 2.07 \text{ Pa}$$

Self-Assessment Test

Questions

1. Indicate whether the following statements are true or false:
   a. The law of the conservation of energy says that all of the energy changes in a system must add up to zero.
   b. The law of the conservation of energy says that no net change, or no creation or destruction of energy, is possible in a system.
   c. The law of the conservation of energy says that any change in heat that is not exactly equal to and opposite a change in work must appear as a change in the total internal energy in a system.
   d. Heat can flow into and out of a system from and to its surroundings. Work can be done by the system on the surroundings, and vice versa. The internal energy of a system can increase or decrease by any of these processes.
   e. For all adiabatic processes between two specified states of a closed system, the net work done is the same regardless of the nature of the closed system and the details of the process.
   f. A closed, unsteady-state system cannot be an isothermal system.
   g. Heat transfer to an unsteady-state, closed system cannot be zero.
   h. A closed system must be adiabatic.

2. Can heat be converted completely into work?

Answers

1. All false except (e)
2. No

Problems

1. A closed system undergoes three successive processes: $Q_1 = +10$ kJ, $Q_2 = +30$ kJ, and $Q_3 = -5$ kJ, respectively. In the first process, $\Delta E = +20$ kJ, and in the third process, $\Delta E = -20$ kJ. What is the work in the second process, and what is the net work output of all three stages if $\Delta E = 0$ for the overall three-stage process?

2. A closed tank contains 20 lb of water. If 200 Btu are added to the water, what is the change in internal energy of the water?

3. When a batch of hot water at 140°F is suddenly well mixed with cold water at 50°F, the water that results is at 110°F. What was the ratio of the hot water to the cold water? You can use the steam tables to get the data.

Answers

1. $W_1 = +10$ kJ; $W_3 = -15$ kJ; $\Delta E_2 = 0$; therefore, $W_2 = -30$ kJ; net work $= -35$ kJ.
2. 200 Btu
3. Energy lost by hot water is equal to energy gained by cold water: $m_h(140 - 110) = m_c(110 - 50)$; $m_h/m_c = 60/30 = 2$.

8.4   Energy Balances for Open, Steady-State Systems

As we mentioned earlier, you will find that the preponderance of industrial processes operate under approximately continuous, open, steady-state conditions. Most processes in the refining and chemical industries are open, steady-state systems. Biological processes are more likely to be closed systems (e.g., batch systems). You will find that continuous processes are most cost-effective in producing high-volume products.

Figure 8.13 shows some examples of open, steady-state processes. In Figure 8.13a, a fuel is burned in a boiler to heat tubes through which water flows and becomes steam. In Figure 8.13b, a dilute liquid feed containing a solute is concentrated to a "thick liquor." Vapor from the liquid is removed overhead. To provide the necessary heat transfer, steam flows through a steam chest (heat exchanger). In Figure 8.13c, a liquid containing a desirable solute is passed through a column countercurrent to an immiscible solvent that favors extracting the solute from the liquid.

Equation (8.4) applies to steady-state, open processes because steady-state means that the final and initial states of the system are the same, $\Delta U_{total} = 0$. In addition, W, ΔKE, and ΔPE are typically quite small compared to the heat transfer Q. Consequently, Equation (8.2) becomes Equation (8.4):

$$\Delta H = Q$$

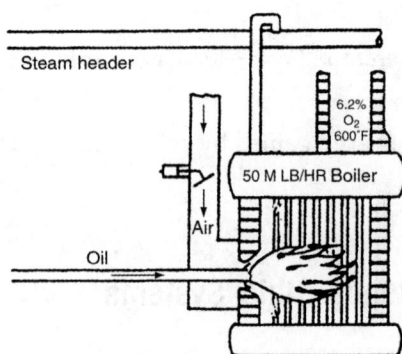

a. Boiler to generate steam

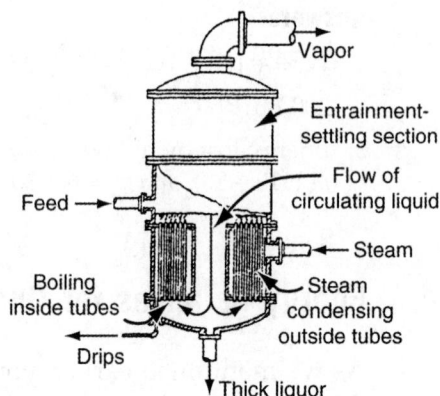

b. Evaporator that concentrates a solute

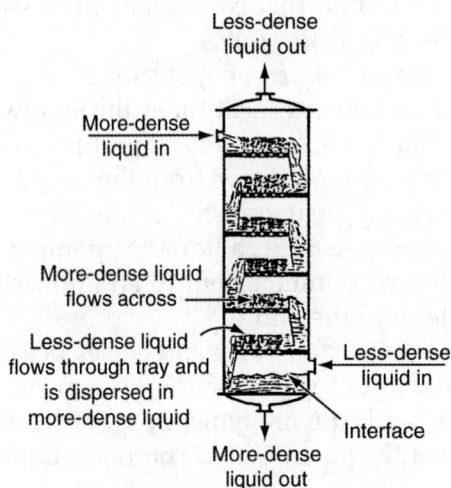

c. Plate extraction column

Figure 8.13   Examples of open, steady-state processes (without reaction occurring)

When are ΔPE and ΔKE negligible? Because the energy terms in the energy balance in most open processes are dominated by Q and ΔH; therefore, ΔPE and ΔKE are not used in Equation (8.4). For example, consider the right side of Equation (8.4). An enthalpy change of 1000 J/kg is really quite small, corresponding in air to a temperature change of about 1 K. For the other terms on the right side of Equation (8.4) to be equivalent to 1000 J:

1. The *PE* change would require 1 kg to go up a distance of 100 m.
2. The *KE* change would require a velocity change from 0 to 45 m/s.

Before we show how to apply energy balances, we need to explain enthalpy and show how to use it. Remember that heat transfer Q was already addressed in the previous section.

8.4.1 Enthalpy

Recall the flow work that was introduce in Section 8.3.2. Instead of treating flow work as work W for open systems, it is combined with the internal energy U to form another variable called the **enthalpy** (pronounced en-THAL-py):

$$H = U + pV \tag{8.12a}$$

where p is the pressure and V is the volume, or on a basis of a unit mass

$$\hat{H} = \hat{U} + p\hat{V} \tag{8.12b}$$

To calculate the specific enthalpy (enthalpy per unit mass), as with the specific internal energy, we use the property that the enthalpy is an exact differential. As you saw for internal energy, the state for the enthalpy for a **single phase** and single component can be completely specified by two intensive variables. We express the enthalpy in terms of the temperature and pressure, which is a more convenient variable than the specific volume. If we let

$$\hat{H} = \hat{H}(T, p)$$

by taking the total derivative of $\hat{H}$, we can form an expression analogous to Equation (8.8):

$$d\hat{H} = \left(\frac{\partial \hat{H}}{\partial T}\right)_p dT + \left(\frac{\partial \hat{H}}{\partial p}\right)_T dp \tag{8.13}$$

By definition, $(\partial \hat{H}/\partial T)_p$ is the heat capacity at constant pressure and is given the special symbol C_p. For most practical purposes, $(\partial \hat{H}/\partial p)_T$ is so small at modest pressures that the second term on the right side of Equation (8.13) can be neglected. Changes in the specific enthalpy can then be calculated by integration of Equation (8.13) as follows:

$$\Delta \hat{H} = \hat{H}_2 - \hat{H}_1 = \int_{H_1}^{H_2} d\hat{H} = \int_{T_1}^{T_2} C_p \, dT \tag{8.14}$$

However, in processes operating at high pressures, the second term on the right side of Equation (8.13) cannot necessarily be neglected but must be evaluated from experimental data. Consult the references at the end of the chapter for details. **One property of ideal gases, but *not* of real gases, to remember is that their enthalpies and internal energies are functions of temperature only** and are not influenced by changes in pressure or specific volume. Also, the relation between C_p and C_v for an ideal gas is $C_v = C_p - R$.

Where can you get values of enthalpies or data to calculate enthalpy values? Here are some sources:

1. Heat capacity values or equations
2. Equations to estimate the enthalpy of a phase transition
3. Tables
4. Enthalpy charts
5. Computer databases

A number of resources are available on the Internet for obtaining the values of enthalpy and other thermophysical properties. Many of them can be accessed without any charge, while a licensing fee is required to access the information contained in others. Some of these resources are listed on page 616. Figure 8.14 shows a listing of the thermophysical property data, which is available from the National Institute of Standards and Technology. You can specify the substance, choose the property units, the type of data, and the standard state condi-tions. Clicking the continue button will bring up a screen that allows you to input the pressure and temperature values, with the software returning the property values immediately.

Enthalpies can also be *estimated* by generalized methods based on the theory of corresponding states or additive bond contributions, but we do not discuss these methods here. Refer instead to the references at the end of the chapter for information.

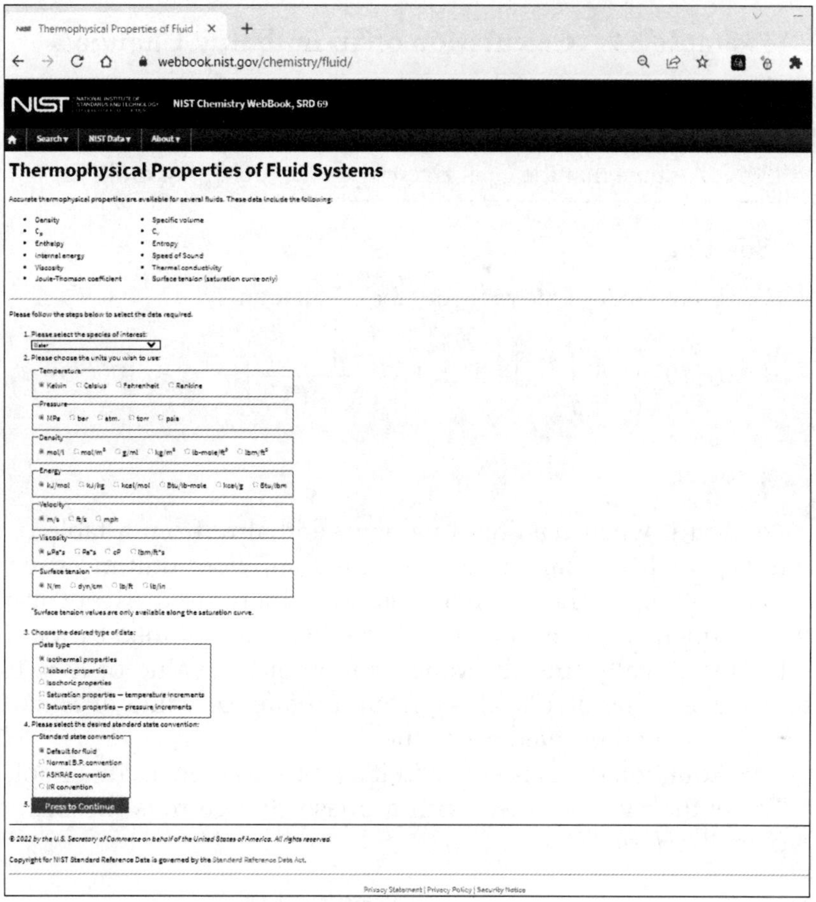

Figure 8.14   Screenshot of NIST webpage

As with internal energy, **enthalpy has no absolute value; only changes in enthalpy can be calculated**. Often you will use a reference set of conditions (perhaps implicitly) in computing enthalpy changes. For example, the reference conditions used in the steam tables are liquid water at 0°C (32°F) and its vapor pressure. This does not mean that the enthalpy is actually zero under these conditions but merely that the enthalpy has arbitrarily been assigned a value of zero at these conditions. In computing enthalpy changes, the reference conditions cancel out, as can be seen from the following:

Initial state of the system (1)   *Final state of the system* (2)

$$\text{Specific enthalpy} = \hat{H}_1 - \hat{H}_{\text{ref}} \qquad \text{Specific enthalpy} = \hat{H}_2 - \hat{H}_{\text{ref}}$$

$$\text{Net specific enthalpy change} = \left(\hat{H}_2 - \hat{H}_{\text{ref}}\right) - \left(\hat{H}_1 - \hat{H}_{\text{ref}}\right) = \hat{H}_2 - \hat{H}_1$$

Example 8.7   Calculation of an Enthalpy Change

Problem Statement

Calculate the enthalpy change for argon for a temperature change from 60°C to 30°C. Assume that the C_p is given by 2.913×10^4 J/(kg mol)(°C).

Solution

Use Equation (8.14) to carry out the calculation:

$$\Delta H = 10\,\text{kg mol} \int_{60°C}^{30°C} \left(2.9 \times 10^4 \frac{\text{J}}{(\text{kg mol})(°C)} \right) dT = 2.9 \times 10^5 \, (30 - 60) = -8.7 \times 10^6 \, \text{J}$$

Be careful when reading a value of enthalpy from a table or chart because the symbol H usually refers to the enthalpy per unit mass (or mole), not the total enthalpy. The symbol $\hat{H}$ is not widely used except in this book, and consequently the meaning of H should be determined by looking at the units located (ideally) in a heading or footnote. A value of $H = 100$ kJ probably means a value of 100 kJ/kg from a reference value of H such as $H = 0$, or possibly another reference value.

Equation (8.14) is *not* valid if a phase change occurs. Look at Figure 8.15. The enthalpy associated with a phase change must be included to get the overall ΔH (or $\Delta \hat{H}$).

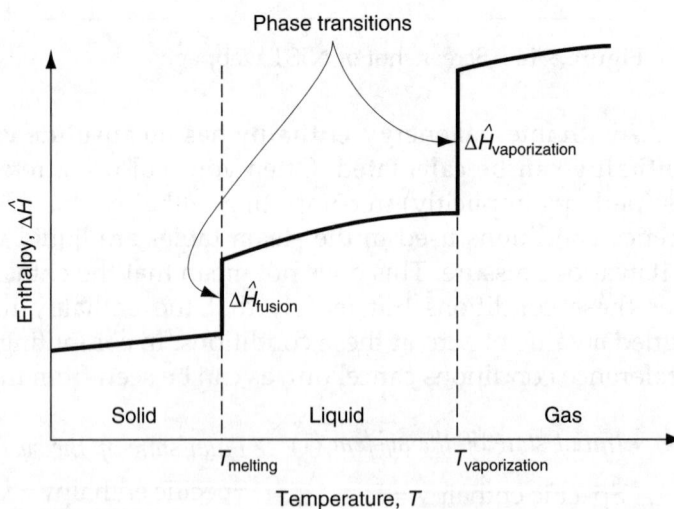

Figure 8.15   The overall enthalpy change includes the sensible heats (the enthalpy changes within a phase) plus the latent heats (the enthalpy changes of the phase transitions).

Do you recall from Chapter 7 that **phase transitions** occur from the solid to the liquid phase, and from the liquid to the gas phase, and vice versa? During these transitions, very large changes in the value of the enthalpy (and internal energy) of a substance occur, changes called **latent heat** changes, because they occur without any noticeable change in temperature. Because of the relatively large enthalpy change associated with a phase transition, it is important to get accurate values of any latent heats involved when applying energy balances. For a single phase, the enthalpy varies as a function of the temperature, as illustrated in Figure 8.15. The enthalpy changes that take place within a single phase are usually called **sensible heat** changes.

The enthalpy changes for the common specific phase transitions are termed **heat of fusion** (for melting), ΔH_{fusion}, and **heat of vaporization** (for vaporization), ΔH_v. The word *heat* has been carried by custom from very old experiments in which enthalpy changes were calculated from experimental data that frequently involved heat transfer. *Enthalpy of fusion* and *vaporization* would be the proper terms, but they are not widely used. The **heat of condensation** is the negative of the heat of vaporization, and the **heat of solidification** is the negative of the heat of fusion. The **heat of sublimation** is the enthalpy change associated with the changing of phase from solid directly to vapor.

The overall specific enthalpy change of a pure substance, as illustrated in Figure 8.15, can be formulated by summing the sensible and latent heats (enthalpies) from the initial state to the final state.

Sensible heat of

Overall enthalpy change   Solid   Melting   Sensible heat of liquid

$$\Delta \hat{H} = \hat{H}(T) - \hat{H}(T_{ref}) = \int_{T_{ref}}^{T_{fusion}} C_{p,solid}\, dT + \Delta \hat{H}_{fusion\ at\ T_{fusion}} + \int_{T_{fusion}}^{T_{vaporization}} C_{p,liquid}\, dT$$

Vaporization   Sensible heat of vapor

$$+ \Delta \hat{H}_{vaporization}\ at\ T_{vap} + \int_{T_{ref}}^{T} C_{p,vapor}\, dT \tag{8.15}$$

The overall enthalpy of a system or stream containing more than one component is the sum of the enthalpies of each component if you ignore the heat (enthalpy change) of mixing (discussed in Chapter 12).

$$H_{tot} = m_1 \hat{H}_1 + m_2 \hat{H}_2 + \ldots + m_n \hat{H}_n$$

Self-Assessment Test

Questions

1. When you read in the steam table in two adjacent columns $\hat{U} = 1022.6$ kJ/kg and $\hat{H} = 1022.6$ kJ/kg, what is the water liquid or vapor?

2. Repeat question 1 for $\hat{U} = 2577.1$ kJ/kg and $\hat{H} = 2770.1$ kJ/kg.

Answers

1. Liquid, because internal energy and enthalpy are the same.

2. Vapor, because the enthalpy is greater than the internal energy.

Problems

1. The enthalpy change of a real gas can be calculated from

$$\hat{H}_2 - \hat{H}_1 = \int_{T_1}^{T_2} C_p \, dT + \int_{P_1}^{P_2} \left[\hat{V} - T\left(\frac{\partial \hat{V}}{\partial T}\right)_p \right] dp$$

What is the enthalpy change for a real gas in (a) an isothermal process, (b) an isochoric (constant volume) process, and (c) an isobaric process?

2. Will the enthalpy change be greater along the path shown by the solid line or the path shown by the dashed line in Figure SAT8.2.5P2?

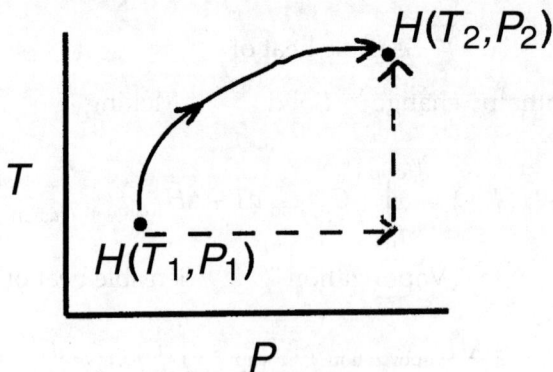

Figure SAT8.2.5P2

Answers

1.

(a) $\hat{H}_2 - \hat{H}_1 = \int_{P_1}^{P_2} \left[\hat{V} - T \left(\frac{\partial \hat{V}}{\partial T} \right)_p \right] dp$

(b) $\hat{H}_2 - \hat{H}_1 = \int_{T_1}^{T_2} C_p \, dT + \int_{P_1}^{P_2} \hat{V} \, dp$

(c) $\hat{H}_2 - \hat{H}_1 = \int_{T_1}^{T_2} C_p \, dT$

2. Both paths will be the same because enthalpy is a state function.

Heat Capacity

Quite a few sources exist for equations expressing C_p as a function of T, and numerous charts exist on which C_p is plotted as a function of temperature. Look at Figure 8.16.

Historically, the integration of heat capacity equations has been used to calculate sensible heat (enthalpy) changes. C_p, of course, is a continuous function of temperature, whereas the enthalpy change may not be. Why? Because of the phase changes that may occur.

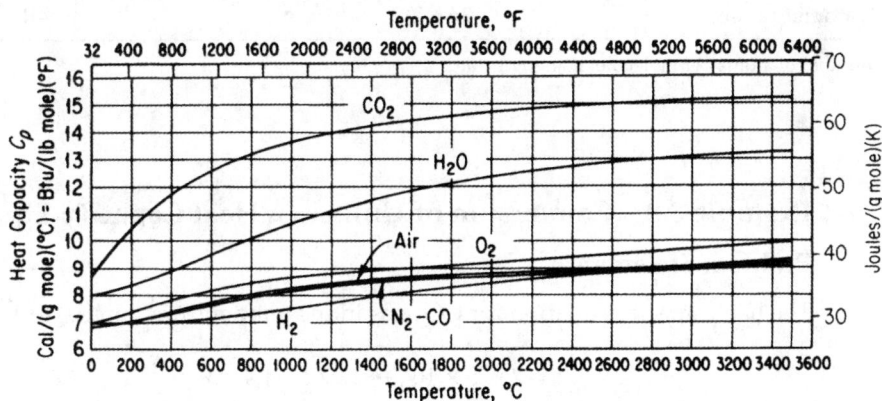

Figure 8.16   Heat capacity curves for the combustion gases at 1 atm

Consequently, it is not possible to have a heat capacity equation for a substance that will go from a low temperature up to any desired temperature.

To give the heat capacity some physical meaning, you can think of C_p as representing the amount of energy required to increase the temperature of a unit mass (or mole) of a substance by 1 degree.

How are the functions such as those in Figure 8.16 determined? What you do is measure experimentally the heat capacity between the temperatures at which the phase transitions occur, and then fit the data with an equation using a built-in function in MATLAB or Python. If you can assume the gas is an ideal gas, the heat capacity at constant pressure is constant even though the temperature varies (examine Table 8.3).

Table 8.3   Heat Capacities of Ideal Gases

Type of Molecule*	Approximate Heat Capacity, C_p	
	High Temperature (Translational, Rotational, and Vibrational Degrees of Freedom)	Room Temperature (Translational and Rotational Degrees of Freedom Only)
Monoatomic	$\dfrac{5}{2}R$	$\dfrac{5}{2}R$
Polyatomic, linear	$\left(3n - \dfrac{3}{2}\right)R$	$\dfrac{7}{2}R$
Polyatomic, nonlinear	$(3n = 2)R$	$4R$

*n, number of atoms per molecule; R, gas constant.

Example 8.8   Conversion of Units in a Heat Capacity Equation

Problem Statement

The heat capacity equation for CO_2 gas in the temperature range 0 to 1500 K is

$$C_p = 2.675 \times 10^4 + 42.27T - 1.425 \times 10^{-2}T^2$$

with Cp expressed in J/(kg mol)(K) and T in kelvin. Convert this equation into a form so that the heat capacity will be expressed over the entire temperature range in Btu/(lb mol)(°F) with T in degrees Fahrenheit.

Solution

Changing a heat capacity equation from one set of units to another is merely a problem in the conversion of units. Each term on the right side of the heat capacity equation must have the same units as the left side of the equation. To avoid confusion in the conversion, you must remember to distinguish between the temperature symbol that in one usage represents temperature and in another usage represents a temperature difference. In the following conversions, we distinguish between the temperature and the temperature difference for clarity, as was done in Chapter 2.

First, multiply each side of the given equation for Cp by appropriate conversion factors to convert J/(kg mol) (Δ K) to Btu/(lb mol) (Δ °F). Multiply the left side by the factor in the square brackets.

$$C_p \frac{J}{(\text{kg mol})(\Delta K)} \times \left[\frac{1\,\text{Btu}}{1055\,\text{J}} \middle| \frac{1\Delta K}{1.8\Delta°R} \middle| \frac{1\Delta°R}{1\Delta°F} \middle| \frac{0.4536\,\text{kg}}{1\,\text{lb}} \right] \to C_p \frac{\text{Btu}}{(\text{lb mol})\,(\Delta°F)}$$

and multiply the right side by the same set of conversion factors.

Next, substitute the relation between the temperature in kelvin and the temperature in degrees Fahrenheit:

$$T_K = \frac{T°_R}{1.8} = \frac{T°_F + 460}{1.8}$$

into the given equation for Cp where T appears.

Finally, carry out all the indicated mathematical operations and consolidate quantities to get

$$C_p \frac{\text{Btu}}{(\text{lb mol})(\Delta°F)} = 8.702 \times 10^{-3} + 4.66 \times 10^{-6}\,T°_F - 1.053 \times 10^{-9}\,T°_F^2$$

When you cannot find a heat capacity for a gas, you can estimate one by using one of the numerous equations that can be found in the supplementary references at the end of this chapter. Estimation of C_p for liquids is fraught with error, but some relationships exist. For aqueous solutions, you can roughly approximate C_p by using the water content only.

Self-Assessment Test

Questions

1. Indicate whether the following statements are true or false:

   a. For a real gas, $\Delta\hat{H} = \int_{T_1}^{T_2} C_p \, dT$ is an exact expression.

   b. For liquids below $T_r = 0.75$ or a solid, $\Delta U \approx \int_{T_1}^{T_2} C_p \, dT$.

   c. For ideal gases near room temperature, $C_p = 5/2\,\mathrm{R}$.

   d. C_v cannot be used to calculate $\Delta\hat{H}$; you have to use C_p.

2. Is the term *specific energy* a better term to use than *specific heat*?

3. What is the heat capacity at constant pressure at room temperature of O_2 if the O_2 is assumed to be an ideal gas?

Answers

1. All false except for (b)

2. Probably, it would be technically be correct, but it is rarely used.

3. 7/2R

Problems

1. Determine the specific enthalpy of liquid water at 400 K and 500 kPa relative to the specific enthalpy value of liquid water at 0°C and 500 kPa using a heat capacity equation. Compare to the value obtained from the steam tables.

2. A problem indicates that the enthalpy of a compound can be predicted by an empirical equation $\hat{H}(J/g) = -30.2 + 4.25T + 0.001T^2$, where T is in kelvin. What is a relation for the heat capacity at constant pressure for the compound?

3. A heat capacity equation in cal/(g mol)(K) for ammonia gas is

$$C_p = 8.4017 + 0.70601 \times 10^{-2} T + 0.10567 \times 10^{-5} T^2 - 1.5981 \times 10^{-9} T^3$$

where T is in degrees Celsius. What are the units of each of the coefficients in the equation?

Answers

1. Assuming that liquid water has a constant C_p of 4.184 kJ/(kg)(°C), the specific enthalpy change is equal to 531 kJ/kg and using the steam tables is equal to 532 kJ/kg.

2. $C_p(J/g) = 4.25 + 0.002T$

3. (a) cal/(g mol)(K); (b) cal/(g mol)(K)(°C); (c) cal/(g mol)(K)(°C)^2; (d) cal/(g mol)(K)(°C)^3

Enthalpies for Phase Transitions

Where can you get values for the enthalpies of phase changes? Values of $\Delta\hat{H}$ for phase change for many common compounds are available from several sources available on Internet, such as those listed on page 616. Other sources of experimental data are cited in reference books listed in the supplementary references at the end of this chapter. You can also estimate values for $\Delta\hat{H}_v$ from one of the relations such as the following three. Use of experimental values for the heat of vaporization is recommended whenever possible.

Chen's equation[2]   An equation that yields values of $\Delta\tilde{H}_v$ (in kilojoules per gram mole) [the overlay tilde (~) on H designates per mole rather than per mass] to within 2% is Chen's equation:

$$\Delta\tilde{H}_v = RT_b \left(\frac{3.978(T_b/T_c) - 3.938 + 1.555\ln p_c}{1.07 - (T_b/T_c)} \right) \tag{8.16}$$

where T_b is the normal boiling point of the liquid in kelvin, T_c is the critical temperature in kelvin, and p_c is the critical pressure in atmospheres.

Riedel's equation[3]

$$\Delta\tilde{H}_v = 1.093R\, T_c \left[\frac{T_b(\ln p_c - 1)}{T_c(0.930 - (T_b/T_c))} \right] \tag{8.17}$$

Watson's equation[2]   Watson found empirically that below the critical temperature, the ratio of two heats of vaporization could be related by

$$\frac{\Delta\tilde{H}_{v2}}{\Delta\tilde{H}_{v1}} = \left(\frac{1 - T_{r2}}{1 - T_{r1}} \right)^{0.38} \tag{8.18}$$

where $\Delta\tilde{H}_{v2}$ = heat of vaporization of a pure liquid at T_1 and $\Delta\tilde{H}_{v1}$ = heat of vaporization of the same liquid at T_2. Yaws[4] lists other values of the exponent for various substances.

---

[2] N. H. Chen, *Ind. Eng. Chem.*, **51**, 1494 (1959).

[3] K. M. Watson, *Ind. Eng. Chem.*, **23**, 360 (1931); 35, 398 (1943).

[4] C. L., Yaws, H. C. Yang, and W. A. Cawley, "Predict Enthalpy of Vaporization," *Hydrocarbon Processing*, 87–90 (June, 1990).

The heat of vaporization at the normal boiling point (1 atm pressure) data for most substances are readily available from many resources, some of which are listed later on page 616. Therefore, it is recommended that when you have no data and need a heat of vaporization for an energy balance calculation, you use the heat of vaporization at the normal boiling point from the physical property software and use Equation (8.18) to correct for temperature.

Example 8.9   Comparison of an Estimate of the Heat of Vaporization with the Experimental Value

Problem Statement

Use Chen's equation to estimate the heat of vaporization of acetone at its normal boiling point and compare your results with the experimental value of 30.2 kJ/g mol listed in Appendix E on the website that accompanies this book.

Solution

The basis is 1 g mol. You have to look up some data for acetone in Appendix E:

$$\text{Normal boiling point:} \qquad 328.2 \text{ K}$$
$$T_c: \qquad 508.0 \text{ K}$$
$$p_c: \qquad 47.0 \text{ atm}$$

The next step is to calculate some of the values of the variables in the estimation equations:

$$\frac{T_b}{T_c} = \frac{329.2}{508.0} = 0.648$$

$$\ln p_c = \ln(47.0) = 3.85$$

From Equation (8.16):

$$\Delta \tilde{H}_v = \frac{8.314 \times 10^{-3} \text{ kJ}}{(\text{g mol})(K)} \bigg| 329.2 \text{ K} \bigg| \frac{[(3.978)(0.648) - 3.938 + (1.555)(3.85)]}{1.07 - 0.648}$$

$$= 30.0 \text{ kJ/g mol (insignificant error)}$$

Self-Assessment Test

Questions

1. Indicate whether the following statements are true or false:
   a. The molar heat of vaporization of water is 40.7 kJ/g mol.
   b. The molar heats of vaporization you look up in a reference book or database come from experimental data.
   c. The molar heat of fusion is the amount of energy necessary to melt or freeze 1 g mol of substance at its melting point.

2. Define (a) heat of vaporization, (b) heat of condensation, and (c) heat of transition.

3. Why do engineers use the term *heat of* for the energy change that occurs in a phase transition rather than the better term *enthalpy change*?

Answers

1. (a) F, because heat of vaporization is a function of temperature; (b) T; (c) F, heat of fusion is for melting

2. See text

3. Tradition

Problems

1. Ethane (C_2H_6) has the heat of vaporization of 14.707 kJ/g mol at 184.6 K. What is the estimated heat of vaporization of ethane at 210 K?

2. At 0°C you melt 315 g of H_2O. What is the energy change corresponding to the process?

3. One hundred grams of H_2O exist in the gas phase at 395 K. How much energy will it take to condense all of the H_2O at 395 K?

Answers

1. By using the Watson equation, 13.445 kJ/g mol (the experimental value is 13.527 kJ/g mol).

2. For water, the $\Delta H_{fusion} = 334$ J/g; energy exchange $= (334)(315) = 105$ kJ

3. From steam table, $\Delta H_{vap}(395K) = 2197$ kJ/kg; energy to condense 100 g $= -219.7$ kJ

Tables and Charts to Retrieve Enthalpy Values

Tables listing smoothed experimental data can cover the values of physical properties well beyond the range valid for a single equation. Because the most commonly measured properties are temperature and pressure, tables of enthalpies (and internal energies) for pure compounds usually are organized in columns and rows, with T and p being the independent variables. If the intervals between table entries are close enough, linear interpolation between entries is reasonably accurate. For example, look at the following calculation for saturated steam in SI units from the steam tables on the website that accompanies this book. The units are kilojoules per kilogram. If you want to calculate the enthalpy change of saturated steam from 305 to 307 K by linear interpolation from a table graduated in 5-degree increments, as explained in Chapter 7 in connection with the steam tables, you would carry out the following computation:

$$\hat{H}_{307} = \hat{H}_{305} + \frac{\hat{H}_{310} - \hat{H}_{305}}{T_{310} - T_{305}}(T_{307} - T_{305}) = 2558.9 + \frac{2567.9 - 2558.9}{310 - 305}(307 - 305)$$

$$= 2562.5$$

Or you can use MATLAB or Python to apply cubic spline interpolation, as shown in Examples 7.5 and 7.6.

Usually, a steam table lists values for the heat of vaporization as well as the enthalpy for liquid water and water vapor. The values of $\Delta \hat{H}_v$ are automatically included in $\hat{H}_g$. Note that the steam tables (and similar tables) include the effects of pressure changes as well as phase transition. **For compressed liquids at values of pressure higher than the saturation pressure, you can use the properties of the saturated liquid at the same temperature as a good approximation if needed.** Look at the 600°F line in Figure 8.17, which is almost vertical. At point B, you could substitute the value of $\hat{H}$ at point A.

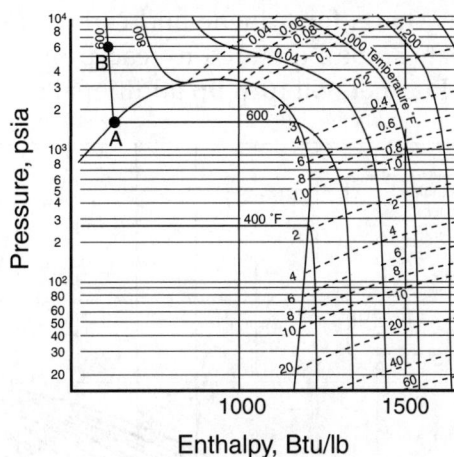

Enthalpy, Btu/lb

Figure 8.17   At higher pressures in which the compound is liquid, the lines of constant temperature (such as 600°F) are almost vertical, and thus the enthalpy of the saturated liquid can be substituted for the actual enthalpy if necessary.

You should look at the tables of enthalpies at 1 atm of selected gases in Appendix F, Tables F.2 through F.6, on the website that accompanies this book. Remember that enthalpy values are all relative to some reference state. What is the reference state for the gases in Table F.6? Did you decide on 273 K (0°C) and 1 atm? Remember, you can calculate enthalpy changes by subtracting the initial enthalpy from the final enthalpy for any two sets of conditions **whatever the reference state**. Thus, despite the reference state for the gas enthalpy tables being slightly different from that of the steam tables, you can choose as a reference state for a problem another temperature, such as the temperature of an entering stream, to serve as a reference for an enthalpy change.

You no doubt have heard the saying "A picture is worth a thousand words." Something similar might be said of two-dimensional charts, such as in Figure 8.17, namely, that you can get an excellent idea of the values of the enthalpy and other properties of a substance in all of the regions displayed in a chart. Although the accuracy of the values read from a chart may be limited (depending on the scale and accuracy of the chart), tracing out various processes on a chart enables you to rapidly visualize and analyze what is taking place. Charts are certainly a simple and quick method for you to get data to compute approximate enthalpy changes. Figure 8.18 for *n*-butane is an example chart. A number of sources of charts are listed in the references at the end of the chapter. Appendix J on the website that accompanies this

book contains charts for toluene and carbon dioxide. The website also has
p-H charts that can be expanded to read property values more accurately. A
search of the Internet will turn up additional charts.

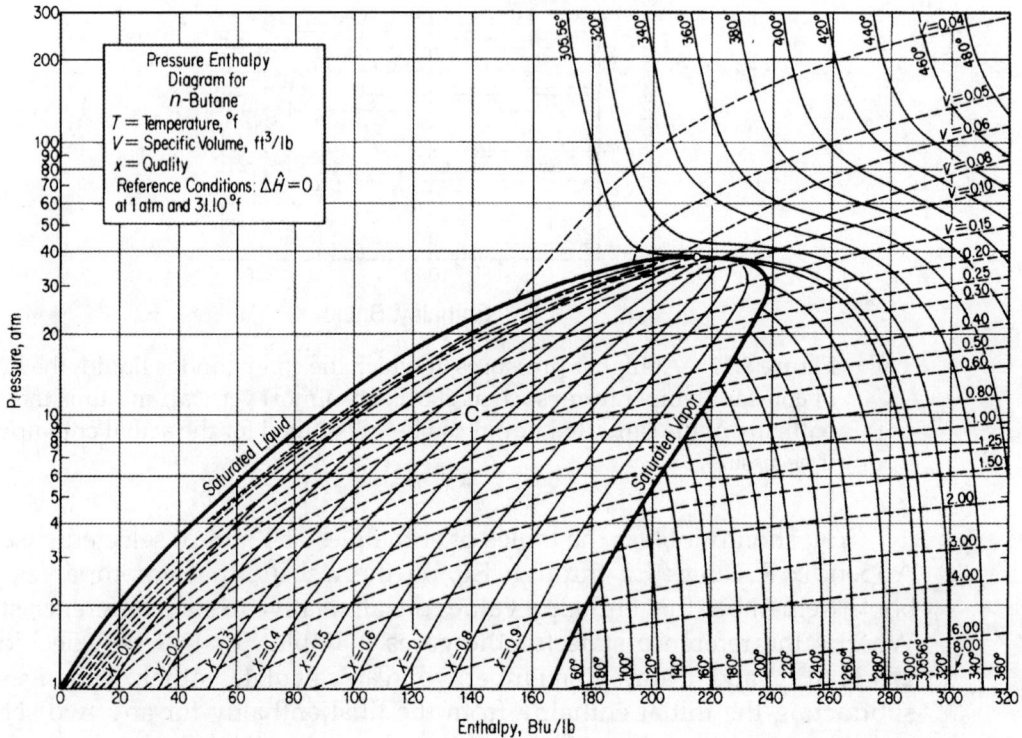

Figure 8.18   A pressure versus enthalpy chart for butane showing lines of
constant temperature and constant specific volume

Charts are drawn with various coordinates, such as p versus $\hat{H}$,
p versus $\hat{V}$, or p versus T. Because a chart has only two dimensions, the coordinates can represent only two variables. The other variables of interest
have to be plotted as lines or curves of constant value across the face of the
chart. Now for some remarks about the two-phase region of a compound,
such as point C in Figure 8.18: You will find that in the two-phase region
only values for the saturated liquid and saturated vapor are listed
in a table. You have to interpolate between these saturated liquid and
vapor values to get properties of vapor-liquid mixtures, as explained in
Chapter 7. Similarly, on a chart with pressure and enthalpy as the axes,
lines of constant specific volume and/or temperature might be drawn as in

Figure 8.18. This situation limits the number of variables that can be presented in tables. How many properties have to be specified for a pure component gas to definitely fix the state of the gas? If you specify two intensive properties for a pure gas, you ensure that all of the other intensive properties will have distinct values, and from the phase rule, any two intensive properties can be chosen at will to ascertain all of the other intensive properties.

Example 8.10   Use of the Pressure-Enthalpy Chart for Butane to Calculate the Enthalpy Difference between Two States

Problem Statement

Calculate $\Delta \hat{H}$, $\Delta \hat{V}$, and ΔT for 1 lb of saturated vapor of n-butane going from 2 atm to 20 atm (saturated).

Solution

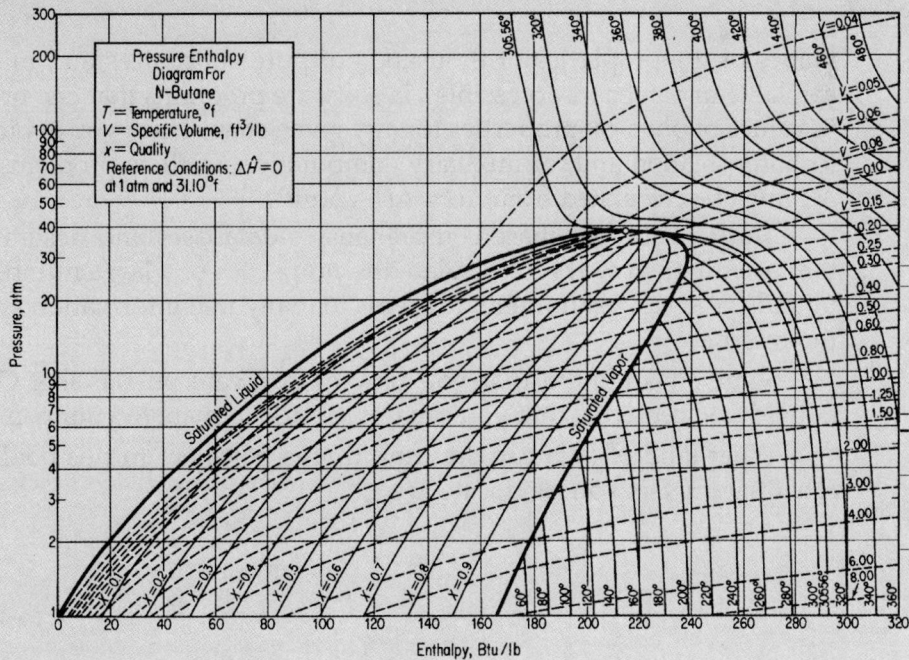

Pressure Enthalpy
Diagram For
N–Butane
T = Temperature, °f
V = Specific Volume, ft³/lb
x = Quality
Reference Conditions: $\Delta \hat{H} = 0$
at 1 atm and 31.10 °f

Figure E8.10

(Continues)

Example 8.10   Use of the Pressure-Enthalpy Chart for Butane to Calculate the Enthalpy Difference between Two States (*Continued*)

Obtain the necessary data from Figure 8.20 or Figure E8.10.

	$\hat{H}$ (Btu/lb)	$\hat{V}$ (ft^3/lb)	T (°F)
Saturated vapor at 2 atm: point A	179	3.00	72
Saturated vapor at 20 atm: point B	233	0.30	239

$$\Delta\hat{H} = 233 - 179 = 54 \text{ Btu/lb}$$
$$\Delta\hat{V} = 0.30 - 3.00 = -2.70 \text{ ft}^3/\text{lb}$$
$$\Delta T = 239 - 72 = 167°\text{F}$$

Computer Databases

Values of the properties of thousands of pure substances and mixtures are available in databases accessible via software programs that can provide sets of values of physical properties for any given state. Thus, you avoid the need for interpolation and/or auxiliary computation, such as problems in which ΔU must be calculated from $\Delta U = \Delta H - \Delta(pV)$.

You can also purchase comprehensive databases and design packages that use different ways to calculate the properties of a large number of compounds. You can get immediate access to many free information systems via the Internet.

Some sites require that you use the software on the site. Others sell individual licenses so that you can download software to your computer.

What kind of data can you find that is pertinent to this book? Among other parameters you can locate are

Boiling point	Heat of condensation
Bubble point and dew point for binary mixtures	Heat of formation
	Heat of vaporization
Critical properties	p, $\hat{V}$, and T values for gases and liquids
Density of liquids	Vapor-liquid equilibria
Heat capacity	Vapor pressure

Some sites are quite specialized, such as those devoted to the properties of petroleum fluids, resins, solvents, and so on. Some sites consist of directories of other sites that pertain to specialized topics such as atomic and molecular data, dielectric constants, ionization energies, refraction indices, refrigerants, safety, and so on. Many sites also have helpful links to other sites.

How valid (consistent, accurate, rectified) are the data you collect from Internet sources? Are the data from a reliable source? Have the data been audited? Are they free of anomalies? What is the uncertainty? Are they easy to use? Such questions are best answered by the footnote found at one data site.

8.4.2   Energy Balance Examples for Open, Steady-State Systems

One step in the strategy for the solution of a problem, a step that becomes a bit more complex when using energy balances than when using strictly material balances, is the analysis of the degrees of freedom, because the intensive variables specifying enthalpy $\hat{H}$ (and internal energy $\hat{U}$), which are intensive variables, are themselves functions of T and p. Consequently, each stream as well as the material in the system can be characterized by its temperature and pressure. Some software programs will allow you to input the values of $\hat{H}$ and p and recover the value of T at the state that is specified by $\hat{H}$ and p. For tables, you have to interpolate. Often, just the temperature of a stream is specified, and the value of the pressure is not mentioned. How can a state be specified by just one variable? In such circumstances, for a saturated liquid you could assume the value of the pressure is the vapor pressure of the liquid. If the liquid or gas is not saturated, you can fall back on assuming 1 atm when no better value is known for the pressure. For two phases, you may have to assume equilibrium exists in the absence of better information to make the degrees of freedom zero in such ill-posed problems.

Figure 8.19 shows a schematic of a heat exchanger in which a cold stream is used to cool a hot stream and a hot stream is used to warm a cold stream.

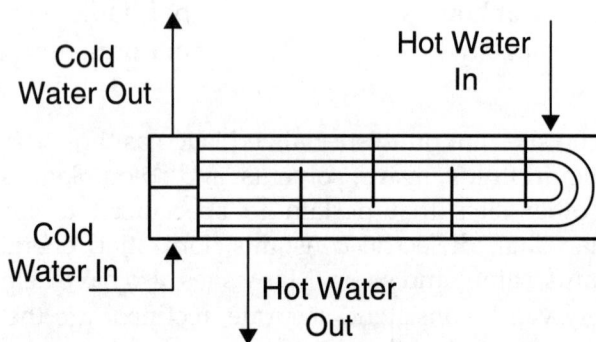

Figure 8.19   Schematic of a heat exchanger

The cold water enters a header that feeds a number of tubes that carry the cold water through the heat exchanger. As the cold water flows through these tubes, heat is transferred through the walls of the tubes from the hot water to the cold water, thus increasing the temperature of the cold water and decreasing the temperature of the hot water. The hot water is added at the opposite end of the exchanger and flows back and forth past the tube bundle due to baffles inside the heat exchanger before exiting the exchanger near the cold water entrance. Assuming that the heat exchanger is properly insulated, the heat lost by the hot water is equal to the heat gained by the cold water, shown graphically in Figure 8.20. Therefore, applying the heat balance equation (i.e., $\Delta H = Q$) to both the hot water system and the cold water system yields

$$\dot{Q} = -F_{hw}C_p\Delta T_{hw}$$

$$\dot{Q} = F_{cw}C_p\Delta T_{cw}$$

where F_{cw} and F_{hw} are the mass flow rate of the cold and hot water, C_p is the heat capacity of water, and ΔT_{cw} and ΔT_{hw} are the temperature changes through the process for the cold and hot water. Note that a minus sign appears in front of the hot water term in the preceding equation because the hot water temperature decreases as it flows through the exchanger.

If we perform an overall balance on the heat exchanger (i.e., $\Delta H = 0$) because the heat exchanger is insulated from its surroundings,

$$F_{cw}C_p\Delta T_{cw} + F_{hw}C_p\Delta T_{hw} = 0$$

Because the heat capacity of the hot water and the cold water are very nearly equal,

$$F_{cw}\Delta T_{cw} = -F_{hw}\Delta T_{hw}$$

From this equation, if the flow rate of the hot and cold water were equal, the magnitude of the temperature change of the hot and cold water would be the same. Furthermore, if the hot water flow rate were twice that of the cold water, the magnitude of the temperature change of the hot water would be half that of the cold water.

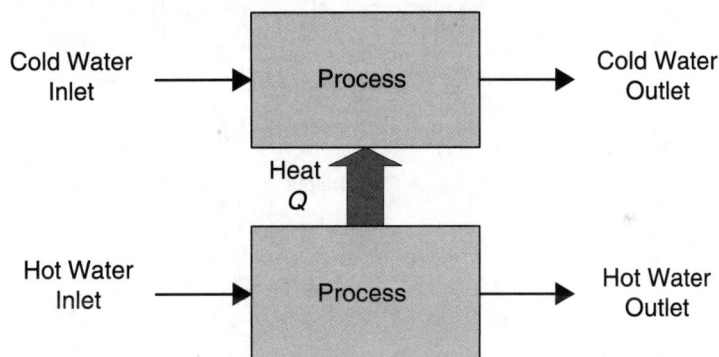

Figure 8.20   A functional diagram of a heat exchanger

Example 8.11   Fluid Warmer

Problem Statement

Figure E8.11a shows a fluid warmer that uses standard IV tubing instead of special disposable equipment. An electrically powered 250 W dry heat warmer supplies the heat transfer to the plastic tubing found in IV tube sets. The IV tubing is easily positioned in the S-shaped channel between the aluminum heating plates.

(Continues)

Example 8.11   Fluid Warmer (*Continued*)

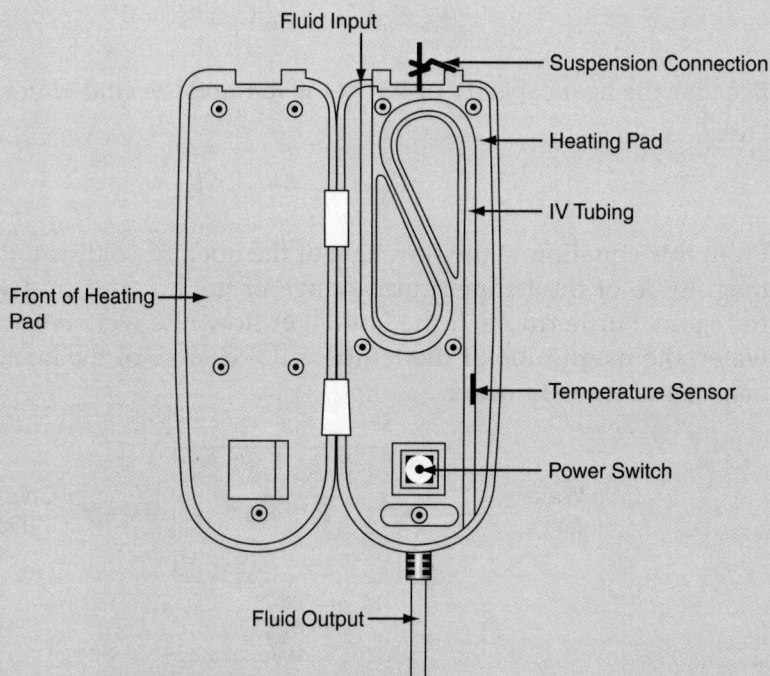

Figure E8.11a

The small size of the device permits a minimal length of tubing between the patient and the unit so that subsequent cooling of the exit fluid is negligible. The device warms up in 2 to 3 min, and then the fluid flows through the device at rates up to 12 cm³/min. Note that the temperature sensor after the fluid has passed through the heating section is used to adjust the power to the heating pad so that the desired temperature of the fluid is attained.

Acyclovir is an antiviral agent used for genital herpes (*Herpes simplex*) and shingles (varicella zoster, the chicken pox virus) by infusion through a vein. The solution flows through the warmer at the rate of 1.67 g/min of infusion solution. The entering fluid is at 24°C and exits at 37°C prior to infusion. The infusion solution in addition to the acyclovir contains 0.45% NaCl and 2.5% glucose.

How many watts must be used by the warmer to warm the solution?

Solution

Steps 1–4

Figure E8.11b is a sketch of the process. Note that the electric work is given as power, watts (joules per second). For simplicity (and in the absence of data!), we use the properties of sugar (glucose) in water for the fluid.

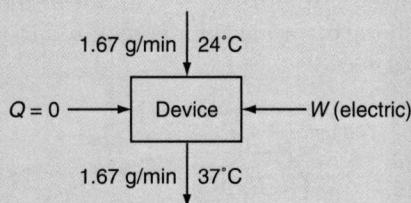

Figure E8.11b

Step 5

Basis: 1 min

Steps 6–8

What should you pick for the system? Choose the warming device. Then the system is open, steady-state (except for a short warmup interval that we will ignore) system. If we consider the entire warming device, electrical work is involved. On the other hand, if we consider the warming device minus the electrical heater, the energy balance would be $\Delta H = Q$, for which Q is the electrical heat provided by the heater:

$$Q = \Delta H_{\text{flow}} = F C_p \left(T_{\text{out}} - T_{\text{in}} \right) = \frac{1.67\,\text{g}}{\text{min}} \left| \frac{4.18\,\text{J}}{(\text{g})(^\circ\text{C})} \right| \frac{17^\circ\text{C}}{} \left| \frac{1\,\text{min}}{60\,\text{s}} \right.$$

$$= 1.98\,\text{J/s} \equiv 1.98\,\text{W}$$

The heat capacity found on the Internet (ignoring the small amount of NaCl) is for 4% sugar solution. The power in W is approximately 2 W.

Example 8.12   Hot Gas Dilution

Problem Statement

In a plant, 10,000 ft^3/hr of a combustion gas at 1000°F and 1 atm are produced from combusting coal containing sulphur. The combustion gas has the following composition: 1.1% SO_2, 14.5% CO_2, 3.4% O_2 and 81.0% N_2 and is to be cooled from 1000°F to 400°F by mixing it with air at 65°F and 760 mm Hg. See Figure E8.12. What volume of air per hour is required to produce an air-gas mixture with a temperature of 400°F?

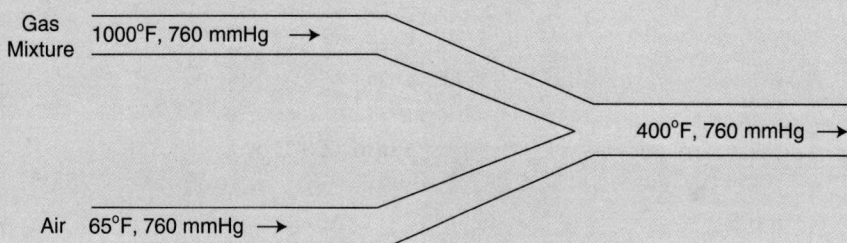

Gas Mixture   1000°F, 760 mmHg →

400°F, 760 mmHg →

Air   65°F, 760 mmHg →

Figure E8.12

Solution

Assuming that the pipes carrying the gas mixture are insulated, the energy balance for this system is $\Delta H = 0$. That is, the thermal energy lost by the gas mixture is gained by the air.

Basis: Flow rate of the combustion gas is 10,000 ft^3/hr.

The first step is to calculate the molar flow rate of the high temperature gas mixture:

$$n_{GM} = \frac{10,000 \text{ ft}^3}{\text{hr}} \left| \frac{\text{lbmol}}{359.05 \text{ ft}^3} \right| \frac{(459.67 + 32)}{(459.67 + 1000)} = 9.381 \text{ lbmol/hr}$$

Now, let's calculate the enthalpy changes for each component on the basis of equation 8.14:

Component	$\Delta H_{sensible}$(J/g mol)	Temperature Changes
SO_2	−16,454.3	1000 °F → 400 °F
CO_2	−15,382.4	1000 °F → 400 °F
O_2	−10,635.4	1000 °F → 400 °F
N_2	−10,308.5	1000 °F → 400 °F
Air	5480.6	65°F → 400 °F

Applying the energy balance yields:

$$0 = \frac{n_{air}\ \text{lbmol}}{\text{hr}} \left| \frac{453.6\ \text{g mol}}{\text{lbmol}} \right| \frac{5480.6\ \text{J}}{\text{g mol}} + \frac{9.381\ \text{lbmol}}{\text{hr}} \left| \frac{453.6\ \text{g mol}}{\text{lbmol}} \right| \times$$

$$\left| [0.011(-16,454.3) + 0.145(-15,382.4) + .034(-10,635.4) + 0.81(-10,308.5)] \frac{\text{J}}{\text{g mol}} \right.$$

Solving for n_{air} yields

$$n_{air} = 19.04\ \text{lbmol/hr}$$

Converting n_{air} to ft^3/hr yields

$$Q_{air} = \frac{19.039\ \text{lbmol}}{\text{hr}} \left| \frac{359.05\ \text{ft}^3}{\text{lbmol}} \right| \frac{(459.67 + 65)°\text{R}}{(459.67 + 32)°\text{R}} = 7295\ \text{ft}^3/\text{hr at 65°F and 1 atm}$$

Example 8.13   Use of Combined Material and Energy Balances

Problem Statement

Figure E8.13 shows a hot gas stream at 500°C being cooled to 300°C by transferring heat to the liquid water that enters at 20°C and exits at 213°C. Assume that the heat exchanger is insulated. The cooling water does not mix with the gas. Calculate the value of the flow rate of the water.

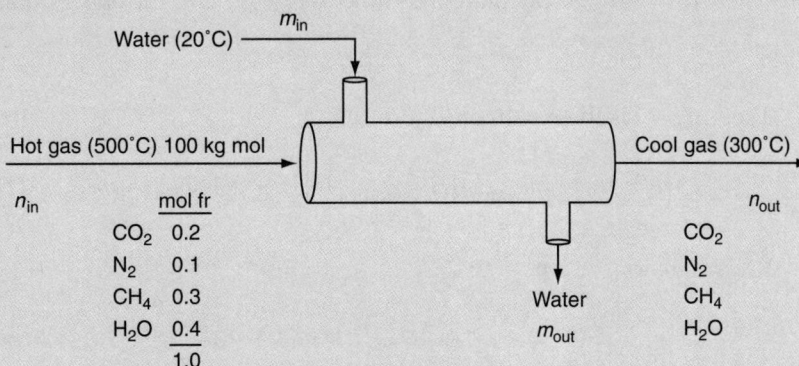

Figure E8.13

(*Continues*)

Example 8.13   Use of Combined Material and Energy Balances (*Continued*)

Solution

Steps 1–4

Figure E8.13 contains all of the data for the process.

Step 5

$$\text{Basis: 100 kg mol of entering gas 1} \equiv \text{min}$$

Steps 6–8

Pick the system as the heat exchanger. Very few unknowns exist for this system if certain easy calculations are made for the material balances:

a. Make material balances in your head to conclude that the exit gas is 100 kg mol and that the kilogram moles of each of the components are known.

b. Make another material balance in your head to conclude that

$$\text{water in = water out} = m_{\text{water}}$$

The result is that after solving the material balances, only one unknown exists, namely, m_{water}. All of the other variables can be assigned known values, but because they are obvious, we will not list them here to save space.

How can you calculate the value of m_{water}? Use the energy balance.

$$\Delta H = Q = 0$$

Then $\Delta H = 0$ is the result, or in full:

$$\left[\left(n_{\text{gas}}\right)\left(\hat{H}_{\text{gasout}}\right)+\left(m_{\text{water}}\right)\left(\hat{H}_{\text{waterout}}\right)\right]-\left[\left(n_{\text{gas}}\right)\left(\hat{H}_{\text{gasin}}\right)+\left(m_{\text{water}}\right)\left(\hat{H}_{\text{waterin}}\right)\right]=0$$

Rearrange to

$$n_{\text{gas}}\Delta\hat{H}_{\text{gas}} - m_{\text{water}}\Delta\hat{H}_{\text{water}} = 0$$

Note that the simplified equation has been written in terms of $\Delta\hat{H}$ rather than $\hat{H}$ to avoid taking into account differences in the reference states for the gas and water.

Where can you get values for $\Delta\hat{H}$ of the gas components and the water for the reduced energy balance? You can assign values to each of the components of the 100 kg mol of gas from data on the website that accompanies this book because you know the temperature and pressure (assume 1 atm) for the entrance and exit streams.

For each of the components of the gas:

$$n_{\text{gas component}}\left(\hat{H}_{\text{gas component at 300°C}} - \hat{H}_{\text{gas component at 500°C}}\right) \equiv n_i\Delta\hat{H}_i$$

Either integrate the heat capacity equations, or if you have access to it, use a software that gives $\Delta\hat{H}_i$ directly for the transition from 500°C to 300°C. You should get the following:

Component	$\Delta\hat{H}_i$ (kJ/kg mol)	n_i (kg mol)	$\Delta H = n_i\,\Delta\hat{H}_i$ (kJ)
CO_2	−9333	20	−186,660
N_2	−6215	10	−62,150
CH_4	−11,307	30	−339,210
H_2O	−7441	40	−297,640
Total		100	$-885,660 = \Delta H_{\text{gas}}$

The next step is easy if you use the steam tables for liquid water. Although the vapor pressure of water and the enthalpy change as the water temperature rises, you can use the value of $\hat{H}_i$ for saturated water without much loss of accuracy.

From the values in the saturated portion of the steam tables on the website:

T (°C)	$\hat{H}$ (kJ/kg)	
20	35.7	$\Big\}\Delta\hat{H} = 350.1$ kJ/kg
213	391.8	

$$m_{\text{water}} = \frac{-\Delta H_{\text{gas}}}{\Delta\hat{H}_{\text{water}}} = \frac{-(-885,660)\ \text{kJ}}{350.1\text{kJ/kg}} = 2487\ \text{kg}$$

$$\dot{m}_{\text{water}} = 2487\ \text{kg/min} \qquad \text{(the basis was 1 min)}$$

Self-Assessment Test

Questions

1. What assumptions are made to attain the energy balance equation $\Delta H = Q$?
2. What does ΔH signify?
3. How is flow work considered for this class of energy balance problems?

Answers

1. Steady-state and open system. W and ΔKE and ΔPE of mass entering and leaving the system are insignificant compared to Q.
2. The enthalpy difference between the outlet streams and the inlet streams
3. Flow work is part of the definition of enthalpy.

Problems

1. Calculate Q for the system shown in Figure SAT 8.4.2P1 in kJ/min.

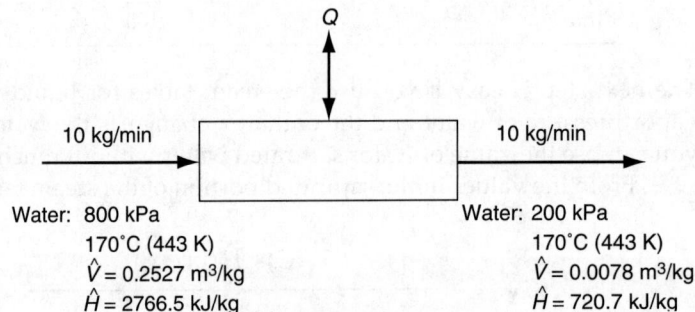

Figure SAT 8.4.2P1

2. Consider a steam heat exchanger in which 500 gallons per minute (gpm) of water are heated from 65°F to 150°F using 50 psig steam. Assume that the steam condenses to saturated liquid at 50 psig. What is the required flow rate of steam for this heat exchanger?

Answers

1. $\hat{H}_{inlet}$ = 2768.7 kJ/kg; $\hat{H}_{outlet}$ = 720.7 kJ/kg (i.e., saturated liquid at 443 K); $Q = 10$ kg/min $(2768.7 - 720.7)$kJ/kg = 20,480 kJ/min

2. $Q = \dot{m}C_p \Delta T = \dfrac{500 \text{ gal}}{\text{min}}\left|\dfrac{\text{ft}^3}{7.481 \text{ gal}}\right|\dfrac{62.4 \text{ lb}_m}{\text{ft}^3}\left|\dfrac{1 \text{ Btu}}{\text{lb}_m - {}^\circ F}\right|\dfrac{85 \text{ }^\circ F}{} = 35,450 \text{ Btu/min};$

   $\Delta H_{vap,stm} = 1059.4 \text{ Btu/lb}_m; \dot{m}_{stm} = 20,480 / 1059.4 = 19.3 \text{ lb}_m/\text{min}$

8.5 Mechanical Energy Balances

Mechanical energy balances apply to steady-state cases in which heat transfer and thermal changes are not important (e.g., fluid flow applications), and the general energy balance becomes

$$W = \Delta E_{conv}$$

For these cases, the following factors are considered: applied work, flow work, changes in kinetic energy, and changes in potential energy. Note that because the thermal changes are negligible, the enthalpy change reduces to the change in flow work, that is, $\Delta(pV)$. Therefore, the mechanical energy balance equation becomes

$$\underset{\text{shaft work}}{W} = \underset{\text{change in kinetic energy}}{\Delta KE} + \underset{\text{change in potential energy}}{\Delta PE} + \underset{\text{change in flow work}}{\Delta(pV)} \qquad (8.19)$$

Note that the change in kinetic energy, potential energy, and flow work are calculated as "out" minus "in." When flow work is considered as work done on the system by the surroundings, it is calculated as flow work in minus flow work out. As a result, the change in the flow work (i.e., out minus in) has a positive sign on the right side of Equation (8.19).

Now we describe kinetic energy and potential energy in more detail before applying the mechanical energy balance to some engineering examples.

8.5.1 Kinetic Energy (*KE*)

Kinetic energy (*KE*) is the energy a system, or some material, possesses because of its velocity relative to the surroundings, which are usually, *but not always*, at rest. The wind, moving automobiles, waterfalls, flowing fluids, and so on, possess kinetic energy. The kinetic energy of a material refers to what is called the *macroscopic kinetic energy*, namely, the energy that is associated with the gross movement (velocity) of the system or material, and not the kinetic energy of the individual molecules, which belongs in the category of internal energy, U, that was discussed earlier.

Do you recall the equation used to calculate the kinetic energy relative to stationary surroundings?

It is

$$KE = \frac{1}{2}mv^2 \qquad (8.20a)$$

The kinetic energy per unit mass (the specific kinetic energy), a state variable, is

$$\widehat{KE} = \frac{1}{2}v^2 \qquad (8.20b)$$

In Equation (8.20a), m refers to the center of mass of the material and v to a suitably averaged velocity of the material. The value of a *change* in the specific kinetic energy ($\Delta \widehat{KE}$) depends only on the inlet and outlet values of the convective flow and the velocity of the material.

Example 8.14   Calculation of the Specific Kinetic Energy for a Flowing Fluid

Problem Statement

Water is pumped from a storage tank through a tube of 3.00 cm inner diameter at the rate of 0.001 m^3/s. See Figure E8.14. What is the specific kinetic energy of the water in the tube?

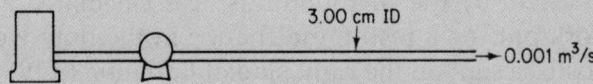

3.00 cm ID

0.001 m³/s

Figure E8.14

Solution

Basis: 0.001 m^3 equivalent to 1 s

Assume that

$$\rho = \frac{1000\text{kg}}{m^3} \text{ and } r = \frac{1}{2}(3.00) = 1.50 \text{ cm}$$

$$v = \frac{0.001 m^3}{s} \left| \frac{}{\pi(1.50)^2 \text{ cm}^2} \right| \left(\frac{100\text{cm}^2}{1m} \right) = 1.415 \text{ m/s}$$

$$\widehat{KE} = \frac{1}{2} \left| \left(\frac{1.415m}{s} \right)^2 \right| \frac{1(N)(s^2)}{1(\text{kg})(m)} \left| \frac{1J}{1(N)(m)} \right| = 1.00 \text{ J/kg}$$

Self-Assessment Test

Questions

1. Can the kinetic energy of a mass be zero if the mass has a velocity, that is, is moving?

2. Temperature is a measure of the average kinetic energy of a material. True or false?

3. The kinetic energy of an automobile going 100 mi/hr is greater than the energy stored in the battery of the automobile (300 Wh). True or false?

Answers

1. No

2. True

3. Assume 3000 lb car; $\Delta KE = 1360$ kJ; 300 Wh = 1080 kJ. Therefore, true, but false for cars <2350 lb.

Problem

1. Calculate the kinetic energy changes in the water that occur when 10,000 lb/hr flow in a pipe that is reduced from 2 in. in diameter to 1 in. in diameter.

Answer

1. $$v_2 = \frac{10,000 \text{ lb}}{\text{hr}} \left| \frac{\text{ft}^3}{62.4 \text{ lb}} \right| \frac{\text{hr}}{3600 \text{ s}} \left| \frac{4}{3.14(1 \text{ in})^2} \right| \frac{144 \text{ in}^2}{\text{ft}^2} = 8.166 \text{ ft/s}; \quad v_1 = 2.052 \text{ ft/s}$$

$$\Delta KE = \frac{1}{2} \left| \frac{10,000 \text{ lb}_m}{\text{hr}} \right| \frac{\text{hr}}{3600 \text{ s}} \left| \frac{(8.166^2 - 2.052^2) \text{ft}^2}{\text{s}^2} \right| \frac{\text{lb}_f\text{-s}^2}{32.2 \text{ lb}_m\text{-ft}} = 2.695 \text{ lb}_m\text{-ft/s}$$

8.5.2   Potential Energy (*PE*)

Potential energy (*PE*) is energy a system possesses because of the force exerted on its mass by a gravitational or electromagnetic field with respect to a reference surface. When an electric car or bus goes uphill, it gains potential energy (Figure 8.21), energy that can be recovered to some extent by regeneration charging the batteries when the vehicle goes down the hill on the other side. You can calculate the potential energy in a gravitational field from

Figure 8.21　Gain of potential energy by an electric automobile going uphill

$$PE = mgh \qquad (8.21a)$$

or the specific potential energy

$$\widehat{PE} = gh \qquad (8.21b)$$

where h is the distance from the reference surface, and where the overlay ($\wedge$) means potential energy per unit mass. The measurement of h is made to the center of mass of a system. Thus, if a ball suspended inside a container somehow is permitted to drop from the top of the container to the bottom, and in the process, it raises the thermal energy of the system slightly, we do not say work is done on the system but instead say that the potential energy of the system is reduced (slightly) because the center of mass changes slightly. The value of a *change* in the specific potential energy, $\Delta\widehat{PE}$, depends only on the location of the inlet and outlet flows in relation to the reference plain, and $\Delta\widehat{PE}$ is a state variable and does not depend on the path followed.

Example 8.15　Calculation of Potential Energy of Water

Problem Statement

Water is pumped from one reservoir to another 300 ft away, as shown in Figure E8.15. The water level in the second reservoir is 40 ft above the water level of the first reservoir. What is the increase in specific potential energy of the water in British thermal units per pound (mass)?

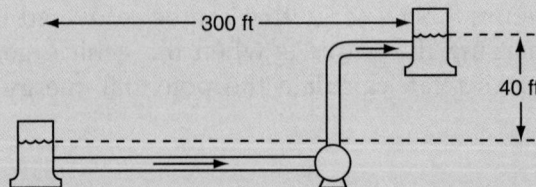

Figure E8.15

> **Solution**
>
> Because you are asked to calculate the potential energy change of 1 lb of water and not of the whole reservoir, you can assume for this problem that the 40 ft difference in height does not change. Think of a Ping-Pong ball riding on top of the water that determines the height difference between the levels in the two reservoirs.
>
> Let the water level in the first reservoir be the reference plane. Then $h = 40$ ft.
>
> $$\Delta \widehat{PE} = \frac{32.2\,\text{ft}}{s^2} \left| (40-0)\,\text{ft} \right| \frac{1(\text{lb}_f)(s^2)}{32.2\,(\text{lb}_m)(\text{ft})} \left| \frac{1\,\text{Btu}}{778.2\,(\text{ft})(\text{lb}_f)} \right| = 0.0514\ \text{Btu/lb}_m$$

Self-Assessment Test

Questions

1. Indicate whether the following statements are true or false:
   a. Potential energy has no unique absolute value.
   b. Potential energy can never be negative.
   c. The attractive and repulsive forces between the molecules in a material contribute to the potential energy of a system.

2. The units of potential energy or kinetic energy in the AE system are (select all of the correct expressions):
   a. $(\text{ft})(\text{lb}_f)$
   b. $(\text{ft})(\text{lb}_m)$
   c. $(\text{ft})(\text{lb}_f)/(\text{lb}_m)$
   d. $(\text{ft})(\text{lb}_m)/(\text{lb}_f)$
   e. $(\text{ft})(\text{lb}_f)/(\text{hr})$
   f. $(\text{ft})(\text{lb}_m)/(\text{hr})$

Answers

1. (a) T; (b) F; (c) F

2. (a); note that (c) represents the specific PE.

Problems

1. A 100 kg ball initially at rest on the top of a 5 m ladder is dropped and hits the ground. With reference to the ground:
   a. What is the initial kinetic and potential energy of the ball?
   b. What is the final kinetic and potential energy of the ball?

   c. What is the change in kinetic and potential energy for the process?

   d. If all of the initial potential energy were somehow converted to heat, how many calories would this amount to? How many British thermal units? How many joules?

2. A 1 kg ball 10 m above the ground is dropped and hits the ground. What is the change in *PE* of the ball?

Solutions

1. (a) $KE = 0$, $PE = (100)(9.8)(5) = 4900$ J; (b) $KE = 0$ J, $PE = 0$; (c) $\Delta KE = 0$, $\Delta PE = -4900$ J; (d) 1171 cal, 4.644 Btu; 4900 J

2. $\Delta PE = mgh = (10 \text{ kg})(9.8 \text{ m/s}^2)(-10 \text{ m}) = -980$ J.

8.5.3 Mechanical Energy Balance Examples

Now that we have defined potential energy and kinetic energy, we can convert Equation (8.19) into a form that can be directly used to solve mechanical energy balance problems:

$$W = \frac{1}{2}\dot{m}\Delta v^2 + \dot{m}g\Delta h + \frac{\dot{m}\Delta p}{\rho}$$

where $\dot{m}$ is the mass flow rate through the system, Δv is change in the velocity of a stream leaving and entering the system, Δh is the change in the elevation of a stream leaving and entering a system, and Δp is the pressure difference between the stream leaving and entering the system. The specific flow work is equal to $\hat{V}p = p/\rho$ so that the total change in flow work is equal to $\dot{m}\Delta p/\rho$. This equation states that the work required is equal to the change in kinetic energy plus potential energy plus the flow work. Dividing through by $\dot{m}$ yields

$$\frac{W}{\dot{m}} = \frac{1}{2}\Delta v^2 + g\Delta h + \frac{\Delta p}{\rho} \qquad (8.22)$$

If no work is added to the system, flow through a system can result from an applied pressure drop across the system (i.e., Δp). In that case, the following equation applies

$$-\frac{\Delta p}{\rho} = \frac{1}{2}\Delta v^2 + g\Delta h$$

Note that because Δp is outlet pressure minus the inlet pressure, when pressure is used to drive flow through a system, Δp is negative, thus the minus sign. This equation is known as **Bernoulli's equation**.

Equation (8.22) assumes a perfectly reversible process without frictional losses (e.g., frictional losses for flow through pipes). Equation (8.22) can be modified by adding a friction loss term E_v; that is,

$$\frac{W}{\dot{m}} = \frac{1}{2}\Delta v^2 + g\Delta h + \frac{\Delta p}{\rho} + E_v$$

E_v depends on the velocity in the lines in the system and the length of the lines. Bird et al.[6] provides formulas for calculating E_v for flow through pipes.

The following examples demonstrate how to apply Equation (8.22).

Example 8.16   Calculation of the Power Needed to Pump Water in an Open, Steady-State System

Problem Statement

Water is pumped from a well (Figure E8.16) in which the water level is a constant 20 ft below the ground level. The water is discharged into a level pipe that is 5 ft above the ground at a rate of 0.50 ft³/s. Assume that negligible heat transfer occurs from the water during its flow. Calculate the electric power required by the pump if it is 100% efficient; you can neglect friction in the pipe and the pump.

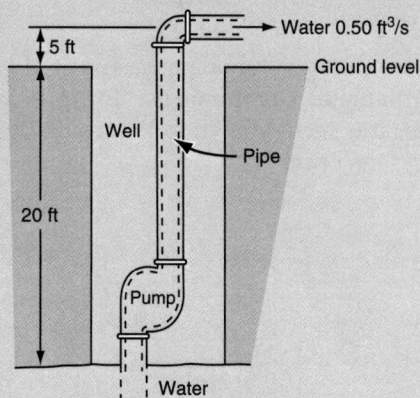

Figure E8.16

(Continues)

---

[6] Bird, R. B., W. E. Stewart, and E. N. Lightfoot. *Transport Phenomena* (New York, Wiley and Sons, 1960), 216–217.

Example 8.16   Calculation of the Power Needed to Pump Water in an Open, Steady-State System (*Continued*)

Solution

Let's pick as the system the pipe from the water level in the well to the place where the water that exits is at 5 ft above the ground so that the pump is included. Let's simplify Equation (8.22). Because this is a steady-state system and the pipe diameter appears to be the same for both the inlet and the exit, $\Delta KE = 0$. Also, the pressure at the exit is equal to the pressure at the inlet. After these simplifications, Equation (8.22) becomes

$$\frac{W}{\dot{m}} = g\Delta h$$

$$W = \frac{0.5\,\text{ft}^3}{\text{s}} \left| \frac{62.4\,\text{lb}_m\,\text{H}_2\text{O}}{\text{ft}^3} \right| \frac{32.2\,\text{ft}}{\text{s}^2} \left| 25\,\text{ft} \right| \frac{\left(\text{s}^2\right)\left(\text{lb}_f\right)}{32.2\,(\text{ft})\,(\text{lb}_m)} \left| \frac{\text{hp-s}}{550\,(\text{lb}_f)\,(\text{ft})} \right| = 1.42\,\text{hp}$$

Because this analysis neglected friction and the efficiency of the pump, it represents an estimate of the required pump size that is less than the actual size required.

Example 8.17   Siphoning Gasoline

Problem Statement

Assume that you want to siphon gasoline out of your gas tank using a hose with a 0.5 in. inside diameter. Determine the flow rate of gasoline in gallons per minute based on the arrangement shown in Figure E8.17 neglecting the friction between the gasoline and the hose.

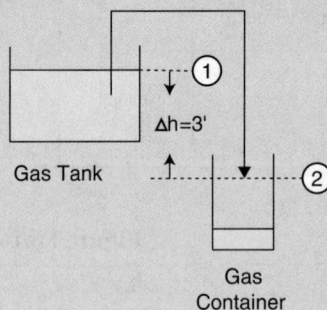

Figure E8.17

Solution

Simplifying Equation (8.22): (1) the pressure at the entrance and exit for this system will both be atmospheric pressure, so the flow work should be zero; and (2) no work is applied to this system. Therefore, the mechanical energy balance equation reduces to

$$0 = \frac{1}{2}\Delta v^2 + g\Delta h$$

Solving for the exit velocity, assuming the entrance velocity (i.e., the velocity of the gasoline level in the gas tank) is zero,

$$v = \sqrt{2g\Delta h} = \sqrt{\frac{2}{}\left|\frac{32.2 \text{ ft}}{s^2}\right|\frac{2 \text{ ft}}{}} = 11.35 \text{ ft/s}$$

Finally, convert the average velocity into the volumetric flow rate using the cross-sectional area:

$$Q = Av = \frac{\pi}{}\left|\frac{(0.5 \text{ in})^2}{4}\right|\frac{11.35 \text{ ft}}{s}\left|\frac{\text{ft}^2}{(12 \text{ in})^2}\right|\frac{7.481 \text{ gal}}{\text{ft}^3}\left|\frac{60 \text{ s}}{\text{min}}\right. = 6.947 \text{ gpm}$$

Example 8.18   Hydroelectric System

Problem Statement

Consider the hydroelectric system shown in Figure E8.18. The Hoover Dam has a power capacity of 2080 MW of electricity. The current water level at the dam is 1070 ft. Assuming 100% efficiency for the hydroelectric process at Hoover Dam, calculate the flow rate of water in kg/s and gpm through this hydroelectric process.

(Continues)

Example 8.18   Hydroelectric System (*Continued*)

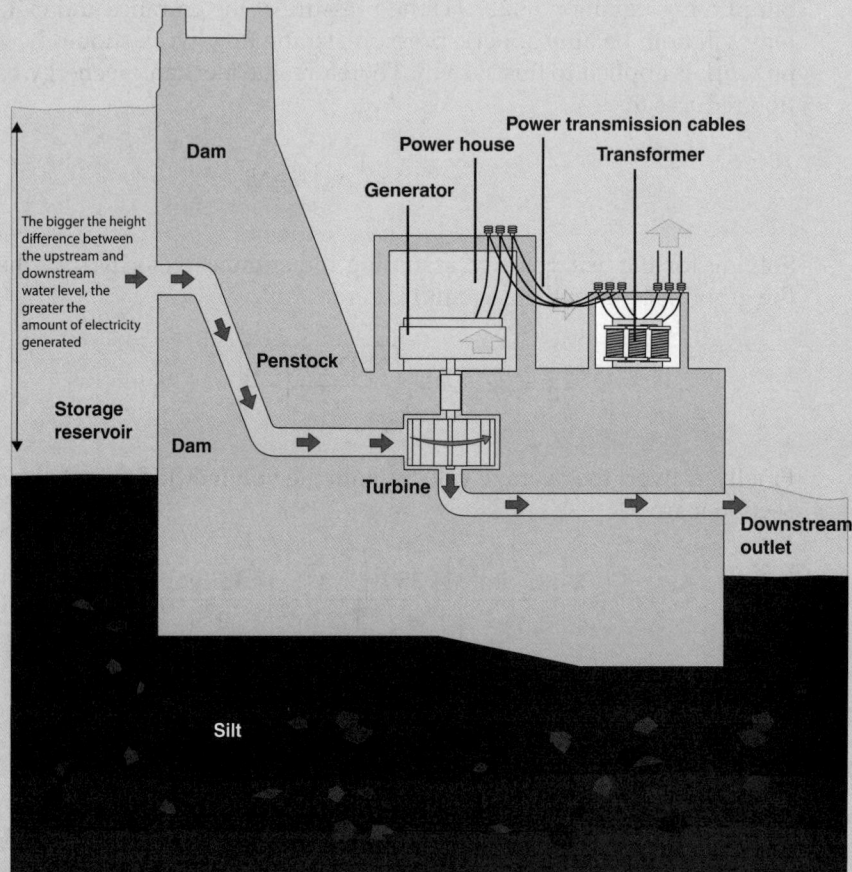

Dam

Power transmission cables

Power house

Transformer

Generator

The bigger the height difference between the upstream and downstream water level, the greater the amount of electricity generated

Penstock

Storage reservoir

Dam

Turbine

Downstream outlet

Silt

Figure E8.18

Solution

Simplifying Equation (8.22), recognizing that the kinetic energy change and the flow work can be ignored, yields

$$\frac{W}{\dot{m}} = g\Delta h$$

Solving for $\dot{m}$ and making the proper unit conversions yields

$$\dot{m} = \frac{2080 \text{ MW}}{1070 \text{ ft}} \left| \frac{s^2}{9.807 \text{ m}} \right| \frac{3.2808 \text{ ft}}{m} \left| \frac{10^6 \text{ W}}{\text{MW}} \right| \frac{\text{kg-m}^2}{\text{W-s}^3} = 6.503 \times 10^5 \text{ kg/s} \left(1.031 \times 10^7 \text{ gpm} \right)$$

Example 8.19 Determine the Flow Rate through a Piping System with a Pump

Problem Statement

Determine the flow rate from tank A to tank B for the system shown in Figure E8.19 in gpm. Assume that the fluid being pumped is water, a 0.25 hp pump is used, and the pipes in this system have a 2 in. inside diameter; neglect the friction in the pipes and the pump.

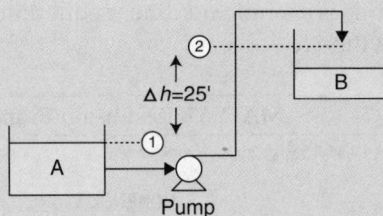

Figure E8.19

Solution

Simplifying Equation (8.22), the pressure at the entrance and exit for this system will both be atmospheric pressure, so the flow work should be zero. Therefore, the mechanical energy balance equation reduces to

$$\frac{W}{\dot{m}} = \frac{1}{2}\Delta v^2 + g\Delta h$$

Note that $\dot{m}$ and v are both depend on the flow rate Q between the tanks: $\dot{m} = Q\rho$ and $v = Q/A$, where A is the cross-sectional areas of the pipe between the tanks. Therefore, the previous equation becomes

$$\frac{W}{Q\rho} = \frac{Q^2}{2A^2} + g\Delta h$$

Because this is a cubic equation in Q, the unknown, we will use MATLAB and Python to solve it, but we must be careful with the units for this problem. Note that each term should have units of ft^2/s^2 to match the units of $g\Delta h$.

$$\left.\frac{W\,\text{hp}}{}\right|\frac{\text{ft}^3}{\rho\,\text{lb}_m}\left|\frac{\text{s}}{Q\,\text{ft}^3}\right|\frac{550\,\text{ft lb}_f}{\text{hp-s}}\left|\frac{32.2\,\text{ft lb}_m}{\text{lb}_f\,\text{s2}}\right| = 17{,}710\frac{W}{\rho Q}\,[=]\text{ft}^2/\text{s}^2$$

$$\left.\frac{2Q^2\,\text{ft}^6}{\text{s}^2}\right|\frac{}{A^2\,\text{ft}^4}\left.\right| = 2Q^2/A^2\,[=]\,\text{ft}^2/\text{s}^2$$

(Continues)

Example 8.19   Determine the Flow Rate through a Piping System with a Pump (*Continued*)

Therefore, the dimensionally consistent equation becomes

$$71,710\frac{W}{Q\rho} = \frac{Q^2}{2A^2} + g\Delta h$$

where W is in horsepower, Q is ft^3/s, ρ is in lb$_m$/ft^3, A is in ft^2, g is in ft/s^2, and Δh is in ft. Because this is a single nonlinear equation, we use the built-in functions in MATLAB and Python to solve it.

MATLAB Results for Example 8.19

```
%%%%%%%%%%%%%%%%%%%%%%%%%%%%%%%%%%%%%%%%%%%%%%%%%%%%%%%%%%%%%%%%%%%
%                       NOMENCLATURE
%
%  x0 - initial guess for the root of the equation
%  xsoln - the root of f(x) determined by function fzero
%%%%%%%%%%%%%%%%%%%%%%%%%%%%%%%%%%%%%%%%%%%%%%%%%%%%%%%%%%%%%%%%%%%
%                         PROGRAM
function Ex8_19_MechEnBal
clear; clc;
x0=1.
W=0.25; den=62.4; A=3.14159*2^2/4/144; g=32.2; Dh=25;
                        % Define anonymous function
fx=@(x)71710/(den*x)-x^2/(2*A)+g*Dh
xsoln=fzero(fx,x0)          % Call built-in function fzero
end

%                       PROGRAM END
%%%%%%%%%%%%%%%%%%%%%%%%%%%%%%%%%%%%%%%%%%%%%%%%%%%%%%%%%%%%%%%%%%%
xsoln =
   0.3151
```

Python Code for Example 8.19

```
Ex8_19 MechEnBal.py
#######################################################################
#                          NOMENCLATURE
#
#  x0 – the initial guess for the root of function f(x)
#  xsol - the root of f(x) determined by function scipy.optimize.newton
#######################################################################
#                            PROGRAM
import numpy as np
import scipy.optimize
#  Define the UD function for the function value
def f(x):
    A, W, den, g, Dh =3.14159*2**2/4/144, 0.25, 62.4, 32.2, 25
    fx=71710*W/(x*den)-x**2/2/A-g*Dh
    return fx
x0=1.                                                                  #
Initial guess for x
#  Apply function scipy.optimize.newton to determine a root of the
   nonlinear equation
xsol=scipy.optimize.newton(f, x0)

#                          PROGRAM END
#######################################################################
```

IPython console:
```
In[1]: runfile(…
In[2]: %precision 4
In[3]: xsol
Out[3]: 0.3151
```

Therefore, the flow rate through this system is calculated to be 0.3151 ft^3/s. This value is based on neglecting friction, but at this flow rate, which corresponds to an average velocity in the pipe of 104 ft/s, friction would actually be a major factor.

Self-Assessment Test

Questions

1. Do turbines and pumps represent examples of unsteady-state systems?

2. Indicate whether the following statements are true or false:
   a. The input stream to a process possesses kinetic energy.
   b. The input stream to a process possesses potential energy.
   c. The input stream to a process possesses internal energy.
   d. The exit stream from the system does flow work.
   e. The shaft work done by a turbine that is rotated by a fluid in a system is positive.

3. From the mechanical energy balance, what can drive flow through a pipe?

Answers

1. No, they usually work in a steady-state fashion.

2. (a) T; (b) possibly T, depending on the location above the reference plane; (c) T; (d) T; (e) F

3. Work, *PE, KE,* or a pressure difference.

Problem

1. Consider a pump that is pumping water uphill for a vertical change of 35 ft. What is the flow rate if a 1.0 hp pump is used?

Answer

1. $\dot{m} = \dfrac{W}{g\Delta h} = \dfrac{1.0\ \text{hp}}{} \left|\dfrac{s^2}{32.2\ \text{ft}}\right| \dfrac{}{35\ \text{ft}} \left|\dfrac{550\ \text{ft-lb}_f}{\text{hp-s}}\right| \dfrac{32.2\ \text{lb}_m\text{-ft}}{\text{lb}_f\text{-s}^2} = 17.08\ \text{lb}_m/\text{s}$

8.6   Energy Balances for Special Cases

The previous sections addressed the most common types of energy balances encountered in the process industries. Nevertheless, sometimes you will encounter a system that does not fit into one of the three categories, and in that case, you should apply the general energy balance equation:

$$\Delta U_{\text{total}} = Q + W - \Delta E_{\text{conv}}$$

remembering that ΔU_{total} contains the internal energy as well as the kinetic and potential energy of the material in the system of interest, and ΔE_{conv} contains the enthalpy as well as the kinetic and potential energy of the material entering and leaving the system. Each of the terms in the general energy balance has been explicitly defined and demonstrated in the previous sections.

The following examples show how to apply the general energy balance for special cases.

Example 8.20   Unsteady-State, Open Process

Problem Statement

For 10 min, 10 lb of water at 35°F flows into a 125 ft³ insulated vessel that initially contains 4 lb of ice at 32°F. To heat and mix the ice and water, 6 lb of steam at 250°F and 20 psia are introduced. What is the final temperature in the vessel after 10 min (assume all the material is well mixed)?

Solution

Steps 1–4

Figure E8.20 is a sketch of the process. Pick the system as the vessel. Let T_2 be the final temperature in the system.

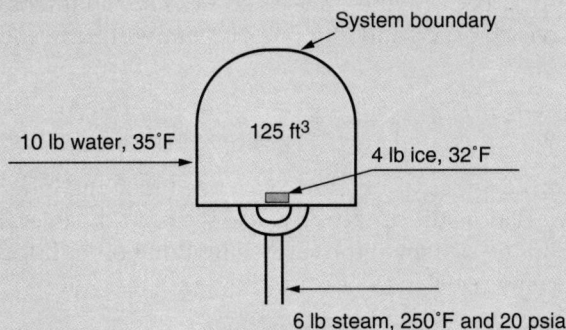

Figure E8.20

Step 5

Pick 10 min as the basis, which is equal to 20 total lb (water plus steam plus ice) in the final state of the system.

(Continues)

Example 8.20   Unsteady-State, Open Process (*Continued*)

	State 1: Initial Conditions in System (Ice)	Water Flow into System	Steam Flow into System	State 2: Known Final Conditions in System
m (lb)	4	10	6	20
T (°F)	32	35	250°F	
$\hat{U}$ (Btu/lb)	−143.6*	3.025	1090.26	
$\hat{H}$ (Btu/lb)	−143.6*	3.025	1167.15	
$\hat{V}$ (ft³/lb)	Ignore	Ignore (0.016)	20.80	6.25
p (psia)	Ignore	Ignore (vapor pressure)	20	

*Heat of fusion

The source of the data are the steam tables, readily available on the Internet. One particularly important piece of information you might normally ignore is $\hat{V}_2 = \hat{V}_{\text{inside, final}} = \dfrac{125 \text{ ft}^3}{20 \text{ lb}} = 6.25 \text{ ft}^3/\text{lb}$. Remember that the values of two intensive variables must be known to fix the final conditions, and $\hat{V}_2$ can thus be one of them. What is another one?

A quick glance at the list of variables whose values have not been assigned indicates that T_2, $\hat{U}_2$, and $\hat{H}_2$ are unknowns. But the situation is not as bad as it might appear because the three unknowns are all intensive variables, they are all related to each other, and any one of them will suffice to fix the final state of the system.

Steps 7–9

To get zero degrees of freedom, you need just one independent equation. Will a material balance suffice? No, because we used the overall material balance to get the 20 lb in the system at the final state. What other information can be involved? Why, the energy balance, of course!

$$\Delta U_{\text{total}} = Q + W - \Delta E_{\text{conv}}$$

To simplify the general energy balance, assume:

1. All the ΔKE and ΔPE terms are zero inside the system and for the flow streams.

2. $Q = 0$ (insulated).

3. $W = 0$.

Then

$$\Delta U = -\Delta E_{conv} = \left(m_2 \hat{U}_2 - m_1 \hat{U}_1 \right) = -\left[0 - \left(m_{\text{water in}} \hat{H}_{\text{water in}} + m_{\text{steam in}} \hat{H}_{\text{steam in}} \right) \right]$$

$$\left[20\,\hat{U}_2 - 4(-143.6) \right] = \left[10(3.025) + 6(1167.15) \right]$$

$$\hat{U}_2 = 6458.75/20 = 322.9 \text{ Btu/lb}$$

Although the values of $\hat{U}_2 = 322.9$ Btu/lb and $\hat{V}_2 = 6.25$ lb/ft³ fix the solution, you still have to obtain the value of T_2. How can you get T_2? Not easily. If you had a way to insert the values of $\hat{U}_2$ and $\hat{V}_2$ into a computer program that would yield T_2, you would be all set. Even a chart of $\hat{V}_2$ versus $\hat{U}_2$ would be of help. Perhaps those tools will be available someday. But at the moment, the best you can do is to look at a chart that shows p, T, $\hat{V}$, and $\hat{H}$ to ascertain the value of T_2, assuming $\hat{U} \cong \hat{H}$, if the solution is in the two-phase region, which it is. The paper steam tables folded in the back of this book have such a chart (in SI units).

Alternatively, you can prepare a small region of your own chart of $\hat{V}_2$ versus $\hat{U}_2$ say, from 280 to 360 for $\hat{U}_2$ in Btu/lb and $\hat{V}_2$ from 3 to 9 ft³/lb. Use the saturated steam tables on the website to calculate at constant pressure values of $\hat{U}_2$ and $\hat{V}_2$ surrounding the point $\hat{V}_2 = 6.25$ lb/ft³ and $\hat{U}_2 = 322.9$ Btu/lb by changing values of the quality of the two-phase mixture x. The lines of p and x will be straight. An answer with a reasonable number of significant figures might require a half hour of your time. The final values of the unknowns are

$$T_2 = 195.6°F$$
$$P_2 = 10.5 \text{ psia}$$
$$x = 17.1\%$$

What if the vessel contained air at atmospheric pressure when the ice was placed in the vessel? How would the solution change?

Example 8.21   Heating of a Biomass

Problem Statement

Steam at 250°C saturated (which is used to preheat a fermentation broth) enters the steam chest. The steam is segregated from the biomass solution in the preheater and is completely condensed in the steam chest. The rate of the heat loss from the surroundings is 1.5 kJ/s. The material to be heated, which enters at 45°C, flows into the process vessel, which initially is at 20°C. If 150 kg of biomass with an average heat capacity of $C_p = 3.26$ J/(g)(K) flows into the process per hour, how many kilograms of steam are needed per hour?

(Continues)

Example 8.21   Heating of a Biomass (*Continued*)

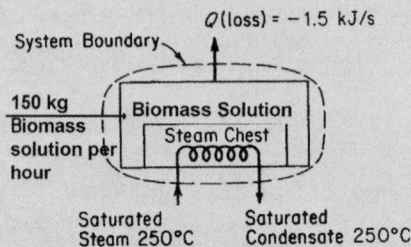

Figure E8.21

Solution

Steps 1–4

Figure E8.21 defines the system and shows the known conditions. If the system is composed of the biomass solution plus the steam chest (a heat exchanger), the process is open and unsteady state because the temperature of the biomass increases as does the mass in the system. Assume that the entering solution causes the biomass to be well mixed.

Step 5

Basis: 1 hr of operation (150 kg of biomass solution heated)

Steps 6–8

The material balance for the steam is in $m_{in} = m_{out} = m$, and m is an unknown. The material balance for the biomass solution is also simple: what flows in stays in the system. Initially, no biomass was in the vessel, and at the end of 1 hr, 150 kg of solution existed in the vessel. As a result, one or more unknowns are associated with the biomass solution.

Next, let's examine the energy balance for the system to ascertain if the balance can be used to solve for m. The general energy balance is

$$\Delta U_{total} = Q + W - \Delta E_{conv} \tag{a}$$

Let's simplify the energy balance:

1. The process is not in the steady state, so $\Delta U_{total} \neq 0$.

2. We can safely assume that $\Delta KE = 0$ and $\Delta PE = 0$ inside the system and for the flow in.

3. $W = 0$.

Consequently, Equation (a) becomes

$$\Delta U = Q - \Delta \left[m\hat{H} \right] \qquad \text{(b)}$$

where ΔU can be calculated from just the change in state of the biomass solution and does not include the water in the steam chest because we will assume that there was no water or steam in the steam chest at the start of the hour and none in the steam chest at the end of the hour.

$$(150 \text{ kg})\left(\hat{U}_{\text{final}}\right) - (0 \text{ kg})\left(\hat{U}_{\text{initial}}\right) = \frac{-1.50 \text{ J}}{\text{s}} \left| \frac{3600 \text{ s}}{1 \text{ hr}} \right| \frac{1 \text{ hr}}{} - m_{\text{steam}}\left(\hat{H}_2 - \hat{H}_1\right) \qquad \text{(c)}$$

Choose as a reference temperature either 0°C or 20°C; the p reference is 1 atm by assumption. Pick $T_{\text{reference}} = 20°C$. Assume $\Delta\hat{U} = \Delta\hat{H}$.

$$\hat{U}_{\text{final}} = \frac{3.26 \text{ kJ}}{(\text{kg})(°C)} \left| \frac{(45-20)°C}{} \right| = 81.5 \text{ kJ/kg}$$

$$\hat{U}_{\text{initial}} = 0 \text{ (Why? 20°C is the reference temperature)}$$

$$\left(\hat{H}_2 - \hat{H}_1\right) = \Delta\hat{H}_{\text{condensation of steam}} = -\Delta\hat{H}_{\text{vaporization}} \text{ at } 250°C = -1701 \text{ kJ/kg}$$

Introduce these values into Equation (c):

$$12{,}225 = -5400 - m(-1701)$$

$$m_{\text{steam}} = 10.4 \text{ kg/hr}$$

Self-Assessment Test

Questions

1. Indicate whether the following statements are true or false:
   a. The shaft work done by a pump and motor located inside the system is positive.
   b. The Δ symbol in an energy balance for a steady-state system refers to the property of the material entering the system minus the property of the material leaving the system.
   c. An input stream to a system does flow work on the system.
   d. The input stream to a system possesses internal energy.
   e. Work done by fluid flowing in a system that drives a turbine coupled with an electric generator is known as *flow work*.

2. What are two circumstances in which you can neglect the heat transfer term in the general energy balance?

3. What term in the general energy balance is always zero for a steady-state process?

4. What intensive variables are usually used to specify the value of the enthalpy?

5. Under what condition can (a) the KE term and (b) the PE term be ignored or deleted from the general energy balance?

6. You read in an engineering book that

$$U_2 - U_1 = Q - W_{nonflow}$$

If you cannot find the notation list in the book, would you agree that the equation represents the energy balance for an unsteady-state, open system?

7. Can a system with no moving parts be treated as an unsteady-state, open system?

Answers

1. (a) If done on the surroundings, F; (b) F, it is out-in; (c) T; (d) T; (e) F

2. Insulated system and no temperature difference between system and the surroundings

3. $\Delta E = 0$

4. T and p

5. (a) No velocity of fluid flow, $KE_{in} = KE_{out}$, or ΔKE, is negligible compared to other terms in the energy balance.

6. Yes, but W is defined to be positive if work is done on the surroundings by the system.

7. Yes

Problems

1. A 3 MW steam-driven turbine operates in the steady state using 2.0 kg/s of steam. The inlet conditions for the steam are $p = 3000$ kPa and 450°C. The outlet conditions are 500 kPa, saturated vapor. The entering velocity of the steam is 250 m/s, and the exit velocity is 40 m/s. What is the heat transfer in kilowatts for the turbine as the system? What fraction of the energy supplied by the steam is generated power?

2. Simplify the general energy balance, Equation (8.2), as much as possible for each of the following circumstances (state which terms can be deleted and why):
   a. The system has no moving parts.
   b. The temperatures of the system and its surroundings are the same.

    c. The velocity of the fluid flowing into the system equals the velocity of the fluid leaving the system.

    d. The fluid exits the system with sufficient velocity so that it can shoot out 3 m.

3. Under what circumstances is $\Delta U = \Delta H$ for an open, unsteady-state system?

4. A compressor fills a tank with air at a service station. The inlet air temperature is 300 K, and the pressure is 100 kPa, the same conditions as for the initial air in the tank with a mass of 0.8 kg. After 1 kg of air is pumped into the tank, the pressure reaches 300 kPa, and the temperature is 400 K. How much heat was added to or removed from the tank during the compression? Hint: Do not forget the initial amount of air in the tank.

Data for the air:

	$\hat{H}(kJ/kg)$	$\hat{U}(kJ/kg)$	$\hat{V}(m^3/kg)$
100 kPa and 300 K	458.85	337.75	0.8497
300 kPa and 400 K	560.51	445.61	0.3830

Answers

1. $\Delta H = -2$ kg/s (3389.1 − 639.8)kJ/kg = −6778 kW; $\Delta KE = -0.5(20$ kg/s)(2502 − 402) m²/s² = −609 kW; $W = -3000$ kW; $W = \Delta H + \Delta KE + Q$; $Q = 3000 - 6778 - 609 = -4387$ kW; fraction = 3000/(6778 + 609) = 0.406

2. (a) No terms dropped; (b) $Q = 0$; (c) $\Delta KE_{conv} = 0$; (d) no terms dropped

3. $\Delta KE = \Delta PE = 0$ in system and for streams entering and leaving the system; $Q = W = 0$

4. $Q = m_{total}\hat{U}(300$ kPa, 400 K) $- m_1\hat{U}(100$ kPa, 300 K) $- m_2\hat{H}(100$ kPa, 400 K) $= 445.61(1.8) - 337.75(.8) - 458.85 = 73.04$ kJ

Summary

This chapter introduced a number of different forms of energy and how to formulate energy balances using them as well as important new terminology. The general energy balance was broken into four types: unsteady-state, closed systems; steady-state, open systems; mechanical energy balances; and general cases. The vast majority of energy balances applied to the process industries fit into one of the first three categories while the last case handles the few exceptions.

Glossary

adiabatic   Describes a system that does not exchange heat with the surroundings during a process.

adiabatic process   A process in which no heat transfer occurs ($Q = 0$).

boundary   Hypothetical perimeter used to define what a system is.

closed system   A system in which mass is not exchanged with the surroundings.

conservation of energy   The total energy of a system plus its surroundings is constant.

electrical work   Work done on or by a system because a voltage difference forces a current to flow.

energy   The capacity to do work or transfer heat.

energy transfer   The movement of energy from one site or state to another.

enthalpy (*H*)   The sum of the variables $U + pV$.

equilibrium state   The properties of a system remain invariant under a balance of potentials.

extensive property   A property that depends on the amount of material present, such as volume or mass.

flow system   See **open system**.

flow work   Work done on a system to put a fluid element into the system, or work done by a system to push a fluid element into the surroundings.

general energy balance   The change of energy inside a system is equal to the net heat and net work interchange with the surroundings plus the net energy transported by flow into and out of the system.

heat (*Q*)   Energy transfer across a system boundary that is caused by a temperature difference (potential) between the system and the surroundings.

heat capacity   Also called the specific heat. One heat capacity, C_v, is defined as the change in internal energy with respect to temperature at constant volume, and a second heat capacity, C_p, is defined as the change in enthalpy with respect to temperature at constant pressure.

heat of condensation   The negative of the heat of vaporization.

heat of fusion   The enthalpy change for the phase transition of melting.

heat of solidification   The negative of the heat of fusion.

heat of sublimation   The enthalpy change for the phase transition of a solid directly to a vapor.

heat of vaporization   The enthalpy change for the phase transition of a liquid to a vapor.

intensive property   A property that is independent of the amount of material present, such as temperature, pressure, or any specific variable (per unit amount of material, mass, or mole).

internal energy (U)   Energy that represents a macroscopic account of all the molecular, atomic, and subatomic energies, all of which follow definite microscopic conservation rules for dynamic systems.

isobaric   Describes a system for which the pressure is invariant during a process.

isochoric   Describes a system for which the volume is invariant during a process.

isothermal   Describes a system in which the temperature is invariant during a process.

kinetic energy (KE)   Energy a system possesses because of its velocity relative to the surroundings.

latent heat   An enthalpy change that involves a phase transition.

nonflow system   See **closed system**.

open system   A system in which interchange of mass occurs with the surroundings.

path variable (function)   A variable (function) whose value depends on how the process takes place, such as heat and work.

phase transition   A change from the solid to the liquid phase, from the liquid to the gas phase, from the solid to the gas phase, or the respective reverses.

point function   See **state variable**.

potential energy (PE)   Energy a system possesses because of the force exerted on its mass by a gravitational or electromagnetic field with respect to a reference surface.

power   Work per unit time.

property   An observable or calculable characteristic of a system, such as temperature or enthalpy.

reference substance plots   Plots used to estimate values for physical properties based on comparison with the same properties of a substance used as a reference.

sensible heat   An enthalpy change that does not involve a phase transition.

shaft work   Work corresponding to a force acting on a shaft to turn it.

single phase   A part of a system that is physically distinct and homogeneous, such as a gas or liquid.

state   Conditions of the system, such as the values of the temperature or pressure.

state variable (function)   A variable (function), also called a point function, whose value depends only on the state of the system and not upon its previous history, such as internal energy.

steady state   The accumulation in a system is zero, the flows in and out are constant, and the properties of the system are invariant.

surroundings   Everything outside a system.

system   The quantity of matter or region of space chosen for analysis.

transient   See **unsteady state**.

unsteady state   Not steady state (see **steady state**).

work (*W*)   Work is a form of energy that represents a transfer of energy between a system and its surroundings. Work cannot be stored.

Supplemental References

Abbott, M. M., and H. C. Van Ness. *Schaum's Outline of Thermodynamics with Chemical Applications*, 2nd ed., McGraw-Hill, New York (1989).

Blatt, F. J. *Modern Physics*, McGraw-Hill, New York (1992).

Boethling, R. S., and D. McKay. *Handbook of Property Estimation Methods for Chemicals*, Lewis Publishers (CRC Press), Boca Raton, FL (2000).

Crawly, G. M. *Energy*, Macmillan, New York (1975).

Daubert, T. E., and R. P. Danner. *Data Compilation of Properties of Pure Compounds*, Parts 1, 2, 3 and 4, Supplements 1 and 2, DIPPR Project, AIChE, New York (1985-1992).

Gallant, R. W., and C. L. Yaws. *Physical Properties of Hydrocarbons*, 4 vols., Gulf Publishing, Houston, TX (1992-1995).

Guvich, L. V., I. V. Veyts, and C. B. Alcock. *Thermodynamic Properties of Individual Substances*, Vol. 3, Parts 1 & 2, Lewis Publishers (CRC Press), Boca Raton, FL (1994).

Levi, B. G. (ed.). *Global Warming, Physics and Facts*, American Institute of Physics, New York (1992).

Pedley, J. B. *Thermochemical Data and Structures of Organic Compounds*, Vol. 1, TRC Data Distribution, Texas A&M University System, College Station, TX (1994).

Poling, B. E., J. M. Prausnitz, and J. P. O'Connell. *The Properties of Gases and Liquids*, 5th ed., McGraw-Hill, New York (2001).

Raznjevic, K. *Handbook of Thermodynamic Tables*, 2nd ed., Begell House, New York (1995).

Stamatoudis, M., and D. Garipis. "Comparison of Generalized Equations of State to Predict Gas-Phase Heat Capacity," *AIChE J.*, **38**, 302-7 (1992).

Vargaftik, N. B., Y. K. Vinogradov, and V. S. Yargin. *Handbook of Physical Properties of Liquids and Gases*, Begell House, New York (1996).

Yaws, C. L. *Handbook of Thermodynamic Diagrams*, Vols. 1, 2, 3 and 4, Gulf Publishing Co., Houston, TX (1996).

Problems

8.1   Terminology Associated with Energy Balances

*8.1.1   Convert 45.0 Btu/lb$_m$ to the following:
   a. cal/kg
   b. J/kg
   c. kWh/kg
   d. (ft)(lbf)/lb$_m$

*8.1.2   Convert the following physical properties of liquid water at 0°C and 1 atm from the given SI units to the equivalent values in the listed AE units:

   a. Heat capacity of 4.184 J/(g)(K)   Btu/(lb)(°F)
   b. Enthalpy of –241.6 J/kg   Btu/lb
   c. Thermal conductivity of 0.59 (kg)(m)/(s^3)(K)   Btu/(ft)(hr)(°F)

*8.1.3   Convert the following quantities as specified:
   a. A rate of heat flow of 6000 Btu/(hr)(ft^2) to cal/(s)(cm^2)
   b. A heat capacity of 2.3 Btu/(lb)(°F) to cal/(g)(°C)
   c. A thermal conductivity of 200 Btu/(hr)(ft)(°F) to J/(s)(cm)(°C)
   d. The gas constant, 10.73 (psia)(ft^3)/(lb mol)(°R) to J/(g mol)(K)

*8.1.4   The energy from the sun incident on the surface of the earth averages 32.0 cal/(min)(cm^2). It has been proposed to use space stations in synchronous orbits 36,000 km from Earth to collect solar energy. How large a collection surface is needed (in square meters) to obtain 10^8 W of electricity (equivalent to a 100 MW power plant)? Assume that 10% of the collected energy is converted to electricity. Is your answer a reasonable size?

*8.1.5   Indicate whether the following statements are true or false:
   a. A simple test to determine whether a property is an extensive property is to split the system in half, and if the value of the sum of the properties in each half is twice the value of the property in one half, then the property is an extensive variable.
   b. The value of a state variable (point function) depends only on its state and not on how the state was reached.
   c. An intensive variable is a variable whose value depends on the amount of mass present.
   d. Temperature, volume, and concentration are intensive variables.
   e. Any variable that is expressed as a "specific"—that is, per unit mass—quantity is deemed to be an intensive variable.

*8.1.6   Indicate whether the following statements are true or false:
   a. An isobaric process is one of constant pressure.
   b. An isothermal process is one of constant temperature.
   c. An isometric process is one of constant volume.
   d. A closed system is one in which mass does not cross the system boundary.
   e. The units of energy in the SI system are joules.
   f. The units of energy in the AE system can be $(ft)(lb_f)/lb_m$.
   g. The difference between an open system and a closed system is that energy transfer can take place between the system and the surroundings in the former but not in the latter.

*8.1.7   Are the following variables intensive or extensive?
   a. Partial pressure
   b. Volume
   c. Specific gravity
   d. Potential energy
   e. Relative saturation
   f. Specific volume
   g. Surface tension
   h. Refractive index

*8.1.8   Classify the following measurable physical characteristics of a gaseous mixture of two components as an intensive property, an extensive property, both, or neither:
   a. Temperature
   b. Pressure
   c. Composition
   d. Mass

**8.1.9 A simplified equation for the heat transfer coefficient from a pipe to air is

$$h = \frac{0.026G^{0.6}}{D^{0.4}}$$

where h = heat transfer coefficient, Btu/(hr)(ft²)(°F)
G = mass rate flow, lbm/(hr)(ft²)
D = outside diameter of the pipe, ft

If h is to be expressed in J/(min)(cm²)(°C) and the units of G and D remain the same, what should the new constant in the equation be in place of 0.026?

**8.1.10 A problem for many people in the United States is excess body weight stored as fat. Many people have tried to capitalize on this problem with fruitless weight-loss schemes. However, since energy is conserved, an energy balance reveals only two real ways to lose weight (other than water loss): (1) reduce the caloric intake and/or (2) increase the caloric expenditure. In answering the following questions, assume that fat contains approximately 7700 kcal/kg (1 kcal is called a "dietetic calorie" in nutrition, or commonly just a "Calorie"):

a. If a normal diet containing 2400 kcal/day is reduced by 500 kcal/day, how many days does it take to lose 1 lb of fat?

b. How many miles would you have to run to lose 1 lb of fat if running at a moderate pace of 12 km/hr expends 400 kJ/km?

c. Suppose that two joggers each run 10 km/day. One runs at a pace of 5 km/hr and the other at 10 km/hr. Which will lose more weight (ignoring water loss)?

*8.1.11 Solar energy has been suggested as a source of renewable energy. If in the desert the direct radiation from the sun (say, for 320 days) is 975 W/m² between 10 AM and 3 PM, and the conversion efficiency to electricity is 21.0%, how many square meters are needed to collect an amount of energy equivalent to the annual U.S. energy consumption of 3×10^{20} J? Is the construction of such an area feasible?

How many tons of coal (of heating value of 10,000 Btu/lb) would be needed to provide the 3×10^{20} J if the efficiency of conversion to electricity was 70%? What fraction of the total U.S. resources of coal (estimated at 1.7×10^{12} tons) is the calculated quantity?

**8.1.12 The thermal conductivity equation for a substance is $k = a + bT$ where

k is in (Btu)/(hr)(ft)(°F)
T is temperature, °F
a, b are constants with appropriate units

Convert this relation to make it possible to introduce the temperature into the modified equation as degrees Celsius and to have the units in which k is expressed to be (J)/(min)(cm)(K). Your answer should be the modified equation. Show all transformations.

**8.1.13   How much weight can you lose by exercise, say, running?

Data: 1 kg fat 5 7700 Calories 5 7700 kcal 5 32,000 J

Running at 5 min/km requires the expenditure of about 400 kJ/km.

**8.1.14   An overweight person decides to lose weight by exercise. Hard aerobic exercise requires 700 W. By exercising for one-quarter of an hour, can the person compensate for a big meal (4000 kJ)?

**8.1.15   A can of soda at room temperature is put into the refrigerator so that it will cool. Would you model the can of soft drink as a closed system or as an open system? Explain.

***8.1.16   Lasers are used in many technologies, but they are fairly large devices even in CD players. A materials scientist has been working on producing a laser from a microchip. He claims that his laser can produce a burst of light with up to 10,000 W, qualifying it for use in eye surgery, satellite communications, and so on. Is it possible to have such a powerful laser in such a small package?

***8.1.17   A letter to the editor of the local newspaper said:

> **Q.** I'm trying to eat a diet that's less than 30 percent fat. I bought some turkey franks advertised as "80 percent fat free." The statistical breakdown on the back of the package shows each frank has 8 grams of fat and 100 calories. If a gram of fat has nine calories, that makes the frank have about 75 percent fat. I'm so confused; is it them or is it me?

How would you answer the question?

**8.1.18   Often in books you read about "heat reservoirs" existing in two bodies at different temperatures. Can this concept be correct?

**8.1.19   An examination question asked: "Is heat conserved?" Sixty percent of the students said no, but 40% said yes. The most common explanation was (a) "Heat is a form of energy and therefore conserved." The next most common was (b) "Heat is a form of energy and therefore is not conserved." Two other common explanations were (c) "Heat is conserved. When something is cooled, it heats something else up. To get heat in the first place, though, you have to use energy. Heat is just one form of energy," and (d) "Yes, heat is transferred from a system to its surroundings, and vice versa. The amount lost by one system equals the amount gained by the surroundings." Explain whether heat is conserved, and criticize each of the four answers.

**8.1.20   Tell what is right and what is wrong with each of these concepts related to heat:
  a. Heat is a substance.
  b. Heat is really not energy.

    c. Heat and cold are the same thing; they represent opposite ends of a continuum.

    d. Heat and temperature are the same thing.

    e. Heat is proportional to temperature.

    f. Heat is not a measurable, quantifiable concept.

    g. Heat is a medium.

    h. Heat flows from one object to another or can be stored.

    i. Heat is produced by burning.

    j. Heat cannot be destroyed.

**8.1.21 This explanation of the operation of a refrigerator was submitted for publication in a professional news magazine:

Recall that in the usual refrigerator, a liquid coolant with a low boiling point vaporizes while at low pressure, absorbing heat from the refrigerator's contents. This heat energy is concentrated by a compressor and the result dissipated in a condenser, with the gas converting to liquid at high pressure. As the liquid passes through an expansion nozzle back into the refrigerator chamber, the cooling cycle begins anew.

Clarify the explanation for the editor of the magazine.

**8.1.22 Indicate whether the following statements are true or false:

    a. In an adiabatic process, no heat transfer occurs.

    b. When a gas is compressed in a cylinder, no heat transfer occurs.

    c. If an insulated room contains an operating freezer, no heat transfer takes place if the room is picked as the system.

    d. Heat and thermal energy are synonymous terms.

    e. Heat and work are the only mechanisms by which energy can be transferred to a closed system.

    f. Light is a form of heat.

    g. You can measure the heat in a system by its temperature.

    h. Heat is a measure of the temperature of a system.

**8.1.23 Indicate whether the following statements are true or false:

    a. In a process in which a pure substance starts at a specified temperature and pressure, goes through several temperature and pressure changes, and then returns to the initial state, $\Delta U = 0$.

    b. The reference enthalpy for the steam tables ($\Delta H = 0$) is at 25°C and 1 atm.

    c. Work can always be calculated as $\Delta(pV)$ for a process going from state 1 to state 2.

    d. An isothermal process is one for which the temperature change is zero.

    e. An adiabatic process is one for which the pressure change is zero.

    f. A closed system is one for which no reaction occurs.

g. An intensive property is a property of material that increases in value as the amount of material increases.

h. Heat is the amount of energy liberated by the reaction within a process.

i. Potential energy is the energy a system has relative to a reference plane.

j. The units of the heat capacity can be (cal)/(g)(°C) or Btu/(lb)(°F), and the numerical value of the heat capacity is the same in each system of units.

8.2 Overview of Types of Energy and Energy Balances

*8.2.1   Consider the following systems:
   a. Open system, steady state
   b. Open system, unsteady state
   c. Closed system, steady state
   d. Closed system, unsteady state

   For which system(s) can energy cross the system boundary?

**8.2.2   Draw a picture of the following processes, draw a boundary for the system, and state for each whether heat transfer, work, a change in internal energy, a change in enthalpy, a change in potential energy, or a change in kinetic energy occurs *inside the system*. Also classify each system as open or closed, and as steady state or unsteady state.
   a. A pump, driven by a motor, pumps water from the first to the third floor of a building at a constant rate and temperature. The system is the pump.
   b. As in a, except the system is the pump and the motor.
   c. A block of ice melts in the sun. The system is the block of ice.
   d. A mixer mixes a polymer into a solvent. The system is the mixer.

**8.2.3   Draw a simple sketch of each of the following processes, and in each, label the system boundary, the system, the surroundings, and the streams of material and energy that cross the system boundary.
   a. Water enters a boiler, is vaporized, and leaves as steam. The energy for vaporization is obtained by combustion of a fuel gas with air outside the boiler surface.
   b. Steam enters a rotary steam turbine and turns a shaft connected to an electric generator. The steam is exhausted at a low pressure from the turbine.
   c. A battery is charged by connecting it to a source of current.

**8.2.4   Explain specifically what the system is for each of the following processes; indicate if any energy transfer takes place by heat or work (use the symbols Q and W, respectively) or if these terms are zero.

    a. A liquid inside a metal can, well insulated on the outside of the can, is shaken very rapidly in a vibrating shaker.

    b. A motor and propeller are used to drive a boat.

    c. Water flows through a pipe at 1.0 m/min, and the temperatures of the water and the air surrounding the pipe are the same.

**8.2.5 Write the simplified energy balances for the following changes:

    a. A fluid flows steadily through a poorly designed coil in which it is heated from 70°F to 250°F. The pressure at the coil inlet is 120 psia and at the coil outlet is 70 psia. The coil is of uniform cross section, and the fluid enters with a velocity of 2 ft/s.

    b. A fluid is expanded through a well-designed adiabatic nozzle from a pressure of 200 psia and a temperature of 650°F to a pressure of 40 psia and a temperature of 350°F. The fluid enters the nozzle with a velocity of 25 ft/s.

    c. A turbine directly connected to an electric generator operates adiabatically. The working fluid enters the turbine at 1400 kPa absolute and 340°C. It leaves the turbine at 275 kPa absolute and at a temperature of 180°C. Entrance and exit velocities are negligible.

    d. The fluid leaving the nozzle of part b is brought to rest by passing it through the blades of an adiabatic turbine rotor, and it leaves the blades at 40 psia and at 400°F.

8.3 Energy Balances for Closed, Unsteady-State Systems

*8.3.1 In a closed system process, 60 Btu of heat are added to the system, and the internal energy of the system increases by 220 Btu. Calculate the work of the process.

*8.3.2 Suppose that a constant force of 40.0 N is exerted to move an object for 6.00 m. What is the work accomplished (on an ideal system) expressed in the following:

    a. joules

    b. $(ft)(lb_f)$

    c. calories

    d. British thermal units

*8.3.3 A gas is contained in a horizontal piston-cylinder apparatus at a pressure of 350 kPa and a volume of 0.02 m^3. Determine the work done by the piston on the gas if the cylinder volume is increased to 0.15 m^3 through heating. Assume the pressure of the gas remains constant throughout the process and that the process is ideal.

*8.3.4 A rigid tank contains air at 400 kPa and 600°C. As a result of heat transfer to the surroundings, the temperature and pressure inside the tank drop to 100°C and 200 kPa, respectively. Calculate the work done during this process.

**8.3.5   Find the value of internal energy for water (relative to the reference state) for the states indicated:
   a. Water at 0.4 MPa, 725°C
   b. Water at 3.0 MPa, 0.01 m³/kg
   c. Water at 1.0 MPa, 100°C

**8.3.6   What is the difference between heat and internal energy?

**8.3.7   A gas is heated at 200 kPa from 300 K to 400 K and then cooled to 350 K. In a different process, the gas is directly heated from 300 K to 350 K at 200 kPa. What difference is there in internal energy and enthalpy changes for the two processes?

**8.3.8   For the systems defined below, state whether Q, W, ΔH, and ΔU are 0, >0, or <0, and compare their relative values if not equal to 0:
   a. An egg (the system) is placed into boiling water.
   b. Gas (the system), initially at equilibrium with its surroundings, is compressed rapidly by a piston in an insulated nonconducting cylinder by an insulated nonconducting piston; give your answer for two cases: (1) before reaching a new equilibrium state and (2) after reaching a new equilibrium state.
   c. A Dewar flask of coffee (the system) is shaken.

**8.3.9   You are asked to calculate the electric power required (in kilowatt-hours) to heat all of an aluminum wire (positioned in a vacuum similar to a lightbulb filament) from 25°C to 660°C (liquid) to be used in a vapor deposition apparatus. The melting point of Al is 660°C. The wire is 2.5 mm in diameter and has a length of 5.5 cm. (The vapor deposition occurs at temperatures in the vicinity of 900°C.) Data: For Al, $C_p = 20.0 + 0.0135T$ where T is in kelvin and C_p is in J/(g mol)(°C). The $\Delta H_{fusion} = 10{,}670$ J/(g mol)(°C) at 660°C. The density of Al is 18.35 g/cm³.

***8.3.10   A horizontal cylinder, closed at one end, is fitted with a movable piston. Originally the cylinder contains 1.2 ft³ of gas at 7.3 atm pressure. If the pressure against the piston face is reduced very slowly to 1 atm, calculate the work done by the gas on the piston, assuming the following relationship to hold for the gas: $pV^{1.3}$ = constant.

***8.3.11   Calculate the change in the internal energy of 1 mol of a monoatomic ideal gas when the temperature goes from 0°C to 50°C.

**8.3.12   Is it possible to compress an ideal gas in a cylinder with a piston isothermally in an adiabatic process? Explain your answer briefly.

**8.3.13   Water is heated in a closed pot on top of a stove while being stirred by a paddle wheel. During the process, 30 kJ of heat are transferred to the water, and 5 kJ of heat are lost to the surrounding air. The work done amounts to 500 J. Determine the final energy of the system if its initial internal energy was 10 kJ.

**8.3.14 A person living in a 4 m × 5 m × 5 m room forgets to turn off a 100 W fan before leaving the room, which is at 100 kPa, 30°C. Will the room be cooler when the person comes back after 5 hr, assuming zero heat transfer? The heat capacity at constant volume for air is 30 kJ/kg mol.

**8.3.15 Two tanks are suspended in a constant temperature bath at 200°F. The first tank contains 1 ft³ of dry saturated steam. The other tank is evacuated. The two tanks are connected. After equilibrium is reached, the pressure in both tanks is 1 psia. Calculate (a) the work done in the process, (b) the heat transfer to the two tanks, (c) the internal energy change of the steam, and (d) the volume of the second tank.

**8.3.16 Two states, 1 and 2, are marked in Figure P8.3.16. Path A is taken from 1 to 2. Two alternative return paths from 2 to 1 are shown: B and C. Two different cycles can now be made up, each going from point 1 to point 2, and then returning to point 1. One cycle is made up from path A and path B, and the other from path A and path C. Are the following equations correct for the cycle 1 to 2 and return?

$$Q_A + Q_B = W_A + W_B \qquad Q_A + Q_C = W_A + W_C$$

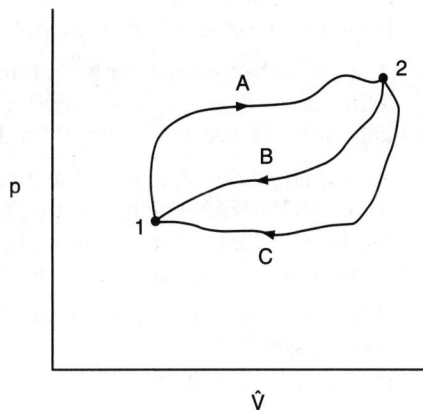

Figure P8.3.16

***8.3.17 You have calculated that the specific enthalpy of 1 kg mol of an ideal gas at 300 kN/m² and 100°C is 6.05×10^5 J/kg mol (with reference to 0°C and 100 kN/m²). What is the specific internal energy of the gas at 300 kPa and 100°C?

***8.3.18 A propane gas tank is filled, closed, and attached to a barbecue grill. After standing for some time, what is the state of the propane inside the tank? What are the temperature and pressure inside the tank? After 80% of the propane in the tank is used, what is the pressure inside the tank after it reaches equilibrium?

****8.3.19

 a. Ten pound moles of an ideal gas are originally in a tank at 100 atm and 40°F. The gas is heated to 440°F. The specific molar enthalpy of the ideal gas is given by the equation $\hat{H} = 300 + 8.00T$, where $\hat{H}$ is in British thermal units per pound mole and T is the temperature in degrees Fahrenheit.

  1. Compute the volume of the container (in cubic feet).
  2. Compute the final pressure of the gas (in atmospheres).
  3. Compute the enthalpy change of the gas.

 b. Use the equation above to develop an equation giving the molar internal energy, in joules per gram mole as a function of temperature, T, in degrees Celsius.

**8.3.20 A national mail-order firm is advertising for $698.99 ("manufacturer's suggested retail price $1,198.00") the Cold Front Portable Air Conditioner, a "freestanding portable unit" that "does not require outside venting." It is intended to be rolled from room to room by the user, who simply plugs it into an ordinary 110 V AC outlet and enjoys the cool air from its "Cold Front." It is claimed to provide 5500 Btu/hr cooling capacity for 695 W power. Is the cooling capacity correct? What is the catch to the advertisement?

**8.3.21 A large high-pressure tank contains 10 kg of steam. In spite of the insulation on the tank, it loses 2050 kJ/hr to the surroundings. How many kilowatts are needed to maintain the steam at 3000 kPa and 600 K?

**8.3.22 An expensive drug is manufactured in a sealed vessel that holds 8 lb of water at 100°F. A 1/4 hp motor stirs the contents of the vessel. What is the rate of heat removal from the vessel in British thermal units per minute to maintain the temperature at 100°F?

**8.3.23 Calculate how much heat is needed to evaporate 1 kg of water in an open vessel if the water starts at 27°C. Use the steam tables. The barometer reads 760 mm Hg.

**8.3.24 Air is being compressed from 100 kPa and 255 K (where it has an enthalpy of 489 kJ/kg) to 1000 kPa and 278 K (where it has an enthalpy of 509 kJ/kg). The exit velocity of the air from the compressor is 60 m/s. What is the power required (in kilowatts) for the compressor if the load is 100 kg/hr of air?

**8.3.25 Write the appropriate simplified energy balances for the following changes; in each case the amount of material to be used as a basis of calculation is 1 lb and the initial condition is 100 psia and 370°F:

 a. The substance, enclosed in a cylinder fitted with a movable frictionless piston, is allowed to expand at constant pressure until its temperature has risen to 550°F.

b. The substance, enclosed in a cylinder fitted with a movable friction-less piston, is kept at constant volume until the temperature has fallen to 250°F.

c. The substance, enclosed in a cylinder fitted with a movable friction-less piston, is compressed adiabatically until its temperature has risen to 550°F.

d. The substance, enclosed in a cylinder fitted with a movable friction-less piston, is compressed at constant temperature until the pressure has risen to 200 psia.

e. The substance is enclosed in a container that is connected to a sec-ond evacuated container of the same volume as the first, there being a closed valve between the two containers. The final condition is reached by opening the valve and allowing the pressures and tem-peratures to equalize adiabatically.

**8.3.26 An insulated, sealed tank that is 2 ft³ in volume holds 8 lb of water at 100°F. A 1/4 hp stirrer mixes the water for 1 hr. What is the fraction vapor at the end of the hour? Assume all of the energy from the stirrer enters the tank.

For this problem, you do not have to get a numerical solution. Instead, list the following in this order:
a. State what the system you select is.
b. Specify open or closed.
c. Draw a picture.
d. Put all of the known or calculated data on the picture in the proper place.
e. Write down the energy balance (use the symbols in the text) and simplify it as much as possible. List each assumption in so doing.
f. Calculate W.
g. List the equations with data introduced that you would use to solve the problem.
h. Explain step by step how to solve the problem (but do not do so).

**8.3.27 One kilogram of gaseous CO_2 at 550 kPa and 25°C was compressed by a piston to 3500 kPa, and in so doing 4.016×10^3 J of work were done on the gas. To keep the container isothermal, the container was cooled by blowing air over fins on the outside of the container. How much heat (in joules) was removed from the system?

**8.3.28 A household freezer is placed inside an insulated sealed room. If the freezer door is left open with the freezer operating, will the tempera-ture of the room increase or decrease? Explain your answer.

***8.3.29 Four kilograms of superheated steam at 700 kPa and 500 K are cooled in a tank to 400 K. Calculate the heat transfer involved.

***8.3.30 A closed vessel having a volume of 100 ft^3 is filled with saturated steam at 265 psia. At some later time, the pressure has fallen to 100 psia due to heat losses from the tank. Assuming that the contents of the tank at 100 psia are in an equilibrium state, how much heat was lost from the tank?

***8.3.31 A large piston in a cylinder does 12,500 (ft)(lb$_f$) of work in compressing 3 ft^3 of air to 25 psia. Five pounds of water in a jacket surrounding the cylinder increased in temperature by 2.3°F during the process. What was the change in the internal energy of the air?

$$C_{p,\text{ water}} = 8.0 \frac{\text{Btu}}{(\text{lb mol})(°F)}$$

***8.3.32 One pound of steam at 130 psia and 600°F is expanded isothermally to 75 psia in a closed system. Thereafter, it is cooled at constant volume to 60 psia. Finally, it is compressed adiabatically back to its original state. For each of the three steps of the process, compute ΔU and ΔH. For each of the three steps, where possible, also calculate Q and W.

****8.3.33 A quantity of an ideal gas goes through the ideal cycle shown in Figure P8.3.33. Calculate:

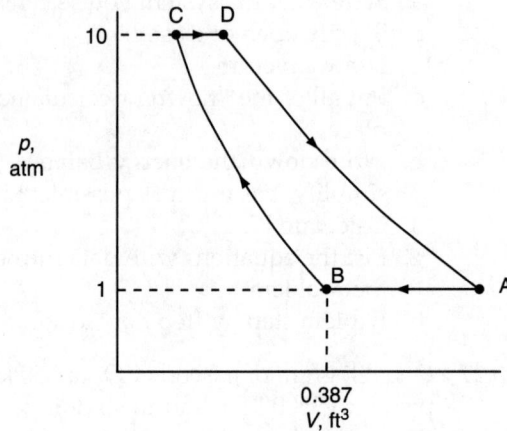

Figure P8.3.33

    a. Pound moles of gas being processed
    b. VD, ft^3
    c. W_{AB}, Btu
    d. W_{BC}, Btu
    e. W_{CD}, Btu
    f. W_{DA}, Btu

     g. W for the cycle, Btu
     h. ΔH for the cycle, Btu
     i. Q for the cycle, Btu

$$\text{Data:}$$
$$T_A = 170°\text{F} \quad T_D = 623°\text{F}$$
$$T_B = 70°\text{F}$$
$$\text{BC} = \text{isothermal process}$$
$$\text{DA} = \text{adiabatic process}$$
$$\text{Assume } C_v = 5 / 2R$$

**8.3.34 Indicate whether the following statements are true or false:
    a. The work done by a constant volume system is always zero.
    b. In a cycle that starts at one state and returns to the same state, the net work done is zero.
    c. Work is the interchange of energy between the system and its surroundings.
    d. Work can always be calculated by the integration of $+pdV$ for a gas.
    e. For a closed system, work is always zero.
    f. When an ideal gas expands in two stages from state one to state two, the first stage being at constant pressure and the second at constant temperature, the work done by the gas is greater during the second stage.
    g. When positive work is done on a system, its surroundings do an equal quantity of negative work, and vice versa.

**8.3.35 A horizontal frictionless piston-cylinder contains 10 lb of liquid water saturated at 320°F. Heat is now transferred to the water until one-half of the water vaporizes. If the piston moves slowly to do work against the surroundings, calculate the work done by the system (the piston-cylinder) during this process.

**8.3.36 Steam is used to cool a polymer reaction. The steam in the steam chest of the apparatus is found to be at 250.5°C and 4000 kPa absolute during a routine measurement at the beginning of the day. At the end of the day, the measurement showed that the temperature was 650°C and the pressure 10,000 kPa absolute. What was the internal energy change of 1 kg of steam in the chest during the day? Obtain your data from the steam tables.

8.4 Energy Balances for Open, Steady-State Systems

*8.4.1 Which is an example of an exothermic phase change: (a) liquid to solid; (b) liquid to gas; (c) solid to liquid; (d) solid to gas?

**8.4.2   Two gram moles of nitrogen are heated from 50°C to 250°C in a cylinder. What is ΔH for the process? The heat capacity equation is

$$C_p = 27.32 + 0.6226 \times 10^{-2}T - 0.0950 \times 10^{-5}T^2$$

where T is in kelvin and C_p is in J/(g mol)(°C).

*8.4.3   Heat is added to a substance at a constant rate and the temperature of the substance remains the same. This substance is (a) solid melting at its melting point; (b) solid below its melting point; (c) liquid above its freezing point; (d) liquid freezing at its freezing point.

**8.4.4   Calculate the heat transfer to the atmosphere per second from a circular pipe, 5 cm in diameter and 100 m long, carrying steam at an average temperature of 120°C if the surroundings are at 20°C. The heat transfer can be estimated from the relation

$$Q = hA\Delta T$$

where $h = 5$ J/(s)(m²)(°C)

      A is the surface area of the pipe.

      ΔT is the temperature difference between the surface of the pipe and ambient conditions.

**8.4.5   Explain why, when a liquid evaporates, the change in enthalpy is greater than the change in internal energy.

**8.4.6   One pound of liquid water is at its boiling point of 575°F. It is then heated at constant pressure to 650°F, then compressed at constant temperature to one-half of its volume (at 650°F), and finally returned to its original state of the boiling point at 575°F. Calculate ΔH and ΔU for the overall process.

**8.4.7   Indicate whether the following statements are true or false:
   a. The enthalpy change of a substance can never be negative.
   b. The enthalpy of steam can never be less than zero.
   c. Both Q and ΔH are state (point) functions.
   d. Internal energy does not have an absolute value.
   e. By definition $U = H - (pV)$.
   f. The work done by a gas expanding into a vacuum is zero.
   g. An intensive property is a property whose value depends on the amount of material present in the system.
   h. The enthalpy change for a system can be calculated by just taking the difference between the final and initial values of the respective enthalpies.
   i. Internal energy has a value of zero at absolute zero.
   j. A batch system and an open system are equivalent terms.

   k. $\left(\dfrac{\partial U}{\partial p}\right)_T$ for an ideal gas 0.

l. Enthalpy is an intensive property.
m. Internal energy is an extensive property.
n. The value of the internal energy for liquid water is about the same as the value for the enthalpy.

**8.4.8   Figure P8.4.8 shows a pure substance that is heated by a constant source of heat supply.

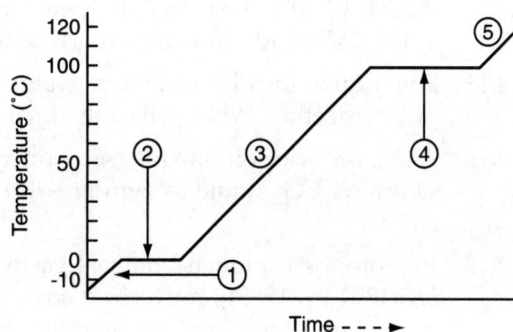

Figure P8.4.8

Use the numbers in the diagram to denote the following stages:
a. Being warmed as a solid
b. Being warmed as a liquid
c. Being warmed as a gas
d. Changing from a solid to a liquid
e. Changing from a liquid to a gas

Also, what is the boiling temperature of the substance? The freezing temperature?

**8.4.9   Estimate ΔH_v for n-heptane at its normal boiling point given $T_b = 98.43°C$, $T_c = 540.2$ K, $p_c = 27$. Use Chen's equation. Calculate the percent error in this value compared to the tabulated value of 31.69 kJ/g mol.

**8.4.10   Estimate the heat capacity of gaseous isobutane at 1000 K and 200 mm Hg by using the Kothari-Doraiswamy relation

$$C_p = A + B \log_{10} T_r$$

from the following experimental data at 200 mm Hg:

C_p [J/(K) (g mol)]	97.3	149.0
Temperature (K)	300	500

The experimental value is 227.6; what is the percentage error in the estimate?

**8.4.11   What is the enthalpy change for acetylene when heated from 37.8°C to 93.3°C?

**8.4.12   A closed vessel contains steam at 1000.0 psia in a 4-to-1 vapor-volume to liquid-volume ratio. What is the steam quality?

**8.4.13   What is the enthalpy change in British thermal units when 1 gal of water is heated from 60°F to 1150°F at 240 psig?

8.4.14   **(8.2.34) What is the enthalpy change that takes place when 3 kg of water at 101.3 kPa and 300 K are vaporized to 15,000 kPa and 800 K?

**8.4.15   Equal quantities by weight of water at +50°C and of ice at –40°C are mixed together. What will be the final temperature of the mixture?

**8.4.16   A chart for carbon dioxide (see Appendix J) shows that the enthalpy of saturated CO_2 liquid is zero at –40°F. Can this be true? Explain your answer.

***8.4.17   In a proposed molten-iron coal gasification process [*Chemical Engineering*, 17 (July, 1985)], pulverized coal of up to 3 mm size is blown into a molten iron bath, and oxygen and steam are blown in from the bottom of the vessel. Materials such as lime for settling the slag, or steam for batch cooling and hydrogen generation, can be injected at the same time. The sulfur in the coal reacts with lime to form calcium sulfide, which dissolves into the slag. The process operates at atmospheric pressure and 1400°C to 1500°C. Under these conditions, coal volatiles escape immediately and are cracked. The carbon conversion rate is said to be above 98%, and the gas is typically 65% to 70% CO, 25% to 35% H_2, and less than 2% CO_2. Sulfur content of the gas is less than 20 ppm. Assume that the product gas is 68% CO, 30% H_2, and 2% CO_2, and calculate the enthalpy change that occurs on the cooling of 1000 m^3 at 1400°C and 1 atm of gaseous product from 1400°C to 25°C and 1 atm. Use the table for the enthalpies of the combustion gases.

***8.4.18   Calculate the enthalpy change (in joules) that occurs when 1 kg of benzene vapor at 150°C and 100 kPa condenses to a solid at –20.0°C and 100 kPa.

***8.4.19   Use the steam tables to answer the following questions:
   a. What is the enthalpy change needed to change 3 lb of liquid water at 32°F to steam at 1 atm and 300°F?
   b. What is the enthalpy change needed to heat 3 lb of water from 60 psia and 32°F to steam at 1 atm and 300°F?
   c. What is the enthalpy change needed to heat 1 lb of water at 60 psia and 40°F to steam at 300°F and 60 psia?
   d. What is the enthalpy change needed to change 1 lb of a water-steam mixture of 60% quality to one of 80% quality if the mixture is at 300°F?

    e. Calculate the ΔH value for an isobaric (constant pressure) change of steam from 120 psia and 500°F to saturated liquid.

    f. Repeat part e for an isothermal change to saturated liquid.

    g. Does an enthalpy change from saturated vapor at 450°F to 210°F and 7 psia represent an enthalpy increase or decrease? A volume increase or decrease?

    h. In what state is water at 40 psia and 267.24°F? At 70 psia and 302°F? At 70 psia and 304°F?

    i. A 2.5 ft^3 tank of water at 160 psia and 363.5°F has how many cubic feet of liquid water in it? Assume that you start with 1 lb of H_2O. Could the tank contain 5 lb of H_2O under these conditions?

    j. What is the volume change when 2 lb of H_2O at 1000 psia and 20°F expand to 245 psia and 460°F?

    k. Ten pounds of wet steam at 100 psia have an enthalpy of 9000 Btu. Find the quality of the wet steam.

***8.4.20 Use the steam tables to calculate the enthalpy change (in joules) of 2 kg mol of steam when heated from 400 K and 100 kPa to 900 K and 100 kPa. Repeat using the table in the text for the enthalpies of combustion gases. Repeat using the heat capacity for steam. Compare your answers. Which is more accurate?

***8.4.21 Use the CO_2 chart in Appendix J for the following calculations:

    a. Four pounds of CO_2 are heated from saturated liquid at 20°F to 600 psia and 180°F.

        1. What is the specific volume of the CO_2 at the final state?

        2. Is the CO_2 in the final state gas, liquid, solid, or a mixture of two or three phases?

    b. The 4 lb of CO_2 are then cooled at 600 psia until the specific volume is 0.07 ft^3/lb.

        1. What is the temperature of the final state?

        2. Is the CO_2 in the final state gas, liquid, solid, or a mixture of two or three phases?

***8.4.22 Calculate the enthalpy change in heating 1 g mol of CO_2 from 50°C to 100°C at 1 atm. Do this problem by three different methods:

    a. Use the heat capacity equation from Appendix F.

    b. Use the CO_2 chart in Appendix J.

    c. Use the table of combustion gases in Appendix E.

***8.4.23 Use the chart for *n*-butane (Figure 8.14) to calculate the enthalpy change for 10 lb of butane going from a volume of 2.5 ft^3 at 360°F to saturated liquid at 10 atm.

****8.4.24   Wet steam flows in a pipe at a pressure of 700 kPa. To check the quality, the wet steam is expanded adiabatically to a pressure of 100 kPa in a separate pipe. A thermocouple inserted into the pipe indicates that the expanded steam has a temperature of 125°C. What was the quality of the wet steam in the pipe prior to expansion?

**8.4.25   Hot reaction products (assume they have the same properties as air) at 1000°F leave a reactor. In order to prevent further reaction, the process is designed to reduce the temperature of the products to 400°F by immediately spraying liquid water into the gas stream. Refer to Figure P8.4.25.

How many pounds of water at 70°F are required per 100 lb of products leaving at 400°F?

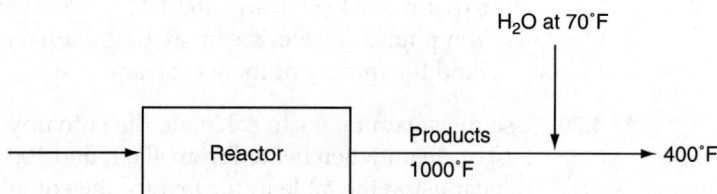

Figure P8.4.25

For this problem, you do not have to get a numerical solution. Instead, list the following in this order:
a. State what the system you select is.
b. Specify open or closed.
c. Draw a picture.
d. Put all the known or calculated data on the picture in the proper places.
e. Write down the material and energy balances (use the symbols in the text) and simplify them as much as possible; list each assumption in so doing.
f. Insert the known data into the simplified equation(s) you would use to solve the problem.

***8.4.26   Air is used to heat an auditorium. The flow rate of the entering air to the heating unit is 150 m^3 per minute at 17°C and 100 kPa. The entering air passes through a heating unit that uses 15 kW for the electric coils. If the heat loss from the heating unit is 200 W, what is the temperature of the exit air?

***8.4.27   Feed-water heaters are used to increase the efficiency of steam power plants. A particular heater is used to preheat 10 kg/s of boiler feed water from 20°C to 188°C at a pressure of 1200 kPa by mixing it with saturated steam bled from a turbine at 1200 kPa and 188°C, as shown in Figure P8.4.27. Although insulated, the heater loses heat at the rate of 50 J/g of exiting mixture. What fraction of the exit stream is steam?

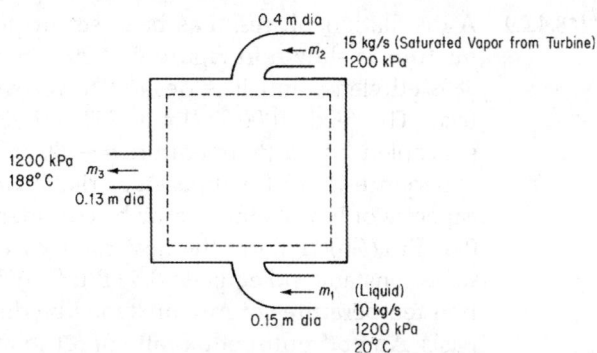

Figure P8.4.27

**** 8.4.28 A process involving catalytic dehydrogenation in the presence of hydrogen is known as *hydroforming*. Toluene, benzene, and other aromatic materials can be economically produced from naphtha feedstocks in this way. After the toluene is separated from the other components, it is condensed and cooled in a process such as the one shown in Figure P8.4.28. For every 100 kg of C charged into the system, 27.5 kg of a vapor mixture of toluene and water (8.1% water) enter the condenser and are condensed by the C stream. Calculate (a) the temperature of the C stream after it leaves the condenser, and (b) the kilograms of cooling water required per hour.

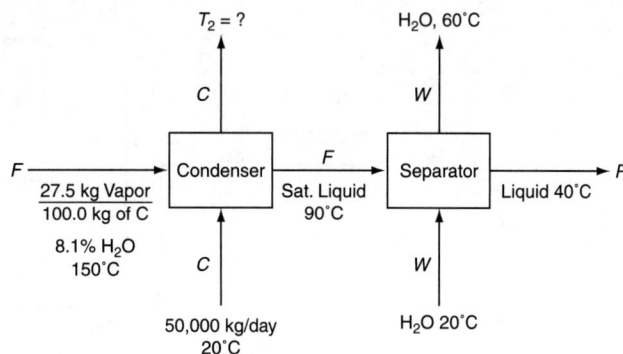

Figure P8.4.28

Additional data:

Stream	C_p[kJ/(kg)(°C)]	B.P. (°C)	(kJ/kg)
$H_2O(l)$	4.2	100	2260
$H_2O(g)$	2.1		
C_7H_8 (l)	1.7	111	230
C_7H_8 (g)	1.3		
C(s)	2.1		

****8.4.29   A distillation process has been set up to separate an ethylene-ethane mixture as shown in Figure P8.4.29. The product stream will consist of 98% ethylene, and it is desired to recover 97% of the ethylene in the feed. The feed, 1000 lb/hr of 35% ethylene, enters the preheater as a subcooled liquid (temperature = −100°F, pressure = 250 psia). The feed experiences a 20°F temperature rise before it enters the still. The heat capacity of liquid ethane may be considered to be constant and equal to 0.65 Btu/(lb)(°F), and the heat capacity of ethylene may be considered to be constant and equal to 0.55 Btu/(lb)(°F). Heat capacities and saturation temperatures of mixtures may be determined on a weight fraction basis. An optimum reflux ratio of 6.1 lb reflux/lb product has been previously determined and will be used. Operating pressure in the still will be 250 psia. Additional data are as follows:

Determine:
   a. The pounds of 30 psig steam required in the reboiler per pound of feed.
   b. The gallons of refrigerant required in the condenser per hour assuming a 25°F rise in the temperature of the refrigerant. Heat capacity is approximately 1.0 Btu/(lb)(°F) and density = 50 lb/ft^3.
   c. The temperature of the bottoms as it enters the preheater.

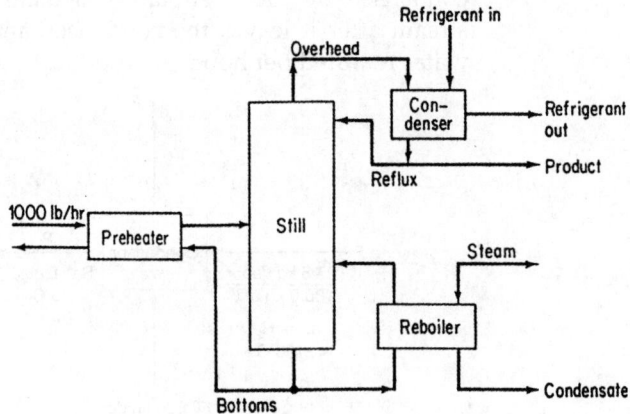

Figure P8.4.29

Component	Temp. Sat. (°F)	Heat of Vaporization (Btu/lb)
C_2H_6	10°	140
C_2H_4	−30°	135

****8.4.30   A boiler house flowsheet for a chemical process plant is shown in Figure P8.4.30. The production rate of 600 psia superheated steam is 100,000 lb/hr. The return condensate flow rate is 50,000 lb/hr. Calculate:

a. The flow rate of 30 psia steam required in the deaerator (lb/hr)
b. The flow rate of makeup feed water (lb/hr)
c. The pump horsepower (hp)
d. The pump electrical consumption if the pump is 55% efficient (kW)
e. Yearly electrical cost to operate the pump (at 0.05/kWh)
f. Yearly electrical savings if the pump could be operated with a 600 psia discharge pressure
g. Heat input to the steam drum (Btu/hr)
h. Heat input to the superheater section (Btu/hr)
i. The amount of 30 psia steam lost to the atmosphere (lb/hr)
j. The amount of 30 psia condensate lost to the atmosphere (lb/hr)

Pressure = 250 psia

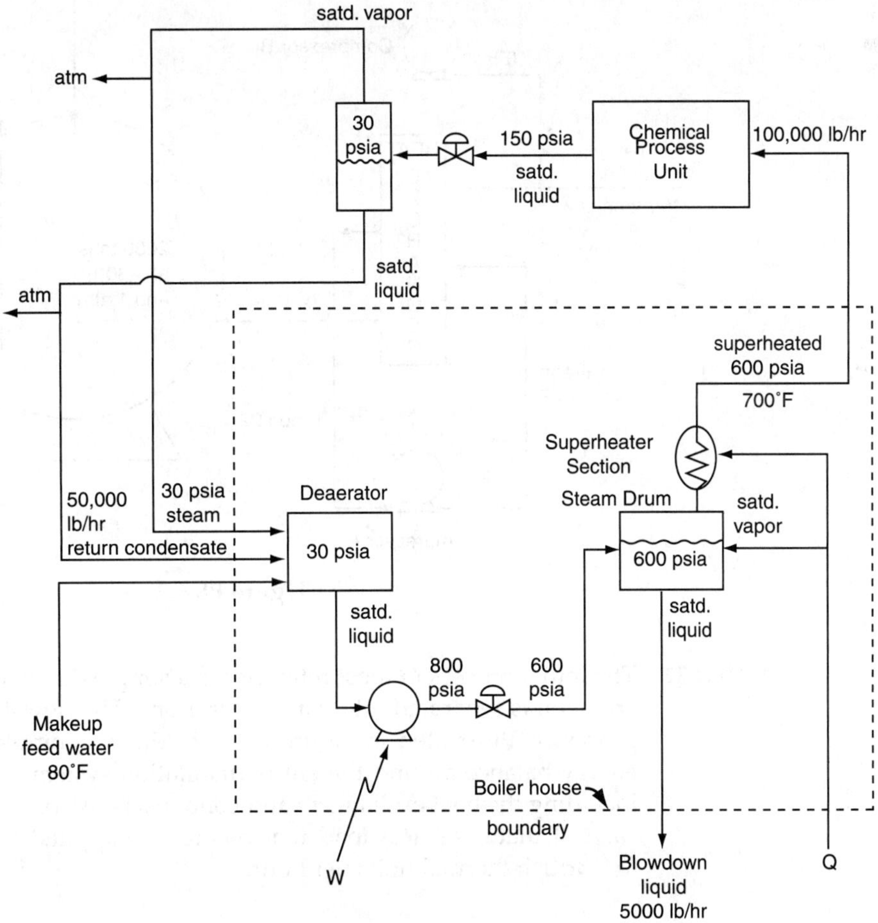

Figure P8.4.30

****8.4.31   A proposal to store Cl_2 as a liquid at atmospheric pressure was recently in the news. The operation is shown in Figure P8.4.31. The normal boiling point of Cl_2 is –30°F. Vapor formed in the storage tank exits through the vent and is compressed to liquid at 0°F and returned to the feed. The vaporization rate is 2.5 tons/day when the sphere is filled to its capacity and the surrounding air temperature is 80°F. If the compressors are driven by electric motors and are about 30% efficient, what is the horsepower input required to make this process successful? Assume lines and heat exchangers are well insulated. Use 8.1 Btu/(lb mol)(°F) for the heat capacity of liquid $\Delta H_{vaporization} = 123.67$ Btu/lb Cl_2.

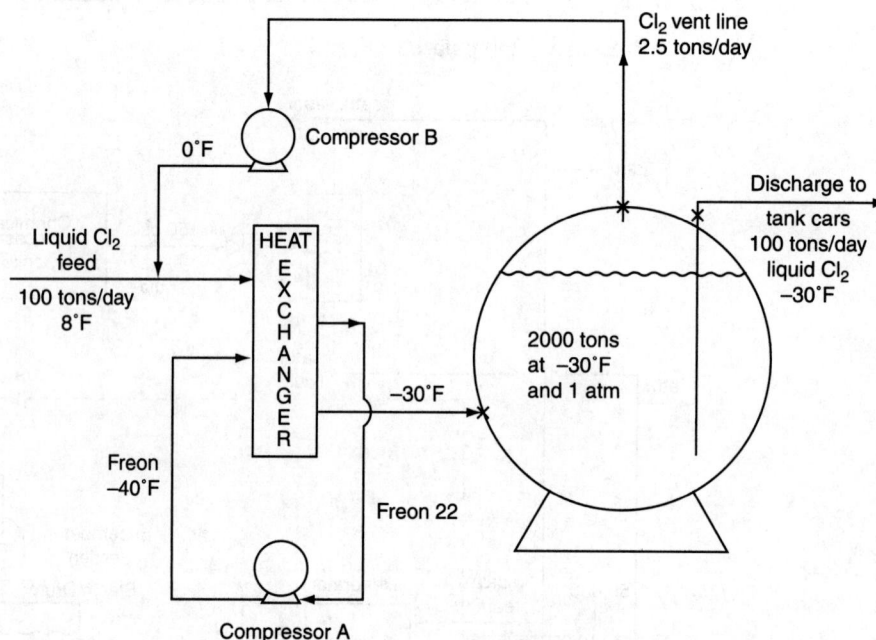

Figure P8.4.31

****8.4.32   The initial process in most refineries is a simple distillation in which the crude oil is separated into various fractions. The flowsheet for one such process is illustrated in Figure P8.4.32. Make a complete material and energy balance around the entire distillation system and for each unit, including the heat exchangers and condensers. Also:

a. Calculate the heat load that has to be supplied by the furnace in British thermal units per hour.

b. Determine the additional heat that would have to be supplied by the furnace if the charge oil were not preheated to 200°F before it entered the furnace.

Do the calculated temperatures of the streams going into storage from the heat exchangers seem reasonable?

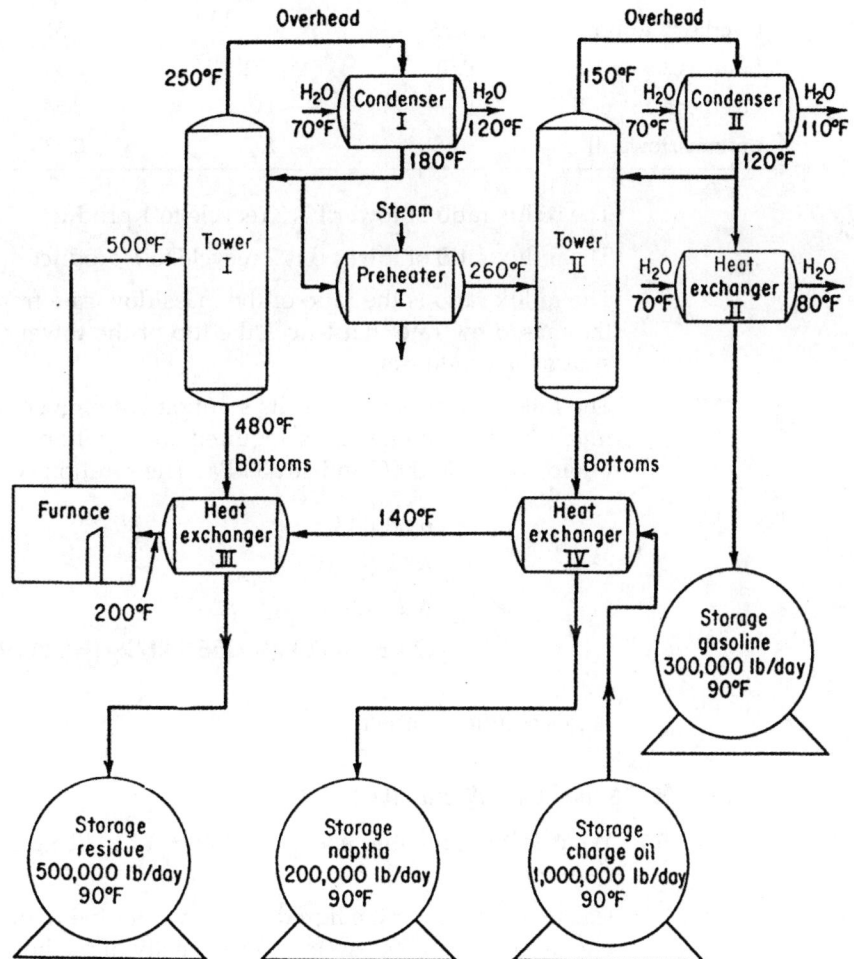

Figure P8.4.32

Additional data:

	Specific Heat of Liquid Btu/(lb)(°F)	Latent Heat of Vaporization Btu/lb	Specific Heat of Vapor Btu/(lb)(°F)	Condensation Temp. °F
Charge oil	0.53	100	0.45	480
Overhead, tower I	0.59	111	0.51	250
Bottoms, tower I	0.51	92	0.42	500
Overhead, tower II	0.63	118	0.58	150
Bottoms, tower II	0.58	107	0.53	260

The reflux ratio of tower I is 3 recycle to 1 product.

The reflux ratio of tower II is 2 recycle to 1 product.

The reflux ratio is the ratio of the mass flow rate from the condenser to the mass flow rate that leaves the top of the tower (the overhead) and enters the condenser.

***8.4.33   The following problem and its solution were given in a textbook: How much heat in kilojoules is required to vaporize 1.00 kg of saturated liquid water at 100°C and 101.3 kPa? The solution is

$$n = 1.00 \text{ kg}$$
$$\Delta E = Q + W + E_{flow}$$
$$\Delta U = Q + W = Q - p\Delta V$$
$$Q = \Delta H = (1 \text{ kg})(2256.9 \text{ kJ/kg}) = 2256.9 \text{ kJ}$$

Is this solution correct?

8.5   Mechanical Energy Balances

*8.5.1   What is the potential energy in joules of a 12 kg mass 25 m above a datum plane?

**8.5.2   Calculate the KE of the liquid flowing in a pipe with a 5 cm inner diameter at the rate of 500 kg/min. The density of the liquid is 1.15 g/cm^3.

**8.5.3   Find the kinetic energy in $(ft)(lb_f)/(lb_m)$ of water moving at the rate of 10 ft/s through a pipe of 2 in. inner diameter.

**8.5.4   In an electric gun, suppose we want the muzzle velocity to be Mach 4, or about 4400 ft/s. This is a reasonable number because at around 3500 ft/s, you begin to get hypervelocity effects, such as straws going through doors. Now the requirement that our slug start from zero and reach 4400 ft/s in 20 ft exactly defines our environment, if we insist that the acceleration is constant, which is a good starting point. By a simple manipulation of Newton's laws, we find that the time of launch is 9 ms. If the slug weighs 1 kilogram, which is much less than a tank shell but quite sufficient to wreck any tank it hits at Mach 4, then the kinetic energy of the bullet is 900,000 J. Given that we have 9 ms to get this energy into the slug, we have to transfer about 100 MW into the bullet during the 9 ms launch time. Is the energy transfer really 100 MW?

***8.5.5 A windmill converts the kinetic energy of the moving air into electrical energy at an efficiency of about 30%, depending on the windmill design and speed of the wind. Estimate the power in kilowatts for a wind flowing perpendicular to a windmill with blades 15 m in diameter when the wind is blowing at 20 mi/hr at 27°C and 1 atm.

***8.5.6 Before it lands, a vehicle returning from space must convert its enormous kinetic energy to heat. To get some idea of what is involved, a vehicle returning from the moon at 25,000 mi/hr can, in converting its kinetic energy, increase the internal energy of the vehicle sufficiently to vaporize it. Obviously, a large part of the total kinetic energy must be transferred from the vehicle. How much kinetic energy does the vehicle have (in British thermal units per pound)? How much energy must be transferred by heat if the vehicle is to heat up by only 20°F/lb?

***8.5.7 The world's largest plant that obtains energy from tidal changes is at Saint-Malo, France. The plant uses both the rising and falling cycles (one period in or out is 6 hr 10 min in duration). The tidal range from low to high is 14 m, and the tidal estuary (the Rance River) is 21 km long with an area of 23 km². Assume that the efficiency of the plant in converting potential to electrical energy is 85%, and estimate the average power produced by the plant. (**Note:** Also assume that after high tide, the plant does not release water until the sea level drops 7 m, and after a low tide it does not permit water to enter the basin until the level outside the basin rises 7 m, and the level differential is maintained during discharge and charge.)

***8.5.8 In one stage of a process for the manufacture of liquid air, air as a gas at 4 atm absolute and 250 K is passed through a long, insulated 3-in.-ID pipe in which the pressure drops 3 psi because of frictional resistance to flow. Near the end of the line, the air is expanded through a valve to 2 atm absolute. State all assumptions.
   a. Compute the temperature of the air just downstream of the valve.
   b. If the air enters the pipe at the rate of 100 lb/hr, compute the velocity just downstream of the valve.

***8.5.9 A turbine that uses steam drives an electric generator. The inlet steam flows through a 10-cm-diameter pipe to the turbine at the rate of 2.5 kg/s at 600°C and 1000 kPa. The exit steam discharges through a 25-cm-diameter pipe at 400°C and 100 kPa. What is the expected power obtained from the turbine if it operates essentially adiabatically?

** 8.5.10 Your company produces small power plants that generate electricity by expanding waste process steam in a turbine. You are asked to study the turbine to determine if it is operating as efficiently as possible. One way to ensure good efficiency is to have the turbine operate adiabatically. Measurements show that for steam at 500°F and 250 psia:

    a. The work output of the turbine is 86.5 hp.

    b. The rate of steam usage is 1000 lb/hr.

    c. The steam leaves the turbine at 14.7 psia and consists of 15% moisture (i.e., liquid H_2O).

Is the turbine operating adiabatically? Support your answer with calculations.

8.6   Energy Balances for Special Cases

***8.6.1 A cylinder contains 1 lb of steam at 600 psia and a temperature of 500°F. It is connected to another equal-size cylinder that is evacuated. A valve between the cylinders is opened. If the steam expands into the empty cylinder, and the final temperature of the steam in both cylinders is 500°F, calculate Q, W, ΔU, and ΔH for the system composed of both cylinders.

***8.6.2 Carbon dioxide cylinders, initially evacuated, are being loaded with CO_2 from a pipeline in which the CO_2 is maintained at 200 psia and 40°F. As soon as the pressure in a cylinder reaches 200 psia, the cylinder is closed and disconnected from the pipeline. If the cylinder has a volume of 3 ft³, and if the heat losses to the surroundings are small, compute (a) the final temperature of the CO_2 in a cylinder, and (b) the number of pounds of CO_2 in a cylinder.

**8.6.3 A vertical cylinder capped by a piston weighing 990 g contains 100 g of air at 1 atm and 25°C. Calculate the maximum possible final elevation of the piston if 100 J of work are used to raise the cylinder and its contents vertically. Assume that all of the work goes into raising the piston.

***8.6.4 Calculate Q, W, ΔU, and ΔH for 1 lb of liquid water which is evaporated at 212°F by (a) a nonflow process and (b) a unsteady-state flow process.

***8.6.5 A cylinder that initially contains nitrogen at 1 atm and 25°C is connected to a high-pressure line of nitrogen at 50 atm and 25°C. When the cylinder pressure reaches 40 atm, the valve on the cylinder is closed. Assume the process is adiabatic, and that nitrogen can be treated as an ideal gas. What is the temperature in the cylinder when the valve is closed? Ideal heat capacities are listed in Table 8.3.

***8.6.6 An insulated tank having a volume of 50 ft³ contains saturated steam at 1 atm. It is connected to a steam line maintained at 50 psia and 291°F. Steam flows slowly into the tank until the pressure reaches 50 psia.

What is the temperature in the tank at that time? Hint: Use the steam tables on the website to make this an easy problem.

***8.6.7 Start with the general energy balance, and simplify it for each of the processes listed below to obtain an energy balance that represents the process. Label each term in the general energy balance by number, and list by their numbers the terms retained or deleted, followed by your explanation. (You do not have to calculate any quantities in this problem.)

  a. One hundred kilograms per second of water are cooled from 100°C to 50°C in a heat exchanger. The heat is used to heat up 250 kg/s of benzene entering at 20°C. Calculate the exit temperature of the benzene.

  b. A feed-water pump in a power generation cycle pumps water at the rate of 500 kg/min from the turbine condensers to the steam generation plant, raising the pressure of the water from 6.5 kPa to 2800 kPa. If the pump operates adiabatically with an overall mechanical efficiency of 50% (including both the pump and its drive motor), calculate the electric power requirement of the pump motor (in kilowatts). The inlet and outlet lines to the pump are of the same diameter. Neglect any rise in temperature across the pump due to friction (i.e., the pump may be considered to operate isothermally).

***8.6.8 It is necessary to evaluate the performance of an evaporator that will be used to concentrate a 5% organic solution. Assume there will be no boiling point rise.

The following information has already been obtained: $U = 300$ Btu/(hr)(ft^2)(°F); $A = 2000$ ft^2; heating steam S is available at 4.5 psia; feed enters at 140°F.

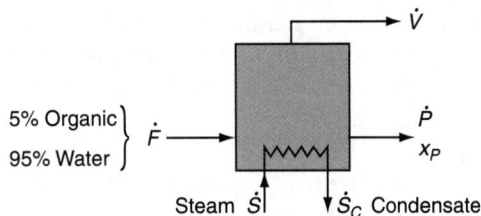

Figure P8.6.8

Must any further measurements be made? The rate of heat transfer from the steam coils to the liquid is $\dot{Q} = UA(T_S - T_V)$.

***8.6.9 A liquid that can be treated as water is being well mixed by a stirrer in a 1 m^3 vessel. The stirrer introduces 300 W of power into the vessel. The heat transfer from the tank to the surroundings is proportional to the temperature difference between the vessel and the surroundings (which are at 20°C). The flow rate of liquid in and out of the tank is 1 kg/min. If the temperature of the inlet liquid is 40°C, what is the temperature of the outlet liquid? The proportionality constant for the heat transfer is 100 W/°C.

***8.6.10 In the vapor-recompression evaporator shown in Figure P8.6.10, the vapor produced on evaporation is compressed to a higher pressure and passed through the heating coil to provide the energy for evaporation. The steam entering the compressor is 98% vapor and 2% liquid, at 10 psia; the steam leaving the compressor is at 50 psia and 400°F; and 6 Btu of heat are lost from the compressor per pound of steam throughput. The condensate leaving the heating coil is at 50 psia, 200°F.

   a. Compute the British thermal units of heat supplied for evaporation in the heating coil per British thermal unit of work needed for compression by the compressor.
   b. If 1,000,000 Btu/hr of heat are to be transferred in the evaporator, what must be the intake capacity of the compressor in cubic feet of wet vapor per minute?

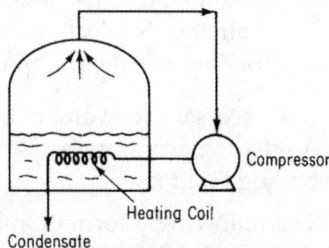

Figure P8.6.10

****8.6.11 A power plant operates as shown in Figure P8.6.11. Assume that the pipes, boiler, and superheater are well lagged (insulated) and that friction can be neglected. Calculate (in British thermal units per pound of steam):
   a. The heat supplied to the boiler
   b. The heat supplied to the superheater
   c. The heat removed in the condenser
   d. The work delivered by the turbine
   e. The work required by the liquid pumps

Also calculate the efficiency of the entire process defined as

$$\frac{\text{Network delivered/lb steam}}{\text{Total heat supplied/lb steam}}$$

If the water rate to the boiler is 2000 lb/hr, what horsepower does the turbine develop? Finally, what suggestion can you offer that will improve the efficiency of the power plant?

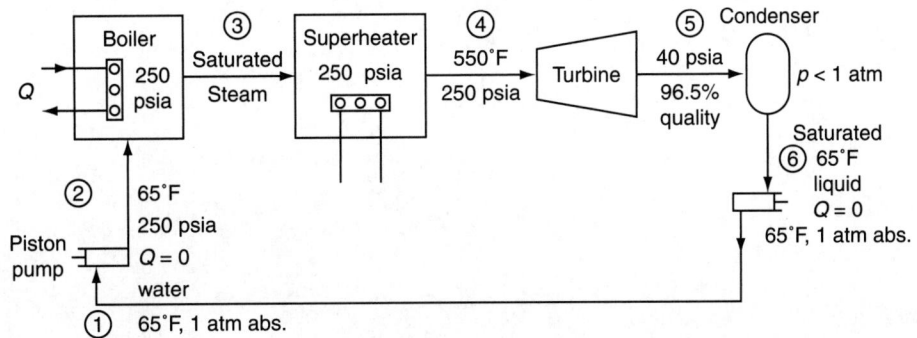

Figure P8.6.11

8.6.12 Simplify the general energy balance so as to represent the process in each of the following cases. Number each term in the general balance, and state why you retained or deleted it.

a. A bomb calorimeter is used to measure the heating value of natural gas. A measured volume of gas is pumped into the bomb. Oxygen is then added to give a total pressure of 10 atm, and the gas mixture is exploded using a hot wire. The resulting heat transfer from the bomb to the surrounding water bath is measured. The final products in the bomb are CO_2 and water.

b. Cogeneration (generation of steam for both power and heating) involves the use of gas turbines or engines as prime movers, with the exhausted steam going to the process to be used as a heat source. A typical installation is shown in Figure P8.6.12.

c. In a mechanical refrigerator, the Freon liquid is expanded through a small insulated orifice so that part of it flashes into vapor. Both the liquid and vapor exit at a lower temperature than the temperature of the liquid entering.

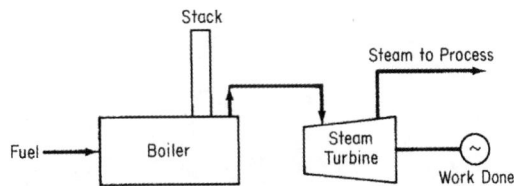

Figure P8.6.12

CHAPTER 9

Energy Balances with Reaction

Chapter Objectives

- Explain the meaning of standard heat (enthalpy) of formation, heat (enthalpy) of reaction, and higher and lower heating values.
- List the standard conventions and reference states used for reactions associated with the standard heat of formation and the heat of combustion.
- Calculate the standard heat of reaction from tabulated standard heats of formation (or combustion) for a given reaction.
- Understand how to combine the heat of formation with sensible heat changes to solve problems involving chemical reactions.
- Solve simple material and energy balance problems involving reactions.

Introduction

Consider the schematic of a reactor shown in Figure 9.1. If the reaction taking place in this reactor liberates thermal energy, the heat transfer fluid (e.g., cooling water) would remove the heat so that the reactor would operate at a desired temperature. Conversely, if the reaction consumes thermal energy, the heat transfer fluid (e.g., steam) would transfer heat to maintain the desired operating temperature. The material that follows demonstrates how to apply energy balances for systems that involve chemical reaction, such as the reactor shown in Figure 9.1. As you will also observe, chemical reactions usually have a dominant effect on the energy balance for a system.

You probably are aware from previous comments that chemical reactions are at the heart of many industrial processes and directly affect the economics of an entire plant. Chemical reactions also are the basis of complex biological systems. We have deferred including the effects of chemical reactions in the discussion of energy balances to this point to avoid confusion. It's time to include them. In this chapter, we explain how you can do it. We describe two closely related accounting procedures that treat the effects of energy generation or consumption due to chemical reaction. In one method, all of the energy effects are consolidated into one term in the energy balance called the *heat of reaction.* In the other, the reaction energy effects are merged with the enthalpies associated with each stream flowing in and out of the system. Both methods are based on a quantity known as the *heat of formation,* which we take up first.

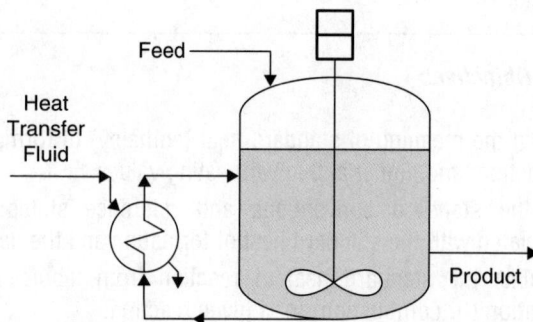

Figure 9.1   Schematic of a reactor

9.1   The Standard Heat (Enthalpy) of Formation

The observed heat transfer that occurs to or from a closed isothermal system in which a reaction takes place represents the energy change associated with the rearrangement of the bonds holding together the atoms of the reacting molecules. For an **exothermic reaction,** heat is removed from the process in order to maintain a fixed system temperature; that is, the energy that is produced by the reaction must be removed to maintain isothermal conditions. The reverse is true of an **endothermic reaction,** in which heat is added to the system.

To include energy changes caused by a reaction in the energy balance, we make use of a quantity called the **standard heat** (really, **enthalpy**) **of formation,** often called just the heat of formation,[1] denoted by the symbol $\Delta \hat{H}_f^o$. The superscript o denotes the **standard state** (reference state) for reaction of 25°C and 1 atm, and the subscript f denotes formation. In this chapter, the overlay caret (^) usually denotes that the value is per mole.

The standard heat of formation is the name given to the special enthalpy change associated with the formation of 1 mol of a compound from its constituent elements and products in their standard state of 25°C and 1 atm. An example is the enthalpy change that occurs for the reaction of carbon and oxygen to form carbon dioxide at 25°C and 1 atm, as indicated in Figure 9.2.

$$C(s) + O_2(g) \rightarrow CO_2(g)$$

Note in the chemical reaction equation that the explicit specification of the states of the compounds is placed in parentheses.

In the reference state, the enthalpies of all of the elements are assigned values of 0. Thus, C and O_2 are assigned 0 values in the reaction shown in Figure 9.2. If you simplify the general energy balance, Equation (8.2), for the isothermal process in Figure 9.2 (steady-state, flow, no *KE* or *PE*, etc.), you get the standard heat of formation of CO_2, calculated from

$$-393.5 \text{ kJ/kg mol } CO_2 = Q = \Delta H = (1)\Delta \hat{H}_{f, CO_2}^o - (1)\Delta \hat{H}_{f, C}^o - (1)\Delta \hat{H}_{f, O_2}^o$$

$$= (1)\Delta \hat{H}_{f, CO_2}^o - 0 - 0 = \Delta \hat{H}_{f, CO_2}^o$$

Because enthalpy $\hat{H}$ is a state variable, any state would do for the standard state of reference, but by convention, the standard state of a substance (both reactants and products) usually is 25°C and 1 atm (absolute) pressure.[2] Fixing a reference state should cause no problem because you are always interested in calculating *enthalpy differences*, so the reference state is eliminated.

---

[1] Historically, the name arose because the changes in enthalpy associated with chemical reactions were generally determined in a device called a calorimeter to which heat is added or removed from the reacting system so as to keep the temperature constant.

[2] The standard pressure is 1 bar (100 kPa) in some tabulations.

$Q = -393.5$ kJ/g mol C

System boundary

1 kg mol C
25°C, 1 atm

Combustion process

1 kg mol CO_2
25°C, 1 atm

1 kg mol O_2
25°C, 1 atm

25°C

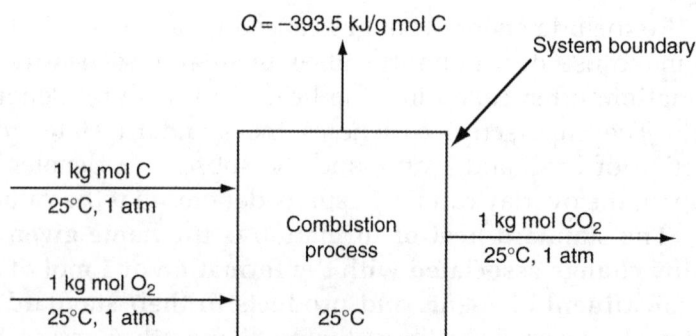

Figure 9.2   The heat transferred from a steady-state combustion process for the reaction $C(s) + O_2(g) \rightarrow CO_2(g)$ with reactants and products at 25°C and 1 atm is the standard heat of formation.

The reaction underlying $\Delta \hat{H}_f^o$ does not necessarily represent a real reaction that would proceed at constant temperature but can be a fictitious process for the formation of a compound from the elements. By defining the **heat of formation as zero in the standard state for each stable element** (e.g., N_2 versus N), it is possible to design a system to express the heats of formation for all *compounds* at 25°C and 1 atm. Appendix C contains standard heats of formation for a number of molecules. Additionally, heats of formation for more than 700 compounds can be obtained using the correlations reported by Professor Yaws.[3] Remember that the values for the **standard heats of formation are negative for exothermic reactions and positive for endothermic reactions.**

For various compounds of interest to bioengineers, $\Delta \hat{H}_f^o$ may not be easy to find. Equation (4.17) shows a general substrate and cellular product that includes C, H, O, and N. To get $\Delta \hat{H}_f^o$ for $C_w H_x O_y N_z$, you have to start with a balanced equation. Use the electron balance technique shown in Chapter 4 to save time. For example, to get $\Delta \hat{H}_f^o$ for $CH_2(NH_3)COOH$ (glycine), the smallest amino acid, write the balanced chemical equation based on the reported heat of combustion of 966.1 kJ/g mol.

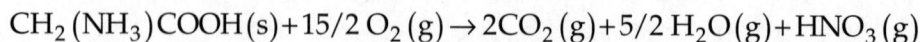

$$CH_2(NH_3)COOH(s) + 15/2\, O_2(g) \rightarrow 2CO_2(g) + 5/2\, H_2O(g) + HNO_3(g)$$

$$-\Delta H_{c,\,glycine\,solid}^o = (2)\Delta H_{f,\,CO_2}^o + (5/2)\Delta H_{f,\,H_2O}^o + (1)\Delta H_{f,\,HNO_3}^o - \Delta H_{f,\,glycine\,solid}^o - \Delta H_{f,\,O_2}^o$$

$$-966.1 = (2)(-393.51) + (5/2)(-241.826) + (1)(-173.23) - \Delta H_{f,\,glycine\,solid}^o - 0$$

$$\Delta H_{f,\,glycine\,solid}^o = -235.98 \text{ kJ/g mol}$$

---

[3] C. L. Yaws, "Correlation for Chemical Compounds," *Chem. Engr.,* **79** (August 15, 1976).

This analysis is based on the assumption that the heat of reaction for this case is relatively small. The reaction would practically be carried out in water because glycine is an odorless white crystal fairly soluble in water; hence, some solvation effects would occur.

Example 9.1   Use of Heat Transfer Measurements to Get a Heat of Formation

Problem Statement

Suppose that you want to find the standard heat of formation of CO from experimental data. Can you prepare pure CO from the reaction of C and O_2 and measure the heat transfer? This would be far too difficult. It would be easier experimentally to find first the heats of the reaction at standard conditions for the two reactions shown in Figure E9.1 (assuming you had some pure CO to start with).

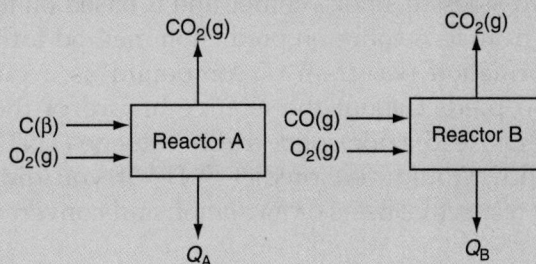

Figure E9.1   Use of two convenient reactions to determine the heat of formation for an inconvenient reaction

---

Solution

$$C(s) + \frac{1}{2}O_2(g) \rightarrow CO(g)$$

$$C(s) + O_2(g) \rightarrow CO_2(g) \quad Q = -393.51 \text{ kJ/g mol C} \equiv \Delta \hat{H}_A \qquad (a)$$

$$CO + \frac{1}{2}O_2(g) \rightarrow CO_2(g) \quad Q = -282.99 \text{ kJ/g mol CO} \equiv \Delta \hat{H}_B \qquad (b)$$

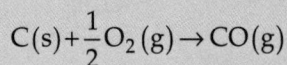

Basis: 1 g mol each of C and CO

(*Continues*)

Example 9.1   Use of Heat Transfer Measurements to Get a Heat of Formation (*Continued*)

According to Hess's law, you subtract reaction (b) from reaction (a), subtract the corresponding $\Delta \hat{H}_i$, and rearrange the compounds to form the desired chemical equation:

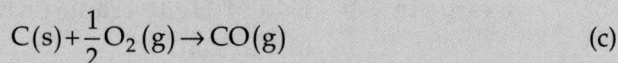

$$C(s)+\frac{1}{2}O_2(g)\rightarrow CO(g) \tag{c}$$

for which the net heat of reaction per gram mole of CO is the heat of formation of CO:

$$\Delta \hat{H}^o_{f,\,CO}=-393.51-(-282.99)=-110.52 \text{ kJ/g mol CO}$$

One hazard assessment of a compound is based on the potential rapid release of energy from it. A common prediction method for such release is to use the heat of formation per gram of compound as a guide. For example, what would you predict about the relative hazard of the following compounds: acetylene gas, lead azide solid, trinitrotoluene (TNT) liquid, and ammonium nitrate solid? Would you predict TNT? If you did, you would be wrong. Check the respective heats of formation and convert to a per gram basis.

Self-Assessment Test

Questions

1. If for the reaction

$$2N(g)\rightarrow N_2(g)$$

   the heat transfer is $Q=-941$ kJ how do you determine the value for the heat of formation of $N_2(g)$?

2. If the reaction for the decomposition of CO

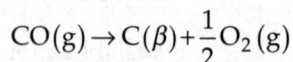

$$CO(g)\rightarrow C(\beta)+\frac{1}{2}O_2(g)$$

   takes place only at high temperature and pressure, how will the value of the standard heat of formation of CO be affected?

3. Will reversing the direction of a reaction equation reverse the sign of the heat of formation of a compound?

Answers

1. The heat of formation of an element in its natural state is defined as zero.

2. It will not be affected because the heat of formation by definition is at 25°C and 1 atm.

3. No, because the heat of formation is defined in a certain direction.

Problems

1. What is the standard heat of formation of HBr(g)?

2. Show that for the process in Figure 9.1, the general energy balance reduces to $Q = \Delta H$. What assumptions do you have to make?

3. Could the heat of formation be calculated from measurements taken in a batch process? If so, show the assumptions and calculations.

4. Calculate the standard heat of formation of CH_4 given the following experimental results at 25°C and 1 atm (Q is for complete reaction):

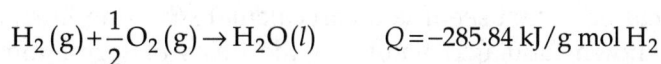

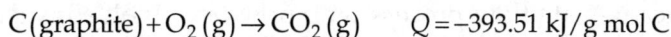

$$H_2(g) + \frac{1}{2}O_2(g) \rightarrow H_2O(l) \qquad Q = -285.84 \text{ kJ/g mol } H_2$$

$$C(graphite) + O_2(g) \rightarrow CO_2(g) \qquad Q = -393.51 \text{ kJ/g mol } C$$

$$CH_4(g) + 2O_2(g) \rightarrow CO_2(g) + 2H_2O(l) \quad Q = -890.36 \text{ kJ/g mol } CH_4$$

Compare your answer with the value found in the table of the heats of formation listed in Appendix C.

Answers

1. 36.4 kJ/g mol HBr

2. Open, steady-state system with negligible change in KE, PE and W.

3. Yes. $\Delta U = \Delta H - \Delta(pV) = Q$ if $W = \Delta KE = \Delta PE = 0$. Also, V is constant and system remains at 1 atm and 25°C. Then, $\Delta H_f^o = Q$

4.
$$2H_2(g) + O_2(g) \rightarrow 2H_2O(l) \qquad Q = 2(-285.84) \text{ kJ/g mol } H_2$$

$$C(graphite) + O_2(g) \rightarrow CO_2(g) \qquad Q = -393.51 \text{ kJ/g mol } C$$

$$CO_2(g) + 2H_2O(l) \rightarrow CH_4(g) + 2O_2(g) \quad Q = 890.36 \text{ kJ/g mol } CH_4$$

Therefore, the heat of formation is −74.83 kJ/g mol and from Appendix C it is −74.84 kJ/g mol.

9.2   The Heat (Enthalpy) of Reaction

As we mentioned previously, one method of including the effect of chemical reaction in the energy balance makes use of the heat of reaction. The **heat of reaction** (which should be but is only rarely called the **enthalpy of reaction**) is the enthalpy change that occurs when reactants at various T and p react to form products at some T and p. The **standard heat of reaction** (ΔH^o_{rxn}) is the name given to the heat of reaction when **stoichiometric quantities of reactants in the standard state** (25°C and 1 atm) **react completely** to produce products in the standard state. Do not confuse the symbol for the standard heat of reaction, (ΔH^o_{rxn}), with the symbol for the more general heat of reaction, ΔH_{rxn}, which applies to a process in which a reaction occurs under any conditions with any amount of reactants and products. The units of ΔH_{rxn} are energy (kilojoules, British thermal units, etc.), and the units of the specific standard heat of reaction, ΔH^o_{rxn}, are energy for one mole of reaction based on the chemical reaction equation, so the result is energy per mole.

You can obtain the heat of reaction from experiments, of course, but it is easier to first see if you can calculate the standard heat of reaction from the known tabulated values of the heats of formation as follows: Consider a steady-state flow process with no work involved, such as the one shown in Figure 9.3 in which benzene (C_6H_6) reacts with the stoichiometric amount of H_2 to produce cyclohexane (C_6H_{12}) in the standard state:

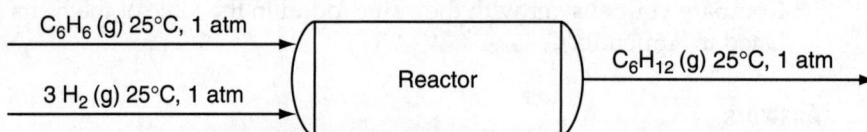

Figure 9.3   Reaction of benzene to form cyclohexane

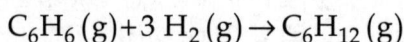

$$C_6H_6\,(g) + 3\,H_2\,(g) \rightarrow C_6H_{12}\,(g)$$

The energy balance for the process reduces to $Q = \Delta H$, where ΔH is by definition the ΔH^o_{rxn} for the specified chemical reaction equation.

Because we adopt for the heat of reaction the same reference conditions (0 enthalpy for the elements at 25°C and 1 atm) as used in defining the heats of formation, the values of the specific enthalpies associated with each species involved in the reaction are just the values of the respective heats of formation. For the process shown in Figure 9.3, the data are as follows:

Compound	Specific Enthalpy $\Delta \hat{H}_f^o$ (kJ/g mol)	Number of Moles
C_6H_6 (g)	82.927	1
H_2 (g)	0	3
C_6H_{12} (g)	−123.1	1

You calculate the standard heat of reaction thus (based on 1 mol of C_6H_6 that reacts):

$$\Delta H_{rxn}^o = n_{C_6H_{12}} \Delta \hat{H}_{f,\,C_6H_{12}}^o - n_{C_6H_6} \Delta \hat{H}_{f,\,C_6H_6}^o - n_{H_2} \Delta \hat{H}_{f,\,H_2}^o$$
$$= (1)(-123.1) - (1)(82.927) - (3)(0) = -206.0 \text{ kJ}$$

or with H_2 as the reference:

$$\Delta H_{rxn}^o = -206.0 \text{ kJ}/3 \text{ g mol } H_2, \text{ hence } \Delta H_{rxn}^o = -68.67 \text{ kJ/g mol } H_2$$

If the reaction equation is multiplied by 2, will the heat of reaction be doubled? Yes. Will the standard heat of reaction also be doubled? No.

In general, for **complete** reaction,

$$\Delta H_{rxn}^o = \left(\overset{\text{Products}}{\underset{i}{\sum} v_i \Delta \hat{H}_{f,i}^o} - \overset{\text{Reactants}}{\underset{i}{\sum} |v_i| \Delta \hat{H}_{f,i}^o} \right) = \overset{\text{All Species}}{\underset{i}{\sum} v_i \Delta \hat{H}_{f,i}^o} \tag{9.1}$$

where v_i is the stoichiometric coefficient in the reaction equation. Keep in mind the sign convention for v_i. Always remember that the "heat of reaction" is actually an **enthalpy change** and not necessarily equivalent to heat transfer to or from the system. If you do not have values for the heats of formation, you can estimate them as described in some of the references listed at the end of this chapter. If the stoichiometry of the reaction is not well known, you may have to carry out an experiment to get the heat of reaction.

Example 9.2   Calculation of the Standard Heat of Reaction from the Standard Heats of Formation

Problem Statement

Calculate ΔH^o_{rxn} for the following reaction of 4 g mol of NH_3 and 5 g mol of O_2:

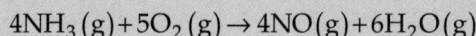

$$4NH_3(g) + 5O_2(g) \rightarrow 4NO(g) + 6H_2O(g)$$

Solution

Basis: 4 g mol of NH_3

Tabulated data:	$NH_3(g)$	$O_2(g)$	$NO(g)$	$H_2O(g)$
$\Delta \hat{H}^o_f$ per mol at 25°C and 1 atm (kJ/g mol)	−46.191	0	+90.374	−241.826

We use Equation (9.1) to calculate ΔH^o_{rxn} (25°C, 1 atm) for 4 g mol of NH_3, assuming complete reaction:

$$\Delta H^o_{rxn} = [4(90.374) + 6(2241.826)] + [(-5)(0) + (-4)(-46.191)] = -904.696 \text{ kJ}$$

Therefore, the heat of reaction per mole of NH_3 is

$$\Delta H^o_{rxn} = \frac{904.646 \text{ kJ}}{4 \text{ g mol } NH_3} = -226.174 \text{ kJ/g mol } NH_3$$

Example 9.3   Green Chemistry: Examining Alternate Processes

Problem Statement

Green chemistry refers to the adoption of chemicals in commercial processes that reduce concern with respect to the environment. An example is the elimination of methyl isocyanate, a very toxic gas, in the production of carbaryl (1-napthalenyl methyl carbamate). In 1984 in Bhopal, India, the accidental release of methyl isocyanate in a residential area led to the death of thousands of people and the injury of many thousands more. The Bhopal process can be represented by the reaction Equations (a) and (b):

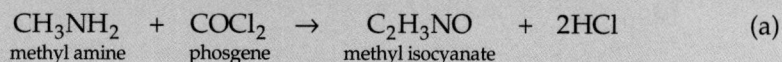

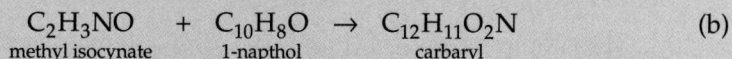

$$CH_3NH_2 \;+\; COCl_2 \;\rightarrow\; C_2H_3NO \;+\; 2HCl \tag{a}$$

methyl amine   phosgene   methyl isocyanate

$$C_2H_3NO \;+\; C_{10}H_8O \;\rightarrow\; C_{12}H_{11}O_2N \tag{b}$$

methyl isocynate   1-napthol   carbaryl

An alternate process eliminating the methyl isocyanate is represented by two other reaction equations:

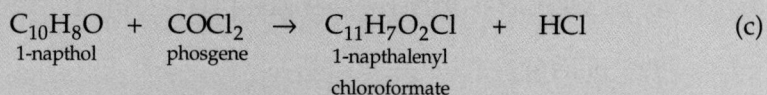

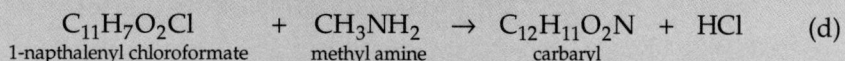

$$C_{10}H_8O \;+\; COCl_2 \;\rightarrow\; C_{11}H_7O_2Cl \;+\; HCl \tag{c}$$

1-napthol   phosgene   1-napthalenyl
chloroformate

$$C_{11}H_7O_2Cl \;+\; CH_3NH_2 \;\rightarrow\; C_{12}H_{11}O_2N \;+\; HCl \tag{d}$$

1-napthalenyl chloroformate   methyl amine   carbaryl

All of the reactions are in the gas phase. Calculate the approximate amount of heat transfer that will be required in each step of each process per gram mole of carbaryl produced overall in the process. Will there be any additional cost of heat transfer using the greener process?

Data: Note that some of the values listed are only estimates. All of the values for $\Delta \hat{H}_f^o$ are in kilojoules per gram mole.

Component	ΔH_f^o (kJ / g mol)
Carbaryl	−26
Hydrogen chloride	−92.311
Methyl amine	−20.0
Methyl isocyanate	-9.0×10^4
1-Napthalenyl chloroformate	−17.9
1-Napthol	30.9
Phosgene	−221.85

(Continues)

> ## Example 9.3   Green Chemistry: Examining Alternate Processes (*Continued*)
>
> **Solution**
>
> Basis: 1 g mol of carbaryl
>
> The simplest analysis is to say $\Delta H_{rxn}^o = Q$ so that the values of the standard heats of reaction will provide a rough measure of the major contributions to the energy balances for the respective processes.
>
> Reaction (a):
>
> $$\Delta H_{rxn}^o = [2(-92.311)+1(-90,000)]-[1(-20.0)+1(-221.85)] \cong -9.0\times10^4 \text{ kJ}$$
>
> Reaction (b):
>
> $$\Delta H_{rxn}^o = [1(-26)]-[1(-90,000)+1(30.9)] \cong \underline{9.0\times10^4 \text{ kJ}}$$
>
> Total ≈ 0 kJ
>
> Reaction (c):
>
> $$\Delta H_{rxn}^o = [1(-17.9)+1(-92.311)]-[1(30.9)+1(-221.85)] = 81.64 \text{ kJ}$$
>
> Reaction (d):
>
> $$\Delta H_{rxn}^o = [1(-26)+1(-92.311)]-[1(-17.9)+1(-20.0)] = \underline{-80.41 \text{ kJ}}$$
>
> Total = 1.23 kJ (≈ 0 kJ)
>
> Both the original process and the suggested process require relatively small heat removal overall, but the Bhopal process requires considerable heat transfer on each of the two stages of the process, with reaction (a) requiring removal and reaction (b) requiring heating. Thus, the capital costs of the Bhopal process could be higher than those of the alternate process.

So far, we have focused on the heat of reaction for complete reaction. What if the reaction is not complete? As mentioned in Chapter 4, for most processes the moles of reactants entering are not in their stoichiometric quantities, and the reaction may not go to completion, so some reactants appear in the products from the reactor. How do you calculate the heat of reaction, ΔH_{rxn} (25°C, 1 atm) (*not* the standard heat of reaction—why?), in the standard state? One way is to start with Equation (9.1) and use the extent of reaction if you know it or can calculate it. For each species associated with the reaction for an unsteady-state, closed system (no flow in or out):

$$n_i^{\text{final}} = n_i^{\text{initial}} + v_i \xi$$

The equivalent for a steady-state flow system is

$$n_i^{\text{in}} = n_i^{\text{out}} + v_i \xi$$

Thus, for a steady-state flow system

$$\Delta H_{\text{rxn}}(25°C,\ 1\ \text{atm}) = \overset{\text{Flow out}}{\sum_i \left(n_i^{\text{in}} + v_i \xi \right) \Delta \hat{H}_{f,i}^{o}} - \overset{\text{Flow in}}{\sum_i \left(n_i^{\text{in}} \right) \Delta \hat{H}_{f,i}^{o}}$$

$$= \xi \sum v_i \Delta \hat{H}_{f,i}^{o} = \xi \Delta H_{\text{rxn}}^{o} \tag{9.2}$$

In Equation (9.2), you use the summations over all of the species associated with the reaction with $n_i = 0$ and $v_i = 0$ for any species not present as a product or reactant.

For example, for the reaction in Figure 9.3, assume the fraction conversion is 0.80 and the limiting reactant is 1 mol of C_6H_6. From Equation (4.10),

$$\xi = \frac{(-f)(n_{\text{limiting reactant}}^{in})}{v_{\text{limiting reactant}}} = \frac{-(0.80)(1)}{-1} = 0.80$$

$$\Delta H_{\text{rxn}}(25°C,\ 1\ \text{atm}) = (0.80)[(1)(-23.1) + (-3)(0) + (-1)(82.927)]$$

$$= -164.8\ \text{kJ}$$

Can you calculate the heat of reaction for a process in which the reactants enter and products exit at temperatures other than 25°C and 1 atm? The answer is yes. Recall that the enthalpy is a state (point) variable. Then you can calculate a change in enthalpy by any path that goes from the initial state to the final state.

Look at Figure 9.4. We want to calculate the ΔH_{rxn} (the enthalpy change) from state 1 to state 2. The value is the same as summing the values of all of the enthalpy changes 1 to 3, 3 to 4, and 4 to 2. The enthalpy change for the reactants and products is the combination of the sensible and latent heat (enthalpy) changes that might be taken from a table, or be calculated using a heat capacity equation for each species. The products are shown leaving the reactor at a common temperature, but each reactant might exist at a separate temperature T_i. Then the sensible heat plus any phase change for each component would be (excluding any mixing effects)

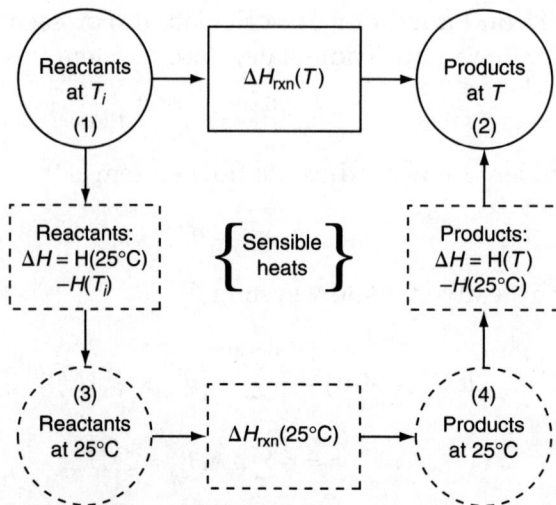

Figure 9.4   Information flow used to calculate the enthalpies constituting ΔH_{rxn} (from state 1 to state 2)

$$H_i(T_i) - H_i(25°C) = n_i \int_{25°C}^{T_i} C_{p,i} \, dT + n_i \Delta \hat{H}_{i,\text{phase change}} \tag{9.3}$$

If you ignore any slight pressure and mixing effects, for a steady-state system, the heat of reaction at a temperature T other than the reference temperature is

$$\Delta H_{\text{rxn}}(T) = \sum_i^{\text{Reactants}} n_i \left[\hat{H}_i(25°C) - \hat{H}_i(T) \right] + \Delta H_{\text{rxn}}(25°C)$$

$$+ \sum_i^{\text{Products}} n_i \left[\hat{H}_i(T) - \hat{H}_i(25°C) \right] \tag{9.4}$$

and in shorter notation

$$\Delta H_{\text{rxn}}(T) = \left[H(T) - H(25°C) \right]_{\text{Products}} - \left[H(T) - H(25°C) \right]_{\text{Reactants}}$$

$$+ \Delta H_{\text{rxn}}(25°C) \tag{9.4a}$$

Next, let's look at how the heat of reaction fits into the general energy balance $\Delta U_{\text{Total}} = 0 = Q + W - \Delta H - \Delta PE - \Delta KE$ for a steady-state flow system.

In the energy balance, the enthalpy change for the compounds involved in the reaction is

$$\Delta H = \left[H(25°C) - H(T^{in})\right]_{\text{Reactants}} + \Delta H_{\text{rxn}}(25°C) + \left[H(T^{out}) - H(25°C)\right]_{\text{Products}} \quad (9.5)$$

Any compounds not involved in the reaction (e.g., N_2, in combustion) can be included in the first and third terms on the right side of the equal sign in Equation (9.5), but they make no contribution to the heat of reaction at 25°C.

The conclusion that you should reach from Equation (9.5) is that when reaction occurs in a system, just one term has to be added to the energy balance, namely, the heat of reaction at 25°C. All of the effects of energy generation or consumption caused by reaction can be lumped into the one quantity $\Delta H_{\text{rxn}}(25°C)$. What would be the analog of Equation (9.5) for an unsteady-state, batch (closed) system? Remember that $\Delta U = \Delta H - \Delta(pV)$.

You have to be a bit careful in calculating the heat of reaction at 25°C to get ΔH for an energy balance because certain conventions exist:

1. The reactants are shown on the left side of the chemical equation, and the products are shown on the right; for example,

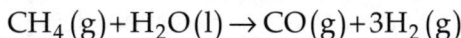

$$CH_4(g) + H_2O(l) \rightarrow CO(g) + 3H_2(g)$$

2. The conditions of phase, temperature, and pressure must be specified unless the last two are standard conditions, as presumed in the preceding equation, when only the phase is required. This is particularly important for a compound such as H_2O, which can exist as more than one phase under common conditions.

3. The amounts of material reacting are assumed to be the quantities shown by the stoichiometry in the chemical equation and the extent of reaction. Thus, if the reaction is

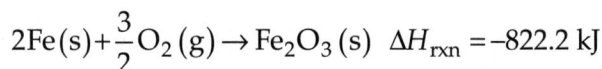

$$2Fe(s) + \frac{3}{2}O_2(g) \rightarrow Fe_2O_3(s) \quad \Delta H_{\text{rxn}} = -822.2 \text{ kJ}$$

the value of –822.2 kJ refers to 2 g mol of Fe(s) reacting and not 1 g mol. The heat of reaction for 1 g mol of Fe(s) would be –411.1 kJ/g mol of Fe.

Example 9.4   Calculation of the Heat of Reaction in a Process in Which the Reactants Enter and the Products Leave at Different Temperatures

Problem Statement

Public concern about the increase in the carbon dioxide in the atmosphere has led to numerous proposals to sequester or eliminate the carbon dioxide. An inventor believes he has developed a new catalyst that can make the gas phase reaction

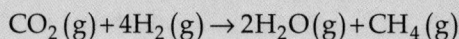

$$CO_2(g) + 4H_2(g) \rightarrow 2H_2O(g) + CH_4(g)$$

proceed with 40% conversion of the CO_2. The source of the hydrogen would be from the electrolysis of water using electricity generated from solar cells. Assume that 1.5 mol of CO_2 enter the reactor at 700°C together with 4 mol of H_2 at 100°C. Determine the heat of reaction if the exit gases leave at 1 atm and 500°C.

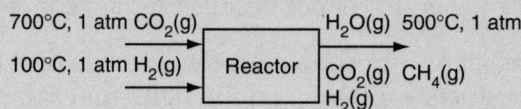

700°C, 1 atm CO_2(g) | Reactor | H_2O(g) 500°C, 1 atm
100°C, 1 atm H_2(g) | | CO_2(g) CH_4(g) H_2(g)

Figure E9.4

Solution

The system is steady-state and open with reaction. Assume 1 atm for the products and reactants. We do not outline each step in the solution to save space. First, complete the material balance, and then the energy balance in three phases: (1) the standard heat of reaction at 25°C, (2) the heat of reaction at 25°C, (3) and the sensible heats.

Basis: 1.5 g mol of CO_2 (g)

The reference temperature is 25°C. The outcome of the material balance and some needed enthalpy data are as follows:

Compounds	g mol In	g mol Out	Fraction Reacted	$\Delta\hat{H}_f^o$ (kJ/g mol) 25°C	$\Delta\hat{H}_{sensible}^o$ (kJ/g mol) 100°C	$\Delta\hat{H}_{sensible}^o$ (kJ/g mol) 500°C	$\Delta\hat{H}_{sensible}^o$ (kJ/g mol) 700°C
CO_2(g)	1.5	0.90	0.40	−393.250		20.996	30.975
H_2(g)	4	1.60	0.60	0	2.123	13.826	
H_2O(g)	2	1.20		−241.835		17.010	
CH_4(g)	1	0.60		−74.848		23.126	

The first step is to calculate the standard heat of reaction at the reference temperature. The reaction is

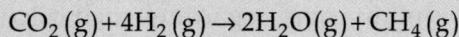

$$CO_2(g) + 4H_2(g) \rightarrow 2H_2O(g) + CH_4(g)$$

The sensible heat data shown in tables below have been obtained through a computer program developed to perform calculations of equation 8.14. Complete reaction is specified for the standard heat of reaction. H_2 is the limiting reactant.

$$\Delta \hat{H}^o_{rxn} = [(1)(-74.848) + (2)(-241.835)]$$
$$- [(1)(-393.250) + (4)(0)] = -165.27 \text{ kJ/g mol } CO_2$$

For 40% conversion of the CO_2 (the excess reactant), the conversion of the limiting reactant (H_2) is 0.60; hence, the extent of reaction is $\xi = \dfrac{(-0.60)(4)}{-4} = 0.60$. Then

$$\Delta H_{rxn}(25°C) = (0.60)(-165.27) = -99.16 \text{ kJ}$$

The next step is to calculate the enthalpy changes (sensible heats) from 25°C to the respective temperatures of the compounds entering and leaving the reactor:

Compound	g mol	T (°C)	$\Delta \hat{H}_{sensible}$ (kJ/g mol)	$\Delta H_{sensible}$ (kJ)
		Sensible Heat In		
$CO_2(g)$	1.5	700	30.975	46.463
$H_2(g)$	4.0	100	2.123	8.492
Total				**54.955**
		Sensible Heat Out		
$CO_2(g)$	0.90	500	20.996	18.896
$H_2(g)$	1.60	500	13.826	28.122
$H_2O(g)$	1.20	500	17.010	20.412
$CH_4(g)$	0.60	500	23.126	13.876
Total				**75.306**

Water is a gas at 500°C, so a phase change is not involved. The heat of reaction is

$$\Delta H_{rxn} = 75.306 - 54.955 + (-99.161) = -78.85 \text{ kJ/g mol } CO_2$$

The standard heat of reaction for biological reactions, such as the growth of yeast with a substrate of glucose, can be calculated using the stoichiometric equation, Equation (4.17), as a guide. However, one factor that you have to keep in mind regarding such calculations is that the substrate may be in solution. You have to find ΔH_f^o for the solution, not the pure solute, or solvent. For example, from F. O. Rossini et al.,[4] the heat of solution of NH_3, often said to be a nitrogen source, in H_2O depends on the concentration of the NH_3 in the water.

Chemical Species	$\Delta \hat{H}_{solution}^o$ (25°C and 1 atm, kJ/g mol)	ΔH_f^o (kJ/g mol)
$NH_3(g)$	0	−67.20
$NH_3(1\ H_2O)$	−7.06	−74.26
$NH_3(5\ H_2O)$	−8.03	−75.25
$NH_3(50\ H_2O)$	−8.25	−75.45
$NH_3(100\ H_2O)$	−8.28	−75.48

The value in parentheses is the total number of moles of H_2O added to 1 mol of $NH_3(g)$. Remember that the pressure of NH_3 over the solution and the pH of the solution vary with its concentration at equilibrium. The heat of formation of NH_3 in water is the sum of the heat of formation of $NH_3(g)$ plus the heat of solution.

If the $\Delta H_{solution}^o$ between different concentrations of solute is small relative to the heat of reaction, you can ignore the heat of solution, but it may be relatively large in some cases. The heats of formation of compounds in solution are available from sources such as *Perry's Chemical Engineers' Handbook*, *CRC Handbook of Chemistry and Physics*, and many others. For many compounds, the values are reported as function of solute concentration, allowing accounting of the heat of solution in calculations for increased accuracy.

Self-Assessment Test

Questions

1. Can the standard heat of reaction ever be positive?

2. Is it correct to calculate a heat of reaction for which the reaction is incomplete?

---

[4] *Selected Values of Chemical Thermodynamic Properties* [National Bureau of Standards Circular 500, USGPU (1952)].

3. How does phase change, such as when water goes from a liquid to a vapor, affect the value of the heat of reaction?

4. What does it mean when the standard heat of reaction is (a) negative and (b) positive?

5. Can you choose a reference state other than 25°C and 1 atm to use in applying the energy balance?

6. What is the difference between the heat of reaction and the standard heat of reaction?

Answers

1. Yes

2. No

3. If the reaction equation is expressed with the proper phases and applied properly, the heat of reaction will properly take into account phase changes.

4. The reaction is (a) exothermic and (b) endothermic.

5. Yes, but it may not prove to be convenient.

6. The standard heat of reaction is the enthalpy change at standard conditions and for complete reaction based on the stoichiometry of the reaction equation. The heat of reaction is the enthalpy change for reaction at a specified temperature and pressure with partial or complete reaction.

Problems

1. Calculate the standard heat of reaction for the following reaction from the heats of formation:

$$C_6H_6 (g) \rightarrow 3C_2H_2 (g)$$

2. Calculate the heat of reaction at 90°C for the Sachse process (in which acetylene is made by partial combustion of LPG):

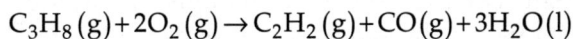

$$C_3H_8 (g) + 2O_2 (g) \rightarrow C_2H_2 (g) + CO(g) + 3H_2O(l)$$

Answers

1. $3(226.75) - 82.927 = 597.323$ kJ/g mol

2. The standard heat of reaction $= 226.75 - 110.52 + 3(-285.84) - (-103.85) = -637.44$ kJ/g mol; reactants sensible heat $= 7.189$; products sensible heat $= 19.48$; heat of reaction at 90°C $= -637.44 - 7.189 + 19.48 = 625.15$ kJ/g mol.

9.3   Integration of Heat of Formation and Sensible Heat

You might have speculated from the tabular formation of the solution in Example 9.3 that one way to simplify the calculation procedure for the overall enthalpy change in a process in which a reaction(s) occurs is to transfer each standard heat of formation from the calculation of the standard heat of reaction and merge it with its respective sensible heat. Also note that the heat of formation and sensible heat are both calculated on a per mol basis. Thus, the heat of reaction does not have to be explicitly calculated. Look at Figure 9.4.

The result is that the enthalpy change term in the energy balance for a steady-state, open system can simplify to just

$$\Delta H = \sum_{\text{outputs}} n_i \Delta \hat{H}_i - \sum_{\text{inputs}} n_i \Delta \hat{H}_i \tag{9.6}$$

because the heats of formation are embedded into the H_is.

Note: All of the compounds in a stream are included in the summation "output" or "input," respectively, whether the compound reacts or not. To be consistent, you can include every compound that exists in the problem in each stream, input or output, if you make its number of moles zero; for example, $n_i = 0$ when the compound is not present in a stream. Although it may seem that you have lost some essential feature in the calculation of ΔH, Figure 9.4 shows that you have simply dispersed the heats of formation into a different sequence in the calculations so that they are no longer aggregated in a term called the *standard heat (enthalpy) of reaction*. If the reaction is not complete, the material balance determines the amount of reactants appearing with the products, and the unreacted reactants become subtracted from the total reactants in the entering compounds if the calculation sequence shown in Figure 9.5 is used for the energy balance. This approach is actually easier to apply and less susceptible to errors than the heat of reaction approach, as the following examples demonstrate.

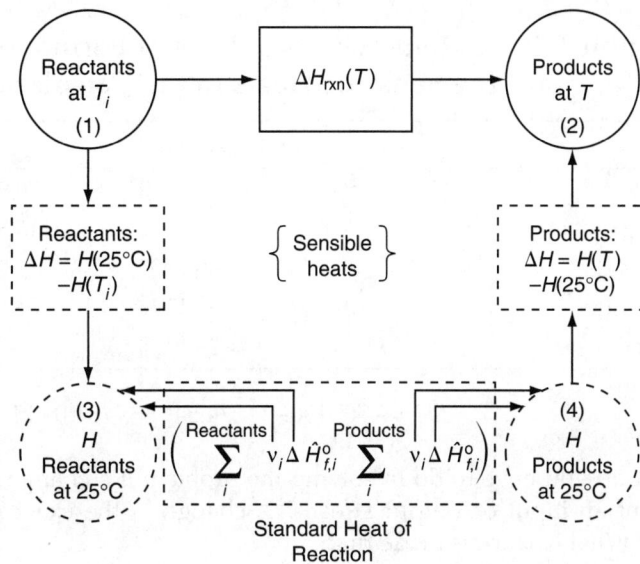

Figure 9.5   Splitting the respective heats of formation from the standard heat of reaction, and merging them with their appropriate sensible heats

---

Example 9.5   Redone with the Heats of Formation Merged with the Sensible Heats in the Calculations

Problem Statement

The problem and all of the data are the same as in Example 9.3. The following table lists the details of the solution.

Solution

			Enthalpy In		
Compound	g mol	T (K)	$\Delta \hat{H}_f^o$ (kJ/g mol)	$\Delta \hat{H}_{sensible}$ (kJ/g mol)	ΔH (kJ)
$CO_2(g)$	1.5	700	−393.250	30.975	−543.413
$H_2(g)$	4.0	100	0	2.123	8.492
Total					**−534.546**

(*Continues*)

Example 9.5   Redone with the Heats of Formation Merged with the Sensible Heats in the Calculations (*Continued*)

			Enthalpy Out		
$CO_2(g)$	0.90	500	−393.250	20.996	−335.029
$H_2(g)$	1.60	500	0	13.826	28.122
$H_2O(g)$	1.20	500	−241.835	17.010	−269.790
$CH_4(g)$	0.60	500	−74.848	23.126	−31.0336
Total					**−613.130**

$$\Delta H = -613.130 - (-534.546) = -78.209 \text{ kJ}$$

What do you have to do in solving the problem if a change occurs in a temperature of an input or output stream? A change in the exact reaction of the CO_2? Hint: What balance is made first?

Frequently Asked Questions

1. Where does the heat of reaction go if you use the tabular solution procedure of Example 9.5 in calculating ΔH? $\Delta H_{rxn}(25°C)$ still exists, but not as an agglomerated term. Its components are decentralized and added to the respective sensible heats of the components. It can still be calculated separately, if needed, as before.

2. How does the fraction conversion or extent of reaction used to modify ΔH^o_{rxn} in Equation (9.2) become involved in calculating ΔH if the reaction is not complete? The preliminary material balance takes care of the amount of a reactant that enters and leaves the process. When the same amount of compound is included in both the output and input—that is, no contribution to reaction occurs—the result is the only contribution to ΔH in the sensible heat and phase changes: $\left(n_i \Delta H^o_{f,i}\right)_{in} - \left(n_i \Delta H^o_{f,i}\right)_{out} = 0$.

3. Do you need to know the specific reaction(s) in the process to use the tabular method in Example 9.5? No. The extent of reaction is automatically taken into account in the material balance (unless it is the unknown), or from process measurements. In biological reactions, the reactions themselves and the extents of reaction may be uncertain, but the overall ΔH_{rxn} can be calculated.

Now that we have described the details of how to calculate the enthalpy change in the energy balance when reaction(s) occurs, we next give some illustrative examples. Here are some typical problems frequently posed for steady-state, open systems:

1. What is the temperature of one stream given data for the other streams?
2. How much heat has to be added to or removed from the process?
3. What is the temperature of the reaction?
4. How much material must be added to or removed from the process to give a specified value of heat transfer?

Example 9.6   A Process in Which More than One Reaction Occurs

Problem Statement

Limestone ($CaCO_3$) is converted to calcium oxide (CaO) in a continuous vertical kiln such as that illustrated in Figure E9.6a. The energy to decompose the limestone is supplied by the combustion of natural gas (CH_4) in direct contact with the limestone using 50% excess air. The $CaCO_3$ enters the process at 25°C, and the CaO exits at 900°C. The CH_4 enters at 25°C, and the product gases exit at 500°C. Calculate the maximum number of pounds of $CaCO_3$ that can be processed per 1000 ft^3 of CH_4 measured at standard conditions (0°C and 1 atm for gases). To simplify the calculations, assume that the heat capacities of $CaCO_3$ (56.0 Btu/(lb mol)(°F)) and CaO (26.7 Btu/(lb mol)(°F)) are constant.

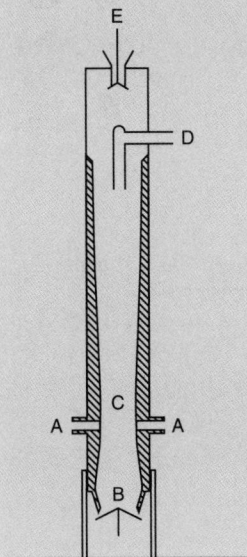

Figure E9.6a   A vertical line kiln C composed of a steel cylinder lined with fire brick approximately 80 ft high and 10 ft in diameter. Fuel is supplied at A, air at B, and limestone ($CaCO_3$) at E. Combustion products and CO_2 exit at D.

(Continues)

Example 9.6   A Process in Which More than One Reaction Occurs (*Continued*)

Solution

Steps 1–3

You have to decide whether to make the calculations using the SI or American Engineering (AE) systems of units (or both). We use the SI system for most of the solutions for convenience, but the calculations can be made just as easily using the AE system. We assume that the entire process occurs at 1 atm (so that H is a function only of temperature) and that the air enters at 25°C. We also have to choose a basis.

Step 5

Let us start with a basis of 1 g mol CH_4 for convenience. Many other bases, of course, could be selected. At the end of the calculations the results can be converted to the requested units.

Steps 1–4

Figure E9.6b is a sketch of the process. Assume that the process is open and steady-state. The assumption of steady-state means that no material was in the kiln at the start or end of the process and ignores a relatively short interval for start-up and shutdown. Your first mission is to solve the material balances.

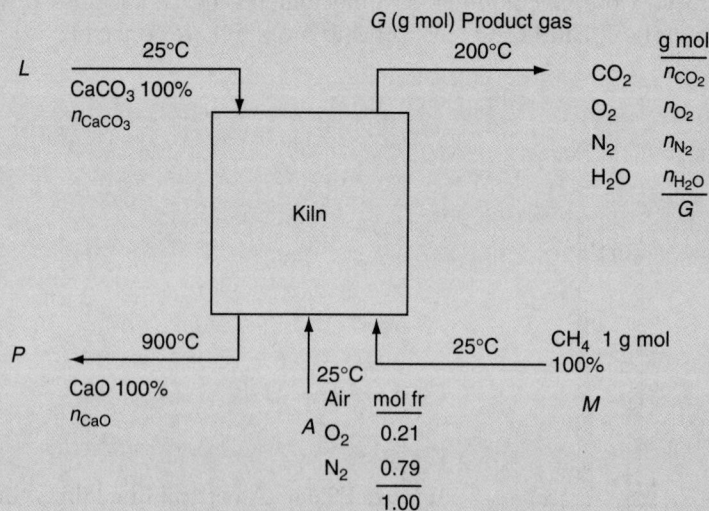

Figure E9.6b

Start by specifying the values of all of the variables you can identify or easily calculate in your head. Be sure to keep track of all of the material balances you use in your preliminary assignments so that you don't find subsequently that you have inadvertently used redundant information.

Steps 3 and 4

To get the values of the entering O_2 and N_2, you can assume that the values cited in the problem statement relate to the following reaction:

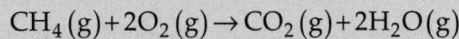

$$CH_4(g) + 2O_2(g) \rightarrow CO_2(g) + 2H_2O(g)$$

Thus, the moles of O_2 and N_2 entering are

1 mol CH_4 requires:	2 g mol O_2
50% excess:	1
Total O_2	$\dfrac{}{3}$
Entering N_2	3 (0.79/0.21) = (11.29) g mol

What other values can be assigned? Look at Figure E9.6b and note that the inlet amount of N_2 equals the outlet N_2 (N_2 doesn't react). Thus, $n_{N_2}^{G} = 11.29$ g mol. Also, all of the H from CH_4 goes into stream G; hence $n_{H_2O}^{G} = 2.0$ g mol. Use the implicit relation of the sum of moles in the stream to get (in gram moles)

$$n_{CaCO_3}^{L} = L \quad A = 14.29$$
$$n_{CaO}^{P} = P \quad M = 1.00$$

From a calcium balance, you know $L = P$. Let's use L as an unknown because that variable has to be used in the final part of the solution.

Steps 6 and 7

The residual variables that can be deemed unknowns are

$$L, \quad G, \quad n_{CO_2}^{G}, \quad n_{O_2}^{G} \quad \text{(Total 4)}$$

and the remaining independent equations that have not been used in assigning values are

Element balances: C, O (H and N were used)	2
Sum of moles: $n_{CO_2}^{G} + n_{O_2}^{G} + 11.29 + 2.00 = G$	1
Total	3

(Continues)

Example 9.6   A Process in Which More than One Reaction
Occurs (*Continued*)

The remaining degrees of freedom are $4 - 3 = 1$.

Note that the specification of complete reaction of the $CaCO_3$ and the CH_4 would be a redundant specification because no $CaCO_3$ or CH_4 was indicated to exit the process (the variables thus are automatically assigned zero values). A specification of less than complete reaction of $CaCO_3$ or CH_4 would be inconsistent with the other information given in the problem.

If you used species balances in the solution of the material balance problem, you would start with 14 variables (L, P, A, M, G, 7 species, and two extents of reaction), assign values to 10 of the variables (the two extents of reaction are unity), the basis is $CH_4 = 1$ mol, the CH_4 reaction gives the H_2O, the $CaCO_3$ reaction gives $L = M$, and the other assignments would be as shown for the element balances, to leave three species balances (O_2, CO_2, $CaCO_3$) available to solve for four unknowns—the same result as obtained from using element balances.

Thus, you need one more independent equation, the energy balance:

$$\Delta H = Q$$

Assume: $Q = 0$ (fairly well-insulated tower by fire brick, and you don't know how to calculate Q):

Conclusion: $\Delta H = 0$.

Remember that H is a function of temperature (p has been assumed to be 1 atm), and note that the temperature is specified for each stream. Consequently, the values of $\hat{H}(T)$ can be looked up and multiplied by the respective number of moles to get $\Delta H(T)$ for each stream (the heat of mixing for a gas is essentially zero).

$$\frac{\text{Out (at 500°C and 900°C)}}{(L)\left(\Delta\hat{H}_{CaO}\right)+\left(n^G_{CO_2}\right)\left(\Delta\hat{H}^G_{CO_2}\right)+\left(n^G_{O_2}\right)\left(\Delta\hat{H}^G_{O_2}\right)+(11.29)\left(\Delta\hat{H}^G_{N_2}\right)+(2)\left(\Delta\hat{H}_{H_2O}\right)}$$

$$-\frac{\text{in(at 25°C)}}{-(1)\left(\Delta\hat{H}_{CH_4}\right)-(3)\left(\Delta\hat{H}_{O_2}\right)-(11.29)\left(\Delta\hat{H}_{N_2}\right)-(L)\left(\Delta\hat{H}_{CaCO_3}\right)}=0$$

Select the reference state as 25°C and 1 atm.

Use a tabular format for efficiency in the calculations (an ideal application for a spreadsheet). A stream ΔH (in joules) is the sum of $\left(\Delta\hat{H}^o_f + \Delta\hat{H}_{\text{sensible}}\right)n_i$ for the respective i compounds.

Compound	Mol	$\Delta \hat{H}_f^o$ (kJ/g mol)	T (°C)	Sensible Heat* (kJ/g mol)	Stream ΔH (kJ)
In					
$CH_4(g)$	1	−49.963	25	0	−49.963
$O_2(g)$	3	0	25	0	0
$N_2(g)$	11.29	0	25	0	0
$CaCO_3(s)$	$n_{CaCO_3}^L = L$	−1206.9	25	0	−1206.9L
Out					
$CaO(s)$	$n_{CaO}^L = P = L$	−635.6	900	(0.062)(900 − 25)	−581.35L
$CO_2(g)$	$n_{CO_2}^G$	−393.25	500	21.425	−371.825 $n_{CO_2}^G$
$O_2(g)$	$n_{O_2}^G$ } $\Sigma = G$	0	500	15.034	15.034
$N_2(g)$	11.29	0	500	14.241	160.781
$H_2O(g)$	2	−241.835	500	17.010	−449.650

*Data from Appendix F, reference No. 1 (Kobe).

We have to solve the two remaining element mass balances (in moles) with the energy balance:

Balance	**In**		**Out**	
C:	$1 + L$	$=$	$n_{CO_2}^G$	(1)
O:	$3L + 2(3)$	$=$	$2n_{CO_2}^G + 2n_{O_2}^G + 2 + L$	(2)

$$\Delta H = 0$$

$$\left(-581.35L - 371.825 n_{CO_2}^G + 15.034 + 160.781 - 449.650\right)$$

$$-(-49.963 - 1206.9L) = 0 \qquad (3)$$

$$L = 2.56 \text{ g mol}$$

On the basis of 1 g mol of CH_4:

$$\frac{2.56 \text{ g mol } CaCO_3}{1 \text{ g mol } CH_4} \left| \frac{100.09 \text{ g } CaCO_3}{1 \text{ g mol } CaCO_3} = \frac{256g \text{ } CaCO_3}{1 \text{ g mol } CH_4} \right.$$

(Continues)

Example 9.6   A Process in Which More than One Reaction Occurs (*Continued*)

To get the ratio asked for (assuming CH_4 is an ideal gas is a good assumption):

$$\frac{1000 \text{ ft}^3 \text{ CH}_4}{} \left| \frac{1 \text{ lb mol CH}_4}{359.05 \text{ ft}^3} \right| \frac{256 \text{ lb CaCO}_3}{1 \text{ lb mol CH}_4} = \frac{713 \text{ lb CaCO}_3}{1000 \text{ ft}^3 \text{ CH}_4 \text{ at S.C.}}$$

If you decided to use the exit gas from the kiln to preheat the entering air, would you increase or decrease the ratio calculated above? Would the exit gases from the entire system, if you include the heat exchange, be at a lower or higher temperature? What would the revised plan do to the sensible heats of the exit gases? Does the tabular format of calculation help you reach an answer?

A term called the **adiabatic reaction (theoretical flame or combustion) temperature** is defined as the temperature obtained from a combustion process when

1. The reaction is carried out under adiabatic conditions; that is, there is no heat interchange between the system in which the reaction occurs and the surroundings.
2. No other effects occur, such as electrical effects, work, ionization, free radical formation, and so on.
3. The limiting reactant reacts completely.

When you calculate the adiabatic reaction temperatures for combustion reactions, you assume complete combustion occurs, but equilibrium considerations may dictate less than complete combustion in practice. For example, the adiabatic flame temperature for the combustion of CH_4 with theoretical air has been calculated to be 2070°C; allowing for incomplete combustion at equilibrium, it would be 1920°C. The actual temperature when measured is 1885°C.

The adiabatic reaction temperature tells you the temperature ceiling of a process. You can do no better, but of course the actual temperature may be less. The adiabatic reaction temperature helps you select the types of materials that must be specified for the equipment in which the reaction is taking place. Chemical combustion with air produces gases with a maximum temperature of roughly 2500 K, but the temperature can be increased to 3000 K with the use of oxygen and more exotic oxidants, and even this value can be exceeded, although handling and safety problems are severe. Applications of such hot gases lie in the preparation of new materials, micromachining, welding using laser beams, and the direct generation of electricity using ionized gases as the driving fluid.

As you have seen, the *open system, steady-state* energy balance with $Q = 0$ reduces to just $\Delta H = 0$. Because you do not know the flame temperature (e.g., the solution) to a problem before you start, the solution will involve trial and error. To find the exit temperature for which $\Delta H = 0$ in the energy balance:

1. Assume a sequence of values of T selected to bracket $\Delta H = 0(+$ and $-)$ for the sum of the enthalpies of the outputs minus the enthalpies of the inputs.

2. Once the bracket is obtained, interpolate within the bracket to get the desired value of T when $\Delta H = 0$.

If you integrate the heat capacity equations from 25°C to T to obtain the sensible heats, ΔH will involve at least a final cubic or quadratic equation for ΔH to be solved for the exit temperature. Make sure you get a unique reasonable solution from the polynomial; more than one solution may exist, as well as negative or complex values.

For an *unsteady-state, closed system* with ΔKE and $\Delta PE = 0$ inside the system and $W = 0$, the energy balance reduces to

$$Q = \Delta U = U_{final} - U_{initial}$$

If you do not have values for $\hat{U}$, you have to calculate Q from

$$Q = \left[H(T) - H(25°C) \right]_{final} - \left[H(T) - H(25°C) \right]_{initial} - \left[(pV)_{final} - (pV)_{initial} \right]$$

The contribution of $\Delta(pV)$ is frequently negligible, so you can just use H in lieu of U, in which case the open and closed system calculations will give the same answer. But if $\Delta(pV)$ is significant, remember that for a constant volume process, $\Delta(pV) = V\Delta p$, and for a constant pressure but expandable closed system, $\Delta(pV) = p\Delta V$.

Example 9.7   Calculation of an Adiabatic Reaction (Flame) Temperature

Problem Statement

Calculate the theoretical flame temperature for CO gas burned at constant pressure with 100% excess air, when the reactants enter at 100°C and 1 atm.

Solution

The solution presentation will be compressed to save space. The system is shown in Figure E9.7. We use data from Appendix C and other resources available on the Internet and literature (correlations reported by Professor Yaws). The process is a steady-state flow system. Ignore any equilibrium effects.

Basis: 1 g mol of CO(g); ref. temp. 25°C and 1 atm

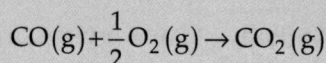

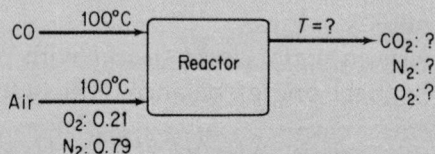

$$CO(g) + \frac{1}{2}O_2(g) \rightarrow CO_2(g)$$

Basis: 1 g mol of CO(g); ref. temp. 25°C and 1 atm

Figure E9.7

The reaction is always assumed to occur with the limiting reactant (CO) reacting completely. The excess air and the N_2 are nonreacting components, but you nevertheless have to calculate their associated sensible heats to obtain the proper total sensible heats (enthalpies) for the process streams. We use the tabular method of bookkeeping used previously for all of the pertinent enthalpy changes for each compound entering and leaving the system. You can solve the material balances (for which the degrees of freedom are zero) independently of the energy balance. A summary of the results of the solution of the material balances is as follows:

Entering Compounds		Exit Compounds	
Component	*g mol*	*Component*	*g mol*
CO(g)	1.00	$CO_2(g)$	1.00
O_2 (req. + xs)0.50 + 0.50	1.00	$O_2(g)$	0.50
N_2	3.76	$N_2(g)$	3.76
(Air = 4.76)			

In the first approach to the solution of the problem, the "sensible heat" (enthalpy) values have been taken from the table of the enthalpy values for the combustion gases in Appendix C. The energy balance (with $Q = 0$) reduces to $\Delta H = 0$. Here are the data needed for the energy balance:

Component	g mol	T (K)	$\Delta \hat{H}_{sens.}$ (J/g mol)	$\Delta \hat{H}_f^o$ (J/g mol)	ΔH (J)
Inputs					
CO	1.00	373	$(2917 - 728)$	$-110{,}520$	$-108{,}331$
O_2	1.0	373	$(2953 - 732)$	0	2221
N_2	3.76	373	$(2914 - 728)$	0	8219
		Total			**−97,891**
Outputs					
Assume $T = 2000$ K:					
CO_2	1.00	2000	$(92{,}466 - 912)$	$-393{,}510$	$-301{,}956$
O_2	0.50	2000	$(59{,}914 - 732)$	0	29,591
N_2	3.76	2000	$(56{,}902 - 728)$	0	211,214
		Total			**−61,151**

$$\Delta H = \Delta H_{outputs} - \Delta H_{inputs} = (-61{,}151) - (-97{,}891) = 36{,}740 > 0$$

Assume $T = 1750$ K:					
CO_2	1.00	1750	$(77{,}455 - 912)$	$-393{,}510$	$-316{,}977$
O_2	0.50	1750	$(50{,}555 - 732)$	0	24,912
N_2	3.76	1750	$(47{,}940 - 728)$	0	177,517
		Total			**−114,548**

$$\Delta H = (-114{,}548) - (-97{,}891) = -16{,}657 < 0$$

Now that $\Delta H = 0$ is bracketed, we can carry out a linear interpolation to find the adiabatic flame temperature (AFT):

$$\text{AFT} = 1750 + \frac{0 - (-16{,}657)}{36{,}740 - (-16{,}657)}(250) = 1750 + 78 = 1828 \text{ K } (1555°\text{C})$$

(Continues)

Example 9.7   Calculation of an Adiabatic Reaction (Flame) Temperature (*Continued*)

An alternate approach to solving this problem would be to develop explicit equations in T to be solved for AFT without trial and error. The difference from the first approach is that you would have to formulate nonlinear polynomial equations in AFT by integrating the heat capacity equations for each compound to obtain the respective sensible heats for each compound. If the heat capacity equations were cubic in T, the integrated equations in AFT would be quadratic. To avoid error, you can take the already-integrated equations for each compound from the website that accompanies this book, introduce them into the energy balance as sensible heats, merge them, and solve the resulting energy balance using an equation solver such as MATLAB or Python (see next example). You can get a rough preliminary solution by first truncating the quartic equations to quadratic equations and solving the latter.

Example 9.8   Adiabatic Flame Temperature using MATLAB and Python

Problem Statement

Determine the adiabatic flame temperature of methane with stochiometric air using MATLAB and Python. The following tables contains the required data:

Component	$\Delta \hat{H}_f^o$ (kJ/g mol)
CH_4	−74.84
O_2	0
N_2	0
CO_2	−393.51
H_2O	−241.826

Component	Cp(T) (J/g mol °C)
N_2	$29.0 + 0.2199T \times 10^{-2} + 0.5723T^2 \times 10^{-5} - 2.871T^3 \times 10^{-7}$
CO_2	$36.11 + 4.233T \times 10^{-2} - 2.887T^2 \times 10^{-5} - 7.464T^3 \times 10^{-7}$
$H_2O(g)$	$33.46 + 0.6880T \times 10^{-2} + 0.7604 T^2 \times 10^{-5} - 3.593T^3 \times 10^{-7}$

Solution

Basis: 1 g mol CH_4

The chemical reaction equation is

$$CH_4 + 2O_2 \rightarrow CO_2 + 2H_2O$$

The material balances for this problem are summarized in the following table:

Component	Moles Initially	Moles after Reaction
CH_4	1	0
O_2	2	0
N_2	7.524	7.524
CO_2	0	1
H_2O	0	2

The combined enthalpies for the initial components are given by

$$\Delta H_{N_2} = 0 \qquad \Delta H_{O_2} = 0 \qquad \Delta H_{CH_4} = \Delta H^o_{f, CH_4}$$

There is no sensible heat because the components are initially at 25°C and 1 atm. The combined enthalpies for the products are given by

$$\Delta H_{N_2} = 0 + n_{N_2} \int_{25°C}^{T} C_{p, N_2} \, dT$$

$$\Delta H_{CO_2} = \Delta H^o_{f, CO_2} + n_{CO_2} \int_{25°C}^{T} C_{p, CO_2} \, dT$$

$$\Delta H_{H_2O} = \Delta H^o_{f, H_2O} + n_{H_2O} \int_{25°C}^{T} C_{p, CO_2} \, dT$$

Because each of the heat capacities have the same form (i.e., $C_p(T) = a + bT + cT^2 + dT^3$), we can analytically integrate the general equation and later change the parameters (i.e., a, b, c, and d) for the particular component. Therefore,

$$\int_{25°C}^{T} C_p \, dT = a(T - 25) + b(T^2 - 25^2)/2 + c(T^3 - 25^3)/3 + d(T^4 - 25^4)/4$$

(Continues)

Example 9.8   Adiabatic Flame Temperature using MATLAB and Python (*Continued*)

The energy balance for this formulation of the problem is simply $\Delta H = 0$. Then substituting yields

$$\Delta H = n_{N_2}\Delta H_{N_2} + n_{CO_2}\Delta H_{CO_2} + n_{H_2O}\Delta H_{H_2O} - n_{CH_4}\Delta H_{CH_4} = 0$$

This is a single nonlinear equation with a single unknown, T. Therefore, we use MATLAB and Python to solve for the adiabatic flame temperature.

MATLAB Solution for Example 9.8

```
%%%%%%%%%%%%%%%%%%%%%%%%%%%%%%%%%%%%%%%%%%%%%%%%%%%%%%%%%%%%%%%%%%%%%%%%%
%
%                          NOMENCLATURE
%
% A(I) - the constant term for the gas heat capacity equation for I
% B(I) - the linear term for the gas heat capacity equation for I
% C(I) - quadratic term for the gas heat capacity equation for I
% D(I) - the cubic term for the gas heat capacity equation for I
% DHCO2 - the combined heat of formation and sensible heat for CO2
% DHf(I) - the heat of formation of component I
% DHN2 - the combined heat of formation and sensible heat for N2
% DHH2O - the combined heat of formation and sensible heat for H2O
% error - the error in the energy balance
% fun - m-file function that calculates the error in the energy balance
% N* - moles of component *
% T - the unknown in the energy balance
% T0 - the initial guess for the adiabatic flame temperature
% T_AFT - the solution for the adiabatic flame temperature
%
%%%%%%%%%%%%%%%%%%%%%%%%%%%%%%%%%%%%%%%%%%%%%%%%%%%%%%%%%%%%%%%%%%%%%%%%%

%                               PROGRAM
function Ex9_8_AdFlmTemp
clear;
T0 = 1000;       % Initial guess for adiabatic flame temp (deg C)
             % Input constants for heat capacity equations
A(1) = 36.11; A(2) = 33.46; A(3) = 29.00;
B(1) = 4.233e-2; B(2) = 0.688e-2; B(3) = 0.2199e-2;
```

```
C(1) = -2.887e-5; C(2) = 0.7604e-5; C(3) = 0.5723e-5;
D(1) = 7.464e-9; D(2) = -3.593e-9; D(3) = -2.871e-9;
NCO2 = 1; NH2O = 2; NN2 = 7.524; NCH4 = 1;
               % Input heats of formation
DHf(1) = -393510; DHf(2) =  -241826; DHf(3) = 0.0; DHfCH4 = -74840;
T_AFT = fzero(@fun,T0)
fprintf('Adiabatic flame temp is %8.2f deg C\n',T_AFT)
% Nested m-file function
    function [error] = fun(T)
          % Combine heat of formation and sensible heat
       DHCO2 = DHf(1)+A(1)*(T-25)+B(1)*(T^2-25^2)/2+ ...
           C(1)*(T^3-25^3)/3+D(1)*(T^4-25^4)/4
       DHH2O = DHf(2)+A(2)*(T-25)+B(2)*(T^2-25^2)/2+ ...
           C(2)*(T^3-25^3)/3+D(2)*(T^4-25^4)/4
       DHN2 = DHf(3)+A(3)*(T-25)+B(3)*(T^2-25^2)/2+ ...
           C(3)*(T^3-25^3)/3+D(3)*(T^4-25^4)/4
       error = NCO2*DHCO2+NH2O*DHH2O+NN2*DHN2-NCH4*DHfCH4
       end
end
%                            PROGRAM END

%%%%%%%%%%%%%%%%%%%%%%%%%%%%%%%%%%%%%%%%%%%%%%%%%%%%%%%%%%%%%%%%%%%%%%%%
>> Ex9_8_AdFlmTemp
Adiabatic flame temp is 2071.91 deg C
```

Description of Program and Results. Note that when an m-file function is used with function `fzero`, the symbol @ must precede the handle for the m-file in the call for function `fzero`. Because ΔH_{N_2}, ΔH_{CO_2} and ΔH_{H_2O} must be recalculated each time a new adiabatic flame temperature is used, an m-file was used instead of an anonymous function.

Python Code for Example 9.8

```
Ex9_8 AdFlmTemp.py
######################################################################

#                          NOMENCLATURE
# A, B, C, D – the heat capacity parameters
# DHf – the heat of formation (J/g mol)
# DH* – the combined heat of formation and sensible heat (J/g mol)
# error – the error in the energy balance equation (J)
```

(Continues)

Example 9.8   Adiabatic Flame Temperature using MATLAB and Python (*Continued*)

```
# N* - the number of moles of component * (g mol)
# T0 - the initial guess for the root of function fun(x)
# TAF - the root of fun(x) determined by function scipy.optimize.newton
########################################################################
#                              PROGRAM
import numpy as np
import scipy.optimize
def fun(T):
    NCO2, NH2O, NN2, NCH4 = 1, 2, 7.524, 1
    A = np.array([36.11, 33.46, 29.00])
    B = np.array([4.233e-2, 0.688e-2, 0.2199e-2])
    C = np.array([-2.887e-5, 0.7604e-5, 0.5723e-5])
    D = np.array([ 7.464e-9, -3.593e-9, -2.871e-9])
            # Input heats of formation
    DHf = np.array([-393510,  -241826, 0.0])
    DHfCH4 = -74840
    DHCO2 = DHf[0]+A[0]*(T-25)+B[0]*(T**2-25**2)/2+C[0]*
            (T**3-25**3)/3+D[0]*(T**4-25**4)/4;
    DHH2O = DHf[1]+A[1]*(T-25)+B[1]*(T**2-25**2)/2+C[1]*
            (T**3-25**3)/3+D[1]*(T**4-25**4)/4;
    DHN2 = DHf[2]+A[2]*(T-25)+B[2]*(T**2-25**2)/2+C[2]*
            (T**3-25**3)/3+D[2]*(T**4-25**4)/4;
    error = NCO2*DHCO2+NH2O*DHH2O+NN2*DHN2-NCH4*DHfCH4;
    return error
T0 = 1000      # Initial guess for adiabatic flame temp (deg C)
    # Input constants for heat capacity equations
TAF = scipy.optimize.newton(fun,T0)
#                            PROGRAM END
########################################################################
```

IPython console:
```
In[1]: runfile(…
In[2]: %precision 2
In[3]: TAF
Out[3]: 2071.91
```

Example 9.9   Production of Citric Acid by a Fungus

Problem Statement

Citric acid ($C_6H_8O_7$) is a well-known compound that occurs in living cells of both plants and animals. The citric acid cycle is a series of chemical reactions occurring in living cells that is essential for the oxidation of glucose, the primary source of energy for the cells. The detailed reaction scheme is far too complicated to show here, but from a macroscopic (overall) viewpoint, for the commercial production of citric acid in a batch (closed) process, three different phases occur for which the stoichiometries are slightly different: Early idiophase (occurs between 80 and 120 hr), initial reaction:

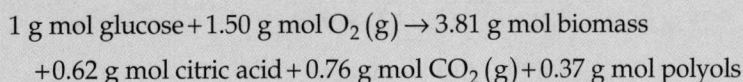

$$1 \text{ g mol glucose} + 1.50 \text{ g mol } O_2\,(g) \rightarrow 3.81 \text{ g mol biomass}$$
$$+ 0.62 \text{ g mol citric acid} + 0.76 \text{ g mol } CO_2\,(g) + 0.37 \text{ g mol polyols}$$

Medium idiophase (occurs between 120 and 180 hr), additional glucose consumed:

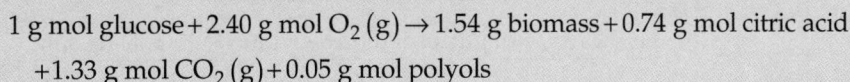

$$1 \text{ g mol glucose} + 2.40 \text{ g mol } O_2\,(g) \rightarrow 1.54 \text{ g biomass} + 0.74 \text{ g mol citric acid}$$
$$+ 1.33 \text{ g mol } CO_2\,(g) + 0.05 \text{ g mol polyols}$$

Late idiophase (occurs between 180 and 220 hr), additional glucose consumed:

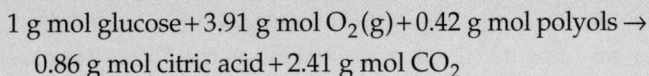

$$1 \text{ g mol glucose} + 3.91 \text{ g mol } O_2\,(g) + 0.42 \text{ g mol polyols} \rightarrow$$
$$0.86 \text{ g mol citric acid} + 2.41 \text{ g mol } CO_2$$

In an aerobic (in the presence of air) batch process, a 30% glucose solution at 25°C is introduced into a fermentation vessel. Citric acid is to be produced by using the fungus *Aspergillus niger*. Stoichiometric sterile air is mixed with the culture solution by a 100 hp aerator. Only 60% overall of the glucose supplied is eventually converted to citric acid. The early phase is run at 32°C, the middle phase at 35°C, and the late phase at 25°C.

Based on the given data, how much net heat has to be added or removed from the fermenter during the production of a batch of 10,000 kg of citric acid?

Solution

The detailed steps of the solution will be merged. Let's omit the inclusion of the respective heats of solution of glucose and citric acid as having minor impact but a complicating role. Because you are interested only in values at the final state and the initial state, the details of the intermediate stage can be ignored for this specific problem (but not for designing the equipment for the dynamic process in which both heating and cooling may be needed). Let citric acid be denoted by CA and the biomass by BM. Given the composition of the biomass (roughly), you can

(Continues)

Example 9.9   Production of Citric Acid by a Fungus (*Continued*)

make a material balance for the overall process in which the overall net reaction of the three steps is

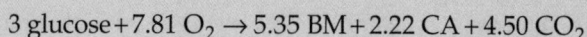

$$3 \text{ glucose} + 7.81 \ O_2 \rightarrow 5.35 \ BM + 2.22 \ CA + 4.50 \ CO_2$$

Assuming that overall conversion is divided equally among the three idiophases.
     CA material balance:

Basis: 10,000 kg of CA produced

$$\frac{10{,}000 \text{ kg CA}}{} \left| \frac{1 \text{ kg mol CA}}{192.12 \text{ kg CA}} \right. = 52.05 \text{ kg mol CA produced}$$

$$\frac{52.05 \text{ kg mol CA}}{} \left| \frac{3 \text{ kg mol glucose}}{2.22 \text{ kg mol CA}} \right| \frac{1.00 \text{ kg mol glucose at start}}{0.60 \text{ kg mol glucose consumed}}$$

$$= 117.23 \text{ kg mol glucose}$$

$$\frac{117.23 \text{ kg mol glucose}}{} \left| \frac{180.16 \text{ kg glucose}}{1 \text{ kg mol glucose}} \right| \frac{1.00 \text{ kg soln}}{0.30 \text{ kg glucose}}$$

$$= 70{,}400 \text{ kg of 30\% solution needed at the beginning of the reaction}$$

Note that the water serves as a culture medium (sort of a constant background) and does not enter the overall stoichiometry. To save space, we show only a summary of the material balances.

Component	Initial (kg mol)	Final (kg mol)
Glucose $(70{,}400)(0.3)/180.16 =$	117.23	46.92
BM $(52.05)(5.35/2.22) =$		125.44
CA		52.05
O_2 $(117.23)(7.8/3) =$	305.03	
CO_2 $(117.23)(4.5/3)(0.60) =$		105.59

We assume that the air (O_2 and N_2) drawn into the system, and the CO_2 and N_2 leaving the system are deemed to be part of the initial state of the system and the final state, respectively, to maintain the presumption of a closed system. The assumption about no flow is not correct, of course, because gas flows in at temperatures not specified and flows out at various temperatures. However, if you were to calculate the difference in the energy associated with the gas flows in an open system, you would find it to be quite insignificant relative to the work done on the system and the heat of reaction.

Energy balance:
     For the closed system,

$$\Delta U = Q + W$$

because the changes in KE and PE inside the system are zero.
The work done is

$$W = \frac{100 \text{ hp}}{} \left| \frac{745.7 \text{ J}}{1(\text{hp})(\text{s})} \right| \frac{220 \text{ hr}}{} \left| \frac{3600 \text{ s}}{1 \text{ hr}} \right| \frac{1 \text{ kJ}}{1000 \text{ J}} = 5.906 \times 10^7 \text{ kJ}$$

Because we do not have a value for U for this system, we have to assume that

$$\Delta U = \Delta H - \Delta(pV) \cong \Delta H \text{ because } \Delta(pV) \text{ is negligible}$$

Then:

The next step is to calculate the enthalpy change. The available data for $\Delta \hat{H}$ are:

Component	MW	ΔH_f^o (kJ/g mol)
d-glucose ($C_6H_{12}O_6$)	180.16	−1266
Citric acid ($C_6H_8O_7$)	192.12	−1544.8
Dry cells (biomass)	28.6	−91.4

The reference temperature will be 25°C. The initial state is 25°C and the final state is also 25°C, so the sensible heats are zero. We omit including the nitrogen in the energy balance because the nitrogen in equals the nitrogen out and the temperature in and out is 25°C.

Component	kg mol	ΔH_f^o (kJ/g mol)	ΔH (kJ)
In			
Glucose	117.32	−1266	$-148{,}530 \times 10^3$
O_2	305.03	0	0
Total			$-148{,}530 \times 10^3$
Out			
Glucose	46.93	−1266	$-59{,}410 \times 10^3$
BM	125.44	−91.4	$-11{,}470 \times 10^3$
CA	52.05	−1544.8	$-80{,}410 \times 10^3$
CO_2	105.59	−393.51	$-41{,}550 \times 10^3$
Total			$-192{,}840 \times 10^3$

$$\Delta H = [(-192{,}840) - (-148{,}530)] \times 10^3 = -44{,}310 \times 10^3 \text{ kJ}$$

$$Q = -4.43 \times 10^7 - 5.91 \times 10^7 = -1.03 \times 10^8 \text{ kJ (heat removed)}$$

Example 9.10 Application of the Energy Balance to a Process Composed of Multiple Units

Problem Statement

Figure E9.10a shows a process in which CO is burned with 80% of the theoretical air in Reactor 1. The combustion gases are used to generate steam, and also to transfer heat to the reactants in Reactor 2. A portion of the combustion gases that are used to heat the reactants in Reactor 2 are recycled. SO_2 is oxidized in Reactor 2. You are asked to calculate the pound moles of CO burned per hour in Reactor 1. Note: The gases involved in the SO_2 oxidation do not come in direct contact with the combustion gas used to heat the SO_2 reactants and products.

Data for Reactor 2 pertaining to the SO_2 oxidation are

Reactants	Mol fr.	T (°F)
SO_2	0.667	77
O_2	0.333	77
	1.00	
Products		
SO_3	0.586	1000
SO_2	0.276	1000
O_2	0.138	1000
	1.000	

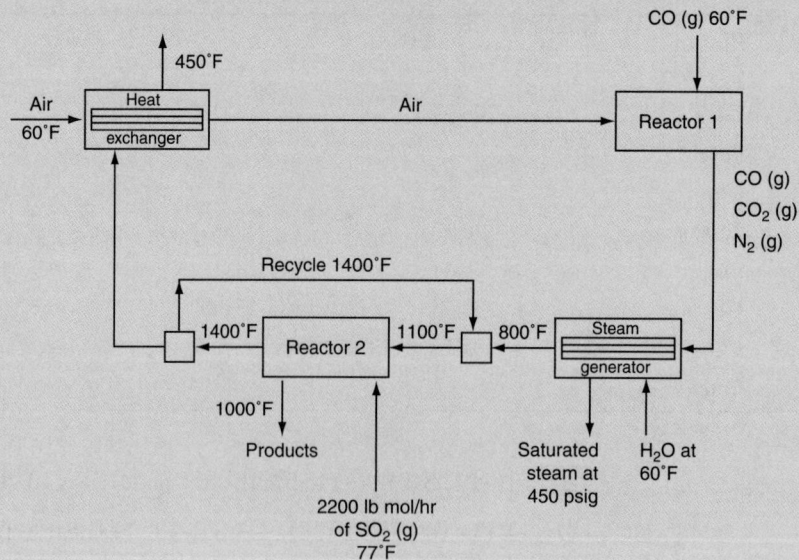

Figure E9.10a

Solution

A number of key decisions must be made before starting any calculations to solve this problem.

1. What units should you use? Although all of the data in the diagram are in the AE system, the convenient enthalpy data are in the SI system; hence, we use the latter. The temperature conversions are

T (°F)	T (K)
77	298
800	700
1000	811
1400	1033

2. What system should you choose to start with in the calculations? Selection of the entire process as the system will lead to about 15 unknowns and require 15 independent equations. It makes sense to look for a system to begin with that involves fewer unknowns and equations. If you examine some of the possible subsystems in the process, the selection of Reactor 2 excluding the local recycle stream proves to be a reasonable choice. It includes SO_2 directly, and the input and output temperatures are known, but we must first determine the composition of the exit stream from Reactor 1.

3. What basis should be used? The given amount of material is the 2200 lb mol/hr of SO_2. A more convenient basis would be 1 g mol of CO entering so that we can first determine the composition of the combustion gas stream leaving Reactor 1. Both bases are suitable we will use the latter, and at the end of the calculations we can convert to a pound mole of CO per hour based on 2200 lb mol/hr of SO_2.

In addition to these three decisions, we have to make some assumptions:

1. The process and its components are continuous, steady-state flow systems; hence, $\Delta U_{total} = 0$ throughout.

2. The pressure everywhere is 1 atm.

3. No heat loss occurs ($Q = 0$ in any energy balances).

4. $\Delta KE = \Delta PE = W = 0$ in any energy balances.

(Continues)

Example 9.10   Application of the Energy Balance to a Process Composed of Multiple Units (*Continued*)

Step 5

$$\text{Basis: 1 g mol of CO entering Reactor 1}$$

Step 3

On the selected basis, you have to calculate the O_2, N_2, CO, and CO_2 in the combustion gases from Reactor 1 that pass through the entire process. No O_2 exists in the combustion products because excess air was not used. Thus, the combustion gases are

$$\text{Basis: 1 g mol CO(g) entering Reactor 1}$$

Reaction:	$CO + 1/2\, O_2 \rightarrow CO_2$
Entering O_2:	$1\,(0.5)(0.8) = 0.4$ g mol
Entering N_2	$0.4\,(79/21) = 1.5$ g mol

Summary of the output of Reactor 1

Component	g mol	Mol fr.
CO	$1(1 - 0.8) = 0.2$	0.08
CO_2	$1(0.8) = 0.8$	0.32
N_2	1.5	0.60
Total	**2.5**	**1.00**

Let's next look at Reactor 2. A figure helps in the analysis; examine Figure E9.10b. The system selected excludes the recycle stream.

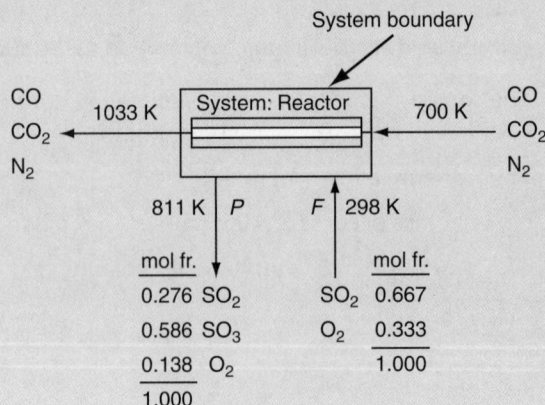

Figure E9.10b

Steps 6 and 7

The degree-of-freedom analysis for Reactor 2 is as follows: We can assign values to the entering and exiting gases in the CO, CO_2, N_2 stream from previous calculations. We know the entering and exit temperatures and pressures. We also know the values of the mole fractions of the input and output streams F and P, so we can calculate the values of $n_{SO_2}^F$, $n_{O_2}^F$, $n_{SO_2}^P$, $n_{O_2}^P$ from the specified mole fractions; for example, $n_{SO_2}^F = 0.667\ F$. The temperatures in and out are specified.

The *unknowns* for the system in Figure E9.10b are $n_{SO_2}^F$, $n_{O_2}^F$, $n_{SO_2}^P$, $n_{SO_3}^P$, $n_{O_2}^P$, ξ, F, and P, a total of eight. How many *independent equations* are there? A typical analysis for the material balances would be as follows:

    a. SO_2, SO_3, and O_2 species balances: 3

    b. Sum of component moles to get P and F (redundant):   0

    c. Specifications ($n_{SO_2}^F$, $n_{O_2}^F$, $n_{SO_2}^P$, $n_{SO_3}^P$, $n_{O_2}^P$):   5

    But wait! The three species balances are

$$SO_2: \quad P(0.276) - F(0.667) = -1(\xi) \tag{a}$$

$$SO_3: \quad P(0.586) - F(0) = 1(\xi) \tag{b}$$

$$O_2: \quad P(0.138) - F(0.333) = -0.5(\xi) \tag{c}$$

Only two are independent. Try to solve Equations (a), (b), and (c), and you will see that this statement is true. Thus, one more equation is needed. What equation can you use? Use an energy balance!

If you carry out the degree-of-freedom analysis using element balances, seven unknowns exist (ξ is not involved), and the equations comprise two element balances plus five specifications.

Steps 8 and 9

To make an energy balance, you have to get information about the heats of formation and the sensible heats using the resources mentioned earlier. With the assumptions made, the energy balance reduces to $\Delta H = 0$.

(Continues)

Example 9.10   Application of the Energy Balance to a Process Composed of Multiple Units (*Continued*)

Data and Calculations for the Energy Balance

Comp	g mol	T (K)	ΔH_f^o (kJ/g mol)	$\Delta \hat{H}_{\text{sensible}}$ (kJ/g mol)	ΔH (kJ)
Out					
$CO(g)$	0.2	1033	−109.054	35.332	−14.744
$CO_2(g)$	0.8	1033	−393.250	35.178	−286.458
$N_2(g)$	1.5	1033	0	22.540	33.810
$SO_2(g)$	$n_{SO_2}^P$	811	−296.855	20.845	$-276.010\, n_{SO_2}^P$
$SO_3(g)$	$n_{SO_3}^P$	811	−395.263	34.302	$-360.961\, n_{SO_3}^P$
$O_2(g)$	$n_{O_2}^P$	811	0	16.313	$16.313\, n_{O_2}^P$
In					
$CO_2(g)$	0.8	700	−393.250	17.753	−300.398
$N_2(g)$	1.5	700	0	11.981	17.972
$SO_2(g)$	$n_{SO_2}^F$	298	−296.855	0	$-296.855\, n_{SO_2}^F$
$O_2(g)$	$n_{O_2}^F$	298	0	0	0

The energy balance is

$$-276.010 n_{SO_2}^P - 360.961 n_{SO_3}^P + 296.855 n_{SO_2}^F + 16.313 n_{O_2}^P = -33.41 n_{SO_2}^F \qquad \text{(d)}$$

Step 9

Substitute for the variables in Equation (d) the following to get Equation (d) in terms of F and P:

$$n_{SO_2}^P = P(0.276) \quad n_{O_2}^P = P(0.138)$$
$$n_{SO_3}^P = P(0.586) \quad n_{SO_2}^F = F(0.667)$$

Solve Equation (d) in terms of F and P together with Equations (a) and (b) (using MATLAB or Python) to get (the n_i are in gram moles):

$$n_{SO_2}^P = 0.312 \quad n_{SO_2}^F = 0.974$$
$$n_{SO_2}^P = 0.662 \quad n_{O_2}^F = 0.486$$
$$n_{O_2}^P = 0.156$$
$$P = 1.33 \qquad \xi = 0.66$$
$$F = 1.46$$

Step 10

Check the solution using the redundant oxygen balance:

$$O: \quad 2n^P_{SO_2} + 3n^P_{SO_3} + 2n^P_{O_2} - 2n^F_{O_2} - 2n^F_{SO_2} = 0$$

$$2(0.312) + 3(0.662) + 2(0.156) - 2(0.486) - 2(0.974) = 0$$

Finally, we have to calculate the pound moles of CO per hour that flow into Reactor 1. Because no loss of combustion gases occurs up to and through Reactor 2, you know (on the basis of 1 g mol of CO entering Reactor 1) that

$$\frac{1 \text{ lb mol CO}}{0.974 \text{ lb mol } SO_2} \left| \frac{2200 \text{ lb mol } SO_2}{\text{hr}} \right. = 2259 \frac{\text{lb mol CO}}{\text{hr}}$$

Although this is all that was requested from the problem statement, the remainder of the unknowns can also be determined sequentially. For example, an energy balance on the air heat exchanger will yield the temperature of the air entering Reactor 1. Then an energy balance on Reactor 1 can be used to calculate the temperature of the combustion gases leaving Reactor 1.

Self-Assessment Test

Questions

1. Explain why omitting the term for the heat of reaction at standard conditions in the energy balance does not prevent you from obtaining the proper value of ΔH to use in the energy balance.

2. What are some of the advantages and disadvantages of calculating the enthalpy change of a compound in the process for use in the energy balance by the method described in Section 9.3 versus that in Section 9.2?

Answers

1. Using the heat of reaction in the energy balance, the heats of formation are lumped together. When the heats of formation are merged with the enthalpies of the sensible heat and phase changes, the components of the heat of reaction are split up and no longer appear as an amalgamated term.

2. Advantages: easier to use for problems in which (1) multiple reactions occur, (2) the reaction equation(s) is unknown, (3) standard software is used. Disadvantages: (1) when the heats of formation of a compound are unknown and cannot be estimated, (2) when only experimental data is available for the reaction enthalpy.

Problems

1. What is the heat transfer to or from a reactor in which methane reacts completely with oxygen to form carbon dioxide gas and water vapor? Base your calculations on a feed of 1 g mol of $CH_4(g)$ at 400 K plus 2 g mol of $O_2(g)$ at 25°C. The exit gases leave at 1000 K. Use the method of this section.

2. Repeat problem 1, but assume that the fractional conversion is only 60%.

3. Calculate the heat added to or removed from a reactor in which stoichiometric amounts of $CO(g)$ and $H_2(g)$ at 400°C react to form $CH_3OH(g)$ (methanol) at 400°C.

Answers

1. $Q = 1(-393.51 + 32.37) + 2(-241.826 + 26.02) - (-74.84) = -717.91$ kJ (removed)

2. $Q = 0.6(-393.51 + 32.37) + 0.6*2(-241.826 + 26.02) + 0.4*2(22.4) + 0.4*(-74.84 + 38.5) - (-74.84) = -397.43$ kJ (removed)

3. $Q = 1(-201.45 + 17.19) - 1(-110.52 + 11.45) - 2(10.84) = -106.87$ kJ (removed)

9.4   The Heat (Enthalpy) of Combustion

An older method of calculating enthalpy changes when chemical reactions occur is via **standard heats (enthalpies) of combustion**, $\Delta \hat{H}_c^o$, which have a different set of reference conditions from the standard heats of formation. The conventions used with the standard heats of combustion are as follows:

1. The compound is oxidized with oxygen or some other substance to the products $CO_2(g)$, $H_2O(l)$, $HCl(aq)$, and so on.

2. The reference conditions are still 25°C and 1 atm.

3. Zero values of $\Delta \hat{H}_c^o$ are assigned to certain of the oxidation products, for example, $CO_2(g)$, $H_2O(l)$, $HCl(aq)$, and to $O_2(g)$ itself.

4. If other oxidizing substances are present, such as S, N_2, or Cl_2, it is necessary to make sure that states of the products are carefully specified and are identical to (or can be transformed into) the final conditions that determine the standard state.

5. Stoichiometric quantities react completely.

The major difference between the standard heats of formation and the standard heats of combustion is item 3 in the preceding list. Certain products have zero values for the heat of combustion, while certain reactants have zero values for reactants.

The heat of combustion has been proposed as one of the criteria to de-termine the ranking of the incinerability of hazardous waste. The rationale is that if a compound has a higher heat of combustion, it can release more en-ergy than other compounds during combustion and would be easier to in-cinerate. Heats of combustion are frequently reported for bioreactions rather than heats of reaction.

How to convert a value of the standard heat of combustion of a com-pound to the corresponding value for the standard heat of formation is shown in the next example.

Example 9.11   Conversion of a Standard Heat of Combustion to the Corresponding Standard Heat of Formation

Problem Statement

The heat of combustion of liquid (>17°C) lactic acid ($C_3H_6O_3$) varies somewhat in the literature. We use −1361 kJ/g mol [*Electronic J. Biotech.*, **7**, No. 2 (2004)]. What is the corresponding standard heat of formation?

Solution

ΔH_f^o is determined by the chemical equation

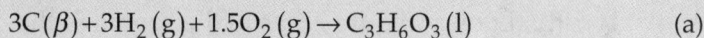

$$3C(\beta) + 3H_2(g) + 1.5O_2(g) \rightarrow C_3H_6O_3(l) \tag{a}$$

The procedure is to start with the chemical equation that gives $\Delta \hat{H}_c^o$ [Equa-tion (b)] and add or subtract other chemical equations whose standard heat(s) of combustion is known so that the final result is the desired Equation (a).

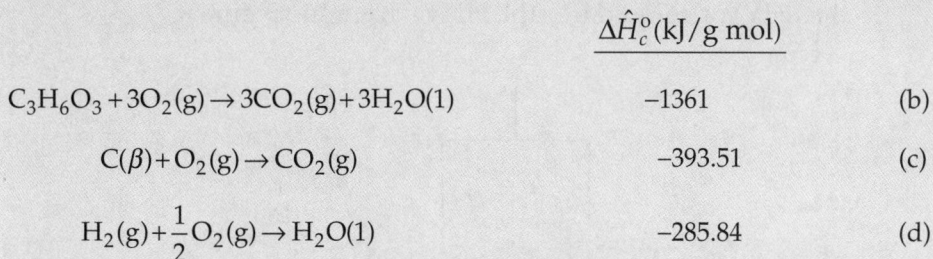

$$\Delta \hat{H}_c^o (\text{kJ/g mol})$$

$$C_3H_6O_3 + 3O_2(g) \rightarrow 3CO_2(g) + 3H_2O(1) \qquad\qquad -1361 \tag{b}$$

$$C(\beta) + O_2(g) \rightarrow CO_2(g) \qquad\qquad -393.51 \tag{c}$$

$$H_2(g) + \frac{1}{2}O_2(g) \rightarrow H_2O(1) \qquad\qquad -285.84 \tag{d}$$

Equation (d) includes a phase change for water (see Figure 9.6).

Equation (a) can be obtained by combining Equations (b), (c), and (d) as follows:

−(b) + 3(c) + 3(d); accordingly, the heats of combustions can also be combined in the same proportions to obtain

$$1361 + 3(-393.51) + 3(-285.84) = \Delta \hat{H}_f^o = -677 \text{ kJ/g mol}$$

$$
\Delta H^{\circ}_{\text{vap}} \text{ at } 25^{\circ}C \\
\text{and 1 atm} \\
= 44{,}000 \text{ J/g mol}
\left\{
\begin{array}{l}
\text{H}_2\text{O(l) } 25^{\circ}C\text{, 1 atm} \qquad\qquad\qquad H_1 = 0 \text{ J/g mol} \\[4pt]
\downarrow \qquad \Delta H_1 \cong 0 \text{ J/g mol} \\[4pt]
\text{H}_2\text{O(l) } 25^{\circ}C\text{, vapor pressure at } 25^{\circ}C \quad H_2 = 0 \text{ J/g mol} \\[4pt]
\downarrow \qquad \Delta H_2 = \Delta H_{\text{vap}} \text{ at the vapor pressure of water} \\[4pt]
\qquad\qquad (p = 3.17 \text{ kPa}) = 44{,}004 \text{ J/g mol} \\[4pt]
\text{H}_2\text{O(g) } 25^{\circ}C\text{, vapor pressure at } 25^{\circ}C \quad H_3 = 44{,}004 \text{ J/g mol} \\[4pt]
\downarrow \qquad \Delta H_3 = -4 \text{ J/g mol} \\[4pt]
\text{H}_2\text{O(g) } 25^{\circ}C\text{, 1 atm} \qquad\qquad\qquad H_4 = 44{,}000 \text{ J/g mol}
\end{array}
\right.
$$

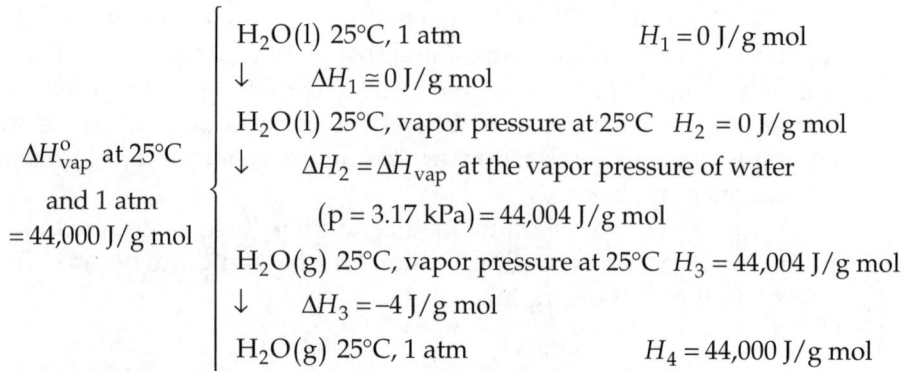

Figure 9.6   The enthalpy change that occurs when $H_2O(l)$ goes from 25°C and 1 atm to $H_2O(g)$ at 25°C and 1 atm. Look at the steam tables on the sheet in the back of the book if the steps are not clear.

The precise heat of vaporization for a compound such as water at 1 atm and 25°C can be calculated as shown in Figure 9.6, but the value of heat of vaporization of a compound at 25°C and the vapor pressure of the compound will suffice for most engineering calculations. Note that the heat of combustion of $H_2O(l)$ is zero, but for $H_2O(g)$ it is –44.00 kJ/g mol. You have to subtract the heat of vaporization of water at 25°C and 1 atm from the value for $H_2O(l)$. (Look at Figure 9.6 as to the source of the $\Delta \hat{H}_{\text{vap}} = 44.00$ kJ/g mol.)

For a fuel such as coal or oil, the negative of the standard heat of combustion is known as the heating value of the fuel. Both a lower (net) heating value (LHV) and a higher (gross) heating value (HHV) exist, depending on whether the water in the combustion products is in the form of a vapor (for the LHV) or a liquid (for the HHV). Examine Figure 9.7.

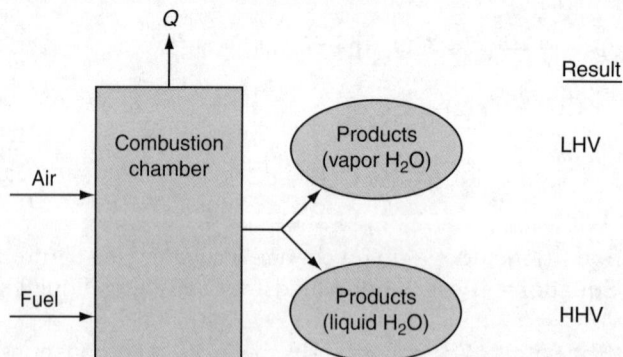

Figure 9.7   The classification of LHV or HHV for a fuel depends on the state of the water exiting from the system.

$$\text{HHV} = \text{LHV} + \left(n_{H_2O(g)}\text{in product}\right)\left(\Delta\hat{H}_{vap} \text{ at } 25°C \text{ and } 1 \text{ atm}\right)$$

You can calculate a standard heat of reaction using the heats of combustion by an equation analogous to Equation (9.1). For **complete combustion:**

$$\Delta H_{rxn}^o\,(25°C) = -\left(\overset{\text{Products}}{\sum} n_i\Delta\hat{H}_{c,i}^o - \overset{\text{Reactants}}{\sum} n_i\Delta\hat{H}_{c,i}^o\right) \qquad (9.7)$$

Note: The minus sign in front of the summation expression occurs because the choice of reference states is zero for the right-hand products of the standard reaction. Refer to Appendix C for values of $\Delta\hat{H}_c^o$.

As an example of the calculation of $\Delta H_{rxn}^o\,(25°C)$ from heat of combustion data, we calculate $\Delta H_{rxn}^o\,(25°C)$ for the reaction (the data are taken from Appendix C in kilojoules per gram mole).

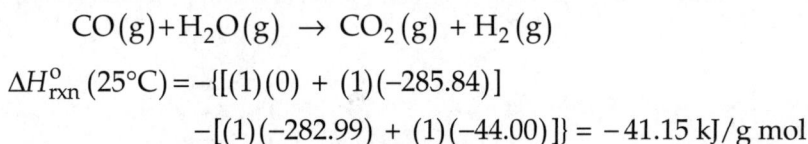

$$CO(g) + H_2O(g) \;\rightarrow\; CO_2(g) + H_2(g)$$

$$\Delta H_{rxn}^o\,(25°C) = -\{[(1)(0)\,+\,(1)(-285.84)]$$
$$-[(1)(-282.99)\,+\,(1)(-44.00)]\} = \,-41.15 \text{ kJ/g mol}$$

Calculation of the heat of reaction for fuels and compounds that have a complicated analysis requires you to use empirical formulas to estimate $\Delta\hat{H}_{rxn}^o$. You can estimate the heating value of a coal within about 3% from the Dulong formula:[5]

Higher heating value (HHV) in Btu per pound

$$= 14{,}544\,C + 62{,}028\left(H - \frac{O}{8}\right) + 4050\,S$$

where C, H, S, and O are the respective *weight* fractions in the fuel.

$$\text{net Btu/lb coal} = \text{gross Btu/lb coal} \;-\; (91.23)(\% \text{ total H by weight})$$

The HHV of fuel oils in British thermal units per pound can be approximated by

$$\text{HHV} = 17{,}887 + 57.5°\text{API} - 102.2\,(\%S).$$

---

[5] H. H. Lowry (ed.), *Chemistry of Coal Utilization*, Wiley, New York (1945), Chapter 4.

Example 9.12 Heating Value of Coal

Problem Statement

Coal gasification consists of the chemical transformation of solid coal into a combustible gas. For many years before the widespread introduction of natural gas, gas generated from coal served as an urban fuel (and also as an illuminant). The heating values of coals differ, but the higher the heating value, the higher the energy value of the gas produced. The analysis of the following coal has a reported heating value of 29,770 kJ/kg as received. Assume that this is the gross heating value at 1 atm and 25°C obtained in an open system.[6] Use the Dulong formula to check the validity of the reported value.

Component	Percent
C	71.0
H_2	5.6
N_2	1.6
Net S	2.7
Ash	6.1
O_2	13.0
Total	**100.0**

Solution

$$\text{HHV} = 14,544\,(0.71) + 62,028\left[(0.056) - \frac{0.130}{8}\right] + 4050\,(0.027) = 12,901 \text{ Btu/lb}$$

Note that 0.056 lb of H_2 is still 0.056 lb of H, and 0.130 lb of O_2 is 0.130 lb of O, because the percent values are the mass of the element no matter how many atoms are joined together as a molecule.

$$\frac{12,901 \text{ Btu}}{\text{lb}} \left|\frac{1 \text{ lb}}{0.454 \text{ kg}}\right| \frac{1.055 \text{ kJ}}{1 \text{ Btu}} = 29,980 \text{ kJ/kg}$$

The two values are quite close.

---

[6] If the reported value was obtained in a closed system, as might be the case, the value reported might be $\Delta \hat{U}$, not $\Delta \hat{H}$.

Example 9.13   Selection of a Fuel to Reduce SO$_2$ Emissions

Problem Statement

SO$_2$ emissions from power plants are the primary source of acid rain. SO$_2$ in the atmosphere is absorbed by tiny water droplets that eventually end up as rain, thus increasing the acidity of lakes and rivers. Consider the two fuels listed in the following table. Determine which fuel would be preferred to provide 10^6 Btu of thermal energy from combustion while minimizing the SO$_2$ emissions. SO$_2$ removal equipment for flue gas can be installed to reduce the SO$_2$ discharge, but at additional cost, bringing into play another important factor in choosing a fuel.

	No. 6 Fuel Oil	No. 2 Fuel Oil
Density (lb/ft^3)	60.2	58.7
Lower heating value (Btu/gal)	155,000	120,000
Carbon (wt %)	87.2	87.3
Hydrogen (wt %)	10.5	12.6
Sulfur (wt %)	0.72	0.62
Ash (wt %)	0.04	

Solution

$$\text{Basis: } 10^6 \text{ Btu from combustion}$$

For No. 6 heating oil:

$$\frac{10^6 \text{ Btu}}{} \left| \frac{1 \text{ gal}}{155,000 \text{ Btu}} \right| \frac{\text{ft}^3}{7.481 \text{ gal}} \left| \frac{60.2 \text{ lb fuel}}{\text{ft}^3} \right| \frac{0.0072 \text{ lb S}}{1 \text{ lb fuel}} = 0.374 \text{ lb S}$$

For No. 2 heating oil:

$$\frac{10^6 \text{ Btu}}{} \left| \frac{1 \text{ gal}}{120,000 \text{ Btu}} \right| \frac{\text{ft}^3}{7.481 \text{ gal}} \left| \frac{58.7 \text{ lb fuel}}{\text{ft}^3} \right| \frac{0.0062 \text{ lb S}}{1 \text{ lb fuel}} = 0.405 \text{ lb S}$$

The No. 6 heating oil should be selected because its combustion will generate less SO$_2$ emissions, even though it has a higher weight percent S due to its larger heating value.

Self-Assessment Test

Questions

1. Explain why, in calculating ΔH using heats of formation, you subtract the ΔH of the inputs from the ΔH of the outputs, whereas in using heat of combustion you subtract the ΔH of the outputs from that of the inputs.

2. Can the HHV ever be the same as the LHV?

3. Can the HHV ever be lower than the LHV?

4. Do you have to use heats of combustion to calculate an HHV or an LHV?

5. Is it true for the reaction $H_2(g) + 1/2O_2(g) \rightarrow H_2(g)$ at 25°C and 1 atm, which has a cited heat of reaction of –241.83 kJ, that if you write the reaction as $2H_2(g) + O_2(g) \rightarrow 2H_2O(g)$, the calculated heat of reaction will be 2(–241.83) kJ?

6. What is the difference in HHV and LHV when CO is burned with O_2 at 25°C and 1 atm?

Answers

1. The reference for the heats of combustion is zero for the right side of the chemical equation, whereas the reference for the heats of formation is zero for the left side of the chemical equation.

2. Yes, if none of the product compounds can undergo a phase transition (e.g., if no water is formed).

3. No

4. No, you can use the heats of formation.

5. If you pick the same basis for the heat of reaction, namely, 1 g mol H_2, it makes no difference. However, if you want the heat of reaction for the chemical reaction equation as written, the heat of reaction for the second reaction equation will be two times the first.

6. None, because water is not involved and CO_2 will not undergo a phase change.

Problems

1. A synthetic gas analyzes 6.1% CO_2, 0.8% C_2H_4, 0.1% O_2, 26.4% CO, 30.2% H_2, 3.8% CH_4, and 32.6% N_2. What is the heating value of the gas in British thermal units per cubic foot measured at 60°F, saturated when the barometer reads 30.0 in. Hg?

2. Calculate the standard heat of reaction using heat of combustion data for the reaction

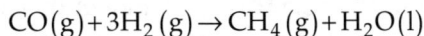

$$CO(g) + 3H_2(g) \rightarrow CH_4(g) + H_2O(l)$$

Answers

1. Basis: 1 ft³; $n = pV/RT = (1\ ft^3)(1.0026\ atm)/((0.7303)(419.6\ °R)) = 0.003272$ lb mol; C_2H_4: 0.003272 lb mol(0.008)(1410.99kJ/g mol)(429.95 Btu-g mol/kJ-lb mol) = 15.88 Btu/ft³; CO: 105.47; H_2: 121.44; CH_4: 47.60; Total = 290.39 Btu/ft³

2. Using heats of combustion, the overall reaction can be represented by the following three combustion reactions:

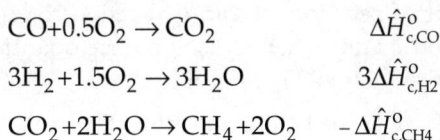

$$CO + 0.5O_2 \rightarrow CO_2 \qquad\qquad \Delta\hat{H}^\circ_{c,CO}$$

$$3H_2 + 1.5O_2 \rightarrow 3H_2O \qquad\qquad 3\Delta\hat{H}^\circ_{c,H2}$$

$$CO_2 + 2H_2O \rightarrow CH_4 + 2O_2 \qquad -\Delta\hat{H}^\circ_{c,CH4}$$

Therefore, the heat of reaction is $-282.99 + 3(-285.84)(-890.4) = -250.11$ kJ/g mol

Summary

This chapter introduced the concept of the heat of formation and showed how it can be used to solve energy balances with reaction using two techniques: (1) the heat of reaction approach or (2) an approach that combines the heat of formation with sensible heat changes. The heat of combustion, which contains the same information as the heat of formation but on a different basis, was presented and used to solve energy balance problems with reaction.

Glossary

endothermic reaction   A reaction for which heat must be added to the system to maintain isothermal conditions.

exothermic reaction   A reaction for which heat must be removed from the system to maintain isothermal conditions.

heating value   The negative of the standard heat of combustion for a fuel such as coal or oil.

heat of reaction   Enthalpy change that is associated with a reaction.

higher (gross) heating value (HHV)   Value of the negative heat of combustion when the product water is a liquid.

lower (net) heating value (LHV)   Value of the negative of the heat of combustion when the product water is a vapor.

reference state   For enthalpy, it is the state at which the enthalpy is zero.

sensible heat   The quantity $(\hat{H}_{T,P} - \hat{H}_{ref}^{o})$ that excludes any phase changes.

standard heat of combustion   Enthalpy change for the oxidation of 1 mol of a compound at 25°C and 1 atm. By definition, the water and carbon dioxide on the right side of the reaction equation are assigned zero values.

standard heat of formation   Enthalpy change for the formation of 1 mol of a compound from its constituent elements at 25°C and 1 atm.

standard heat of reaction   Heat of reaction for components in the standard state (25°C and 1 atm) when stoichiometric quantities of reactants react completely to give products at 25°C and 1 atm.

standard state   For the heat of reaction, 25°C and 1 atm.

Supplemental References

Benedek, P., and F. Olti. *Computer Aided Chemical Thermodynamics of Gases and Liquids Theory, Model and Programs*, Wiley-Interscience, New York (1985).

Benson, S. W., et al. "Additivity Rules for the Estimation of Thermophysical Properties," *Chem Rev.*, **69**, 279 (1969) [cited in Reid, R. C., et al. *The Properties of Gases and Liquids*, 4th ed., McGraw-Hill, New York (1987)].

Danner, R. P., and T. E. Daubert. "Manual for Predicting Chemical Process Design Data," *Documentation Manual*, Chapter 11 "Combustion," American Institute of Chemical Engineers, New York (1987).

Daubert, T. E., et al. *Physical and Thermodynamic Properties of Pure Chemicals: Data Compilation*, American Institute of Chemical Engineers, New York, published by Taylor and Francis, Bristol, PA, periodically.

Garvin, J. "Calculate Heats of Combustion for Organics," *Chem. Eng. Progress*, 43–45 (May, 1998).

Kraushaar, J. J. *Energy and Problems of a Technical Society*, John Wiley, New York (1988).

Poling, B. E., J. M. Prausnitz, and J. P. O'Connell. *The Properties of Gases and Liquids*, 5th ed., McGraw-Hill, New York (2001).

Rosenberg, P. *Alternative Energy Handbook*, Association of Energy Engineers, Lilbum, GA (1988).

Seaton, W. H., and B. K. Harrison. "A New General Method for Estimation of Heats of Combustion for Hazard Evaluation," *Journal of Loss Prevention*, **3**, 311–20 (1990).

Vaillencourt, R. *Simple Solutions to Energy Calculations*, Association of Energy Engineers, Lilbum, GA (1988).

Problems

9.1   The Standard Heat (Enthalpy) of Formation

*9.1.1   Which of the following heats of formation would indicate an endothermic reaction?

    a. −32.5 kJ

    b. 32.5 kJ

    c. −82 kJ

    d. 82 kJ

    e. Both a and c

    f. Both b and d

*9.1.2   Which of the following changes of phase is exothermic?

    a. Gas to liquid

    b. Solid to liquid

    c. Solid to gas

    d. Liquid to gas

*9.1.3   Exothermic reactions are usually self-sustaining because of which of the following explanations?

    a. Exothermic reactions usually require low activation energies.

    b. Exothermic reactions usually require high activation energies.

    c. The energy released is sufficient to maintain the reaction.

    d. The products contain more potential energy than the reactants.

**9.1.4   How would you determine the heat of formation of gaseous fluorine at 25°C and 1 atm?

**9.1.5   Determine the standard heat of formation for FeO(s) given the following values for the heats of reaction at 25°C and 1 atm for the following reactions:

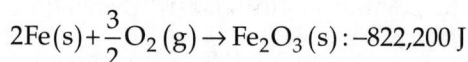

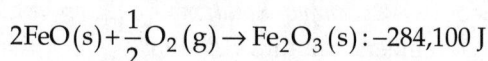

$$2Fe(s) + \frac{3}{2}O_2(g) \rightarrow Fe_2O_3(s) : -822,200 \text{ J}$$

$$2FeO(s) + \frac{1}{2}O_2(g) \rightarrow Fe_2O_3(s) : -284,100 \text{ J}$$

***9.1.6   The following enthalpy changes are known for reactions at 25°C and 1 atm:

No.				$\Delta H°$ (kJ/g mol)
1	$C_3H_6(g) + H_2(g)$	$\rightarrow$	$C_3H_8(g)$	−123.8
2	$C_3H_8(g) + 5O_2(g)$	$\rightarrow$	$3CO_2(g) + 4H_2O(l)$	−2220.0
3	$H_2(g) + \frac{1}{2}O_2(g)$	$\rightarrow$	$H_2O(l)$	−285.8
4	$H_2O(l)$	$\rightarrow$	$H_2O(g)$	43.9
5	$C(\text{diamond}) + O_2(g)$	$\rightarrow$	$CO_2(g)$	−395.4
6	$C(\text{graphite}) + H_2(g)$	$\rightarrow$	$CO_2(g)$	−393.5

Calculate the heat of formation of propylene (C_3H_6).

***9.1.7   In a fluidized bed gasification system, you are asked to find out the heat of formation of a solid sludge of the composition (formula) C_5H_2 from the following data:

			ΔH (kJ/g mol)
$C(s) + \frac{1}{2}O_2(g)$	$\rightarrow$	$CO(g)$	−110.4 kJ/g mol C
$C(s) + O_2(g)$	$\rightarrow$	$CO_2(g)$	−394.1 kJ/g mol C
$H_2(g) + \frac{1}{2}O_2(g)$	$\rightarrow$	$H_2O(g)$	−241.826 kJ/g mol H_2
$CO(g) + \frac{1}{2}O_2(g)$	$\rightarrow$	$CO_2(g)$	−283.7 kJ/g mol CO
$H_2O(g) + CO(g)$	$\rightarrow$	$H_2O(g) + CO_2(g)$	−38.4 kJ/g mol H_2O
$C_5H_2(s) + 5\frac{1}{2}O_2(g)$	$\rightarrow$	$5CO_2(g) + H_2O(l)$	−2110.5 kJ/g mol C_5H_2
ΔH vaporization H_2O at 25°C			+43.911 kJ/g mol H_2O_2

***9.1.8   Look up the heat of formation of (a) liquid ammonia, (b) formaldehyde gas, (c) acetaldehyde liquid. If you can't find the value, get data to calculate it.

9.2 The Heat (Enthalpy) of Reaction

*9.2.1 Calculate the standard (25°C and 1 atm) heat of reaction per gram mole of the first reactant on the left side of the reaction equation for the following reactions:

    a. $NH_3(g) + HCl(g) \rightarrow NH_4Cl(s)$

    b. $CH_4(g) + 2O_2(g) \rightarrow CO_2(g) + 2H_2O(l)$

    c. $C_6H_{12}(g) \rightarrow C_6H_6(l) + 3H_2(g)$

*9.2.2 Calculate the standard heat of reaction for the following reactions:

    a. $CO_2(g) + H_2(g) \rightarrow CO(g) + H_2O(l)$

    b. $2CaO(s) + 2MgO(s) + 4H_2O(l) \rightarrow 2Ca(OH)_2(s) + 2Mg(OH)_2(s)$

    c. $Na_2SO_4(s) + C(s) \rightarrow Na_2SO_3(s) + CO(g)$

    d. $NaCl(s) + H_2SO_4(l) \rightarrow NaHSO_4(s) + HCl(g)$

    e. $4NaCl(s) + 2SO_2(g) + 2H_2O(l) + O_2(g) \rightarrow 2Na_2SO_4(s) + 4HCl(g)$

    f. $SO_2(g) + \frac{1}{2}O_2(g) + H_2O(l) \rightarrow H_2SO_4(l)$

    g. $N_2(g) + O_2(g) \rightarrow 2NO(g)$

    h. $Na_2CO_3(s) + 2Na_2S(s) + 4SO_2(g) \rightarrow 3Na_2S_2O_3(s) + CO_2(g)$

    i. $CS_2(l) + 2Cl_2(g) \rightarrow S_2Cl_2(l) + CCl_2(l)$

    j. $\underset{\text{ethylene}}{C_2H_4(g)} + HCl(g) \rightarrow \underset{\text{ethyl chloride}}{CH_3CH_2Cl(g)}$

    k. $\underset{\text{methyl alcohol}}{CH_3OH(g)} + \frac{1}{2}O_2(g) \rightarrow \underset{\text{formaldehyde}}{H_2CO(g)} + H_2O(g)$

    l. $\underset{\text{acetylene}}{C_2H_2(g)} + H_2O(l) \rightarrow \underset{\text{acetaldehyde}}{CH_3CHO(l)}$

    m. $\underset{\text{butane}}{n\text{-}C_4H_{10}(g)} \rightarrow \underset{\text{ethylene}}{C_2H_4(g)} + \underset{\text{ethance}}{C_2H_6(g)}$

9.2.3 J. D. Park et al. [*JACS* **72, 331-3 (1950)] determined the heat of hydrobromination of propene (propylene) and cyclopropane. For hydrobromination (addition of HBr) of propene to 2-bromopropane, they found that $\Delta H = -84,441\ J/g\ mol$. The heat of hydrogenation of propene to propane is $\Delta H = -126,000\ J/g\ mol$. N.B.S. Circ. 500 gives the heat of formation of HBr(g) as $-36,233\ J/g\ mol$ when the bromine is liquid, and the heat of vaporization of bromine as $30,710\ J/g\ mol$. Calculate the heat of bromination of propane using gaseous bromine to form 2-bromopropane using these data.

**9.2.4   In the reaction

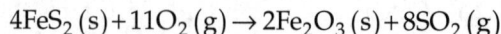

$$4FeS_2\,(s) + 11O_2\,(g) \rightarrow 2Fe_2O_3\,(s) + 8SO_2\,(g)$$

the conversion of $FeS_2(s)$ to $Fe_2O_3(s)$ is only 80% complete. If the standard heat of reaction for the reaction is calculated to be -567.4 kJ/g mol FeS_2 (s), what value of ΔH^o_{rxn} will you use per kilogram of FeS_2 burned in an energy balance?

9.2.5   M. Beck et al. [*Can J. Ch.E.*, **64, 553 (1986)] described the use of immobilized enzymes (E) in a bioreactor to convert glucose (G) to fructose (F):

$$G + E \leftrightarrow EG \leftrightarrow E + F$$

At equilibrium the overall reaction can be considered to be $G + E \leftrightarrow E + F$.

Glucose	**Fructose**
CHO	CH$_2$OH
$\|$	$\|$
CHOH	C = O
$\|$	$\|$
CHOH $\Rightarrow$	CHOH
$\|$	$\|$
CHOH	CHOH
$\|$	$\|$
CHOH	CHOH
$\|$	$\|$
CH$_2$OH	CH$_2$OH
$\Delta \hat{H}^o_f : 0.990 \times 10^9$ J/g mol	$\Delta \hat{H}^o_f : 1.040 \times 10^9$ J/g mol

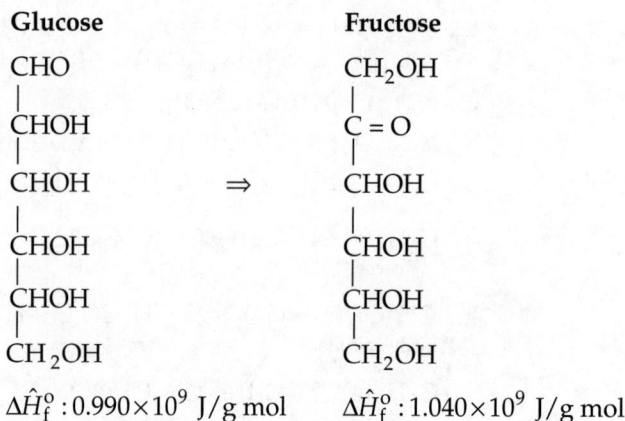

The fractional conversion is a function of the flow rate through the reactor and the size of the reactor, but for a flow rate of 3×10^{-3} m/s and a bed height of 0.44 m, the fractional conversion on a pass through the reactor was 0.48. Calculate the heat of reaction at 25°C per mole of G converted.

**9.2.6   A consulting laboratory is called upon to determine the heat of reaction at 25°C of a natural gas in which the combustible is entirely methane. They do not have a Sargent flow calorimeter but do have a Parr bomb calorimeter. They pump a measured volume of the natural gas into the Parr bomb, add oxygen to give a total pressure of 1000 kPa, and explode the gas-O_2 mixture with a hot wire. From the data, they calculate that the gas has a heating value of 39.97 kJ/m^3. Should they report this value as the heat of reaction? Explain. What value should they report?

**9.2.7   Is it possible to calculate the heat of reaction by using property tables that are prepared using different reference states? How?

**9.2.8 A fat is a glycerol molecule bonded to a combination of fatty acids or hydrocarbon chains. Usually, the glycerol bonds to three fatty acids, forming a triglyceride. A fat that has no double bonds between the carbon atoms in the fatty acid is said to be saturated.

How are fats treated by the human digestive system? They are first enzymatically broken down into smaller units, fatty acids and glycerol. This is called digestion and occurs in the intestine or cellularly by lysosomes. Next, enzymes remove two carbons at a time from the carboxyl end of the chain, a process that produces molecules of acetyl CoA, NADH, and $FADH_2$. The acetyl CoA is a high-energy molecule that is then treated by the citric acid cycle which oxidizes it to CO_2 and H_2O.

Calculate the heat of reaction when tristearin is used in the body.

Data:	$\Delta \hat{H}_f^o$ (kJ / g mol)
Stearic acid (s) $(C_{18}H_{36}O_2)$, a fatty acid	−964.3
Glycerol (1) $(C_3H_8O_3)$	−159.16
Tristearin(s) $(C_{63}H_{132}O_{15})$ a triglyceride	−3.820

**9.2.9 Answer the following questions briefly (in no more than three sentences):
   a. Does the addition of an inert diluent to the reactants entering an exothermic process increase, decrease, or make no change in the heat transfer to or from the process?
   b. If the reaction in a process is incomplete, what is the effect on the value of the standard heat of reaction? Does it go up, go down, or remain the same?
   c. Consider the reaction $H_2\,(g) + \frac{1}{2}O_2\,(g) \rightarrow H_2O(g)$. Is the heat of reaction with the reactants entering and the products leaving at 500 K higher, lower, or the same as the standard heat of reaction?

**9.2.10 Compute the heat of reaction at 600 K for the following reaction:

$$S(l) + O_2\,(g) \rightarrow SO_2\,(g)$$

**9.2.11 Calculate the heat of reaction at 500°C for the decomposition of propane:

$$C_3H_8 \rightarrow C_2H_2 + CH_4 + H_2$$

**9.2.12 Calculate the heat of reaction of the following reactions at the stated temperature:

   a. $\underset{\text{methyl alcohol}}{CH_3OH(g)} + \frac{1}{2}O_2(g) \xrightarrow{\;200°C\;} \underset{\text{formaldehyde}}{H_2CO(g)} + H_2O(g)$

   b. $SO_2(g) + \frac{1}{2}O_2(g) \xrightarrow{\;300°C\;} SO_3(g)$

**9.2.13   In a new process for the recovery of tin from low-grade ores, it is desired to oxidize stannous oxide, SnO, to stannic oxide, SnO_2, which is then soluble in a caustic solution. What is the heat of reaction at 90°C and 1 atm for the reaction $SnO + \frac{1}{2}O_2 \rightarrow SnO_2$? Data are as follows:

	$\Delta \hat{H}_f^\circ$ (kJ / g mol)	C_p [J / (g mol)(K)], T in K
SnO	−283.3	$39.33 + 15.15 \times 10^{-3}\, T$
SnO_2	−577.8	$73.89 + 10.04 \times 10^{-3}\, T - \dfrac{2.16 \times 10^4}{T^2}$

**9.2.14   One hundred gram moles of CO at 300°C are burned with 100 g mol of O_2, which is at 100°C; the exit gases leave at 400°C. What is the heat transfer to or from the system in kilojoules?

**9.2.15   The composition of a strain of yeast cells is determined to be $C_{3.92}H_{6.5}O_{1.94}$, and the heat of combustion was found to be −1518 kJ/g mol of yeast. Calculate the heat of formation of 100 g of the yeast.

***9.2.16   Physicians measure the metabolic rate of conversion of foodstuffs in the body by using tables that list the liters of O_2 consumed per gram of foodstuff. For a simple case, suppose that glucose reacts

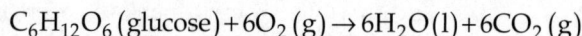

$$C_6H_{12}O_6 \text{(glucose)} + 6O_2 \text{(g)} \rightarrow 6H_2O \text{(l)} + 6CO_2 \text{(g)}$$

How many liters of O_2 would be measured for the reaction of 1 g of glucose (alone) if the conversion were 90% complete in your body? How many kilojoules per gram of glucose would be produced in the body?

Data: $\Delta \hat{H}_f^\circ$ of glucose is −1260 kJ/g mol of glucose. Ignore the fact that your body is at 37°C and assume it is at 25°C.

***9.2.17   Calculate the heat of reaction in the standard state for 1 mol of C_3H_8(l) for the following reaction from the given data:

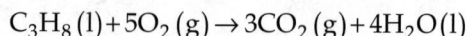

$$C_3H_8 \text{(l)} + 5O_2 \text{(g)} \rightarrow 3CO_2 \text{(g)} + 4H_2O \text{(l)}$$

Compound	$-\Delta \hat{H}_f^\circ$ (kcal/g mol)	Vaporization at 25° C (kcal/g mol)
C_3H_8(g)	24.820	3.823
CO_2(g)	94.052	1.263
H_2O(g)	57.798	10.519

***9.2.18   Hydrogen is used in many industrial processes, such as the production of ammonia for fertilizer. Hydrogen also has been considered to have a potential as an energy source because its combustion yields a clean product and it is easily stored in the form of a metal hydride. Thermo-chemical cycles (a series of reactions resulting in a recycle of some of the reactants) can be used in the production of hydrogen from an abundant natural compound water. One process involving a series of five steps is outlined in Figure P9.2.18. State assumptions about the states of the compounds.

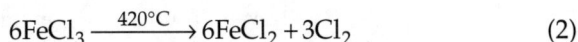

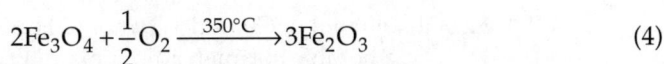

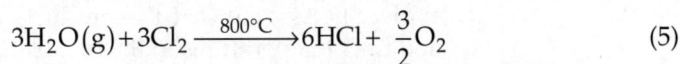

$$3Fe_2O_3 + 18HCl \xrightarrow{120°C} 6FeCl_3 + 9H_2O(g) \tag{1}$$

$$6FeCl_3 \xrightarrow{420°C} 6FeCl_2 + 3Cl_2 \tag{2}$$

$$6FeCl_2 + 8H_2O(g) \xrightarrow{650°C} 2Fe_3O_4 + 12HCl + 2H_2 \tag{3}$$

$$2Fe_3O_4 + \frac{1}{2}O_2 \xrightarrow{350°C} 3Fe_2O_3 \tag{4}$$

$$3H_2O(g) + 3Cl_2 \xrightarrow{800°C} 6HCl + \frac{3}{2}O_2 \tag{5}$$

a. Calculate the standard heat of reaction for each step.
b. What is the overall reaction? What is its standard heat of reaction?

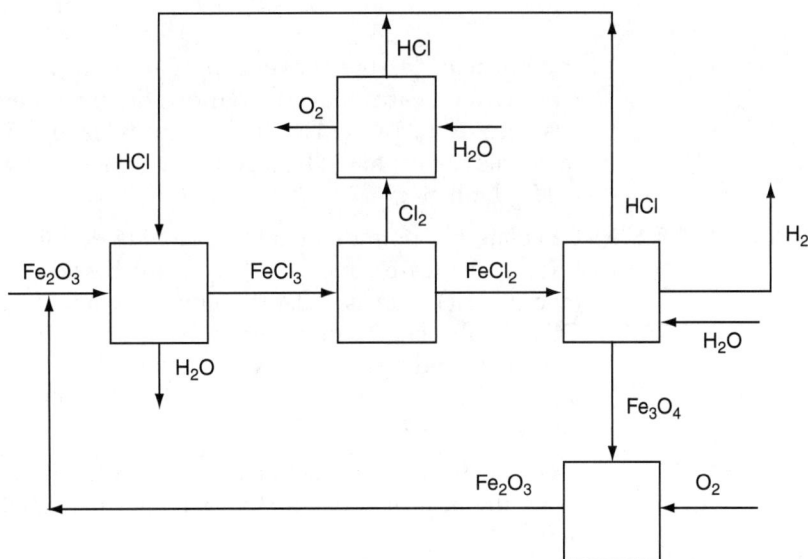

Figure P9.2.18

***9.2.19** About 30% of crude oil is processed eventually into automobile gaso-line. As petroleum prices rise and resources dwindle, alternatives must be found. However, automobile engines can be tuned so that they will run on simple alcohols. Methanol and ethanol can be derived from coal or plant biomass, respectively. While the alcohols produce fewer pollutants than gasoline, they also reduce the travel radius of a tank of fuel. What percent increase in the size of a fuel tank is needed to give an equivalent travel radius if gasoline is replaced by alcohol? Base your calculations on 40 kJ/g of gasoline having a specific gravity of 0.84 and with the product water as a gas. Make the calculations for (a) methanol; (b) ethanol.

***9.2.20 Calculate the heat transfer for the following reaction if it takes place at constant volume and a constant temperature of 25°C:

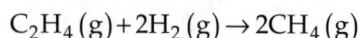

$$C_2H_4(g) + 2H_2(g) \rightarrow 2CH_4(g)$$

***9.2.21 If 1 lb mol of Cu and 1 lb mol of H_2SO_4 (100%) react together completely in a bomb calorimeter, how many British thermal units are absorbed (or evolved)? Assume that the products are $H_2(g)$ and $CuSO_4$. The initial and final temperatures in the bomb are 25°C.

***9.2.22 Calculate the standard heat of reaction for the conversion of cyclohexane to benzene:

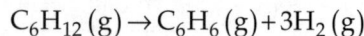

$$C_6H_{12}(g) \rightarrow C_6H_6(g) + 3H_2(g)$$

If the reactor for the conversion operates at 70% conversion of C_6H_{12}, what is the heat removed from or added to the reactor if (a) the exit gases leave at 25°C and (b) the exit gases leave at 300°C and the entering materials consist of C_6H_{12} together with one-half mole of N_2 per mole of C_6H_{12}, both at 25°C?

***9.2.23 Calculate the pounds of carbon dioxide emitted per gallon of fuel for three fuels: (a) ethanol (C_2H_5OH), (b) benzene (C_6H_6), and (c) isooctane (C_8H_{18}), that is, 2,2,4-trimethyl pentane. Compare the pounds of CO_2 per British thermal unit and the British thermal units per gallon of fuel for the respective reactions with the stoichiometric quantity of air entering the process at 100°C and the other compounds entering and leaving at 77°F.

***9.2.24 Calculate the heat of reaction at standard conditions for 1 g mol of $H_2(g)$ using the heat of combustion data, and then calculate ΔH_{rxn} for $H_2(g)$ at 0°C.

***9.2.25 Yeast cells can be grown in a culture in which glucose is the sole source of carbon to produce cells that are up to 50% by weight of the glucose consumed. Assume the following chemical reaction equation represents the process:

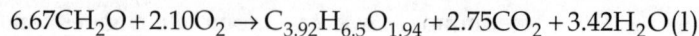

$$6.67CH_2O + 2.10O_2 \rightarrow C_{3.92}H_{6.5}O_{1.94} + 2.75CO_2 + 3.42H_2O(l)$$

The formula for glucose is $C_6H_{12}O_6$; hence CH_2O is directly proportional to 1 mol of glucose. Given the following data for the heat of formation:

	$\Delta \hat{H}_f^o$ (kJ / g mol)	MW
Dry cells ($C_{3.92}H_{6.5}O_{1.94}$)	−1517	84.58
Glucose (CH_2O)	−2817	30.02

Calculate the standard heat of reaction per 100 g of dry cells.

9.3   Integration of Heat of Formation and Sensible Heat

***9.3.1   A synthesis gas at 500°C that analyzes 6.4% CO_2, 0.2% O_2, 40.0% CO, 50.8% H_2, and the balance N_2 is burned with 40% dry excess air which is at 25°C. The composition of the flue gas which is at 720°C is 13.0% CO_2, 14.3% H_2O, 67.6% N_2, and 5.1% O_2. What was the heat transfer to or from the combustion process?

***9.3.2   Dry coke composed of 4% inert solids (ash), 90% carbon, and 6% hydrogen at 40°C is burned in a furnace with dry air at 40°C. The solid refuse at 200°C that leaves the furnace contains 10% carbon and 90% inert, and no hydrogen. The ash does not react. The Orsat analysis of the flue gas which is at 1100°C gives 13.9% CO_2, 0.8% CO, 4.3% O_2, and 81.0% N_2. Calculate the heat transfer to or from the process. Use a constant C_p for the inert of 8.5 J/g.

**** 9.3.3   An eight-room house requires 200,000 Btu per day to maintain the interior at 68°F. How much calcium chloride hexahydrate must be used to store the energy collected by a solar collector for one day of heating? The storage process involves heating the salt from 68°F to 86°F and then converting the hexahydrate to dihydrate and gaseous water:

$$CaCl_2 \cdot 6H_2O(s) \rightarrow CaCl_2 \cdot 2H_2O(s) + 4H_2O(g)$$

The water from the dehydration is evaporated during the process. Use the following data:

	$\Delta \hat{H}_f^o$ (kJ/g mol)	C_p (J/(g)(°C))
$CaCl_2 \cdot 6H_2O(s)$	−2607.89	1.34
$CaCl_2 \cdot 2H_2O(s)$	−1402.90	0.97

9.4   The Heat (Enthalpy) of Combustion

**9.4.1 Estimate the higher heating value (HHV) and lower heating value (LHV) of the following fuels in British thermal units per pound:
  a. Coal with the analysis C (80%), H (0.3%), O (0.5%), S (0.6%), and ash (18.60%)
  b. Fuel oil that is 30°API and contains 12.05% H and 0.5% S

**9.4.2 Is the higher heating value of a fuel ever equal to the lower heating value? Explain.

**9.4.3 Find the higher (gross) heating value of $H_2(g)$ at 0°C.

**9.4.4 The chemist for a gas company finds a gas analyzes 9.2% CO_2, 0.4% C_2H_4, 20.9% CO, 15.6% H_2, 1.9% CH_4, and 52.0% N_2. What should the chemist report as the gross heating value of the gas?

***9.4.5 What is the higher heating value of 1 m^3 of n-propylbenzene measured at 25°C and 1 atm and with a relative humidity of 40%?

***9.4.6 An off-gas from a crude oil topping plant has the following composition:

Component	Vol. %
Methane	88
Ethane	6
Propane	4
Butane	2

  a. Calculate the higher heating value on the following bases: (1) Btu per pound, (2) Btu per lb mole, and (3) Btu per cubic foot of off-gas measured at 60°F and 760 mm Hg.
  b. Calculate the lower heating value on the same three bases indicated in part a.

**9.4.7 The label on a 43 g High Energy ("Start the day with High Energy") bar states that the bar contains 10 g of fat, 28 g of carbohydrate, and 4 g of protein. The label also says that the bar has 200 calories per serving (the serving size is 1 bar). Does this information agree with the information about the contents of the bar?

Data:

Component	$\Delta \hat{H}_c^\circ$ (kJ/g)
Carbohydrate	17.1
Fat	39.5
Protein	14

**9.4.8 Under what circumstances would the heat of formation and the heat of combustion have the same value?

***9.4.9 One of the ways to destroy chlorinated hydrocarbons in waste streams is to add a fuel (here toluene waste) and burn the mixture. In a test of the combustion apparatus, 1200 lb of a liquid mixture composed of 0.0487% hexachloroethane (C_2Cl_6), 0.0503% tetrachloroethylene (C_2Cl_4), 0.2952% chlorobenzene (C_6H_5Cl), and the balance toluene was burned completely with air. What was the HHV of the mixture calculated using data for the heats of combustion in British thermal units per pound? Compare the resulting value with the one obtained from the Dulong formula. The observed value was 15,200 Btu/lb. What is one major problem with the incineration process described?

**9.4.10 Gasohol is a mixture of ethanol and gasoline used to increase the oxygen content of fuels and thus reduce pollutants from automobile exhaust. What is the heat of reaction for 1 kg of the mixture calculated using heat of combustion data for a fuel composed of 10% ethanol and the rest octane? How much is the heat of reaction reduced by adding the 10% ethanol to the octane?

PART V

COMBINED MATERIAL AND ENERGY BALANCES

CHAPTER 10

Humidity (Psychrometric) Charts

Chapter Objectives

- Define and understand humidity, dry-bulb temperature, wet-bulb temperature, humidity charts, moist volume, and adiabatic cooling line.
- Use the humidity chart to determine the properties of moist air.
- Calculate enthalpy changes and solve heating and cooling problems involving moist air.

Introduction

Cooling towers are the method of choice for removing excess thermal energy for large-scale processing systems. Cooling towers contact the heated water with ambient air, evaporating a portion of the hot water into the air and thus cooling the water. The cooling water is then used throughout the process to remove heat from process streams (e.g., condensers on distillation columns). Figure 10.1 shows the elements of a cooling tower typically used by the process industries. The contacting of the air with the hot water in the "exchange surface" in Figure 10.1 evaporates a relatively small portion of water into the air, which in turn reduces the temperature of the hot water. The material in this chapter helps you to understand and analytically represent the behavior of cooling towers and other processes involving moist air.

The humidification and drying of air with water is a commonly encountered system. Therefore, instead of applying material and energy balances for these systems, engineers can use psychometric charts developed specifically for air-water systems at 1 atm pressure to represent humidification and drying for these systems. We introduce some new terminology specific to this field, present psychometric charts, and demonstrate how to use psychometric charts to apply material and energy balances.

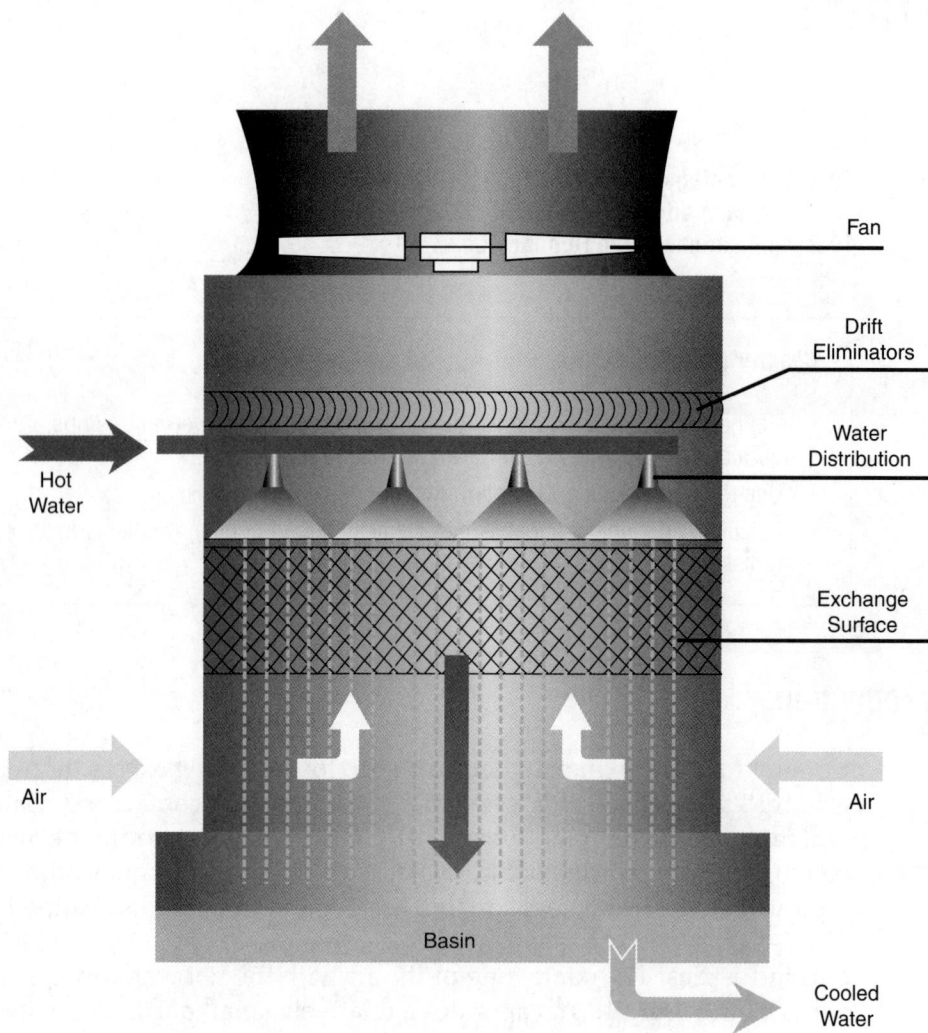

Figure 10.1   Elements of a cooling tower

10.1   Terminology

Following are new terms that relate to humidification and drying.

a.   **Relative saturation (relative humidity)** is defined as

$$RS = \frac{p_{vapor}}{p^*} = \text{relative saturation} \qquad (10.1)$$

where p_{vapor} = partial pressure of the water vapor in the gas mixture.
    p^* = partial pressure of the water vapor in the gas mixture if the gas is saturated at the given temperature of the mixture (i.e., the vapor pressure of water). Then, for brevity, if the subscript 1 denotes the water,

$$RS = \frac{p_1}{p_1^*} = \frac{p_1/p_{tot}}{p_1^*/p_{tot}} = \frac{V_1/V_{tot}}{V_{satd}/V_{tot}} = \frac{n_t}{n_{satd}} = \frac{mass_1}{mass_{satd}} \qquad (10.2)$$

You can see that relative saturation, in effect, represents the fractional approach to total saturation. If you listen to the radio or TV and hear the announcer say that the temperature is 30°C (86°F) and the relative humidity is 60%, the announcer implies that

$$\frac{p_{H_2O}}{p_{H_2O}^*}(100) = \%RH = 60$$

with both the p_{H_2O} and the $p_{H_2O}^*$ being measured at 30°C. Zero percent relative saturation means no water vapor in the gas. What does 100% relative saturation mean? It means that the partial pressure of the water vapor in the gas is the same as the vapor pressure of water (i.e., the gas is saturated with water).

b.   **Humidity** H (specific humidity) is the mass (in pounds or kilograms) of water vapor per unit mass (in pounds or kilograms) of bone-dry air (some texts use moles of water vapor per mole of dry air as the humidity):

$$H = \frac{m_{H_2O}}{m_{dry\ air}} = \frac{18 p_{H_2O}}{29\left(p_{total} - p_{H_2O}\right)} = \frac{18 n_{H_2O}}{29\left(n_{total} - n_{H_2O}\right)} \qquad (10.3)$$

c.   **Dry-bulb temperature** (T_{DB}) is the ordinary temperature you always have been using for a gas in degrees Fahrenheit or Celsius (or Rankine or kelvin).

d. **Wet-bulb temperature** (T_{WB}) is something new. Suppose that you put a wick, or porous cotton cloth, on the mercury bulb of a thermometer and wet the wick. Next you either (1) whirl the thermometer in the air, as in Figure 10.2 (this apparatus is called a *sling psychrometer* when a wet-bulb and a dry-bulb thermometer are mounted together), or (2) set up a fan to blow rapidly on the bulb at a high linear velocity. What happens to the temperature recorded by the wet-bulb thermometer?

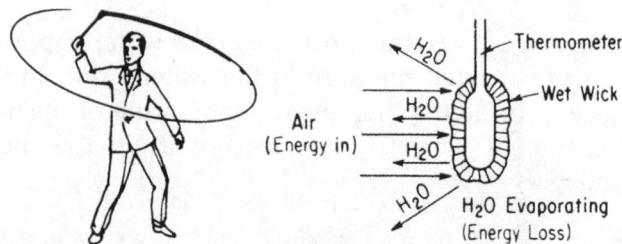

Figure 10.2 Wet-bulb temperature obtained with a sling psychrometer

As the water from the wick evaporates, the wick cools down and continues to cool until the steady-state rate of energy transferred to the wick by the air blowing on it equals the steady-state rate of loss of energy caused by the water evaporating from the wick. We say that the temperature of the bulb when the water in the wet wick is at equilibrium with the water vapor in the air is the wet-bulb temperature. (Of course, if water continues to evaporate, it eventually will disappear, and the wick temperature will rise to the dry-bulb temperature.) The equilibrium temperature for this process will lie on the 100% relative humidity curve (saturated-air curve). Look at Figure 10.3.

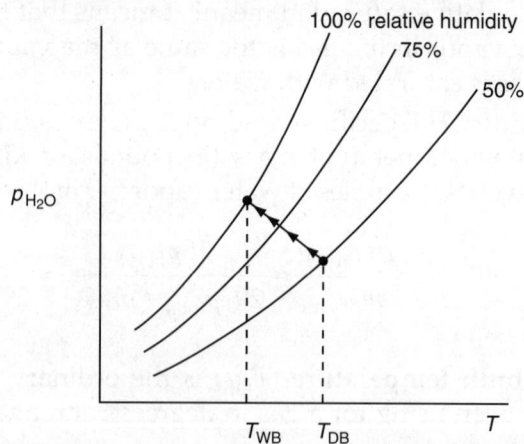

Figure 10.3 Evaporative cooling of the wick initially at T_{DB} causes a thermometer with a moist wick at equilibrium to reach the wet-bulb temperature, T_{WB}.

Example 10.1   Application of Relative Humidity to Calculate the Dew Point

Problem Statement

The weather report on the radio this morning was that the temperature this afternoon would reach 94°F, the relative humidity would be 43%, the barometer was 29.67 in. Hg, partly cloudy to clear, with the wind from SSE at 8 mi/hr. How many pounds of water vapor would be in 1 mi^3 of afternoon air? What would be the dew point of this air?

Solution

The vapor pressure of water at 94°F is 1.61 in. Hg. You can calculate the partial pressure of the water vapor in the air from the given percent relative humidity:

$$p_{H_2O} = (1.61 \text{ in. Hg})(0.43) = 0.692 \text{ in. Hg}$$

Basis: 1 mi^3 of water vapor at 94°F and 0.692 in. Hg

$$\frac{1 \text{ mi}^3}{} \left| \left(\frac{5280 \text{ ft}}{1 \text{ mi}} \right)^3 \right| \frac{492°R}{555°R} \left| \frac{0.692 \text{ in. Hg}}{29.92 \text{ in. Hg}} \right| \frac{1 \text{ lb mol}}{359 \text{ ft}^3} \left| \frac{18 \text{ lb H}_2\text{O}}{1 \text{ lb mol}} \right.$$

$$= 1.52 \times 10^8 \text{ lb H}_2\text{O vapor}$$

Now, the dew point is the temperature at which the water vapor in the air will first condense on cooling at *constant pressure and composition*. As the gas is cooled, you can see from Equation (10.1) that the relative humidity increases because the partial pressure of the water vapor is constant, while the vapor pressure of water decreases with temperature. When the relative humidity reaches 100%,

$$100 \frac{p_{H_2O}}{p^*_{H_2O}} = 100\% \text{ or } p_{H_2O} = p^*_{H_2O} = 0.692 \text{ in. Hg}$$

the water vapor will start to condense. From the steam tables, you can see that this corresponds to a temperature of about 68°F to 69°F.

Self-Assessment Test

Questions

1. On a copy of the *p-H* chart for water, plot the locus of (a) where the dry-bulb temperatures can be located and (b) where the wet-bulb temperatures can be located.

2. Would the dew point temperatures be the locus of the wet-bulb temperatures?

3. Can psychrometric charts exist for mixtures other than water-air?

4. What is the difference between the wet- and dry-bulb temperatures?

5. Can the wet-bulb temperature ever be higher than the dry-bulb temperature?

Answers

1. See Figure 10.3.

2. Yes

3. Yes

4. The wet-bulb temperature is the temperature of the evaporating water that is in equilibrium with the temperature of the ambient air, which is known as the dry-bulb temperature.

5. No

Problems

1. Apply the Gibbs phase rule to an air-water vapor mixture. How many degrees of freedom exist (for the intensive variables)? If the pressure on the mixture is fixed, how many degrees of freedom result?

2. Prepare a chart in which the vertical axis is the humidity (Equation (10.1) in kilograms of H_2O per kilogram of dry air) and the horizontal axis is the temperature (in degrees Celsius). On the chart, plot the curve of 100% relative humidity for water.

Answers

1. Degrees of freedom $= 2 - P + C = 2 - 1 + 2 = 3$. For a fixed pressure, DOF $= 2$.

2. See Figure 10.4.

10.2   The Humidity (Psychrometric) Chart

The **humidity chart,** more formally known as the **psychrometric chart,** relates the various parameters involved in making combined material and energy balances for moist air. In the Carrier chart, the type of chart we will discuss, the vertical axis (usually placed on the right-hand side) is the (specific) humidity, and the horizontal axis is the dry-bulb temperature. Examine Figure 10.4. On this chart, we want to construct several other lines and curves featuring different parameters.

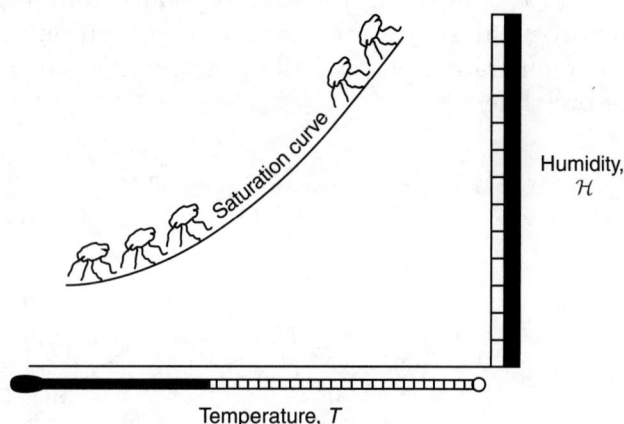

Figure 10.4   Major coordinates of the humidity chart

We start with two of the new concepts and relations you read about in Section 10.1.

10.2.1   Wet-Bulb Line (Equation)

The equation for the **wet-bulb** lines is based on a number of assumptions, a detailed discussion of which is beyond the scope of this book. Nevertheless, as mentioned in Section 10.1, the idea of the wet-bulb temperature is based on the equilibrium between the *rates* of energy transfer to the bulb and the evaporation of water. The fundamental idea is that a large amount of air is brought into contact with a little bit of water and that, presumably, the evaporation of the water leaves the temperature and humidity of the air unchanged. Only the temperature of the water changes. The equation for the wet-bulb line evolves from an energy balance that equates the heat transfer to the water to the heat of vaporization of the water. Figure 10.5 shows several such lines on the psychrometric chart for air-water. The plot of H_{WB} versus T_{WB}, the wet-bulb line, is approximately straight and has a negative slope.

What use is the wet-bulb line (equation)? You require two pieces of information to fix a state (point A in the humidity chart). One piece of information can be T_{DB}, H_{DB}, RS, and so on. The other can be T_{WB}. Where do you locate the value of T_{WB} on the humidity chart? The Carrier chart lists values of T_{WB} along the saturation curve (such as point B). You can also find the same values by projecting vertically upward from a temperature (point C) on the horizontal axis to the saturation curve (point B). If the process is a wet-bulb temperature process, all of the possible states of the process fall on the wet-bulb line.

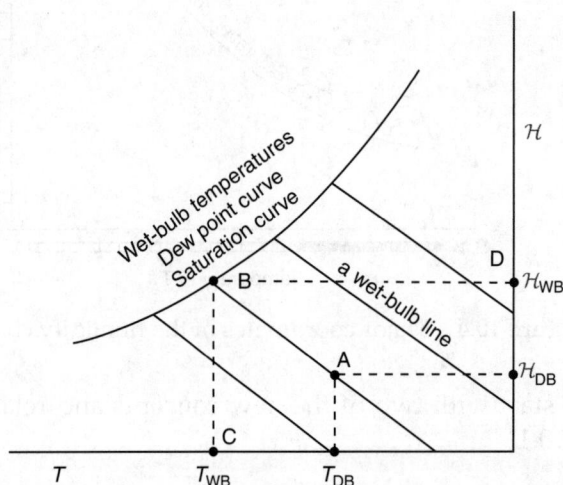

Figure 10.5   Representation of the wet-bulb process on an **H-T** chart

For example, for a known wet-bulb process, given an initial T_{DB} (along the horizontal axis) and H_{DB} (along the vertical axis), the combination fixes a point on the graph (point A). You can follow (up to the left) the wet-bulb process (line) to the saturation curve (point B). Then project vertically downward onto the temperature axis to get the related T_{WB} (point C), and project horizontally to the right to the related H_{WB} (point D). All points along the wet-bulb line are fixed by just one additional piece of information.

10.2.2   Adiabatic Cooling Line (Equation)

Another type of process of some importance occurs when **adiabatic cooling** or **humidification** takes place between air and water that is recycled, as illustrated in Figure 10.6. In this process, the air is both cooled and humidified (its water content rises) while a little bit of the recirculated water is evaporated. Thus, makeup water is added. At **equilibrium,** in the steady state, the temperature of the exit air is the same as the temperature of the water, and the exit air is saturated at this temperature. By making an overall energy balance around the process (with $Q = 0$), you can obtain the equation for the adiabatic cooling of the air.

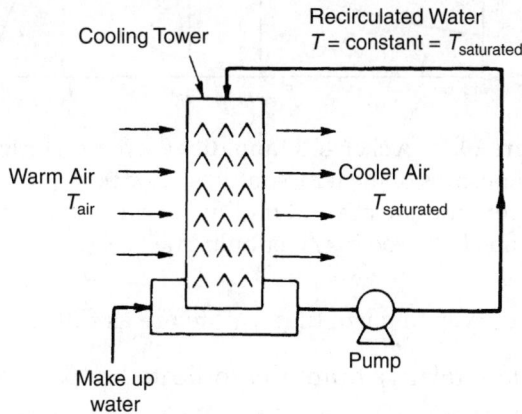

Figure 10.6   Adiabatic humidification with recycle of water

It turns out that the wet-bulb process equation, *for water only,* is essentially the same relation as the adiabatic cooling equation. Thus, you have the nice feature that two processes can be represented by the same set of lines. The remarks we made previously about locating parameters of moist air on the humidity chart for the wet-bulb process apply equally to the adiabatic humidification process. For a detailed discussion of the uniqueness of this coincidence, consult any of the references at the end of the chapter. For most other substances besides water, the respective equations will have different slopes.

In addition to the wet-bulb line being identical to the adiabatic cooling line, let's look at some of the other details found on a humidity chart in which the vertical axis is H and the horizontal axis is T. Look at Figure 10.7.

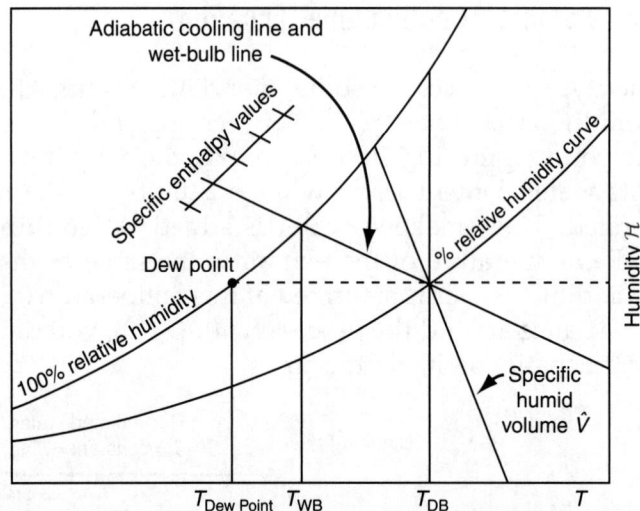

Figure 10.7   A skeleton humidity (psychrometric) chart showing typical relationships of the temperatures, dew point, wet- and dry-bulb temperatures, relative humidity, specific humid volume, humidity enthalpy, and adiabatic cooling/wet-bulb line

Lines and curves to particularly note are as follows:

1. Constant relative humidity indicated in percent

2. Constant moist volume (specific humid volume)

3. Adiabatic cooling lines that are the same (for water vapor only) as the wet-bulb or psychrometric lines

4. The 100% relative humidity curve (i.e., saturated-air curve)

5. The specific enthalpy per mass of dry air (not air plus water vapor) for a saturated air-water vapor mixture:

$$\Delta\hat{H} = \Delta\hat{H}_{air} + \Delta\hat{H}_{H_2O\ vapor}\,(H) \tag{10.4}$$

Enthalpy adjustments for air that is less than saturated (identified by a minus sign) are shown on the chart itself by a series of curves.

Figures 10.8a and 10.8b are reproductions of the Carrier psychrometric charts.

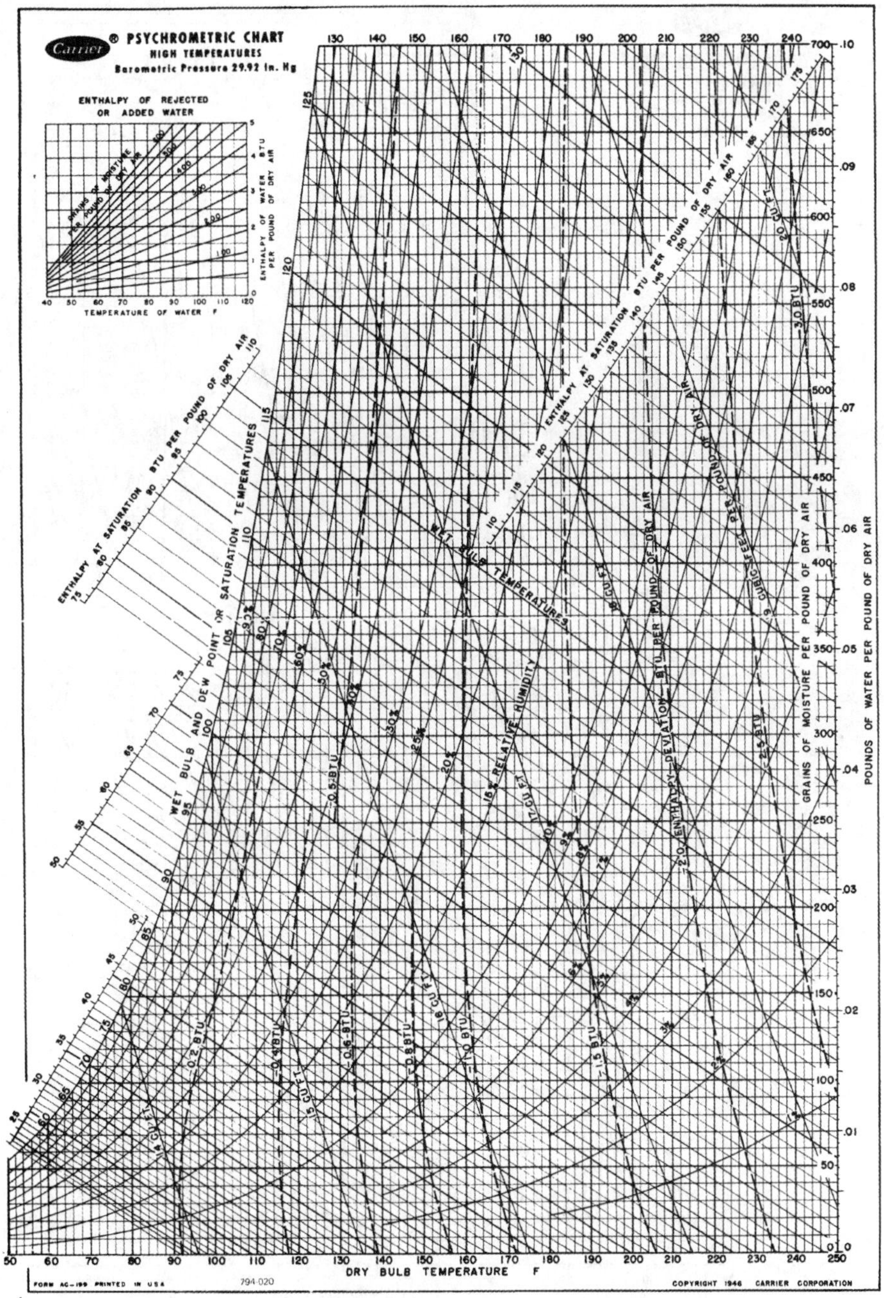

Figure 10.8a   Psychrometric chart in American Engineering units

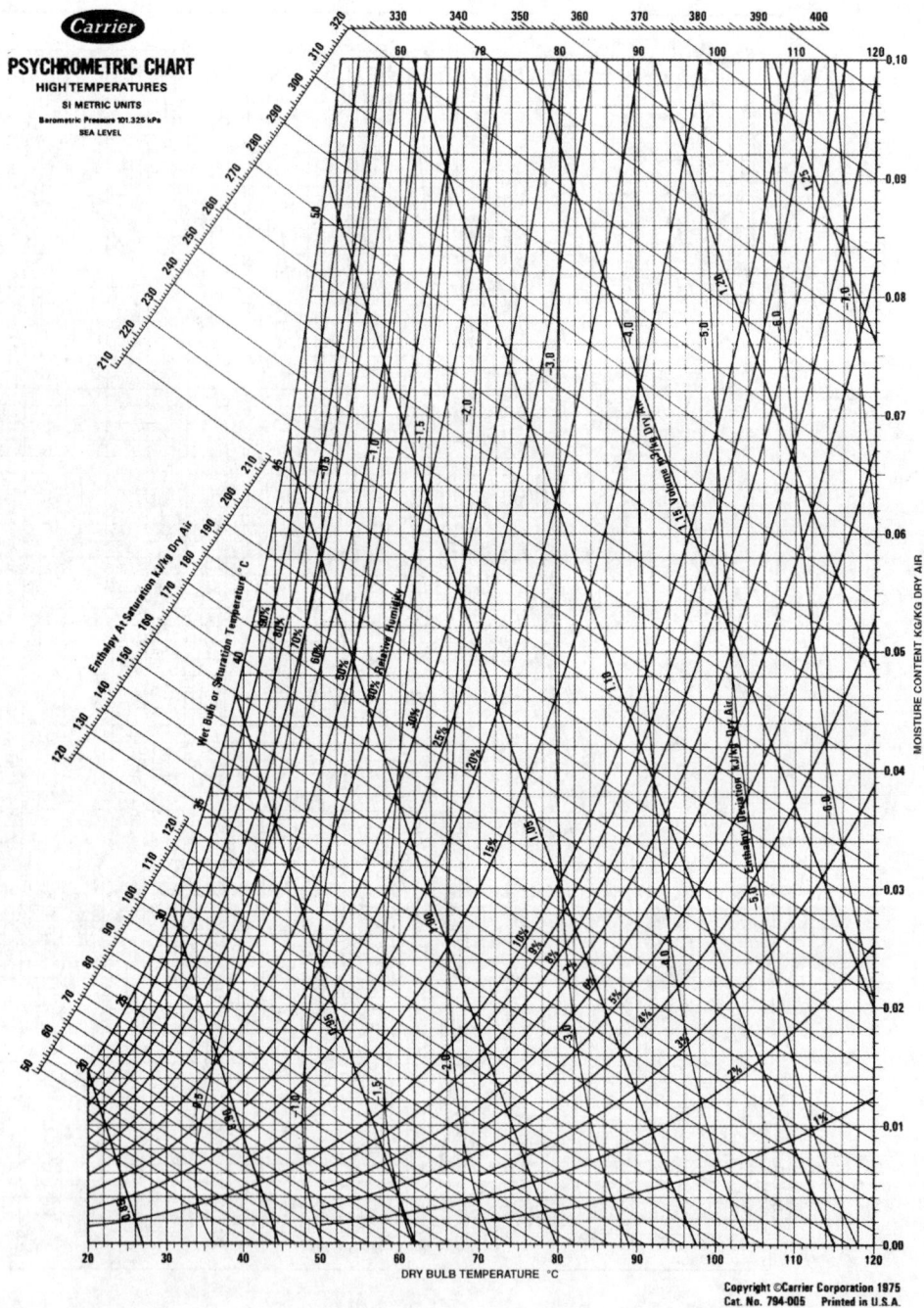

Figure 10.8b Psychrometric chart in SI units

If you analyze the degrees of freedom for the intensive variables via the phase rule, you find

$$F = 2 - P + C = 2 - 1 + 2 = 3$$

Is this a paradox in view of our previous remarks stating that only two parameters are required to fix a point on the humidity chart? No. Basis for the chart is that the *pressure is fixed* at 1 atm. Consequently, $F = 2$, and specification of any two values of the various variables shown on the chart fix a specific point. Values for all of the other variables are correspondingly fixed.

The adiabatic cooling lines are lines of almost constant enthalpy for the entering air-water mixture, and you can use them as such without much error (1% or 2%). Follow the line on the chart up to the left where the values for the enthalpies of saturated air appear. If you want to correct a saturated enthalpy value for the deviation that exists for a less-than-saturated air-water vapor mixture, you can employ the enthalpy deviation lines that appear on the chart, which can be used as illustrated in the following examples.

Any process that is not a wet-bulb process or an adiabatic process with recirculated water can be analyzed using the usual material and energy balances, taking the basic data for the calculations from the humidity charts (or equations). If there is any increase or decrease in the moisture content of the air in a psychrometric process, the small enthalpy effect of the moisture added to the air or lost by the air may be included in the energy balance for the process to make it more exact, as illustrated in the examples.

Tables of the properties shown in the humidity charts exist if you need better accuracy than can be obtained via the charts, and computer programs that you can find on the Internet can be used to retrieve psychrometric data. Although we have been discussing humidity charts exclusively, charts can be prepared for mixtures of any two substances in the vapor phase, such as CCl_4 and air or acetone and nitrogen.

Now let's look at an example of getting data from the humidity chart.

Example 10.2   Determining Properties of Moist Air from the Humidity Chart

Problem Statement

List all of the properties you can find on the humidity chart in American Engineering (AE) units for moist air at a dry-bulb temperature of 90°F and a wet-bulb temperature of 70°F. It is reasonable to assume that someone could measure the dry-bulb temperature using a typical mercury-in-glass thermometer and the wet-bulb temperature using a sling psychrometer.

Solution

A diagram helps to explain the various properties that can be obtained from the humidity chart. See Figure E10.2.

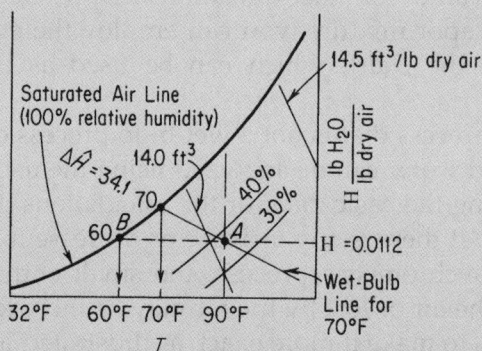

Figure E10.2

You can find the location of point A for 90°F DB (dry-bulb) and 70°F WB (wet-bulb) by following a vertical line from $T_{DB} = 90°F$ until it crosses the wet-bulb line for 70°F. You can locate the end of a wet-bulb line by searching along the 100% relative humidity curve (saturation curve) until the label for the saturation temperature of 70°F is reached. Alternatively, you can proceed up a vertical line at 70°F until it intersects with the 100% humidity line.

Next, from the wet-bulb temperature of 70°F, follow the adiabatic cooling line (which is the same as the wet-bulb temperature line on the humidity chart) down to the right until it intersects the 90°F DB line. Now that point A has been fixed, you can read the other properties of the moist air from the chart.

a. **Dew point:** When the air at A is cooled at constant pressure (and in effect at *constant humidity*), it eventually reaches a temperature at which the moisture begins to condense. This process is represented by a horizontal line, a constant humidity line, on the humidity chart, and the dew point is located at B, or about 60°F.

b. **Relative humidity:** By interpolating between the 40% RH and 30% RH lines, you can find that point A is at about 37% RH.

c. **Humidity (H):** You can read the humidity from the right-hand ordinate as 0.0112 lb H_2O/lb of dry air.

d. **Humid volume:** By interpolation again between the 14.0 ft^3/lb and the 14.5 ft^3/lb lines, you can find the humid volume to be 14.1 ft^3/lb of dry air.

e. **Enthalpy:** The enthalpy value for saturated air with a wet-bulb temperature of 70°F is $\Delta\hat{H} = 34.1$ Btu/lb of dry air (a more accurate value can be obtained from psychrometric tables if needed). The enthalpy deviation shown by the dashed curves in Figure 10.7a (not shown in Figure E10.1) for less than saturated air is about −0.2 Btu/lb of dry air; consequently, the actual enthalpy of air at 37% RH is $34.0 - 0.2 = 33.8$ Btu/lb of dry air.

Frequently Asked Questions

1. For what range of temperatures can a humidity chart be applied? Air conditioning applications range from about −10°C to 50°C. Outside this range of temperatures (and for other pressures), applications may require amended charts.

2. Can air and water vapor be treated as ideal gases in that range? Yes, with an error less than 0.2%.

3. Will the enthalpy of the water vapor be a function of the temperature only? Yes, because it is assumed to be an ideal gas.

4. How can you achieve greater accuracy than the values read from the humidity chart? Use the equations listed in Sections 10.1 and 10.2, or use a computer program.

Self-Assessment Test

Questions

1. What properties of an air-water vapor mixture are displayed on a humidity chart?

2. Explain what happens to both the air and water vapor when an adiabatic cooling process occurs.

3. Wet towels are to be dried in a laundry by blowing hot air over them. Can the Carrier humidity chart be used to help design the dryer?

4. Can a humidity chart use values per mole rather than values per mass?

Answers

1. Moisture content (i.e., pound of water per pound of dry air), dry-bulb temperature, wet-bulb temperature, enthalpy at saturation, cubic feet per pound of dry air, relative humidity

2. For adiabatic cooling, water is evaporated into the air, thus cooling the mixture.

3. Yes

4. Yes

Problems

1. Air at a dry-bulb temperature of 200°F has a humidity of 0.20 mol H_2O/mol of dry air.
   a. What is its dew point?
   b. If the air is cooled to 150°F, what is its dew point?

2. Air at a dry-bulb temperature of 71°C has a wet-bulb temperature of 52°C.
   a. What is its percentage relative humidity?
   b. If this air is passed through a washer-cooler, what would be the lowest temperature to which the air could be cooled without using refrigerated water?

3. Estimate for air at 70°C dry-bulb temperature, 1 atm, and 15% relative humidity:
   a. kg H_2O/kg of dry air
   b. m^3/kg of dry air
   c. dew point (in °C)

Solutions

1. a. 0.2 lb mol water/lb mol of BDA = 0.0124 lb water/lb BDA; from psychrometric chart, dew point = 63°F; b. same

2. a. 39% *RH*; b. 52°C

3. a. 0.03 kg H_2O/kg BDA; b. 16.25 ft^3/lb BDA = 1.015 m^3/kg BDA; c. dew point = 31.7°C

10.3   Applications of the Humidity Chart

Quite a few industrial processes exist for which you can involve the properties found on a humidity chart, including

- **Drying:** Dry air enters, absorbs moisture from a material, and moist air leaves the process.
- **Humidification:** Liquid water is vaporized into moist air.
- **Combustion:** Moist air enters a process, and additional water is added to the moist air from the combustion products.
- **Air conditioning:** Moist air is cooled.
- **Condensation:** Moist air is cooled to the saturation temperature, and additional cooling condenses water from the moist air.

Example 10. 3   Heating at Constant Humidity in a Home Furnace

Problem Statement

Moist air at 38°C and 49% RH is heated in your home furnace to 86°C. How much heat has to be added per cubic meter of initial moist air, and what is the final dew point of the air?

Solution

As shown in Figure E10.3, the process has an input state at point A at $T_{DB} = 38°C$ and 49% RH, and an output state at point B, which is located at the intersection of a horizontal line of constant humidity with the vertical line up from $T_{BD} = 86°C$. The dew point is unchanged in this process because the humidity is unchanged and is located at C at 24.8°C.

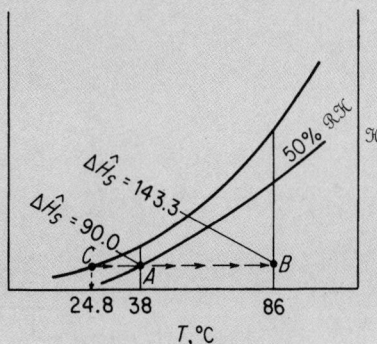

Figure E10.3

(Continues)

Example 10. 3   Heating at Constant Humidity in a Home Furnace (*Continued*)

The enthalpy values are as follows (all in kilojoules per kilogram of dry air):

Point	$\Delta \hat{H}_{said}$	δH	$\Delta \hat{H}_{actual}$
A	90.0	−0.5	89.5
B	143.3	−3.3	140.0

The value for the reduction of $\Delta \hat{H}_{actual}$ is obtained from the enthalpy deviation lines (not shown in Figure E10.2) whose values are printed about a quarter of the way down from the top of the chart. Also, at A, the volume of the moist air is 0.91 m³/kg of dry air. Consequently, the heat added is (the energy balance reduces to $Q = \Delta \hat{H}$) 140.0 − 89.5 = 50.5 kJ/kg of dry air.

$$\frac{50.5 \text{ kJ}}{\text{kg dry air}} \left| \frac{1 \text{ kg dry air}}{0.91 \text{ m}^3} \right. = 55.5 \text{ kJ/m}^3 \text{ initial moist air}$$

Example 10.4   Cooling and Humidification Using a Water Spray

Problem Statement

One way of adding moisture to air is by passing it through water sprays or air washers. See Figure E10.4a. Normally, the water used is recirculated rather than wasted. Then, in the steady state, the water is at its adiabatic saturation temperature, which is the same as the wet-bulb temperature. The air passing through the washer is cooled, and if the contact time between the air and the water is long enough, the air will be at the wet-bulb temperature also.

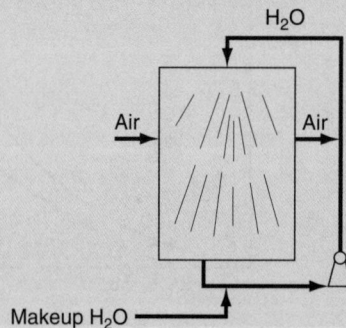

Figure E10.4a

However, we shall assume that the washer is small enough so that the air does not reach the wet-bulb temperature; instead, the following conditions prevail:

	T_{DB}(°C)	T_{WB}(°C)
Entering air:	40	22
Exit air:	27	

Find the moisture added in kilograms per kilogram of dry air going through the humidifier.

Solution

The whole process is assumed to be adiabatic. As shown in Figure E10.4b, the process inlet conditions are at A. The outlet state is at B, which occurs at the intersection of the adiabatic cooling line (the same as the wet-bulb line) with the vertical line at $T_{DB} = 27$°C. The wet-bulb temperature remains constant at 22°C. Humidity values (from Figure 10.8b) are

$$H_B = 0.0145 \quad H_A = 0.0093 \quad \text{where} \quad H\left(\frac{\text{kg H}_2\text{O}}{\text{kg dry air}}\right)$$

$$\text{Difference: } 0.0052 \frac{\text{kg H}_2\text{O}}{\text{kg dry air}} \text{ added}$$

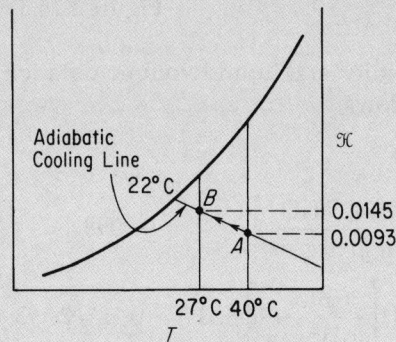

Figure E10.4b

Example 10.5   Combined Material and Energy Balances for a Cooling Tower

Problem Statement

You have been asked to redesign a water-cooling tower that has a blower with a capacity of 8.30×10^6 ft³/hr of moist air. The moist air enters at 80°F and a wet-bulb temperature of 65°F. The exit air is to leave at 95°F and 90°F wet-bulb. How much water can be cooled in pounds per hour if the water to be cooled enters the tower at 120°F, leaves the tower at 90°F, and is not recycled?

Solution

Figure E10.5 shows the process and the corresponding states on the humidity chart.

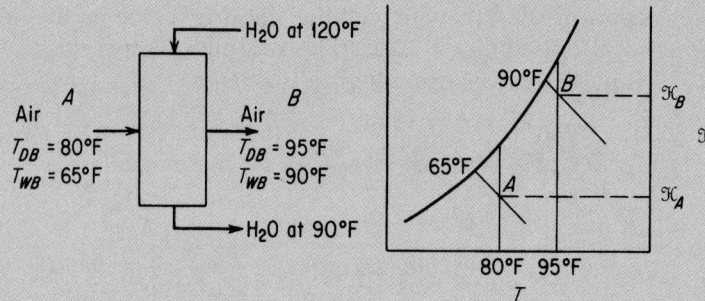

Figure E10.5

Enthalpy, humidity, and humid volume data for the air taken from the humidity chart are as follows:

	A	**B**
$H\left(\dfrac{\text{lb H}_2\text{O}}{\text{lb dry air}}\right)$	0.0098	0.0297
$\Delta \hat{H}\left(\dfrac{\text{Btu}}{\text{lb dry air}}\right)$	$30.05 - 0.12 = 29.93$	$55.93 - 0.10 = 55.83$
$\hat{V}\left(\dfrac{\text{ft}^3}{\text{lb dry air}}\right)$	13.82	14.65

The cooling water exit rate can be obtained from an energy balance around the process.

$$\text{Basis: } 8.30 \times 10^6 \text{ ft}^3/\text{hr of moist air} \equiv 1 \text{ hr}$$

The entering air is

$$\frac{8.30 \times 10^6 \text{ ft}^3}{} \left| \frac{1 \text{ lb dry air}}{13.82 \text{ ft}^3} = 6.01 \times 10^5 \text{ lb dry air}\right.$$

The relative enthalpy of the entering water stream per pound is (the reference temperature for the water stream is 32°F and 1 atm)

$$\Delta \hat{H} = C_{p_{H_2O}} \Delta T = (120 - 32)(1) = 88 \text{ Btu/lb } H_2O$$

and that of the exit stream is $90 - 32 = 58$ Btu/lb H_2O. (The value from the steam tables at 120°F for liquid water of 87.92 Btu/lb H_2O is slightly different because it represents water at its vapor pressure, 1.69 psia, based on reference conditions of 32°F and liquid water at its vapor pressure.) Any other reference datum could be used instead of 32°F for the liquid water. For example, if you chose 90°F, one water stream would not have to be taken into account because its relative enthalpy would be zero. In any case, the enthalpies of the reference state will cancel when you calculate enthalpy differences.

The transfer of water to the air is

$$0.0297 - 0.0098 = 0.0199 \text{ lb } H_2O/\text{lb dry air}$$

a. Material balance for the liquid water stream: Let W = pounds of H_2O entering the tower in the water stream per pound of dry air. Then

$$W - 0.0199 = \text{lb } H_2O \text{ leaving tower in the water stream per lb dry air}$$

b. Material balance for the dry air:

$$6.01 \times 10^5 \text{ lb in} = 6.01 \times 10^5 \text{ lb dry air out}$$

(Continues)

Example 10.5   Combined Material and Energy Balances for a Cooling Tower (*Continued*)

c. Energy balance (enthalpy balance) around the entire process (although the reference temperature for the moist air, 0°F, is not the same as that for liquid water, 32°F, the reference enthalpies cancel in the calculations, as mentioned previously):

Air and water in air entering   **Water stream entering**

$$\frac{29.93 \text{ Btu}}{1 \text{ lb dry air}}\bigg|6.01 \times 10^5 \text{ lb dry air} + \frac{88 \text{ Btu}}{1 \text{ lb H}_2\text{O}}\bigg|\frac{W \text{ lb H}_2\text{O}}{1 \text{ lb dry air}}\bigg|6.01 \times 10^5 \text{ lb dry air} =$$

Air and water in air leaving   **Water stream leaving**

$$\frac{55.83 \text{ Btu}}{\text{lb dry air}}\bigg|6.01 \times 10^5 \text{ lb dry air} + \frac{58 \text{ Btu}}{\text{lb H}_2\text{O}}\bigg|\frac{(W-0.0199)1 \text{ lb H}_2\text{O}}{\text{lb dry air}}\bigg|6.01 \times 10^5 \text{ lb dry air}$$

Simplifying this expression:

$$29.93 + 88W = 55.83 + 58(W - 0.0199)$$

$$W = 0.825 \text{ lb H}_2\text{O/lb dry air}$$

$$W - 0.0199 = 0.805 \text{ lb H}_2\text{O/lb dry air}$$

The total water leaving the tower is

$$\frac{0.805 \text{ lb H}_2\text{O}}{\text{lb dry air}}\bigg|\frac{6.01 \times 10^5 \text{ lb dry air}}{\text{hr}} = 4.83 \times 10^5 \text{ lb/hr}$$

Self-Assessment Test

Questions

1. In combustion calculations with air in previous chapters, you usually neglected the moisture in the air. Is this assumption reasonable?

2. Is the use of the humidity charts restricted to adiabatic cooling or wet-bulb processes?

Answers

1. Usually, it is okay, but in a flue gas, for example, you want to make sure that the condensation in the exit duct does not occur. For such a case, the water entering with the air may be important to include in the calculations.

2. No

Problem

1. A process that takes moisture out of the air by passing the air through water sprays sounds peculiar but is perfectly practical as long as the water temperature is below the dew point of the air. Equipment such as that shown in Figure SAT10.3P1 would do the trick. If the entering air has a dew point of 70°F and is at 40% *RH*, how much heat has to be removed by the cooler, and how much water vapor is removed, if the exit air is at 56°F with a dew point of 54°F?

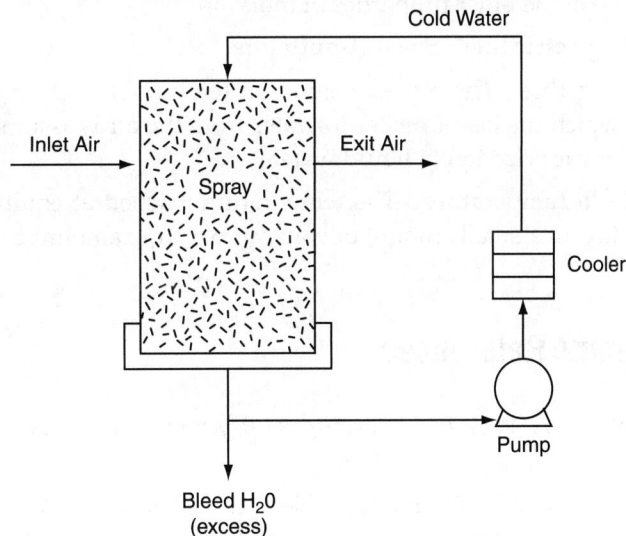

Figure SAT10.3P1

Answer

1. Inlet air humidity of 0.087 lb/lb BDA and enthalpy of 301 Btu/lb BDA; exit air humidity of 0.10 lb/lb BDA and enthalpy of 313 Btu/lb BDA; therefore, net water makeup is 0.013 lb/lb BDA, and heat removed is 13 Btu/lb BDA.

Summary

In this chapter, we described the structure of and information obtained from humidity charts and showed how they can be used in material and energy balance calculations.

Glossary

dry-bulb temperature   The usual temperature of a gas or liquid.

humid heat   The heat capacity of an air-water vapor mixture per mass of bone-dry air.

humidity chart   See **psychrometric chart.**

humid volume   The volume of air including the water vapor per mass of bone-dry air.

psychrometric chart   A chart showing the humidity versus temperature along with all of the other properties of moist air.

psychrometric line   See **wet-bulb line.**

wet-bulb line   The representation on the humidity chart of the energy balance in which the heat transfer to water from the air is assumed to equal the enthalpy of vaporization of liquid water.

wet-bulb temperature   The temperature reached at equilibrium for the vaporization of a small amount of water into a large amount of air.

Supplementary References

Barenbrug, A. W. T. *Psychrometry and Psychrometric Charts*, 3d ed., Chamber of Mines of South Africa (1991).

Bullock, C. E. "Psychrometric Tables," in *ASHRAE Handbook Product Directory*, paper No. 6, American Society of Heating, Refrigeration, and Air Conditioning Engineers, Atlanta (1977).

McCabe, W. L., J. C. Smith, and P. Harriott. *Unit Operations of Chemical Engineering*, McGraw-Hill, New York (2001).

McMillan, H. K., and J. Kim. "A Computer Program to Determine Thermodynamic Properties of Moist Air," *Access*, **36** (January, 1986).

Shallcross, D. C. *Handbook of Psychrometric Charts—Humidity Diagrams for Engineers.* Kluwer Academic Publishers (1997).

Treybal, R. E. *Mass Transfer Operations*, 3d ed., McGraw-Hill, New York (1980).

Wilhelm, L. R. "Numerical Calculation of Psychrometric Properties in SI Units," *Trans. ASAE*, **19**, 318 (1976).

Problems

Section 10.1   Terminology

*10.1.1   If a gas at 60.0°C and 101.6 kPa absolute has a molar humidity of 0.030, determine (a) the relative humidity and (b) the dew point of the gas (in degrees Celsius).

**10.1.2   What is the relative humidity of 28.0 m³ of wet air at 27.0°C that is found to contain 0.636 kg of water vapor?

**10.1.3   Air at 80°F and 1 atm has a dew point of 40°F. What is the relative humidity of this air? If the air is compressed to 2 atm and 58°F, what is the relative humidity of the resulting air?

**10.1.4   If a gas at 140°F and 30 in. Hg absolute has a molar humidity of 0.03 mole of H_2O per mole of dry air, calculate (a) the relative humidity (percent) and (b) the dew point of the gas (in degrees Fahrenheit).

**10.1.5   A wet gas at 30°C and 100.0 kPa with a relative humidity of 75.0% was compressed to 275 kPa, then cooled to 20°C. How many cubic meters of the original gas were compressed if 0.341 kg of condensate (water) was removed from the separator that was connected to the cooler?

***10.1.6   A constant volume bomb contains air at 66°F and 21.2 psia. One pound of liquid water is introduced into the bomb. The bomb is then heated to a constant temperature of 180°F. After equilibrium is reached, the pressure in the bomb is 33.0 psia. The vapor pressure of water at 180°F is 7.51 psia.
   a. Did all of the water evaporate?
   b. Compute the volume of the bomb in cubic feet.
   c. Compute the humidity of the air in the bomb at the final conditions in pounds of water per pound of air.

***10.1.7   A dryer must remove 200 kg of H_2O per hour from a certain material. Air at 22°C and 50% relative humidity enters the dryer and leaves at 72°C and 80% relative humidity. What is the weight (in kilograms) of bone-dry air used per hour? The barometer reads 103.0 kPa.

***10.1.8   Thermal pollution is the introduction of waste heat into the environment in such a way as to adversely affect environmental quality. Most thermal pollution results from the discharge of cooling water into the surroundings. It has been suggested that power plants use cooling towers and recycle water rather than dump water into streams and rivers. In a proposed cooling tower, air enters and passes through baffles over which warm water from the heat exchanger falls. The air enters at a temperature of 80°F and leaves at a temperature of 70°F. The partial pressure of the water vapor in the air entering is 5 mm Hg, and the partial pressure of the water vapor in the air leaving the tower is 18 mm Hg. The total pressure is 740 mm Hg. Calculate:

a. The relative humidity of the air-water vapor mixture entering and of the mixture leaving the tower
b. The percentage composition by volume of the moist air entering and of that leaving
c. The percentage composition by weight of the moist air entering and of that leaving
d. The percent humidity of the moist air entering and of that leaving
e. The pounds of water vapor per 1000 ft^3 of mixture entering and of that leaving
f. The pounds of water vapor per 1000 ft^3 of vapor-free air entering and of that leaving
g. The weight of water evaporated if 800,000 ft^3 of air (at 740 mm and 80°F) enters the cooling tower per day

Section 10.2   The Humidity (Psychrometric) Chart

*10.2.1   Autumn air in the deserts of the southwestern United States during the day is typically moderately hot and dry. If a dry-bulb temperature of 27°C and a wet-bulb temperature of 17°C are measured for the air at noon:
a. What is the dew point?
b. What is the percent relative humidity?
c. What is the humidity?

*10.2.2   Under what conditions are the dry-bulb, wet-bulb, and dew point temperatures equal?

*10.2.3   Explain how you locate the dew point on a humidity chart for a given state of moist air.

*10.2.4   Under what conditions can the dry-bulb and wet-bulb temperatures be equal?

*10.2.5   Under what conditions are the adiabatic saturation temperature and the wet-bulb temperature the same?

*10.2.6   Moist air has a humidity of 0.020 kg of H_2O/kg of air. The humid volume is 0.90 m³/kg of air.
   a. What is the dew point?
   b. What is the percent relative humidity?

*10.2.7   Use the humidity chart to estimate the kilograms of water vapor per kilogram of dry air when the dry-bulb temperature is 30°C and the relative humidity is 65%.

*10.2.8   Urea is produced in cells as a product of protein metabolism. From the cells, it flows through the circulatory system, is extracted in the kidneys, and is excreted in the urine. In an experiment, the urea was separated from the urine using ethyl alcohol and dried in a stream of carbon dioxide. The gas analysis at 40°C and 100 kPa was 10% alcohol by volume and the rest carbon dioxide. Determine (a) the grams of alcohol per gram of CO_2 and (b) the percent relative saturation.

*10.2.9   What is the difference between the constant enthalpy and constant wet-bulb temperature lines on the humidity chart?

*10.2.10   What does a home air conditioning unit do besides cool the air in the house?

*10.2.11   Moist air at 100 kPa, a dry-bulb temperature of 90°C, and a wet-bulb temperature of 46°C is enclosed in a rigid container. The container and its contents are cooled to 43°C.
   a. What is the molar humidity of the cooled moist air?
   b. What is the final total pressure in atmospheres in the container?
   c. What is the dew point in degrees Celsius of the cooled moist air?

*10.2.12   Humid air at 1 atm has a dry-bulb temperature of 180°F and a wet-bulb temperature of 120°F. The air is then cooled at 1 atm to a dry-bulb temperature of 115°F. Calculate the enthalpy change per pound of dry air.

*10.2.13   What is the lowest temperature that air can attain in an evaporative cooler if it enters at 1 atm, 29°C, and 40% relative humidity?

**10.2.14 The air supply for a dryer has a dry-bulb temperature of 32°C and a wet-bulb temperature of 25.5°C. It is heated to 90°C by coils and blown into the dryer. In the dryer, it cools along an adiabatic cooling line as it picks up moisture from the dehydrating material and leaves the dryer fully saturated.
   a. What is the dew point of the initial air?
   b. What is its humidity?
   c. What is its percent relative humidity?
   d. How much heat is needed to heat 100 m³ of initial air to 90°C?
   e. How much water will be evaporated per 100 m³ of air entering the dryer?
   f. At what temperature does the air leave the dryer?

Section 10.3 Applications of the Humidity Chart

**10.3.1 A rotary dryer operating at atmospheric pressure dries 10 tons/day of wet grain at 70°F, from a moisture content of 10% to 1% moisture. The airflow is countercurrent to the flow of grain, enters at 225°F dry-bulb and 110°F wet-bulb temperature, and leaves at 125°F dry-bulb. See Figure P10.3.1. Determine (a) the humidity of the entering and leaving air if the latter is saturated, (b) the water removal in pounds per hour, (c) the daily product output in pounds per day, (d) the heat input to the dryer.

  Assume that there is no heat loss from the dryer, that the grain is discharged at 110°F, and that its specific heat is 0.18.

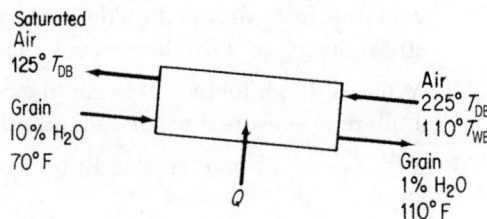

Saturated Air
125° T_{DB}
Grain
10% H_2O
70° F

Air
225° T_{DE}
110° T_{WE}
Grain
1% H_2O
110° F

Q

Figure P10.3.1

**10.3.2 A stream of warm air with a dry-bulb temperature of 40°C and a wet-bulb temperature of 32°C is mixed adiabatically with a stream of saturated cool air at 18°C. The dry air mass flow rates of the warm and cool airstreams are 8 and 6 kg/s, respectively. Assuming a total pressure of 1 atm, determine (a) the temperature, (b) the specific humidity, and (c) the relative humidity of the mixture.

**10.3.3 Temperatures (in degrees Fahrenheit) taken around a forced-draft cooling tower are as follows:

	In	Out
Air	85	90
Water	102	89

The wet-bulb temperature of the entering air is 77°F. Assuming that the air leaving the tower is saturated, calculate (a) the humidity of the entering air, (b) the pounds of dry air through the tower per pound of water into the tower, and (c) the percentage of water vaporized in passing through the tower.

10.3.4 A person uses energy simply by breathing. For example, suppose that you breathe in and out at the rate of 5.0 L/min and that air with a relative humidity of 30% is inhaled at 25°C. You exhale at 37°C, saturated. What must be the heat transfer to the lungs from the blood system in kilojoules per hour to maintain these conditions? The barometer reads 97 kPa.

***10.3.5** A dryer produces 180 kg/hr of a product containing 8% water from a feed stream that contains 1.25 g of water per gram of dry material. The air enters the dryer at 100°C dry-bulb and a wet-bulb temperature of 38°C; the exit air leaves at 53°C dry-bulb and 60% relative humidity. Part of the exit air is mixed with the fresh air supplied at 21°C, 52% relative humidity, as shown in Figure P10.3.5. Calculate the air and heat supplied to the heater, neglecting any heat lost by radiation, used in heating the conveyor trays, and so forth. The specific heat of the product is 0.18.

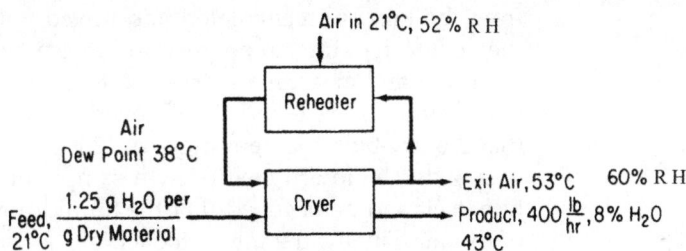

Figure P10.3.5

***10.3.6** Air, dry-bulb 38°C, wet-bulb 27°C, is scrubbed with water to remove dust. The water is maintained at 24°C. Assume that the time of contact is sufficient to reach complete equilibrium between air and water. The air is then heated to 93°C by passing it over steam coils. It is then used in an adiabatic rotary dryer from which it issues at 49°C. It may be assumed that the material to be dried enters and leaves at 46°C. The material loses 0.05 kg H_2O per kilogram of product. The total product is 1000 kg/hr.
   a. What is the humidity (1) of the initial air? (2) after the water sprays? (3) after reheating? (4) leaving the dryer?
   b. What is the percent humidity at each of the points in part a?
   c. What is the total weight of dry air used per hour?
   d. What is the total volume of air leaving the dryer?
   e. What is the total amount of heat supplied to the cycle in joules per hour?

***10.3.7   In the final stages of the industrial production of penicillin, air enters a dryer having a dry-bulb temperature of 34°C and a wet-bulb temperature of 17°C. The moist air flows over the penicillin at 1 atm at the rate of 4500 m³/hr. The air exits at 34°C. The penicillin feed enters at 34°C with a moisture content of 80% and exits at 50%.

   a. How much water is evaporated from the penicillin per hour?

   b. What is the enthalpy change in kilojoules per hour of the air from inlet to exit?

   c. How many kilograms of the entering penicillin are dried per hour?

Assume that the properties of the wet penicillin are those of the water content of the mixture.

***10.3.8   A plant has a waste product that is too wet to burn for disposal. To reduce the water content from 63.4% to 22.7%, it is passed through a rotary kiln dryer. Prior to flowing through the kiln, the air is preheated in a heater by steam coils. To conserve energy, part of the exit gases from the kiln are recirculated and mixed with the heated airstream as they enter the kiln. An engineering intern stood by the entrance to the air heater and measured a dry-bulb temperature of 80°F and a wet-bulb temperature of 54°F and then moved to the exit of the kiln and found that the dry-bulb temperature was 120°F and the wet-bulb temperature was 94°F. The intern next drew a sample of the moist air entering the kiln itself and determined that the humidity was 0.0075 lb water per lb of dry air. Finally, the intern looked at the weather data on the TV and saw that the barometer reading was 29.92 in. Hg.

Calculate (a) the percentage of the air leaving the kiln that is recirculated; (b) the pounds of inlet air per ton of dried waste; (c) the cubic feet of moist air leaving the kiln per ton of dried waste.

****10.3.9   Air is used for cooling purposes in a plant process. The scheme shown in Figure P10.3.9 has been proposed as a way to cool the air and continuously reuse it. In the process, the air is heated and its water content remains constant. It is estimated that the air leaving the process will be at 234°F regardless of the entering temperature or the throughput rate. The air must remove 425,000 Btu/hr from the process. Water for the spray tower is available at 100°F. This tower operates adiabatically with the air leaving saturated. Determine the following: (a) the wet-bulb temperature and percent relative humidity of the air leaving the process; (b) the temperature and dew point of the air leaving the cooler; (c) the psychrometric chart for this cycle; (d) the air circulation rate needed in pounds bone-dry air (BDA) per hour.

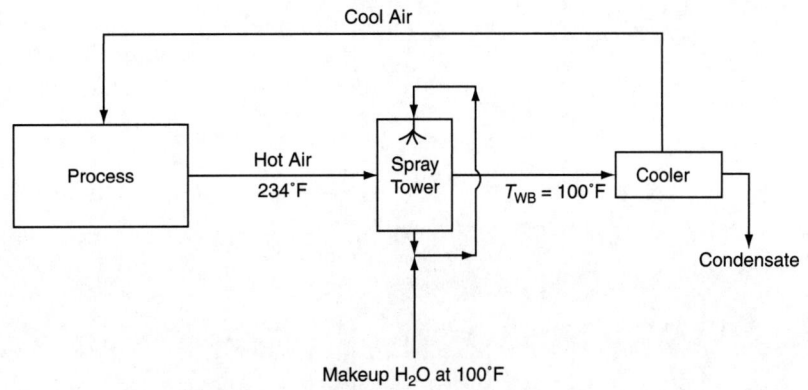

Figure P10.3.9

CHAPTER 11

Unsteady-State Material and Energy Balances

Chapter Objectives

- Be able to develop differential equations that describe the dynamic behavior of unsteady-state macroscopic systems.
- Be able to use MATLAB or Python to numerically solve differential equations that represent unsteady-state systems.

Introduction

Consider the batch reactor shown in Figure 11.1. Initially, the feed is added to the reactor (Figure 11.1a). During the reaction phase (Figure 11.1b), the reactor is heated to the operating temperature where the reaction occurs. After the reaction is complete, the product is removed from the reactor (Figure 11.1.c) for further processing to recover the desired product(s). Figure 11.2 shows the reactant concentration (C_A) in the reactor during the reaction phase for a first-order reaction. Note that C_A decreases exponentially during the reaction phase for this case. This chapter demonstrates how to develop the differential equations that describe how C_A varies with time and how to solve these equations numerically.

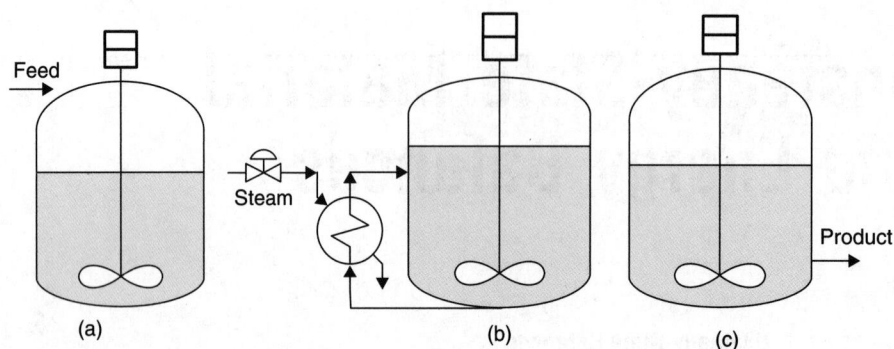

Figure 11.1   The three stages of a batch reactor: (a) the feed charging stage, (b) the reaction phase, and (c) the product removal phase

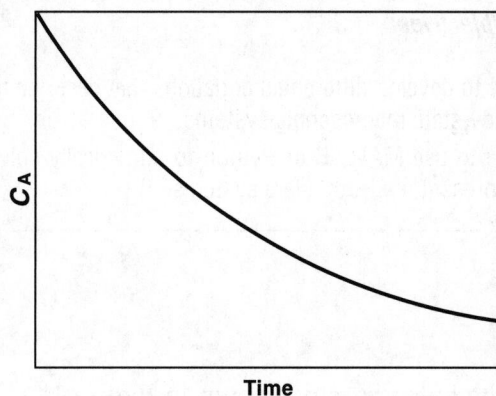

Figure 11.2   The time behavior of the reactant concentration for a batch reactor

The material and energy balances presented previously in this text were based on a steady-state analysis and were macroscopic in approach. This chapter continues to use a macroscopic approach, but develops approaches that describe the unsteady-state behavior of these systems. That is, we previously applied a steady-state analysis to batch processes and developed a solution at the end of a period of time. Now, we develop a solution for batch system and other unsteady-state systems that provides the dependent variable (e.g., the composition and/or the temperature) as a function of time.

Let's focus our attention on unsteady-state processes in which the value of the state (dependent variable) as a function of time is of interest. Recall that the term *unsteady state* refers to processes in which quantities or operating conditions within the system change with time. These systems are sometimes said to be in the transient state. The unsteady state can be somewhat more complicated than the steady state, and, in general, problems involving unsteady-state processes can be more difficult to formulate and solve than those involving steady-state processes. However, a wide variety of important industrial problems fall into this category, such as the start-up of a process, batch heating or reactions, the change from one set of operating conditions to another, and the perturbations that develop as process conditions fluctuate.

11.1   Unsteady-State Balances

Unsteady-state balance equations are similar to steady-state balance equations except that the unsteady-state equation has an additional term: accumulation. Accumulation is equal to the increase or decrease of the balance item in the system. For example, for a material balance based on moles, accumulation is equal to the increase or the decrease in the number of moles in the system. For an energy balance, the accumulation is equal to the increase or decrease in the thermal energy in the system.

11.1.1   Unsteady-State Material Balances

The previous balance equations have been based on steady-state operation. For example, Equation (4.7) is steady-state component mole balance equation considering chemical reactions:

$$
\begin{Bmatrix} \text{moles of } i \\ \text{leaving} \\ \text{the system} \end{Bmatrix} = \begin{Bmatrix} \text{moles of } i \\ \text{entering} \\ \text{the system} \end{Bmatrix} + \begin{Bmatrix} \text{moles of } i \\ \text{generated} \\ \text{by reaction} \end{Bmatrix} - \begin{Bmatrix} \text{moles of } i \\ \text{consumed} \\ \text{by reaction} \end{Bmatrix} \tag{4.7}
$$

When we consider an unsteady-state mole balance, we must add an additional term: the accumulation term. The accumulation term is equal to the increase or decrease in the moles of a component in the system under consideration. Therefore, the unsteady-state component balance equation becomes

$$
\begin{Bmatrix} \text{accumulation} \\ \text{of moles of } i \\ \text{in the system} \end{Bmatrix} = \begin{Bmatrix} \text{moles of } i \\ \text{entering} \\ \text{the system} \end{Bmatrix} + \begin{Bmatrix} \text{moles of } i \\ \text{generated} \\ \text{by reaction} \end{Bmatrix} - \begin{Bmatrix} \text{moles of } i \\ \text{consumed} \\ \text{by reaction} \end{Bmatrix} - \begin{Bmatrix} \text{moles of } i \\ \text{leaving} \\ \text{the system} \end{Bmatrix} \tag{11.1}
$$

This equation can be understood by considering how each term on the right side of the equation affects the number of moles of component i in the system. That is, the flow of component i into the system and the generation by reaction both cause the amount of component i in the system to increase; thus, both terms have a positive sign. Conversely, the flows leaving the system and the consumption by reaction both cause the amount of component i in the system to decrease; thus, both have a negative sign.

If we apply Equation (11.1) for the total mass in the system, the following equation results:

$$M(t_{\text{final}}) - M(t_{\text{initial}}) = F^{\text{in}}(t_{\text{final}} - t_{\text{initial}}) - F^{\text{out}}(t_{\text{final}} - t_{\text{initial}})$$

where M is the mass inside the system, t_{final} is the time at the end of the time interval, t_{initial} is the time at the beginning of the time interval, F^{in} is the mass flow rate entering the system, and F^{out} is the mass flow rate leaving the system. If we consider an infinitely small time interval ($dt = t_{\text{final}} - t_{\text{initial}}$), the previous equation becomes

$$M(t_{\text{initial}} + dt) - M(t_{\text{initial}}) = dM = F^{\text{in}}dt - F^{\text{out}}dt$$

Dividing this equation by dt and taking the limit as dt approaches zero yields the unsteady-state mass balance equation:

$$\frac{dM}{dt} = F^{\text{in}} - F^{\text{out}} \tag{11.2}$$

which is an ordinary differential equation (ODE). Note that the units of each term in Equation (11.2) are mass or moles per unit time. The next section demonstrates how to numerically solve this type of differential equation using MATLAB and Python.

If we apply Equation (11.1) for a component A, the following equation results:

$$n_{\text{A}}(t_{\text{final}}) - n_{\text{A}}(t_{\text{initial}}) = F_{\text{A}}^{\text{in}}(t_{\text{final}} - t_{\text{initial}}) + v_{A}rV_{\text{r}}(t_{\text{final}} - t_{\text{initial}}) - F_{\text{A}}^{\text{out}}(t_{\text{final}} - t_{\text{initial}})$$

where n_{A} is the total number of moles of component A in the system, F_{A}^{in} is the molar flow rate of A entering the system, $F_{\text{A}}^{\text{out}}$ is the molar flow rate of A leaving the system, r is the reaction rate involving A, v_{A} is the stoichiometric coefficient for A in the reaction, and V is the volume of the system.

If we consider an infinitely small time interval (i.e., $dt = t_{\text{final}} - t_{\text{initial}}$), the previous equation becomes

$$\frac{dn_{\text{A}}}{dt} = F_{\text{A}}^{\text{in}} + v_{A}r(C_{\text{A}})V - F_{\text{A}}^{\text{out}} \tag{11.3}$$

We can express the previous equation in terms of the concentration of A (C_A) using the relationships $n_A = C_A V$ and $F_A = C_A Q$, where V is the volume of the system and Q is the volumetric flow rate. Alternatively, Equation (11.3) can be expressed in terms of mass or mole fractions using the relationships $n_A = x_A M$, where M is the total mass or moles in the system and x_A is the mass or mole fraction of A, respectively; and F_A is equal to $x_A F$, where F is the mass or molar flow rate of the inlet or outlet stream.

Example 11.1   Tank Level

Problem Statement

Consider the level in a tank, shown in Figure E11.1, which has material added to the tank and removed from the tank. Develop a differential equation (i.e., ODE) that describes the level in the tank as a function of time. Assume that at time equal to zero, the level is equal to L_0, F_{in} and F_{out} are expressed in mass per time, the cross-sectional area of the tank is equal to A_c, and the density of the liquid is equal to ρ.

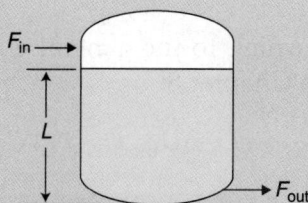

Figure E11.1   Level in a tank

Solution

Applying Equation (11.2) to this problem yields

$$\frac{dM}{dt} = \frac{d(A_c L \rho)}{dt} = F_{in} - F_{out}$$

Simplifying yields

$$A_c \rho \frac{dL}{dt} = F_{in} - F_{out} \qquad \text{at } t = 0 \quad L = L_0$$

This differential equation along with the initial conditions can be used to determine the level in the tank as a function of time.

Example 11.2   Batch Reactor

Problem Statement

Develop an ODE that represents a batch reactor (Figure 11.1b) in which the reaction $A \rightarrow B$ occurs with a reaction rate equal to kC_A^2 and the volume of feed in the reactor is equal to V_r. Initially, the concentration of A in the reactor is equal to C_{A0}.

Solution

Applying Equation (11.3), recognizing that there is no flow into or out of the reactor yields

$$V_r \frac{dC_A}{dt} = -kC_A^2 V_r \qquad \text{at } t = 0 \ \ C_A = C_{A0}$$

11.1.2 Unsteady-State Energy Balances

Let's begin by returning to the general energy balance equation—Equation (8.2)—presented in Chapter 8:

$$\Delta U_{total} = Q + W - \Delta E_{conv}$$

The dynamics of fluid flow systems and gas-piston systems are generally quite fast, whereas the dynamics of open systems are usually much slower. Usually, systems with very fast dynamics are assumed to operate at steady-state conditions, and open systems, which exhibit much slower dynamics, require dynamic modeling in order to represent their unsteady-state behavior. Therefore, the general energy balance equation reduces to

$$\Delta U = Q - \Delta H$$

The term ΔU represents the accumulation of thermal energy inside the system during a specified time period and is given by $MC_v(T_{final} - T_{initial})$, where M is the mass inside the system, which is assumed here to be constant for simplicity; C_v is the heat capacity at constant volume of the material inside the system; T_{final} is the temperature of the material inside the system at the end of the specified time period; and $T_{initial}$ is the temperature of the material inside the system at the beginning of the time period. Substituting into the previous equation yields

$$MC_v(T_{\text{final}} - T_{\text{initial}}) = Q - \Delta H = Q - FC_p(T - T_{in})(t_{\text{final}} - t_{\text{initial}})$$

where Q is the total thermal energy transferred to the system during the time period, F is the flow rate of the material entering the system, T is the temperature of the material inside the system, and T_{in} is the temperature of the material entering the system so that $FC_p(T - T_{\text{in}})(t_{\text{final}} - t_{\text{initial}})$ represents the total thermal energy added to the system by convection during the time period ($t_{\text{final}} - t_{\text{initial}}$). This equation assumes that M is constant; otherwise, you will need to develop an ODE for M, and the previous equation will also change. Because M is assumed to be constant, the inlet flow (F) is equal to the outlet flow (F), and the temperature of the outlet flow is set equal to the temperature of the system, T.

Now, if we apply this equation using an infinitely small time period, dt (i.e., $t_{\text{final}} - t_{\text{initial}}$), the following equation results:

$$MC_p dT = \dot{Q}dt - FC_p(T - T_{\text{in}})dt$$

Finally, if we take the limit as $dt \to 0$, the unsteady-state energy balance equation results,

$$MC_v \frac{dT}{dt} = \dot{Q} - FC_p(T - T_{\text{in}}) \tag{11.4}$$

which is an ordinary differential equation (ODE). The next section demonstrates how to numerically solve this type of differential equation using MATLAB and Python.

If a chemical reaction occurs in the system, the chemical reaction will either add thermal energy to the system (exothermic reaction) or consume energy (endothermic reaction), and the following equation results:

$$MC_p \frac{dT}{dt} = \dot{Q} - FC_p(T - T_{\text{in}}) - rV\Delta H_{\text{rxn}} \tag{11.5}$$

where r is the reaction rate, V is the volume of the system, and ΔH_{rxn} is the heat of reaction. Note the negative sign on the last term results because exothermic reactions have a negative ΔH_{rxn}. Again, we are assuming that M is constant. Because r is normally a function of the composition of the reactant, Equation (11.5) cannot be solved by itself. It must be solved in conjunction with Equation (11.3). That is, the energy balance equation is coupled with the material balance equation (see Example 11.4).

Example 11.3   Thermal Mixing Tank

Problem Statement

Develop an ODE that describes the thermal mixing tank shown in Figure E11.3. In this system, stream F_1 is combined with stream F_2 in a mixing tank. Assume that the product flow rate is equal to the sum of F_1 and F_2, M is the mass of the liquid in the tank, the temperature of the liquid in the mixing tank is initially T_0, and the mixing tank is adiabatic. F_1 and F_2 are expressed in mass per unit time.

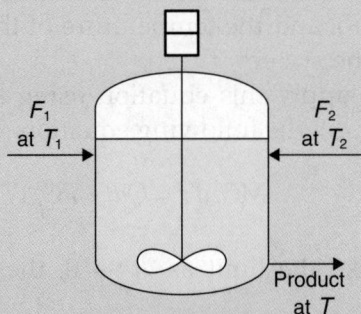

Figure E11.3   Thermal mixing tank

Solution

Applying Equation (11.4) to this problem yields

$$MC_v \frac{dT}{dt} = -F_1 C_p (T - T_1) - F_2 C_p (T - T_2)$$

Note that because there are two feed streams, the last term in Equation (11.4) is applied twice: once for each feed stream. Because we are dealing with liquid streams and liquid in the mixing tank, it is appropriate to assume that C_v is equal to C_p. Therefore, the ODE becomes

$$M \frac{dT}{dt} = -F_1 (T - T_1) - F_2 (T - T_2) \qquad \text{at } t = 0 \quad T = T_0$$

Example 11.4   Adiabatic Exothermic Batch Reactor

Problem Statement

Develop the differential equations that describe an adiabatic exothermic batch reactor in which the reaction rate is represented by

$$r = k_0 \exp(-E_r/RT)C_A$$

Solution

The application of Equation (11.3) to this system yields

$$\frac{dn_A}{dt} = V\frac{dC_A}{dt} = -Vk_0 \exp(-E_r/RT)C_A$$

And the application of Equation (11.5) yields

$$MC_p\frac{dT}{dt} = \rho VC_p\frac{dT}{dt} = -r\, V\Delta H_{rxn} = -V\Delta H_{rxn}k_0 \exp(-E_r/RT)C_A$$

where ρ is the molar density of the liquid in the reactor. Simplifying these equations yields

$$\frac{dC_A}{dt} = -k_0 \exp(-E_r/RT)C_A \qquad \text{at } t=0 \quad C_A = C_{A0}$$

$$\rho C_p\frac{dT}{dt} = -\Delta H_{rxn}k_0 \exp(-E_r/RT)C_A \qquad \text{at } t=0 \quad T=T_0$$

Note that the first ODE, which defines C_A, is also a function of T, and the second ODE, which defines T, is also a function of C_A. Therefore, this set of differential equations is referred to as a coupled set of ODEs and thus must be solved simultaneously.

Self-Assessment Test

Questions

1. Is time the independent or dependent variable in macroscopic unsteady-state equations?

2. For an unsteady-state material balance, what would accumulation represent?

3. What is the difference between a steady-state and an unsteady-state balance?

4. What information is required in order to solve an ODE problem?

Answers

1. Time is the independent variable.

2. Accumulation for a material balance would be an increase or decrease in the moles or mass in the system.

3. The unsteady-state balance is the same as a steady-state balance except that it also includes accumulation. Note that if you set the accumulation term for an unsteady-state equation equal to zero, the result is a steady-state balance.

4. The initial conditions for the dependent variable

Problem

1. Consider an unsteady-state balance on a mixing tank that has one stream entering it (i.e., F_1) and two streams leaving (F_2 and F_3). Write the ODE that represents the overall unsteady-state material balance.

Answer

1. $\dfrac{dM}{dt} = F_1 - F_2 - F_3$

11.2 Numerical Integration of ODEs

This section considers the numerical solution of first-order ODEs, also known as **initial value problems** (IVPs). An ODE is a differential equation that contains derivatives with respect to only one independent variable. A single first-order ODE has the following general form:

$$\frac{dy}{dx} = f(x, y) \text{ where } y(x_0) = y_0$$

where y is the dependent variable, x is the independent variable, and $y(x_0)$ is the initial conditions for the dependent variable. For certain simple problems, it may be possible to separate $f(x,y)$; that is,

$$\frac{dy}{dx} = f(x, y) = g(x)h(y)$$

In that case, an analytical solution to the IVP may be possible. For example,

$$\frac{dy}{h(y)} = g(x)dx$$

Then

$$\int_{y_0}^{y} \frac{dy}{h(y)} = \int_{x_0}^{x} g(x)\,dx$$

However, most industrial problems do not lend themselves to this approach and thus must be solved numerically.

The easiest way to numerically solve an IVP is the Euler method. Using the Euler method, you calculate the slope [i.e., $f(x,y)$] at the starting point using the ODE. Then choose an increment in x (i.e., Δx) and multiply it by the slope to calculate the corresponding change in y:

$$\Delta y = f(x,y)\Delta x$$

Then

$$y_1 = y_0 + \Delta y = y_0 + \Delta x f(x_0, y_0)$$

Likewise,

$$y_2 = y_1 + \Delta x f(x_1, y_1)$$

The general recursion relation for the Euler method is

$$y_{i+1} = y_i + \Delta x f(x_i, y_i)$$

While the Euler method is simple to apply, it is not very accurate. The Runge-Kutta method is more complicated but much more accurate.

Runge Kutta (RK) Method. The fourth-order RK method uses four slope evaluations (k_1 to k_4) to approximate Δy:

$$y_{i+1} = y_i + \frac{\Delta x}{6}(k_1 + 2k_2 + 2k_3 + k_4) \tag{11.6}$$

where

$$k_1 = f(x,y) \qquad k_2 = f(x + \tfrac{1}{2}\Delta x, y + \tfrac{1}{2}k_1\Delta x) \qquad k_3 = f(x + \tfrac{1}{2}\Delta x, y + \tfrac{1}{2}k_2\Delta x)$$
$$k_4 = f(x + \Delta x, y + k_3\Delta x)$$

then

$$\Delta y = \frac{\Delta x}{6}(k_1 + 2k_2 + 2k_3 + k_4)$$

Therefore, the effective slope over Δx for the fourth-order RK method is equal to $(k_1 + 2k_2 + 2k_3 + k_4)/6$.

Figures 11.3 a–d present a graphical representation of this fourth-order RK method. Note that k_1 is used to estimate y at $x = x_0 + \Delta x/2$, which in turn is used to calculate k_2. Similarly, k_2 is used to calculate k_3, and k_3 is used to calculate k_4. Finally, the k values are used to estimate the slope of y over the interval $(x_0, x_0 + \Delta x)$ by applying Equation 11.5.

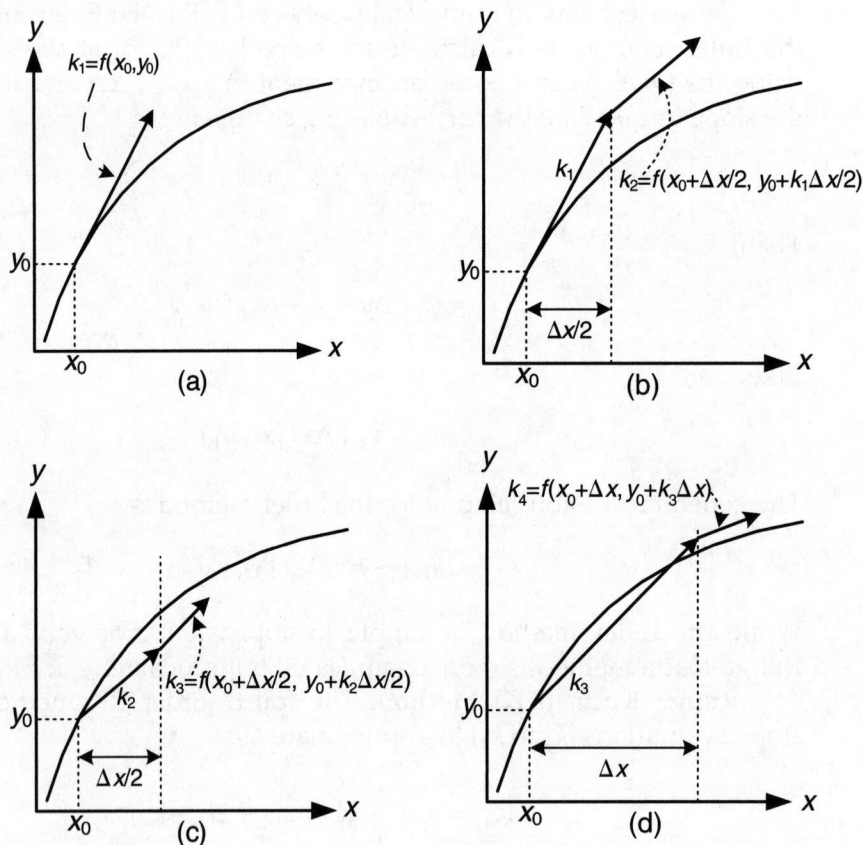

Figure 11.3　Graphical representation of the Runge-Kutta method

11.2.1　MATLAB

MATLAB offers a number of ODE integrators based on a variety of integration formulas. The recommended built-in function for ODE integration is `ode45`, which offers a fourth-order RK method.

MATLAB's ode45 is an adjustable step size fourth-order RK integrator. The step size is adjusted by using a fourth-order and a fifth-order RK integrator at each step to select the appropriate step size to meet the error criterion. The results from the fifth-order RK are used as the exact solution to determine when the integration has met the accuracy requirements, and the results of the fourth- and fifth-order RK integrators are used to estimate the step size for the next step. The call statement for ode45 is

$$[\mathbf{t},\mathbf{Y}] = \text{ode45}(\text{FunctionName},[t_0,dt,t_f], [\mathbf{y}_0],\text{options})$$

where **t** is a column vector of values of the independent variable between t_0 and t_f, **Y** is a matrix of dependent variable values (i.e., a data structure) calculated by the integrator corresponding to the values of the independent variables in vector **t**, FunctionName is the name of the function that is used to calculate the derivatives of the dependent variable with respect to the independent variable, $[t_0,dt,t_f]$ is the range of integration of the ODEs from the initial conditions t_0 to the final conditions t_f using increments of dt, and $[\mathbf{y}_0]$ is a column vector of the initial conditions for the dependent variable. An alternate form for calling ode45 involves using tspan instead of the terms in the brackets with the initial and final values of the independent variable. When using tspan, it must be defined before the call to the ODE integrator; for example,

$$\text{tspan} = t_0 \text{:} dt \text{:} t_f$$

where t_0, dt, and t_f are defined above.

Example 11.5   Numerical Solution of an IVP Using MATLAB

Problem Statement

Numerically solve the following IVP using MATLAB's ode45 function from $x = 0$ to $x = 0.6$:

$$\frac{dy}{dx} = y\exp(yx) \qquad y(0) = 1$$

(Continues)

Example 11.5   Numerical Solution of an IVP Using MATLAB (*Continued*)

MATLAB Code for Example 11.5

```
%%%%%%%%%%%%%%%%%%%%%%%%%%%%%%%%%%%%%%%%%%%%%%%%%%%%%%%%%%%%%%%%%%%%
%                        NOMENCLATURE
%
% dydx - the vector of functions of the derivative of y with respect
%        to x
% soln - the solution matrix for the ODE problem
% x - a vector containing the values of the independent variable
% x0 - the initial value of x (0)
% xf - the final value of x (0.6)
% y - a vector containing the values of the dependent variable
% y0 - the initial value of y (1)
%%%%%%%%%%%%%%%%%%%%%%%%%%%%%%%%%%%%%%%%%%%%%%%%%%%%%%%%%%%%%%%%%%%%
%                          PROGRAM
function Ex11_5_ode45
clear; clc;
x0=0; xf=0.6; y0=[1];           % Input problems specifications
soln=ode45(@f,[x0 xf],y0);      % Call ode45 and store solution in soln
x=linspace(0,xf,100);           % Generate the values of x
y=deval(soln,x,1);              % Retrieve value of y from soln
%% Output Plot %%
axes('FontSize',20);
plot(x,y,'k-','LineWidth',2.5)
xlabel('x','FontSize',20)
ylabel('y','FontSize',20)
grid on;
title('MATLAB Solution','FontSize',24);
end
%
% User-specified function for dydx for each dependent variable
%
function dydx=f(x,y)
                                % Specify dydx in column vector form
dydx=[y*exp(y*x)];
end
%                        PROGRAM END
%%%%%%%%%%%%%%%%%%%%%%%%%%%%%%%%%%%%%%%%%%%%%%%%%%%%%%%%%%%%%%%%%%%%
```

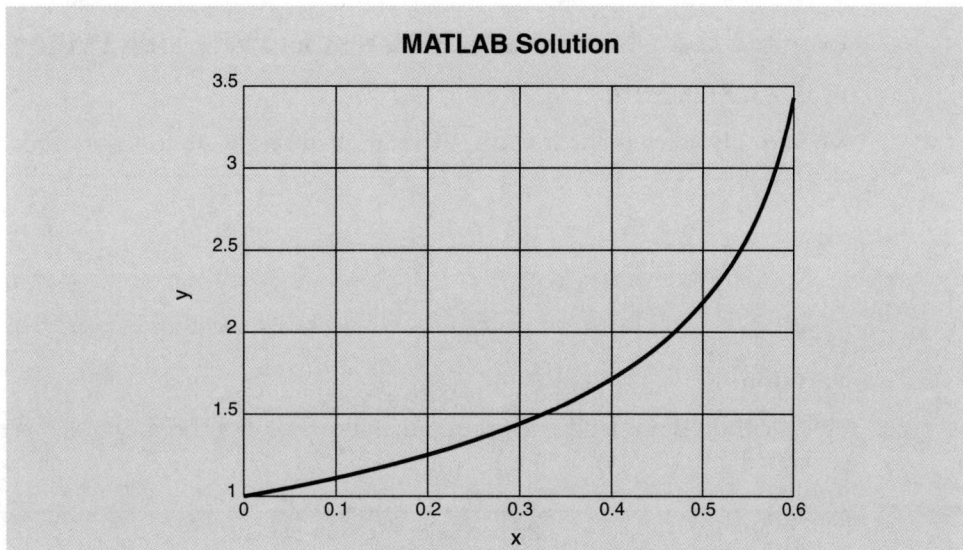

Figure E11.5

Description of Program and Results. The ODE is specified in a user-defined function that is referenced in the call to function ode45. After the solution is generated, a number of values of x are generated using function linspace. The function deval is used to retrieve the values of y corresponding to each generated value of x. Finally, the results are plotted (Figure E11.5).

11.2.2   Python

Python offers a number of ODE integrators based on a variety of integration formulas. The recommended built-in function for ODE integration is RK45. The call statement to apply RK45 is given by

```
soln=scipy.integrate.solve_ivp(fun, tspan, y0,
          method='RK45', t_eval=tvalues)
```

where **fun** is the name of the user-defined function that calculates dy/dx from the values of y and x, **tspan** is a tuple that contains the initial and final value of x, **y0** is a vector that contains the initial value(s) of the dependent variable(s), **method** defines the integrator that will be applied (i.e., RK45), and **t-eval** is a vector of values of x for which the corresponding values of y will be determined.

Example 11.6   Numerical Solution of an IVP Using Python

Problem Statement

Numerically solve the following IVP using Python's RK45 function from $x = 0$ to $x = 0.6$:

$$\frac{dy}{dx} = y\exp(yx) \qquad y(0) = 1$$

Solution

This ODE does not lend itself to an analytical solution because $f(x,y)$ cannot be separated into $g(x)h(y)$.

Python Code for Example 11.6

```
Ex11_6 High Order ODE.py
#
#                        NOMENCLATURE
#
# dydx – the vector of functions of the derivative of y with respect to x
# soln – the solution matrix for the ODE problem
# time – the values of the independent variable used for the plot of the
#         results
# tspan – a tuple that contains the initial and final condition for the
#          independent variable ([0,0.6])
# tvalues – the values for y that will be used to plot the solution
# x – a vector containing the values of the independent variable
# x0 – the initial value of x (0)
# xf – the final value of x (0.6)
# y – a vector containing the values of the dependent variable
# y0 – the initial value of y (1)
#
#                        PROGRAM
import scipy.integrate
import numpy as np
import matplotlib.pyplot as plt
```

```
#
# The user-defined function for dydx for each dependent variable
#
def fun(x, y):
    dy1=y[0]*np.exp(y[0]*x)
    return [dy1]
#  Define input variables
xspan = [0.0,0.6]
y0=[1]
xvalues=np.linspace(0, 0.6, 100)
#
# Apply scipy.integrate.solve_ivp with method equal to RK45
#
soln=scipy.integrate.solve_ivp(fun, xspan, y0, method='RK45',
t_eval=xvalues)
#
# Plot solution
#
x=soln.t                # Retrieve the independent variable vector from
                          soln matrix

y1=soln.y[0]            # Retrieve y1 from soln matrix
plt.plot(x, y1, 'k-', linewidth=2)
plt.axis([0, 0.6, 1.0, 3.5])
plt.title('Python Solution', fontsize=20)
plt.xlabel('x ', fontsize=14)
plt.ylabel('y', fontsize=14)
plt.show
plt.savefig("FigE11.6 ODE integrate.jpg")
#
#                        PROGRAM END
####################################################################
```

(Continues)

Example 11.6   Numerical Solution of an IVP Using Python (*Continued*)

Python Solution

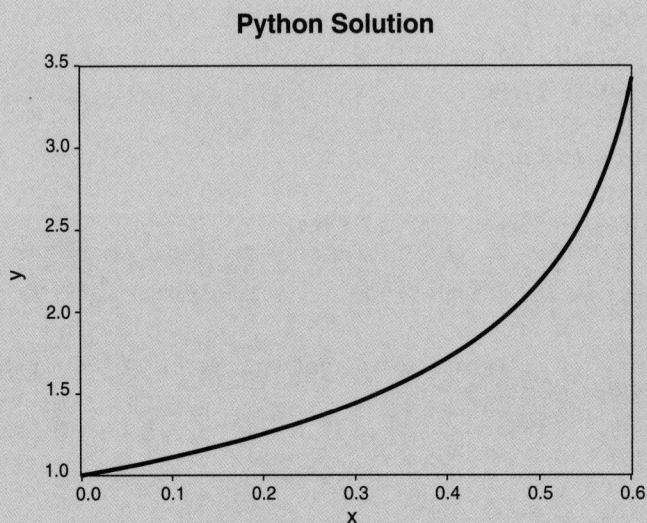

Figure E11.6

Description of Program and Results. The ODE is specified in a user-defined function that is referenced in the call to function `scipy.integrate.solve_ivp`. After the solution is generated, a number of values of x are generated using function `np.linspace`. The values of x and y are then retrieved from `soln`. Finally, the results are plotted (Figure E11.6).

Self-Assessment Test

Questions

11.2.1 For what conditions is it possible to develop an analytical solution for an IVP?

11.2.2 Why is the Runge-Kutta method preferred over the Euler method?

11.2.3 How many function evaluations [i.e., $f(x,y)$] are required for one step for the RK method?

Answers

11.2.1   When it is possible to factor $f(x,y)$ into $h(x)g(y)$

11.2.2   Because the RK method is much more accurate than the Euler method

11.2.3   Four

Problem

11.2.1   Using the fourth-order RK method, determine $y(0.1)$ for the following IVP using $\Delta t = 0.1$:

$$\frac{dy}{dt} = (t^2 - 1)y \qquad y(0) = 1$$

Answer

11.2.1   Applying the fourth-order RK yields

$$k_1 = [(0)^2 - 1]y(0) = -1; \quad k_2 = [(0.05)^2 - 1](y(0) + 0.05k_1) = -0.9476;$$

$$k_3 = [(0.05)^2 - 1](y(0) + 0.05k_2) = -0.9502; \quad k_4 = [(0.1)^2 - 1](y(0) + 0.1k_3) = -0.8959$$

$$y(0.1) = y(0) + \frac{0.1}{6}[-1 + 2(-0.9476) + 2(-0.9502) - 0.8959] = 1 - 0.09485 = 0.9051$$

11.3   Examples

Example 11.7   Numerical Solution of Thermal Mixer Case

Problem Statement

Consider the thermal mixer introduced in Example 11.3:

$$M\frac{dT}{dt} = -F_1(T - T_1) - F_2(T - T_2) \qquad \text{at } t = 0 \quad T = T_0$$

(Continues)

Example 11.7   Numerical Solution of Thermal Mixer Case (*Continued*)

with the following values for its parameters:

- F_1 – mass flow rate of stream 1 (5 kg/s)
- F_2 – mass flow rate of stream 2 (5 kg/s)
- M – mass of liquid in the mixer (100 kg)
- T – the temperature of the mixed liquid (°C)
- T_0 – the initial temperature of the mixed liquid in the mixing tank (50°C)
- T_1 – temperature of stream 1 (25°C)
- T_2 – temperature of stream 2 (75°C)
- t – time (s)

At time equal to 10 s, the flow rate of F_1 is changed from 5 to 4 kg/s. By integrating the ODE for this problem, determine the temperature from 0 to 75 s using MATLAB and Python. Note that the products of $F_1 \times T$ and $T_2 \times T$ render this ODE nonlinear and inseparable, and thus an analytical solution is not available.

Solution

```
                         MATLAB Code for Example 11.7
%%%%%%%%%%%%%%%%%%%%%%%%%%%%%%%%%%%%%%%%%%%%%%%%%%%%%%%%%%%%%%%%%%%%%%%%%%%

%                            NOMENCLATURE
%
% dTdt - the vector of functions of the derivative of T with
%        respect to t
% F1 - the flow rate of stream 1 (5 kg/s)
% F2 - the flow rate of stream 2 (initially 5 kg/s, t>10, 4 kg/s)
% soln - the solution matrix for the ODE problem
% M - the mass of liquid in the mixer (100 kg)
% t - a vector containing the values of the independent variable
% t0 - the initial value of t(0)
% tf - the final value of t (75 s)
% T - a vector containing the values of the dependent variable
% T0 - the initial value of T (50ºC)
% T1 - the temperature of stream 1 (25ºC)
% T2 - the temperature of stream 2 (75ºC)
%
%%%%%%%%%%%%%%%%%%%%%%%%%%%%%%%%%%%%%%%%%%%%%%%%%%%%%%%%%%%%%%%%%%%%%%%%%%%
```

```
%                                PROGRAM
function Ex11_7 MATLAB
clear; clc;
t0=0; tf=75;                      % Input problems specifications
soln=ode45(@f,[x0 xf],y0);        % Call ode45 & store solution in soln
T1=25; T2=75; F1=5; F2=5;         % Specify problem parameters
T0=50; M=100;                     % Specify problem parameters
t=linspace(0,tf,100);             % Generate the values of x
T=deval(soln,t,1);                % Retrieve value of y from soln
%% Output Plot %%
axes('FontSize',20);
plot(t,T,'k-','LineWidth',2.5)
xlabel('t (s)','FontSize',20)
ylabel('T (oC)','FontSize',20)
grid on;
title('MATLAB Solution','FontSize',24);
%
% User-specified function for dTdt
%
function dTdt=f(t,T)
if t>=10
    F2=4
end
dTdt=[(-F1*(T-T1)-F2*(T-T2))/M]
end
end
%                                PROGRAM END
%%%%%%%%%%%%%%%%%%%%%%%%%%%%%%%%%%%%%%%%%%%%%%%%%%%%%%%%%%%%%%%%%%
```

Description of Program and Results. The ODE is specified in a user-defined function that is referenced in the call to function ode45. Note that an if statement is used to change F_2 from 5 to 4 at $t = 10$ s. Finally, the results are plotted (Figure E11.7a). The small hump in the plot centered around $t = 10$ s is not real but was caused by the filtering function used by MATLAB to plot the data. That is, the actual data should be $T = 50°C$ at $t = 10$ s and then should abruptly begin to decrease for $t > 10$ s, but the filtering used by MATLAB to plot the data caused the hump in an effort to smooth out this abrupt change.

(Continues)

Example 11.7   Numerical Solution of Thermal Mixer Case (*Continued*)

MATLAB Solution

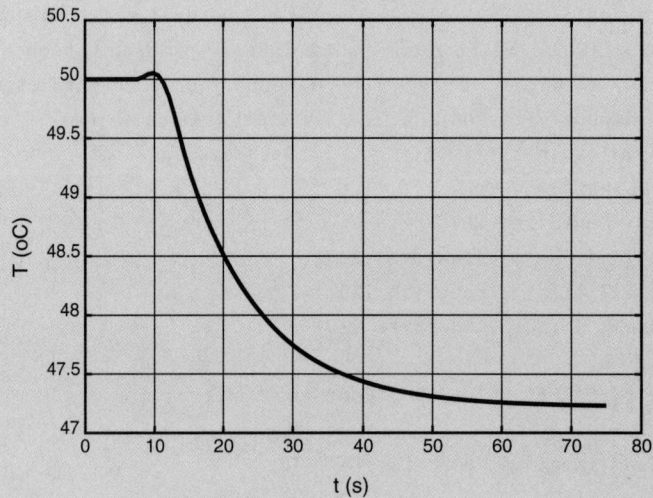

Figure E11.7a

Python Code for Example 11.7

```
Ex11_7 High Order ODE.py

#                     NOMENCLATURE
#
#  dTdt – the vector of functions of the derivative of T with
#        respect to t
#  F1 – the flow rate of stream 1 (5 kg/s)
#  F2 – the flow rate of stream 2 (initially 5 kg/s, t>10, 4 kg/s)
#  soln – the solution matrix for the ODE problem
#  M – the mass of liquid in the mixer (100 kg)
#  t – a vector containing the values of the independent variable
#  t0 – the initial value of t(0)
#  tf – the final value of t (75 s)
#  T – a vector containing the values of the dependent variable
#  T0 – the initial value of T (50oC)
#  T1 – the temperature of stream 1 (25oC)
#  T2 – the temperature of stream 2 (75oC)

#
```

```
#                                    PROGRAM
import scipy.integrate
import numpy as np
import matplotlib.pyplot as plt
#
# The user-defined function for dydx for each dependent variable
#
def fun(t, T) :
    T1, T2, F1, F2, M = 25, 75, 5, 5, 100
    if t>=10 :
        F2=4
    dy1=(-F1*(T[0]-T1)-F2*(T[0]-T2))/M
    return [dy1]
#   Define input variables
xspan = [0.0,75]
T0=[50]
xvalues=np.linspace(0, 75, 100)
#
#   Apply scipy.integrate.solve_ivp with method equal to RK45
#
soln=scipy.integrate.solve_ivp(fun, xspan, T0, method='RK45',
t_eval=xvalues)
#
#   Plot solution
#
time=soln.t                 #  Retrieve the independent variable vector from
                               soln matrix
Tsoln=soln.y[0]             #  Retrieve y1 from soln matrix
plt.plot(time, Tsoln, 'k-', linewidth=2)
plt.axis([0, 75, 47, 50.5])
plt.title('Python Solution', fontsize=20)
plt.xlabel('t (sec) ', fontsize=14)
plt.ylabel('T (oC)', fontsize=14)
plt.show
plt.savefig(" FigE11_7 Python.jpg ")
#                                    PROGRAM END
```

(*Continues*)

Example 11.7   Numerical Solution of Thermal Mixer Case (*Continued*)

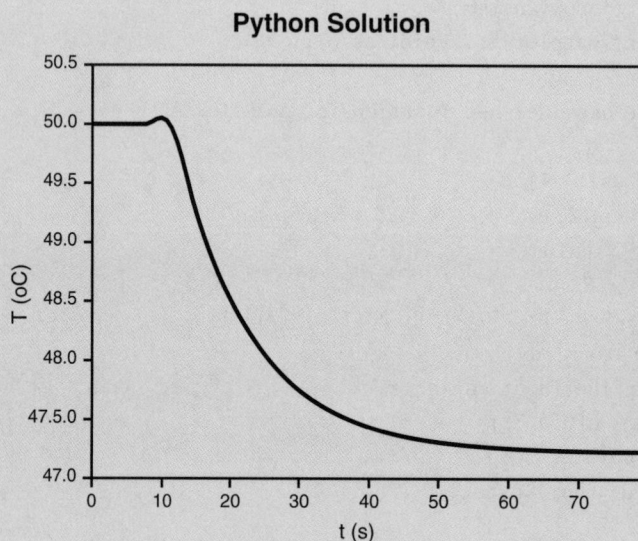

Figure E11.7b

Description of Program and Results. The ODE is specified in a user-defined function that is referenced in the call to function `scipy.intergrate.solve_ivp`. The results are then plotted (Figure E11.7b). Note that an `if` statement is used to change F_2 from 5 to 4 at $t = 10$ s. The small hump in the plot centered around $t = 10$ s is not real but was caused by the filtering function used to plot the data. That is, the actual data should be $T = 50°C$ at $t = 10$ s and then should abruptly begin to decrease for $t > 10$ s, but the filtering used by Python to plot the data caused the hump in an effort to smooth out this abrupt change.

Example 11.8   Material Balances for a Batch Distillation Process

Problem Statement

Consider the batch distillation process shown in Figure E11.8 for the separation of propane from butane. Initially, the system is filled with 10 kg mole of a mixture of propane and butane that is 50% propane and 50% butane under a pressure of 10 atm. Determine the time required to reduce the propane concentration in the batch distillation unit to 30% and the corresponding amount and composition of the distillate product. Assume that the distillate rate is constant at 1 kg mol per hour.

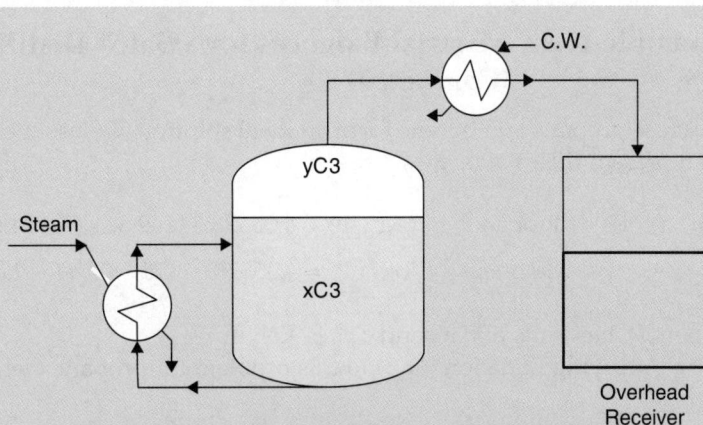

Figure E11.8a   Schematic of a batch distillation unit

Solution

To solve this problem, we must be able to calculate the composition of the vapor stream leaving the batch distillation unit. As the composition in the batch distillation unit changes from 50% propane to 30% propane, both the bubble point temperature and the composition of the distillate will change. Therefore, as the composition changes, we must first calculate bubble point temperature. We use the computer code presented in Chapter 7, Example 7.11, to determine the bubble point temperature as the composition in the batch distillation unit changes from 50% to 30% propane. After the bubble point temperature is determined at each point in time, the composition of the propane in the vapor stream is given by

$$y_{C3} = \frac{x_{C3} P_{C3}^0 (T_{BP})}{P_{Tot}}$$

where y_{C3} is the composition of propane in the vapor stream, x_{C3} is the composition of propane in the liquid in the batch distillation unit, $P_{C3}^0(T_{BP})$ is the vapor pressure of propane at the bubble point temperature (T_{BP}), and P_{Tot} is the total pressure in the batch distillation unit.

(Continues)

Example 11.8   Material Balances for a Batch Distillation Process (*Continued*)

Because propane and butane form an ideal solution, we use the Antoine constants to represent their vapor pressures:[1]

$$\text{Propane:} \quad \log_{10} P^0 = 4.53678 - 1149.36 / (T + 24.906)$$

$$\text{Butane:} \quad \log_{10} P^0 = 4.35576 - 1175.581 / (T - 2.071)$$

where P^0 has units of bars and T is in kelvin.

Applying Equation (11.3) to this problem for propane yields

$$\frac{dn_{C3}}{dt} = \frac{d(x_{C3}M)}{dt} = M\frac{dx_{C3}}{dt} + x_{C3}\frac{dM}{dt} = M\frac{dx_{C3}}{dt} - x_{C3} = -F_{C3}^{out} = -1y_{C3}$$

where y_{C3} is the composition of propane in the vapor leaving the batch distillation unit. Simplifying the previous equation,

$$M(t)\frac{dx_{C3}}{dt} = -y_{C3} + x_{C3} \quad x_{C3}(t=0) = 0.5 \quad M(t) = 10 - t \quad t[=]\text{hr}$$

A trial-and-error procedure is used to determine the time to attain a 30% propane concentration; that is, the length of time for integrating the model equations is adjusted until a 30% propane concentration is obtained. Once the time to $x_{C3} = 0.3$ is determined, an overall material balance can be applied to determine the amount and composition in the overhead receiver.

MATLAB Code for Example 11.8

```
%%%%%%%%%%%%%%%%%%%%%%%%%%%%%%%%%%%%%%%%%%%%%%%%%%%%%%%%%%%%%%%%%%%%%%

%                        NOMENCLATURE

%
%  dxC3dt - the derivative of xC3 with respect to t
%  M - the moles of liquid in the batch unit (initially 10 kg mol)
%  PoC3 - the partial pressure of C3 at the bubble point (atm)
%  Ptot - the operating pressure of the still (10 atm)
%  soln - the solution matrix for the ODE problem
%  t - a vector containing the values of the independent variable
```

---

[1] NIST database from the National Institute of Standards and Technology website.

```
%  t0 - the initial value of t(0)
%  tf - the final value of t (4.949 h)
%  T - a vector containing the values of the dependent variable
%  xC30 - the initial value of xC3 (0.5)
%  xC3 - the mole fraction of C3 in the still
%  yC3 - the mole fraction of C3 in the vapor leaving the still
%
%%%%%%%%%%%%%%%%%%%%%%%%%%%%%%%%%%%%%%%%%%%%%%%%%%%%%%%%%%%%%%%%%%%%%%%
%                                    PROGRAM
function Ex11_8 Batch Distillation
clear; clc;
t0=0; tf=4.949; xC30=0.5;            % Input problems specifications
soln=ode45(@fode,[t0 tf],xC30);      % Call ode45 and store solution
t=linspace(0,tf,100);                % Generate the values of x
T=deval(soln,t);                     % Retrieve value of y from soln
%% Output Plot %%
axes('FontSize',20) ;
plot(t,T,'k-','LineWidth',2.5)
xlabel('t (h)','FontSize',20)
ylabel('xC3','FontSize',20)
grid on;
title('MATLAB Solution','FontSize',24);
%
% User-specified function for dxC3dt
%
function [dxc3dt]=fode(t,xC3)
[yC3]=BubPt(xC3);                    % Call bubble point function for yC3
M=10-t;                              % Calculate current mass in still
dxC3dt=[(-yC3+xC3)/M] ;
end
% Function for bubble point and yC3 calculation
function [yC3]=BubPt(xC3)
Ptot=10.;
TBP=fzero(@TBPt,350)                 % Bubble point calculation
PoC3=10^(4.53678-1149.36/(24.906+TBP))/1.013;
yC3=xC3*PoC3/Ptot                    % Calculate yC3 from Bub Pt
% Function for error in bubble point calculation
```

(Continues)

Example 11.8   Material Balances for a Batch Distillation Process (*Continued*)

```
function [er]=TBPt(TBP)
er=10*1.013-10^(4.53678-1149.36/(24.906+TBP))*xC3-10^(4.35576-… 1175.581/
(-2.071+TBP))*(1-xC3);
end
end
end

%                              PROGRAM END

%%%%%%%%%%%%%%%%%%%%%%%%%%%%%%%%%%%%%%%%%%%%%%%%%%%%%%%%%%%%%%%%%%%%%%%
```

Description of Program and Results. The ODE is specified in a user-defined function fode that is referenced in the call to function ode45. Before function fode can calculate dc3dt, the composition of the vapor (i.e., yC3) must be calculated by function BubPt. Function BubPt uses function fzero to calculate the bubble point temperature (TBP), and TBP is used to calculate yC3 by zeroing function er. Note that functions BubPt and TBPt are nested so that function TBPt will have the current value of xC3. By trial and error, the time required to reduce xC3 to 0.3 was found to be 4.949 h (plotted on Figure E11.8b). An overall steady-state material balance indicates that the amount of liquid in the receiver at the end of the run was 4.949 kg mol with a mole fraction of propane equal to 0.704.

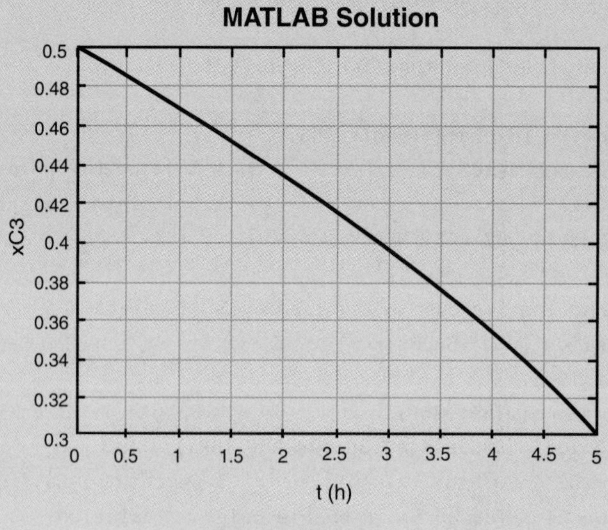

Figure E11.8b

Python Code for Example 11.8

```
#                               NOMENCLATURE
#
#  dxC3dt - the derivative of xC3 with respect to t
#  M - the moles of liquid in the batch unit (initially 10 kg mol)
#  PoC3 - the partial pressure of C3 at the bubble point (atm)
#  Ptot - the operating pressure of the still (10 atm)
#  soln - the solution matrix for the ODE problem
#  t - a vector containing the values of the independent variable (h)
#  t0 - the initial value of t(0)
#  tf - the final value of t (4.949 h)
#  xC30 - the initial value of xC3 (0.5)
#  xC3 - the mole fraction of C3 in the still
#  yC3 - the mole fraction of C3 in the vapor leaving the still
#  y1 - a vector containing the values of the dependent variable
#
#                               PROGRAM
import numpy as np
import matplotlib.pyplot as plt
import scipy.optimize
#
# The user-defined function for error in bubble point calculation
#
def er(T,xC3):
    fx=10*1.013-10.**(4.53678-1149.36/(24.906+T))*xC3 -10.
    **(4.35574-1175.581/(-2.071+T))*(1.-xC3)
    return fx
#
# The user-defined function for calculation of yC3
#
def BPT(xC3):
    Ptot=10.
    TBP=scipy.optimize.newton(er, 350, args=(xC3,))
    PoC3=10.**(4.53678-1149.36/(24.906+TBP))/1.013
    yC3=xC3*PoC3/Ptot
    return yC3
```

(Continues)

Example 11.8   Material Balances for a Batch Distillation Process (*Continued*)

```
#
# The user-defined function for dxC3dt
#
def fode(t, xC3):
    M=10-t
    yC3=BPT(xC3)
    dxC3dt=(-yC3+xC3)/M
    return [dxC3dt]
#  Define input variables
xspan = [0.0, 4.949]
xC30=[0.5]
xvalues=np.linspace(0, 4.949, 100)
#
#  Apply scipy.integrate.solve_ivp with method equal to RK45
#
soln=scipy.integrate.solve_ivp(fode, xspan, xC30, method='RK45',
t_eval=xvalues)
#
#  Plot solution
#
time=soln.t                # Retrieve the independent variable vector
from soln matrix
y1=soln.y[0]               # Retrieve y1 from soln matrix
plt.plot(time, y1, 'k-', linewidth=2)
plt.axis([0, 4.949, 0.3, 0.5])
plt.title('Python Solution', fontsize=20)
plt.xlabel('Time (h) ', fontsize=14)
plt.ylabel('xC3', fontsize=14)
plt.show
plt.savefig("FigE11_7 Batch Distill.jpg")

#                          PROGRAM END
#############################################################################
```

Description of Program and Results. The ODE is specified in a user-defined function fode that is referenced in the call to function scipy.integrate.solve_ivp. Before function fode can calculate dc3dt, the composition of the vapor (i.e., yC3) must be calculated by function BPT. Function BPT uses function **scipy.optimize. newton** to calculate the bubble point temperature (TBP) by zeroing function er, and then TBP is used to calculate yC3. Note that function er requires the value of xC3; therefore, the value of xC3 is supplied to function er through function **scipy. optimize.newton** using args as an argument to the call to scipy.optimize. newton. By trial and error, the time required to reduce xC3 to 0.3 was found to be 4.949 h (plotted in Figure E11.8c). An overall steady-state material balance indicates that the amount of liquid in the receiver at the end of the run was 4.949 kg mol with a mole fraction of propane equal to 0.704.

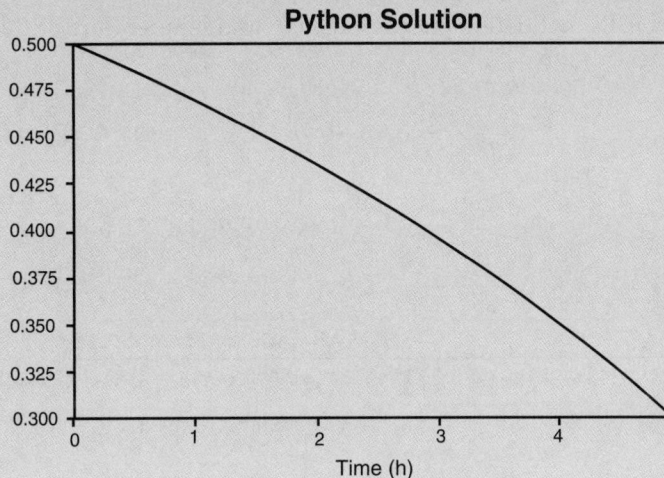

Figure E11.8c

Example 11.9   Numerical Solution of the Adiabatic Exothermic Batch Reactor

Problem Statement

Using MATLAB and Python, develop numerical solutions for the coupled set of ODEs developed for an adiabatic exothermic batch reactor in Example 11.4 using the following data:

- C_{A0} – initial reactant concentration (1 g mol/L)
- T_0 – initial temperature of the reactor (300 K)

(Continues)

Example 11.9   Numerical Solution of the Adiabatic Exothermic Batch Reactor (*Continued*)

- k_0 – pre-exponential rate constant (0.1)
- E_r/R – normalized activation energy for the reaction (900 K)
- C_p – heat capacity of the material in the batch reactor (4.18 J/g mol-K)
- ΔH_{rxn} – heat of reaction (–4000 J/g mol)
- ρ – molar density of the reaction mixture (50 g mol/L)

Solution

The ODEs that define the composition and temperature of this adiabatic exothermic reactor are

$$\frac{dC_A}{dt} = -k_0 \exp(-E_r/RT)C_A \qquad \text{at } t = 0 \quad C_A = C_{A0}$$

$$\rho C_p \frac{dT}{dt} = -\Delta H_{rxn} k_0 \exp(-E_r/RT)C_A \quad \text{at } t = 0 \quad T = T_0$$

MATLAB Code for Example 11.9

```
%%%%%%%%%%%%%%%%%%%%%%%%%%%%%%%%%%%%%%%%%%%%%%%%%%%%%%%%%%%%%%%%%%%%%%%
%                        NOMENCLATURE
%
% dydx - the vector of functions of the derivative of y with respect
%        to x
% soln - the solution matrix for the ODE problem
% x - a vector containing the values of the independent variable
% x0 - the initial value of x (0)
% xf - the final value of x (500)
% y - a vector containing the values of the dependent variable
% y0 - the initial value of y (1, 300)
%
%%%%%%%%%%%%%%%%%%%%%%%%%%%%%%%%%%%%%%%%%%%%%%%%%%%%%%%%%%%%%%%%%%%%%%%
%                          PROGRAM
function Ex11_9
clear; clc;
```

```
x0=0; xf=500; y0=[1, 300]';      % Input problems specifications
soln=ode45(@f,[x0 xf],y0);       % Call ode45 and store solution in soln
x=linspace(0,xf,100)             % Generate the values of x
y1=deval(soln,x,1);              % Retrieve value of y from soln
%% Output Plot %%
axes('FontSize',20);
plot(x,y1,'k-','LineWidth',2.5)
xlabel('t (sec)','FontSize',20)
ylabel('CA (gmol/L)','FontSize',20)
grid on;
title('MATLAB Solution','FontSize',24);
y2=deval(soln,x,2);              % Retrieve value of y from soln
%% Output Plot %%
axes('FontSize',20);
plot(x,y2,'k-','LineWidth',2.5)
xlabel('t (sec)','FontSize',20)
ylabel('T(K)','FontSize',20)
grid on;
title('MATLAB Solution','FontSize',24);
end
%
% User-specified function for dydx for each dependent variable
%
function dydx=f(x,y)
                                 % Specify dydx in column vector for
k0=1.E-1; ER=900; Cp=4.18; den=50; DHrxn=-4000;
dydx=[-k0*y(1)*exp(-ER/y(2)), -DHrxn*k0*y(1)*exp(-ER/y(2))/Cp/den]';
end
%                                PROGRAM END
%%%%%%%%%%%%%%%%%%%%%%%%%%%%%%%%%%%%%%%%%%%%%%%%%%%%%%%%%%%%%%%%%%%%%%
```

Description of Program and Results. The ODE is specified in a user-defined function f that is referenced in the call to function ode45. Most of the data for this problem is specified in the user-defined function f. Note that the reactant concentration decreased dramatically, while the reactor temperature increased by 18° K (Figure E11.9a).

(Continues)

Example 11.9   Numerical Solution of the Adiabatic Exothermic Batch Reactor (*Continued*)

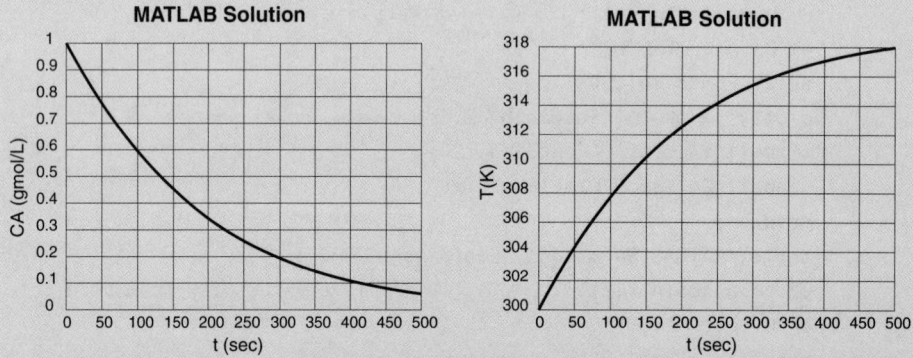

Figure E11.9a

Python Code for Example 11.9

```
#############################################################################

#                          NOMENCLATURE

#

#  dydx – the vector of functions of the derivative of y with respect to x
#  soln – the solution matrix for the ODE problem
#  time – the values of the independent variable used for the plot of the
#          results
#  tspan – a tuple that contains the initial and final condition for the
#           independent variable ([0, 150])
#  tvalues – the values for y that will be used to plot the solution
#  x – a vector containing the values of the independent variable
#  x0 – the initial value of x (0)
#  xf – the final value of x (500)
#  y – a vector containing the values of the dependent variable
#  y0 – the initial value of y , i.e., [1, 300]
#

#############################################################################

#                            PROGRAM
import scipy.integrate
import numpy as np
import matplotlib.pyplot as plt
```

```
#
# The user-defined function for dydx for each dependent variable
#
def fun(x, y):
    k0, ER, Cp, den, DHrxn = 0.1, 900, 4.18, 50, -4000
    dy1=-k0*y[0]*np.exp(-ER/y[1])
    dy2=-DHrxn*k0*y[0]*np.exp(-ER/y[1])/Cp/den
    return [dy1, dy2]
# Define input variables
xspan = [0.0,500]
y0=[1, 300]
xvalues=np.linspace(0, 500, 100)
#
# Apply scipy.integrate.solve_ivp with method equal to RK45
#
soln=scipy.integrate.solve_ivp(fun, xspan, y0, method='RK45',
t_eval=xvalues)
#
# Plot solution CA
#
time=soln.t               # Retrieve the independent variable vector
                               from soln matrix
y1=soln.y[0]              # Retrieve y1 from soln matrix
plt.plot(time, soln.y[0], 'k-', linewidth=2)
plt.axis([0, 500, 0, 1])
plt.title('Python Solution', fontsize=20)
plt.xlabel('Time (sec)', fontsize=14)
plt.ylabel('CA', fontsize=14)
plt.show
plt.savefig("FigE11.9 CA.jpg")
y1=soln.y[0]                 # Retrieve y1 from soln matrix
#
# Plot solution T
#
plt.plot(time, soln.y[1], 'k-', linewidth=2)
plt.axis([0, 500, 300, 318])
plt.title('Python Solution', fontsize=20)
```

(Continues)

Example 11.9 Numerical Solution of the Adiabatic Exothermic Batch Reactor (*Continued*)

```
plt.xlabel('Time (sec)', fontsize=14)
plt.ylabel('T (K)', fontsize=14)
plt.show
plt.savefig("FigE11.9 T.jpg")
#

#                          PROGRAM  END

###############################################################################
```

Description of Program and Results. The ODE is specified in a user-defined function fode that is referenced in the call to function scipy.integrate.solve_ivp. Most of the data for this problem is specified in the user-defined function fun. Note that the reactant concentration decreased dramatically, while the reactor temperature increased by 18° K (Figure E11.9b).

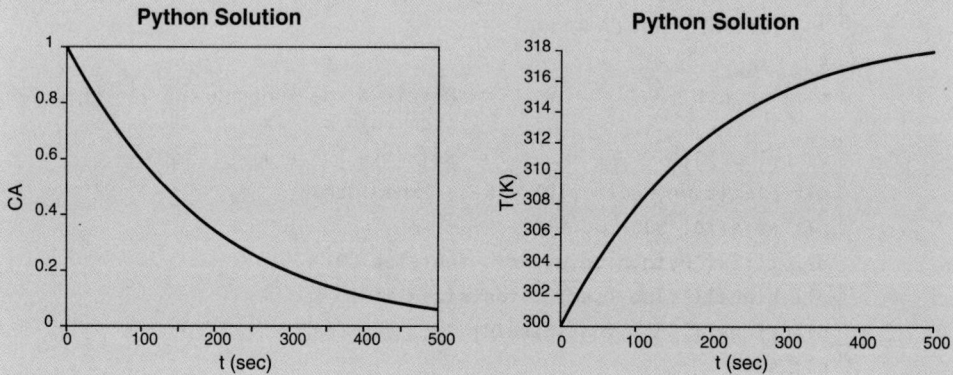

Figure E11.9b

Supplementary References

Clements, W. C. Unsteady-State Balances, AIChE Chemi Series No. F5.6, American Institute of Chemical Engineers, New York (1986).

Himmelblau, D. M., and K. B. Bischoff. Process Analysis and Simulation, Swift Publishing Co., Austin, TX (1980).

Porter, R. L., Unsteady-State Balances—Solution Techniques for Ordinary Differential Equations, AIChE Chemi Series No. F5.5, American Institute of Chemical Engineers, New York (1986).

Riggs, J. B., Computational Methods for Chemical Engineers, Ferret Publishing, Austin, Texas (2020).

Glossary

initial value problems (IVPs)   Problems involving a system of ordinary differential equations where values of all dependent variables are specified at the same value of the independent variable.

Runge-Kutta Method   A technique to solve IVPs by approximating the slope(s) of a function(s) for an interval through assigning weights to the slope(s) of the function(s) at several locations within the interval.

Problems

11.3.1   A tank containing 100 kg of a 60% brine (60% salt) is filled with a 10% salt solution at the rate of 10 kg/min. Solution is removed from the tank at the rate of 15 kg/min. Assuming complete mixing, find the kilograms of salt in the tank after 10 min.

11.3.2   A defective tank of 1500 ft^3 volume containing 100% propane gas (at 1 atm) is to be flushed with air (at 1 atm) until the propane concentration reduces to less than 1%. At that concentration of propane, the defect can be repaired by welding. If the flow rate of air into the tank is 30 ft^3/min, for how many minutes must the tank be flushed out? Assume that the flushing operation is conducted so that the gas in the tank is well mixed.

11.3.3   A 2% uranium oxide slurry (2 lb UO_2/100 lb H_2O) flows into a 100 gal tank at the rate of 2 gal/min. The tank initially contains 500 lb of H_2O and no UO_2. The slurry is well mixed and flows out at the same rate at which it enters. Find the concentration of slurry in the tank at the end of 1 hr.

11.3.4   The catalyst in a fluidized-bed reactor of 200 m^3 volume is to be regenerated by contact with a hydrogen stream. Before the hydrogen can be introduced in the reactor, the O_2 content of the air in the reactor must be reduced to 0.1%. If pure N_2 can be fed into the reactor at the rate of 20 m^3/min, for how long should the reactor be purged with N_2? Assure that the catalyst solids occupy 6% of the reactor volume and that the gases are well mixed.

11.3.5   An advertising firm wants to get a special inflated sign out of a warehouse. The sign is 20 ft in diameter and is filled with H_2 at 15 psig. Unfortunately, the door frame to the warehouse permits only 19 ft to pass. The maximum rate of H_2 that can be safely vented from the balloon is 5 ft^3/min (measured at room conditions). How long will it take to get the sign small enough to just pass through the door?

   a. First assume that the pressure inside the balloon is constant so that the flow rate is constant.

   b. Then assume the amount of H_2 escaping is proportional to the volume of the balloon and initially is 5 ft^3/min. Could a solution to this problem be obtained if the amount of escaping H_2 were proportional to the pressure difference inside and outside the balloon?

11.3.6   A plant at Canso, Nova Scotia, makes fish-protein concentrate (FPC). It takes 6.6 kg of whole fish to make 1 kg of FPC, and therein is the problem—to make money, the plant must operate most of the year. One of the operating problems is the drying of the FPC. It dries in the fluidized dryer at a rate approximately proportional to its moisture content. If a given batch of FPC loses one-half of its initial moisture in the first 15 min, how long will it take to remove 90% of the water in the batch of FPC?

11.3.7   Water flows from a conical tank at the rate of $0.020(2 + h^2)$ m^3/min, as shown in Figure P11.3.7. If the tank is initially full, how long will it take for 75% of the water to flow out of the tank? What is the flow rate at that time?

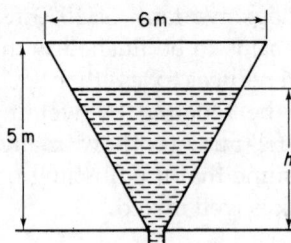

Figure P11.3.7

11.3.8   A sewage disposal plant has a big concrete holding tank of 100,000 gal capacity. It is three-fourths full of liquid to start with and contains 60,000 lb of organic material in suspension. Water runs into the holding tank at the rate of 20,000 gal/hr, and the solution leaves at the rate of 15,000 gal/hr. How much organic material is in the tank at the end of 3 hr?

11.3.9   Suppose that in problem 11.3.8 the bottom of the tank is covered with sludge (precipitated organic material) and that the stirring of the tank causes the sludge to go into suspension at a rate proportional to the difference between the concentration of sludge in the tank at any time and 10 lb of sludge/gal. If no organic material were present, the sludge would go into suspension at the rate of 0.05 lb/(min)(gal solution) when 75,000 gal of solution are in the tank. How much organic material is in the tank at the end of 3 hr?

11.3.10 In a chemical reaction, the products X and Y are formed according to the equation

$$C \rightarrow X + Y$$

The rate at which each of these products is being formed is proportional to the amount of C present. Initially, $C = 1$, $X = 0$, $Y = 0$. Find the time for the amount of X to equal the amount of C.

11.3.11 A tank is filled with water. At a given instant, two orifices in the side of the tank are opened to discharge the water. The water at the start is 3 m deep, and one orifice is 2 m below the top while the other one is 2.5 m below the top. The coefficient of discharge of each orifice is known to be 0.61. The tank is a vertical right circular cylinder 2 m in diameter. The upper and lower orifices are 5 and 10 cm in diameter, respectively. How much time will be required for the tank to be drained so that the water level is at a depth of 1.5 m?

11.3.12 Suppose that you have two tanks in series, as diagrammed in Figure P11.3.12. The volume of liquid in each tank remains constant because of the design of the overflow lines. Assume that each tank is filled with a solution containing 10 lb of A, and that each tank contains 100 gal of aqueous solution of A. If freshwater enters at the rate of 10 gal/hr, what is the concentration of A in each tank at the end of 3 hr? Assume complete mixing in each tank and ignore any change of volume with concentration.

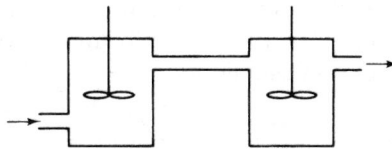

Figure P11.3.12

11.3.13 A well-mixed tank has a maximum capacity of 100 gal and it is initially half full. The discharge pipe at the bottom is very long, and thus it offers resistance to the flow of water through it. The force that causes the water to flow is the height of the water in the tank, and in fact, that flow is just proportional to the height. Since the height is proportional to the total volume of water in the tank, the volumetric flow rate of water out, q_o, is

$$q_o = kV$$

The flow rate of water into the tank, q_i, is constant. Use the information given in Figure P11.3.13 to decide whether the amount of water in the tank increases, decreases, or remains the same. If it changes, how much time is required to completely empty or fill the tank, as the case may be?

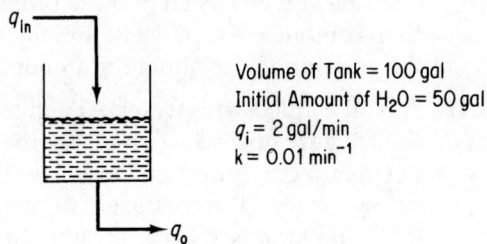

Volume of Tank = 100 gal
Initial Amount of H_2O = 50 gal
q_i = 2 gal/min
k = 0.01 min^{-1}

Figure P11.3.13

11.3.14 A stream containing a radioactive fission product with a decay constant of 0.01 hr^{-1} (i.e., $dn/dt = 0.01n$) is run into a holding tank at the rate of 100 gal/hr for 24 hr. Then the stream is shut off for 24 hr. If the initial concentration of the fission product was 10 mg/L and the tank volume is constant at 10,000 gal of solution (owing to an overflow line), what is the concentration of fission product (a) at the end of the first 24-hr period and (b) at the end of the second 24-hr period? What is the maximum concentration of fission product? Assume complete mixing in the tank.

11.3.15 A radioactive waste that contains 1500 ppm of ^{92}Sr is pumped at the rate of 1.5×10^{-3} m^3/min into a holding tank that contains 0.40 m^3. ^{92}Sr decays as follows:

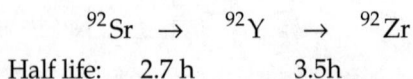

$$^{92}Sr \quad \rightarrow \quad ^{92}Y \quad \rightarrow \quad ^{92}Zr$$

Half life: 2.7 h   3.5h

If the tank contains clear water initially and the solution runs out at the rate of 1.5×10^{-3} m^3/min, assuming perfect mixing:

    a. What is the concentration of ^{92}Sr, ^{92}Y, and ^{92}Zr after 1 day?

    b. What is the equilibrium concentration of ^{92}Sr and ^{92}Y in the tank? The rate of decay of such isotopes is $dN/dt = -\lambda N$, where $\lambda = 0.693/t_{1/2}$ and the half-life is $t_{1/2}$. $N =$ moles.

11.3.16 A tank contains 3 m^3 of pure oxygen at atmospheric pressure. Air is slowly pumped into the tank and mixes uniformly with the contents, an equal volume of which is forced out of the tank. What is the concentration of oxygen in the tank after 9 m^3 of air has been admitted?

11.3.17 Suppose that an organic compound decomposes as follows: $C_6H_{12} \rightarrow C_4H_8 + C_2H_4$. If 1 mol of C_6H_{12}, but no C_4H_8 and C_2H_4, exists at $t = 0$ set up equations showing the moles of C_4H_8 and C_2H_4 as a function of time. The rates of formation of C_4H_8 and C_2H_4 are proportional to the number of moles of C_6H_{12} present.

11.3.18 A large tank is connected to a smaller tank by means of a valve. The large tank contains N_2 at 690 kPa, and the small tank is evacuated. If the valve leaks between the two tanks and the rate of leakage of gas is proportional to the pressure difference between the two tanks (p1 – p2), how long does it take for the pressure in the small tank to be one-half its final value? The instantaneous initial flow rate with the small tank evacuated is 0.091 kg mol/hr.

	Tank 1	Tank 2
Initial Pressure (kPa)	700	0
Volume (m^3)	30	15

Assume that the temperature in both tanks is held constant and is 20°C.

11.3.19 The following chain reactions take place in a constant-volume batch tank:

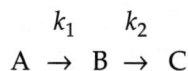

$$A \xrightarrow{k_1} B \xrightarrow{k_2} C$$

Each reaction is first order and irreversible. If the initial concentration of A is C_{A0} and if only A is present initially, find an expression for C_B as a function of time. Under what conditions will the concentration of B be dependent primarily on the rate of reaction of A?

11.3.20 Consider the following chemical reactions in a constant-volume batch tank: All of the indicated reactions are first order. The initial concentration of A is C_{A0}, and nothing else is present at that time. Determine the concentrations of A, B, and C as functions of time.

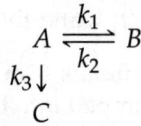

$$A \underset{k_2}{\overset{k_1}{\rightleftharpoons}} B$$
$$k_3 \downarrow$$
$$C$$

Figure P11.3.20

11.3.21 Tanks A, B, and C are each filled with 1000 gal of water. See Figure P11.3.21. Workers have instructions to dissolve 2000 lb of salt in each tank. By mistake, 3000 lb are dissolved in each of tanks A and C and none in B. You wish to bring all of the compositions to within 5% of the specified 2 lb/gal. If the units are connected A—B—C—A by three 50 gal/min (gpm) pumps:
  a. Express the concentrations C_A, C_B, and C_C in terms of t (time).
  b. Find the shortest time at which all concentrations are within the specified range. Assume the tanks are all well mixed.

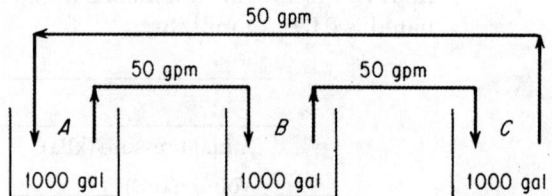

Figure P11.3.21

11.3.22 Determine the time required to heat a 10,000 lb batch of liquid from 60°F to 120°F using an external, counterflow heat exchanger having an area of 300 ft². Water at 180°F is used as the heating medium and flows at a rate of 6000 lb/hr. An overall heat transfer coefficient of 40 Btu/(hr) (ft²)(°F) may be assumed; use Newton's law of heating. The liquid is circulated at a rate of 6000 lb/hr, and the specific heat of the liquid is the same as that of water (1.0). Assume that the residence time of the liquid in the external heat exchanger is very small and that there is essentially no holdup of liquid in this circuit.

11.3.23 A ground material is to be suspended in water and heated in preparation for a chemical reaction. It is desired to carry out the mixing and heating simultaneously in a tank equipped with an agitator and a steam coil. The cold liquid and solid are to be added continuously, and the heated suspension will be withdrawn at the same rate. One method of operation for starting up is to (1) fill the tank initially with water and solid in the proper proportions, (2) start the agitator, (3) introduce freshwater and solid in the proper proportions and simultaneously begin to withdraw the suspension for reaction, and (4) turn on the steam. An estimate is needed of the time required, after the steam is turned on, for the temperature of the effluent suspension to reach a certain elevated temperature.

a. Using the nomenclature given below, formulate a differential equation for this process. Integrate the equation to obtain n as a function of B and ϕ (see nomenclature).

b. Calculate the time required for the effluent temperature to reach 180°F if the initial contents of the tank and the inflow are both at 120°F and the steam temperature is 220°F. The surface area for heat transfer is 23.9 ft^2, and the heat transfer coefficient is 100 Btu/(hr)(ft^2)(°F). The tank contains 6000 lb, and the rate of flow of both streams is 1200 lb/hr. In the proportions used, the specific heat of the suspension may be assumed to be 1.00. If the area available for heat transfer is doubled, how will the time required be affected? Why is the time with the larger area less than half that obtained previously? The heat transferred is $Q = UA(T - T_S)$.

Nomenclature

W – weight of tank contents, lb
G – rate of flow of suspension, lb/hr
T_S – temperature of steam, °F
T – temperature in tank at any instant, perfect mixing assumed, °F
T_0 – temperature of suspension introduced into tank; also, initial temperature of tank contents, °F
U – heat transfer coefficient, Btu/(hr)(ft2)(°F)
A – area of heat transfer surface, ft^2
C_p – specific heat of suspension, Btu/(lb)(°F)
t – time elapsed from the instant the steam is turned on, hr
n – dimensionless time, Gt/W
B – dimensionless ratio, UA/GC_p
ϕ – dimensionless temperature (relative approach to the steam temperature) $(T - T_0)/(T_S - T_0)$

11.3.24 Consider a well-agitated cylindrical tank in which the heat transfer surface is in the form of a coil that is distributed uniformly from the bottom of the tank to the top of the tank. The tank itself is completely insulated. Liquid is introduced into the tank at a uniform rate, starting with no liquid in the tank, and the steam is turned on at the instant that liquid flows into the tank.

      a. Using the nomenclature of problem 11.3.23, formulate a differential equation for this process. Integrate the differential equation to obtain an equation for ϕ as a function of B and f, where f = fraction filled = W/W_{filled}.

      b. If the heat transfer surface consists of a coil of 10 turns of 1-in.-OD tubing 4 ft in diameter, the feed rate is 1200 lb/hr, the heat capacity of the liquid is 1.0 Btu/(lb)(°F), the heat transfer coefficient is 100 Btu/(hr)(°F)(ft²) of covered area, the steam temperature is 200°F, and the temperature of the liquid introduced into the tank is 70°F, what is the temperature in the tank when it is completely full? What is the temperature when the tank is half full? The heat transfer is given by $Q = UA(T - T_S)$.

11.3.25 A cylindrical tank 5 ft in diameter and 5 ft high is full of water at 70°F. The water is to be heated by means of a steam jacket around the sides only. The steam temperature is 230°F, and the overall coefficient of heat transfer is constant at 40 Btu/(hr)(ft²)(°F). Use Newton's law of cooling (heating) to estimate the heat transfer. Neglecting the heat losses from the top and the bottom, calculate the time necessary to raise the temperature of the tank contents to 170°F. Repeat, taking the heat losses from the top and the bottom into account. The air temperature around the tank is 70°F, and the overall coefficient of heat transfer for both the top and the bottom is constant at 10 Btu/(hr)(ft²)(°F).

CHAPTER 12

Heats of Solution and Mixing

Your objectives in studying this chapter are to be able to

1. Distinguish between ideal solutions and real solutions
2. Understand how energy changes occur on mixing
3. Distinguish between integral heat of solution, differential heat of solution, heat of solution at infinite dilution, and heat of solution in the standard state
4. Calculate the heat of mixing, or the heat of dissolution, at standard conditions given the moles of the materials forming the mixture and experimental data
5. Calculate the standard integral heat of solution
6. Apply an energy balance to problems in which the heat of mixing is significant
7. Use an enthalpy-concentration chart in solving material and energy balances

---

We have deferred consideration of the heat of solution/mixing until this chapter to avoid making the presentation of energy balances more complicated. But, in many processes, the energy involved when solution/mixing occurs is too large to be ignored. Recall from chemistry the warning "Never add water to sulfuric acid, but slowly add the acid to water."

Looking Ahead

In this chapter we explain how mixing and solution of one component in another to form a binary solution affect the energy balance. We focus only on binary mixtures here. You will learn several new terms and related calculations that affect the enthalpy terms in the general energy balance.

12.1   Heats of Solution, Dissolution, and Mixing

Up to this chapter we have assumed that when a stream consists of several components, the total properties of the stream are the appropriately weighted sum of the properties of the individual components. For such **ideal solutions,** we could write down for the heat capacity of an ideal mixture, for example,

$$C_{p\,\text{mixture}} = x_A C_{p_A} + x_B C_{p_B} + x_C C_{p_C} + \cdots$$

or, for the enthalpy,

$$\Delta \hat{H}_{\text{mixture}} = x_A \Delta \hat{H}_A + x_B \Delta \hat{H}_B + x_C \Delta \hat{H}_C + \cdots$$

In particular, mixtures of gases have been treated as ideal solutions.

However, you must take into consideration other types of mixtures. You can prepare various kinds of binary solutions or mixtures:

  a. gas-gas
  b. gas-liquid
  c. gas-solid
  d. liquid-liquid
  e. liquid-solid
  f. solid-solid

You can ignore the energy changes that occur on mixing for cases a, c, and f. They are negligible. The other mixtures constitute **real solutions.** When a gaseous or solid **solute** (the compound to be dissolved) is mixed with a liquid **solvent** (the compound in which the solute is dissolved), the energy effect that occurs is referred to as the **heat** (really **enthalpy**) **of solution.** When a liquid is mixed with a liquid, the energy effect is called the **heat (enthalpy) of mixing.** The negative of the heat of solution or mixing is the **heat (enthalpy) of dissolution.**

The heat of solution can be positive (endothermic) or negative (exothermic). Examine Figure 12.1, which shows the relative enthalpy values of a mixture of a fluorocarbon (C_6F_6) in benzene (C_6H_6).

You can treat heats of solution/mixing in the same way you treat chemical reactions. In the energy balance, you can (a) merge the heats of solution/mixing of the compounds in the system with the heats of formation, or (b) consolidate the effects of the heats of solution/mixing in one lumped term analogous to a heat of reaction term. For example, let us represent the solution of 1 g mol of HCl(g) into 5 g mol of by the following chemical equation:

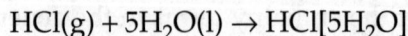

$$HCl(g) + 5H_2O(l) \rightarrow HCl[5H_2O]$$

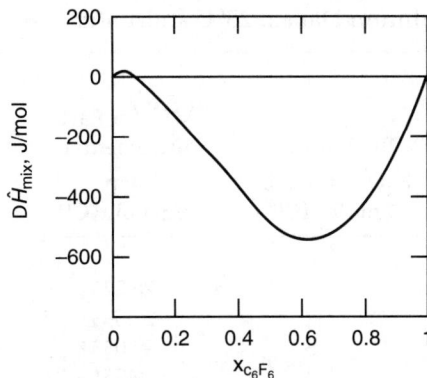

Figure 12.1   Relative enthalpy change on mixing C_6F_6 in C_6H_6 at 25°C

If you carry out experiments to measure the heat transfer from an apparatus at a constant 25°C and 1 atm by successively adding water to HCl, and arrange the experiments so that the energy balance reduces to $Q = \Delta H$, then the values of ΔH would be the tabulation in the third column of Table 12.1. (The values incorporate a slight adjustment in the measured values of Q at the vapor pressure of the solution to adjust them to 1 atm, the standard state.) If you cumulate each incremental change in ΔH, you would obtain the fourth column in Table 12.1. Appendix I on the CD that accompanies this book contains other tables listing the heats of solution for common compounds. Table 12.1 shows that there are actually two concepts that incorporate the name "heat of solution":

a. The **incremental (differential) heat of solution,** column 3
b. The **integral heat of solution,** column 4—the heat of solution for the combination of 1 mole of HCl(g) with n moles of $H_2O(l)$

Usually "heat of solution" refers to concept b, and **the enthalpy change is stated per mole of solute.**

Figure 12.2 is a plot of the values listed in column 4 of Table 12.1. The asymptotic value of the heat of solution of HCl dissolved in an infinite amount of water is known as the **heat of solution at infinite dilution** (–75, 144 J/g mol HCl).

If you want to calculate the heat of formation of any of the solutions of HCl(g) in all you have to do is add the heat of solution to the heat of formation of HCl(g), as shown in column 5 in Table 12.1

$$\Delta \hat{H}^{\circ}_{f,\text{ solution}} = \Delta \hat{H}^{\circ}_{f,\text{ solute}} + \Delta \hat{H}^{\circ}_{\text{solution}} \qquad (12.1)$$

where $\Delta \hat{H}^{\circ}_{\text{solution}}$ is the integral heat of solution at standard conditions per mole of HCl, and $\Delta \hat{H}^{\circ}_{f,\text{ solution}}$ is the heat of formation of the solution itself per

Table 12.1 Heat of Solution Data at 25°C and 1 atm

Composition	Total Moles H_2O Added to 1 mole HCl	$\Delta\hat{H}^o$ for Each Incremental Step (J/g mol HCl)	Integral Heat of Solution (Cumulative $\Delta\hat{H}^o$) (J/g mol HCl)	Heat of Formation $\Delta\hat{H}_f^0$ (J/g mol HCl)
HCl(g)	0			−92,311
HCl[$1H_2O$(aq)]	1	−26,225	−26,225	−118,536
HCl[$2H_2O$(aq)]	2	−22,593	−48,818	−141,129
HCl[$3H_2O$(aq)]	3	−8033	−56,851	−149,161
HCl[$4H_2O$(aq)]	4	−4351	−61,202	−153,513
HCl[$5H_2O$(aq)]	5	−2845	−64,047	−156,358
HCl[$8H_2O$(aq)]	8	−4184	−68,231	−160,542
HCl[$10H_2O$(aq)]	10	−1255	−69,486	−161,797
HCl[$15H_2O$(aq)]	15	−1503	−70,989	−163,300
HCl[$25H_2O$(aq)]	25	−1276	−72,265	−164,576
HCl[$50H_2O$(aq)]	50	−1013	−73,278	−165,589
HCl[$100H_2O$(aq)]	100	−569	−73,847	−166,158
HCl[$200H_2O$(aq)]	200	−356	−74,203	−166,514
HCl[$500H_2O$(aq)]	500	−318	−74,521	−166,832
HCl[$1000H_2O$(aq)]	1000	−163	−74,684	−166,995
HCl[$50,000H_2O$(aq)]	50,000	−146	−75,077	−167,388
HCl[∞H_2O]		−67	−75,144	−167,455

SOURCE: *National Bureau of Standards Circular 500*, U.S. Government Printing Office, Washington, DC (1952).

mole of HCl. It is important to remember that **the heat of formation of the H_2O does not enter into the calculation** in Equation (12.1); it is defined as zero for the process of solution. Tables in reference books usually list data for the heats of formation of solutions in the standard state rather than the heats of solution themselves. In the following processes and examples, we assume that the systems are open, steady-state, or, if closed, that the accumulation term in the energy balance is $\Delta U = \Delta H$ so that we discuss only enthalpies.

You can treat a solution as a single compound in making calculations in an energy balance by using the property that enthalpies of solutions are state variables. One convenient procedure is to merge the heats of solution of a compound with the heats of formation, as indicated in column 5 of Table 12.1

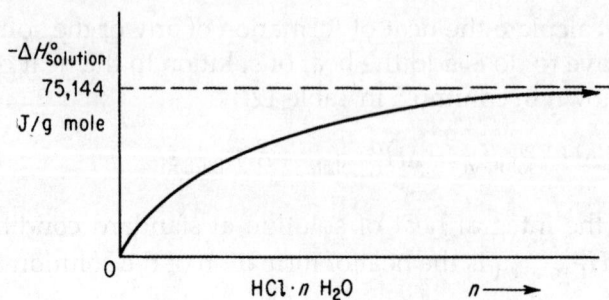

Figure 12.2 Integral heat of solution of HCl in water

and by Equation (12.1). Then the specific enthalpy of a solution relative to S.C. would be

$$\hat{H}_{solution}(T) = \Delta\hat{H}^{o}_{f,\ solution} + [\hat{H}(T) - \hat{H}(\ S.C.\ )]_{solution} \qquad (12.2)$$

where the term in the brackets represents the sensible heat of the solution itself (any phase change is unlikely).

For example, suppose you want to find the $\Delta\hat{H}^{o}_{solution}$ that occurs at S.C. when a solution of 1 g mol of HCl dissolved in 1 g mol of H_2O is placed in an infinite amount of water. Use the data listed in column 5 of Table 12.1, and subtract the enthalpy of the initial state from the enthalpy of the final state as follows:

$$(-167{,}455) - (-118{,}536) = -48{,}911 \text{ J/g mol HCl}$$

As another example, consider a process in which a dilute solution of HCl is to be concentrated. Because enthalpy changes for heats of solution are state variables, you can easily look up the heats of formation of HCl solutions at their respective concentrations. Then you can calculate the enthalpy change between the final and initial states for a closed process, or the enthalpy change between the output and the input for a flow process. Thus, if you mix 1 mole of $HCl[15H_2O]$ and 1 mole of $HCl[5H_2O]$, you obtain 2 moles of $HCl[10H_2O]$, and the total enthalpy change at 25°C and 1 atm is

$$\Delta H^{o} = [2(-69{,}486)] - [1(-70{,}989) + 1(-64{,}047)]$$
$$= -3936 \text{ J}$$

You would have to remove 3936 J to keep the temperature of the final mixture at 25°C.

We should mention that the solution of a hydrated salt such as $CaCl_2 \cdot 6H_2O$ requires a little care in the calculations. If you mix the hydrated salt with water or a $CaCl_2$ solution, the procedure to calculate the enthalpy change is as follows: You first have to decompose (melt) the hydrated salt into a solid and water. Then you dissolve the total salt available into the total water available after the melting and solution. For example, from the data in problem 12.3, if 1 g mol of $Na_2CO_3 \cdot 7H_2O$ is dissolved in 8 g mol of the resulting solution contains 1 g mol of Na_2CO_3 and 15 g mol of H_2O. The melting step involves the following enthalpy change:

$$Na_2CO_3 \cdot 7H_2O(s) \quad \rightarrow \quad Na_2CO_3(s) \quad + \quad 7H_2O(l)$$

$$\Delta\hat{H}^{o}_{f}(kJ/gmol): \qquad -3201.18 \qquad\qquad -1130.92 \qquad -285.840$$

$$\Delta H^{o} = [7(-285.840) + 1(-1130.92)] - [1(-3201.18)] = +69.38 \text{ kJ}$$

The solution step is

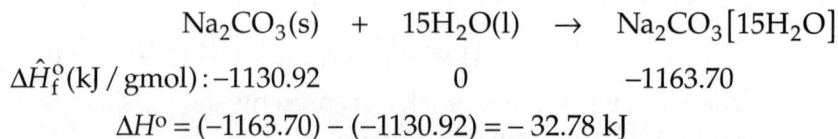

$$Na_2CO_3(s) \quad + \quad 15H_2O(l) \quad \rightarrow \quad Na_2CO_3[15H_2O]$$

$$\Delta\hat{H}_f^o(kJ/gmol): -1130.92 \qquad\qquad 0 \qquad\qquad -1163.70$$

$$\Delta H^o = (-1163.70) - (-1130.92) = -32.78 \text{ kJ}$$

The overall enthalpy change is $69.38 - 32.78 = 36.60$ kJ.

Example 12.1   Application of Heat of Solution Data

You are asked to prepare an ammonium hydroxide solution at 77°F by dissolving gaseous NH_3 in water. Calculate (a) the amount of cooling needed in British thermal units to prepare a 3.0% solution containing 1 lb mol of NH_3, and (b) the amount of cooling needed in British thermal units to prepare 100 gal of a solution of 32.0% NH_3.

  Data: The following heat of solution data have been taken from *NBS circular 500*.

Composition	State	$-\Delta\hat{H}_f^o$ (Btu/lb mol)	$-\Delta\hat{H}_{soln}^o$ (Btu/lb mol)
	g	19,900	0
1H₂O	aq	32,600	12,700
2H₂O	aq	33,600	13,700
3H₂O	aq	34,000	14,100
4H₂O	aq	34,200	14,300
5H₂O	aq	34,350	14,450
10H₂O	aq	34,600	14,700
20H₂O	aq	34,700	14,800
30H₂O	aq	34,700	14,800
40H₂O	aq	34,700	14,800
50H₂O	aq	34,750	14,850
100H₂O	aq	34,750	14,850
200H₂O	aq	34,800	14,900
∞ H₂O	aq	34,800	14,900

Solution

The solution will be presented in abbreviated form.
Reference temperature: 77°F

a. Basis: 1 lb mol $NH_3 \equiv 17$ lb NH_3

$$\text{wt \% NH}_3 = \frac{\text{lb NH}_3}{\text{lb H}_2\text{O} + \text{lb NH}_3}(100)$$

$$3 = \frac{17(100)}{17 + m_{H_2O}} \qquad m_{H_2O} = 550 \text{ lb or about 30 lb mol H}_2\text{O}$$

(Continues)

Example 12.1 Application of Heat of Solution Data (*Continued*)

From the table above, the $\Delta \hat{H}^{\circ}_{soln} = -14,800$ Btu/lb mol NH_3, which is equal to 14,800 Btu removed from the system.

b. Basis: 100 gal solution

From Lange's *Handbook of Chemistry*, a 32.0% solution of NH_3 has the following properties:

	NH_3	H_2O
Sp.gr.:	0.889	1.003

The density of the 32.0% solution in pounds per 100 gal is

$$\frac{(0.889)(62.4)(1.003)(100)}{7.48} = 744 \text{lb} / 100 \text{gal}$$

$$\frac{744(0.32)}{17} = 14.0 \text{ lb mol } NH_3/100 \text{ gal solution}$$

Basis: 1 lb mol NH_3

$$32.0 = \frac{1700}{17 + m_{H_2O}} \quad m_{H_2O} = 36 \text{ lb and } n_{H_2O} = 2 \text{ lb mol}$$

$$\left. \begin{array}{c} \text{Cooling req'd} \\ \text{Btu/100 gal} \end{array} \right\} = \left(\frac{\text{lbmolNH}_3}{100 \text{gal}} \right) \left(-\Delta H^{\circ}_{soln} \frac{\text{Btu}}{\text{lbmolNH}_3} \right)$$

$$\left. \begin{array}{c} \text{Cooling req'd} \\ \text{for 100 gal} \\ 32.0\% NH_3 \text{ soln} \end{array} \right\} = (14.0)(-13,700) = -191,000 \text{Btu} / 100 \text{ gal soln (heat removed)}$$

Self-Assessment Test

Questions

1. Define and show on a sketch (a) the integral heat of solution, and (b) the differential heat of solution.

2. Indicate whether the following statements are true or false:
   a. Heats of reaction and heats of solution represent the same physical phenomena.
   b. All mixtures have significant heats of solution.
   c. The heat of mixing at infinite dilution involves an infinite amount of solvent.
   d. Heats of solution can be positive and negative.
   e. A gas mixture is usually an ideal solution.

3. a. What is the reference state for H_2O in the table for the heat of solution of HCl?
   b. What is the value of the enthalpy of the H_2O in the reference state?

4. Repeat question 3 for HCl.

Problems

1. Calculate the heat of solution at standard conditions when 1 mol of a solution of 20 mol % HCl is mixed with 1 mol of a solution of 25 mol % HCl.

2. How much heat has to be added to a solution of 1 g mol of HCl in 10 g mol of H_2O to concentrate the solution to 1 g mol of HCl in 4 g mol of H_2O?

3. The heat of formation of H_2SO_4 is -811.319 kJ/g mol H_2SO_4. What is the heat of formation per gram mole of H_2SO_4 of a solution of 20% sulfuric acid?

Thought Problems

1. A tanker truck of hydrochloric acid was inadvertently unloaded into a large storage tank used for sulfuric acid. After about one-half of the 3000 gal load had been discharged, a violent explosion occurred, breaking the inlet and outlet lines and buckling the tank. What might be the cause of the explosion?

2. A concentrated solution (73%) of sodium hydroxide was stored in a vessel. Under normal operations, the solution was forced out by air pressure as needed. When application of air pressure did not work, apparently due to solidification of the caustic solution, water was poured through a manhole to dilute the caustic and free up the pressure line. An explosion took place and splashed caustic out of the manhole 15 ft into the air. What caused the incident?

Discussion Problem

A significant amount of energy is liberated when freshwater and saltwater are mixed. It has been calculated that the dilution of a cubic meter of freshwater per second in a large volume of seawater dissipates roughly 2.3 MW of power. If this energy could be put to use rather than heating the ocean, it is estimated that the potential of the flow of the Columbia River would yield 15,000 MW. The technology to collect such potential has been proposed, namely, to use a selective membrane that lets certain molecules through but holds others back. Instead of separating water from saltwater by imposing an electric potential on the membrane as in desalinization, the idea is to reverse the process and mix freshwater with saltwater to generate an electric current. The membranes are arranged so that positive ions flow in one direction and negative ions in the other direction. What do you think of the proposal?

12.2   Introducing the Effects of Mixing into the Energy Balance

In Section 12.1 we restricted the discussion and examples to the standard state (25°C and 1 atm). In this section we proceed with what happens when the temperatures of the inlet and outlet streams differ from 25°C for a binary mixture in an open, steady-state process. (For a closed system the initial and final states of the internal energy would be involved rather than the stream flows.) You can treat problems involving the heat of solution/mixing in exactly the same way that you can treat problems involving reaction. The heat of solution/mixing is analogous to the heat of reaction in the energy balance. You can carry out the needed calculations by

a. Associating heats of formation of the compounds and solutions with each of the respective compounds and solutions, or

b. Computing the overall lumped heat of solution at the reference state,

and for either option calculating the sensible heats (and phase change effects) for the compounds and solutions from the reference state.

The next example shows the detailed procedure.

Example 12.2   Application of Heat of Solution Data

Hydrochloric acid is an important industrial chemical. To make aqueous solutions of it in a commercial grade (known as *muriatic acid*), purified HCl(g) is absorbed in water in a tantalum absorber in a steady-state continuous process. How much heat must be removed from the absorber by the cooling water per 100 kg of product if hot HCl(g) at 120°C is fed into water in the absorber as shown in Figure E12.2? The feed water can be assumed to be at 25°C, and the exit product HCl(aq) is 25% HCl (by weight) at 35°C. The cooling water does not mix with the HCl solution.

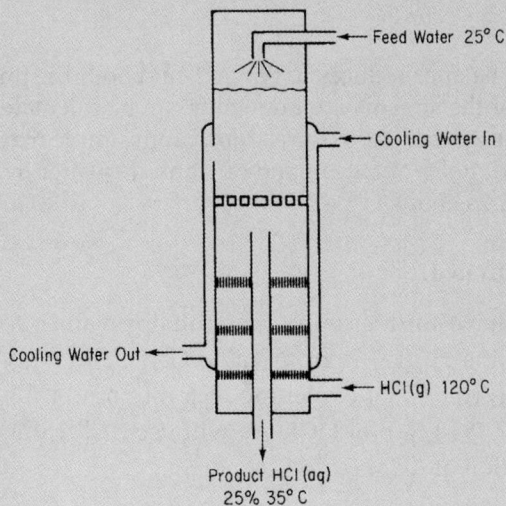

Feed Water 25° C

Cooling Water In

Cooling Water Out

HCl(g) 120° C

Product HCl (aq)
25% 35° C

Figure E12.2

Solution

Steps 1–4

You need to convert the process data to moles of HCl to be able to use the data in Table 12.1. Consequently, we will first convert the product into moles of HCl and moles of H_2O.

Table E12.2a

Component	kg	Mol. wt.	kg mol	Mole Fraction
HCl	25	36.37	0.685	0.141
H_2O	75	18.02	4.163	0.859
Total	100		4.848	1.000

The mole ratio of H_2O to HCl is $4.163/0.685 = 6.077$.

Step 5

The system will be the HCl and water (not including the cooling water).

Basis: 100 kg of product

Ref. temperature: 25°C

Steps 6 and 7

The energy balance reduces to $Q = \Delta H$, and both the initial and final enthalpies of all of the streams are known or can be calculated directly; hence the problem has zero degrees of freedom. From simple material balances the kilograms and moles of HCl in and out, and the water in and out, are as listed in Table E12.2a above.

Step 3 (continued)

Next, you have to determine the enthalpy values for the streams. Data are: C_p for the HCl(g) is from Table G.1; C_p for the product is approximately 2.7 J/(g)(°C) equivalent to 55.6 J/(g mol) (°C); $\Delta \hat{H}_f^o$ for HCl · 6.077 $H_2O \cong -157{,}753$ J/g mol HCl. We will use $\Delta \hat{H}_f^o$ values for each stream in the calculation of ΔH.

(Continues)

Example 12.2   Application of Heat of Solution Data (*Continued*)

Steps 8 and 9

Table E12.2b

Stream	g mol	$T(°C)$	$\Delta \hat{H}_f^o$ (J/g mol HCl)	$\Delta \hat{H}_{sensible}$ (J/g mol)
OUT				
HCl (aq)	4.848*	35	−157,753	$\int_{25°C}^{35°C} (2.7)dT$
IN				
$H_2O(l)$	4.163	25	−	
HCl(g)	0.685	120	−92,311	$\int_{25°C}^{120°C} \left(29.13 - 0.134 \times 10^{-2} T\right)dT$ $= 2758$

*HCl=0.685

$$Q = \Delta H_{out} - \Delta H_{in}$$

$$\underset{Out}{=[0.685(-157,753)+4.848(27)]} - \underset{in}{[0+0.685(-92,311)+0.685(2758)]}$$

$$= -46,586 J$$

If you use heat of solution values, the calculation is (from Table 12.1 the heat of solution is −65,442 J/g mol HCl for the ratio of $HCl/H_2O = 6.077$)

$$Q = \Delta H_{out, \, sensible} - \Delta H_{in, \, sensible} + \Delta H_{solution}$$
$$= (4.848)\,(27) - [0.685(2753) + 0] + (0.685)\,(-65,442)$$
$$= -46,586 \, J \quad \text{as expected}$$

In a process simulation code, table lookup or equations would be used to calculate the heats of formation at various temperatures (and pressures) other than 25°C (and 1 atm). The details would be buried in the computer code. You can better understand what the calculations for an energy balance involve if you use a graph—at the expense of some accuracy—instead of equations.

A convenient graphical way to represent enthalpy data for binary solutions is via an **enthalpy-concentration diagram.** Enthalpy-concentration diagrams (*H-x*) are plots of specific enthalpy versus concentration (usually mass or mole fraction) with temperature as a parameter. Figure 12.3 illustrates one such plot. If available,[1] such charts are useful in making combined material

---

[1]For a literature survey as of 1957, see Robert Lemlich, Chad Gottschlich, and Ronald Hoke, *Chem. Eng. Data Ser.*, **2**, 32 (1957). Additional references: for CCl_4, see M. M. Krishniah et al, *J. Chem. Eng. Data,* **10**, 117 (1965); for EtOH-EtAc, see Robert Lemlich, Chad Gottschlich, and Ronald Hoke, *Br. Chem. Eng.*, **10**, 703 (1965); for methanol-toluene, see C. A. Plank and D. E. Burke, *Hydrocarbon Process*, **45**, No. 8, 167 (1966); for acetone-isopropanol, see S. N. Balasubramanian, *Br. Chem. Eng.*, **12**, 1231 (1967); for acetonitrile-water-ethanol, see Reddy and Murti, ibid., **13**, 1443 (1968); for alcohol-aliphatics, see Reddy and Murti, ibid., **16**, 1036 (1971); and for H_2SO_4, see D. D. Huxtable and D. R. Poole, *Proc. Int. Solar Energy Soc.*, Winnipeg, August 15, 1976, **8**, 178 (1977). For more recent sources search the Internet.

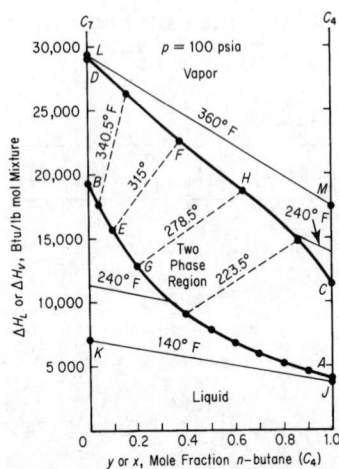

Figure 12.3 Enthalpy concentration of n-butane-n-heptane at 100 psia. Curve *DFHC* is the saturated vapor; curve *BEGA* is the saturated liquid. The dashed lines are equilibrium tie lines connecting y, the mole fraction of C_4 in the vapor, and x, the mole fraction of C_4 in the liquid, at the same temperature.

and energy balance calculations in distillation, crystallization, and all sorts of mixing and separation problems. You will find a few examples of enthalpy-concentration charts in Appendix J.

As you might expect, the preparation of an enthalpy-concentration chart requires numerous calculations and valid enthalpy or heat capacity data for solutions of various concentrations. Refer to *Unit Operations of Chemical Engineering* [W. L. McCabe and J. C. Smith, 3rd ed., McGraw-Hill, New York (1976)] for instructions if you have to prepare such a chart. In the next example we show how to use an *H-x* chart.

Example 12.3   Application of an Enthalpy-Concentration Chart

Six hundred pounds of 10% NaOH per hour at 200°F are added to 400 lb/hr of 50% NaOH at the boiling point in an insulated vessel. Calculate the following:

a. The final temperature of the exit solution
b. The final concentration of the exit solution
c. The pounds of water evaporated per hour during the process

---

Solution

You can use the steam tables and the NaOH-H_2O enthalpy-concentration chart in Appendix J as your sources of data. What are the reference conditions for the chart? The reference conditions for the latter chart are $\Delta \hat{H}_f^\circ = 0$ at 32°F for pure liquid water, an infinitely dilute solution of NaOH. Pure caustic has an enthalpy at 68°F of 455 Btu/lb above this datum. Treat the process as a flow process even if is not. The energy balance reduces to $\Delta H = 0$. Basis: 1000 lb of final solution = 1 hr.

(Continues)

Example 12.3   Application of an Enthalpy-Concentration Chart (*Continued*)

You can write the following material balances:

Component	10% solution	+	50% solution	=	Final solution	wt%
NaOH	60		200		260	26
H_2O	540		200		740	74
Total	600		400		1000	100

Enthalpy data from the *H-x* chart

$$\Delta \hat{H}(\text{Btu / lb}): \quad \frac{10\% \text{ solution}}{152} \quad \frac{50\% \text{ solution}}{290}$$

The energy balance is

10% solution		50% solution		Final solution
600(152)	+	400(290)	=	ΔH
91,200	+	116,000	=	207,200

Note that the enthalpy of the 50% solution at its boiling point is taken from the bubble point at $\omega_{NaOH} = 0.50$. The enthalpy per pound of the final solution is

$$\frac{207,200 \text{Btu}}{1000 \text{lb}} = 207 \text{ Btu/lb}$$

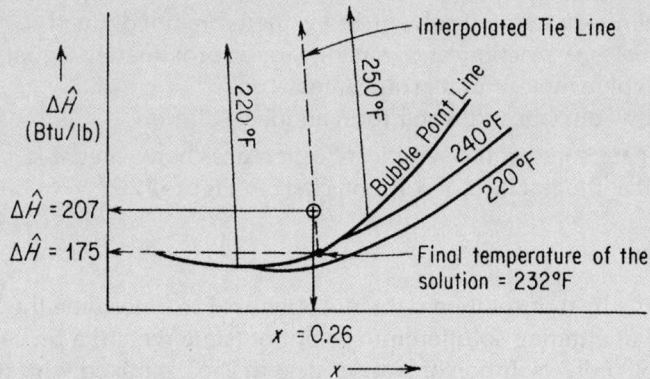

Figure E12.3

On the enthalpy-concentration chart for NaOH-H_2O, for a 26% NaOH solution with an enthalpy of 207 Btu/lb, you would find that only a two-phase mixture of (1) saturated H_2O vapor and (2) NaOH-H_2O solution at the boiling point could exist. To get the fraction H_2O vapor, you have to make an additional energy (enthalpy) balance. By interpolation, draw the tie line through the point $x = 0.26$, $H = 207$ (make it parallel to the 220°F and 250°F tie lines). The final temperature of the tie line appears from Figure E12.3 to be 232°F;

(*Continues*)

Example 12.3   Application of an Enthalpy-Concentration Chart (*Continued*)

the enthalpy of the liquid at the bubble point at this temperature is about 175 Btu/lb. The enthalpy of the saturated water vapor (no NaOH is in the vapor phase) from the steam tables at 232°F is 1158 Btu/lb. Let $x =$ pounds of H_2O evaporated.

Basis: 1000 lb of final solution

$$x(1158) + (1000 - x) 175 = 1000 (207.2)$$

$$x = 32.8 \text{ of } H_2O \text{ evaporated/hr}$$

Self-Assessment Test

Questions

1. Is a gas mixture an ideal solution?
2. Give **(a)** two examples of exothermic mixing of two liquids and **(b)** two examples of endothermic mixing based on your experience.
3. Indicate whether the following statements are true or false:
   a. Heat of mixing at infinite dilution involves mixing of 1 mole of solute and an infinite amount of solvent and is therefore not defined.
   b. Heats of reaction and mixing are approximately equal because they both involve molecular rearrangement.
   c. Ethyl alcohol and water form an ideal solution.
4. What are some of the significant differences between the *H-x* chart in Figure 12.3 and the integral heat of solution chart in Figure 12.2?

Problems

1. Use the heat of solution data in Appendix I to determine the heat transferred per mole of entering solution into or out of (state which) a process in which 2 g mol of a 50 mol % solution of sulfuric acid at 25°C is mixed with water at 25°C to produce a solution at 25°C containing a mole ratio of 10 H_2O to 1 H_2SO_4.
2. Calculate the heat that must be added or removed per ton of 50 wt % H_2SO_4 produced by the process shown in Figure SAT12.2P2.

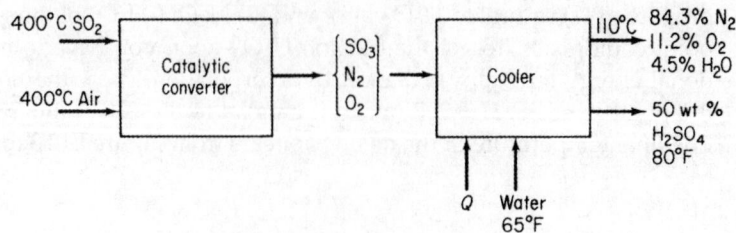

Figure SAT12.2P2

3. For the sulfuric acid–water system, what are the phase(s), composition(s), and enthalpy(ies) existing at and $T = 260°F$?

4. Estimate the heat of vaporization of an ethanol-water mixture at 1 atm and an ethanol mass fraction of 0.50 from the enthalpy-concentration chart in Appendix J.

Thought Problems

1. Because the driver was in a hurry, a tank truck of concentrated 6000 gal of hydrochloric acid was unloaded by mistake into a large storage tank used for concentrated sulfuric acid. After about one-half of the load had been discharged, a violent explosion took place, buckling the tank and breaking the inlet and outlet lines. What was the cause of the explosion?

2. A second example of an accident occurred in a tank as shown in Figure SAT12.2TP2. A problem occurred in discharging the NaOH (caustic) solution by air pressure because the end of the dip pipe was broken. The manhole cover was removed, and a scraper rod was lowered into the liquid to remove the broken part of the dip pipe. An immediate explosion occurred that shot NaOH solution 30 ft into the air, killing one man. What was the cause of the explosion?

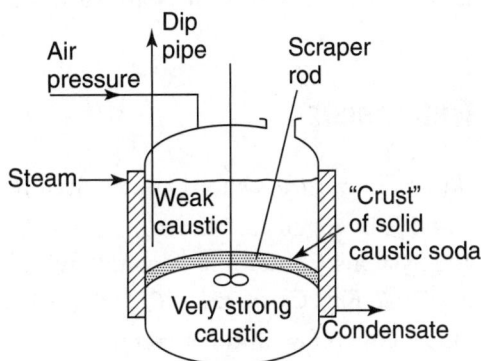

Figure SAT12.2TP2

Looking Back

In this chapter we described how to carry out energy balances when enthalpy changes occur because of the heat of solution or mixing. We also described how to use binary enthalpy-concentration charts.

Glossary

enthalpy-concentration diagram   A convenient graphical way to represent enthalpy data for binary solutions that features a plot of specific enthalpy versus concentration (usually mass or mole fraction) with temperature as a parameter.

heat (enthalpy) of dissolution   The negative of the heat of solution or mixing.

heat (enthalpy) of mixing   The enthalpy change that occurs when a liquid is mixed with a liquid.

heat (enthalpy) of solution   The enthalpy change that occurs when a solute is mixed with a solvent.

heat of solution at infinite dilution   The asymptotic value of the heat of solution of 1 mole of solute dissolved in an infinite amount of solvent.

ideal solution   A solution composed of several components in which a property is the weighted sum of the properties of the individual components.

incremental (differential) heat of solution   The derivative of the integral heat of solution curve.

integral heat of solution   The heat of solution for the combination of 1 mole of solute with n moles of solvent.

real solution   Nonideal solution.

solute   The compound that is dissolved in a solvent.

solvent   The compound in which the solute is dissolved.

Supplementary References

Berry, R. S. *Physical Chemistry,* Oxford University Press, Oxford, UK (2000).

Brandani, V., and F. Evangelista. "Correlation and Predication of Enthalpies of Mixing for Systems Containing Alcohols with UNIQUAC Associated-Solution Theory," *Ind. Eng. Chem. Res.,* **26,** 2423 (1987).

Christensen, C., J. Gmehling, and P. Rasmussen. *Heats of Mixing Data Collection,* Dechema, Frankfurt (1984).

Christensen, J. J., R. W. Hanks, and R. M. Izatt. *Handbook of Heats of Mixing,* John Wiley, New York (1982).

Dan, D., and D. P. Tassios. "Prediction of Enthalpies of Mixing with a UNIFAC Model," *Ind. Eng. Chem. Process Des. Develop.,* **25,** 22 (1986).

Johnson, J. E., and D. J. Morgan. "Graphical Techniques for Process Engineering," *Chem. Eng.,* 72 (July 8, 1985).

Sandler, S. I. *Chemical and Engineering Thermodynamics,* 3rd ed., John Wiley, New York (1998).

Smith, J. M., H. C. Van Ness, and M. M. Abbot. *Introduction to Chemical Engineering Thermodynamics,* 5th ed., McGraw-Hill, New York (1998).

Problems

*12.1 If the heat of formation of LiCl(s) is –408.78 kJ/kg LiCl, calculate the heat of formation of LiCl in 10 mol of water. The heat of solution is –32.84 kJ/kg LiCl.

**12.2 Home ice cream makers use a mixture of 3 parts of ice to 1 part of salt (NaCl) to freeze the ice cream. Why does this process work?

Based on the following data, is salt the best compound to use (with the amount of water shown) for a freezing mixture?

Compound	$\Delta \hat{H}_f^o$ (kJ / gmol)
H_2O(s)	–290.852
(l)	–285.840
(g)	–241.826
NaCl(s)	–411.00
NaCl[$10H_2O$]	–408.99
NH_4Cl(s)	–315.43
NH_4Cl [$10H_2O$]	–301.33
$CaCl_2$(s)	–794.97
$CaCl_2$ [$10H_2O$]	–860.03
$CaCl_2 \cdot 6H_2O$	–2586.5
$CaCl_2$ [$6H_2O$]	–854.4

**12.3 Based on the following data for the solution of Na_2CO_3 at S.C.:

Compound	$\Delta \hat{H}_f^o$ (kJ / gmol)
Na_2CO_3(s)	–1130.92
in 15 mol H_2O	–1163.70
20	–1162.78
25	–1161.98
40	–1160.72
75	–1158.01
100	–1157.17
200	–1155.50
400	–1154.39
$Na_2CO_3 \cdot H_2O$(s)	–1430.09
$Na_2 CO_3 \cdot 7H_2O$(s)	–3201.18
$Na_2CO_3 \cdot 10H_2O$(s)	–4081.9

a. Draw a standard integral heat of solution curve for sodium carbonate dissolved in water showing the heat of solution in kilojoules per mole of sodium carbonate versus mole of water.

b. If 143 kg of sodium carbonate are added to 180 kg of water, what would be the approximate final temperature of the solution if the mixing were adiabatic?

12.4   a. From the data below, plot the enthalpy of 1 *mole of solution* at 27°C as a
function of the weight percent HNO_3. Use as reference states liquid water
at 0°C and liquid HNO_3 at 0°C. You can assume that C_p for H_2O is 75 J/
(g mol)(°C) and for HNO_3, 125 J/(g mol)(°C).

$-\Delta H_{soln}$ at 27°C (J/g mol HNO_3)	Moles H_2O Added to 1 g mol HNO_3
0	0.0
3350	0.1
5440	0.2
6900	0.3
8370	0.5
10,880	0.67
14,230	1.0
17,150	1.5
20,290	2.0
24,060	3.0
25,940	4.0
27,820	5.0
30,540	10.0
31,170	20.0

b. Compute the energy absorbed or evolved at 27°C on making a solution of
4 moles of HNO_3 and 4 moles of water by mixing a solution of 33 1/3 mol
% acid with one of 60 mol % acid.

12.5   *National Bureau of Standards Circular 500* gives the following data for calcium
chloride (mol. wt. 111) and water:

Formula	State	$-\Delta H_f$ at 25°C (kcal/g mol)
H_2O	Liquid	68.317
	Gas	57.798
$CaCl_2$	Crystal	190.0
	in 25 mol	
	of H_2O	208.51
	50	208.86
	100	209.06
	200	209.20
	500	209.30
	1000	209.41
	5000	209.60
	∞	209.82
$CaCl_2 \cdot H_2O$	Crystal	265.1
$CaCl_2 \cdot 2H_2O$	Crystal	335.5
$CaCl_2 \cdot 4H_2O$	Crystal	480.2
$CaCl_2 \cdot 6H_2O$	Crystal	623.15

Calculate the following:
a. The energy evolved when 1 lb mol of $CaCl_2$ is made into a 20% solution at 77°F
b. The heat of hydration of $CaCl_2 \cdot 2H_2O$ to the hexahydrate
c. The energy evolved when a 20% solution containing 1 lb mol of $CaCl_2$ is diluted with water to 5% at 77°F

*12.6 For each of the following processes that occur in an open, steady-state system, calculate the heat transfer to or from the system if it is isothermal:
a. 1000 g of O_2 are mixed with 1000 g of CO_2.
b. 900 kg of water are mixed with 63 kg of nitric acid.

Data at S.C.:

Compound	$\hat{H}_f^o$ (kJ/kg)
HNO_3 (liquid)	−173.234
In 1 g mol H_2O	−186.347
$2H_2O$	−193.318
$5H_2O$	−201.962
$10H_2O$	−205.014
$50H_2O$	−205.978
$100H_2O$	−205.983

**12.7 An insulated closed tank contains 250 kg of a 20% solution of sulfuric acid at its boiling point. One hundred pounds of a 98% solution of sulfuric acid are carefully added to the original solution with stirring. What are the final temperature, composition, and weight of solution in the tank? Use Appendix J.

*12.8 An insulated tank contains 500 kg of a solution of 20% sulfuric acid at 340 K. To this solution are added 300 kg of a 96% solution of sulfuric acid at 310 K. To heat the solution, 100 kg of superheated steam are introduced at 1 atm and 400 K. What is the final temperature in the solution, and what are the concentrations of sulfuric acid and water in the final solution?

**12.9 A vessel contains 100 g of an NH_4OH-H_2O liquid mixture at 1 atm that is 15.0% by weight NH_4OH. Just enough aqueous H_2SO_4 is added to the vessel from an H_2SO_4 liquid mixture at 25°C and 1 atm (25.0 mol % H_2SO_4) so that the reaction to $(NH_4)_2SO_4$ is complete. After the reaction, the products are at 25°C and 1 atm. How much heat (in joules) is absorbed or evolved by this process? It may be assumed that the final volume of the products is equal to the sum of the volumes of the two initial mixtures.

***12.10 An ammonium hydroxide solution is to be prepared at 77°F by dissolving gaseous NH_3 in water. Prepare charts showing
a. The amount of cooling needed (in British thermal units) to prepare a solution containing 1 lb mol of NH_3 at any concentration desired
b. The amount of cooling needed (in British thermal units) to prepare 100 gal of a solution of any concentration up to 35% NH_3
If a 10.5% NH_3 solution is made up without cooling, at what temperature will the solution be after mixing?

*12.11   An evaporator at atmospheric pressure is designed to concentrate 10,000 lb/hr of a 10% NaOH solution at 70°F into a 40% solution. The steam pressure inside the steam chest is 40 psig. Determine the pounds of steam needed per hour if the exit strong caustic preheats the entering weak caustic in a heat exchanger, leaving the heat exchanger at 100°F.

**12.12   A 50% by weight sulfuric acid solution is to be made by mixing the following:
   a. Ice at 32°F
   b. 80% H_2SO_4 at 100°F
   c. 20% H_2SO_4 at 100°F
   How much of each must be added to make 1000 lb of the 50% solution with a final temperature of 100°F if the mixing is adiabatic?

*12.13   Saturated steam at 300°F is blown continuously into a tank of 30% H_2SO_4 at 70°F. What is the highest concentration of liquid H_2SO_4 that can result from this process?

**12.14   One thousand pounds of 10% NaOH solution at 100°F are to be fortified to 30% NaOH by adding 73% NaOH at 200°F. How much 73% solution must be used? How much cooling must be provided so that the final temperature will be 70°F?

12.15*   A mixture of ammonia and water in the vapor phase, saturated at 250 psia and containing 80% by weight ammonia, is passed through a condenser at a rate of 10,000 lb/hr. Heat is removed from the mixture at the rate of 5,800,000 Btu/hr while the mixture passes through a cooler. The mixture is then expanded to a pressure of 100 psia and passes into a separator. A flowsheet of the process is given in Figure P12.15. If the heat loss from the equipment to the surroundings is neglected, determine the composition of the liquid leaving the separator by material and energy balances.

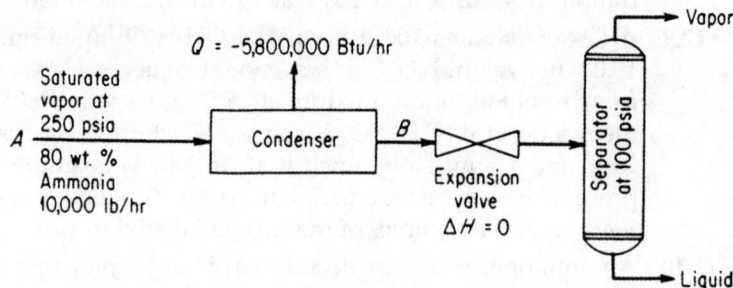

Figure P12.15

CHAPTER 13

Liquids and Gases in Equilibrium with Solids

Your objectives in studying this chapter are to be able to

1. Predict adsorption of gases and liquids on solids at equilibrium
2. Determine the values of coefficients in adsorption equilibrium relations from physical measurements

---

In Chapter 8 you read about equilibria involving gas and liquids. Equilibria between fluids and solids are also important—they certainly cannot be ignored.

Looking Ahead

In this chapter we discuss the adsorption of gases and liquids on solids when the system is at *equilibrium*. You will learn what some of the relations are that are used to predict the amounts of absorption, and what kinds of data are collected to get the values of the coefficients in the relations.

13.1 Main Concepts

Many important processes involve the adsorption of gases or liquids on solids. Some examples for liquids are

- Decolorizing, drying, or "degumming" of petroleum fractions
- Odor, taste, and color removal from municipal water supplies
- Decolorizing of vegetable and animal oils, and of crude sugar syrups
- Clarification of beverages and pharmaceutical preparations
- Purification of process effluents and gases for pollution control

- Solvent recovery from air such as in removing evaporated dry-cleaning solvents
- Dehydration of gases
- Odor and toxic gas removal from air or vent gases
- Separation of rare gases at low temperatures
- Impurity removal from air prior to low-temperature fractionation
- Storage of hydrogen

What goes on in adsorption? **Adsorption** is a physical phenomenon that occurs when gas or liquid molecules are brought into contact with a solid surface, the **adsorbent.** Some of the molecules may condense (the **adsorbate**) on the exterior surface and in the cracks and pores of the solid. If interaction between the solid and condensed molecules is relatively weak, the process is called physical adsorption; if the interaction is strong (similar to a chemical reaction), it is called **chemisorption,** or activated adsorption. We focus on equilibria in physical adsorption in this chapter.

Equilibrium adsorption is analogous to the gas-liquid and liquid-liquid equilibria described in previous chapters. Picture a small section of adsorbent surface. As soon as molecules come near the surface, some condense on the surface. A typical molecule will reside on the adsorbent for some finite time before it acquires sufficient energy to leave. Given sufficient time, an equilibrium state will be reached: The number of molecules leaving the surface will just equal the number arriving.

The number of molecules on the surface at equilibrium is a function of the (1) nature of the solid adsorbent, (2) nature of the molecule being adsorbed (the adsorbate), (3) temperature of the system, and (4) concentration of the adsorbate over the adsorbent surface. Numerous theories have been proposed to relate the amount of fluid adsorbed to the amount of adsorbent. The various theories lead to equations that represent the equilibrium state of the adsorption system. Theoretical models hypothesize various physical conditions such as (1) a solid surface on which there is a monomolecular layer of molecules, or (2) a multimolecular layer, or (3) capillary condensation.

Figure 13.1 illustrates typical simple equilibrium adsorption isotherms for the adsorption of water vapor on Type 5A molecular sieves at various temperatures. Note how the amount of gas adsorbed decreases as the temperature increases.

No single relation can represent all of the myriad types of equilibrium data found in practice, as indicated by the variety of curves for isothermal adsorption called **adsorption isotherms,** shown in Figure 13.2.

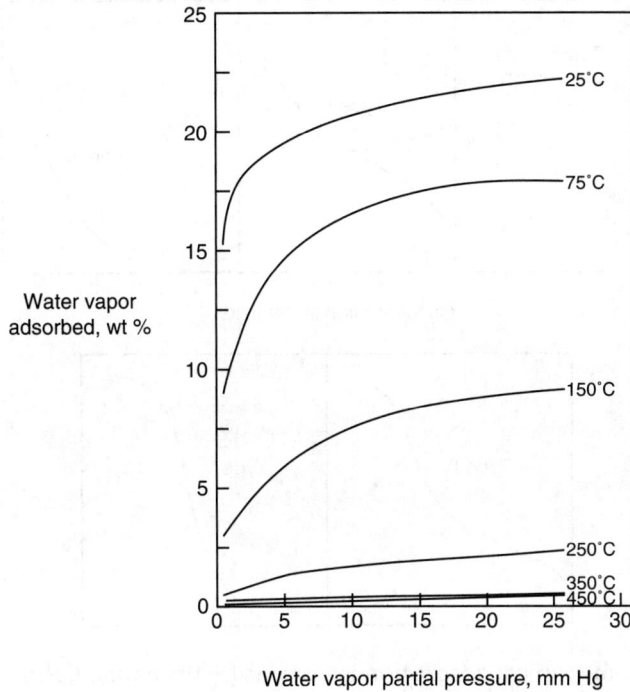

Figure 13.1   Equilibrium adsorption of water on Type 5A molecular sieves

Two of the simpler equilibrium relations for physical adsorption of Type I in Figure 13.2 relate the amount adsorbed for a single component (the adsorbate) on the adsorbent as a function of the adsorbate partial pressure in the gas phase, or the concentration in the liquid phase, at some temperature. These equations apply for low concentrations at equilibrium.

Freundlich Isotherm

For a gas: *For a liquid:*

$$y = k_1 p^{n_1}$$ $$y = k_2 x^{n_2}$$

where p is the partial pressure of the adsorbate in the gas phase

  y is the cumulative mass of solute or gas adsorbed (adsorbate) per mass of adsorbent in the solid phase

  x is the mass of solute per mass of solution in the liquid phase

  n_i and k_i are coefficients

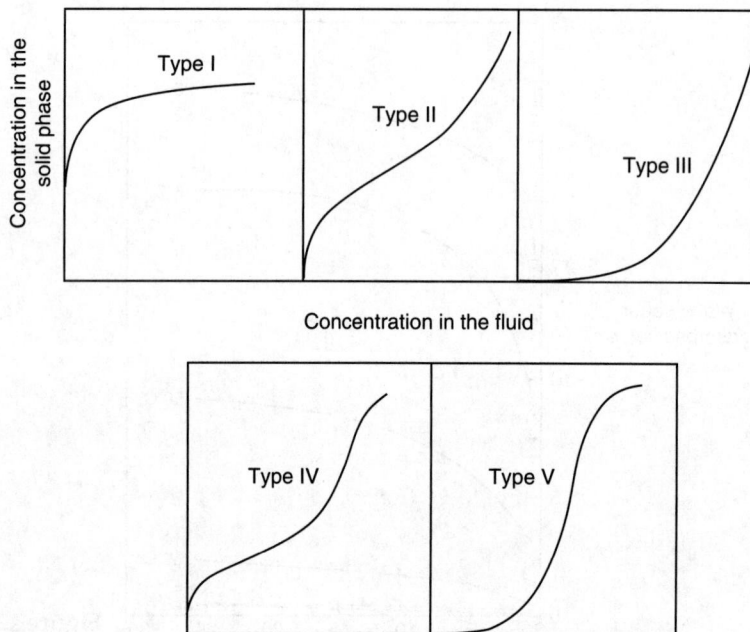

Figure 13.2 Some typical adsorption isotherms [S. Brunauer, *J. Am. Chem. Soc.*, **62**, 1723 (1940.)]

Langmuir Isotherm

For a gas:

$$y = \frac{(1 + ak_3)p}{1 + k_3 p}$$

For a liquid:

$$y = \frac{(1 + bk_4)x}{1 + k_4 x}$$

The respective equations correspond to monolayer adsorption. Numerous other relations have been developed to match the particular shapes of the adsorption equilibrium data shown in Figure 13.2. Refer to the references at the end of this chapter.

You can obtain values for the coefficients in the equilibrium isotherms by fitting them to experimental data, as shown in Example 13.1. The expressions $(1 + ak_3)$ and $(1 + bk_4)$ can be treated as single coefficients in the fitting.

Example 13.1   Fitting Adsorption Isotherms to Experimental Data

The following data for the adsorption of CO_2 on 5A molecular sieves at 298 K have been taken from the PhD dissertation *Periodic Countercurrent Operation of Pressure-Swing Adsorption Processes Applied to Gas Separations* by Carol L. Blaney [University of Delaware (1985), p. 131].

p_{CO_2} (mm Hg)	y (g adsorbed/g sieves)
0	0
25	6.69×10^{-2}
50	9.24×10^{-2}
100	0.108
200	0.114
400	0.127
760	0.137

The estimated coefficients in the respective isotherms obtained by using the regression function in Polymath 5 were:

Freundlich Isotherm			Langmuir Isotherm		
Model: $y + ap^n$		$R^2 = 0.993$	**Model:** $ap/(1 + bp)$		$R^2 = 0.981$
Variable	Initial guess	Final value	Variable	Initial guess	Final value
a	1	0.0448	a	100	0.0052
n	0.5	0.1729	b	1	0.0386

The parameter R^2 is a measure of the degree of fit; $R^2 = 1$ is a perfect fit. Figure E13.1 shows the data and the adsorption isotherms.

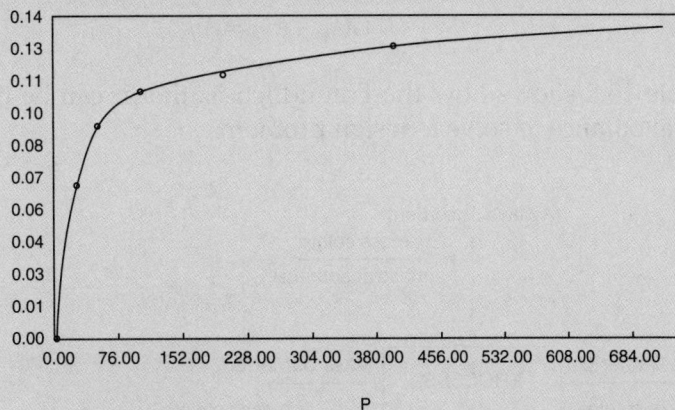

Figure E13.1   The Freundlich and Langmuir isotherms coincide for the adsorption of CO_2 on 5A molecular sieves; O are the experimental points.

A type of constant that is used in environmental studies is the soil sorption partition coefficient, K_{oc}. It characterizes the partitioning of a compound between the solid and liquid phases in soil, and is used to determine the mobility of a compound in soil:

$$K_{oc} = \frac{\dfrac{\mu g \text{ of compound adsorbed}}{g \text{ of organic carbon in the soil}}}{\dfrac{\mu g \text{ of compound in the liquid phase}}{mL \text{ liquid in the liquid phase}}}$$

How do you use the information about adsorption in practice? One way is to combine it with material balances to help design adsorption equipment. Examine Figure 13.3 in which a carrier G (gas or liquid) is contacted with an adsorbent A (solid). Assume that the vessel is well mixed and that the exit streams are in equilibrium with each other.

The mass of G is the mass of the carrier excluding the solute, and the mass of A is also solute free. Often we assume the solution entering is dilute enough so that G is equivalent to the total flow. The material balance corresponding to the process in Figure 13.3 is (note that x and y are *not* mass fractions in what follows)

$$\overset{\text{In}}{} \qquad \overset{\text{Out}}{}$$
$$G(x_0 - x_1) = A(y_1 - y_0) \tag{13.1}$$

Usually the A entering is free of solute so that $y_0 = 0$, and thus the material balance becomes

$$G(x_0 - x_1) = Ay_1 \tag{13.2}$$

Example 13.2 shows how the Freundlich isotherm can be combined with a material balance to solve a design problem.

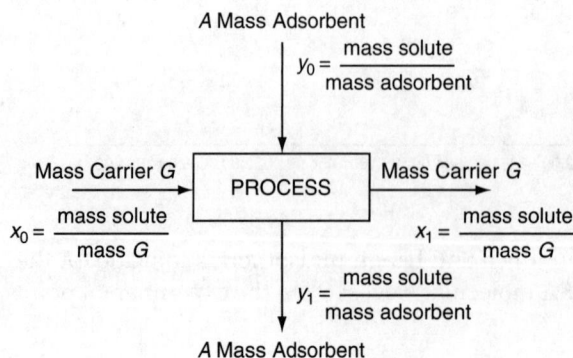

Figure 13.3   The process of adsorption with the notation

Example 13.2   Combination of an Adsorption Isotherm with a Material Balance

You are asked to calculate the minimum mass of activated carbon required to reduce a contaminating solute in a fermentation system from $x_0 = 19.2$ g solute/L solution (for simplicity the value can be treated as grams of solute per 1000 g solution) to $x_1 = 1.4$ g solute/L solution for a 1 L batch of solution by adsorption on activated charcoal in a well-mixed vessel so that the output products are in equilibrium. To solve this problem you collect the following liquid-solid equilibria data. Note that the third column contains calculated, not measured, values.

	Measured at equilibrium	
Cumulative g Carbon Added per 1000 g Solution	x g Solute / 1000 g Solution	y (Calculated) g Solute Adsorbed / g Carbon
0	19.2	—
0.01	17.2	(19.2–17.2)/0.01 =200
0.04	12.6	(19.2–12.6)/0.04 =165
0.08	8.6	(19.2–8.6)/0.08 = 133
0.20	3.4	(19.2–3.4)/0.20 = 79
0.40	1.4	(19.2–1.4)/0.40 = 45

Solution

Steps 1–4

If you fit the data using Polymath, the Freundlich isotherm proves to be

$$y = 37.919\, x^{0.583} \qquad R^2 = 0.999 \qquad\qquad \text{(a)}$$

and the Langmuir isotherm is

$$y = (29.698x) / (1 + 0.0955x) \qquad R^2 = 0.987 \qquad\qquad \text{(b)}$$

Figure E13.2, the process diagram, shows the flows in and out of the system. From a different viewpoint, the process is equivalent to a batch process in which the materials are put into an empty vessel at the start and removed at the end of the process. Let G be the grams of solution. Because the solution is so dilute, G is essentially equivalent to the grams of solute-free solvent. If you wanted to, you could calculate the ratios of the solute to the solute-free solvent coming in and out of the process, but we will not do so.

Step 5

Let the basis be 1000 g solution (1 L).

(Continues)

Example 13.2   Combination of an Adsorption Isotherm with a Material Balance (*Continued*)

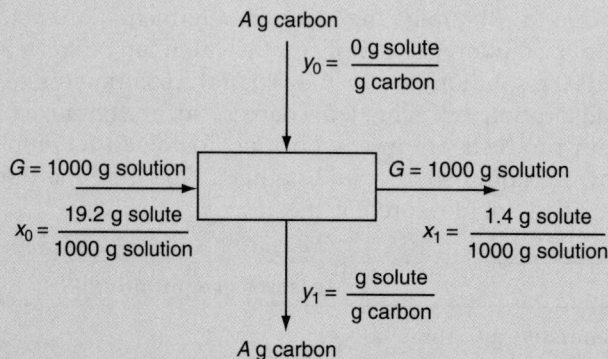

Figure E13.2

Steps 6–9

Two independent equations are involved in the solution, an equilibrium relation and the material balance. Two unknowns exist, y_1 and the ratio A/G.

Combine Equation (a) with Equation (13.2) to get

$$\frac{A}{G} = \frac{x_0 - x_1}{y_1} = \frac{(19.2 - 1.4)\dfrac{\text{g solute}}{1000\text{g solution}}}{37.919(1.4)^{0.583}\dfrac{\text{g solute}}{\text{g carbon}}} = \frac{17.8}{46.1} = 0.39\frac{\text{g carbon}}{1000\text{g solution}}$$

Hence, 0.39 g carbon is required.

Step 10

You can check the answer by using the Langmuir isotherm together with the material balance:

$$\frac{A}{G} = \frac{19.2 - 1.4}{Y_1}$$

$$y_1 = \frac{29.698(1.4)}{1 + 0.0955(1.4)} \quad \text{or } y_1 = 36.67$$

hence

$$\frac{A}{G} = \frac{17.8}{36.67} = 0.48 \text{ g carbon}/1000 \text{ g solution}$$

Because $y_1 = 46.1$ from the Freundlich isotherm is closer to the experimental value of 45 than is 36.7 from the Langmuir isotherm, the value of 0.39 g as the answer would be preferred. Of course, more carbon will have to be used in practice because to reach equilibrium takes a very long time.

Can you now calculate how much carbon would be required to remove all of the polluting solute?

Self-Assessment Test

Questions

1. Would condensation of water below its dew point on charcoal be considered adsorption?
2. Would the conversion of solid NaOH by HCl gas to NaCl be considered adsorption?
3. Would the drying of moist air by charcoal be considered adsorption?
4. Can more than one component be removed simultaneously from a gas or liquid by a solid adsorbent?

Problems

1. One hundred pounds per minute of moist air at 70°F that has a humidity of 0.01 lb H_2O/lb dry air is dehumidified to 0.002 lb H_2O/lb dry air in a dryer. If 7 lb/min of dry silica gel enter the dryer at 70°F, how many pounds of H_2O leave the dryer per pound of silica gel entering?

2. The dissertation of Blaney (see Example 13.1) contains the following data for the adsorption of N_2 on activated coconut charcoal at 298 K:

P (mm Hg)	y (g mol N_2/g Charcoal) $\times 10^4$
0	0
26.8	0.1562
45.93	0.2945
67.72	0.4149
88.21	0.5354
108.05	0.6360
127.91	0.7406
147.42	0.8477
166.77	0.9458
246.86	1.3786
369.93	1.9808
428.79	2.2396
492.32	2.5430
587.86	2.9713
688.05	3.3996
761.10	3.6851

Determine the coefficients in the Langmuir relation $y = k_0 k_1 p/1 + k_1 p$.

3. Water contains organic color, which is to be extracted with alum and lime. Five parts of alum and lime per million parts of water will reduce the color to 25% of the original color, and 10 parts will reduce the color to 3.5%.

Estimate how much alum and lime as parts per million are required to reduce the color to 0.5% of the original color.

Thought Problems

1. Why might a solid adsorbent be used rather than distillation to separate a liquid mixture?
2. How might you regenerate an adsorbent used for the separation of a mixture of (a) gases and (b) liquids?

Glossary

adsorbent   A solid surface on which gas or liquid molecules condense to form a film.

adsorption   The physical process that occurs when gas or liquid molecules are brought into contact with a solid surface and condense on the surface.

adsorption isotherm   The mathematical or experimental relation between the amount a single component adsorbed (the adsorbate) on the adsorbent, and the bulk amount of the adsorbate in a different phase expressed in terms of the partial pressure in the gas phase, or the concentration in the liquid phase, at some temperature.

chemisorption   Adsorption when interaction between the solid and the condensed molecules is relatively strong as contrasted with physical adsorption.

Freundlich isotherm   Mathematical relation for adsorption that takes place at equilibrium.

Langmuir isotherm   Mathematical relation for adsorption that takes place at equilibrium.

Supplementary References

Basmadjian, D. *The Little Adsorption Book: A Practical Guide for Engineers and Scientists,* CRC Press, Boca Raton, FL (1996).

De Nevers, N. *Physical and Chemical Equilibrium for Chemical Engineers,* Wiley-Interscience, New York (2002).

Masel, R. I. *Principles of Adsorption and Reaction on Solid Surfaces,* Wiley-Interscience, New York (1996).

Thomas, W. J. *Adsorption Technology and Design,* Butterworth-Heinemann, Oxford, UK (1998).

Toth, J. *Adsorption,* Marcel Dekker, New York (2002).

Problems

*13.1 The adsorption of sulfur dioxide by polymer pellets at 0°C is listed below. Using these data, determine the Langmuir constants and the Freundlich constants.

p_{SO_2} (mm Hg)	Uptake (mg mol/g)
5	1.75
10	2.20
15	2.40
20	2.62
30	2.75
40	2.85
50	3.00
60	3.05
70	3.12

Do these two isotherms fit the data well?

*13.2 Emmett studied the adsorption of argon on 0.606 g of silica gel at −183°C. From the following data, calculate the Freundlich and Langmuir constants:

p (mm Hg)	Uptake (cm³ at S.C.)
78.46	55.03
176.92	72.73
224.62	80.00
378.46	106.67
432.31	117.58
513.38	138.18
584.62	166.06

Do these two isotherms fit the data well?

*13.3 The adsorption of ethane on 5A molecular sieves was studied by Glessner and Myers (1969) at 35°C. Using the data given below, determine (a) if the Langmuir equation can be used to model the data and (b) if the Freundlich equation can be used to model the data.

p (mm Hg)	Uptake (cm³ at S.C./g)
0.17	0.059
0.95	0.318
5.57	1.638
12.09	3.613
111.32	24.236
220.87	34.278
300.05	38.340
401.25	41.779
500.18	44.037
602.74	45.693

*13.4 In the isothermal adsorption of a mixture of butanol-2 (component 1) and t-amyl alcohol (component 2) on porous activated carbon, the following relation represented the data within an accuracy of ±2%:

$$c_{s1} = \frac{1.06 c_{f1}^{1.217}}{c_{f1}^{0.812} + 0.626 c_{f2}^{0.764}}$$

where c_s = concentration of solute in the solid phase, g/cm^3

c_f = concentration of the solute in the fluid phase, g/cm^3

What would the Freundlich equation be for butanol-2 as a pure liquid?

*13.5 A solution containing a trace amount of chloroform (12 mg/L) is to be reduced to 1 $\mu g/L$ of CCl_4 using activated charcoal as the adsorbent. The Freundlich equation that applies to this system is $y = k\,C^n$ where C is in milligrams adsorbate per liter of solution after processing, and y is the cumulative milligrams of adsorbate per gram of adsorbent. $k = 100$ and $n = 0.30$ at the temperature of the process. How much activated charcoal is required per liter of solution?

**13.6 One hundred pounds per minute of moist air that includes 1 lb of water vapor flow through a dehumidifier. Dry silica is fed to the process and flows counter-current to the air. The water content of the exit air is reduced to 10^{-3} lb of water per pound of dry air. The water in the exit silica gel is in equilibrium with the water in the entering air. What is the required flow rate of the silica gel per minute? The equilibrium relation that applies to this process is $y = 0.0375 p^{0.65}$ where y is the grams of water per gram of silica gel and p is the partial pressure of the water in the entering air in millimeters of Hg.

**13.7 The compound 1, 1-dichloroethane (DE) in air at 40°C and with a dew point of 12°C is to be removed with 2 kg/min of activated carbon. The entering activated carbon contains 50 g DE per kilogram of DE-free activated carbon and leaves the process with a DE content of 300 g of DE per kilogram of DE-free activated carbon. Calculate the concentration of DE in the exit air, assuming that the DE in the exit air is in equilibrium with the exit activated carbon. The barometer is 770 mm of Hg. The equilibrium relation for the DE is $y = 0.089 p^{0.51}$ where y is the grams of DE adsorbed per gram of activated charcoal, and p is the partial pressure in the gas phase of the DE in millimeters of Hg. How many gram moles of bone-dry air pass through the process per minute?

*13.8 The partition of a compound between soil and water can be estimated by using a soil adsorption coefficient K_{OC}

$$K_{OC} = \frac{\dfrac{\text{Mass of compound adsorbed in } \mu g}{\text{Mass of organic carbon in the soil in g}}}{\dfrac{\text{Mass of compound in the liquid in } \mu g}{\text{Volume of the liquid in mL}}}$$

Values of K_{OC} for low adsorption are in the range $2.5 > \log_{10}(K_{OC}) > 1.5$.

How much nitrobenzene will be adsorbed at equilibrium from 1 L of a saturated solution in water at 20°C by 1 kg of soil that has a carbon content of 11%? Data: The solubility of nitrobenzene in water is 0.19 g/100 g water at 20°C, and $\log_{10}(K_{OC}) = 2.27$.

CHAPTER 14

Solving Material and Energy Balances Using Process Simulators (Flowsheeting Codes)

> **Your objectives in studying this chapter are to be able to**
>
> 1. Understand the differences between equation-based and modular-based flowsheeting
> 2. Understand how material and energy balances are formulated for equation- and modular-based flowsheeting codes

Looking Ahead

In this chapter we survey process simulators (flowsheeting codes) that are used in industrial practice to solve material and energy balances.

14.1 Main Concepts

As explained in Chapter 6, a plant **flowsheet** such as the simple diagram in Figure 14.1 mirrors the stream network and equipment arrangement in a process. Once the process flowsheet is specified, or during its formulation, the solution of the appropriate process material and energy balances is referred to as **process simulation** or **flowsheeting,** and the computer code used in the solution is known as a **process simulator** or **flowsheeting code.** Codes for both steady-state and dynamic processes exist. The essential problem in flowsheeting is to solve (satisfy) a large set of linear and nonlinear equations and constraints to an acceptable degree of precision. Such a program can also, at the same time, determine the size of equipment and piping, evaluate costs, and optimize performance. Figure 14.2 shows the information flow that occurs in a process simulator.

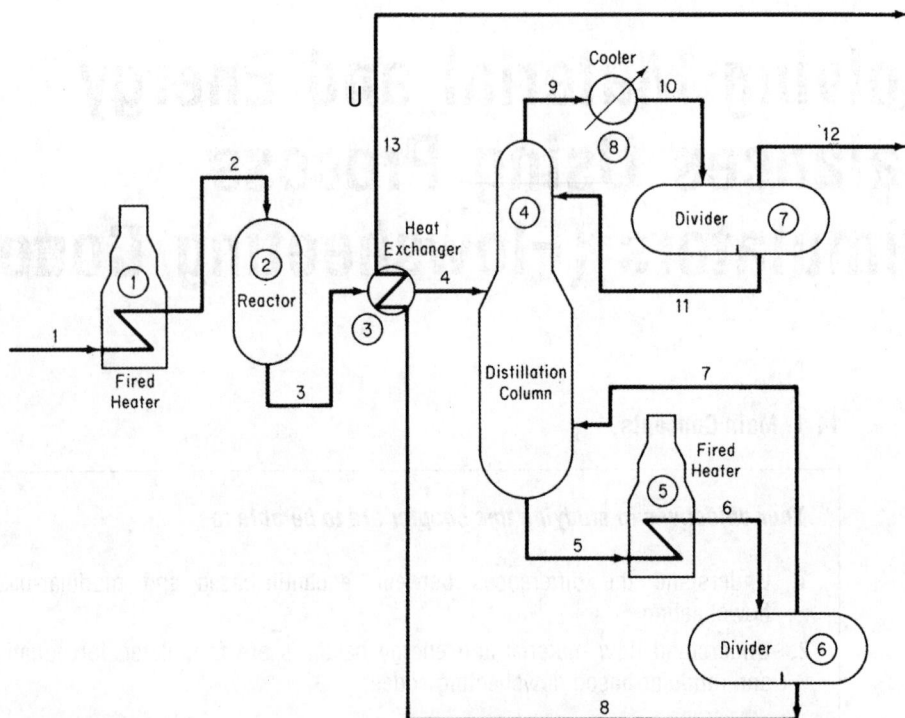

Figure 14.1 Hypothetical process flowsheet showing the materials flow in a process that includes reaction. The encircled numbers denote the unit labels, and the other numbers label the interconnecting streams.

The software must facilitate the transfer of information between equipment and streams, have access to a reliable database, and be flexible enough to accommodate equipment specifications provided by the user to supplement the library of programs that come with the code. Fundamental to all flowsheeting codes is the calculation of mass and energy balances for the entire process. Valid inputs to the material and energy balance phase of the calculations for the flowsheet must be defined in sufficient detail to determine all of the intermediate and product streams and the unit performance variables for all units.

Frequently, process plants contain recycle streams and control loops, and the solution for the stream properties requires iterative calculations. Thus, efficient numerical methods must be used. In addition, appropriate physical properties and thermodynamic data have to be retrieved from a database. Finally, a master program must exist that links all of the building blocks, physical property data, thermodynamic calculations, subroutines,

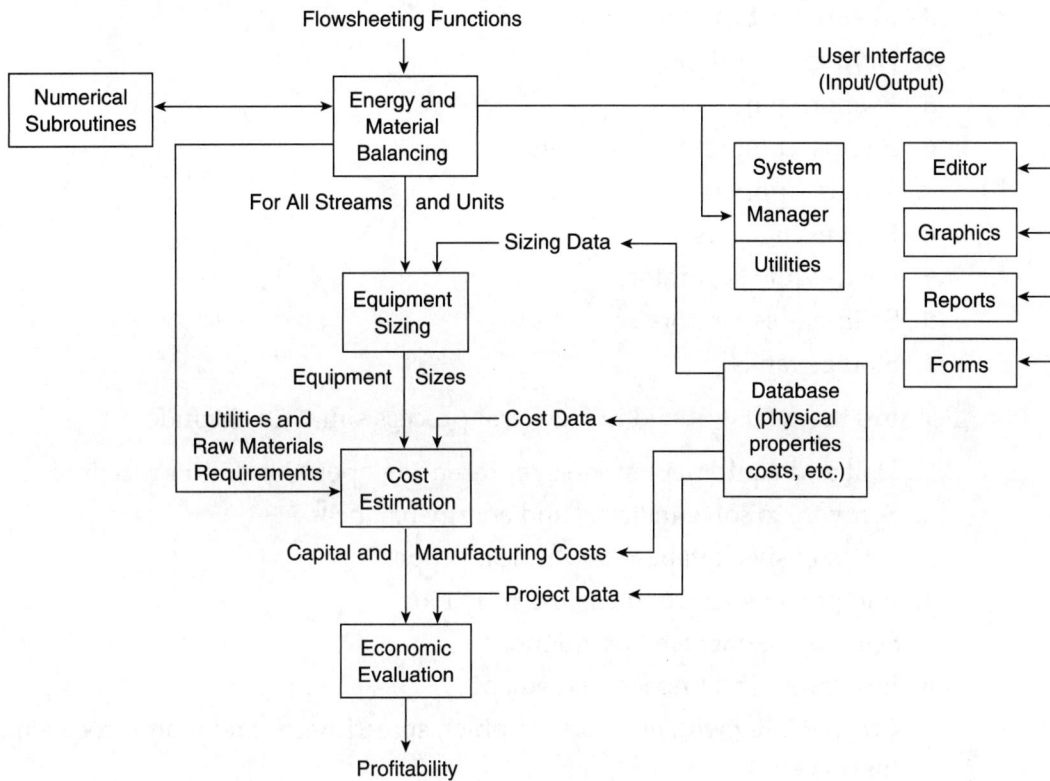

Figure 14.2   Information flow in a typical flowsheeting code

and numerical subroutines, and that also supervises the information flow. You will find that optimization and economic analysis are often the ultimate goals in the use of flowsheeting codes.

Other specific applications include

1. Steady- and unsteady-state simulation to help improve and verify the design of a process and examine complicated or dangerous designs
2. Training of operators
3. Data acquisition and reconciliation
4. Process control, monitoring, diagnostics, and troubleshooting
5. Optimization of process performance
6. Management of information
7. Safety analysis

Typical unit process models found in process simulators for both steady-state and unsteady-state operations include

1. Reactors of various kinds
2. Phase separation equipment

3. Ion exchange and absorption
4. Drying
5. Evaporation
6. Pumps, compressors, blowers
7. Mixers, splitters
8. Heat exchangers
9. Solid-liquid separators
10. Solid-gas separators
11. Storage tanks

Features that you will find in a general process simulator include

1. Unit and equipment models representing operations and procedures
2. Software to solve material and energy balances
3. An extensive database of physical properties
4. Equipment sizing and costing functions
5. Scheduling of batch operations
6. Environmental impact assessment
7. Compatibility with auxiliary graphics, spreadsheets, and word-processing functions
8. Ability to import and export data

Table 14.1 lists some commercial process simulators.

From the viewpoint of a user of a process simulator code you should realize:

Table 14.1 Vendors of Commercial Process Simulators

Name of Program	Source
ABACUSS II	MIT, Cambridge, MA
Aspen Engineering Suite (AES)	Aspen Technology, Cambridge, MA
CHEMCAD	Chemstations, Houston, TX
DESIGN II	WinSim, Houston, TX
D-SPICE	Fantoft Process Technologies, Houston, TX
HYSIM, HYSYS	Hyprotech, Calgary, Alberta
PRO/II, PROTISS	Simulation Sciences, Fullerton, CA
ProSim	Bryan Research and Engineering, Bryan, TX
SPEEDUP	Aspen Technology Corp., Cambridge, MA
SuperPro Designer	Intelligen, Scotch Plains, NJ

1. Several levels of analysis can be carried out beyond just solving material balances, including solving material plus energy balances, determining equipment sizing, profitability analysis, and much more. Crude, approximate flowsheets are usually studied before fully detailed flowsheets.

2. The results obtained by simulation rest heavily on the type and validity of the choices you make in the selection of the physical property package to be used.

3. You have to realize that the basic function of the process simulator is to solve equations. In spite of the progress made in equation solvers in the last 50 years, the information structure you introduce into the code may yield erroneous or no results. Check essential results by hand. Limits introduced on the range of variables must be valid.

4. A learning curve exists in using a process simulator so that initially a simple problem may take hours to solve, whereas as your familiarity with the simulator increases, it may take only minutes to solve the same problem.

5. GIGO (Garbage In Garbage Out). You have to take care to put appropriate data and connections between units into the data files for the code. Some diagnostics are provided, but they cannot troubleshoot all of your blunders.

Two extremes exist in process simulator software. At one extreme, the entire set of equations (and inequalities) representing the process is written down, including the material and energy balances, the stream connections, and the relations representing the equipment functions. This representation is known as the **equation-oriented method** of flowsheeting. The equations can be solved in a sequential fashion analogous to the modular representation described below, or simultaneously by Newton's method (or the equivalent), or by employing sparse matrix techniques to reduce the extent of matrix manipulations; you can find specific details in the references at the end of this chapter.

At the other extreme, the process can be represented by a collection of modules (**the modular method of flowsheeting**) in which the equations (and other information) representing each subsystem or piece of equipment are collected together and coded so that the module may be used in isolation from the rest of the flowsheet and hence is portable from one flowsheet to another, or can be used repeatedly in a given flowsheet. A module is a model of an individual element in a flowsheet (such as a reactor) that can be coded, analyzed, debugged, and interpreted by itself. In its usual formulation it is an input-output model; that is, given the input values, the module calculates the output values, but the reverse calculation is not feasible. Units represented solely by equations sometimes can yield inputs given the outputs. Some modular-based software codes such as Aspen Plus integrate equations with modules to speed up the calculations.

Another classification of flowsheeting codes focuses on how the equations or modules are solved. One treatment is to solve the equations or modules sequentially, and the other is to solve them simultaneously. Either the program and/or the user must select the decision variables for recycle and provide estimates of certain stream values to make sure that convergence of the calculations occurs, especially in a process with many recycle streams.

A third classification of flowsheeting codes is whether they solve steady-state or dynamic problems. We consider only the former here.

We will review equation-based process simulators first, although historically modular-based codes were developed first, because they are much closer to the techniques described up to this point in this book, and then turn to consideration of modular-based flowsheeting.

14.1.1 Equation-Based Process Simulation

Sets of linear and/or nonlinear equations can be solved simultaneously using an appropriate computer code. Equation-based flowsheeting codes have some advantages in that the physical property data needed to obtain values for the coefficients in the equations are transparently transmitted from a database at the proper time in the sequence of calculations. Figure 14.3 shows the information flow corresponding to the flowsheet in Figure 14.1.

Figure 14.4 is a set of equations that represents the basic operation of a flash drum.

The interconnections between the unit modules may represent information flow as well as material and energy flow. In the mathematical representation of

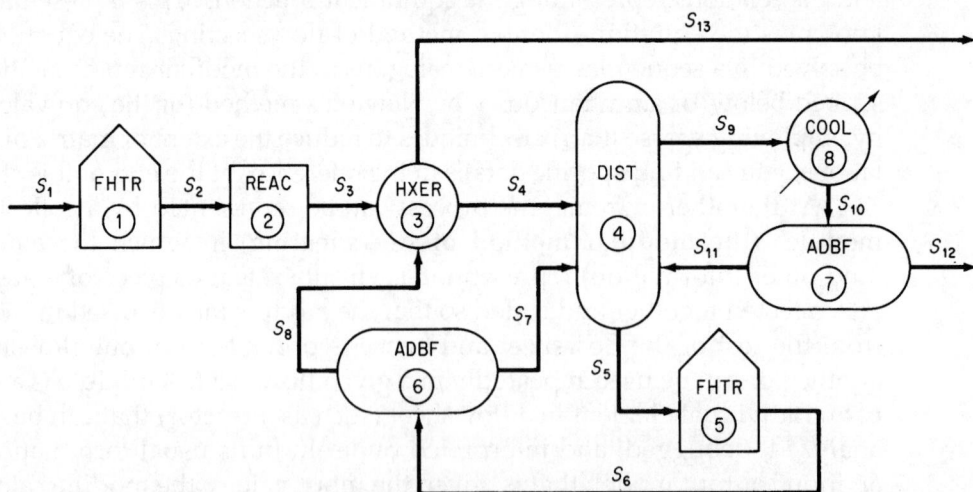

Figure 14.3 Information flowsheet for the hypothetical process in Figure 14.1 (S stands for stream; the module or computer code number is encircled.)

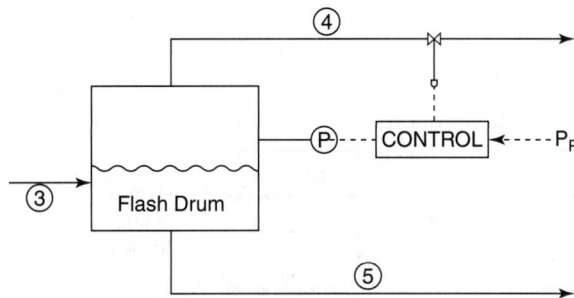

Material balances:

$$x_{A3}F_3 - y_{A4}F_4 - x_{A5}F_5 = 0$$

$$x_{P3}F_3 - y_{P4}F_4 - x_{P5}F_5 = 0$$

$$x_{G3}F_3 - y_{G4}F_4 - x_{G5}F_5 = 0$$

$$y_{A4} + y_{P4} + y_{G4} = 1$$

$$x_{A5} + x_{P5} + x_{G5} = 1$$

Equilibrium relations:

$$T_4 = T_5$$

$$y_{A4} = K_A x_{A5}$$

$$y_{P4} = K_P x_{P5}$$

$$y_{G4} = K_G x_{G5}$$

where $\kappa_i = p_i^*(T_4)/p_F \quad (i = A, P, G)$

Energy balance:

$$F_5(x_{A3}C_A + x_{P3}C_P + x_{G3}C_G)T_3 = F_5(x_{A5}C_A + x_{P5}C_P + x_{G5}C_G)T_5$$

$$+ F_4\left[(y_{A4}C_A + y_{P4}C_P + y_{G4}C_G)T_4 + y_{A4}\lambda_A + y_{P4}\lambda_P + y_{G4}\lambda_G\right]$$

Figure 14.4   A set of a linear and two nonlinear equations representing a system of three components, *A*, *P*, and *E*, passing through a flash drum

the plant, the interconnection equations are the material and energy balance flows between model subsystems. Equations for models such as mixing, reaction, heat exchange, and so on must also be listed so that they can be entered into the computer code used to solve the equation. Figure 14.5 (and Table 14.2) lists the common types of equations that might be used for a single subsystem.

In general, similar process units repeatedly occur in a plant and can be represented by the same set of equations that differ only in the names of variables, the number of terms in the summations, and the values of any coefficients in the equations.

<u>System diagram</u>

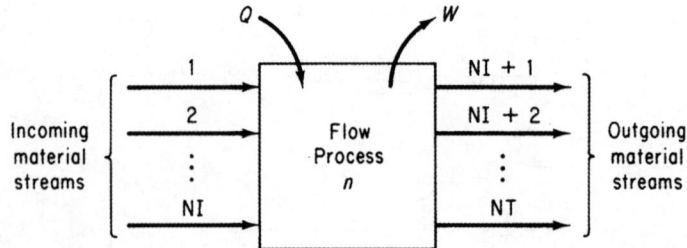

<u>Total mass balance (or mole balance) without reaction)</u>

$$\sum_{i=1}^{NI} F_i = \sum_{i=NI+1}^{NT} F_i$$

<u>Energy balance</u>

$$\sum_{i=1}^{NI} F_i H_i + Q_n - W_{s,n} = \sum_{i=N+1}^{NI} F_i H_i$$

<u>Vapor-liquid equilibrium distribution</u>

$$y_j = K_j x_j \quad \text{for} \quad j=1,2,\dots,NC$$

<u>Equilibrium vaporization coefficients</u>

$$K_j = K\left(T_i, P_i, \bar{W}_i\right) \quad j=1,2, \text{ for } \dots,,NC$$

<u>Total mole balance (with reaction)</u>

$$\sum_{i=1}^{NI} F_i + \sum_{l=1}^{NI} R_l \left[\sum_{j=1}^{NI} V_{j,l}\right] = \sum_{i=NI+1}^{NI} F_i$$

<u>Component mole balances (with reaction)</u>

$$\sum_{i=1}^{NI} F_i w_{i,j} + \sum_{j=1}^{NR} V_{j,l} R_l = \sum_{i=NI+1}^{NT} F_i w_{i,j} \quad \text{for } j=1,2,\dots,NC$$

<u>Molar atom balances</u>

$$\sum_{i=1}^{NI} F_i \left[\sum_{j=1}^{NC} w_{i,j} a_{j,k}\right] = \sum_{i=NI+1}^{NI} F_i \left[\sum_{j=1}^{NC} w_{i,j} a_{j,k}\right] \quad \text{for } k=1,2,\dots,NE$$

<u>Mechanical energy balance</u>

$$\sum_{i=1}^{NI} (K_i + P_i) + \sum_{i=1}^{NI} \int_{P_{1,i}}^{P_{2,i}} V_i dp_i = \sum_{i=NI+1}^{NI} (K_i + P_i) + \sum_{i=NI+1}^{NI} \int_{P_{1,i}}^{P_{2,i}} V_i dP_i + W_{s,n} + E_{v,n}$$

<u>Component mass or mole balances</u>

$$\sum_{i=1}^{NI} F_i w_{i,j} = \sum_{i=NT+1}^{NT} F_i w_{i,j}$$

$$\text{for } j=1,2,\dots,NC$$

<u>Summation of mole or mass fractions</u>

$$\sum_{j=1}^{NC} w_{i,j} = 1.0 \; \text{ for } i=1,2,\dots,NI$$

<u>Physical property functions</u>

$$H_i = H_{VL}\left(T_i, P_i, \bar{W}_i\right)$$

$$S_i = S_{VL}\left(T_i, P_i, \bar{W}_i\right) \quad i=1,2,\dots,NI$$

Figure 14.5 Generic equations for a steady-state, open system

Table 14.2   Notation for Figure 14.5

$a_{j,k}$	Number of atoms of the kth chemical element in the jth component
F_i	Total flow rate of the ith stream
H_i	Relative enthalpy of the ith stream
K_j	Vaporation coefficient of the jth component
NC	Number of chemical components (compounds)
NE	Number of chemical elements
NI	Number of incoming material streams
NR	Number of chemical reactions
NT	Total number of material streams
p_i	Pressure of the ith stream
Q_n	Heat transfer for the nth process unit
R_l	Reaction expression for the lth chemical reaction
T_i	Temperature of the ith stream
$V_{j,l}$	Stoichiometric coefficient of the jth component in the lth chemical reaction
$w_{i,j}$	Fractional composition (mass of mole) of the jth component in the ith stream
$\overline{W}_i$	Average composition in the ith stream
$W_{s,n}$	Work for the nth process unit
x_j	Mole fraction of component j in the liquid
y_j	Mole fraction of component j in the vapor

Equation-based codes can be formulated to include inequality constraints along with the equations. Such constraints might be of the form $a_1x_1 + a_2x_2 + \ldots \le b$ and might arise from such factors as

1. Conditions imposed in linearizing any nonlinear equations
2. Process limits for temperature, pressure, concentration
3. Requirements that variables be in a certain order
4. Requirements that variables be positive or integer

As you can see from Figures 14.4 and 14.5, if all of the equations for the material and energy balances plus the phase and chemical equilibrium relationships plus the thermodynamic and kinetic relations are combined, they form a huge, sparse (few variables in any equation) array. The set of equations can be partitioned into subsets of equations that cannot further be decomposed and have to be solved simultaneously. Two important aspects of solving the sets of nonlinear equations in flowsheeting codes, both equation-based and modular, are (1) the procedure for establishing the precedence order in solving the equations, and (2) the treatment of recycle (feedback) of information, material, and/or energy. Details of how to accommodate these important issues efficiently can be found in the references at the end of this chapter.

Whatever the process simulator used to solve material and energy balance problems, you must provide certain input information to the code in an acceptable format. All flowsheeting codes require that you convert the information in the flowsheet (see Figure 14.1) and the information flowsheet as illustrated in Figure 14.3, or something equivalent. In the information flowsheet, you use the name of the mathematical model (or the subroutine in modular-based flowsheeting) that will be used for the calculations instead of the name of the process unit.

Once the information flowsheet is set up, the determination of the process topology is easy; that is, you can immediately write down the stream interconnections between the modules (or subroutines) that have to be included in the input data set. For Figure 14.3 the matrix of stream connections (the **process matrix**) is (a negative sign designates an exit stream):

Unit	Associated Streams				
1	1	−2			
2	2	−3			
3	3	8	−4	−13	
4	4	7	11	−9	−5
5	5	−6			
6	6	−8	−7		
7	10	−11	−12		
8	9	−10			

14.1.2 Modular-Based Process Simulators

Because plants are composed of various unit operations (such as distillation, heat transfer, and so on) and unit processes (such as alkylation, hydrogenation, and so on), chemical engineers historically developed representations of each of these units or processes as self-contained modules. Each module (refer to Figure 14.6) might be composed of equations, equipment sizes, material and energy balance relations, component flow rates, and the temperatures, pressures, and phase conditions of each stream that enters and leaves the physical equipment represented by the module.

Figure 14.7 shows a flash module and the computer code that yields an output for a given input.

Values of certain parameters and variables determine the capital and operating costs for the units. Of course, the interconnections set up for the modules must be such that information can be transferred from module to module concerning the streams, compositions, flow rates, coefficients, and so on. In other words, the modules constitute a set of building blocks that can be arranged in general ways to represent any process.

The sequential modular method of flowsheeting is the one most commonly encountered in commercial computer software. A module exists for each process unit in the information flowsheet. Given the values of each input

A MODULE

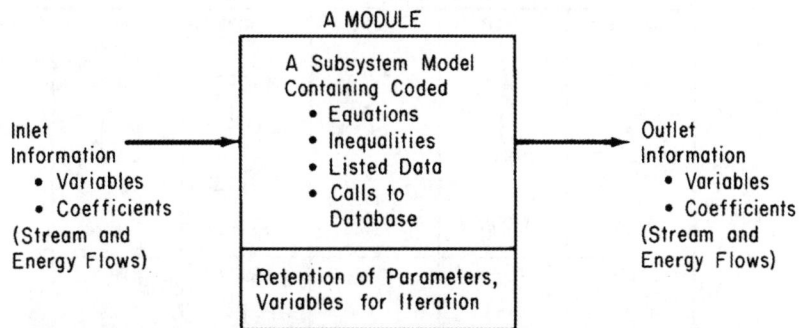

Figure 14.6 A typical process module showing the necessary interconnections of information

stream composition, flow rate, temperature, pressure, enthalpy, and the equipment parameters, the output of a module can become the input stream to another module for which the calculations can then proceed, and so on, until the material and energy balances are resolved for the entire process. Modules are portable. By portable we mean that a subroutine corresponding to a module can be assembled as an element of a large group of subroutines and successfully represent a certain type of equipment in any process. Figure 14.8 shows icons for typical standardized unit operation modules.

Other modules take care of equipment sizing and cost estimation, perform numerical calculations, handle recycle calculations (described in more detail below), optimize, and serve as controllers (executives for the whole set of modules so that they function in the proper sequence). Internally, a very simple module might just be a table lookup program. However, most modules consist of Fortran or C subroutines that execute a sequence of calculations. Subroutines may consist of hundreds to thousands of lines of code.

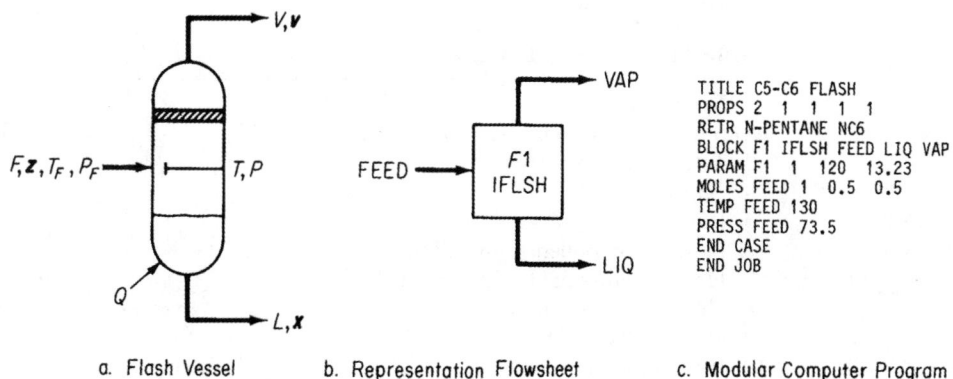

```
TITLE C5-C6 FLASH
PROPS 2  1  1  1  1
RETR N-PENTANE NC6
BLOCK F1 IFLSH FEED LIQ VAP
PARAM F1  1   120  13.23
MOLES FEED 1  0.5  0.5
TEMP FEED 130
PRESS FEED 73.5
END CASE
END JOB
```

a. Flash Vessel   b. Representation Flowsheet   c. Modular Computer Program

Figure 14.7 A module that represents a flash unit [From J. D. Seader, W. D. Seider, and A. C. Pauls, *Flowtran Simulation: An Introduction*, CACHE, Austin, TX (1987)]

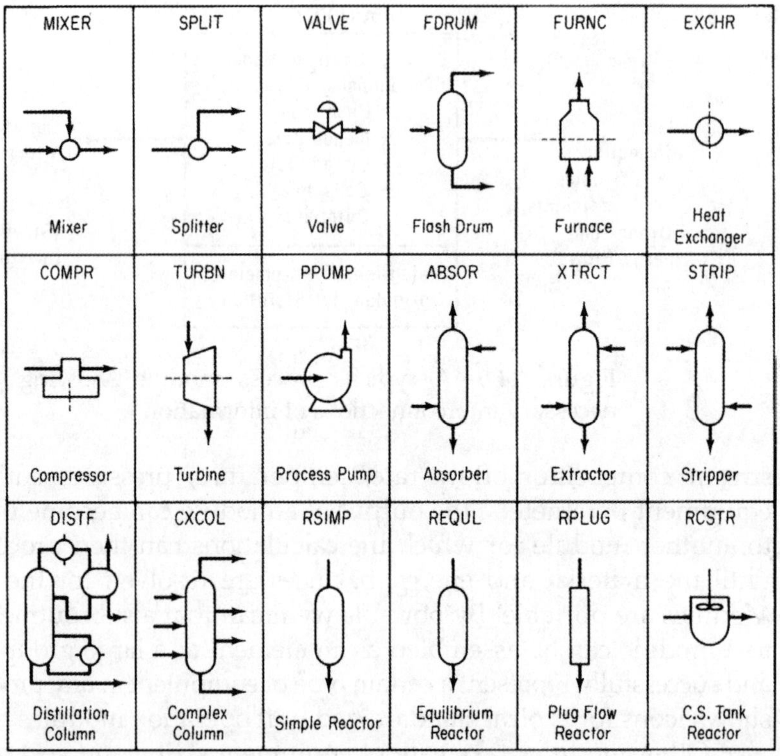

Figure 14.8 Typical process modules used in sequential modular-based flowsheeting codes with their subroutine names

Information flows between modules via the material streams. Associated with each stream is an ordered list of numbers that characterize the stream. Table 14.3 lists a typical set of parameters associated with a stream. The presentation of the results of simulations also follows the same format as shown in Table 14.3.

Table 14.3   Stream Parameters

1. Stream number*
2. Stream flag (designates type of stream)
3. Total flow, lb mol/hr
4. Temperature, °F
5. Pressure, psia
6. Flow of component 1, lb mol/hr
7. Flow of component 2, lb mol/hr
8. Flow of component 3, lb mol/hr
9. Molecular weight
10. Vapor fraction
11. Enthalpy
12. Sensitivity

*Corresponds to an arbitrary numbering scheme used on the information flowsheet.

As a user of a modular-based code, you have to provide

1. The process topology
2. Input stream information, including physical properties and connections
3. Design parameters needed in the modules and equipment specifications
4. Convergence criteria

In addition, you sometimes may have to insert a preferred calculation order for the modules. When economic evaluation and optimization are being carried out, you must also provide cost data and optimization criteria.

Modular-based flowsheeting exhibits several advantages in design. The flowsheet architecture is easily understood because it closely follows the process flowsheet. Individual modules can easily be added and removed from the computer package. Furthermore, new modules may be added to or removed from the flowsheet without affecting other modules. Modules at two different levels of accuracy can be substituted for one another.

Modular-based flowsheeting also has certain drawbacks:

1. The output of one module is input to another. The input and output variables in a computer module are fixed so you cannot arbitrarily introduce an output and generate an input as sometimes can be done in equation-based code.
2. The modules require extra computer time to generate reasonably accurate derivatives or their substitutes, especially if a module contains tables, functions with discrete variables, discontinuities, and so on. Perturbation of the input to a module is the primary way in which a finite-difference substitute for a derivative can be generated.
3. The modules may require a fixed precedence order of solution—that is, the output of one module must become the input of another—hence convergence may be slower than in an equation-solving method, and the computational costs may be high.
4. To specify a parameter in a module as a decision variable in the design of a plant, you have to place a control block around the module and adjust the parameter such that design specifications are met. This arrangement creates a loop. If the values of many design variables are to be determined, you might end up with several nested loops of calculation (which do, however, enhance stability). A similar arrangement must be used if you want to impose constraints.
5. Conditions imposed on a process (or a set of equations for that matter) may cause the unit physical states to move from two-phase to single-phase operation, or the reverse. (This situation is true of equation-based codes as well.) You have to foresee and accommodate such changes in state.

An engineer can usually carry out the partitioning and nesting and determine the computational sequence for a flowsheet by inspection if the flowsheet is not too complicated. In some codes, the user supplies the computational sequence as input. Other codes determine the sequence automatically. In Aspen, for example, the code is capable of determining the entire computational sequence, but the user can supply as many specifications as desired, up to and including the complete computational sequence. Consult one of the supplementary references at the end of this chapter for detailed information on optimal ways of using simulator techniques beyond our scope in this text.

Once the sequence of calculation codes is specified, everything is in order for the solution of material and energy balances. All that has to be done is to calculate the correct values for the stream flow rates and their properties. To execute the calculations, various numerical algorithms can be selected by the user or determined by the simulator. The results can be displayed as tables, graphs, charts, and so on.

Looking Back

In this chapter we described the two main ways of solving the material and energy balances in process simulators: using equation-based and modular-based computer software.

Discussion Question

A number of articles have been written on the subject of paper versus polystyrene as materials for disposable cups. Set up the flowsheets for the production of each, and include all of the quantitative and qualitative factors, both positive and negative, for the production from basic raw materials to the final product. Indicate what material and energy balances are needed, and, if possible, collect data so that they can be solved. Summarize the material and energy usage in the manufacture of a cup.

Supplementary References

American Institute of Chemical Engineers. *CEP Software Directory*, AIChE, New York (issued annually on the Web).

Benyaha, F. "Flowsheeting Packages: Reliable or Fictitious Process Models?" *Transactions Inst. Chemical Engineering*, **78A,** 840–44 (2000).

Bequette, B. W. *Process Dynamics: Modeling, Analysis, and Simulation,* Prentice Hall, Upper Saddle River, NJ (1998).

Biegler, L. T., I. E. Grossmann, and A. W. Westerberg. *Systematic Methods of Chemical Process Design,* Prentice Hall, Upper Saddle River, NJ (1997).

Canfield, F. B., and P. K. Nair. "The Key of Computed Integrated Processing," in *Proceed. ESCAPE-1,* Elsinore, Denmark (May, 1992).

Chen, H. S., and M. A. Stadtherr. "A Simultaneous-Modular Approach to Process Flowsheeting and Optimization: I. Theory and Implementation," *AIChE J.,* **30** (1984).

Clark, G., D. Rossiter, and P. W. H. Chung. "Intelligent Modeling Interface for Dynamic Process Simulators," *Transactions Inst. Chemical Engineering,* **78A,** 823–39 (2000).

Gallun, S. E., R. H. Luecke, D. E. Scott, and A. M. Morshedi. "Use Open Equations for Better Models," *Hydrocarbon Processing,* **78** (July, 1992).

Lewin, D. R., et al. *Using Process Simulators in Chemical Engineering: A Multimedia Guide for the Core Curriculum,* John Wiley, New York (2001).

Mah, R. S. H. *Chemical Process Structures and Information Flows,* Butterworths, Seven Oaks, UK (1990).

Seider, W. D., J. D. Seader, and D. R. Lewin. *Process Design Principles,* John Wiley, New York (1999).

Slyberg, O., N. W. Wild, and H. A. Simons. *Introduction to Process Simulation,* 2nd ed., TAPPI Press, Atlanta (1992).

Thome, B. (ed.). *Systems Engineering—Principles and Practice of Computer-Based Systems Engineering,* John Wiley, New York (1993).

Turton, R., R. C. Bailie, W. B. Whiting, and J. A. Shaeiwitz. *Analysis, Synthesis, and Design of Chemical Processes,* Prentice Hall, Upper Saddle River, NJ (1998).

Westerberg, A. W., H. P. Hutchinson, R. L. Motard, and P. Winter. *Process Flowsheeting,* Cambridge University Press, Cambridge, UK (1979).

Problems

14.1 In petroleum refining, lubricating oil is treated with sulfuric acid to remove unsaturated compounds, and after settling, the oil and acid layers are separated. The acid layer is added to water and heated to separate the sulfuric acid from the sludge contained in it. The dilute sulfuric acid, now 20% H_2SO_4 at 82°C, is fed to a Simonson-Mantius evaporator, which is supplied with saturated steam at 400 kPa gauge to lead coils submerged in the acid, and the condensate leaves at the saturation temperature. A vacuum is maintained at 4.0 kPa by means of a barometric leg. The acid is concentrated to 80% H_2SO_4; the boiling point at 4.0 kPa is 121°C. How many kilograms of acid can be concentrated per 1000 kg of steam condensed?

14.2 You are asked to perform a feasibility study on a continuous stirred tank reactor shown in Figure P14.2 (which is presently idle) to determine if it can be used for the second-order reaction

$$2A \rightarrow B + C$$

Since the reaction is exothermic, a cooling jacket will be used to control the reactor temperature. The total amount of heat transfer may be calculated from an overall heat transfer coefficient (U) by the equation

$$\dot{Q} = UA\,\Delta T$$

where $\dot{Q}$ = total rate of heat transfer from the reactants to the water jacket in the steady state

U = empirical coefficient

A = area of transfer

ΔT = temperature difference (here $T_4 - T_2$)

Some of the energy released by the reaction will appear as sensible heat in stream F_2, and some concern exists as to whether the fixed flow rates will be sufficient to keep the fluids from boiling while still obtaining good conversion. Feed data are as follows:

Component	Feed Rate (lb mol/hr)	C_p [Btu/(lb mol)(°F)]	MW
A	214.58	41.4	46
B	23.0	68.4	76
C	0.0	4.4	16

The consumption rate of A may be expressed as

$$-2k(C_A)^2 V_R$$

where

$$C_A = \frac{(F_{1,A})(\rho)}{\sum (F_{1,i})(MW_i)} = \text{concentration of } A, \text{lbmol}/\text{ft}^3$$

$$k = k_0 \exp\left(\frac{-E}{RT}\right)$$

k_0, E, R are constants, and T is the absolute temperature.

Solve for the temperatures of the exit streams and the product composition of the steady-state reactor using the following data:

Fixed parameters

$$\text{Reactor volume} = V_R = 13.3 \text{ ft}^3$$
$$\text{Heat transfer area} = A = 29.9 \text{ ft}^2$$
$$\text{Heat transfer coefficient} = U = 74.5 \text{ Btu}/(\text{hr})(\text{ft}^2)(°\text{F})$$

Variable input

$$\text{Reactant feed rate} = F_i \text{ (see table above)}$$
$$\text{Reactant feed temperature} = T_1 = 80°\text{F}$$
$$\text{Water feed rate} = F_3 = 247.7 \text{ lb mol/hr water}$$
$$\text{Water feed temperature} = T_3 = 75°\text{F}$$

Physical and thermodynamic data

$$\text{Reaction rate constant} = k_0 = 34 \text{ ft}^3/(\text{lb mol})/(\text{hr})$$
$$\text{Activation energy/gas constant} = E/R = 1000°\text{R}$$

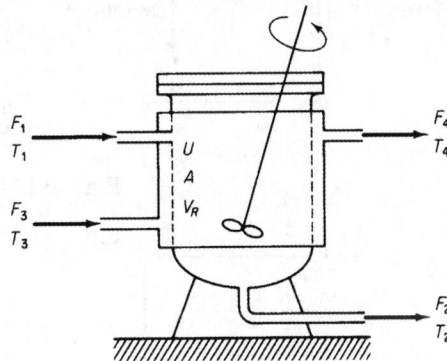

Figure P14.2

$$\text{Heat of reaction} = \Delta H = -5000 \text{ Btu/lb mol } A$$
$$\text{Heat capacity of water} = C_{pw} = 18 \text{ Btu/(lb mol)(°F)}$$
$$\text{Product component density} = \rho = 55 \text{ lb/ft}^3$$

The densities of each of the product components are essentially the same. Assume that the reactor contents are perfectly mixed as well as the water in the jacket, and that the respective exit stream temperatures are the same as the reactor contents or jacket contents.

14.3 The stream flows for a plant are shown in Figure P14.3. Write the material and energy balances for the system and calculate the unknown quantities in the diagram (A to F). There are two main levels of stream flow: 600 psig and 50 psig. Use the steam tables for the enthalpies.

14.4 Figure P14.4 shows a calciner and the process data. The fuel is natural gas. How can the energy efficiency of this process be improved by process modification? Suggest at least two ways based on the assumption that the supply conditions of the air and fuel remain fixed (but these streams can possibly be passed through heat exchangers). Show all calculations.

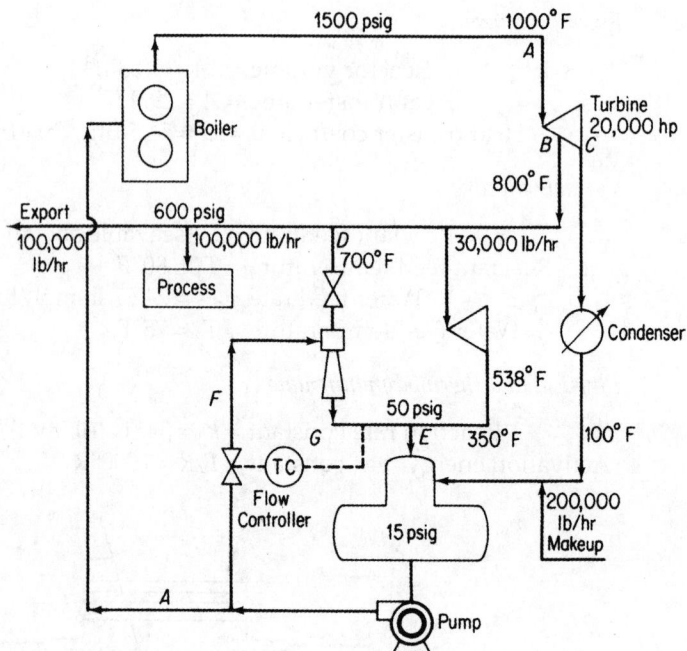

Figure P14.3

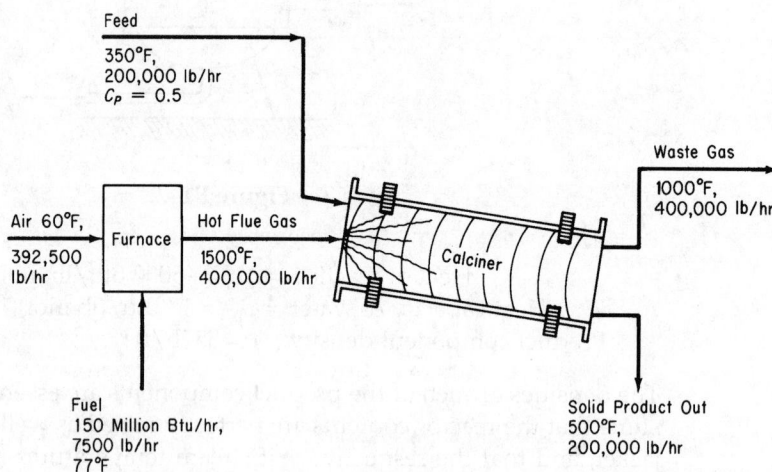

Figure P14.4

14.5 Limestone ($CaCO_3$) is converted into CaO in a continuous vertical kiln (see Figure P14.5). Heat is supplied by combustion of natural gas (CH_4) in direct contact with the limestone using 50% excess air. Determine the kilograms of $CaCO_3$ that can be processed per kilogram of natural gas. Assume that the following average heat capacities apply:

$$C_p \text{ of } CaCO_3 = 234 \text{ J}/(\text{g mol})(°C)$$
$$C_p \text{ of } CaO \ = 111 \text{ J}/(\text{g mol})(°C)$$

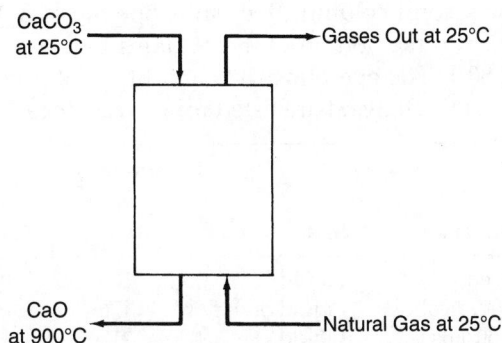

Figure P14.5

14.6 A feed stream of 16,000 lb/hr of 7% by weight NaCl solution is concentrated to a 40% by weight solution in an evaporator. The feed enters the evaporator, where it is heated to 180°F. The water vapor from the solution and the concentrated solution leave at 180°F. Steam at the rate of 15,000 lb/hr enters at 230°F and leaves as condensate at 230°F. See Figure P14.6.

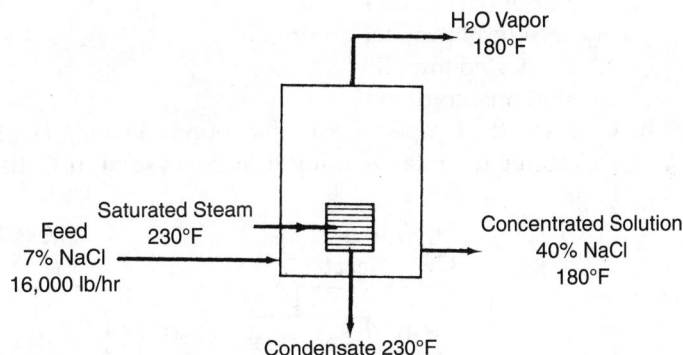

Figure P14.6

a. What is the temperature of the feed as it enters the evaporator?
b. What weight of 40% NaCl is produced per hour?
   Assume that the following data apply:

$$\text{Average } C_p \text{ 7\% NaCl soln:} \quad 0.92 \text{ Btu/(lb)(°F)}$$
$$\text{Average } C_p \text{ 40\% NaCl soln:} \quad 0.85 \text{ Btu/(lb)(°F)}$$
$$\Delta \hat{H}_{vap} \text{ of } H_2O \text{ at } 180°F = 990 \text{ Btu/lb}$$
$$\Delta \hat{H}_{vap} \text{ of } H_2O \text{ at } 230°F = 959 \text{ Btu/lb}$$

14.7 The Blue Ribbon Sour Mash Company plans to make commercial alcohol by a process shown in Figure P14.7. Grain mash is fed through a heat exchanger where it is heated to 170°F. The alcohol is removed as 60% by weight alcohol from the first fractionating column; the bottoms contain no alcohol. The 60% alcohol is further fractionated to 95% alcohol and essentially pure water in the second column. Both stills operate at a 3:1 reflux ratio and heat is supplied to the bottom of the columns by steam. Condenser water is obtainable at 80°F. The operating data and physical properties of the streams have been accumulated and are listed for convenience:

Stream	State	Boiling Point (°F)	C_p[Btu/(lb)(°F)] Liquid	Vapor	Heat of Vaporization (Btu/lb)
Feed	Liquid	170	0.96	—	950
60% alcohol	Liquid or vapor	176	0.85	0.56	675
Bottoms I	Liquid	212	1.00	0.50	970
95% alcohol	Liquid or vapor	172	0.72	0.48	650
Bottoms II	Liquid	212	1.0	0.50	970

Prepare the material balances for the process, calculate the precedence order for solution, and

a. Determine the weight of the following streams per hour:
  1. Overhead product, column I
  2. Reflux, column I
  3. Bottoms, column I
  4. Overhead product, column II
  5. Reflux, column II
  6. Bottoms, column II
b. Calculate the temperature of the bottoms leaving Heat Exchanger III.
c. Determine the total heat input to the system in British thermal units per hour.

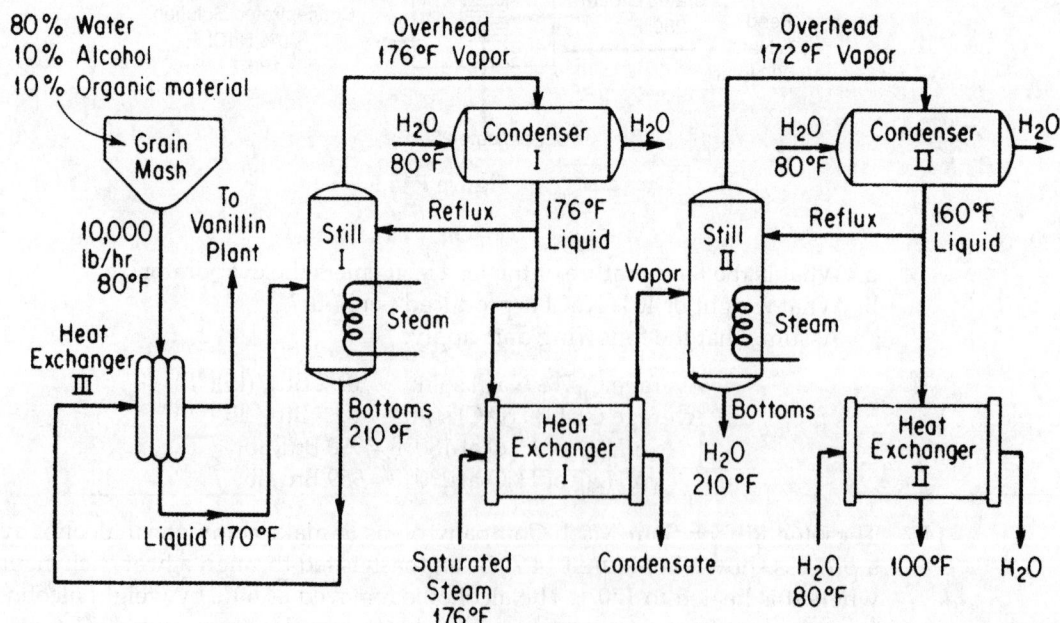

Figure P14.7

d. Calculate the water requirements for each condenser and Heat Exchanger II in gallons per hour if the maximum exit temperature of water from this equipment is 130°F.

14.8 Toluene, manufactured by the conversion of n-heptane with a Cr_2O_3-on-Al_2O_3 catalyst

$$CH_3CH_2CH_2CH_2CH_2CH_2CH_3 \rightarrow C_6H_5CH_3 + 4H_2$$

by the method of hydroforming, is recovered by use of a solvent. See Figure P14.8 for the process and conditions.

The yield of toluene is 15 mol % based on the n-heptane charged to the reactor. Assume that 10 kg of solvent are used per kilogram of toluene in the extractors.

a. Calculate how much heat has to be added to or removed from the catalytic reactor to make it isothermal at 425°C.
b. Find the temperature of the n-heptane and solvent stream leaving the mixersettlers if both streams are at the same temperature.
c. Find the temperature of the solvent stream after it leaves the heat exchanger.
d. Calculate the heat duty of the fractionating column in kilojoules per kilogram of n-heptane feed to the process.

	$-\Delta H_f^{\circ*}$ (kJ/g mol)	$C_p[J/(g)(°C)]$		$\Delta H_{vaporization}$ (kJ/kg)	Boiling Point (K)
		Liquid	Vapor		
Toulene[†]	12.00	2.22	2.30	364	383.8
n-Heptane	−224.4	2.13	1.88	318	371.6
Solvent	—	1.67	2.51	—	434

*As liquids.
[†]The heat of solution of toluene in the solvent is −23 J/g toluene.

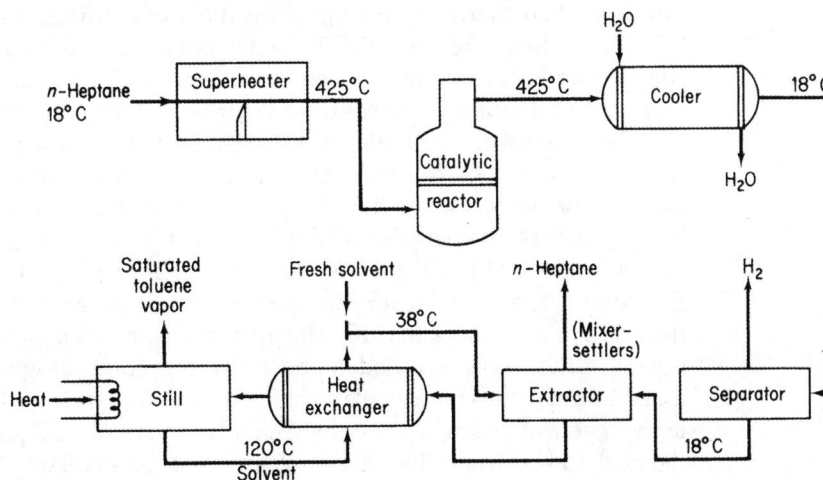

Figure P14.8

14.9 One hundred thousand pounds of a mixture of 50% benzene, 40% toluene, and 10% o-xylene is separated every day in a distillation-fractionation plant as shown on the flowsheet for Figure P14.9.

	Boiling Point (°C)	C_p Liquid [cal/(g)(°C)]	Latent Heat of Vap. (cal/g)	C_p Vapor [cal/(g)(°C)]
Benzene	80	0.44	94.2	0.28
Toluene	109	0.48	86.5	0.30
o-Xylene	143	0.48	81.0	0.32
Charge	90	0.46	88.0	0.29
Overhead T_I	80	0.45	93.2	0.285
Residue T_I	120	0.48	83.0	0.31
Residue T_{II}	413	0.48	81.5	0.32

The reflux ratio for Tower I is 6:1; the reflux ratio for Tower II is 4:1; the charge to Tower I is liquid; the chart to Tower II is liquid. Compute:

a. The temperature of the mixture at the outlet of the heat exchanger (marked as T^*)
b. The British thermal units supplied by the steam reboiler in each column
c. The quantity of cooling water required in gallons per day for the whole plant
d. The energy balance around Tower I

14.10 Sulfur dioxide emissions from coal-burning power plants cause serious atmospheric pollution in the eastern and midwestern portions of the United States. Unfortunately, the supply of low-sulfur coal is insufficient to meet the demand. Processes presently under consideration to alleviate the situation include coal gasification followed by desulfurization and stack gas cleaning. One of the more promising stack gas cleaning processes involves reacting SO_2 and O_2 in the stack gas with a solid metal oxide sorbent to give the metal sulfate, and then thermally regenerating the sorbent and absorbing the result SO_3 to produce sulfuric acid. Recent laboratory experiments indicate that sorption and regeneration can be carried out with several metal oxides, but no pilot or full-scale processes have yet been put into operation.

You are asked to provide a preliminary design for a process that will remove 95% of the SO_2 from the stack gas of a 1000 MW power plant. Some data are given below and in the flow diagram of the process (Figure P14.10). The sorbent consists of fine particles of a dispersion of 30% by weight CuO in a matrix of inert porous Al_2O_3. This solid reacts in the fluidized-bed absorber at 315°C. Exit solid is sent to the regenerator, where SO_3 is evolved at 700°C, converting the $CuSO_4$ present back to CuO. The fractional conversion of CuO to $CuSO_4$ that occurs in the sorber is called α and is an important design variable. You are asked to carry out your calculations for $\alpha = 0.2$, 0.5, and 0.8. The SO_3 produced in the regenerator is swept out by recirculating air. The SO_3-laden air is sent to the acid tower, where the SO_3 is absorbed in recirculating sulfuric acid and oleum, part of which is withdrawn as salable by-products. You will notice that the sorber, regenerator, and perhaps the acid tower are adiabatic; their

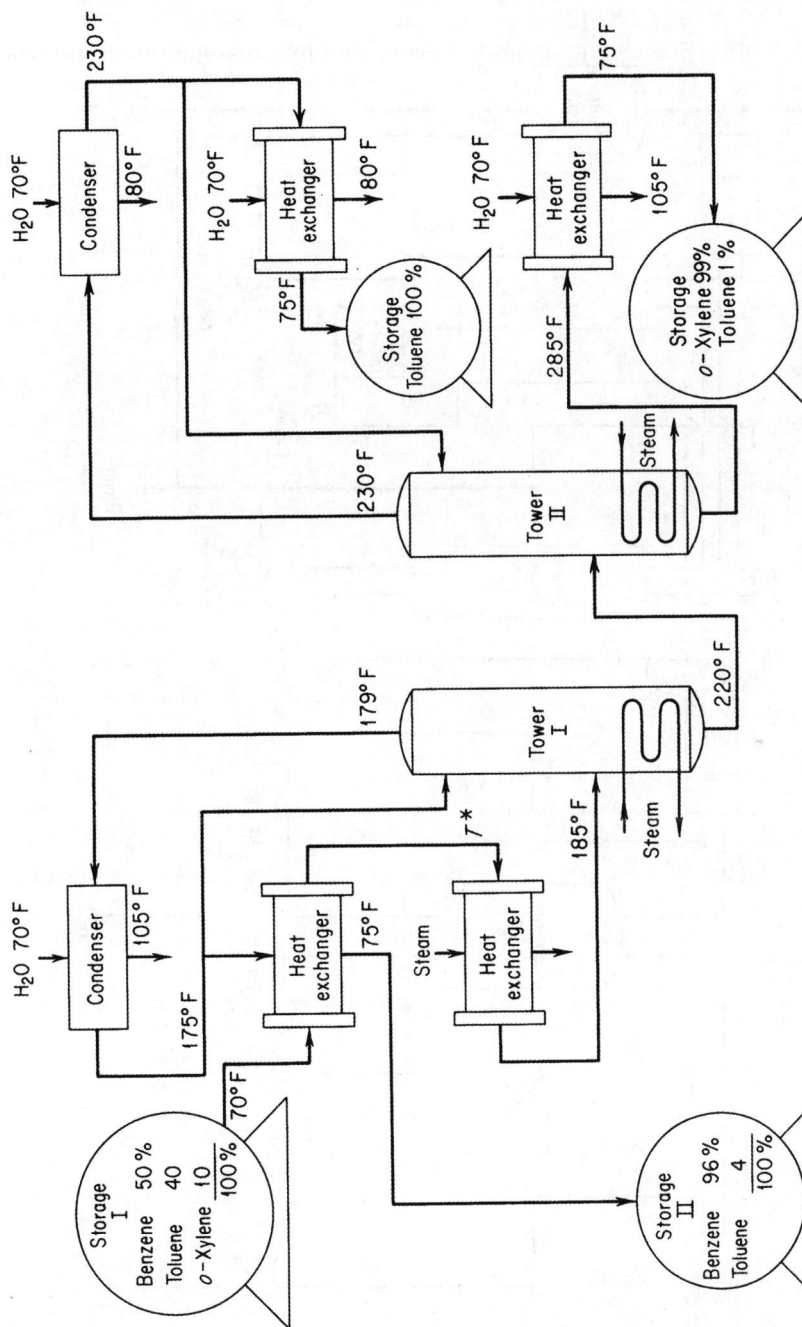

Figure P14.9

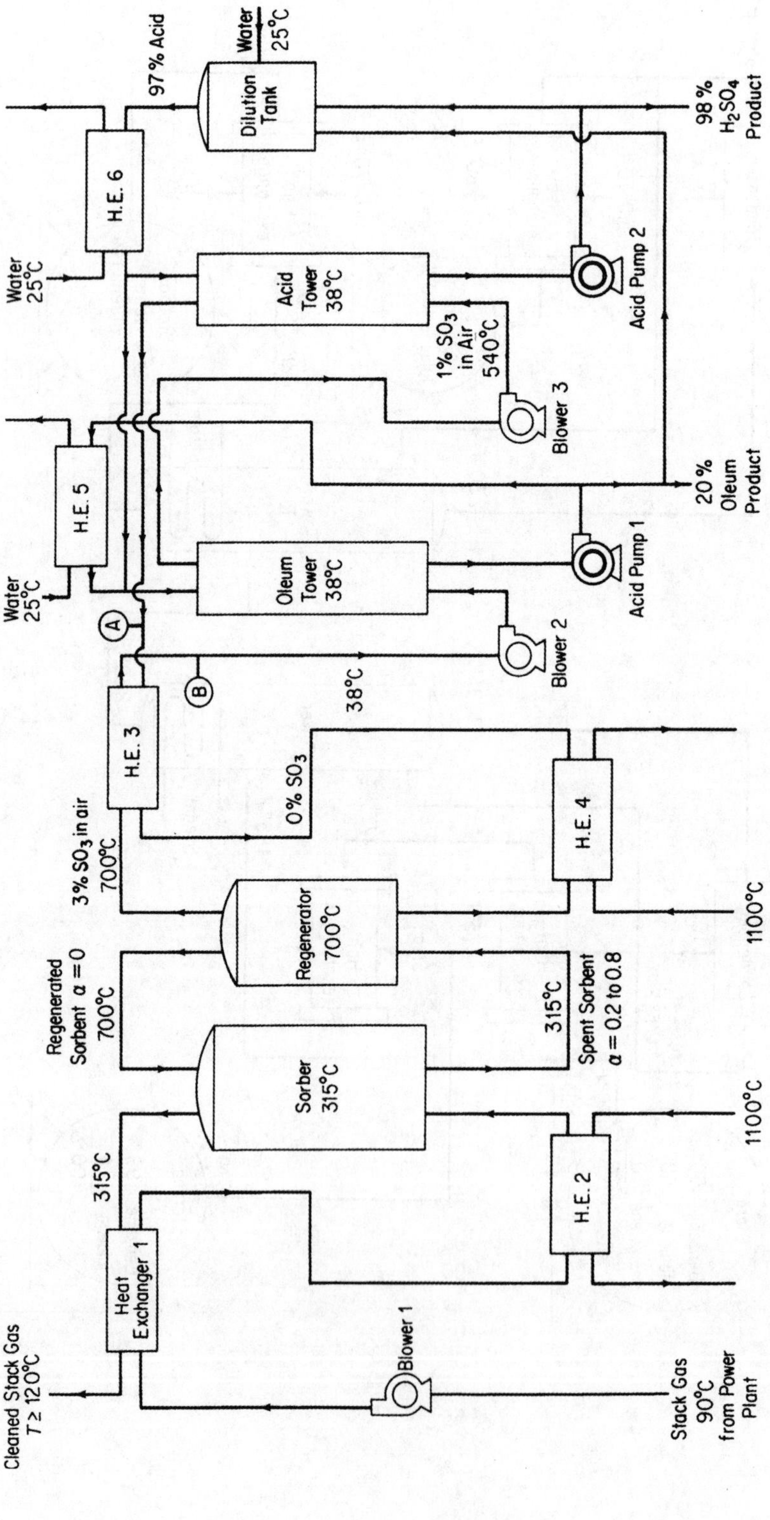

Figure P14.10

temperatures are adjusted by heat exchange with incoming streams. Some of the heat exchangers (nos. 1 and 3) recover heat by countercurrent exchange between the feed and exit streams. Additional heat is provided by withdrawing flue gas from the power plant at any desired high temperature up to 1100°C and then returning it at a lower temperature. Cooling is provided by water at 25°C. As a general rule, the temperature difference across heat-exchanger walls separating the two streams should average about 28°C. The nominal operating pressure of the whole process is 10 kPa. The three blowers provide 6 kPa additional head for the pressure losses in the equipment, and the acid pumps have a discharge pressure of 90 kPa gauge. You are asked to write the material and energy balances and some equipment specifications as follows:

a. Sorber, regenerator, and acid tower. Determine the flow rate, composition, and temperature of all streams entering and leaving.
b. Heat exchangers. Determine the heat load and flow rates, temperatures, and enthalpies of all streams.
c. Blowers. Determine the flow rate and theoretical horsepower.
d. Acid pump. Determine the flow rate and theoretical horsepower.

Use SI units. Also, use a basis of 100 kg of coal burned for all of your calculations; then convert to the operating basis at the end of the calculations.

Power plant operation. The power plant burns 340 metric tons/hr of coal having the analysis given below. The coal is burned with 18% excess air, based on complete combustion to CO_2, H_2O, and SO_2. In the combustion only the ash and nitrogen are left unburned; all the ash has been removed from the stack gas.

Element	Wt %
C	76.6
H	5.2
O	6.2
S	2.3
N	1.6
Ash	8.1

Data on Solids (Units of C_p are J/(g mol)(K); units of H are kJ/g mol)

T(K)	Al_2O_3 C_p	Al_2O_3 $H_T - H_{298}$	CuO C_p	CuO $H_T - H_{298}$	$CuSO_4$ C_p	$CuSO_4$ $H_T - H_{298}$
298	79.04	0.00	42.13	0.00	98.9	0.00
400	96.19	9.00	47.03	4.56	114.9	10.92
500	106.10	19.16	50.04	9.41	127.2	23.05
600	112.5	30.08	52.30	14.56	136.3	36.23
700	117.0	41.59	54.31	19.87	142.9	50.25
800	120.3	53.47	56.19	25.40	147.7	64.77
900	122.8	65.65	58.03	31.13	151.0	79.71
1000	124.7	77.99	59.87	37.03	153.8	94.98

14.11 When coal is distilled by heating without contact with air, a wide variety of solid, liquid, and gaseous products of commercial importance are produced, as well as some significant air pollutants. The nature and amounts of the products produced depend on the temperature used in the decomposition and the type of coal. At low temperatures (400°C to 750°C) the yield of synthetic gas is small relative to the yield of liquid products, whereas at high temperatures (above 900°C) the reverse is true. For the typical process flowsheet, shown in Figure P14.11:

a. How many tons of the various products are being produced?

b. Make an energy balance around the primary distillation tower and benzol tower.

c. How much (in pounds) of 40% NaOH solution is used per day for the purification of the phenol?

d. How much 50% H_2SO_4 is used per day in the pyridine purification?

e. What weight of Na_2SO_4 is produced per day by the plant?

f. How many cubic feet of gas per day are produced? What percent of the gas (volume) is needed for the ovens?

Products Produced per Ton of Coal Charged	Mean C_p Liquid (cal/g)	Mean C_p Vapor (cal/g)	Mean C_p Solid (cal/g)	Melting Point (°C)	Boiling Point (°C)
Synthetic gas–10,000 ft³					
(555 Btu/ft³)					
$(NH_4)_2SO_4$, 22 lb					
Benzol, 15 lb	0.50	0.30	—	—	60
Toluol, 5 lb	0.53	0.35	—	—	109.6
Pyridine, 3 lb	0.41	0.28	—	—	114.1
Phenol, 5 lb	0.56	0.45	—	—	182.2
Naphthalene, 7 lb	0.40	0.35	0.281	80.2	218
			$+0.00111\ T_{°F}$		
Cresols, 20 lb	0.55	0.50	—	—	202
Pitch, 40 lb	0.65	0.60	—	—	400
Coke, 1500 lb	—	—	0.35		—

	ΔH_{vap} (cal/g)	ΔH_{fusion} (cal/g)
Benzol	97.5	—
Toluol	86.53	—
Pyridine	107.36	—
Phenol	90.0	—
Naphthalene	75.5	35.6
Cresols	100.6	—
Pitch	120	—

14.12 A gas consisting of 95 mol % hydrogen and 5 mol % methane at 100°F and 30 psia is to be compressed to 569 psia at a rate of 440 lb mol/hr. A two-stage compressor system has been proposed with intermediate cooling of the gas to 100°F via a heat exchanger. See Figure P14.12. The pressure drop in the heat

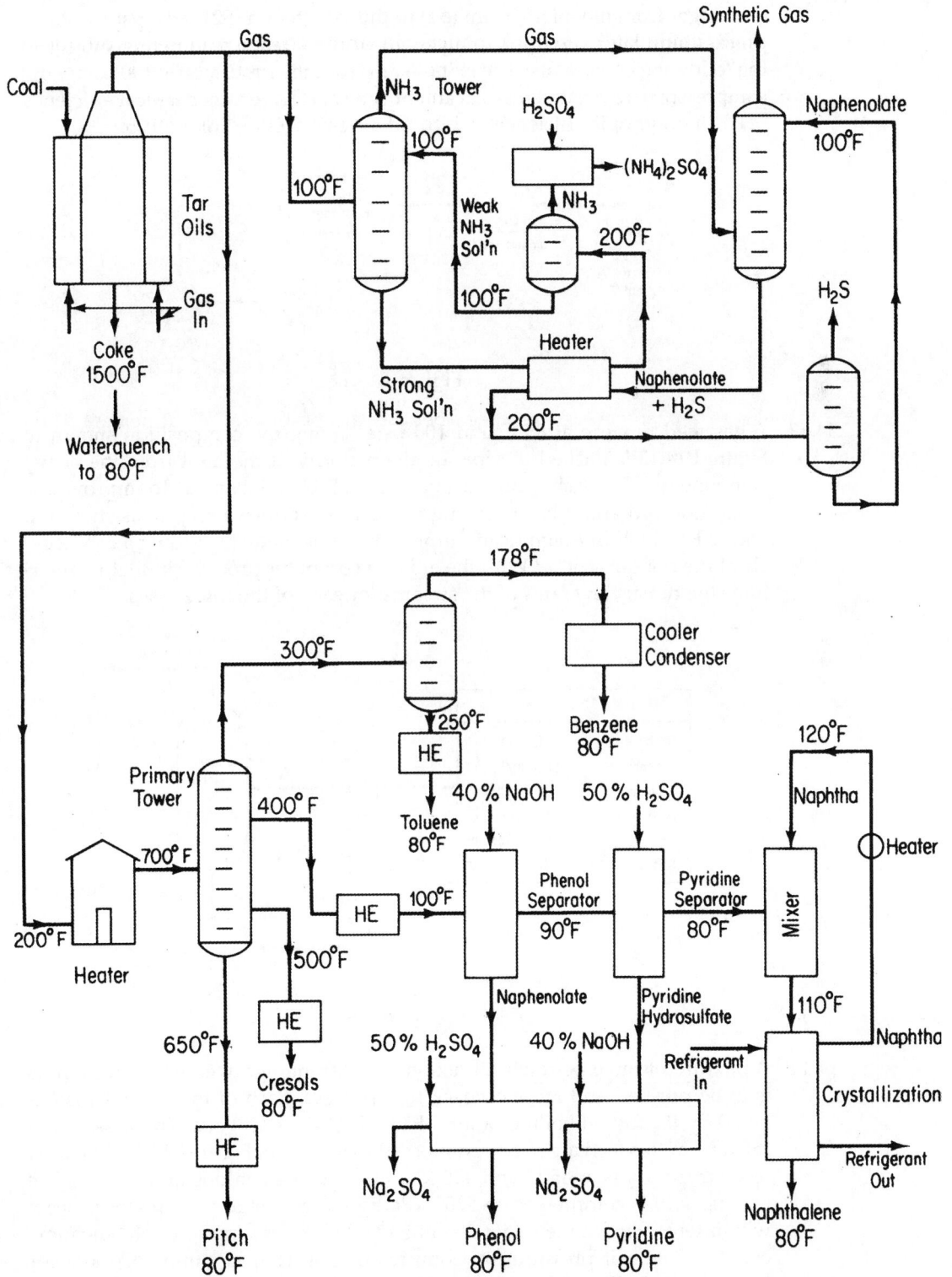

Figure P14.11

exchanger from the inlet stream (S1) to the exit stream (S2) is 2.0 psia. Using a process simulator program, analyze all of the stream parameters subject to the following constraints: The exit stream from the first stage is 100 psia; both compressors are positive-displacement type and have a mechanical efficiency of 0.8, a polytropic efficiency of 1.2, and a clearance fraction of 0.05.

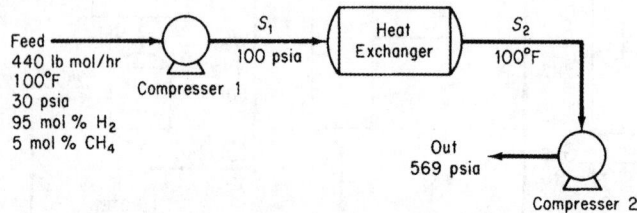

Figure P14.12

14.13 A gas feed mixture at 85°C and 100 psia having the composition shown in Figure P14.13 is flashed to separate the majority of the light from the heavy components. The flash chamber operates at 5°C and 25 psia. To improve the separation process, it has been suggested to introduce a recycle as shown in Figure P14.13. Will a significant improvement be made by adding a 25% recycle of the bottoms? 50%? With the aid of a computer process simulator, determine the molar flow rates of the streams for each of the three cases.

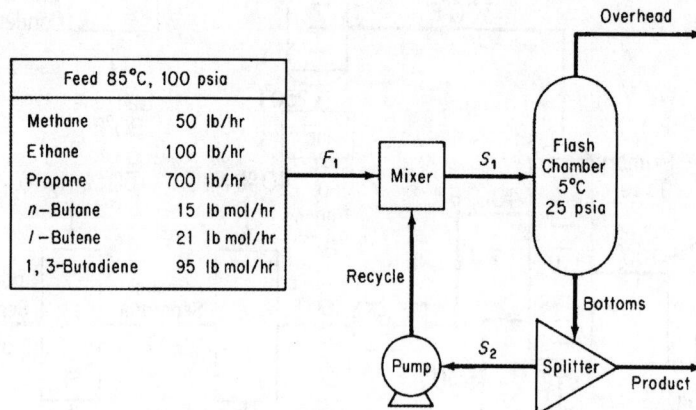

Figure P14.13

14.14 A mixture of three petroleum fractions containing lightweight hydrocarbons is to be purified and recycled back to a process. Each of the fractions is denoted by its normal boiling point: BP135, BP260, and BP500. The gases separated from this feed are to be compressed as shown in Figure P14.14. The inlet feed stream (1) is at 45°C and 450 kPa and has the composition shown. The exit gas (10) is compressed to 6200 kPa by a three-stage compressor process with intercooling of the vapor streams to 60°C by passing through a heat exchanger. The exit pressure for Compressor 1 is 1100 kPa and 2600 kPa for Compressor 2. The efficiencies for Compressors 1, 2, and 3, with reference to an adiabatic compression, are 78%, 75%, and 72%, respectively. Any liquid fraction drawn off from a separator is recycled to the previous stage. Estimate the heat duty (in kilojoules per hour) of the heat exchangers and the various stream compositions (in kilogram moles per hour) for the system. Note that the separators may be considered as adiabatic flash tanks in which the pressure decrease is zero. This problem has been formulated from *Application Briefs of Process*, the user manual for the computer simulation software package of Simulation Science, Inc.

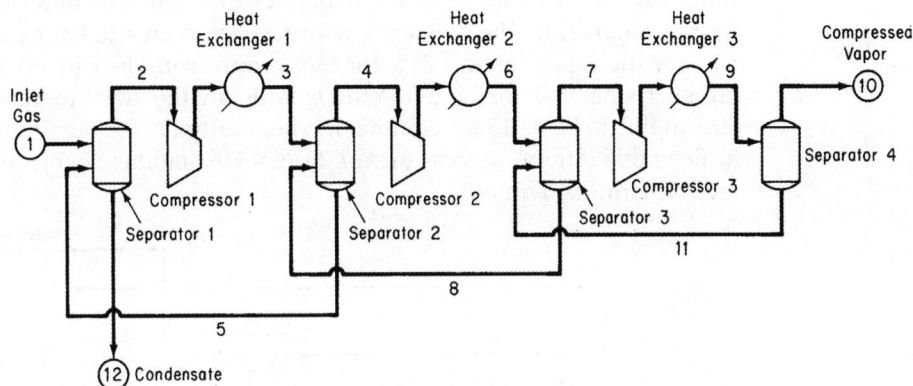

Figure P14.14

Component	kg mol/hr	MW	Sp.Gr.	Normal Boiling Point (°C)
Nitrogen	181			
Carbon dioxide	1920			
Methane	14515			
Ethane	9072			
Propane	7260			
Isobutane	770			
n-Butane	2810			
Isopentane	953			
n-Pentane	1633			
Hexane	1542			
BP135	11975	120	0.757	135
BP260	9072	200	0.836	260
BP500	9072	500	0.950	500

14.15 A demethanizer tower is used in a refinery to separate natural gas from a light hydrocarbon gas mixture stream (1) having the composition listed below. However, initial calculations show that there is considerable energy wastage in the process. A proposed improved system is outlined in Figure P14.15. Calculate the temperature (°F), pressure (psig), and composition (lb mol/hr) of all of the process streams in the proposed system.

Inlet gas at 120°F and 588 psig, stream 1, is cooled in the tube side of a gas-gas heat exchanger by passing the tower overhead, stream 8, through the shell side. The temperature difference between the exit streams 2 and 10 of the heat exchanger is to be 10°F. Note that the pressure drop through the tube side is 10 psia and 5 psia on the shell side. The feed stream (2) is then passed through a chiller in which the temperature drops to −84°F and a pressure loss of 5 psi results. An adiabatic flash separator is used to separate the partially condensed vapor from the remaining gas. The vapor then passes through an expander turbine and is fed to the first tray of the tower at 125 psig. The liquid stream (5) is passed through a valve, reducing the pressure to that of the third tray on the lower side. The expander transfers 90% of its energy output to the compressor. The efficiency with respect to an adiabatic compression is 80% for the expander and 75% for the compressor. The process requirements are such that the methane-to-ethane ratio in the demethanizer liquids in stream 9 is to be 0.015 by volume; the heat duty on the reboiler is variable to achieve this ratio. A process rate of 23.06×10^6 standard cubic feet per day of feed stream 1 is required.

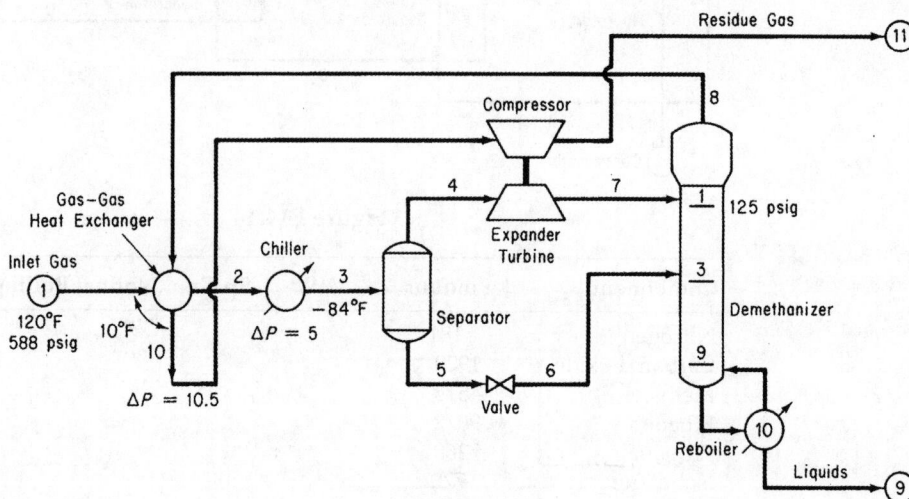

Figure P14.15

Component	Mol %
Nitrogen	7.91
Methane	73.05
Ethane	7.68
Propane	5.69
Isopropane	0.99
n-Butane	2.44
Isopentane	0.69
n-Pentane	0.82
C_6	0.42
C_7	0.31
Total	100.00

The tower has 10 trays, including the reboiler. Note: To reduce the number of trials, the composition of stream 3 may be referenced to stream 1, and if the exit stream of the chiller is given a dummy symbol, the calculation sequence can begin at the separator, thus eliminating the recycle loop.

Carry out the solution of the material and energy balances for the flowsheet in Figure P14.15, determine the component and total mole flows, and determine the enthalpy flows for each stream. Also find the heat duty of each heat exchanger.

This problem has been formulated from *Application Briefs of Process,* the user manual for the computer simulation software package of Simulation Sciences, Inc.

14.16 Determine the values of the unknown quantities in Figure P14.16 by solving the following set of linear material and energy balances that represent the steam balance:

a. $181.60 - x_3 - 132.57 - x_4 - x_5 = -y_1 - y_2 + y_5 + y_4 = 5.1$

b. $1.17x_3 - x_6 = 0$

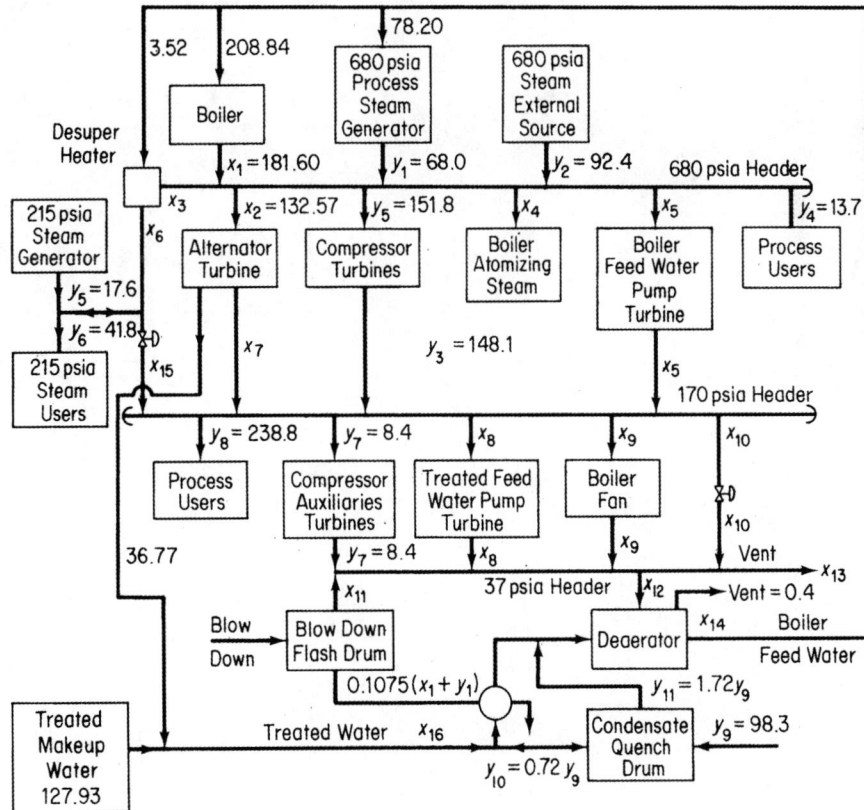

Figure P14.16

c. $132.57 - 0.745x_7 = 61.2$

d. $x_5 + x_7 - x_8 - x_9 - x_{10} + x_5 = y_7 + y_8 = y_3 = 99.1$

e. $x_8 + x_9 + x_{10} + x_{11} - x_{12} - x_{13} = -y_7 = -8.4$

f. $x_6 - x_{15} = y_{12} = y_5 = 24.2$

g. $-1.15(181.60) + x_3 - x_6 + x_{12} + x_{16} = 1.15y_1 - y_9 + 0.4 = -19.7$

h. $181.60 - 4.594x_{12} - 0.11x_{16} = -y_1 + 1.0235y_9 + 2.45 = 35.05$

i. $-0.0423(181.60) + x_{11} = 0.0423y_1 = 2.88$

j. $-0.016(181.60) + x_4 = 0$

k. $x_8 - 0.0147x_{16} = 0$

l. $x_5 - 0.07x_{14} = 0$

m. $-0.0805(181.60) + x_9 = 0$

n. $x_{12} - x_{14} + x_{16} = 0.4 - y_9 = -97.9$

There are four levels of steam: 680, 215, 170, and 37 psia. The 14 x_i, $i = 3, \ldots,$ 16, are the unknowns and the y_i are given parameters for the system. Both x_i and y_i have the units of 10^3 lb/hr.